THE HANDBOOK OF

MARKET DESIGN

THE HANDBOOK OF

MARKET

DESIGN

Edited by

NIR VULKAN,

ALVIN E. ROTH,

and

ZVIKA NEEMAN

OXFORD

UNIVERSITY PRESS

OXFORD
UNIVERSITY PRESS

Great Clarendon Street, Oxford, OX2 6DP,
United Kingdom

Oxford University Press is a department of the University of Oxford.
It furthers the University's objective of excellence in research, scholarship,
and education by publishing worldwide. Oxford is a registered trade mark of
Oxford University Press in the UK and in certain other countries

© Oxford University Press 2013

The moral rights of the authors have been asserted

First Edition published in 2013

Impression: 1

Published in the United States of America by Oxford University Press
198 Madison Avenue, New York, NY 10016, United States of America

British Library Cataloguing in Publication Data
Data available

Library of Congress Control Number: 2013944501

ISBN 978-0-19-957051-5

Printed and bound by
CPI Group (UK) Ltd, Croydon, CR0 4YY

We would like to dedicate this book to Fiona, Norette and Tom Vulkan; Emilie, Aaron and Ben Roth; and Haya and Uri Neeman.

CONTENTS

PART I GENERAL PRINCIPLES

PART II CASES
SECTION II.A MATCHING MARKETS

PART III EXPERIMENTS

PART IV COMPETING DESIGNS

List of Figures

List of Tables

List of Contributors

Atila Abdulkadiroğlu is Professor in Economics at Duke University. He taught at Northwestern University and Columbia University before coming to Duke. He received his PhD in Economics at the University of Rochester. He has consulted school districts in redesigning student assignment systems, including Boston (MA), Chicago (IL), Denver (CO), New Orleans (LA), and New York City (NY). He is a recipient of an Alfred P. Sloan Research Fellowship and a National Science Foundation CAREER award. He serves as an Editor-in-Chief of *Review of Economic Design* and on the board of the Institute for Innovation in Public School Choice.

Lawrence M. Ausubel is Professor of Economics at the University of Maryland. He has published widely on auctions, bargaining, the credit card market, and other aspects of industrial organization and financial markets. He has been awarded fifteen US patents relating to auction methodology and he has several other patents pending. He received his AB in Mathematics from Princeton University, his MS in Mathematics from Stanford University, his MLS in Legal Studies from Stanford Law School, and his PhD in Economics from Stanford University. He is also chairman of Power Auctions LLC and president of Market Design Inc.

Sarbartha Bandyopadhyay manages the servers and database systems for EconJob-Market.org. He is also the maintainer of the EJM codebase. He is the Co-President of Technoluddites Inc. and a developer for Editorial Express, Conference Maker and HeadHunter.

Gary E. Bolton is the O.P. Jindal Chair of Management Measurement Economics in the Jindal School of Management, University of Texas at Dallas, and is Director of the Laboratory of Behavioral Operations and Economics. He studies economic and business decision-making and strategic games, with special interest in bargaining, cooperation, reputation building, social utility, and strategic learning.

Andrew Byde is Head of Research at Acunu Ltd, a database startup, where his research focuses on algorithms and data structures for data storage systems. Prior to joining Acunu, he spent ten years at Hewlett-Packard laboratories, publishing in diverse areas of computer science, including autonomous agents, biologically inspired computing and market-based design. He held a Royal Society Industry Fellowship to the University of Southampton for two years. His PhD is in Mathematics, from Stanford University.

Peter Cramton is Professor of Economics at the University of Maryland. Since 1983, he has conducted widely cited research on auction theory and practice. The main focus is the design of auctions for many related items. Applications include auctions for radio spectrum, electricity, financial securities, diamonds, and timber. He has introduced innovative market designs in many industries. He has advised numerous governments on market design and has advised dozens of bidders in major auction markets. He received his BS in Engineering from Cornell University and his PhD in Business from Stanford University.

Robert Day is currently an Associate Professor of Operations and Information Management at the University of Connecticut. He received his PhD in Applied Mathematics with a concentration in Operations Research from the University of Maryland in 2004. His dissertation, which focused on combinatorial auctions, received INFORMS' Dantzig dissertation award in 2005. He continues to study combinatorial auctions and other related applications, including markets for grid computing, and the scheduling of operating-room resources in order to improve hospital efficiency. Further, he has recently consulted on the design of combinatorial auctions for spectrum licenses held in the UK and other countries.

Samuel Dinkin is Senior Auction Consultant at Power Auctions LLC, where he helps governments and companies design, implement, and participate in high-stakes auctions on six continents. Since 1995, he has designed and conducted over fifty auctions totaling over $100 billion in transactions, including auction rules for the world's four largest multi-unit electricity auctions and first multi-unit telecom auctions for seven countries. He is a seventh-plateau IBM master inventor. He captained the US contract bridge team, winning a silver medal in the 2009 world championships. He received his BS in economics from Caltech and his MA and PhD from the University of Arizona.

Martin Dufwenberg got his PhD in Uppsala in 1995; he worked at universities in Tilburg, Uppsala, Stockholm, Bonn, and is currently affiliated with the University of Arizona, the University of Gothenburg and Bocconi University. His research uses game theory and experiments to incorporate insights from psychology into economic analysis.

Benjamin Edelman is an Associate Professor at the Harvard Business School. His research explores the public and private forces shaping Internet architecture and business opportunities, including online advertising, consumer protection, and regulation. He is a member of the Massachusetts Bar. His writings are available at <http://www.benedelman.org>.

Aytek Erdil is a Lecturer in Economics at the University of Cambridge, and is a fellow of King's College, Cambridge. He received his PhD from the University of Chicago, and held postdoctoral fellowships at Harvard Business School and University of Oxford, prior to his current position.

Haluk Ergin is Associate Professor in the Department of Economics at the University of California, Berkeley.

Emel Filiz-Ozbay is Assistant Professor of Economics at the University of Maryland. She received her PhD in Economics from Columbia University in 2007 and joined the University of Maryland faculty in the same year. She is interested in experimental and behavioral economics, decision theory, industrial organization, and market design. She has conducted several auction and market design experiments to better understand the performance of different mechanisms. Her research also addresses the behavioral biases of agents in their decision-making, and how those biases vary between agents and depending on the environment. She teaches courses in microeconomics and contract theory.

Joshua S. Gans is a Professor of Strategic Management and holder of the Jeffrey S. Skoll Chair of Technical Innovation and Entrepreneurship at the Rotman School of Management, University of Toronto. While his research interests are varied, he has developed specialties in the nature of technological competition and innovation, economic growth, publishing economics, industrial organization and regulatory economics. In 2007, he was awarded the Economic Society of Australia's Young Economist Award, and in 2008 was elected as a Fellow of the Academy of Social Sciences, Australia.

Uri Gneezy is The Epstein/Atkinson Chair in Behavioral Economics and Professor of Economics & Strategy at Rady School of Management, University of California San Diego. As a researcher, his focus is on putting behavioral economics to work in the real world, where theory can meet application. He is looking for basic research as well as more applied approaches to such topics as incentives-based interventions to increase good habits and decrease bad ones, "pay-what-you-want" pricing, and the detrimental effects of small and large incentives. In addition to the traditional laboratory and field studies, he is currently working with several firms, conducting experiments in which basic findings from behavioral economics are used to help companies achieve their traditional goals in non-traditional ways.

Ernan Haruvy is an Associate Professor in Marketing at the University of Texas at Dallas. He earned his PhD in Economics in 1999 from the University of Texas at Austin. His main interests are in market design applications, including auctions, procurement, matching, learning, e-commerce, and software markets. His main methodological tools come from behavioral and experimental economics. He publishes in various disciplines with publications in journals such as *American Economic Review*, *Marketing Science*, *Journal of Marketing Research* and *Journal of Finance*.

Aviad Heifetz is a Professor of Economics at the Economics and Management Department of the Open University of Israel, where he served as department chair, 2006–09. He is the author of the textbook *Game Theory: Interactive Strategies in Economics and Management* (Cambridge University Press, 2012). His research in game theory and economic theory provided insights into the evolution of preferences, market design, bargaining, competitive economies with asymmetric information, and interactive epistemology. He serves on the editorial boards of *Games and Economic Behavior, International Journal of Game Theory*, and *Mathematical Social Sciences*.

Nathaniel Higgins is an Economist at the Economic Research Service, United States Department of Agriculture (USDA), and Adjunct Professor in the Department of International Economics at the Johns Hopkins School of Advanced International Studies. He uses experimental and computational economics to study problems in market design, especially problems related to the design of USDA conservation programs. He has published articles on the design of auctions, behavioral economics, and commodities prices, and consulted in the design of auctions for spectrum and airport landing slots. He received his BA in Mathematics and Economics from Ithaca College and his PhD in Agricultural and Resource Economics from the University of Maryland.

Fedor Iskhakov is Senior Research Fellow at ARC Centre of Excellence in Population Ageing Research at University of New South Wales. He received his PhD in Economics from the University of Oslo, Norway, in 2009 and Candidate of Sciences degree from St Petersburg State University, Russia, in 2006. Iskhakov's PhD dissertation, "A dynamic structural analysis of health and retirement," was awarded the HM King of Norway golden medal (H.M. Kongens gullmedalje) as best research in social sciences among young researchers in Norway in 2008.

Terence Johnson is Assistant Professor of Economics at the University of Notre Dame. He received his PhD from the University of Maryland in 2011, specializing in microeconomic theory and industrial organization.

Elena Katok is Ashbel Smith Professor at the Naveen Jundal School of Management, at the University of Texas at Dallas. She has co-authored a number of scholarly articles in behavioral operations management, focusing on using laboratory experiments to test game-theoretic models of contracting and competitive procurement. She has been a member of INFORMS since 1995, and was the winner of the 2000 Franz Edelman competition. She is one of the organizers of the annual Behavioral Operations Management conference, a Department Editor for Behavioral Operations, and a Deputy Editor at the *Production and Operation Management (POM) Journal*, and the President of the INFORMS Section for Behavioral Operations Management.

Alon Klement writes on various subjects in the fields of Civil Procedure, Law and Economics and Law and Social Norms. He teaches at the Radzyner School of Law, in the Interdisciplinary Center, Herzliya. In recent years he has visited and taught at Columbia University and Boston University in the US, and at the University of Bologna, in Italy. He earned his LLB in Law and BA in Economics from Tel Aviv University. After practicing law for several years in a private law firm in Israel he went to Harvard, where he earned his SJD degree.

Paul Klemperer is the Edgeworth Professor of Economics at Oxford University. He has advised numerous governments, including devising the UK government's 3G mobile-phone license auction, which raised £22.5 billion, and assisting the US Treasury in the financial crisis. He has also developed new auction designs; his most recent innovation—the product-mix auction—is regularly used by the Bank of England. He co-invented the concept of "strategic complements;" developed the "supply function" analysis of electricity markets, and the theory of consumer switching costs; and has applied techniques from auction theory in a range of other economic contexts, from finance to political economy.

Soohyung Lee is an Assistant Professor at University of Maryland. She received her PhD from Stanford University and BA from Seoul National University. Prior to starting her PhD program, she served the Ministry of Strategy and Finance in Korea, as a Deputy Director from 1999 to 2002. Her research interests broadly lie in applied econometrics and market design.

David McArthur is an Economist at the Federal Reserve Board of Governors. His research interests are information economics, networks, and industrial organization. His PhD is in Economics from the University of Maryland.

Paul Milgrom is the Leonard and Shirley Ely Professor of Humanities and Sciences at Stanford University and a member of the National Academy of Sciences and American Academy of Arts and Sciences. He has published widely on the subject of auctions and market design, including thirty published articles, a book (*Putting Auction Theory to Work*) and several issued and pending patents. He is inventor or co-inventor of some of the leading auction methods used for large auctions today, including the simultaneous multiple-round auction, the clock-proxy auction, the core-selecting auction, and the assignment auction. He is the founder and chairman of Auctionomics and leads its team of economists assisting the FCC in designing and implementing its "incentive auction" to buy television broadcast rights and repurpose the spectrum for wireless broadband services.

Zvika Neeman is Professor of Economics at the Berglas School of Economics at Tel Aviv University. He is a microeconomic and game theorist with diverse interests who specializes in mechanism design. Before joining Tel Aviv University, he held positions at Boston University and at the Hebrew University of Jerusalem. He received his PhD from Northwestern University in 1995.

Axel Ockenfels is Professor of Economics at the University of Cologne, Director of the Cologne Laboratory of Economic Research, and Coordinator of the DFG research group "Design & Behavior." He publishes in leading journals in economics, but also in business administration, information systems, psychology, and sociology, as well as in application-oriented outlets. Ockenfels is a Member of the Berlin-Brandenburgische and of the North Rhine-Westphalian Academy of Sciences, the Academic Advisory Board at the Federal Ministry of Economics and Technology, and the Scientific Advisory Board of the University of Cologne. In 2005 he received the Gottfried Wilhelm Leibniz Prize of the German Science Foundation.

Erkut Y. Ozbay is Assistant Professor of Economics at the University of Maryland. He received his PhD in Economics from New York University in 2007 and joined the faculty at the University of Maryland in the same year. He is also the Director of the Experimental Economics Laboratory at the University of Maryland. His research interests are experimental economics and theory. His research mainly focuses on understanding how economic agents make decisions when they are faced with different types of uncertainty and how their behavior is affected by their experience, concerns, and the change of their understanding of the economic problem they are facing.

Michael Peters is Professor of Economics at the University of British Columbia since 2003 and a Fellow of the Econometric Society. His research focuses on search and competing mechanisms.

Chris Preist is Reader in Sustainability and Computer Systems at University of Bristol Computer Science Department. He is also a member of the Cabot Institute and the Systems Centre. His research interests include life cycle analysis of digital services with particular reference to the digital transformation of the news and media industry. Prior to joining Bristol, he was a master scientist at HP Labs, Bristol. He was the technical lead on the EU Framework 5 Semantic Web-based Web Services project, coordinating input from over twenty researchers across eight institutions, and chair of the Scientific Advisory Board of the Framework 6 DIP project. His work in the application of artificial intelligence techniques to automated diagnosis led to the deployment of several systems within HP manufacturing facilities and the development of an award-winning product (Agilent Fault Detective). He has a degree in Mathematics from Warwick University, and a PhD in the Semantics of Logic Programming from Imperial College, London.

Ashok Rai is a development economist who works on microfinance. His field research has taken him to Bangladesh, Colombia, India, and Kenya. Rai has a PhD from the University of Chicago and an undergraduate degree from Stanford. He is currently an Associate Professor at Williams College in the United States, and a Member of the Courant Center at the University of Göttingen in Germany.

Alvin E. Roth received his BS from Columbia University in 1971 and PhD from Stanford University in 1974. He taught at the University of Illinois, 1974–82, at the University of Pittsburgh, 1982–98, at Harvard University, 1998–2012, and now teaches at Stanford University. He shared the 2012 Nobel Memorial Prize in Economics for his work in market design.

John Rust is Professor of Economics at Georgetown University. He was previously a Professor of Economics at University of Maryland, Yale University, and University of Wisconsin. He received his PhD from MIT in 1983, specializing in applied econometrics and computational economics. He is a co-founder of EconJobMarket.org, and Technoluddites, Inc., which provides web-based software to assist academics in publishing, holding conferences, and evaluation of job candidates. He received the Ragnar Frisch

Medal from the Econometric Society in 1992 for his 1987 *Econometrica* paper "Optimal Replacement of GMC Bus Engines: An Empirical Model of Harold Zurcher."

Tuomas Sandholm is Professor in the Computer Science Department at Carnegie Mellon University. He has published over 450 papers on market design and other topics in computer science, operations research, and game theory. He holds sixteen patents on market design. He is best known for his work on combinatorial auctions. Applications include sourcing, TV and Internet display advertising, sponsored search, and radio spectrum. He is Founder, President, and CEO of Optimized Markets, Inc. Previously he was Founder, Chairman, and CTO/Chief Scientist of CombineNet, Inc. His technology also runs the US-wide kidney exchange. He serves as the design consultant of Baidu's sponsored search auctions. He has also consulted for Yahoo!, Netcycler, Google, and many other companies and government institutions on market design. He is recipient of the NSF Career Award, the inaugural ACM Autonomous Agents Research Award, the Alfred P. Sloan Foundation Fellowship, the Carnegie Science Center Award for Excellence, and the Computers and Thought Award. He is Fellow of the ACM and AAAI.

Ella Segev is a researcher in the Department of Industrial Engineering and Management at Ben-Gurion University, Beer Sheva, Israel. Her research interests include auction theory, bargaining theory, and contests. She has a PhD from Tel Aviv University and was a research scholar at the Institute for Advanced Study in Princeton, USA. She has published papers in journals such as *International Economic Review, Games and Economic Behavior*, and *Public Choice*, among others.

Tomas Sjöström did his undergraduate studies in Stockholm and received a PhD from the University of Rochester. He taught at Harvard and Penn State before moving to Rutgers in 2004, where he is currently Professor of Economics. His interests include mechanism design, theories of conflict, and neuroeconomics.

Tayfun Sönmez is a Professor at Boston College, Department of Economics. He received his PhD in Economics in 1995 from the University of Rochester.

Scott Stern is School of Management Distinguished Professor and Chair of the Technological Innovation, Entrepreneurship and Strategic Management Group at the Sloan School at MIT. He explores how innovation—the production and distribution of ideas—differs from more traditional economic goods, and the implications of these differences for entrepreneurship, business strategy, and public policy. He received his PhD from Stanford University, and he is the Director of the Innovation Policy Working Group at the National Bureau of Economic Research. In 2005 he was awarded the Kauffman Prize Medal for Distinguished Research in Entrepreneurship.

Andrew Stocking is a market design economist at the US Congressional Budget Office (CBO). He provides analysis of market rules and their effect on expected outcomes for markets, which include environmental cap-and-trade programs, spectrum auctions, Medicare auctions, oil and gas markets, and financial markets. Prior to working at the

CBO, he designed and worked with online advertising auctions, charitable fundraising markets, airport auctions for landing slots, and international telecom auctions. He has published several papers on the unintended consequences of market rules. He holds a BS in chemical engineering and an MS in environmental engineering, both from Stanford University, and a PhD in resource economics from the University of Maryland.

Eric Talley is the Rosalinde and Arthur Gilbert Professor in Law, Business and the Economy at the University of California Berkeley (Boalt Hall) School of Law. His research focuses on the intersection corporate law, firm governance and financial economics. He holds a bachelor's degree from UC San Diego, a PhD in economics from Stanford, and a JD also from Stanford, where he was articles editor for the *Stanford Law Review*. He has previously held permanent or visiting appointments at the University of Southern California, the University of Chicago, Harvard University, Georgetown University, the California Institute of Technology, Stanford University, the University of Sydney (Australia), and the University of Miami. He is a frequent commentator on the radio show *Marketplace*, and often speaks to corporate boards and regulators on issues pertaining to fiduciary duties, governance, and corporate finance.

M. Utku Ünver is a Professor at Boston College, Department of Economics. He received his PhD in Economics in 2000 from the University of Pittsburgh.

Nir Vulkan is Economics Professor at the Said Business School and a Fellow of Worcester College, both at Oxford University. He is the authors of dozens of articles on market design and the book *The Economics of e-Commerce* (Princeton University Press, 2003). He has worked with many software and e-commerce companies designing markets mainly on the Internet, which are used by humans and software agents. His algorithms for automated trading have been used by hedge funds to trade futures in markets all over the world.

Joel Watson is Professor of Economics at the University of California, San Diego. His research mainly addresses how contractual relationships are formed and managed, and the role of institutions, using game-theoretic models. He authored a popular textbook on game theory (*Strategy: An Introduction to Game Theory*). He co-founded and serves as the CEO of Econ Job Market Inc., a non-profit charitable (501c3) corporation that manages application materials in the economics PhD job market. He obtained his BA from UCSD and his PhD from Stanford's Graduate School of Business. He was a Prize Research Fellow at Oxford's Nuffield College.

John Watson is Director of Analytics for the Institute of Evidence-Based Change, Encinitas, and technologist for Watson Education. He designs data collection and analysis applications primarily in the education field, oversees the largest K-20 student records database in California, and conducts research in the area of intelligent data systems. He holds a patent for mobile data-environment technologies, and has contributed to a half-dozen technology-related inventions. He received a joint doctoral degree from Claremont Graduate University and San Diego State University. He also received degrees from University of California at San Diego (BA) and San Diego State University (MA).

Robert Wilson is an Emeritus Professor at the Stanford Business School. His main research interest is game theory, but he has worked on market designs since the early 1970s, including government auctions of exploration leases and spectrum licenses, and wholesale markets for natural gas and for electric power, reserves, and capacity. His book *Nonlinear Pricing*, in print since 1993, won the Melamed Prize of the Chicago Business School. He is a Fellow of the Econometric Society and the American Economic Association, a member of the National Academy of Sciences, and on the board of Market Design Inc.

Hadas Yafe is CEO of GreenHands.

INTRODUCTION

NIR VULKAN, ALVIN E. ROTH,
AND ZVIKA NEEMAN

"MARKET design" is the term used to refer to a growing body of work that might also be called *microeconomic engineering* and to the theoretical and empirical research that supports this effort and is motivated by it.

Economists often look at markets as given, trying to make predictions about who will do what and what will happen in these markets. Market design, in contrast, does not take markets as given; instead, it combines insights from economic and game theory together with common sense and lessons learned from empirical work and experimental analysis to aid in the design and implementation of actual markets. In recent years the field has grown dramatically—partly because of the successful wave of spectrum auctions in the US and in Europe, partly because of the clearinghouses and other marketplaces which have been designed by a number of prominent economists, and partly because of the increased use of the Internet as the platform over which markets are designed and run. There are now a large number of applications and a growing theoretical literature, which this book surveys.

Market design is both a science and an art. It is a science in that it applies the formal tools of game theory and mechanism design and it is an art because practical design often calls for decisions that are beyond the reliable scientific knowledge of the field, and because the participants in these markets are often different than they are modeled by these theories. Nevertheless, as the book demonstrates, lessons can be learned from successful and unsuccessful market designs which can be transferred to new and different environments.

In this book we attempt to bring together the latest research and provide a relatively comprehensive description of applied market design as it has taken place around the world over the last two decades or so. In particular we survey many matching markets: These are environments where there is a need to match large two-sided populations of agents such as medical residents and hospitals, law clerks and judges, or patients and kidney donors, to one another. Experience shows that if the arranged match is not appropriately stable, then participants will try to transact outside of the indicated marketplace, and the market will unravel leading to very inefficient results. We also survey a number of applications related to electronic markets and e-commerce: The Internet is now the preferred platform for many markets and this raises some interesting issues, such as the impact of automation (for example you use a software agent to bid in an Internet auction). Also related is the resulting competition between exchanges— since anyone can access the Internet anywhere in the world, the geographic location of a market is less relevant and participants now often face a real choice of trading mechanisms which they can use. While many of the chapters in the book consider a single marketplace that has established such a dominant share of the market that most participants have no other desirable choice (e.g. medical residents), a number of chapters in this book consider the implications to market designers of the fact that participants have a choice.

Market design involves the specification of detailed rules, which are typically analyzed using what used to be called "noncooperative" game theory. The analysis focuses on the incentives for individual behavior in the particular environment considered and its consequences. Specific environments and problems can be very different from one another, and, as we'll see, details and differences can be of huge importance in practical design. But there are also some general themes beginning to emerge from all this detail and diversity, and it will help to keep some of these in mind.

Specifically, a marketplace or the setting in which market design is performed, is part of a broader economic environment in which potential participants also have other choices to make, which may be less well known and harder to model. That is, a marketplace being designed or studied is typically part of a larger game that cannot be modeled in detail with the same confidence as the marketplace. So, to work well and attract wide participation, it may be desirable for marketplaces to promote outcomes that are in the *core* of the larger game, so that there don't exist any coalitions that might prefer to transact outside of the marketplace, instead of participating in it.[1]

A related, less formal organizing theme is that, if a marketplace is to be successful, the rules and behavior in the marketplace, together with the (unmodeled) opportunities and behavior outside the marketplace, have to form an equilibrium in which, given how the marketplace works, it makes sense for participants to enter it and participate. In this respect, experience suggests we can start to diagnose whether a marketplace is working well or badly, by examining how well it provides *thickness*, deals with *congestion*, and makes it *safe* and *simple* to participate (cf. Roth, Chapter 1).

---

[1] The core and various related notions of stability not only capture a very general notion of what constitutes a competitive outcome, they also apply to the less detailed models of what used to be called "cooperative" game theory, and in doing so tell us something about the options that may be available to coalitions of players even when we don't know their strategies in detail. This is why the former distinction between cooperative and noncooperative game theory is not very useful in market design; both perspectives are employed together, to answer different kinds of question and to deal with different kinds of design constraint.

A market provides thickness when it makes many potential transactions available at the same time, so that relevant offers can be compared. (Availability in this sense has a big information component; offers must be available in a way that allows comparison.)

A market is congested if there is insufficient time or resources to fully evaluate all the potentially available transactions. Sometimes this will involve the physical resources needed to carry out transactions (e.g. they may be time consuming, and other possibilities may disappear while a transaction is being attempted), but it can also involve the information needed to make the comparisons among alternative transactions that are needed to choose among them. Congestion is thus a particular problem of thick markets with many quite heterogeneous matching opportunities, and one task of an effective market is to deal with congestion in a way that allows the potential benefits of thickness to be achieved.[2]

To be thick, a marketplace must also make it safe to participate, at least relative to transacting outside the marketplace. Depending on the information and sophistication of the participants, safety may also involve what kinds of strategies the rules of the marketplace require participants to be able to execute, and how sensitive it is to how well others execute their strategies. This is one of the ways in which market design differs most clearly from the theoretical literature on mechanism design, in which different mechanisms are compared by comparing their equilibria. In practical markets, particularly new ones in which all participants will begin without experience, the risks to participants out of equilibrium must also be considered, and so designers often analyze "worst cases" as well as equilibria. Unlike the presumptions made in the literature on theoretical mechanism design and implementation, market designers never know the whole game and therefore need to be cognizant of the fact that their design is one piece of a larger game. Market designers typically do not try to design a market *all* of whose equilibria accomplish something, but rather try to design a marketplace with a good equilibrium, and then try to achieve that equilibrium. If unanticipated behavior develops, the market can be modified, for example with appeals processes, or with making bidders use dropdown menus instead of typing in their own bids, and so on.

This brings us to simplicity, which involves both the market rules themselves, and the kind of behavior they elicit. Simplicity of rules is sometimes discussed under the heading of "transparency," which also involves participants being able to audit the outcome and verify that the rules were followed. But rules may be simple and transparent yet require complex strategizing by the participants. Strategic complexity is often the more important issue, since it may affect both participation in the market, for example if implementing good strategies is costly, and market performance, by leading to mistakes and misjudgments. And the risk associated with such mistakes and misjudgments may also deter participation.

---

[2] Congestion sometimes manifests itself as coordination failure, and so signaling and other attempts to facilitate sorting are one way to deal with it. Another reaction to congestion is unraveling, i.e. starting to transact before the opening of the marketplace, and therefore often not participating in the thick market.

This volume includes chapters that provide a conceptualization of new markets or marketplaces and other designs, together with chapters that describe the adoption and implementation of specific designs (and their subsequent adjustments in light of experience), as well as the theoretical and empirical questions raised in the process. We begin with three chapters that discuss general principles in market design: Al Roth's chapter reviews some of the markets that he, his students, and colleagues have designed, and draws general conclusions from these; Gary Bolton's chapter describes how to stress test models in the lab; and Paul Klemperer's explains how to sensibly use economic theory to create good designs, and he demonstrates how using too much theory can be bad.[3]

Part II is the main part of the book and it provides many cases and applications of market design, some that have been running for years, and some that are still in very early stages. Part II is subdivided into sections on matching markets, auctions, e-commerce applications, and law design (a small section).

Part III focuses on market design experiments, and finally Part IV discusses the implications for market design when there is competition between markets.

---

[3] Klemperer's chapter focuses on the design of large-scale auctions. However, we believe his advice is very relevant to all kinds of market design.

PART I

GENERAL PRINCIPLES

CHAPTER 1

....

WHAT HAVE WE LEARNED
FROM MARKET DESIGN?

....

ALVIN E. ROTH[1]

INTRODUCTION

....

IN the centennial issue of the *Economic Journal*, I wrote (about game theory) that

> the real test of our success will be not merely how well we understand the general
> principles that govern economic interactions, but how well we can bring this knowl-
> edge to bear on practical questions of microeconomic engineering. (Roth, 1991a)

Since then, economists have gained significant experience in practical market design.
One thing we learn from this experience is that transactions and institutions matter at
a level of detail that economists have not often had to deal with, and, in this respect, all
markets are different. But there are also general lessons. The present chapter considers
some ways in which markets succeed and fail, by looking at some common patterns we
see of market failures, and how they have been fixed.

This is a big subject, and I will only scratch the surface, by concentrating on markets
my colleagues and I helped design in the last few years. My focus will be different than in
Roth (2002), where I discussed some lessons learned in the 1990s. The relevant parts of

---

[1] The first part of this chapter was prepared to accompany the Hahn Lecture I delivered at the Royal
Economic Society meetings, on April 11, 2007, and was published as Roth (2008a). The present chapter
extends the 2008 paper with a Postscript to bring it up to date, and to include some details appropriate
to this *Handbook*. I have also updated references and added some footnotes to the first part of the
chapter, but otherwise it remains essentially as published in 2008. One reason for keeping this format,
with a distinct Postscript to bring it up to date is that it will become clear that some of the developments
anticipated in the 2008 paper have been realized in the intervening years. The work I report here is a
joint effort of many colleagues and coauthors. I pay particular attention here to work with Atila
Abdulkadiroğlu, Muriel Niederle, Parag Pathak, Tayfun Sönmez, and Utku Ünver. I've also benefited
from many conversations on this topic with Paul Milgrom (including two years teaching together a
course on market design). In the Postscript I also report on work done with Itai Ashlagi. This work has
been supported by grants from the NSF to the NBER.

that discussion, which I will review briefly in the next section, gathered evidence from a variety of labor market clearinghouses to determine properties of successful clearinghouses, motivated by the redesign of the clearinghouse for new American doctors (Roth and Peranson, 1999). Other big market design lessons from the 1990s concern the design of auctions for the sale of radio spectrum and electricity; see for example Cramton (1997), Milgrom (2000), Wilson (2002), and, particularly, Milgrom (2004).[2]

As we have dealt with more market failures, it has become clear that the histories of the American and British markets for new doctors, and the market failures that led to their reorganization into clearinghouses, are far from unique. Other markets have failed for similar reasons, and some have been fixed in similar ways. I'll discuss common market failures we have seen in recent work on more senior medical labor markets, and also on allocation procedures that do not use prices, for school choice in New York City and Boston, and for the allocation of live-donor kidneys for transplantation. These problems were fixed by the design of appropriate clearinghouses. I will also discuss the North American labor market for new economists, in which related problems are addressed by marketplace mechanisms that leave the market relatively decentralized.

The histories of these markets suggest a number of tasks that markets and allocation systems need to accomplish to perform well. The failure to do these things causes problems that may require changes in how the marketplace is organized. I will argue that, to work well, marketplaces need to

1. provide *thickness*—that is, they need to attract a sufficient proportion of potential market participants to come together ready to transact with one another;
2. overcome the *congestion* that thickness can bring, by providing enough time, or by making transactions fast enough, so that market participants can consider enough alternative possible transactions to arrive at satisfactory ones;
3. make it *safe* to participate in the market as simply as possible
   a. as opposed to transacting outside the marketplace, or
   b. as opposed to engaging in strategic behavior that reduces overall welfare.

I will also remark in passing on some other lessons we have started to learn, namely that

4. some kinds of transactions are *repugnant*, and this can be an important constraint on market design.

And, on a methodological note,

5. experiments can play a role, in diagnosing and understanding market failures and successes, in testing new designs, and in communicating results to policy makers.

---

[2] Following that literature to the present would involve looking into modern designs for package auctions; see for example Cramton et al. (2006), and Milgrom (2007).

The chapter is organized as follows. The following section will describe some of the relevant history of markets for new doctors, which at different periods had to deal with each of the problems of maintaining thickness, dealing with congestion, and making it safe to participate straightforwardly in the market. In the subsequent sections I'll discuss markets in which these problems showed up in different ways.

The third section will review the recent design of regional kidney exchanges in the United States, in which the initial problem was establishing thickness, but in which problems of congestion, and, lately, making it safe for transplant centers to participate, have arisen. This is also the market most shaped by the fact that many people find some kinds of transactions repugnant. In particular, buying and selling kidneys for transplantation is illegal in most countries. So, unlike the several labor markets I discuss in this chapter, this market operates entirely without money, which will cast into clear focus how the "double coincidence of wants" problems that are most often solved with money can be addressed with computer technology (and will highlight why these problems are difficult to solve even with money, in markets like labor markets in which transactions are heterogeneous).

The fourth section will review the design of the school choice systems for New York City high schools (in which congestion was the immediate problem to be solved), and the design of the new public school choice system in Boston, in which making it safe to participate straightforwardly was the main issue. These allocation systems also operate without money.

The fifth section will discuss recent changes in the market for American gastroenterologists, who wished to adopt the kind of clearinghouse organization already in place for younger doctors, but who were confronted with some difficulties in making it safe for everyone to change simultaneously from one market organization to another. This involved making changes in the rules of the decentralized market that would precede any clearinghouse even once it was adopted.

This will bring us naturally to a discussion of changes recently made in the decentralized market for new economists in the United States.

Markets for new doctors in the United States, Canada, and Britain[3]

The first job American doctors take after graduating from medical school is called a residency. These jobs are a big part of hospitals' labor force, a critical part of physicians' graduate education, and a substantial influence on their future careers. From 1900 to 1945, one way that hospitals competed for new residents was to try to hire them earlier than other hospitals. This moved the date of appointment earlier, first slowly and then

---

[3] The history of the American medical market given here is extracted from more detailed accounts in Roth (1984, 2003, 2007).

quickly, until by 1945 residents were sometimes being hired almost two years before they would graduate from medical school and begin work.

When I studied this in Roth (1984) it was the first market in which I had seen this kind of "unraveling" of appointment dates, but today we know that unraveling is a common and costly form of market failure. What we see when we study markets in the process of unraveling is that offers not only come increasingly early, but also become dispersed in time and of increasingly short duration. So not only are decisions being made early (before uncertainty is resolved about workers' preferences or abilities), but also quickly, with applicants having to respond to offers before they can learn what other offers might be forthcoming.[4] Efforts to prevent unraveling are venerable; for example, Roth and Xing (1994) quote Salzman (1931) on laws in various English market from the 13th century concerning "forestalling" a market by transacting before goods could be offered in the market.[5]

In 1945, American medical schools agreed not to release information about students before a specified date. This helped control the date of the market, but a new problem emerged: hospitals found that if some of the first offers they made were rejected after a period of deliberation, the candidates to whom they wished to make their next offers had often already accepted other positions. This led hospitals to make exploding offers to which candidates had to reply immediately, before they could learn what other offers might be available, and led to a chaotic market that shortened in duration from year to year, and resulted not only in missed agreements but also in broken ones. This kind of congestion also has since been seen in other markets, and in the extreme form it took in the American medical market by the late 1940s it also constitutes a form of market failure (cf. Roth and Xing, 1997, and Avery et al., 2007, for detailed accounts of congestion in labor markets in psychology and law).

---

[4] On the costs of such unraveling in some markets for which unusually good data have been available, see Niederle and Roth (2003b) on the market for gastroenterology fellows, and Fréchette et al. (2007) on the market for post-season college football bowls. For some other recent unraveled markets, see Avery et al. (2003) on college admissions; and Avery et al. (2001) on appellate court clerks. For a line of work giving theoretical insight into some possible causes of unraveling, see Li and Rosen (1998), Li and Suen (2000), Suen (2000), and Damiano et al. (2005).

[5] "Thus at Norwich no one might forestall provisions by buying, or paying 'earnest money' for them before the Cathedral bell had rung for the mass of the Blessed Virgin; at Berwick-on-Tweed no one was to buy salmon between sunset and sunrise, or wool and hides except at the market-cross between 9 and 12; and at Salisbury persons bringing victuals into the city were not to sell them before broad day." Unraveling could be in space as well as in time. Salzman also reports (p. 132) that under medieval law markets could be prevented from being established too near to an existing market, and also, for markets on rivers, nearer to the sea. "Besides injury through mere proximity, and anticipation in time, there might be damage due to interception of traffic...." Such interception was more usual in the case of waterborne traffic. In 1233 Eve de Braose complained that Richard fitz-Stephen had raised a market at Dartmouth to the injury of hers at Totnes, as ships which ought to come to Totnes were stopped at Dartmouth and paid customs there. No decision was reached, and eight years later Eve's husband, William de Cantelupe, brought a similar suit against Richard's son Gilbert. The latter pleaded that his market was on Wednesday and that at Totnes on Saturday; but the jury said that the market at Dartmouth was to the injury of Totnes, because Dartmouth lies between it and the sea, so that ships touched there and paid toll instead of going to Totnes; and also that cattle and sheep which used to be taken to Totnes market were now sold at Dartmouth; the market at Dartmouth was therefore disallowed.

Faced with a market that was working very badly, the various American medical associations (of hospitals, students, and schools) agreed to employ a centralized clearinghouse to coordinate the market. After students had applied to residency programs and been interviewed, instead of having hospitals make individual offers to which students had to respond immediately, students and residency programs would instead be invited to submit rank order lists to indicate their preferences. That is, hospitals (residency programs) would rank the students they had interviewed, students would rank the hospitals (residency programs) at which they had been interviewed, and a centralized clearinghouse—a matching mechanism—would be employed to produce a matching from the preference lists. Today this centralized clearinghouse is called the National Resident Matching Program (NRMP).

Roth (1984) showed that the algorithm adopted in 1952 produced a matching of students to residency programs that is *stable* in the sense defined by Gale and Shapley (1962), namely that, in terms of the submitted rank order lists, there was never a student and a residency program that were not matched to each other but would have mutually preferred to have been matched to each other than to (one of) their assigned match(es). However, changes in the market over the years made this more challenging.

For example, one change in the market had to do with the growing number of married couples graduating from American medical schools and wishing to be matched to jobs in the same vicinity. This hadn't been a problem in the 1950s, when virtually all medical students were men. Similarly, the changing nature of medical specialization sometimes produced situations in which a student needed to be simultaneously matched to two positions. Roth (1984) showed that these kinds of changes can sometimes make it impossible to find a stable matching, and, indeed, an early attempt to deal with couples in a way that did not result in a stable matching had made it difficult to attract high levels of participation by couples in the clearinghouse.

In 1995, I was invited to direct the redesign of the medical match, in response to a crisis in confidence that had developed regarding its ability to continue to serve the medical market, and whether it appropriately served student interests. A critical question was to what extent the stability of the outcome was important to the success of the clearinghouse. Some of the evidence came from the experience of British medical markets. Roth (1990, 1991b) had studied the clearinghouses that had been tried in the various regions of the British National Health Service (NHS) after those markets unraveled in the 1960s. A Royal Commission had recommended that clearinghouses be established on the American model, but since the American medical literature didn't describe in detail how the clearinghouse worked, each region of the NHS adopted a different algorithm for turning rank order lists into matches, and the unstable mechanisms had largely failed and been abandoned, while the stable mechanisms succeeded and survived.[6]

---

[6] The effects of instability were different in Britain than in the US, because positions in Britain were assigned by the National Health Service, and so students were not in a position to receive other offers (and decline the positions they were matched to) as they were in the US. Instead, in Britain, students and potential employers acted in advance of unstable clearinghouses. For example, Roth

Of course, there are other differences between regions of the British NHS than how they organized their medical clearinghouses, so there was also room for controlled experiments in the laboratory on the effects of stable and unstable clearinghouses. Kagel and Roth (2000) report a laboratory experiment that compared the stable clearinghouse adopted in Edinburgh with the unstable one adopted in Newcastle, and showed that, holding all else constant, the difference in how the two clearinghouses were organized was sufficient to account for the success of the Edinburgh clearinghouse and the failure of the unstable one in Newcastle.

Roth and Peranson (1999) report on the new clearinghouse algorithm that we designed for the NRMP, which aims to always produce a stable matching. It does so in a way that makes it safe for students and hospitals to reveal their preferences.[7] The new algorithm has been used by the NRMP since 1998, and has subsequently been adopted by over three dozen labor market clearinghouses. The empirical evidence that has developed in use is that the set of stable matchings is very seldom empty.

An interesting historical note is that the use of stable clearinghouses has been explicitly recognized as part of a pro-competitive market mechanism in American law. This came about because in 2002, sixteen law firms representing three former medical residents brought a class-action antitrust suit challenging the use of the matching system for medical residents. The theory of the suit was that the matching system was a conspiracy to hold down wages for residents and fellows, in violation of the Sherman Antitrust Act. Niederle and Roth (2003a) observed that, empirically, the wages of medical specialties with and without centralized matching in fact do not differ.[8] The case was dismissed after the US Congress passed new legislation in 2004 (contained in Public Law 108–218)

(1991) reports that in Newcastle and Birmingham it became common for students and consultants (employers) to reach agreement in advance of the match, and then submit only each other's name on their rank order lists.

[7] Abstracting somewhat from the complexities of the actual market, the Roth–Peranson algorithm is a modified student-proposing deferred acceptance algorithm (Gale and Shapley, 1962; see also Roth, 2008b). In simple markets, this makes it a dominant strategy for students to state their true preferences (see Roth, 1982a, 1985; Roth and Sotomayor, 1990). Although it cannot be made a dominant strategy for residency programs to state their true preferences (Roth, 1985; Sönmez, 1997), the fact that the medical market is large turns out to make it very unlikely that residency programs can do any better than to state their true preferences. This was shown empirically in Roth and Peranson (1999), and has more recently been explained theoretically by Immorlica and Mahdian (2005) and Kojima and Pathak (2009).

[8] Bulow and Levin (2006) sketch a simple model of one-to-one matching in which a centralized clearinghouse, by enforcing impersonal wages (i.e. the same wage for any successful applicant), could cause downward pressure on wages (see also Kamecke, 1998). Subsequent analysis suggests more skepticism about any downward wage effects in actual medical labor markets. See, for example, Kojima (2007), who shows that the Bulow-Levin results don't follow in a model in which hospitals can employ more than one worker, and Niederle (2007), who shows that the results don't follow in a model that includes some of the options that the medical match actually offers patients. Crawford (2008) considers how the deferred acceptance algorithm of Kelso and Crawford (1982) could be adapted to adjust personal wages in a centralized clearinghouse (see also Artemov, 2008).

noting that the medical match is a pro-competitive market mechanism, not a conspiracy in restraint of trade. This reflected modern research on the market failures that preceded the adoption of the first medical clearinghouse in the 1950s, which brings us back to the main subject of the present chapter.[9]

To summarize, the study and design of a range of clearinghouses in the 1980s and 1990s made it clear that producing a stable matching is an important contributor to the success of a labor clearinghouse. For the purposes of the present chapter, note that such a clearinghouse can persistently attract the participation of a high proportion of the potential participants, and when it does so it solves the problem of establishing a thick market. A computerized clearinghouse like those in use for medical labor markets also solves the congestion problem, since all the operations of the clearinghouse can be conducted essentially simultaneously, in that the outcome is determined only after the clearinghouse has cleared the market. And, as mentioned briefly, these clearinghouses can be designed to make it safe for participants to reveal their true preferences, without running a risk that by doing so they will receive a worse outcome than if they had behaved strategically and stated some other preferences.

In the following sections, we'll see more about how the failure to perform these tasks can cause markets to fail.

---

[9] See Roth (2003). The law states in part: "Congress makes the following findings: For over 50 years, most United States medical school seniors and the large majority of graduate medical education programs (popularly known as 'residency programs') have chosen to use a matching program to match medical students with residency programs to which they have applied.... Before such matching programs were instituted, medical students often felt pressure, at an unreasonably early stage of their medical education, to seek admission to, and accept offers from, residency programs. As a result, medical students often made binding commitments before they were in a position to make an informed decision about a medical specialty or a residency program and before residency programs could make an informed assessment of students' qualifications. This situation was inefficient, chaotic, and unfair and it often led to placements that did not serve the interests of either medical students or residency programs. The original matching program, now operated by the independent non-profit National Resident Matching Program and popularly known as 'the Match', was developed and implemented more than 50 years ago in response to widespread student complaints about the prior process.... The Match uses a computerized mathematical algorithm ... to analyze the preferences of students and residency programs and match students with their highest preferences from among the available positions in residency programs that listed them. Students thus obtain a residency position in the most highly ranked program on their list that has ranked them sufficiently high among its preferences.... Antitrust lawsuits challenging the matching process, regardless of their merit or lack thereof, have the potential to undermine this highly efficient, pro-competitive, and long-standing process. The costs of defending such litigation would divert the scarce resources of our country's teaching hospitals and medical schools from their crucial missions of patient care, physician training, and medical research. In addition, such costs may lead to abandonment of the matching process, which has effectively served the interests of medical students, teaching hospitals, and patients for over half a century.... It is the purpose of this section to-confirm that the antitrust laws do not prohibit sponsoring, conducting, or participating in a graduate medical education residency matching program, or agreeing to do so; and ensure that those who sponsor, conduct or participate in such matching programs are not subjected to the burden and expense of defending against litigation that challenges such matching programs under the antitrust laws."

KIDNEY EXCHANGE

Kidney transplantation is the treatment of choice for end-stage renal disease, but there is a grave shortage of transplantable kidneys. In the United States there are over 70,000 patients on the waiting list for cadaver kidneys, but in 2006 fewer than 11,000 transplants of cadaver kidneys were performed. In the same year, around 5,000 patients either died while on the waiting list or were removed from the list as "Too Sick to Transplant." This situation is far from unique to the United States: In the UK at the end of 2006 there were over 6,000 people on the waiting list for cadaver kidneys, and only 1,240 such transplants were performed that year. [10]

Because healthy people have two kidneys, and can remain healthy with just one, it is also possible for a healthy person to donate a kidney, and a live-donor kidney has a greater chance of long-term success than does one from a deceased donor. However, good health and goodwill are not sufficient for a donor to be able to give a kidney to a particular patient: the patient and donor may be biologically incompatible because of blood type, or because the patient's immune system has already produced antibodies to some of the donor's proteins. In the United States in 2006 there were 6,428 transplants of kidneys from living donors (in the UK there were 590).

The total supply of transplantable kidneys (from deceased and living donors) clearly falls far short of the demand. But it is illegal in almost all countries to buy or sell kidneys for transplantation. This legislation is the expression of the fact that many people find the prospect of such a monetized market highly repugnant (see Roth, 2007).

So, while a number of economists have devoted themselves to the task of repealing or relaxing laws against compensating organ donors (see e.g. Becker and Elias, 2007, and the discussion of Elias and Roth, 2007), another task that faces a market designer is how to increase the number of transplants subject to existing constraints, including those that forbid monetary incentives.

It turns out that, prior to 2004, in just a very few cases, incompatible patient–donor pairs and their surgeons had managed to arrange an *exchange* of donor kidneys (sometimes called "paired donation"), when the patient in each of two incompatible patient–donor pairs was compatible with the donor in the other pair, so that each patient received a kidney from the other's donor. Sometimes a different kind of exchange had also been accomplished, called a *list exchange*, in which a patient's incompatible donor donated a kidney to someone who (by virtue of waiting a long time) had high priority on the waiting list for a cadaver kidney, and in return the donor's intended patient received high priority to receive the next compatible cadaver kidney that became available. Prior

---

[10] For US data see <http://www.optn.org/data> (accessed August 13, 2007; website since moved to <http://optn.transplant.hrsa.gov>). For UK data, see <http://www.uktransplant.org.uk/ukt/statistics/calendar_year_statistics/pdf/yearly_statistics_2006.pdf> (accessed August 13, 2007). As I update this in 2012, the number of US patients waiting for cadaver kidneys has risen to over 90,000, while in 2011 there were just barely over 11,000 transplants from cadaver kidneys (so the waiting list has grown considerably while the number of deceased donors has not).

to December 2004 only five exchanges had been accomplished at the fourteen transplant centers in New England. Some exchanges had also been accomplished at Johns Hopkins in Baltimore, and among transplant centers in Ohio. So, these forms of exchange were feasible and non-repugnant. [11] Why had so very few happened?

One big reason had to do with the (lack of) thickness of the market, i.e. the size of the pool of incompatible patient–donor pairs who might be candidates for exchange. When a kidney patient brought a potential donor to his or her doctor to be tested for compatibility, donors who were found to be incompatible with their patient were mostly just sent home. They were not patients themselves, and often no medical record at all was retained to indicate that they might be available. And in any event, medical privacy laws made these potential donors' medical information unavailable.

Roth et al. (2004a) showed that, in principle, a substantial increase in the number of transplants could be anticipated from an appropriately designed clearinghouse that assembled a database of incompatible patient–donor pairs. That paper considered exchanges with no restrictions on their size, and allowed list exchange to be integrated with exchange among incompatible patient–donor pairs. That is, exchanges could be a cycle of incompatible patient–donor pairs of any size such that the donor in the first pair donated a kidney to the patient in the second, the second pair donated to the third, and so on, until the cycle closed with the last pair donating to the first. And pairs that would have been interested in a list exchange in which they donated a kidney in exchange for high priority on the cadaver waiting list could be integrated with the exchange pool by having them donate to another incompatible pair in a chain that would end with donation to the waiting list.

We sent copies of that paper to many kidney surgeons, and one of them, Frank Delmonico (the medical director of the New England Organ Bank), came to lunch to pursue the conversation. Out of that conversation, which grew to include many others (and led to modifications of our original proposals), came the New England Program for Kidney Exchange, which unites the fourteen kidney transplant centers in New England to allow incompatible patient–donor pairs from anywhere in the region to find exchanges with other such pairs.

For incentive and other reasons, all such exchanges have been done simultaneously, to avoid the possibility of a donor becoming unwilling or unable to donate a kidney after that donor's intended patient has already received a kidney from another patient's donor. So, one form that congestion takes in organizing kidney exchanges is that multiple operating rooms and surgical teams have to be assembled. (A simultaneous exchange between two pairs requires four operating rooms and surgical teams, two for the nephrectomies that remove the donor kidneys, and two for the transplantations that immediately follow. An exchange involving three pairs involves six operating rooms and teams, etc.) Roth et al. (2004a) noted that large exchanges would arise relatively infrequently, but could pose logistical difficulties.

[11] See Rapoport (1986), Ross et al. (1997), Ross and Woodle (2000), for some early discussion of the possibility of kidney exchange, and Delmonico (2004), and Montgomery et al. (2005) for some early reports of successful exchanges.

These logistical difficulties loomed large in our early discussions with surgeons, and out of those discussions came the analysis in Roth et al. (2005a) of how kidney exchanges might be organized if only two-way exchanges were feasible. The problem of two-way exchanges can be modeled as a classic problem in graph theory, and, subject to the constraint that exchanges involve no more than two pairs, efficient outcomes with good incentive properties can be found in computationally efficient ways. When the New England Program for Kidney Exchange was founded in 2004 (Roth et al., 2005b), it used the matching software that had had been developed to run the simulations in Roth et al. (2005a,b), and it initially attempted only two-way matches (while keeping track of the potential three-way matches that were missed). This was also the case when Sönmez, Ünver and I started running matches for the Ohio-based consortium of transplant centers that eventually became the Alliance for Paired Donation.[12]

However, some transplants are lost that could have been accomplished if three-way exchanges were available. In Saidman et al. (2006) and in Roth et al. (2007), we showed that to get close to the efficient number of transplants, the infrastructure to perform both two-way and three-way exchanges would have to be developed, but that once the population of available patient–donor pairs was large enough, few transplants would be missed if exchanges among more than three pairs remained difficult to accomplish. Both the New England Program for Kidney Exchange and the Alliance for Paired Donation have since taken steps to be able to accommodate three-way as well as two-way exchanges. Being able to deal with the (six operating room) congestion required to accomplish three-way exchanges has the effect of making the market thicker, since it creates more exchange possibilities.

As noted above, another way to make the market thicker is to integrate exchange between pairs with list exchange, so that exchange chains can be considered, as well as cycles. This applies as well to how the growing numbers of non-directed (altruistic) donors are used. A non-directed (ND) donor is someone who wishes to donate a kidney without having a particular patient in mind (and whose donor kidney therefore does not require another donor kidney in exchange). The traditional way to utilize such ND donors was to have them donate to someone on the cadaver waiting list. But as exchanges have started to operate, it has now become practical to have the ND donor donate to some pair that is willing to exchange a kidney, and have that pair donate to someone on the cadaver waiting list. Roth et al. (2006) report on how and why such exchanges are now done in New England. As in traditional exchange, all surgeries are conducted simultaneously, so there are logistical limits on how long a chain is feasible.

---

[12] The New England Program for Kidney Exchange has since integrated our software into theirs, and conducts its own matches. The Alliance for Paired Donation originally used our software, and as the size of the exchange pool grew, the integer programming algorithms were written in software that can handle much larger numbers of pairs (Abraham et al., 2007). The papers by Roth et al. (2005a,b) were also widely distributed to transplant centers (as working papers in 2004). The active transplant program at Johns Hopkins has also begun to use software similar in design to that in Roth et al. (2004b, 2005a) to optimize pairwise matches (see Segev et al., 2005).

But we noted that, when a chain is initiated by a ND donor, it might be possible to relax the constraints that all parts of the exchange be simultaneous, since

> If something goes wrong in subsequent transplants and the whole ND-chain cannot be completed, the worst outcome will be no donated kidney being sent to the waitlist and the ND donation would entirely benefit the KPD [kidney exchange] pool. (Roth et al., 2006, p. 2704)

That is, if a conventional exchange were done in a non-simultaneous way, and if the exchange broke down after some patient–donor pair had donated a kidney but before they had received one, then that pair would not only have lost the promised transplant, but also have lost a healthy kidney. In particular, the patient would no longer be in position to exchange with other incompatible patient–donor pairs. But in a chain that begins with a ND donor, if the exchange breaks down before the donation to some patient–donor pair has been made (because the previous donor in the chain becomes unwilling or unable to donate), then the pair loses the promised transplant, but is no worse off than they were before the exchange was planned, and in particular they can still exchange with other pairs in the future. So, while a non-simultaneous ND chain of donations could create an incentive to break the chain, the costs of a breach would be less than in a pure exchange, and so the benefits (in terms of longer chains) are worth exploring. The first such non-simultaneous "never ending" altruistic donor (NEAD) chain was begun by the Alliance for Paired Donation in July 2007. A week after the first patient was transplanted from an altruistic (ND) donor, her husband donated a kidney to another patient, whose mother later donated her kidney to a third patient, whose daughter donated (simultaneously) to a fourth patient, whose sister is, as I write, now waiting to donate to another patient whose incompatible donor will be willing to "pass it forward" (Rees et al., 2009a). [13]

To summarize the progress to date, the big problem facing kidney exchange prior to 2004 was the lack of thickness in the market, so that incompatible patient–donor pairs were left in the difficult search for what Jevons (1876) famously described as a double coincidence of wants (Roth et al., 2007). By building a database of incompatible patient–donor pairs and their relevant medical data, it became possible to arrange more transplants, using a clearinghouse to maximize the number (or to achieve some quality- or priority-adjusted number) of transplants subject to various constraints. The state of the art now involves both two-way and three-way cyclical exchanges and a variety of chains, either ending with a donation to someone on the cadaver waiting list or beginning with an altruistic ND donor, or both. While large simultaneous exchanges remain logistically infeasible, the fact that almost all efficient exchanges can be accomplished in cycles of no more than three pairs, together with clearinghouse technology that can efficiently

---

[13] Increasing the number of patients who benefit from the altruism of a ND donor may also increase the willingness of such donors to come forward. After publicity of the first NEAD chain on ABC *World News Tonight*, July 26, 2007 (see <http://utoledo.edu/utcommcenter/kidney>), the Alliance for Paired Donation has had over 100 registrations on its website of people who are offering to be altruistic living ND donors (Rees, personal communication).

find such sets of exchanges, substantially reduces the problem of congestion in carrying out exchanges. And, for chains that begin with ND donors, the early evidence is that some relaxation of the incentive constraint that all surgeries be simultaneous seems to be possible.[14]

There remain some challenges to further advancing kidney exchange that are also related to thickness, congestion, and incentives.

Some patients have many antibodies, so that they will need very many possible donors to find one who is compatible. For that reason and others, it is unlikely that purely regional exchanges, such as presently exist, will provide adequate thickness for all the gains from exchange to be realized. Legislation has recently been passed in the US House and Senate to remove a potential legal obstacle to a national kidney exchange.[15] Aside from expanding kidney exchange to national scale, another way to increase the thickness of the market would be to make kidney exchange available not just to incompatible patient–donor pairs, but also to those who are compatible but might nevertheless benefit from exchange.[16]

While some of the congestion in terms of actually conducting transplants has been addressed, there is still congestion associated with the time it takes to test for immuno-logical incompatibility between patients and donors who (based on available tests) are matched to be part of an exchange. That is, antibody production can vary over time, and so a patient and donor who appear to be compatible in the database may not in fact be. Because it now sometimes takes weeks to establish this, during which time other exchanges may go forward, some exchanges are missed that could have been accomplished if the tests for compatibility were done more quickly, so that the overall pattern of exchanges could have been adjusted.

And as regional exchanges have grown to include multiple transplant centers, a new issue has come to the fore concerning how kidney exchange should be organized to give transplant centers the incentive to inform the central exchange of all of their incompati-ble patient–donor pairs. Consider a situation in which transplant center A has two pairs who are mutually compatible, so that it could perform an in-house exchange between

[14] The Postscript describes how non-simultaneous chains have indeed come to play a very large role in kidney exchange.

[15] The proposed bill (HR 710, introduced on January 29, 2007 and passed in the House on March 7, 2007, and S 487, introduced on February 1, 2007 and passed in the Senate February 15, 2007) is "To amend the National Organ Transplant Act to clarify that kidney paired donations shall not be considered to involve the transfer of a human organ for valuable consideration." Kidney exchange is also being organized in the UK; see <http://www.uktransplant.org.uk/ukt/about_transplants/organ_allocation/kidney_(renal)/living_donation/paired_donation_matching_scheme.jsp>. The first British exchange was carried out on July 4, 2007 (see the BBC report at <http://news.bbc.co.uk/1/hi/health/7025448.stm>.

[16] For example, a compatible middle-aged patient-donor pair, and an incompatible patient-donor pair in which the donor is a twenty-five-year-old athlete could both benefit from exchange. Aside from increasing the number of pairs available for exchange, this would also relieve the present shortage of donors with blood type O in the kidney exchange pool, caused by the fact that O donors are only rarely incompatible with their intended recipient. Simulations on the robust effects of adding compatible patient-donor pairs to the exchange pool are found in Roth et al. (2004a, 2005b), and in Gentry et al. (2007).

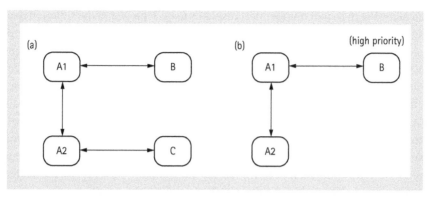

FIGURE 1.1. Potential kidney exchanges between patient–donor pairs at multiple centers. Double-headed arrows indicate that the connected pairs are compatible for exchange, i.e. the patient in one pair is compatible with the donor in the other. Pairs A1 and A2 are both from transplant center A; pairs B and C are from different transplant centers. Transplant center A, which sees only its own pairs, can conduct an exchange among its pairs A1 and A2 since they are compatible, and, if it does so, this will be the only exchange, resulting in two transplants. However, if in Figure 1a transplant center A makes its pairs available for exchange with other centers, then the exchanges will be A1 with B and A2 with C, resulting in four transplants. However, in Figure 1b the suggested exchange might be A1 with B, which would leave the patient in A2 without a transplant. Faced with this possibility (and not knowing if the situation is as in 1a or 1b) transplant center A might choose to transplant A1 and A2 by itself, without informing the central exchange.

these two pairs. If the mutual compatibilities are as shown in Figure 1.1a, then if these two pairs exchange with each other, only those two transplants will be accomplished. If instead the pairs from transplant center A were matched with the pairs from the other centers, as shown in Figure 1.1a, four transplants could be accomplished (via exchanges of pair A1 with pair B, and pair A2 with C).

But, note that if the situation had been that of Figure 1.1b, then transplant center A runs the risk that if it informs the central exchange of its pairs, then the recommended exchange will be between A1 and B, since B has high priority (e.g. B is a child). This would mean that pair A2 did not get a kidney, as they would have if A1 and A2 had exchanged in-house. So, the situation facing transplant center A, not knowing what pairs will be put forward for exchange by the other transplant centers, is that it can assure itself of doing two transplants for its patients in pairs A1 and A2, but it is not guaranteed two transplants if it makes the pairs available for exchange and the situation is as in Figure 1.1b. If this causes transplant centers to withhold those pairs they can transplant by themselves, then a loss to society results where the situation is as in Figure 1.1a. (In fact, if transplant centers withhold those pairs they can exchange in-house, then primarily hard-to-match pairs will be offered for exchange, and the loss will be considerable.)

One remedy is to organize the kidney exchange clearinghouse in a way that guarantees center A that any pairs it could exchange in-house will receive transplants. This would allow the maximal number of transplants to be achieved in the situation depicted in Figure 1.1a, and it would mean that in the situation depicted in Figure 1.1b the exchange between A1 and A2 would be made (and so the high-priority pair B would not participate in exchange, just as they would not have if pairs A1 and A2 had not been put forward). This is a bit of a hard discussion to have with surgeons, who find it repugnant that, for example, the child patient in pair B would receive lower priority than pairs A1 and A2 just because of the accident that they were mutually compatible and were being treated at the same transplant center. (Needless to say, if transplant center A withholds its pairs and transplants them in-house, they effectively have higher priority than pair B, even if no central decision to that effect has been made.) But this is an issue that will have to be resolved, because the full participation of all transplant centers substantially increases the efficiency of exchange.

Note that, despite all the detailed technical particulars that surround the establishment of kidney exchange programs, and despite the absence of money in the kidney exchange market, we can recognize some of the basic lessons of market design that were also present in designing labor market clearinghouses. The first issue was making the market thick, by establishing a database of patient–donor pairs available to participate in exchange. Then issues of congestion had to be dealt with, so that the clearinghouse could identify exchanges involving sufficiently few pairs (initially two, now three) for transplants to be done simultaneously. Simultaneity is related to making sure that everyone involved in an exchange never has an incentive not to go forward with it, but as exchanges have grown to include multiple transplant centers, there are also incentive issues to be resolved in making it safe for a transplant center to enroll all of its eligible pairs in the central exchange.

School choice

Another important class of allocation problems in which no money changes hands is the assignment of children to big-city public schools, based both on the preferences of students and their families, and on the preferences of schools, or on city priorities. Because public school students must use whatever system local authorities establish, establishing a thick market is not the main problem facing such systems. (Although how well a school choice system works may influence how many children ultimately attend city schools.) But how well a school choice system works still has to do with how effectively it deals with congestion, and how safe it makes it for families to straightforwardly reveal their preferences.

My colleagues and I were invited to help design the current New York City (NYC) high-school choice program, chiefly because of problems the old decentralized system had in dealing with congestion. In Boston we were invited to help design the current school choice system because the old system, which was itself a centralized clearinghouse, did not make it safe for families to state their preferences.[17] In both Boston and NYC the newly designed systems incorporate clearinghouses to which students (and, in NYC, schools) submit preferences. Although another alternative was considered in Boston, both Boston and NYC adopted clearinghouses similar to the kinds of stable clearinghouses used in medical labor markets (powered by a student-proposing deferred acceptance algorithm), adapted to the local situations. For my purpose in the present chapter, I'll skip any detailed discussion of the clearinghouse designs, except to note that they make it safe for students and families to submit their true preferences. Instead, I'll describe briefly what made the prior school choice systems congested or risky.[18]

In NYC, well over 90,000 students a year must be assigned to over 500 high-school programs. Under the old system, students were asked to fill out a rank order list of up to five programs. These lists were then copied and sent to the schools. Subject to various constraints, schools could decide which of their applicants to accept, waitlist, or reject. Each applicant received a letter from the NYC Department of Education with the decisions of the schools to which she or he had applied, and applicants were allowed to accept no more than one offer, and one waitlist. This process was repeated: after the responses to the first letter were received, schools with vacant positions could make new offers, and after replies to the second letter were received, a third letter with new offers was sent. Students not assigned after the third step were assigned to their zoned schools, or assigned via an administrative process. There was an appeals process, and an "over the counter" process for assigning students who had changed addresses, or were otherwise unassigned before school began.

Three rounds of processing applications to no more than five out of more than 500 programs by almost 100,000 students was insufficient to allocate all the students. That is, this process suffered from congestion (in precisely the sense explored in Roth and Xing, 1997): not enough offers and acceptances could be made to clear the market. Only about 50,000 students received offers initially, about 17,000 of whom received multiple offers. And when the process concluded, approximately 30,000 students had been assigned to a school that was nowhere on their choice list.

Three features of this process particularly motivated NYC Department of Education's desire for a new matching system. First were the approximately 30,000 students not assigned to a school they had chosen. Second, students and their families had to be strategic in their choices. Students who had a substantial chance of being rejected by their true first-choice school had to think about the risk of listing it first, since, if one of their lower-choice schools took students' rankings into account in deciding on admissions, they might have done better to list it first. (More on this in a

---

[17] The invitation to meet with Boston Public Schools came after a newspaper story recounted the difficulties with the Boston system, as described in Abdulkadiroğlu and Sönmez (2003). For subsequent explorations of the old Boston system, see Chen and Sönmez (2006), Ergin and Sönmez (2006), Pathak and Sönmez (2008), and Abdulkadiroğlu et al. (2007).

[18] The description of the situation in NYC is from Abdulkadiroğlu et al. (2005a); for Boston see Abdulkadiroğlu and Sönmez (2003), Abdulkadiroğlu et al. (2005b, 2007).

moment, in the discussion of Boston schools.) Finally, the many unmatched students, plus those who may not have indicated their true preferences (and the consequent instability of the resulting matching) gave schools an incentive to be strategic: a substantial number of schools managed to conceal capacity from the central administration, thus preserving places that could be filled later with students unhappy with their assignments.

As soon as NYC adopted a stable clearinghouse for high-school matching (in 2003, for students entering high school in 2004), the congestion problem was solved; only about 3,000 students a year have had to be assigned administratively since then, down from 30,000 (and many of these are students who for one reason or another fail to submit preference lists). In addition, in the first three years of operation, schools learned that it was no longer profitable to withhold capacity, and the resulting increase in the availability of places in desirable schools resulted in a larger number of students receiving their first choices, second choices, and so forth from year to year. Finally, as submitted rank order lists have begun to more reliably reflect true preferences, these have begun to be used as data for the politically complex process of closing or reforming undesirable schools (Abdulkadiroğlu et al., 2005a, 2009).

In Boston, the problem was different. The old school choice system there made it risky for parents to indicate their true first-choice school if it was not their local school. The old system was simple in conception: parents ranked schools, and the algorithm tried to give as many families as possible their first-choice school. Where the capacity of a school was less than the number of students who ranked it first, priority was given to students who had siblings in the school, or who lived within walking distance, or, finally, who had been assigned a good lottery number. After these assignments were made, the algorithm tried to match as many remaining students as possible with their second-choice school, and so on. The difficulty facing families was that, if they ranked a popular school first and weren't assigned to it, they might find that by the time they were considered for their second-choice school, it was already filled with people who had ranked it first. So, a family who had a high priority for their second-choice school (e.g. because they lived close to it), and could have been assigned to it if they had ranked it first, might no longer be able to get in if they ranked it second.

As a consequence, many families were faced with difficult strategic decisions, and some families devoted considerable resources to gathering relevant information about the capacities of schools, how many siblings would be enrolling in kindergarten, etc. Other families were oblivious to the strategic difficulties, and sometimes suffered the consequences; if they listed popular schools for which they had low priority, they were often assigned to schools they liked very little.

In Boston, the individual schools are not actors in the school choice process, and so there was a wider variety of mechanisms to choose from than in New York. My colleagues and I recommended two possibilities that were *strategy-proof* (in the sense that they make it a dominant strategy for students and families to submit their true preferences), and which thus would make it safe for students to submit their true preferences

(Abdulkadiroğlu et al., 2005b, 2007).[19] This proved to be decisive in persuading the Boston School Committee to adopt a new algorithm. Then Superintendent of Schools, Thomas Payzant, wrote, in a 2005 memo to the School Committee:

> The most compelling argument for moving to a new algorithm is to enable families to list their true choices of schools without jeopardizing their chances of being assigned to any school by doing so.

Superintendent Payzant further wrote:

> A strategy-proof algorithm levels the playing field by diminishing the harm done to parents who do not strategize or do not strategize well.

Making the school choice system safe to participate in was critical in the decision of Boston public schools to move from a clearinghouse that was not strategy-proof to one that was. Different issues of safety were critical in the market for gastroenterologists, discussed next.

GASTROENTEROLOGISTS[20]

An American medical graduate who wishes to become a gastroenterologist first completes three years of residency in internal medicine, and then applies for a job as a fellow in gastroenterology, a subspecialty of internal medicine.[21] The market for gastroenterology fellows was organized via a stable labor market clearinghouse (a "match") from 1986 through the late 1990s, after which the match was abandoned (following an unexpected shock to the supply and demand for positions in 1996; see McKinney et al., 2005). This provided an opportunity to observe the unraveling of a market as it took place. From the late 1990s until 2006, offers of positions were made increasingly far in advance of employment (moving back to almost two years in advance, so that candidates were often being interviewed early in their second year of residency). Offers also became dispersed in time, and short in duration, so that candidates faced a thin market. One consequence was that the market became much more local than it had been, with gastroenterology fellows more likely to be recruited at the same hospital at which they had worked as a resident (Niederle and Roth, 2003b; Niederle et al., 2006).

Faced with these problems, the various professional organizations involved in the market for gastroenterology fellows agreed to try to resume using a centralized

---

[19] In addition to the student-proposing deferred acceptance algorithm that was ultimately adopted, we proposed a variation of the "top trading cycles" algorithm originally explored by Shapley and Scarf (1974), which was shown to be strategy-proof by Roth (1982b), and which was extended, and explored in a school choice context, by Abdulkadiroğlu and Sönmez (1999, 2003).

[20] A much more thorough treatment of the material in this section is given in Niederle and Roth (2009b).

[21] The American system of residents and fellows is similar but not precisely parallel to the system in the UK of house officers and registrars, which has also recently faced some problems of market design.

clearinghouse, to be operated one year in advance of employment. However, this raised the question of how to make it safe for program directors and applicants to wait for the clearinghouse, which would operate almost a year later than hiring had been accomplished in the immediate past. Program directors who wanted to wait for the match worried that if their competitors made early offers, then applicants would lose confidence that the match would work and consequently would accept those early offers. That is, in the first year of a match, applicants might not yet feel safe to reject an early offer in order to wait for the match. Program directors who worried about their competitors might thus be more inclined to make early offers themselves.

The gastroenterology organizations did not feel able to directly influence the hiring behavior of programs that might not wish to wait for the match. Consequently we recommended that policies be adopted that would allow applicants who wished to wait for the match to more effectively deal with early offers themselves (Niederle et al., 2006). We modeled our recommendation on the policies in place in the American market for graduate school admission. In this market, a policy (adopted by the large majority of universities) states that offers of admission and financial support to graduate students should remain open until April 15.

> Students are under no obligation to respond to offers of financial support prior to April 15; earlier deadlines for acceptance of such offers violate the intent of this Resolution. In those instances in which a student accepts an offer before April 15, and subsequently desires to withdraw that acceptance, the student may submit in writing a resignation of the appointment at any time through April 15. However, an acceptance given or left in force after April 15 commits the student not to accept another offer without first obtaining a written release from the institution to which a commitment has been made. Similarly, an offer by an institution after April 15 is conditional on presentation by the student of the written release from any previously accepted offer. It is further agreed by the institutions and organizations subscribing to the above Resolution that a copy of this Resolution should accompany every scholarship, fellowship, traineeship, and assistantship offer." (See <http://www.cgsnet.org/april-15-resolution)>

This of course makes early exploding offers much less profitable. A program that might be inclined to insist on an against-the-rules early response is discouraged from doing so, because they can't "lock up" a student to whom they make such an offer, because accepting such an offer does not prevent the student from later receiving and accepting a preferred offer.[22]

A modified version of this policy was adopted by all four major gastroenterology professional organizations, the American Gastroenterological Association (AGA), the American College of Gastroenterology (ACG), the American Society for

---

[22] Niederle and Roth (2009a) study in the laboratory the impact of the rules that govern the types of offers that can be made (with or without a very short deadline) and whether applicants can change their minds after accepting an early offer. In the uncongested laboratory environments we studied, eliminating the possibility of making exploding offers, or making early acceptances non-binding, prevents the markets from operating inefficiently early.

Gastrointestinal Endoscopy (ASGE), and the American Association for the Study of Liver Diseases (AASLD), regarding offers made before the (new) match. The resolution states, in part:

> The general spirit of this resolution is that each applicant should have an opportunity to consider all programs before making a decision and be able to participate in the Match. ... It therefore seeks to create rules that give both programs and applicants the confidence that applicants and positions will remain available to be filled through the Match and not withdrawn in advance of it. This resolution addresses the issue that some applicants may be persuaded or coerced to make commitments prior to, or outside of, the Match. ... Any applicant may participate in the matching process ... by ... resigning the accepted position if he/she wishes to submit a rank order list of programs. ... The spirit of this resolution is to make it unprofitable for program directors to press applicants to accept early offers, and to give applicants an opportunity to consider all offers. ...

The gastroenterology match for 2007 fellows was held on June 21, 2006, and succeeded in attracting 121 of the 154 eligible fellowship programs (79%). Of the positions offered in the match, 98% were filled through the match, and so it appears that the gastroenterology community succeeded in making it safe to participate in the match, and thus in changing the timing and thickness of the market, while using a clearinghouse to avoid congestion.

The policies adopted by gastroenterologists prior to their match make clear that market design in this case consists not only of the "hardware" of a centralized clearinghouse, but also of the rules and understandings that constitute elements of "market culture." This leads us naturally to consider how issues of timing, thickness, and congestion are addressed in a market that operates without any centralized clearinghouse.

MARKET FOR NEW ECONOMISTS

The North American market for new PhDs in economics is a fairly decentralized one, with some centralized marketplace institutions, most of them established by the American Economics Association (AEA).[23] Some of these institutions are of long standing, while others have only recently been established. Since 2005 the AEA has had an Ad Hoc Committee on the Job Market, charged with considering ways in which the market for economists might be facilitated.[24]

---

[23] This is not a closed market, as economics departments outside North America also hire in this market, and as American economics departments and other employers often hire economists educated elsewhere. But a large part of the market involves new American PhDs looking for academic positions at American colleges and universities. See Cawley (2006) for a description of the market aimed at giving advice to participants, and Siegfried and Stock (2004) for some descriptive statistics.

[24] At the time of writing its members were Alvin E. Roth (chair), John Cawley, Philip Levine, Muriel Niederle, and John Siegfried, and the committee had received assistance from Peter Coles, Ben Greiner, and Jenna Kutz.

Roughly speaking, the main part of this market begins each year in the early fall, when economics departments advertise for positions. Positions may be advertised in many ways, but a fairly complete picture of the academic part of the market can be obtained from the AEA's monthly publication *Job Openings for Economists* (JOE), which provides a central location for employers to advertise and for job seekers to see who is hiring (<http://www.aeaweb.org/joe>). Graduate students nearing completion of their PhDs answer the ads by sending applications, which are followed by letters of reference, most typically from their faculty advisors.[25]

Departments often receive several hundred applications (because it is easy for applicants to apply to many schools), and junior recruiting committees work through the late fall to read applications, papers, and letters, and to seek information through informal networks of colleagues, to identify small subsets of applicants they will invite for half-hour preliminary interviews at the annual AEA meeting in early January. This is part of a very large annual set of meetings, of the Allied Social Science Associations (ASSA), which consist of the AEA and almost fifty smaller associations. Departments reserve suites for interviewing candidates at the meeting hotels, and young economists in new suits commute up and down the elevators, from one interview to another, while recruiting teams interview candidates one after the other, trading off with their colleagues throughout long days. While the interviews in hotel suites are normally prearranged in December, the meetings also host a spot market, in a large hall full of tables, at which both academic and non-academic employers can arrange at the last minute to meet with candidates. The spot market is called the Illinois Skills Match (because it is organized in conjunction with the Illinois Department of Employment Security).

These meetings make the early part of the market thick, by providing an easy way for departments to quickly meet lots of candidates, and by allowing candidates to efficiently introduce themselves to many departments. This largely controls the starting time of the market.[26] Although a small amount of interviewing goes on beforehand, it is quite rare to hear of departments that make offers before the meetings, and even rarer to hear of departments pressing candidates for replies before the meetings.[27]

---

[25] These applications are usually sent through the mail, but now often also via email and on webpages set up to receive them. Applicants typically apply to departments individually, by sending a letter accompanied by their curriculum vitae and job market paper(s) and followed by their letters of reference. Departments also put together "packages" of their graduating students who are on the market, consisting of curricula vitae, job market papers, and letters of reference, and these are sent by mail and/or posted on department websites (without the letters of reference). In 2007 a private organization, EconJobMarket.org, offered itself as a central repository of applications and letters of reference on the web. The European Economics Association in collaboration with the Asociación Española de Economía has initiated a similar repository at <http://jobmarketeconomist.com>.

[26] The situation is different in Europe, for example, where hiring is more dispersed in time. In an attempt to help create a thicker European market, the Royal Economic Society held a "PhD presentations event" for the first time in late January 2006. Felli and Sutton (2006) remark that "The issue of timing, unsurprisingly, attracted strong comment...."

[27] While the large-scale interviewing at the annual meetings has not been plagued by gradual unraveling, some parts of the market have broken off. In the 1950s, for example, the American Marketing Association used to conduct job market meetings at the time of the ASSA meetings, but for a

But while the preliminary interviewing part of the market is thick, it is congested. A dedicated recruiting committee might be able to interview thirty candidates, but not a hundred, and hence can meet only a small fraction of the available applicants. Thus the decision of whom to interview at the meetings is an important one, and for all but elite schools a strategic one as well. That is, while a few departments at the top of the pecking order can simply interview the candidates they like best, a lower-ranked department that uses all its interview slots to interview the same candidates who are interviewed by the elite schools is likely to find that it cannot convert its initial interviews into new faculty hires. Thus most schools have to give at least some thought not only to how much they like each candidate, but to how likely it is that they can successfully hire that candidate. This problem is only made more difficult by the fact that students can easily apply for many positions, so the act of sending an application does not itself send a strong signal of how interested the candidate might be. The problem may be particularly acute for schools in somewhat special situations, such as liberal arts colleges, or British and other non-American universities in which English is the language of instruction, since these may be concerned that some students who strongly prefer positions at North American research universities may apply to them only as insurance.

Following the January meetings, the market moves into a less organized phase, in which departments invite candidates for "flyouts," day-long campus visits during which the candidate will make a presentation and meet a substantial portion of the department faculty and perhaps a dean. Here, too, the market is congested, and departments can fly out only a small subset of the candidates they have interviewed at the meetings, because of the costs of various sorts.[28] This part of the market is less well coordinated in time: some departments host flyouts in January, while others wait until later. Some departments try to complete all their flyouts before making any offers, while others make offers while still interviewing. And some departments make offers that come with moderate deadlines of two weeks or so, which may nevertheless force candidates to reply to an offer before knowing what other offers might be forthcoming.[29]

By late March, the market starts to become thin. For example, a department that interviewed twenty people at the meetings, invited six for flyouts, made offers to two, and was rejected by both, may find that it is now difficult to assess which candidates it did not interview may still be on the market. Similarly, candidates whose interviews

long time it has held its job market in August, a year before employment will begin, with the result that assistant professors of marketing are often hired before having made as much progress on their dissertations as is the case for economists (Roth and Xing, 1994).

[28] These costs arise not only because budgets for airfares and hotels may be limited, but also because faculties quickly become fatigued after too many seminars and recruiting dinners.

[29] In 2002 and 2003 Georg Weizsacker, Muriel Niederle, Dorothea Kubler, and I conducted surveys of economics departments regarding their hiring practices, asking in particular about what kinds of deadlines, if any, they tended to give when they made offers to junior candidates. Loosely speaking, the results suggested that departments that were large, rich, and elite often did not give any deadlines (and sometimes were able to make all the offers they wanted to make in parallel, so that they would not necessarily make new offers upon receiving rejections). Less well endowed departments often gave candidates deadlines, although some were in a position to extend the deadline for candidates who seemed interested but needed more time.

and flyouts did not result in job offers may find it difficult to know which depart-ments are still actively searching. To make the late part of the market thicker, the first thing our AEA job market committee did was to institute a "scramble" web-page through which departments with unfilled positions and applicants still on the market could identify each other (see *Guide to the Economics Job Market Scramble* at <http://www.aeaweb.org/joe/scramble/guide.pdf>). For simplicity, the scramble web-page was passive (i.e. it didn't provide messaging or matching facilities): it simply announced the availability of any applicant or department who chose to register. The scramble webpage operated for the first time in the latter part of the 2005–06 job market, when it was open for registrants between March 15 and 20, and was used by 70 employers and 518 applicants (of whom only about half were new, 2006 PhDs). It was open only briefly, so that its information provided a snapshot of the late market, which didn't have to be maintained to prevent the information from becoming stale.

The following year our committee sought to alleviate some of the congestion sur-rounding the selection of interview candidates at the January meetings, by introducing a signaling mechanism through which applicants could have the AEA transmit to no more than two departments a signal indicating their interest in an interview at the meetings. The idea was that, by limiting applicants to two signals, each signal would have some information value that might not be contained merely in the act of sending a department an application, and that this information might be helpful in averting coor-dination failures.[30] The signaling mechanism operated for the first time in December 2006, and about 1,000 people used it to send signals. [31]

---

[30] For a simple conceptual example of how a limited number of signals can improve welfare, consider a market with two applicants and two employers, in which there is only time for each employer to make one offer, and each applicant can take at most one position. Even if employers and applicants wish only to find a match, and have no preference with whom they match, there is a chance for signals to improve welfare by reducing the likelihood of coordination failure. In the absence of signals, there is a symmetric equilibrium in which each firm makes an offer to each worker with equal probability, and at this equilibrium, half the time one worker receives two offers, and so one worker and one employer remain unmatched. If the workers are each permitted to send one signal beforehand, and if each worker sends a signal to each firm with equal probability, then if firms adopt the strategy of making an offer to an applicant who sends them a signal, the chance of coordination failure is reduced from one-half to one-quarter. If workers have preferences over firms, the welfare gains from reducing coordination failure can be even larger. For recent treatments of signaling and coordination, see Coles et al. (forthcoming), Lee and Schwarz (2007a,b), Lien (2007), and Stack (2007). See also Abdulkadiroğlu et al. (2011), who discuss allowing applicants to influence tie-breaking by signaling their preferences in a centralized clearinghouse that uses a deferred acceptance algorithm.

[31] The document "Signaling for Interviews in the Economics Job Market," at <http://www.aeaweb.org/joe/signal/signaling.pdf> includes the following advice:

"**Advice to Departments**: Applicants can only send two signals, so if a department *doesn't* get a signal from some applicant, that fact contains almost no information. (See advice to applicants, below, which suggests how applicants might use their signals). But because applicants can send only two signals, the signals a department *does* receive convey valuable information about the candidate's interest." "A department that has more applicants than it can interview can use the signals to help break ties for interview slots, for instance. Similarly, a department that receives applications from some candidates who it thinks are unlikely to really be interested (but might be submitting many applications out of excessive risk aversion) can be reassured of the candidate's interest if the department receives one of the

Both the scramble and the signaling facility attracted many users, although it will take some time to assess their performance. Like the JOE and the January meetings, they are marketplace institutions that attempt to help the market provide thickness and deal with congestion.

DISCUSSION

In the tradition of market design, I have concentrated on the details of particular markets, from medical residents and fellows to economists, and from kidney exchange to school choice. But, despite their very different details, these markets, like others, struggle to provide thickness, to deal with the resulting congestion, and to make it safe and relatively simple to participate. While the importance of thick markets has been understood by economists for a long time, my impression is that issues of congestion, safety, and simplicity were somewhat obscured when the prototypical market was thought of as a market for a homogeneous commodity.[32]

Thickness in a market has many of the properties of a public good, so it is not surprising that it may be hard to provide it efficiently, and that free riders have to be resisted, whether in modern markets with a tendency to unravel, or in medieval markets with rules against "forestalling." Notice that providing thickness blurs the distinction between centralized and decentralized markets, since marketplaces—from traditional farmers' markets, to the AEA job market meetings, to the New York Stock Exchange—provide thickness by bringing many participants to a central place. The possibility of having the market perform other centralized services, as clearinghouses or signaling mechanisms do, has only grown now that such central places can also be electronic, on the Internet or elsewhere. And issues of thickness become if anything more important when there are network externalities or other economies of scope.[33]

---

candidate's two signals. A department that receives a signal from a candidate will likely find it useful to open that candidate's dossier and take one more look, keeping in mind that the candidate thought it worthwhile to send one of his two signals to the department."

"**Advice to Applicants**: The two signals should **not** be thought of as indicating your top two choices. Instead, you should think about which two departments that you are interested in would be likely to interview you if they receive your signal, but not otherwise (see advice to departments, above). You might therefore want to send a signal to a department that you like but that might otherwise doubt whether they are likely to be able to hire you. Or, you might want to send a signal to a department that you think might be getting many applications from candidates somewhat similar to you, and a signal of your particular interest would help them to break ties. You might send your signals to departments to whom you don't have other good ways of signaling your interest."

[32] Establishing thickness, in contrast, is a central concern even in financial markets; see for example the market design ("market microstructure") discussions of how markets are organized at their daily openings and closings, such as Biais et al. (1999) on the opening call auction in the Paris Bourse and Kandel et al. (2007) on the closing call auctions in the Borsa Italiana and elsewhere.

[33] Thickness has received renewed attention in the context of software and other "platforms" that serve some of the functions of marketplaces, such as credit cards, which require large numbers of both

Congestion is especially a problem in markets in which transactions are heterogeneous, and offers cannot be made to the whole market. If transactions take even a short time to complete, but offers must be addressed to particular participants (as in offers of a job, or to purchase a house), then someone who makes an offer runs the risk that other opportunities may disappear while the offer is being considered. And even financial markets (in which offers can be addressed to the whole market) experience congestion on days with unusually heavy trading and large price movements, when prices may change significantly while an order is being processed, and some orders may not be able to be processed at all. As we have seen, when individual participants are faced with congestion, they may react in ways that damage other properties of the market, for example if they try to gain time by transacting before others.[34]

Safety and simplicity may constrain some markets differently than others. Parents engaged in school choice may need more of both than, say, bidders in very-high-value auctions of the sort that allow auction experts to be hired as consultants. But even in billion-dollar spectrum auctions, there are concerns that risks to bidders may deter entry, or that unmanageable complexity in formulating bids and assessing opportunities at each stage may excessively slow the auction.[35] Somewhere in between, insider trading laws with criminal penalties help make financial markets safe for non-insiders to participate. And if it is risky to participate in the market, individual participants may try to manage their risk in ways that damage the market as a whole, such as when transplant centers withhold patients from exchange, or employers make exploding offers before applicants can assess the market, or otherwise try to prevent their trading counterparties from being able to receive other offers.[36]

In closing, market design teaches us both about the details of market institutions and about the general tasks markets have to perform. Regarding details, the word "design" in "market design" is not only a verb, but also a noun, so economists can help to design some markets, and profitably study the design of others. And I have argued in this chapter that among the general tasks markets have to perform, difficulties in providing

consumers and merchants; see for example Evans and Schmalensee (1999) and Evans et al. (2006); and see Rochet and Tirole (2006), who concentrate on how the price structure for different sides of the market may be an important design feature.

[34] The fact that transactions take time may in some markets instead inspire participants to try to transact very late, near the market close, if that will leave other participants with too little time to react. See for example the discussion of very late bids ("sniping") on eBay auctions in Roth and Ockenfels (2002), and Ariely et al. (2005).

[35] Bidder safety lies behind discussions both of the "winner's curse" and collusion (cf. Kagel and Levin 2002; Klemperer, 2004), as well as of the "exposure problem" that faces bidders who wish to assemble a package of licenses in auctions that do not allow package bidding (see e.g. Milgrom, 2007). And simplicity of the auction format has been addressed in experiments prior to the conduct of some (U.S.) Federal Communications Commission (FCC) auctions (see e.g. Plott, 1997). Experiments have multiple uses in market design, not only for investigation of basic phenomena, and small-scale testing of new designs, but also in the considerable amount of explanation, communication, and persuasion that must take place before designs can be adopted in practice.

[36] For example, Roth and Xing (1994) report that in 1989 some Japanese companies scheduled recruiting meetings on the day an important civil service exam was being given, to prevent their candidates from also applying for government positions.

thickness, dealing with congestion, and making participation safe and simple are often at the root of market failures that call for new market designs.

I closed my 1991 *Economic Journal* article (quoted in the introduction) on a cautiously optimistic note that, as a profession, we would rise to the challenge of market design, and that doing so would teach us important lessons about the functioning of markets and economic institutions. I remain optimistic on both counts.

POSTSCRIPT 2012: WHAT HAVE WE LEARNED FROM MARKET DESIGN LATELY?[37]

The design of new marketplaces raises new theoretical questions, which sometimes lead to progress in economic theory. Also, after a market has been designed, adopted, and implemented, it is useful to monitor how things are going, to find out if there are problems that still need to be addressed. In this update, I'll briefly point to developments of each of these kinds since the publication of Roth (2008a), "What have we learned from market design?" I'll again discuss theoretical results only informally, to avoid having to introduce the full apparatus of notation and technical assumptions. And while I will try to separate "theoretical" and "operational" issues for clarity, what will really become clear is how closely theoretical and operational issues are intertwined in practical market design.

In Roth (2008a) I described how marketplace design often involves attracting enough participants to make a market thick, dealing with the congestion that can result from attracting many participants, and making participation in the market safe and simple. Accomplishing these tasks requires us to consider, among other things, the strategy sets of the participants, the behavior elicited by possible market designs, and the stability of the resulting outcomes (see e.g. Roth, 2002; Roth and Sotomayor, 1990). To bring theory to bear on a practical problem, we need to create a simple model that allows these issues to be addressed. In what follows, I'll discuss how sometimes an initially useful simple model becomes less useful as the marketplace changes, or as new problems have to be addressed, and how this feeds back to modifications of the original model, and to new theory developed with the help of the new models.

School choice

Theoretical issues

School assignment systems face different problems in different cities. In NYC, high-school assignment had a strong resemblance to the problems facing labor markets for

---

[37] An earlier update, in Spanish, appeared in Roth (2011).

medical school graduates. In both cases, a large number of people have to be matched with a large number of positions at around the same time. And in both cases, the "positions" are in fact strategic players: NYC high-school principals, like directors of medical residency programs, have preferences over whom they match with, and have some strategic flexibility in meeting their goals. So it made sense to think of the NYC high-school assignment process as a two-sided matching market that needed to reach a stable matching—one in which no student and school would prefer to be matched to one another than to accept their assigned matches—in order to damp down some of the strategic behavior that made it hard for the system to work well. And in NYC, as in the medical residency match, there were compelling reasons to choose the applicant-optimal stable matching mechanism—implemented via a student-proposing deferred acceptance algorithm—that makes it safe for applicants to reveal their true preferences.

However, there is an important difference between labor markets and school choice. In a labor market like the one for medical graduates, assuming that the parties have strict preferences (and requiring the graduates to rank order them) probably doesn't introduce much distortion into the market. But in a school choice setting, schools in many cases have (and are often required to have) very large indifference classes, i.e. very many students between whom they can't distinguish. So the question of tie-breaking arises: when there are enough places in a given school to admit only some of a group of otherwise equivalent students, who should get the available seats?

How to do tie-breaking was one of the first questions we confronted in the design of the NYC high-school match, and we had to make some choices among ways to break ties by lottery. In particular, we considered whether to give each student a single number to be used for tie-breaking at every school (single tie-breaking), or to assign numbers to each student at each school (multiple tie-breaking). Computations with simulated and then actual submitted preferences indicated that single tie-breaking had superior welfare properties. Subsequent theoretical and empirical work has clarified the issues involved in tie-breaking. A simple example with just one-to-one matching is all that will be needed to explain, but first it will be helpful to look at how the deferred acceptance algorithm works. (For a description of how the algorithm is adapted to the complexities of the NYC school system, see Abdulkadiroğlu et al., 2009.)

The basic deferred acceptance algorithm with tie-breaking proceeds as follows:

- Step 0.0: Students and schools *privately*[38] submit preferences (and school preferences may have ties, i.e. schools may be indifferent between some students).

---

[38] One feature of the old NYC high-school assignment process was that schools saw how students ranked them, and quite a few schools would only admit students who had ranked them first. Of course, if in the new system schools had still been permitted to see students' rank order lists, even a student-proposing deferred acceptance algorithm would not be strategy-proof. The proof that the student-proposing deferred acceptance algorithm makes it a dominant strategy for students to state their true preferences incorporates the assumption that preference lists are private, through the assumption that the strategy sets available to the players consist of preference lists as a function (only) of own preferences, so that schools' strategies do not include the possibility of making their preference list contingent on the preference lists submitted by students (see Roth, 1982).

- Step 0.1: Arbitrarily break all ties in preferences.
- Step 1: Each student "applies" to her or his first choice. Each school tentatively assigns its seats to its applicants one at a time in their priority order. Any remaining applicants are rejected.

...

- Step k: Each student who was rejected in the previous step applies to her or his next choice if one remains. Each school considers the students it has been holding together with its new applicants and tentatively assigns its seats to these students one at a time in priority order. Any remaining applicants are rejected.
- The algorithm terminates when no student application is rejected, and each student is assigned her or his final tentative assignment.

Notice that—just as Gale and Shapley (1962) showed—the matching produced in this way is stable, not just with respect to the strict preferences that follow step 0.1, but with respect to the underlying preferences elicited from the parties, which may have contained indifferences. That is, there can't be a "blocking pair," a student and a school, not matched to one another, who would prefer to be. The reason is that, if a student prefers some school to the one she was matched with in the algorithm, she must have already applied to that school and been rejected. This applies to the original preferences too, which may not be strict, since tie-breaking just introduces more blocking pairs; so any matching that is stable with respect to artificially strict preferences is also stable with respect to the original preferences. But those additional blocking pairs are constraints, and these additional constraints can harm welfare. A simple 1–1 ("marriage market") matching example is sufficient to see what's going on.

Example 1 (*Tie-breaking can be inefficient*). *Let $M = \{m_1, m_2, m_3\}$ and $W = \{w_1, w_2, w_3\}$ be the sets of students and schools respectively, with preferences given by*:

$$P(m_1) = w_2, w_1, w_3 \quad P(w_1) = [m_1, m_2, m_3]$$
$$P(m_2) = w_1, w_2, w_3 \quad P(w_2) = m_3, m_1, m_2$$
$$P(m_3) = w_1, w_2, w_3 \quad P(w_3) = m_1, m_2, m_3$$

The brackets around w_1's preferences indicate that w_1 is indifferent between any of $[m_1, m_2, m_3]$ while, in this example, everyone else has strict preferences. Since there is only one place at w_1, but w_1 is the first choice of two students (m_2 and m_3), some tie-breaking rule must be used.

Suppose, at step 0 of the deferred acceptance algorithm, the ties in w_1's preferences are broken so as to produce the (artificial) strict preference $P(w_1) = m_1, m_2, m_3$. The deferred acceptance algorithm operating on the artificial strict preferences produces $\mu_M = [(m_1, w_1); (m_2, w_3); (m_3, w_2)]$, at which m_1 and m_3 each receive their second choice (while m_2 receives his last choice). But note that the matching $\mu = [(m_1, w_2); (m_2, w_3); (m_3, w_1)]$ is Pareto superior for the students, as m_1 and m_3 each receive their first choice, so they are both strictly better off than at μ_M, and m_2 is not worse off. If the preferences of school

w_1 were in fact strict, the matching μ would be unstable, because m_2 and w_1 would be a blocking pair. But w_1 doesn't really prefer m_2 to m_3; in fact, μ is stable with respect to the original, non-strict preferences. The pair (w_1, m_2) is not a blocking pair for μ, and only appeared to be in the deferred acceptance algorithm because of the arbitrary ways in which ties were broken to make w_1's preferences look strict.

So, there are costs to arbitrary or random tie-breaking. Erdil and Ergin (2006, 2007), Abdulkadiroğlu et al. (2009), and Kesten (2010) explore this from different angles.[39] Kesten notes that students are collectively better off at μ than at μ_M in example 1 because, in the deferred acceptance algorithm, m_2's attempt to match with w_1 harms m_1 and m_3 without helping m_2. Kesten defines an *efficiency-adjusted deferred acceptance mechanism* that produces μ in example 1 by disallowing the blocking pair (w_1, m_2) via a definition of "reasonable fairness" that generalizes stable matchings. But he shows that there is no mechanism that is Pareto efficient, reasonably fair, and strategy-proof.

To understand Erdil and Ergin's approach, note that the Pareto improvement from μ_M to μ in example 1 comes from an exchange of positions between m_1 and m_3. This exchange doesn't introduce any new blocking pairs, since, among those who would like to change their positions, m_1 and m_3 are among the most preferred candidates of w_1 and w_2. Since there weren't any blocking pairs to the initial matching, this exchange can occur without creating any new blocking pairs.

Formally, Erdil and Ergin define a *stable improvement cycle* starting from some stable matching to be a cycle of students who each prefer the school that the next student in the cycle is matched to, and each of whom is one of the school's most preferred candidates among the students who prefer that school to their current match. They prove the following theorem.

Theorem 15 (*Erdil and Ergin, 2007*). *If μ is a stable matching that is Pareto dominated (from the point of view of students) by another stable matching, then there is a stable improvement cycle starting from μ.*

This implies that there is a computationally efficient algorithm that produces stable matchings that are Pareto optimal with respect to students. The initial step of the algorithm is a student-proposing deferred acceptance algorithm with arbitrary tie-breaking of non-strict preferences by schools. The output of this process (i.e. the student optimal stable matching of the market with artificially strict preferences) is then improved by finding and satisfying stable improvement cycles, until no more remain. Erdil and Ergin show, however, that this algorithm is not strategy-proof; that is, unlike the student-proposing deferred acceptance algorithm, this deferred acceptance plus stable improvement cycle algorithm doesn't make it a dominant strategy for students to

---

[39] In the computer science literature there has been a focus on the *computational* costs of non-strict preferences, which adds to the computational complexity of some calculations (but not others) (see e.g. Irving, 1994; Irving et al., 2008). When preferences aren't strict, not all stable matchings will have the same number of matched people, and Manlove et al. (2002) show that the problem of finding a maximal stable matching is NP hard.

state their true preferences. They show in fact that no mechanism that always produces a stable matching that is Pareto optimal for the students can be strategy-proof.

Abdulkadiroğlu et al. (2009) establish that no mechanism (stable or not, and Pareto optimal or not) that is better for students than the student-proposing deferred acceptance algorithm with tie breaking can be strategy-proof. Following the design of the New York and Boston school choice mechanisms, define a *tie-breaking rule* T to be an ordering of students that is applied to any school's preferences to produce a strict order of students within each of the school's indifference classes (that is, when a school is indifferent between two students, the tie-breaking rule determines which is preferred in the school's artificial strict preferences). *Deferred acceptance with tie breaking rule T is* then simply the deferred acceptance algorithm operating on the strict preferences that result when T is applied to schools' preferences. One mechanism *dominates* another if, for every profile of preferences, the first mechanism produces a matching that is at least as good for every student as the matching produced by the second mechanism, and for some preference profiles the first mechanism produces a matching that is preferred by some students.

Theorem 16 (*Abdulkadiroğlu et al., 2009*). *For any tie-breaking rule T, there is no individually rational mechanism that is strategy-proof for every student and that dominates student-proposing deferred acceptance with tie-breaking rule T.*

But Abdulkadiroğlu, Pathak, and Roth also analyze the preferences submitted in recent NYC high-school matches (under a deferred acceptance with a tie-breaking mechanism) and find that, *if* the preferences elicited from the strategy-proof mechanism could have been elicited by a stable improvement cycle mechanism, then about 1,500 out of about 90,000 NYC students could have gotten a more preferred high school. (In contrast, the same exercise with the preferences submitted in the Boston school choice system yields almost no improvements.) So a number of open questions remain, among them, what accounts for the difference between NYC and Boston, and to what extent could the apparent welfare gains in NYC actually be captured? The potential problem is that, when popular schools are known, it's not so hard to find manipulations of stable improvement cycle mechanisms (which give families the incentive to rank popular schools more highly than in their true preferences, because of the possibility of using them as endowments from which to trade in the improvement cycles). Azevedo and Leshno (2010) show by example that at equilibrium such manipulations could sometimes be welfare decreasing compared to the (non-Pareto optimal) outcome of the deferred acceptance algorithm with tie-breaking.[40]

So far I have been speaking of tie-breaking when a school is indifferent among a group of students only some of whom can be admitted. Students being indifferent among

---

[40] There has been a blossoming of new theory on school choice, including reconsideration of some of the virtues of the Boston algorithm, new hybrid mechanisms, and experiments. See for example Abdulkadiroğlu et al. (2010, 2011), Calsamiglia et al. (2010), Featherstone and Niederle (2010), Haeringer and Klijn (2009), Kojima and Unver (2010), and Mirrales (2009).

schools arose in a different way, because different seats in the same school (which are indistinguishable from the point of view of students) may be allocated according to different priority rules. We encountered this in New York because some schools, called Educational Option schools, are required to allocate half of their seats randomly, while the other half can be allocated according to the school's preferences. We also encountered it in Boston, where some schools use a "walk zone" priority for only half their seats. In each case, we created two "virtual schools" to which students could be admitted, one of which used each relevant priority rule. This is what introduced indifference in student preferences: each student was indifferent between a place in either of the virtual schools corresponding to a particular real school. But how these ties were broken could have consequences. So, for example, as reported in Abdulkadiroğlu et al. (2005a), the design decision we made in New York was that "If a student ranked an EdOpt school, this was treated in the algorithm as a preference for one of the random slots first, followed by a preference for one of the slots determined by the school's preferences." This was welfare improving for schools, since it meant that random slots would fill up before slots governed by the school's preferences, so a desirable student who happened to be admitted to a random slot would allow an additional preferred student to be admitted. However, other, more flexible rules can be considered. Kominers and Sönmez (2012) explore this issue with care, and reveal some subtle issues in the underlying theory.

New operational issues

One of the problems facing the old NYC school assignment system was congestion, caused in part by the time required for students who had received multiple offers to make a decision and allow waiting lists to move. In Boston, in contrast, the old school assignment system wasn't congested; it already used a centralized, computerized clearinghouse to give just one offer per student. Its problems arose from the way in which the assignment was made. However, as new kinds of public/private schools emerged, such as charter schools, Boston school choice has become something of a hybrid system, in which students get a single offer from the public school system but may get parallel offers from charter schools. Consequently, there is now some congestion and delay in processing waiting lists until these students choose which school to attend. Since the charter schools admit by lottery, this problem could easily be solved by including them in the centralized clearinghouse.

This is a problem we can hope to address from the outset as school choice technology continues to spread to other cities. Neil Dorosin, one of the NYC Department of Education administrators with whom we worked on the implementation of their high-school choice process, subsequently founded the non-profit Institute for Innovation in Public School Choice (IIPSC). With technical support from Abdulkadiroğlu, Pathak, and myself, IIPSC helped introduce new school choice systems in Denver and New Orleans. Denver uses a deferred acceptance algorithm, while in the Recovery School District in New Orleans the matching of children to schools in 2013 was due to be done by a version of a top trading cycles algorithm, along the lines discussed as a possibility for

Boston in Abdulkadiroğlu et al. (2005). The New Orleans school choice system includes charter schools (but not yet all of its schools).

Medical labor markets

Theoretical issues

One of the longstanding empirical mysteries regarding the medical labor market clearinghouse is why it works as well as it does in connection with helping couples find pairs of jobs. The story actually began sometime in the 1970s, when for the first time the percentage of women medical graduates from US medical schools rose above 10% (it is now around 50%). With this rise in women doctors came a growing number of graduating doctors who were married to each other, and who wished to find two residency positions in the same location. Many of these couples started to defect from the match. As noted in Roth (1984), not only does the deferred acceptance algorithm not produce a matching that is stable when couples are present (even when couples are allowed to state preferences over *pairs* of positions), but when couples are present it is possible that no stable matching exists. The following simple example from Klaus and Klijn (2005) makes this clear. This version is from Roth (2008b).

Example 2. *Market with one couple and no stable matchings (Klaus and Klijn 2005): Let c=(s1,s2) be a couple, and suppose there is another (single) student, s3, and two hospitals, h1 and h2. Suppose that the acceptable matches for each agent, in order of preference, are given by*

$$c: (h_1, h_2);^{41} \quad s_3: h_1, h_2$$
$$h_1: s_1, s_3; \quad h_2: s_3, s_2$$

Then no individually rational matching μ (i.e. no μ that matches agents only to acceptable mates) is stable. We consider two cases, depending on whether the couple is matched or unmatched.

Case 1: $\mu(c) = (h_1, h_2)$. Then s_3 is unmatched, and he and h_2 can block μ, because h_2 prefers s_3 to $\mu(h_2)=s_2$.

Case 2: $\mu(c) = c$ (unmatched). If $\mu(s_3) = h_1$, then (c, h_1, h_2) blocks μ. If $\mu(s_3) = h_2$ or $\mu(s_3) = s_3$ (unmatched), then (s_3, h_1) blocks μ.

The new algorithm designed for the National Resident Matching Program by Roth and Peranson (1999) allows couples to state preferences over pairs of positions, and

---

[41] Couple c submits a preference list over pairs of positions, and specifies that only a single pair, h1 for student s1 and h2 for student s2, is acceptable. Otherwise couple c prefers to remain unmatched. For a couple, this could make perfect sense, if for example h1 and h2 are in a different city than the couple now resides, and they will move only if they find two good jobs.

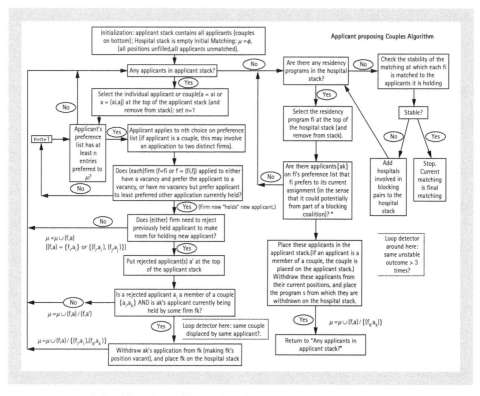

FIGURE 1.2. High-level flowchart of the Roth and Peranson (1999) applicant-proposing deferred acceptance algorithm with couples.

seeks to find a stable matching (see Figure 1.2).[42] The left side of the flow chart describes a fairly standard deferred acceptance algorithm with applicants proposing, much like the basic deferred acceptance algorithm described above in connection with school choice. However, because some applicants are couples who submit preferences over pairs of positions, it may be that a member of a couple sometimes needs to be *withdrawn* from a tentative assignment without having been displaced by a preferred applicant, something that never happens when all applicants are single. This occurs when one member of a couple is displaced by a preferred applicant, so the couple has to apply to another pair of positions, necessitating the withdrawal of the other couple member from the residency program that is holding his or her application. Since that residency program may have rejected other applicants in order to hold this one, this withdrawal may create blocking pairs. Therefore the right side of the flowchart describes an algorithm that tries to repair any blocking pairs that may have arisen in this way. Of course, the

---

[42] The flowchart of the Roth-Peranson algorithm in Figure 1.2 was prepared for an early draft of Roth and Peranson (1999), but was removed in the editorial process, so it is published for the first time here (although it has been available on the Internet for some years in the lecture notes for my market design classes).

algorithm may cycle and fail to find a stable matching, as it must when there is no stable matching, for instance.

The empirical puzzle is why it almost never fails to find a stable matching, in the several dozen annual labor markets in which it has now been employed for over a decade (see Roth, 2008b, for a recent list). Some insight into this, reported in Kojima et al. (2010), connects the success in finding stable matchings that include couples to other recent results about the behavior of large markets.

Roth and Peranson (1999) initiated a line of investigation into large markets by showing computationally that if, as a market gets large, the number of places that a given applicant interviews (and hence the size of his rank order list) does not grow, then the set of stable matchings becomes small (when preferences are strict). Immorlica and Mahdian (2005) showed analytically that in a one-to-one marriage model with uncorrelated preferences, the set of people who are matched to different mates at different stable matchings grows small as the market grows large in this way, and that therefore the opportunities for profitable manipulation grow small. Kojima and Pathak (2009) substantially extend this result to the case of many-to-one matching, in which opportunities for employers to profitably manipulate can occur even when there is a unique stable matching, and in which employers can manipulate capacities as well as preferences. They show that as the size of a market grows towards infinity in an appropriate way, the proportion of employers who might profit from (any combination of) preference or capacity manipulation goes to zero in the worker-proposing deferred acceptance algorithm. Ashlagi et al. (2013) showed that small sets of stable matchings may be typical of large markets. Kojima et al. (2010) showed that when couples are present, if the market grows large in a sufficiently regular way that makes couples a small part of the market, then the probability that a stable matching exists converges to one. That is, in big enough markets with not too many couples we should not be surprised that the algorithm succeeds in finding a stable matching so regularly (see also Ashlagi et al., 2010).

A key element of the proofs is that if the market is large, but no applicant can apply to more than a small fraction of positions, then, even though there may be more applicants than positions, it is a high-probability event that there will be a large number of hospitals with vacant positions after the centralized clearinghouse has found a stable matching. This result is of interest independently from helping in the proofs of the results described above: it means that stable clearinghouses are likely to leave both people unmatched and positions unfilled, even when the market grows very large. Most clearinghouses presently have a secondary, post-match market, often called a "scramble," at which these unmatched people and positions can find one another. The newly developing theory of large markets suggests that post-match marketplaces will continue to be important in markets in which stable centralized clearinghouses are used.

Operational issues

While there has been theoretical progress on managing post-match scrambles, some of this has yet to make its way into practice. In 2012 the National Resident Matching

Program introduced a formal scramble mechanism, called the Supplemental Offer and Acceptance Program. It appears to rely on punishments and sanctions to incentivize orderly participation, and my colleagues and I have expressed some reservations that this will be an effective design for the long term (Coles et al., 2010b).

The clearinghouse for gastroenterology fellowship positions discussed in the first part of this chapter seems to have established itself as a reliable marketplace; in the (2006) match for 2007 positions, 283 positions were offered and 585 applicants applied, of whom 276 were matched. In the match for 2011 positions, 383 positions were offered to 642 applicants, of whom 362 were matched (Proctor et al., 2011). This suggests that the policies adopted to decrease the frequency and effectiveness of exploding offers have been effective (see also Niederle and Roth, 2009a,b).[43] However Proctor et al. (2011) note that there are some warning signs that thickness may be difficult to maintain in the small part of the market that involves research positions. They observe that "the competition for these increasingly scarce, well-qualified, research-track applicants has become fierce, and the authors are aware of several examples during the last application cycle of candidates interested in research being offered fellowship positions outside the Match."

Kidney transplantation

The theoretical and operational issues in kidney exchange are too intertwined for me to try to separate them here. Perhaps the most dramatic recent change in kidney exchange is that, following the publication of Rees et al.'s (2009a) report on the first non-simultaneous extended altruistic donor (NEAD) chain in the *New England Journal of Medicine*, there has been a small explosion of such chains, not only by established exchange networks, but also by transplant centers of all sorts around the United States. See for example the various chains reported at <http://marketdesigner. blogspot.com/search/label/chains>, or the more detailed report of chains conducted by the Alliance for Paired Donation (APD) in Rees et al. (2010). Simulations by Ashlagi et al. (2011a,b) using clinical data from the APD suggest that such chains can play an important role in increasing the number of live donor transplants, and recent theoretical progress has been made in understanding this in Ashlagi et al. (2012) (see also Ashlagi and Roth, 2012; and Dickerson et al., 2012).

The passage into law of what became the 'Charlie W. Norwood Living Organ Donation Act' (Public Law 110–144, 100th Congress) in December 2007 has set in motion plans that may eventually become a national kidney exchange network, but this is still moving slowly, and the issues involved with providing the right incentives for transplant centers to fully participate have not yet been resolved. Indeed, when I discussed this incentive problem in Roth (2008a) it looked like a problem that would become

---

[43] The job market for some other medical subspecialties continues to unravel, and orthopedic surgeons have recently taken steps to organize a centralized match (see Harner et al., 2008).

significant in the future, and today it has become a big issue. Ashlagi and Roth (2011) introduce a random graph model to explore some of these incentive issues in large markets, and show that the cost of making it safe for hospitals to participate fully is low, while the cost of failing to do so could be large if that causes hospitals to match their own internal patient–donor pairs when they can, rather than making them available for more efficient exchanges. That is, guaranteeing hospitals that patients whom they can transplant internally will receive transplants will not be too costly in terms of the overall number of transplants that can be accomplished in large markets. Among the easy-to-match pairs that hospitals withhold are those who are compatible, so that the donor can give directly to the intended recipient, even though such pairs might receive a better-matched kidney through exchange. The inclusion of compatible pairs would greatly increase the efficiency of kidney exchange, in no small part because it would ease the shortage of blood type O donors (see e.g. Roth et al., 2005; and Sönmez and Ünver, 2011; and see also Ünver, 2010, for a discussion of dynamic kidney exchange in large markets). But in the meantime, kidney exchange networks are seeing a disproportionate percentage of hard-to-match pairs, and Ashlagi et al. (2012) use models of sparse random graphs to suggest that this is costly in terms of lost transplants, and that it also accounts for why long ND donor chains have become so useful.

While kidney exchange is growing quickly[44] it still accounts for only a very small fraction of the number of transplants, and the growth is not yet enough to halt the growth of the waiting list for deceased-donor kidneys. (By early 2012 more than 90,000 candidates were registered on the kidney transplant waiting list in the United States.) This has led to continued discussion about ways to recruit more donors, and to continued interest in assessing views on whether kidneys might, in an appropriately regulated environment, under some circumstances be bought and sold, or whether donors could in some way be compensated. The whole question of compensation for donors remains an extremely sensitive subject.

For example, two recent surveys published in the surgical literature showed that public opinion and patient opinion both reflect a willingness to consider payment for organs (Leider and Roth, 2010; and Herold, 2010 respectively). However, the journal that published those surveys also published an editorial (Segev and Gentry, 2010) expressing the opinion that it was a waste of resources even considering the opinions of anyone other than physicians, and expressing the view that physicians were unalterably opposed to any change from current law prohibiting any "valuable consideration" for transplant organs. (This view of physician opinion seems not to be quite accurate, based on available surveys of physician opinion, and on the letters to the editor the journal

---

[44] See Wallis et al. (2011), with the caveat that the UNOS data on kidney exchange and ND donation appears to be incomplete, and may substantially underestimate the kidney exchange transplants to date, for instance because an initially ND donation may be recorded as a directed donation. The data collected by the US Department of Health and Human Services (Health Resources and Services Administration) at <http://optn.transplant.hrsa.gov> are incomplete and ambiguous, but suggest that between 367 and 636 transplants from exchange were reported to it in 2010, compared to between 228 and 441 in 2008, and between 34 and 121 in 2004. (The larger numbers come from including categories that today may include kidney exchange, but almost certainly did not in 2004.)

received in reply to what seems to be a fringe view.) Nevertheless, it is an indication that this remains a controversial subject, with views ranging widely, from those who might contemplate a fairly unregulated market (cf. Becker and Elias, 2007), to those who favor a moderately regulated market like the one in Iran (described in Fatemi, 2010), to those who would consider less direct forms of donor compensation (cf. Satel, 2009), to those, like the editorialists mentioned above, who consider the issue to be beyond discussion except insofar as it impacts physicians.

The continued shortage of kidneys (and other organs) for transplant therefore under-lines the importance of continuing to try to expand deceased donation. Kessler and Roth (2012) report on possibilities of increasing donation by changing organ allocation policy to give increased priority to people who have been long-time registered donors. (This is an element of Singapore's organ allocation policy, and lately also Israel's policy.)

Economists and lawyers: two markets worth watching

Coles et al. (2010a) describe the recent experience of the market for new PhD economists with the newly instituted "pre-market" signaling mechanism, and "post-market" scramble. From 2006 through 2009, the number of candidates who used the signaling mechanism remained roughly constant at around 1,000 per year. The evi-dence is suggestive if not conclusive that judicious signaling increases the probability of receiving an interview. The pattern of signals suggests something about what might constitute "judicious" signaling; when one compares the reputational ranks of the school a student is graduating from and those he signals to, very few signals are sent from lower-ranking to higher-ranking schools. It appears that the signals play a coordination role in ameliorating congestion, with signals distributed across a very broad range of schools. Some new theory of "preference signaling" motivated by this market is presented in Coles et al. (forthcoming).

Participation in the post-market "scramble" has been more variable, with from 70 to 100 positions listed in each of the years 2006–10. It appears that at least 10% of these positions are filled each year through contacts made in the scramble.

Further developments in the market for new PhD economists will provide an ongoing window into the possibilities of dealing with congestion through signaling in a decen-tralized market, and in achieving thickness in the aftermarket.

A window of a different kind is being provided by several of the markets for new law graduates in the United States, which continue to suffer from problems related to the timing of transactions. The market for federal court clerks now appears to be nearing the end of the latest attempt to enforce a set of dates before which applications, interviews, and offers will not be made. Avery et al. (2007) already reported a high level of cheating in that market, as judges accepted applications, conducted interviews, and made offers before the designated dates. Roth and Xing (1994) reported on various ways that markets could fail through the unraveling of appointment dates, but the markets for lawyers have frequently offered the opportunity to observe new failures of this kind. Presently the market for new associates at large law firms is also unraveling (see Roth, 2012).

Conclusions

The new marketplace designs reported in Roth (2008a), for labor markets, for schools, and for kidney exchange, have continued to operate effectively. However, in each of these domains, unsolved operational problems remain. In school choice, integrating standard public schools with other options such as charter schools in a single clearinghouse will help to avoid congestion. In kidney exchange, making it safe for hospitals to enroll all of their appropriate patient–donor pairs will help establish thickness and increase the number of transplants. In labor markets, it may be necessary to pay special attention to submarkets such as medical fellows interested in research.

These examples illustrate how market design, and the close attention it demands to the details of how particular markets operate, raises new theoretical questions about how markets work, and how market failures can be avoided and repaired. Holmstrom et al. (2002) quote Robert Wilson (1993) on this: "for the theorist, the problems encountered by practitioners provide a wealth of topics."

REFERENCES

110th Congress, Public Law 110–144, Charlie W. Norwood Living Organ Donation Act, December 21, 2007, <http://frwebgate.access.gpo.gov/cgi-bin/getdoc.cgi?dbname=110_cong_ public_laws&docid=f:publ144.110.pdf>.

Abdulkadiroğlu, A. and Sönmez, T. (1999) "House allocation with existing tenants," *Journal of Economic Theory*, 88: 233–60.

———— (2003) "School choice: a mechanism design approach," *American Economic Review*, 93(3): 729–47.

———— Pathak, P. A. and Roth, A. E. (2005a) "The New York City high school match," *American Economic Review, Papers and Proceedings*, 95(2): 364–7.

———— ———— Sönmez, T. (2005b) "The Boston public school match," *American Economic Review Papers and Proceedings*, 95(2): 368–71.

———— ———— ———— (2007) "Changing the Boston school choice mechanism: strategy-proofness as equal access," NBER Working Paper No. 11965.

———— ———— ———— (2009) "Strategy-proofness versus efficiency in matching with indifferences: redesigning the NYC high school match," *American Economic Review*, 99(5): 1954–78.

———— Che, Y.-K., and Yasuda, Y. (2010) "Expanding 'choice' in school choice," working paper.

———— ———— ———— (2011) "Resolving conflicting preferences in school choice: the 'Boston' mechanism reconsidered," *American Economic Review*, 101(1): 399–410.

Abraham, D., Blum, A. and Sandholm, T. (2007) "Clearing algorithms for barter exchange markets: enabling nationwide kidney exchanges," In *Proceedings of the ACM Conference on Electronic Commerce* (EC).

Ariely, D., Ockenfels, A. and Roth, A. E. (2005) "An experimental analysis of ending rules in internet auctions," *Rand Journal of Economics*, 36(4): 891–908.

Artemov, G. (2008) "Matching and price competition: would personalized prices help?" *International Journal of Game Theory*, 36(3): 321–31.

Ashlagi, I. and Roth, A. E. (2011) "Individual rationality and participation in large scale, multi-hospital kidney exchange," Working Paper.

—— —— (2012) "New challenges in multi-hospital kidney exchange," *American Economic Review: Papers and Proceedings*, 102(3): 354–9.

—— Braverman, M. and Hassidim, A. (2010) "Matching with couples in large markets revisited," unpublished mimeo, MIT Sloan School.

—— Gilchrist, D. S., Roth, A. E. and Rees, M. A. (2011a) "Nonsimultaneous chains and dominos in kidney paired donation—revisited," *American Journal of Transplantation*, 11(5): 984–94.

—— —— —— —— (2011b) "NEAD chains in transplantation," *American Journal of Transplantation*, 11: 2780–1.

—— Gamarnik, D., Rees, M. and Roth, A. E. (2012) "The need for (long) chains in kidney exchange," Working Paper.

—— Kanoria, K. and Leshno, J. D. (2013) "Unbalanced Random Matching Markets," Working Paper.

Avery, C., Jolls, C., Posner, R. A. and Roth, A. E. (2001) "The market for federal judicial law clerks," *University of Chicago Law Review*, 68: 793–902.

—— Fairbanks, A. and Zeckhauser, R. (2003) *The Early Admissions Game: Joining the Elite*, Harvard University Press.

—— Jolls, C., Posner, R. A. and Roth, A. E. (2007) "The new market for federal judicial law clerks," *University of Chicago Law Review*, 74: 447–86.

Azevedo, E. M. and Leshno, J. D. (2010) "Can we make school choice more efficient? An incentives approach", draft, Harvard University.

Becker, G. S. and Elías, J. J. (2007) "Introducing incentives in the market for live and cadaveric organ donations," *Journal of Economic Perspectives*, 21(3): 3–24.

Biais, B., Hillion, P. and Spatt, C. (1999) "Price discovery and learning during the preopening period in the Paris Bourse," *Journal of Political Economy*, 107: 1218–48.

Bulow, J. and Levin, J. (2006) "Matching and price competition," *American Economic Review*, 96(3): 652–68.

Calsamiglia, C., Haeringer, G. and Klijn, F. (2010) "Constrained school choice: an experimental study," *American Economic Review*, 100(4): 1860–74.

Cawley, J. (2006) "A guide (and advice) for economists on the U.S. junior academic job market," October <http://www.aeaweb.org/joe/articles/2006/cawley.pdf>.

Chen, Y. and Sönmez, T. (2006) "School choice: an experimental study," *Journal of Economic Theory*, 127: 2002–231.

Coles, P. A., Cawley, J. H., Levine, P. B., Niederle, M., Roth, A. E. and Siegfried, J. J. (2010a) "The job market for new economists: a market design perspective," *Journal of Economic Perspectives*, 24(4): 187–206.

—— Featherstone, C. R., Hatfield, J. W., Kojima, F., Kominers, S. D., Niederle, M., Pathak, P. A. and Roth, A. E. (2010b) "Comment on the NRMP's "Supplemental Offer and Acceptance Program" proposed to replace the post-match scramble," <http://kuznets.fas.harvard.edu/~aroth/papers/NRMP%20comment.pdf>.

Coles, P., Kushnir, A. and Niederle, M. (forthcoming) "Preference signaling in matching markets," *American Economic Journal: Microeconomics*.

Cramton, P. (1997) "The FCC spectrum auctions: an early assessment," *Journal of Economics and Management Strategy*, 6(3): 431–95.

_____ Shoham, Y. and Steinberg, R. (eds) (2006) *Combinatorial Auctions*, MIT Press.

Crawford, V. P. (2008) "The flexible-salary match: a proposal to increase the salary flexibility of the National Resident Matching Program," *Journal of Economic Behavior and Organization*, 66: 149–60.

Damiano, E., Li, J. and Suen, W. (2005) "Unraveling of dynamic sorting," *Review of Economic Studies*, 72: 1057–76.

Delmonico, F. L. (2004) "Exchanging kidneys: advances in living-donor transplantation," *New England Journal of Medicine*, 350(18): 1812–14.

Dickerson, J. P., Procaccia, A. D. and Sandholm, T. (2012) "Optimizing kidney exchange with transplant chains: theory and reality," Working Paper.

Elias, J. J. and Roth, A. E. (2007) "Econ one on one: kidney transplantation," WSJ online, November 13 <http://online.wsj.com/public/article/SB118901049137818211.html?mod=todays_free_feature>.

Erdil, A. and Ergin, H. (2007) "What's the matter with tie-breaking? Improving efficiency in school choice," *American Economic Review*, 98(3): 669–89.

Ergin, H. and Sönmez, T. (2006) "Games of school choice under the Boston mechanism," *Journal of Public Economics*, 90: 215–37.

Evans, D. S. and Schmalensee, R. (with D.S. Evans) (1999) *Paying with Plastic: The Digital Revolution in Buying and Borrowing*, MIT Press.

_____ Hagiu, A. and Schmalensee, R. (2006) *Invisible Engines: How Software Platforms Drive Innovation and Transform Industries*, MIT Press.

Fatemi, F. (2010) "The regulated market for kidneys in Iran," Sharif University of Technology, <http://gsme.sharif.edu/~ffatemi/Research/Kidney_Market-Farshad_Fatemi-11Jan2010.pdf>.

Featherstone, C. and Niederle, M. (2010) "Ex ante efficiency in school choice mechanisms: an experimental investigation," draft.

Felli, L. and Sutton, J. (2006) "The Royal Economic Society's first PhD presentations event," <http://econ.lse.ac.uk/news/openfiles/js_RES_LSE_job_market_report.pdf>.

Fréchette, G., Roth, A. E. and Ünver, M. U. (2007) "Unraveling yields inefficient matchings: evidence from post-season college football bowls," *Rand Journal of Economics*, 38(4): 967–82.

Gale, D. and Shapley, L. (1962) "College admissions and the stability of marriage," *American Mathematical Monthly*, 69: 9–15.

Gentry, S. E., Segev, D. L., Simmerling, M. and Montgomery, R. A. (2007) "Expanding kidney paired donation through participation by compatible pairs," *American Journal of Transplantation*, 7: 2361–70.

Grosskopf, B. and Roth, A. E. (2009) "If you are offered the right of first refusal, should you accept? An INVESTIGATION OF CONTRACT DEsign," *Games and Economic Behavior*, Special Issue in Honor of Martin Shubik, 65 (January): 176–204.

Haeringer, G. and Klijn, F. (2009) "Constrained school choice," *Journal of Economic Theory*, 144(5): 1921–47.

Hanto, R. L., Roth, A. E., Ünver, M. U. and Delmonico, F. L. (2010) "New sources in living kidney donation," in D. McKay (ed.), *Kidney Transplantation: A Guide to the Care of Transplant Recipients*, Springer, pp. 103–17.

_____ Saidman, S. L., Roth, A. E. and Delmonico, F. L. (2010) "The evolution of a successful kidney paired donation program," XXIII International Congress of The Transplantation Society, August 16, Vancouver.

Harner, C. D., Ranawat, A. S., Niederle, M., Roth, A. E., Stern, P. J., Hurwitz, S. R., Levine, W., DeRosa, G. P. and Hu, S. S. (2008) "Current state of fellowship hiring: Is a universal match necessary? Is it possible?" *Journal of Bone and Joint Surgery*, 90: 1375–84.

Herold, D. K. (2010) "Patient willingness to pay for a kidney for transplantation," *American Journal of Transplantation*, 10: 1394–400.

Holmstrom, B., Milgrom, P. and Roth, A. E. (2002) "Introduction to 'Game theory in the tradition of Bob Wilson'," in B. Holmstrom, P. Milgrom and A. E. Roth (eds), *Game Theory in the Tradition of Bob Wilson*, Berkeley Electronic Press <http://www.bepress.com/wilson>.

Immorlica, N. and Mahdian, M. (2005) "Marriage, honesty, and stability," *SODA*: 53–62.

Irving, R. W. (1994) "Stable marriage and indifference," *Discrete Applied Mathematics*, 48: 261–72.

_____ Manlove, D. F. and Scott, S. (2008) "The stable marriage problem with master preference lists," *Discrete Applied Mathematics*, 156: 2959–27.

Kessler, J. B. and Roth, A. E. (forthcoming) "Organ allocation policy and the decision to donate," *American Economic Review*.

Jevons, W. S. (1876) *Money and the Mechanism of Exchange*, D. Appleton and Company.

Kagel, J. H. and Levin, D. (2002) *Common Value Auctions and the Winner's Curse*, Princeton University Press.

_____ and Roth, A. E. (2000) "The dynamics of reorganization in matching markets: a laboratory experiment motivated by a natural experiment," *Quarterly Journal of Economics*, 115(1): 201–35.

Kamecke, U. (1998) "Wage formation in a centralized matching market," *International Economic Review*, 39(1): 33–53.

Kandel, E., Rindi, B. and Bosetti, L. (2007) "The effect of a closing call auction on market quality and trading strategies," Working Paper (Closing Call Auction in the Borsa Italiana).

Kelso, A. S. and Crawford, V. P. (1982) "Job matching, coalition formation, and gross substitutes," *Econometrica*, 50(6): 1483–504.

Kesten, O. (2010) "School choice with consent," *Quarterly Journal of Economics* 125(3): 1297–348.

Klaus, B. and Klijn, F. (2005) "Stable matchings and preferences of couples," *Journal of Economic Theory*, 121(1): 75–106.

Klemperer, P. (2004) *Auctions: Theory and Practice. The Toulouse Lectures in Economics*, Princeton University Press.

Kojima, F. (2007) "Matching and price competition: comment", *American Economic Review*, 97(3): 1027–31.

_____ and Pathak, P. A. (2009) "Incentives and stability in large two-sided matching markets," *American Economic Review*, 99(3): 608–27.

_____ and Unver, M. U. (2010) "The 'Boston' school-choice mechanism," Working Paper, Boston College, February.

_____ Pathak, P. A. and Roth, A. E. (2012) "Matching with couples: stability and incentives in large markets," April 2010, revised September.

Kominers, S. D. and Sönmez, T. (2012) "Designing for diversity in matching," Working Paper, September <http://www.scottkom.com/articles/Kominers_Sonmez_Designing_for_Diversity_in_Matching.pdf>.

Lee, R. S. and Schwarz, M. (2007a) "Interviewing in two-sided matching markets," NBER Working Paper 14922.

_____ _____ (2007b) "Signaling preferences in interviewing markets," in P. Cramton, R. Müller, E. Tardos and M. Tennenholtz (eds), *Computational Social Systems and the Internet*, no. 07271 in Dagstuhl Seminar Proceedings, Dagstuhl, Germany.

Leider, S. and Roth, A. E. (2010) "Kidneys for sale: who disapproves, and why?" *American Journal of Transplantation*, 10: 1221–7.

Li, H. and Rosen, S. (1998) "Unraveling in matching markets." *American Economic Review*, 88: 371–87.

_____ and Suen, W. (2000) "Risk sharing, sorting, and early contracting," *Journal of Political Economy*, 108: 1058–91.

Lien, Y. (2007) "Application choices and college rankings," Working Paper, Stanford University.

Manlove, D. F., Irving, R. W., Iwama, K., Miyazaki, S. and Morita, Y. (2002) "Hard variants of stable marriage," *Theoretical Computer Science*, 276: 261–79.

McKinney, C. N., Niederle, M. and Roth, A. E. (2005) "The collapse of a medical labor clearinghouse (and why such failures are rare)," *American Economic* Review, 95(3): 878–89.

Milgrom, P. (2000) "Putting auction theory to work: the simultaneous ascending auction," *Journal of Political Economy*, 108(2): 245–72.

_____ (2004) *Putting Auction Theory to Work*, Cambridge University Press.

_____ (2007) "Package auctions and package exchanges," *Econometrica*, 75(4): 935–66.

Mirrales, A. (2009) "School choice: the case for the Boston mechanism," Boston University.

Montgomery, R. A., Zachary, A. A., Ratner, L. E., Segev, D. L., Hiller, J. M., Houp, J., Cooper, M., et al. (2005) "Clinical results from transplanting incompatible live kidney donor/recipient pairs using kidney paired donation," *Journal of the American Medical Association*, 294(13): 1655–63.

Niederle, M. (2007) "Competitive wages in a match with ordered contracts," *American Economic Review*, 97(5): 1957–69.

_____ and Roth, A. E. (2003a) "Relationship between wages and presence of a match in medical fellowships," *Journal of the American Medical Association*, 290(9): 1153–4.

_____ _____ (2003b) "Unraveling reduces mobility in a labor market: gastroenterology with and without a centralized match," *Journal of Political Economy*, 111(6): 1342–52.

_____ _____ (2004) "The gastroenterology fellowship match: how it failed, and why it could succeed once again," *Gastroenterology*, 127: 658–66.

_____ _____ (2005) "The gastroenterology fellowship market: should there be a match?" *American Economic Review: Papers and Proceedings*, 95(2): 372–5.

_____ _____ (2009a) "Market culture: how rules governing exploding offers affect market performance," *American Economic Journal: Microeconomics*, 1(2): 199–219.

_____ _____ (2009b) "The effects of a centralized clearinghouse on job placement, wages, and hiring practices," in D. Autor (ed.), *Labor Market Intermediation*, University of Chicago Press, pp. 273–306.

_____ Proctor, D. D. and Roth, A. E. (2006) "What will be needed for the new GI fellowship match to succeed?" *Gastroenterology*, 130: 218–24.

_____ _____ _____ (2008) "The gastroenterology fellowship match – the first two years," *Gastroenterology*, 135(2): 344–6.

Pathak, P. and Sönmez, T. (2008) "Leveling the playing field: sincere and strategic players in the Boston mechanism," *American Economic Review*, 98(4): 1636–52.

Payzant, T. W. (2005) "Student assignment mechanics: algorithm update and discussion," memorandum to the Boston School Committee, May 25 <http://boston.k12.ma.us/assignment/faq5-25-05.doc>.

Plott, C. R. (1997) "Laboratory experimental testbeds: application to the PCS auction," *Journal of Economics and Management Strategy*, 6(3): 605–38.

Proctor, D. D., Decross, A. J., Willis, C. E., Jones, T. N. and Pardi, D. S. (2011) "The match: five years later," *Gastroenterology*, 140(1): 15–18.

Rapaport, F. T. (1986) "The case for a living emotionally related international kidney donor exchange registry," *Transplantation Proceedings*, 18: 5–9.

Rees, M. A., Kopke, J. E., Pelletier, R. P., Segev, D. L., Rutter, M. E., Fabrega, A. J., Rogers, J., Pankewycz, O. G., Hiller, J., Roth, A. E., Sandholm, T., Ünver, U. and Montgomery, R. A. (2009a) "A non-simultaneous extended altruistic donor chain," *New England Journal of Medicine*, 360(11): 1096–101.

Rees, M., Kopke, J., Pelletier, R., Segev, D., Fabrega, A., Rogers, J., Pankewycz, O., Hiller, J., Roth, A., Sandholm, T., Unver, M. U., Nibhunupudy, B., Bowers, V., Van Buren, C. and Montgomery, R. (2009b) "Four never-ending altruistic donor chains," *American Journal of Transplantation*, 9 (suppl. 2): 389.

Rees, M. A., Kopke, J. E., Pelletier, R. P., Segev, D. L. Fabrega, A. J., Rogers, J., Pankewycz, O. G., Roth, A. E., Taber, T. E., Ünver, M. U., Nibhunubpudy, B., Leichtman, A. B., VanBuren, C. T., Young, C. J., Gallay, B. J. and Montgomery, R. A. (2010) "Nine non-simultaneous extended altruistic donor (NEAD) chains," XXIII International Congress of The Transplantation Society, August 15–19, Vancouver <http://kuznets.fas.harvard.edu/~aroth/papers/Rees%20et%20al.%20ITC%202010%20NEAD%20Chain%20Poster.pdf>.

Rochet, J-C. and Tirole, J. (2006) "Two-sided markets: a progress report", *RAND Journal of Economics*, 35(3): 645–67.

Ross, L. F. and Woodle, E. S. (2000) "Ethical issues in increasing living kidney donations by expanding kidney paired exchange programs," *Transplantation*, 69: 1539–43.

_____ Rubin, D. T., Siegler, M., Josephson, M. A., Thistlethwaite, J. R., Jr and Woodle, E. S. (1997) "Ethics of a paired-kidney-exchange program," *New England Journal of Medicine*, 336: 1752–5.

Roth, A. E. (1982a) "The economics of matching: stability and incentives," *Mathematics of Operations Research*, 7: 617–28.

_____ (1982b) "Incentive compatibility in a market with indivisible goods," *Economics Letters*, 9: 127–32.

_____ (1984) "The evolution of the labor market for medical interns and residents: a case study in game theory," *Journal of Political Economy*, 92: 991–1016.

_____ (1985) "The college admissions problem is not equivalent to the marriage problem," *Journal of Economic Theory*, 36: 277–88.

_____ (1990) "New physicians: a natural experiment in market organization," *Science*, 250: 1524–8.

_____ (1991a) "Game theory as a part of empirical economics," *Economic Journal*, 101: 107–14.

_____ (1991b) "A natural experiment in the organization of entry level labor markets: regional markets for new physicians and surgeons in the U.K.," *American Economic Review*, 81: 415–40.

_____ (2002) "The economist as engineer: game theory, experimental economics and computation as tools of design economics," *Econometrica*, 70(4): 1341–78.

_____ (2003) "The origins, history, and design of the resident match," *Journal of the American Medical Association*, 289(7): 909–12.

_____ (2007) "Repugnance as a constraint on markets," NBER Working Paper 12702, November, *Journal of Economic Perspectives*, 21(3): 37–58.

_____ (2008a) "What have we learned from market design?" *Economic Journal*, 118: 285–310.

_____ (2008b) "Deferred acceptance algorithms: history, theory, practice, and open questions," *International Journal of Game Theory*, Special Issue in Honor of David Gale on his 85th birthday, 36: 537–69.

_____ (2011) "¿Qué hemos aprendido del diseño de mercados?" *El Trimestre Económico*, 78(2): 259–314.

_____ (2012) "Marketplace institutions related to the timing of transactions: reply to Priest (2010)," *Journal of Labor Economics*, 30(2): 479–94.

_____ and Ockenfels, A. (2002) "Last-minute bidding and the rules for ending second-price auctions: evidence from eBay and Amazon auctions on the internet," *American Economic Review*, 92(4): 1093–103.

_____ and Peranson, E. (1999) "The redesign of the matching market for American physicians: some engineering aspects of economic design," *American Economic Review*, 89(4): 748–80.

_____ and Sotomayor, M. (1990) *Two-Sided Matching: A Study in Game-Theoretic Modeling and Analysis*, Econometric Society Monograph Series, Cambridge University Press.

_____ and Xing, X. (1994) "Jumping the gun: imperfections and institutions related to the timing of market transactions," *American Economic Review*, 84: 992–1044.

_____ and Xing, X. (1997) "Turnaround Times and Bottlenecks in Market Clearing: Decentralized Matching in the Market for Clinical Psychologists," *Journal of Political Economy*, 105: 284–329.

_____ Sönmez, T. and Ünver, M. U. (2004a) "Kidney exchange," *Quarterly Journal of Economics*, 119(2): 457–88.

_____ _____ _____ (2004b) "Pairwise kidney exchange," NBER Working Paper w10698.

_____ _____ _____ (2005a) "Pairwise kidney exchange," *Journal of Economic Theory*, 125(2): 151–88.

_____ _____ _____ (2005b) "A kidney exchange clearinghouse in New England," *American Economic Review: Papers and Proceedings*, 95(2): 376–80.

_____ _____ _____ Delmonico, F. L. and Saidman, S. L. (2006) "Utilizing list exchange and undirected good samaritan donation through 'chain' paired kidney donations," *American Journal of Transplantation*, 6(11): 2694–705.

_____ _____ _____ (2007) "Efficient kidney exchange: coincidence of wants in markets with compatibility-based preferences," *American Economic Review*, 97(3): 828–51.

Saidman, S. L., Roth, A. E., Sönmez, T., Ünver, M. U. and Delmonico, F. L. (2006) "Increasing the opportunity of live kidney donation by matching for two and three way exchanges," *Transplantation*, 81(5): 773–82.

Salzman, L. F. (1931) *English Trade in the Middle Ages*, Clarendon.

Satel, S. (ed.) (2009) *When Altruism Isn't Enough: The Case for Compensating Kidney Donors*, AEI Press.

Segev, D. L. and Gentry, S. E. (2010) "Kidneys for sale: whose attitudes matter?" *American Journal of Transplantation*, 10: 1113–14.

_____ _____ Warren, D. S., Reeb, B. and Montgomery, R. A. (2005) "Kidney paired donation and optimizing the use of live donor organs," *Journal of the American Medical Association*, 293(15): 1883–90.

Shapley, L. S. and Scarf, H. (1974) "On cores and indivisibility," *Journal of Mathematical Economics*, 1: 23–8.

Siegfried, J. J. and Stock, W. A. (2004) "The labor market for new Ph.D. economists in 2002," *American Economic Review: Papers and Proceedings*, 94(2): 272–85.

Sönmez, T. (1997) "Manipulation via capacities in two-sided matching markets," *Journal of Economic Theory*, 77(1): 197–204.

_____ and Ünver, M. U. (2011) "Altruistic kidney exchange," Unpublished Working Paper.

Stack, J. N. (2007) "Three essays in applied economics," Harvard University, PhD dissertation.

Suen, W. (2000) "A competitive theory of equilibrium and disequilibrium unravelling in two-sided matching," *Rand Journal of Economics*, 31: 101–20.

Ünver, M. U. (2010) "Dynamic kidney exchange," *Review of Economic Studies*, 77(1): 372–414.

Wallis, C. B., Samy, K. P., Roth, A. E. and Rees, M. A. (2011) "Kidney paired donation," *Nephrology Dialysis Transplantation*, 26(7): 2091–9.

Wilson, R. B. (1993) *Nonlinear Pricing*, Oxford University Press.

_____ (2002) "Architecture of power markets," *Econometrica*, 70(4): 1299–340.

CHAPTER 2

NOT UP TO STANDARD: STRESS TESTING MARKET DESIGNS FOR MISBEHAVIOR

GARY E. BOLTON

INTRODUCTION

A good market design is a robust market design. Market incentives and transaction rules need to be arranged so that the market outcomes we aim for are resilient to gaming. Most of the chapters in this handbook deal with this critical point. Yet other elements of human behavior can challenge a market's resilience. These involve the complexities of human objectives and judgment. Importantly, the theories of strategic behavior that we currently rely on to guard against gaming make strong assumptions about the objectives and the rational judgment of market participants. For the purpose of the exposition, I call these assumptions the "behavioral standards." Some deviations from the behavioral standards are well documented, others not so much. This then raises the question of how we can discover the sometimes hard-to-anticipate "misbehavior" that can sabotage an otherwise attractive design idea.

In this chapter I discuss recent work on two kinds of market design problems, one dealing with the complexity of human objectives and the other with the bounds of rational judgment. While both problems are reported in the literature, my focus will be somewhat different here. The three specific points I wish to illustrate are these:

First, a behavioral standard that well approximates behavior in one sphere of the market may be inadequate along other dimensions of the same market. Many markets are highly price competitive. From this observation it is tempting to conclude that trader objectives in such markets are highly self-interested, the usual behavioral standard. We know, however, that price-competitive behavior is consistent with other preference structures, such as social preferences for reciprocity and fairness (Cooper and Kagel, forthcoming). Moreover, after a deal is struck in a multilateral price competition, the transaction must be executed in a bilateral buyer–seller relationship. It is precisely these kinds of setting where social preferences are not so easily ignored. I elaborate in the following section.

Second, people can deviate from the benchmark assumptions in diverse ways, with more heterogeneity across individuals than the behavioral standard anticipates. The challenge for market design, therefore, is to write rules that are robust against a range of misbehavior. I illustrate this point in the third section.

Third, while some misbehavior is understood well enough to be anticipated, some other misbehavior is not. As a consequence, fully vetting a new design is necessarily an engineering exercise, one that, particularly when the market design is new, is well suited to laboratory stress testing. The laboratory models employed for these tests may or may not line up squarely with established theoretical models. This can happen because market design can take us into institutional mechanisms, where theory is less developed but might nevertheless be critical to the success or failure of the design.

Reciprocal feedback and trust on eBay

The eBay marketplace and feedback system misbehavior

The eBay marketplace is highly price competitive, something easily explained by appealing to standard benchmark, self-interested behavior. But this will not take you far in understanding the workings of eBay's feedback system, the trader rating system that promotes trust and trustworthiness on the site. In many respects, the system is successful. For instance, many (but not all) studies find that feedback has positive value for the market, as indicated by positive correlations between the feedback score of a seller and the revenue and the probability of sale—see for example Bajari and Hortaçsu (2003, 2004), Ba and Pavlou (2002), Dellarocas (2004), and Houser and Wooders (2005).

Yet there are also problems involving misbehavior, one of which was the subject of a market design study undertaken and reported by Bolton et al. (forthcoming). The misbehavior involved sellers retaliating for a buyer's negative review by giving the buyer a negative review. In some cases, the motive for this behavior appears to have been getting the buyer to withdraw their negative feedback. But other cases appear best described as a reciprocal response, perhaps with some sort of social preference motive. At the time of the study there was a good deal of evidence that buyers knew about seller retaliatory behavior even if they had not experienced it (for instance, seller retaliation was widely discussed on eBay chat sites); many buyers would not report an unhappy experience with a seller in order to avoid the risk of retaliation (Dellarocas and Wood, 2008). As a consequence, feedback given on the site was too positive relative to the true mix of satisfactory and unsatisfactory trades. Most importantly, some sellers had undeservedly high ratings, making it hard for buyers to gauge the true risk of entering into a trade. This version of the "lemons" problem was thought to diminish market efficiency.

Two proposed solutions, one modeled, one not

The first of the two proposed solutions in the literature is a straightforward market design fix to the problem eBay's feedback system, I will call it the "conventional system," was experiencing (see for example Güth et al., 2007; Reichling, 2004; Klein et al., 2007). The proposal made the important observation that it was the feedback timing convention on the site that enabled retaliatory behavior, the convention being that feedback was posted immediately after it was given. This allowed a seller who suspected a buyer would give negative feedback to withhold his own feedback, the implicit threat being retaliation. The new design proposal would make the system double blind: feedback would be revealed simultaneously so that a trader could not condition his feedback on the feedback of that of his transaction partner's. Retaliation, in response to a bad feedback score, would no longer be possible.

The proposal is appealingly straightforward. There are, however, two potential problems with this approach. First, the hard-close feedback period in a double-blind feedback system should be long enough so that it does not interfere with the natural flow of the transaction. To close a deal, traders must have time for payment to clear, for goods to be received, and for any resulting problems to be straightened out; forcing feedback prior to the close of the deal would be self-defeating to the system. We calculated that any deadline of less than thirty days would unduly interfere with the natural flow of transactions and, in this regard, a sixty-day deadline would be better. The problem is that such a deadline can be gamed. A transaction partner expecting negative feedback from his counterpart has an incentive to delay feedback as long as possible. If negative feedback enters the system only very late, a fraudulent seller might have disappointed many other buyers, who otherwise could have been warned.

The second problem has to do with the negative influence the double-blind system might have on the frequency with which feedback is given. Here we need to understand something of the objective behind giving feedback. First, it is not easily explained in terms of strict self-interest: feedback information is largely for public benefit, helping all traders to manage the risks involved in trusting unknown transaction partners. Yet in our data about 70% of the traders, sellers and buyers alike, leave feedback. Moreover, there is a pronounced reciprocal tendency to giving feedback, one that goes beyond the tendency for sellers to retaliate for negative buyer feedback. If feedback were given independently among trading partners, one would expect the percentage of transactions for which both partners give feedback to be 70%×70% = 49%. Yet, in our data-set, mutual feedback is given much more often, about 64% of the time. Sellers have an interest in receiving positive feedback and there is anecdotal evidence that they use this reciprocal tendency to their advantage, by noting to an obviously pleased buyer that the seller gave him or her positive feedback, hoping this will trigger the buyer to reciprocate. Getting legitimately satisfactory trades reported is not only good for the seller but for the system as a whole.[1] A double-blind system would obstruct this kind of reciprocal trigger. The worry, then, is that double-blind feedback would lower reporting frequencies and in a way that would bias reported feedback in the negative direction. This could be bad for the larger marketplace, in that buyers, particularly new buyers considering using the site, would see an unduly biased picture of overall seller performance. There was also evidence for this concern: "Rent-a-coder," a site where software coders bid for contracts offered by software buyers, transitioned to a double-blind feedback system. Data we collected before and after the transition showed a drop in the frequency of giving feedback.

The second proposal is a design put together, in part, to respond to the potential problems of the double-blind proposal. Under this proposal, the system of posting feedback immediately would continue but would be supplemented with an option giving only the buyer an opportunity to leave additional feedback, blind to the seller. The system would also permit fine-tuning the details of the new feedback so as to provide more information on sellers than the conventional system relating issues buyers are known to be concerned with, such as shipping speed and accuracy of the description of the good. For this reason the proposal was known as the *DSR* system, where DSR stands for "detailed seller ratings." A possible negative consequence is that the conventional and DSR feedback given to sellers might diverge, with unhappy buyers giving positive conventional feedback to avoid seller retaliation, and then being truthful with the (blind) DSR score. This might not be a problem for experienced traders, who would know to pay exclusive attention to DSR scores. But it might also make it harder and more costly for new eBay traders to learn how to interpret reputation profiles. For some traders, the inconsistency might damage the institutional credibility of the feedback system.

Importantly, the DSR system addresses the two potential problems with the double-blind system. Maintaining the conventional system permits sellers involved in smooth transactions to continue to trigger reciprocal responses from their satisfied buyers, so that the misbehavior that is good for the market can continue. By the same token, posting conventional feedback without delay permits buyers with major grievances an outlet to immediately alert other buyers about the problem seller. The DSR feedback then allows buyers to make more nuanced, perhaps critical statements about seller performance. At least this was the hope.

---

[1] In theory, a feedback system can be successful only if negative experiences are reported. Perhaps the most persuasive evidence that positive feedback is important on eBay is how eager sellers are to receive positive feedback.

Stress testing a laboratory model

We stress tested the two proposed designs against potential misbehavior using a laboratory experiment. What field evidence there was for the performance of the double-blind system came from Internet markets that differed in scope and institutional detail from eBay. There was no meaningful field evidence for the DSR system, testifying to its novelty. The laboratory experiment was designed as a level playing field for comparing the performance of the competing designs. The control the laboratory affords also helps us to identify the role of reciprocal behavior in the context of giving feedback, and to establish causal relationships between feedback and market performance (for example relating to efficiency).

It is useful to think of the experiment (laboratory test instrument together with the test subject decisions) as a model. As with any model, we sacrifice some details of the real world in order to gain clarity. In this case, we want a clean look at how each proposal interacts with feedback (mis)behavior and subsequently influences market performance. We then need to model the market-making mechanism as well as the feedback mechanism. We modeled the market mechanism as a private-value second-price auction, as eBay is a second-price auction. The details of this mechanism (action space, draw of private valuations, etc.) closely parallel laboratory experiments designed to test second-price auction theory. Modeling the feedback system required a different wellspring, if only because the misbehavior in question, seller retaliatory feedback, is not theoretically well understood. The design we adopted, a simple stage procedure for giving feedback, reflects the need to capture the essential differences among the three feedback systems we tested (the conventional system as baseline plus the two proposed new systems) as well as the need to examine potential side problems, particularly changes in overall feedback frequency or divergence between conventional and DSR feedback scores.

The results from the experiment were stark and easily described. Both double-blind and DSR feedback systems reduce seller feedback retaliation and improve the efficiency of the market, and in quantitatively similar ways. The double-blind system, however, exhibited a feedback frequency lower than that of the conventional system baseline, while the DSR system showed no difference. DSR scores deviated from conventional scores in the DSR treatment, but not by much. Based in part on our findings, eBay implemented the DSR system during 2007. Preliminary data from the site showed the system to be working much as we would expect from the laboratory model.

For the present purposes, this example illustrates two important points. First, motives that explain behavior well in one facet of the market, in this case price behavior, do not necessarily fit well with other facets of the market. It is difficult to explain the reciprocal nature of trader feedback, or, indeed, the fact that traders take the time to give feedback at all, in terms of the standard benchmark of rational self-interest. Second, even though we do not fully understand the objectives behind reciprocal behavior, we can capture it in a laboratory model and stress test market designs intended to curb the misbehavior in a fairly nuanced way. The DSR system curves the undesirable misbehavior involving seller retaliation while allowing the misbehavior that is good for the market to go forward.

HETEROGENEOUS MISBEHAVIOR IN THE NEWSVENDOR PROBLEM

The pull-to-center effect

The newsvendor problem was first studied by Arrow et al. (1951). It remains today a fundamental building block for models of inventory management in the face of stochastic demand (Porteus, 1990), and at a broader level for models of supply chain systems (Cachon, 2002). The newsvendor's problem is that he must stock his entire inventory prior to the selling season, knowing only the stochastic distribution from which the quantity demanded will be drawn. Order too little, and he loses sales; order too much, and he must dispose of the excess stock at a loss. The optimal solution for a risk-neutral newsvendor is to stock up to the point where the expected loss of (over)stocking one more newspaper is equal to the expected loss of (under)stocking one less.

Schweitzer and Cachon (2000) conducted the first laboratory study of the newsvendor problem. An important feature of their design was treatments that examined both a high-safety and a low-safety stock version of the game in which the optimum inventory order was above (below) average demand. The game was repeated and subjects were provided feedback on realized demand and profitability at the end of each round. The data showed a pull-to-center effect in both kinds of conditions; that is, newsvendors on average tended to order away from the expect profit-maximizing order and towards the average demand. This is important because, as Schweitzer and Cachon show, the pattern is inconsistent with any expected utility profile, while prospect theory is consistent with some but not all of the pattern. So pull-to-center qualifies as a form of misbehavior.

It is a pattern of misbehavior that has proven remarkably robust. The bias persists for a variety of demand distributions (Benzion et al., 2008), with substantial task repetition, and even when descriptive statistics on the performance are provided to subjects (Bolton and Katok, 2008). More frequent feedback can actually degrade performance (Lurie and Swaminathan, 2009; Katok and Davis, 2008).

There is as yet no commonly agreed explanation for the pull-to-center bias, although several candidates have emerged. Schweitzer and Cachon (2000) offer two explanations consistent with their data. One is anchoring and insufficient adjustment (Tversky and Kahneman, 1974), the anchoring in this case being the mean demand. The other is minimization of the ex-post inventory error. Recent learning models rely on adaptive behavior (Bostian et al., 2008), computational errors (Su, 2008), an overconfidence bias (Ren and Croson, 2012), or limitations on cognitive reflection (Moritz et al., 2011). Ho et al. (2010) posit psychological costs associated with leftovers and stockouts.

The limitations of information, experience, and training

A natural reaction to these findings is to wonder how robust they are to the student subject pool used in the aforementioned studies. Perhaps managers with experience in procurement would decide more optimally. Perhaps better information or training would lead subjects to perform more optimally.

An experiment by Bolton et al. (2012) explores these issues. The experiment samples three experiential groups: freshman business students who have had no course in operations management; graduate business students who have had at least one undergraduate course in operations management and so have likely been exposed to the newsvendor problem; and working managers with practical experience in newsvendor-type procurement. Classroom instruction on the newsvendor problem exposes students to the broad principles underlying inventory control. Actual procurement experience provides intensive exposure to practical inventory problems. Procurement managers are also subject to market selection pressure. So we might expect managers and students to approach the newsvendor problem differently.

Since the ability to handle information is critical here, the experiment exposed subjects to varying levels of information and task training. Most of the previous studies provided subjects with the information about the demand distribution. In the first phase of this experiment, subjects were provided with only historical information about demand, the kind of information condition that managers often face in the field. In the second phase, information on the demand distribution was provided. In the third phase, information regarding the expected profit from orders was given. The information provided in either the second or the third phase is sufficient to identify the expected profit-maximizing order, but in the latter case less deduction is required. Thus the experiment allows a comparison of analytical sophistication across the subject groups.

In addition to the basic briefing, some subjects received a sixty-minute video lecture immediately before the game. The lecture explained in detail the rationale behind the optimal order-quantity calculation and informed the subjects that people often have a tendency to order toward the mean demand and explained why that is wrong. This on-the-spot training is more immediate than the classroom experience and provides more rationale than does a simple presentation of the expected profit statistics.

The main finding of the study is that manager decisions exhibit the same pull-to-center effect as do both groups of students in the study (similar to that in previous studies). As analytical information about the demand distribution and expected profits is introduced, orders adjust toward the expected profit-maximizing quantity, but not all the way. The student group with an operations management background best utilized this information. All three groups benefited substantially from on-the-spot training, and in fact performed approximately the same.

These findings suggest that experience has limited value as a corrective to newsvendor misbehavior. Experienced managers exhibit a similar bias as do the students. The fact that students with an operations management background handle analytical information better than the other groups suggests that classroom education provides important insight into the process behind the newsvendor solution. Consistent with this observation, training has a strongly positive effect on performance, particularly when it is coupled with an operations management background. However, the fact that all groups perform better and about the same with the addition of training than without suggests that time lags, too, play an important role in the effectiveness of classroom education. And the fact that theoretically redundant information on expected profit significantly improves performance across all subject pools suggests that overcoming the computational problems involved in the newsvendor problem is a challenge even with education and training. For these reasons, it seems sensible to look for market design fixes to newsvendor misbehavior.

Stress testing design cures: action restrictions and incentives

One way one might attempt to curve the misbehavior is to put in place an institutional rule that discourages suboptimal decisions. The challenge here is that the misbehavior takes a variety of forms. Indeed, the lack of a commonly agreed explanation for the pull-to-center effect is arguably due at least in part to the heterogeneity of behavioral patterns observed among newsvendor subjects. Bolton and Katok (2008) clustered individual newsvendors into categories of search (mis)behavior, as behavior consistent with: the gambler's fallacy, based on a fallacious belief that independent draws are either positively correlated (as with the "hot hand" fallacy in basketball) or negatively correlated (e.g. believing a number on the roulette wheel is "due"); choices not statistically different from random; (mostly) optimum ordering behavior; or demand matching behavior. For the baseline newsvendor treatment in their experiment, about two-thirds of the subjects either correspond to the gambler's fallacy (about 40%) or have a modal order of the average demand (25%). About 30% have a modal order that is the optimum order. Choices of about 5% are not distinguishable from random.

On the surface, the misbehavior is diverse, yet underneath is a uniting pattern: the "law of small numbers," a tendency to believe that statistically (too) small samples are representative (Tversky and Kahneman, 1971). In fact, in the Bolton and Katok study, for newsvendors not classified as optimum, the average sample run for a single order was 2.4, with a median and mode of just 1. The uninformative nature of this kind of cursory sampling might explain why so many newsvendors move so little from the initial anchor of ordering average demand.

Bolton and Katok studied an institutional rule that attacks the law of small numbers. In one treatment, newsvendors were restricted to ordering a standing (fixed) quantity for a sequence of ten demand periods. As a point of comparison, we also ran a treatment in which newsvendors order for one demand period at a time but receive, prior to ordering, a statistical analysis of order profitability, including the expected profitability. This manipulation permits a test of whether it is the restriction on ordering behavior that is critical to behavior or whether the additional information the subjects gain from the extended sampling is an adequate explanation.

The data showed that the additional information had but a marginal effect on order-ing decisions. In contrast, the ten-demand-period restriction had a strong effect on the pattern of individual ordering, effectively doubling the amount of optimal ordering (about 60%), while wiping out the negative correlation category of misbehavior as well as anchoring on average demand misbehavior. All in all, restricting newsvendors to longer-term sampling of an order was an effective way to encourage more optimal ordering.

Becker-Peth et al. (2013) take a different approach to discouraging misbehavior by tailoring the costs associated with over- and understocking. They begin by constructing a behavioral model that supposes newsvendors are influenced by loss aversion and anchoring toward the mean. They fit the model, at both the aggregate and the individual level, to decisions laboratory subjects make over a variety of parameterizations of the newsvendor problem. The same subjects then play a new series of games, this time with parameterizations modified to account for the psychological biases identified by the model. The aim here is to provide subjects with incentives that nudge them away from misbehavior and toward optimum behavior.

The behavior in the experiment shows an improved fit with optimum behavior. Indi-vidual parameterizations work better than the aggregate model. A quote for the Becker-Peth et al. paper serves well as a summary to this section:

> There are a number of managerial implications from our research. It shows that people respond irrationally to supply contracts, but that their responses can be reasonable well predicted. Contract designers who are aware of this can use this knowledge in contract negotiations. For instance, if a buyer is reluctant to accept a contract with a low wholesale price and low buyback price, the contract designer might consider offering a higher buyback price and simultaneously increase the wholesale price. Our research indicates that such a contract would be preferred by many buyers. However, there are also buyers who prefer the opposite and the task of the contract designer is to classify the buyer. Because people's behavioral preferences differ, we cannot provide recommendations that hold universally. However, we can provide the general recommendation to realize that people often value different income streams differently, that they frame a contract, and that they place a different value on gains than on losses, information that can be valuable in contract design.

CONCLUSION

One of the comparative advantages of market design as an empirical research method is documenting anomalies of apparent functional importance. Focusing on the price-setting function of markets, it is easy dismiss the human propensity of reciprocity. Self-interest works just fine to explain what we observe. But when confronted with the feedback system, the trust backbone of the market whose very existence is doubtful on the basis of self-interest, the importance of understanding reciprocal behavior becomes more pressing.

The same is true for newsvendor decision errors, which at first might seem best explained as random noise but on closer inspection are biased toward inefficiency. An important role for market design then is to find market rules to move decisions toward more efficient outcomes.

The lab can serve as an important tool for stress testing a market design against misbehavior. Competing design proposals can be tightly manipulated for clear comparison. And the lab is relatively cheap. The newsvendor experiments on procurement managers reported here were collected on site, over the Internet. Given the difficulties in forecasting human (mis)behavior, a new market design might be tested in stages, starting first in a lab setting (cheapest, little risk), proceeding to small-scale field tests (more expensive, somewhat more risk), and then proceeding to a broad implementation (most expensive, with the most risks).

REFERENCES

Arrow, K. J., Harris, T. and Marschak, J. (1951) "Optimal inventory policy," *Econometrica*, 19(3): 250–72.

Ba, S. and Pavlou, P. (2002) "Evidence of the effect of trust building technology in electronic markets: price premiums and buyer behavior," *MIS Quarterly*, 26(3): 243–68.

Bajari, P. and Hortaçsu, A. (2003) "The winner's curse, reserve prices and endogenous entry: empirical insights from eBay auctions," *Rand Journal of Economics*, 34(2): 329–55.

———— (2004) "Economic insights from Internet auctions," *Journal of Economic Literature*, 42(2): 457–86.

Becker-Peth, M., Katok, E. and Thonemann, U. W. (2013) "Designing contracts for irrational but predictable newsvendor," Working Paper.

Benzion, U., Cohen, Y., Peled, R. and Shavit, T. (2008) "Decision-making and the newsvendor problem—an experimental study," *Journal of the Operational Research Society*, 59: 1281–7.

Bolton, G. E. and Katok, E. (2008) "Learning-by-doing in the newsvendor problem: a laboratory investigation of the role of experience and feedback," *Manufacturing and Services Operations Management*, 10: 519–38.

——— Greiner, B. and Ockenfels, A. (2012) "Engineering trust: reciprocity in the production of reputation information," *Management Science*.

——— Ockenfels, A. and Thonemann, U. W. (forthcoming) "Managers and students as newsvendors," *Management Science*.

Bostian, A., Holt, C. and Smith, A. (2008) "The newsvendor 'pull-to-center effect': adaptive learning in a laboratory experiment," *Manufacturing and Service Operations Management*, 10(4): 590–608.

Cachon, G. P. (2002) "Supply chain coordination with contracts," in S. Graves and T. de Kok (eds), *Handbook in OR & MS, Supply Chain Management*, Elsevier, pp. 229–339.

Cooper, D. and Kagel, J. (forthcoming) "Other-regarding preferences," in J. Kagel and A. Roth (eds), *The Handbook of Experimental Economics*, vol. 2.

Dellarocas, C. (2004) "Building trust on-line: the design of robust reputation mechanisms for online trading communities," in G. Doukidis, N. Mylonopoulos and N. Pouloudi (eds), *Social and Economic Transformation in the Digital Era*, Idea Group Publishing.

——— and Wood, C. A. (2008) "The sound of silence in online feedback: estimating trading risks in the presence of reporting bias," *Management Science*, 54(3): 460–76.

Güth, W., Mengel, F. and Ockenfels, A. (2007) "An evolutionary analysis of buyer insurance and seller reputation in online markets," *Theory and Decision*, 63: 265–82.

Ho, T., Lim, N. and Cui, T. (2010) "Reference dependence in multilocation newsvendor models: a structural analysis," *Management Science*, 56(11): 1891–910.

Houser, D. and Wooders, J. (2005) "Reputation in auctions: theory and evidence from eBay," *Journal of Economics and Management Strategy*, 15(2): 353–69.

Katok, E. D. T. and Davis, A. (2008) "Inventory service-level agreements as coordination mechanisms: the effect of review periods," *Manufacturing and Service Operations Management*, 10(4): 609–24.

Klein, T. J., Lambertz, C., Spagnolo, G. and Stahl, K. O. (2006) "Last minute feedback," CEPR Discussion Papers 5693, C.E.P.R. Discussion Papers.

Lurie, N. H. and Swaminathan, J. M. (2009) "Is timely information always better? The effect of feedback frequency on decision making," *Organizational Behavior and Human Decision Processes*, 108(2): 315–29.

Moritz, B., Hill, A. V. and Donohue, K. (2013) "Individual differences in the newsvendor problem: behavior and cognitive reflection," *Journal of Operations Management*, 31(1–2): 72–85.

Porteus, E. L. (1990) "Stochastic inventory theory," in D. P. Heyman and M. J. Sobel (eds), *Handbook in OR & MS*, Elsevier, vol. 2, pp. 605–52.

Reichling, F. (2004) "Effects of reputation mechanisms on fraud prevention in eBay auctions," Working Paper, Stanford University.

Ren, Y. and Croson, R. T. A. (2012) "Explaining biased newsvendor orders: an experimental study," Working Paper, University of Texas, Dallas.

Schweitzer, M. E. and Cachon, G. P. (2000) "Decision bias in the newsvendor problem with known demand distribution: experimental evidence," *Management Science*, 46: 404–20.

Su, X. (2008) Bounded rationality in newsvendor models. *Manufacturing and Service Operations Management*, 10(4), 566–589.

Tversky, A. and Kahneman, D. (1971) "The belief in the law of small numbers," *Psychological Bulletin*, 76: 105–110.

———— (1974) "Judgment under uncertainty: heuristics and biases," *Science*, 185: 1124–31.

CHAPTER 3

··

USING AND ABUSING
AUCTION THEORY

··

PAUL KLEMPERER[1]

INTRODUCTION

··

FOR half a century or more after the publication of his *Principles* (1890), it was routinely asserted of economic ideas that "they're all in Marshall." Of course, that is no longer true of the theory itself. But Marshall was also very concerned with applying economics, and when we think about how to use the theory, the example that Marshall set still remains a valuable guide. In this chapter, therefore, I want to use some of Marshall's views, and my own experience in auction design, to discuss the use (and abuse) of economic theory.[2]

[1] This chapter was originally published in the *Journal of the European Economic Association* (2003), 1(2–3): 272–300. It is reproduced here with the kind permission of the European Economic Association and the MIT Press. It was improved by an enormous number of helpful comments from Tony Atkinson, Sushil Bikhchandani, Erik Eyster, Nils-Henrik von der Fehr, Tim Harford, Michael Landsberger, Kristen Mertz, Meg Meyer, Paul Milgrom, David Myatt, Marco Pagnozzi, Rob Porter, Kevin Roberts, Mike Rothschild, Peter Temin, Chris Wallace, Mike Waterson, and many others. I advised the UK government on the design of its '3G' mobile-phone auction, and I was a member of the UK Competition Commission from 2001 to 2005, but the views expressed in this paper are mine alone. I do not intend to suggest that any of the behaviour discussed below violates any applicable rules or laws.

[2] This chapter was the text of the 2002 Alfred Marshall Lecture of the European Economic Association, given at its Annual Congress, in Venice. I gave a similar lecture at the 2002 Colin Clark Lecture of the Econometric Society, presented to its Annual Australasian Meeting. Like Marshall, Clark was very involved in practical economic policy-making. He stressed the importance of quantification of empirical facts, which, I argue here, is often underemphasized by modern economic theorists. Similar material also formed the core of the biennial 2002 Lim Tay Boh Lecture in Singapore. Lim was another very distinguished economist (and Vice-Chancellor of the National University of Singapore), who also made significant contributions to policy as an advisor to the Singapore government. Finally, some of these ideas were presented in the Keynote Address to the 2002 Portuguese Economic Association's meeting. I am very grateful to all those audiences for helpful comments.

Although the most elegant mathematical theory is often the most influential, it may not be the most useful for practical problems. Marshall (1906) famously stated that "a good mathematical theorem dealing with economic hypotheses [is] very unlikely to be good economics," and continued by asserting a series of rules: "(1) translate [mathematics] into English; (2) then illustrate by examples that are important in real life; (3) burn the mathematics; (4) if you can't succeed in 2, burn 1"! Certainly this view now seems extreme, but it is salutary to be reminded that good mathematics need not necessarily be good economics. To slightly update Marshall's rules, if we can't (1) offer credible intuition and (2) supply empirical (or perhaps case-study or experimental) evidence, we should (4) be cautious about applying the theory in practice.[3]

Furthermore, when economics is applied to policy, proposals need to be robust to the political context in which they are intended to operate. Too many economists excuse their practical failure by saying "the politicians (or bureaucrats) didn't do exactly what I recommended." Just as medical practitioners must allow for the fact that their patients may not take all the pills they prescribe, or follow all the advice they are given, so economics practitioners need to foresee political and administrative pressures and make their plans robust to changes that politicians, bureaucrats, and lobbyists are likely to impose. And in framing proposals, economists must recognize that policies that seem identical, or almost identical, to them may seem very different to politicians, and vice versa.

Some academics also need to widen the scope of their analyzes beyond the confines of their models, which, while elegant, are often short on real-world detail. Marshall always emphasized the importance of a deep "historical knowledge of any area being investigated and referred again and again to the complexity of economic problems and the naivety of simple hypotheses."[4] Employing "know it all" consultants with narrowly focused theories instead of experienced people with a good knowledge of the wider context can sometimes lead to disaster.

One might think these lessons scarcely needed stating—and Marshall certainly understood them very well—but the sorry history of "expert" advice in some recent auctions shows that they bear repetition. So although the lessons are general ones, I will illustrate them using auctions and auction theory. Auction theory is often held up as a triumph of the application of economic theory to economic practice, but it has not, in truth, been an unalloyed success. For example, while the European and Asian 3G spectrum auctions famously raised over €100 billion in total revenues, Hong Kong's, Austria's, the Netherlands', and Switzerland's auctions, among others, were catastrophically badly run, yielding only a quarter or less of the per capita revenues earned

---

[3] I *mean* cautious about the theory. Not dismissive of it. And (3) seems a self-evident mistake, if only because of the need for efficient communication among, and for the education of, economists, let alone the possibilities for further useful development of the mathematics.

[4] Sills (1968, p. 28). An attractively written appreciation of Marshall and his work is in Keynes (1933).

elsewhere—and economic theorists deserve some of the blame.[5,6] Hong Kong's auction, for example, was superficially well designed, but not robust to relatively slight political interference, which should perhaps have been anticipated. Several countries' academic advisors failed to recognize the importance of the interaction between different countries' auction processes, and bidders advised by experts in auction theory who ignored (or were ignorant of) their clients' histories pursued strategies that cost them billions of euros. Many of these failures could have been avoided if the lessons had been learned to pay more attention to elementary theory, to the wider context of the auctions, and to political pressures—and to pay less attention to sophisticated mathematical theory.[7]

Of course, mathematical theory, even when it has no direct practical application, is not merely beautiful. It can clarify the central features of a problem, provide useful benchmarks and starting points for analysis, and—especially—show the deep relationships between problems that are superficially unconnected. Thus, for example, the sophisticated tools of auction theory that have sometimes been abused in practical contexts turn out to have valuable applications to problems that, at first blush, do not look like auctions.

The following section briefly discusses what is often taken to be the "standard auction theory," before discussing its real relevance. The three sections after that illustrate the abuse of the theory using examples from the Asian and European 3G auctions, and discuss the broader lessons that can be drawn from these misapplications. The third section is in large part based on Klemperer (2000b, 2002a–d), where many additional details can be found—and this section may be skipped by readers familiar with that material—but the other sections make different points using additional examples. The sixth section illustrates how the same concepts that are abused can have surprisingly valuable uses in different contexts. The seventh section concludes.

---

[5] We take the governments' desire for high revenue as given, and ask how well the auctions met this objective. While an efficient allocation of licenses was most governments' first priority, there is no clear evidence of any differences between the efficiencies of different countries' allocations, so revenues were seen as the measure of success. Binmore and Klemperer (2002, section 2) argue that governments were correct to make revenue a priority because of the substantial deadweight losses of raising government funds by alternative means, and because the revenues were one-time sunk costs for firms so should be expected to have only limited effects on firms' subsequent investment and pricing behavior.

[6] The six European auctions in the year 2000 yielded, per capita, €100 (Austria), €615 (Germany), €240 (Italy), €170 (Netherlands), €20 (Switzerland), and €650 (UK) for very similar properties. True, valuations fell during the year as the stock markets also fell, but Klemperer (2002a) details a variety of evidence that valuations ranged from €300 to €700 per capita in all of these auctions. Klemperer (2002a) gives a full description of all nine west European 3G auctions.

[7] Another topical example of overemphasis on sophisticated theory at the expense of elementary theory is European merger policy's heavy focus on the "coordinated" effects that may be facilitated by a merger (and about which we have learnt from repeated game theory) and, at the time of writing, relative lack of concern about the more straightforward "unilateral" effects of mergers (which can be understood using much simpler static game theory). (As a former UK Competition Commissioner, I stress that this criticism does not apply to UK policy!)

THE RECEIVED AUCTION THEORY

The core result that everyone who studies auction theory learns is the remarkable *revenue equivalence theorem* (RET).[8] This tells us, subject to some reasonable-sounding conditions, that all the standard (and many non-standard) auction mechanisms are equally profitable for the seller, and that buyers are also indifferent between all these mechanisms.

If that were all there was to it, auction design would be of no interest. But of course the RET rests on a number of assumptions. Probably the most influential piece of auction theory apart from those associated with the RET is Milgrom and Weber's (1982) remarkable paper—it is surely no coincidence that this is also perhaps the most elegant piece of auction theory apart from the RET. Milgrom and Weber's seminal analysis relaxes the assumption that bidders have independent private information about the value of the object for sale, and instead assumes bidders' private information is *affiliated*. This is similar to assuming positive correlation,[9] and under this assumption they show that ordinary ascending auctions are more profitable than standard (first-price) sealed-bid auctions, in expectation.

Milgrom and Weber's beautiful work is undoubtedly an important piece of economic theory and it has been enormously influential.[10] As a result, many economists leave graduate school "knowing" two things about auctions: first, that if bidders' information is independent, then all auctions are equally good; and second, that if information is affiliated (which is generally the plausible case), then the ascending auction maximizes the seller's revenue.[11]

But is this correct?

---

[8] The RET is due in an early form to Vickrey (1961), and in its full glory to Myerson (1981), Riley and Samuelson (1981), and others. A typical statement is: "Assume each of a given number of risk-neutral potential buyers has a privately known signal about the value of an object, independently drawn from a common, strictly increasing, atomless distribution. Then any auction mechanism in which (1) the object always goes to the buyer with the highest signal, and (2) any bidder with the lowest feasible signal expects zero surplus, yields the same expected revenue (and results in each bidder making the same expected payment as a function of her signal)."

Klemperer (1999a) gives an elementary introduction to auction theory, including a simple exposition, and further discussion, of the RET. See also Klemperer (2004a).

[9] Affiliation is actually a stronger assumption, but it is probably typically approximately satisfied.

[10] Not only is the concept of affiliation important in applications well beyond auction theory (see the section "Using economic theory") but this paper was also critical to the development of auction theory, in that it introduced and analyzed a general model including both private and common value components.

[11] Or, to take just one very typical example from a current academic article, "The one useful thing that our single unit auction theory can tell us is that when bidders' [signals] are affiliated…the English [that is, ascending] auction should be expected to raise the most revenue," (Klemperer 2003a).

Relevance of the received theory

Marshall's (updated) tests are a good place to start. The value of empirical evidence needs no defense, while examining the plausibility of an intuition helps check whether an economic model provides a useful caricature of the real world, or misleads us by absurdly exaggerating particular features of it. [12]

The intuition behind the exact RET result cannot, to my knowledge, be explained in words that are both accurate and comprehensible to lay people. Anyone with the technical skill to understand any verbal explanation would probably do so by translating the words back into the mathematical argument. But it is easier to defend the weaker claim that it is ambiguous which of the two most common auction forms is superior: it *is* easy to explain that participants in a sealed-bid auction shade their bids below their values (unlike in an ascending auction), but that the winner determines the price (unlike in an ascending auction), so it is not hard to be convincing that there is no clear reason why either auction should be more profitable than the other. This is not quite the same as arguing that the standard auction forms are approximately similarly profitable, but the approximate validity of the RET (under its key assumptions) in fact seems consistent with the available evidence. (Some would say that the mere fact that both the ascending auction and the sealed-bid auction are commonly observed in practice is evidence that neither is always superior.) So the "approximate RET" seems a reasonable claim in practice, and it then follows that issues assumed away by the RET's assumptions should be looked at to choose between the standard auction forms. These issues should include not just those made explicitly in the statement of the theorem (for example bidders are symmetric and risk-neutral), but also those that are implicit (for example bidders share common priors and play non-cooperative Nash equilibrium) or semi-implicit (for example the number and types of bidders are independent of the auction form).

However, as already noted, much attention has focused on just one of the RET's assumptions, namely independence of the bidders' information, and the theoretical result that if information is non-independent (affiliated), then ascending auctions are more profitable than first-price sealed-bid auctions. There is no very compelling intuition for this result. The verbal explanations that are given are unconvincing and/or misleading, or worse. The most commonly given "explanation" is that ascending auctions allow bidders to be more aggressive, because their "winner's curses" are reduced, [13] but this argument is plain wrong: the winner's curse is a feature only of common-value auctions, but common values are neither necessary nor sufficient for the result. [14]

---

[12] Whether the intuition need be non-mathematical, or even comprehensible to lay people, depends on the context, but we can surely have greater confidence in predicting agents' actions when the agents concerned understand the logic behind them, especially when there are few opportunities for learning.

[13] The "winner's curse" reflects the fact that winning an auction suggests one's opponents have pessimistic views about the value of the prize, and bidders must take this into account by bidding more conservatively than otherwise.

[14] The result applies with affiliated private values, in which bidders' values are unaffected by others' information, so there is no winner's curse, and the result does not apply to independent-signal

A better explanation of the theoretical result is that bidders' profits derive from their private information, and the auctioneer can profit by reducing that private information. [15] An ascending auction reveals the information of bidders who drop out early, so partially reveals the winner's information (if bidders' information is correlated), and uses that information to set the price (through the runner-up's bid), whereas the price paid in a sealed-bid auction cannot use that information. Since the ascending and sealed-bid auctions are revenue-equivalent absent any correlation (that is, with independent signals), and provided the runner-up's bid responds to the additional information that an ascending auction reveals in the appropriate way (which it does when information is affiliated), this effect makes the ascending auction the more profitable. Of course, this argument is obviously still incomplete, [16, 17] and even if it were fully convincing, it would depend on the *exact* RET applying—which seems a very strong claim.

Furthermore, before relying on any theory mattering in practice, we need to ask: what is the likely order of magnitude of the effect? In fact, numerical analysis suggests the effects of affiliation are often tiny, even when bidders who exactly fit the assumptions of the theory compute their bids exactly using the theory. Riley and Li (1997) analyze

---

common-value auctions, which do suffer from the winner's curse. (Where there is a winner's curse, the "theory" behind the argument is that bidders' private information can be inferred from the points at which they drop out of an ascending auction, so less "bad news" is discovered at the moment of winning than is discovered in winning a sealed-bid auction, so bidders can bid more aggressively in an ascending auction. But this assumes that bidders' more aggressive bidding more than compensates for the reduced winner's curse in an ascending auction—in independent-signal common-value auctions it exactly compensates, which is why there is no net effect, as the RET proves.) In fact, many experimental and empirical studies suggest bidders fail to fully account for winner's curse effects, so these effects may in practice make sealed-bid auctions more profitable than ascending auctions!

[15] Absent private information, the auctioneer would sell to the bidder with the highest expected valuation at that expected valuation, and bidders would earn no rents. The more general result that, on average, the selling price is increased by having it depend on as much information as possible about the value of the good, is Milgrom and Weber's (1982, 2000) linkage principle. However, in more recent work, Perry and Reny (1999) show that the principle applies less generally (even in theory) than was thought.

[16] Revealing more information clearly need not necessarily reduce bidders' profits (if bidders' information is negatively correlated, the contrary is typically true); the conditions that make the ascending price respond correctly to the additional information revealed are quite subtle, and nor does the argument say anything about how affiliation affects sealed bids. Indeed, there are simple and not unnatural examples with the "wrong kind" of *positive* correlation in which the ranking of auctions' revenues is reversed (see Bulow and Klemperer, forthcoming), and Perry and Reny (1999) also show the trickiness of the argument by demonstrating that the result holds only for single-unit auctions. A more complete verbal argument for the theoretical result is given in Klemperer (1999a, appendix C), but it is very hard (certainly for the layman).

[17] Another loose intuition is that in an ascending auction each bidder acts as if he is competing against an opponent with the same valuation. But in a sealed-bid auction a bidder must outbid those with lower valuations. With independent valuations, the RET applies. But if valuations are affiliated, a lower-valuation bidder has a more conservative estimate of his opponent's valuation and therefore bids more conservatively. So a bidder in a sealed-bid auction attempting to outbid lower-valuation bidders will bid more conservatively as well. But this argument also rests on the RET applying exactly, and even so several steps are either far from compelling (for example, the optimal bid against a more conservative opponent is not always to be more conservative), or very non-transparent.

equilibrium in a natural class of examples and show that the revenue difference between ascending and first-price auctions is very small unless the information is very strongly affiliated: when bidders' values are jointly normally distributed, bidders' expected rents are about 10% (20%) higher in a sealed-bid auction than in an ascending auction even for correlation coefficients as high as 0.3 (0.5). So these results suggest affiliation could explain why a 3G spectrum auction earned, for example €640 rather than €650 per capita when bidders' valuations were €700 per capita. But the actual range was from just €20 (*twenty*) to €650 per capita! Riley and Li also find that even with very strong affiliation, other effects, such as those of asymmetry, are more important and often reverse the effects of affiliation, even taking the numbers of bidders, non-cooperative behaviour, common priors, and so on, as given.[18] This kind of quantitative analysis surely deserves more attention than economists often give it.

Finally, all the previous discussion is in the context of single-unit auctions. Perry and Reny (1999) show that the result about affiliation does not hold—even in theory—in multi-unit auctions.[19]

Given all this, it is unsurprising that there is no empirical evidence (that I am aware of) that argues that the affiliation effect is important.[20, 21]

---

[18] An easier numerical example than Riley and Li's assumes bidder i's value is $v_i = \theta + t_i$, in which θ and the t_i's are independent and uniform on $[0,1]$, and i knows only v_i. With two bidders, expected revenue is $14/18$ in a first-price sealed-bid auction and $15/18$ in an ascending auction, so bidder rents are $7/18$ and $6/18$ respectively (though with n bidders of whom $n/2$ each win a single object, as $n \to \infty$ bidder rents are 42% higher in the sealed-bid auction). With very extreme affiliation, an auctioneer's profits may be more sensitive to the auction form. Modifying the previous example so that there are two bidders who have completely diffuse priors for θ, bidder rents are 50% higher in a first-price sealed-bid auction than in an ascending auction (see Klemperer, 1999a, appendix D), and Riley and Li's example yields a similar result for correlation coefficients around 0.9 (when bidder rents are anyway small). These examples assume private values. Auctioneers' profits may also be more sensitive to auction form with common values and, in the previous extreme-affiliation model with diffuse priors on θ, if bidders' signals are v_i and the true common value is θ, bidders' rents are twice as high in the sealed-bid auction as in the ascending auction. But, with common values, small asymmetries between bidders are *very* much more important than affiliation (see Klemperer, 1998; Bulow and Klemperer, 2002). Moreover, we will see that other effects also seem to have been quantitatively much more important in practice than affiliation is even in any of these theoretical examples.

[19] The RET, also, only generalizes to a limited extent to multi-unit auctions.

[20] For example, empirical evidence about timber sales suggests rough revenue equivalence, or even that the sealed-bid auction raises more revenue given the number of bidders (Hansen, 1986; Mead and Schneipp, 1989; Paarsch, 1991; Rothkopf and Engelbrecht-Wiggans, 1993; Haile, 1996) though information is probably affiliated. The experimental evidence (see Kagel and Roth, 1995; Levin et al. (1996) is also inconclusive about whether affiliation causes any difference between the revenues from ascending and sealed-bid auctions.

[21] Like Marshall, Colin Clark (1940) emphasized the importance of quantification and real-world facts (see note 2), writing "I have ... left my former colleagues in the English universities ... with dismay at their continued preference for the theoretical ... approach to economic problems. Not one in a hundred ... seems to understand [the need for] the testing of conclusions against ... observed facts.... The result is a vast output of literature of which, it is safe to say, scarcely a syllable will be read in fifty years' time." I think he would be pleased that an academic from an English university is quoting his syllables well over fifty years after he wrote them.

So there seems no strong argument to expect affiliation to matter much in most practical applications; independence is not the assumption of the RET that most needs relaxing.

The theory that really matters most for auction design is just the very elementary undergraduate economics of relaxing the implicit and semi-implicit assumptions of the RET about (fixed) entry and (lack of) collusion.[22] The intuitions are (as Marshall says they should be) easy to explain—we will see that it is clear that bidders are likely to understand and therefore to follow the undergraduate theory. By contrast, the intuition for affiliation gives no sense of how bidders should compute their bids, and the calculations required to do so optimally require considerable mathematical sophistication and are sensitive to the precise assumptions bidders make about the "prior" distributions from which their and others' private information is drawn. Of course, this does not mean agents cannot intuitively make approximately optimal decisions (Machlup, 1946; Friedman, 1953), and individual agents need not understand the intuitions behind equilibrium group outcomes. But we can be more confident in predicting that agents will make decisions whose logic is very clear, especially in one-off events, as many auctions are.

Not surprisingly, practical examples of the undergraduate theory are easy to give (as Marshall also insists). But there is no elegant theory applying to the specific context of auctions; such theory is unnecessary since the basic point is that the main concerns in auctions are just the same as in other economic markets, so much of the same theory applies (see later). Furthermore, some of the key concerns are especially prominent when the assumption of symmetry is dropped, and models with asymmetries are often inelegant.

So graduate students are taught the elegant mathematics of affiliation and whenever, and wherever, I give a seminar about auctions in practice,[23] I am asked a question along the lines of "Haven't Milgrom and Weber shown that ascending auctions raise the most revenue, so why consider other alternatives?" This is true of seminars to academics. It is even more true of seminars to policy-makers. Thus, although a little knowledge of economic theory is a good thing, too much knowledge can sometimes be a dangerous thing. Moreover, the extraordinary influence of the concept of affiliation is only the most important example of this. I give a further illustration, involving overattention to some of my own work, in the next subsection. In short, a little graduate education in auction theory can often distract attention from the straightforward "undergraduate" issues that really matter.[24]

[22] See Klemperer (2002b). Risk aversion and asymmetries (even absent entry issues) also arguably matter more than affiliation (and usually have the opposite effect). It is striking that Maskin and Riley's (1984, 2000) important papers on these topics (see also Matthews, 1983) failed to have the same broad impact as Milgrom and Weber's work on affiliation.

[23] I have done this in over twenty countries on five continents.

[24] True, the generally accepted notion of the "received auction theory" is changing and so is the auction theory that is emphasized in graduate programs. And recent auctions research has been heavily influenced by practical problems. But it will probably remain true that the elegance of a theory will remain an important determinant of its practical influence.

THE ELEMENTARY ECONOMIC THEORY
THAT MATTERS

What really matter in practical auction design are attractiveness to entry and robustness against collusion—just as in ordinary industrial markets.[25] Since I have repeatedly argued this, much of the material of this section is drawn from Klemperer (2000b, 2002a,b) and any reader familiar with these papers may wish to skip to the following section.

Entry

The received theory described above takes the number of bidders as given. But the profitability of an auction depends crucially on the number of bidders who participate, and different auctions vary enormously in their attractiveness to entry; participating in an auction can be a costly exercise that bidders will undertake only if they feel they have realistic chances of winning. In an ascending auction a stronger bidder can always top any bid that a weaker bidder makes, and knowing this the weaker bidder may not enter the auction in the first place—which may then allow the stronger bidder to win at a very low price. In a first-price sealed-bid auction, by contrast, a weaker bidder may win at a price that the stronger bidder could have beaten, but didn't because the stronger bidder may risk trying to win at a lower price and can't change his bid later. So more bidders may enter a first-price sealed-bid auction.[26]

The intuition is very clear, and there is little need for sophisticated theory. Perhaps because of this, or because the argument depends on asymmetries between bidders so any theory is likely to be inelegant, theory has largely ignored the point. Vickrey's (1961) classic paper contains an example (relegated to an appendix, and often overlooked) which illustrates the basic point that the player who actually has the lower value may win a first-price sealed-bid auction in Nash equilibrium, but that this cannot happen in an ascending auction (with private values). But little has been said since.

---

[25] Of course, auction theorists have not altogether ignored these issues—but the emphasis on them has been far less. The literature on collusion includes Robinson (1985), Cramton et al. (1987), Graham and Marshall (1987), Milgrom (1987), Hendricks and Porter (1989), Graham et al. (1990), Mailath and Zemsky (1991), McAfee and McMillan (1992), Menezes (1996), Weber (1997), Engelbrecht-Wiggans and Kahn (2005), Ausubel and Schwartz (1999), Brusco and Lopomo (2002a), Hendricks et al. (1999), Cramton and Schwartz (2000). That on entry includes Matthews (1984), Engelbrecht-Wiggans (1987, 1993), McAfee and McMillan (1987, 1988), Harstad (1990), Levin and Smith (1994), Bulow and Klemperer (1996), Menezes and Monteiro (2000), Persico (2000), Klemperer (1998), Gilbert and Klemperer (2000). See also Klemperer (1999a, 2000a, 2004a,b, 2005, 2008).

[26] The point is similar to the industrial organization point that because a Bertrand market is more competitive than a Cournot market for any given number of firms, the Bertrand market may attract less entry, so the Cournot market may be more competitive if the number of firms is endogenous.

In fact, some of what has been written about attracting entry provides a further illustration of the potentially perverse impact of sophisticated theory. Although the point that weaker bidders are unlikely to win ascending auctions, and may therefore not enter them, is very general, some work—including Klemperer (1998)[27]—has emphasized that the argument is especially compelling for 'almost-common-value' auctions, and this work may have had the unintended side effect of linking the entry concern to common values in some people's minds;[28] I have heard economists who know the latter work all too well say that because an auction does not involve common values, there is no entry problem![29] To the extent that the almost-common values theory (which is both of more limited application, and also assumes quite sophisticated reasoning by bidders) has distracted attention from the more general point, this is another example of excessive focus on sophisticated theory at the expense of more elementary, but more crucial, theory.

There is an additional important reason why a first-price sealed-bid auction may be more attractive to entrants: bidders in a sealed-bid auction may be much less certain about opponents' strategies, and the advantage of stronger players may therefore be less pronounced, than standard equilibrium theory predicts. The reason is that, in practice, players are not likely to share common priors about distributions of valuations and, even if they do, they may not play Nash equilibrium strategies (that is, a sealed-bid auction induces "strategic uncertainty"). So even if players were in fact ex ante symmetric (that is, their private information is drawn from identical distributions), the lower-value player might win a first-price sealed-bid auction, but would never win an ascending auction (in which bidders' strategies are very straightforward and predictable). When players are not symmetric, Nash equilibrium theory predicts that a weaker player will sometimes beat a stronger player in a sealed-bid auction, but I conjecture that strategic uncertainty and the absence of common priors make this outcome even more likely than Nash equilibrium predicts. Since this point is very hard for standard economic theory to capture, it has largely been passed over. But it reinforces the point that a sealed-bid auction is in many circumstances more likely than an ascending auction to attract entry, and this will often have a substantial effect on the relative profitabilities of the auctions.

The 3G auctions provide good examples of oversensitivity to the significance of information revelation and affiliation at the expense of insensitivity to the more important issue of entry. For example, the Netherlands sold five 3G licenses in a context in which there were also exactly five incumbent mobile-phone operators that were the natural winners, leaving no room for any entrant. (For competition-policy reasons, bidders were permitted to win no more than one license each.) The problem of attracting enough entry to have a competitive auction should therefore have been uppermost in planners'

[27] See also Bikhchandani (1988), Bulow et al. (1999), and Bulow and Klemperer (2002).

[28] In spite of the fact that I have made the point that the argument applies more broadly in, for example, Klemperer (1999b, 2002b). See also Gilbert and Klemperer (2000).

[29] Similarly, others have asserted that the reason the UK planned to include a sealed-bid component in its 3G design if only four licenses were available for sale (see below) was because the auction designers (who included me) thought the auction was almost-common values—but publicly available government documents show that we did not think this was likely.

minds. But the planners seem instead to have been seduced by the fact that ascending auctions raise (a little) extra revenue because of affiliation and also increase the likelihood of an efficient allocation to those with the highest valuations.[30] The planners were probably also influenced by the fact that previous spectrum auctions in the US and the UK had used ascending designs,[31] even though they had usually done so in contexts in which entry was less of a concern, and even though some US auctions did suffer from entry problems. The result of the Netherlands auction was both predictable and predicted—see, for example, Maasland (2000) and Klemperer (2000b), quoted in the Dutch press prior to the auction. There was no serious entrant.[32] Revenue was less than a third of what had been predicted and barely a quarter of the per capita amounts raised in the immediately preceding and immediately subsequent 3G auctions (in the UK and Germany respectively). The resulting furor in the press led to a parliamentary inquiry.

By contrast, when Denmark faced a very similar situation in its 3G auctions in late 2001—four licenses for sale and four incumbents—its primary concern was to encourage entry.[33] The designers had both observed the Netherlands fiasco, and also read Klemperer (2000b). It chose a sealed-bid design (a "4th price" auction) and had a resounding success. A serious entrant bid, and revenue far exceeded expectations and was more than twice the levels achieved by any of the other three European 3G auctions (Switzerland, Belgium, and Greece) that took place since late 2000.

The academics who designed the UK sale (which was held prior to the Netherlands and Danish auctions) also thought much harder about entry into their 3G auction.[34]

---

[30] It seems unlikely that the efficiency of the Netherlands auction was much improved by the ascending design.

[31] We discuss the UK design below. The design of the US auctions, according to McMillan (1994, pp. 151-2), who was a consultant to the US government, was largely determined by faith in the linkage principle and hence in the revenue advantages of an ascending auction in the presence of affiliation; the economic theorists advising the government judged other potential problems with the ascending design "to be outweighed by the bidders' ability to learn from other bids in the auction" (McMillan, 1994; see also Perry and Reny, 1999). Efficiency was also a concern in the design of the US auctions.

[32] There was one entrant which probably did not seriously expect to win a license in an ascending auction—indeed, it argued strongly prior to the auction that an ascending auction gave it very little chance and, more generally, reduced the likelihood of entry into the auction. Perhaps it competed in the hope of being bought off by an incumbent by, for example, gaining access rights to an incumbent's network, in return for its quitting the auction early. The Netherlands government should be very grateful that this entrant competed for as long as it did! See Klemperer (2002a) and van Damme (2002) for details.

[33] Attracting entry was an even more severe problem in late 2001 than in early summer 2000 when the Netherlands auction was held. The dotcom boom was over, European telecoms' stock prices at the time of the Danish auction were just one-third the levels they were at in the Dutch auction, and the prospects for 3G were much dimmer than they had seemed previously.

[34] I was the principal auction theorist advising the Radiocommunications Agency, which designed and ran the UK auction. Ken Binmore had a leading role, including also supervising experiments testing the proposed designs. Other academic advisors included Tilman Börgers, Jeremy Bulow, Philippe Jehiel, and Joe Swierzbinksi. Ken Binmore subsequently advised the Danish government on its very successful auction. The views expressed in this paper are mine alone.

The UK had four incumbent operators, and when design work began it was unclear how many licenses it would be possible to offer, given the technological constraints. We realized that if there were just four licenses available it would be hard to persuade a non-incumbent to enter, so we planned in that case to use a design including a sealed-bid component (an "Anglo-Dutch" design) to encourage entry. In the event, five licenses were available so, given the UK context, we switched to an ascending auction, since there was considerable uncertainty about who the fifth strongest bidder would be (we ran the world's first 3G auction in part to ensure this—see the section "Understanding the wider context").[35] Thirteen bidders entered, ensuring a highly competitive auction which resulted in the highest per capita revenue among all the European and Asian 3G auctions.

Collusion

The received auction theory also assumes bidders play non-cooperatively in Nash equilibrium. We have already discussed how Nash equilibrium may be a poor prediction because of "strategic uncertainty" and the failure of the common priors assumption, but a more fundamental problem is that players may behave collusively rather than non-cooperatively. In particular, a standard ascending auction—especially a multi-unit ascending auction—often satisfies *all* the conditions that elementary economic theory tells us are important for facilitating collusion, even without any possibility of interaction or discussion among bidders beyond the information communicated in their bids.

For example, Waterson's (1984) standard industrial organization textbook lists five questions that must be answered affirmatively for firms to be able to support collusion in an ordinary industrial market: (1) Can firms easily identify efficient divisions of the market? (2) Can firms easily agree on a division? (3) Can firms easily detect defection from any agreement? (4) Can firms credibly punish any observed defection? (5) Can firms deter non-participants in the agreement from entering the industry? In a multi-unit ascending auction: (1) the objects for sale are well defined, so firms can see how to share the collusive 'pie' among them (by contrast with the problem of sharing an industrial market whose definition may not be obvious); (2) bids can be used to signal proposals about how the division should be made and to signal agreement; (3) firms'

---

[35] With five licenses, the licenses would be of unequal size, which argued for an ascending design. Note that in some contexts an ascending design may promote entry. For example, when Peter Cramton, Eric Maskin, and I advised the UK government on the design of its March 2002 auction of reductions in greenhouse gas emissions, we recommended an ascending design to encourage the entry of small bidders for whom working out how to bid sensibly in a discriminatory sealed-bid auction might have been prohibitively costly. (Strictly speaking, the auction was a descending one, since the auction was a reverse auction in which firms were bidding to sell emissions reductions to the government. But this is equivalent to an ascending design for a standard auction to sell permits.) (Larry Ausubel and Jeremy Bulow were also involved in the implementation of this design.)

pricing (that is, bidding) is immediately and perfectly observable, so defection from any collusive agreement is immediately detected; (4) the threat of punishment for defection from the agreement is highly credible, since punishment is quick and easy and often costless to the punisher in a multi-object auction in which a player has the ability to raise the price only on objects that the defector will win;[36] and (5) we have already argued that entry in an ascending auction may be hard.

So, collusion in an ascending auction seems much easier to sustain than in an "ordinary" industrial market, and it should therefore be no surprise that ascending auctions provide some particularly clear examples of collusion, as we illustrate below.

By contrast, a first-price sealed-bid auction is usually much more robust against collusion: bidders cannot "exchange views" through their bids, or observe opponents' bids until after the auction is over, or punish defection from any agreement during the course of the auction, or easily deter entry. But, perhaps because auction theorists have little that is new or exciting to say about collusion, too little attention has been given to this elementary issue in practical applications.

In the Austrian 3G auction, for example, twelve identical blocks of spectrum were sold to six bidders in a simultaneous ascending auction (bidders were allowed to win multiple blocks each). No one was in the least surprised when the bidding stopped just above the low reserve price, with each bidder winning two blocks, at perhaps one-third the price that bidders valued them at.[37] Clearly, the effect of "collusion" (whether explicit and illegal, or tacit and possibly legal) on revenues is first order.

Another elegant example of bidders' ability to "collude" is provided by the 1999 German DCS-1800 auction in which ten blocks of spectrum were sold by ascending auction, with the rule that any new bid on a block had to exceed the previous high bid by at least 10 percent.[38] There were just two credible bidders, the two largest German mobile-phone companies, T-Mobil and Mannesman, and Mannesman's first bids were DM18.18 million per MHz on blocks 1–5 and DM20 million per MHz on blocks 6–10. T-Mobil—which bid even less in the first round—later said, "There were no agreements with Mannesman. But [we] interpreted Mannesman's first bid as an offer" (Stuewe, 1999, p. 13). The point is that 18.18 plus a 10 percent raise equals 20.00. It seems T-Mobil

---

[36] For example, in a multi-license US spectrum auction in 1996–7, US West was competing vigorously with McLeod for lot number 378—a license in Rochester, Minnesota. Although most bids in the auction had been in exact thousands of dollars, US West bid $313,378 and $62,378 for two licenses in Iowa in which it had earlier shown no interest, overbidding McLeod, which had seemed to be the uncontested high-bidder for these licenses. McLeod got the point that it was being punished for competing in Rochester, and dropped out of that market. Since McLeod made subsequent higher bids on the Iowa licenses, the "punishment" bids cost US West nothing (Cramton and Schwartz, 2000).

[37] Although it did not require rocket science to determine the obvious way to divide twelve among six, the largest incumbent, Telekom Austria, probably assisted the coordination when it announced in advance of the auction that it "would be satisfied with just two of the 12 blocks of frequency on offer" and "if the [five other bidders] behaved similarly it should be possible to get the frequencies on sensible terms," but "it would bid for a third frequency block if one of its rivals did" (Crossland, 2000).

[38] Unlike my other examples this was not a 3G auction; however, it is highly relevant to the German 3G auction which we will discuss.

understood that if it bid DM20 million per MHz on blocks 1–5, but did not bid again on blocks 6–10, the two companies would then live and let live, with neither company challenging the other on the other's half. Exactly that happened. So the auction closed after just two rounds, with each of the bidders acquiring half the blocks for the same low price, which was a small fraction of the valuations that the bidders actually placed on the blocks.[39]

This example makes another important point. The elementary theory that tells us that "collusion" is easy in this context is important. The reader may think it obvious that bidders can "collude" in the setting described, but that is because the reader has been exposed to elementary undergraduate economic theory. This point was beautifully illustrated by the behavior of the subjects in an experiment that was specifically designed to advise one of the bidders in this auction by mimicking its setting and rules: the experimental subjects completely failed to achieve the low-price "collusive" outcome that was achieved in practice. Instead, "in [all] the [experimental] sessions the bidding was very competitive. Subjects went for all ten units in the beginning, and typically reduced their bidding rights only when the budget limit forced them to do so" (Abbink et al., 2002). So the elementary economic theory of collusion which makes it plain, by contrast, that the "collusive" outcome that actually arose was to be expected from more sophisticated players does matter—and I feel confident that the very distinguished economists who ran the experiments advised their bidder more on the basis of the elementary theory than on the basis of the experiments.[40]

Both the UK's and Denmark's academic advisors gave considerable thought to preventing collusion. Denmark, for example, not only ran a sealed-bid auction, but also allowed bidders to submit multiple bids at multiple locations, with the rule that only the highest bid made by any bidder would count, and also arranged for phony bids to be submitted—the idea was that bidders could not (illegally) agree to observe each other's bids without fear that their partners in collusion would double-cross them, and nor could bidders observe who had made bids, or how many had been made.[41]

[39] See Jehiel and Moldovanu (2001) and Grimm et al. (2003). Grimm et al. argue that this outcome was a non-cooperative Nash equilibrium of the fully specified game. This is similar to the familiar industrial organization point that oligopolistic outcomes that we call "collusive" may be Nash equilibria of repeated oligopoly games. But our focus is on whether outcomes look like competitive, non-cooperative, behavior in the simple analyzes that are often made, not on whether or not they can be justified as Nash equilibria in more sophisticated models.

[40] Abbink et al. write "The lessons learnt from the experiments are complemented by theoretical strategic considerations." Indeed, auctions policy advice should always, if possible, be informed by both theory and experiments.

[41] In the UK's ascending auction, the fact that bidders were each restricted to winning at most a single object, out of just five objects, ruled out tacit collusion to divide the spoils (provided that there were more than five bidders). More important, the large number of bidders expected (because the UK ran Europe's first 3G auction—see the section "Understanding the wider context") also made explicit (illegal) collusion much less likely (see Klemperer, 2002a), and the fact that the UK retained the right to cancel the auction in some circumstances also reduced bidders' incentive to collude.

ROBUSTNESS TO POLITICAL PRESSURES

To be effective, economic advice must also be sensitive to the organizational and political context; it is important to be realistic about how advice will be acted on. Economic advisors commonly explain a policy failure with the excuse that "it would have been okay if they had followed our advice." But medical practitioners are expected to take account of the fact that patients will not follow their every instruction.[42] Why should economic practitioners be different? Maybe it should be regarded as economic malpractice to give advice that will actually make matters worse if it is not followed exactly.

For example, the economic theorists advising the Swiss government on its 3G auction favored a multi-unit ascending auction, apparently arguing along the standard received-auction-theory lines that this was best for both efficiency and revenue. But they recognized the dangers of such an auction encouraging "collusive" behavior and deterring entry, and the advisors therefore also proposed setting a high reserve price. This would not only directly limit the potential revenue losses from collusion and/or inadequate entry but, importantly, also reduce the likelihood of collusion. With a high reserve price, bidders are relatively more likely to prefer to raise the price to attempt to drive their rivals out altogether than to collude with them at the reserve price—see Klemperer (2002b) and Brusco and Lopomo (2002b).

But high reserve prices are often unpopular with politicians and bureaucrats who—even if they have the information to set them sensibly—are often reluctant to run even a tiny risk of not selling the objects, which outcome they fear would be seen as "a failure."

The upshot was that no serious reserve was set. Through exit, joint venture, and possibly—it was rumored—collusion,[43] the number of bidders shrank to equal the number of licenses available, so the remaining bidders had to pay only the trivial reserve price that had been fixed. (Firms were allowed to win just a single license each.) The outcome was met with jubilation by the bidders and their shareholders; per capita revenues were easily the lowest of any of the nine western European 3G auctions, and less

---

[42] Doctors are trained to recognize that some types of patient may not take all prescribed medicines or return for follow-up treatment. Pharmaceutical companies have developed one-dose regimens that are often more expensive or less effective than multiple-dose treatments, but that overcome these specific problems. For example, the treatment of chlamydial infection by a single dose of azithromycin is much more expensive and no more effective than a seven-day course of doxycycline; there is a short (two-month) course of preventive therapy for tuberculosis that is both more expensive, and seems to have more problems with side effects, than the longer six-month course; and the abridged regimen for HIV-positive women who are pregnant (to prevent perinatal transmission) is less effective than the longer, more extensive treatment.

[43] Two bidders merged the day before the auction was to begin, and a total of five bidders quit in the last four days before the auction. At least one bidder had quit earlier after hearing from its bidding consultants that because it was a weaker bidder it had very little chance of winning an ascending auction. Furthermore, the regulator investigated rumors that Deutsche Telekom agreed not to participate in the auction in return for subsequently being able to buy into one of the winners.

than one-thirtieth of what the government had been hoping for.[44] Perhaps an ascending auction together with a carefully chosen reserve price was a reasonable choice. But an ascending auction with only a trivial reserve price was a disaster, and the economic-theorist advisors should have been more realistic that this was a likely outcome of their advice.[45]

Economic similarity ≠ political similarity

Hong Kong's auction was another case where designers should perhaps have anticipated the political response to their advice. The Hong Kong auction's designers, like Denmark's, had observed the Netherlands fiasco (and had also read Klemperer, 2000b). So they were keen to use a sealed-bid design, given Hong Kong's situation.[46] Specifically, they favored a "fourth-price" sealed-bid design so that all four winners (there were four licenses and firms could win just one license each) would pay the same fourth-highest bid—charging winners different amounts for identical properties might both be awkward and lead to cautious bidding by managements who did not want to risk the embarrassment of paying more than their rivals.[47]

---

[44] In fact, when the denouement of the auction had become clear, the Swiss government tried to cancel it and rerun it with different rules. But in contrast to the UK auction (see note 41), the designers had omitted to allow themselves that possibility. The final revenues were €20 per capita, compared to analysts' estimates of €400–600 per capita in the week before the auction was due to begin. Meeks (2001) shows the jumps in Swisscom's share price around the auction are highly statistically significant and, controlling for general market movements, correspond to the market believing that bidders paid several hundred euros per capita less in the auction than was earlier anticipated.

[45] I am not arguing that an ascending auction plus reserve price is always bad advice, or even that it was necessarily poor advice here. But advisors must make it very clear if success depends on a whole package being adopted, and should think carefully about the likely implementation of their proposals. Greece and Belgium did set reserve prices that seem to have been carefully thought out, but they were perhaps encouraged to do so by the example of the Swiss auction, and also of the Italian and Austrian auctions, which also had reserve prices that were clearly too low, even if not as low as Switzerland's.

[46] In Hong Kong, unlike in the Netherlands and Denmark, there were actually more incumbents than licenses. But not all Hong Kong's incumbents were thought strong. Furthermore, it is much more attractive for strong firms to form joint ventures or collude with their closest rivals prior to a standard ascending auction (when the strengthened combined bidder discourages entry) than prior to a standard sealed-bid auction (when reducing two strong bidders to one may attract entry). So even though the difference in strength between the likely winners and the also-rans seemed less dramatic in Hong Kong than in the Netherlands and Denmark, a standard ascending auction still seemed problematic. So there was a very serious concern—well justified as it turned out—that a standard ascending auction would yield no more bidders than licenses.

[47] In a simple model, if a winning bidder suffers "embarrassment costs", which are an increasing function of the difference between his payment and the lowest winning payment, then bidders are no worse off in expectation than in an auction which induces no embarrassment costs, but the auctioneer suffers. This is a consequence of the revenue equivalence theorem: under its assumptions, mechanisms that induce embarrassment costs cannot affect bidders' utilities (it is irrelevant to the bidders whether the "embarrassment costs" are received by the auctioneer or are social waste), so, in equilibrium, winning bidders' expected payments are lower by the expected embarrassment costs they suffer. See Klemperer (2004a, part I).

However, the designers were also afraid that if the public could observe the top three bids after the auction, then if these were very different from the price that the firms actually paid (the fourth highest bid), the government would be criticized for selling the licenses for less than the firms had shown themselves willing to pay. Of course, such criticism would be ill informed, but it could still be damaging, because even well intentioned commentators find it hard to explain to the general public that requiring firms to pay their own bids would result in firms bidding differently. Thus far, nothing was different from the situation in Denmark. However, whereas the Danish government simply followed the advice it was given to keep all the bids secret and reveal only the price paid, the Hong Kong government felt it could not do this.

Openness and transparency of government was a big political issue in the wake of Hong Kong's return to Chinese rule, and it was feared that secrecy would be impossible to maintain. The advisors therefore proposed to run an auction that was *strategically equivalent* (that is, has an identical game-theoretic structure and therefore should induce identical behavior) to a fourth-price auction, but that did not reveal the three high bids to *anyone*.[48] To achieve this, an ascending auction would be run for the four identical licenses, but dropouts would be kept secret and the price would continue to rise until the point at which the number of players remaining dropped from four to *three*. At this point the last four (including the firm that had just dropped out) would pay the last price at which four players remained in the bidding. Since nothing was revealed to any player until the auction was over, no player had any decision to make except to choose a single dropout price, in the knowledge that if its price was among the top four then it would pay the fourth-highest dropout price; that is, the situation was identical from the firm's viewpoint to choosing a single bid in a fourth-price sealed-bid auction. But, unlike in Denmark, no one would ever see the "bids" planned by the top three winners (and since these bids would never even have been placed, very little credibility would have attached to reports of them).

However, although the proposed auction was mathematically (that is, strategically) equivalent to a sealed-bid auction, its verbal description was very different. The stronger incumbents lobbied vigorously for a "small change" to the design—that the price be determined when the numbers dropped from five to four, rather than from four to three.

This is the "standard" way of running an ascending auction, and it recreates the standard problem that entry is deterred because strong players can bid aggressively in the knowledge that the winners will pay only a loser's bid (the fifth bid) and not have to pay one of the winners' bids.

Revealingly, one of the strong players that, it is said, lobbied so strongly for changing the proposal was at the same time a weaker player (a potential entrant) in the Danish market and, it is said, professed itself entirely happy with the fourth-price sealed-bid rules for *that* market.

---

[48] I had no direct involvement with this auction but, embarrassingly, I am told this "solution" was found in a footnote to Klemperer (2000b) that pointed out this method of running a strategically equivalent auction to the uniform fourth-price auction, and that it might (sometimes) be more politically acceptable. See also Binmore and Klemperer (2002).

The lobbyists' arguments that their suggested change was "small" and made the auction more "standard," and also that it was "unfair" to have the bidders continue to "bid against themselves" when there were just four left, were politically salient points, even though they are irrelevant or meaningless from a strictly game-theoretic viewpoint.[49] Since the academic consultants who proposed the original design had very little influence at the higher political levels at which the final decision was taken, and since perhaps not all the ultimate decision-makers understood—or wanted to understand—the full significance of the change, the government gave way and made it.[50]

The result? Just the four strongest bidders entered and paid the reserve price—a major disappointment for the government, and yielding perhaps one-third to one-half the revenue that had been anticipated (allowing for market conditions). Whether other potential bidders gave up altogether, or whether they made collusive agreements with stronger bidders not to enter (as was rumored in the press), is unknown. But what is certain is that the design finally chosen made entry much harder and collusion much easier.

It is not clear what the economic theorists advising should have recommended. Perhaps they should have stuck to a (fourth-price) sealed-bid auction run in the standard way, but used computer technology that could determine the price to be paid while making it impossible for anyone other than the bidders to know the other bids made.

The moral, however, is clear. Auction designs that seem similar to economic theorists may seem very different to politicians, bureaucrats, and the public, and vice versa. And political and lobbying pressures need to be predicted and planned for in advance.

When the designers of the UK 3G auction proposed a design—the Anglo-Dutch— that was very unattractive to the incumbent operators, it probably helped that two alternative versions of the design were initially offered. While the incumbent operators hated the overall design and lobbied furiously against it,[51] they also had strong preferences between its two versions, and much of their lobbying effort therefore focused on the choice between them. When the government selected the version the operators preferred (the designers actually preferred this version too) the operators felt they had got a part of what they had asked for, and it proved politically possible for the government to stick to the Anglo-Dutch design until the circumstances changed radically.[52]

Another notorious "political failure" was the design of the 1998 Netherlands 2G spectrum auction. The Commission of the European Union (EU) objected to the

[49] The lobbyists also successfully ridiculed the original design, calling it the "dark auction," arguing that it "perversely" hid information when "everyone knows that transparent markets are more efficient," and claiming it was an "unfair tax" since bidders "paid more than if they had all the information," (Klemperer 2003a).

[50] The highly sophisticated security arrangements that had been made to ensure secrecy of the dropouts (removal of bidding teams to separate top-secret locations in army camps and so on) were not altered even though they had become much less relevant; there was no need to lobby against these.

[51] It is rumored that a single bidder's budget for economic advice for lobbying against the design exceeded the UK government's expenditure on economic advice during the entire three-year design process; the lobbying effort included hiring two Nobel Prize winners in the hope of finding arguments against the design. See Binmore and Klemperer (2002) for details of the two versions of the design.

[52] When it became possible to offer an additional fifth license in the UK the design changed—as had been planned for this circumstance—to a pure ascending one (see the section "Entry").

Netherlands government's rules for the auction shortly before the (EU-imposed) deadline for the allocation of the licenses. The rules were therefore quickly rewritten by a high-ranking civil servant on a Friday afternoon. The result was an auction that sold similar properties at prices that differed by a factor of about two, and almost certainly allocated the licenses inefficiently.[53]

Economists are now waking up to the importance of these issues: Wilson (2002) addresses political constraints in the design of auction markets for electricity, and Roth (2002) also discusses political aspects of market design. But the politics of design remains under-studied by economic theorists, and underappreciated by them in their role as practitioners.

Understanding the wider context

Any consultant new to a situation must beware of overlooking issues that are well understood by those with more experience of the environment. The danger is perhaps particularly acute for economic theorists who are used to seeing the world through models that, while very elegant, are often lacking in real-world detail and context.

The German 3G auction illustrates the importance of the wider context. As we described in the section "Collusion," in Germany's 1999 DCS-1800 auction Mannesman used its bids to signal to T-Mobil how the two firms should divide the blocks between them and end the auction at a comparatively low price. T-Mobil then cut back its demand in exactly the way Mannesman suggested, and Mannesman followed through with its half of the "bargain" by also cutting back its demand, so the auction ended with the two firms winning similar amounts of spectrum very cheaply.

It seems that Mannesman used the same advisors in the 3G auction that it had used in the GSM auction. Although the rules for the 3G auction were not identical, it was another simultaneous ascending auction in which individual bidders were permitted to win multiple blocks. After the number of bidders had fallen to six, competing for a total of twelve blocks, and when it was clear that the other four bidders would be content with two blocks each, Mannesman apparently signaled to T-Mobil to cut back its demand to just two blocks.[54] If T-Mobil and Mannesman had both done this, the auction would have ended at modest prices. Instead, T-Mobil seemingly ignored Mannesman's signals,

---

[53] See van Damme (1999). This auction also illustrates the potential importance of bidders' errors: although high stakes were involved (the revenues were over €800 million) it seems that the outcome, and perhaps also the efficiency of the license allocation, was critically affected by a bidder unintentionally losing its eligibility to bid on additional properties later in the auction; it has been suggested (van Damme, 1999) that the bidder's behavior can be explained only by the fact that it happened on "Carnival Monday", a day of celebrations and drinking in the south of the Netherlands, where the bidder was based! (The German 3G auction described later in the chapter provides another example of the large role that bidder error can play.)

[54] According to the *Financial Times*, "One operator has privately admitted to altering the last digit of its bid . . . to signal to other participants that it was willing to accept a small licence" (November 3, 2000, p. 21).

and drove up the total price by €15 billion before cutting back demand. Once T-Mobil did cut back its demand, Mannesman followed, so the auction ended with the allocation that Mannesman had originally signaled but with each of the six firms paying an additional €2.5 billion!

It seems that Mannesman's advisors saw the GSM auction as a template for the 3G auction; they took the view that, following previous practice, Mannesman would signal when to reduce demand, T-Mobil would acquiesce, and Mannesman would then follow through on its half of the bargain.[55] The bargain would be enforced by firms not wishing to jeopardize their future cooperation in subsequent auctions (including 3G auctions in other countries) and in negotiating with regulators, and so on—and the short-run advantage that could be gained by failing to cooperate was anyway probably small (see Klemperer, 2002c). But given their expectation that T-Mobil would cut back demand first, Mannesman's advisors were unwilling to reduce demand when T-Mobil did not.

Clearly, T-Mobil's advisors saw things differently. It seems that its main advisors had not been involved in the GSM auction and the example of the previous auction was certainly not in the forefront of their minds. Instead, they mistrusted Mannesman's intentions, and were very unwilling to cut back demand without proof that Mannesman had already done so. True, the 3G auction was a much more complicated game than the GSM auction because of the other parties involved, and Klemperer (2002c) discusses other factors that may have contributed to the firms' failure to reduce demand.[56] But T-Mobil's refusal to cut back demand very likely stemmed partly from viewing the 3G auction in a different, and narrower, context than Mannesman did.

Just as previous auctions within any country might have been an important part of the wider context, auctions in other countries were also relevant parts of the broader environment: the sequencing of the 3G auctions across countries was crucial. Countries that auctioned earlier had more entrants, because weaker bidders had not yet worked out that they were weaker and quit the auctions, because stronger bidders had not yet worked out how and with whom to do joint ventures, and because complementarities between the values of licenses in different countries reinforced these effects—the number of entrants in the nine western European auctions were (in order) 13, 6, 7, 6, 6, 4, 3, 3, and 5 respectively.[57] Countries that auctioned earlier also suffered less from "collusive" behavior, because bidders had had less practice in learning how best to play the game. For example, when the Austrian 3G auction followed the German 3G auction that we

---

[55] It seems that another reason why Mannesman expected the firms to coordinate by T-Mobil reducing demand first in response to Mannesman's signals was that Mannesman saw itself as the leading firm in the market. However, T-Mobil may not have seen Mannesman as the leading firm—the two firms were closely matched—and this seems to have contributed to the problem.

[56] In particular, the firms might have been concerned about their relative performances. See also Grimm et al. (2002), Jehiel and Moldovanu (2003), and Ewerhart and Moldovanu (2002).

[57] Furthermore, the number (6) achieved in the second auction (Netherlands) was perhaps lowered by the peculiarly incompetent design; the number (5) achieved in the last auction (Denmark) was raised by its design, which was very skilful except in its timing (see the section "Entry"). Of course, other factors, in particular the fall in the telecoms stock price index, may have contributed to the fall in the number of entrants.

have just described, using almost the same design, all the bidders very quickly saw the mutual advantage of coordinating a demand reduction (see the section "Collusion").[58]

The UK government's advisors anticipated this pattern of declining competition, and chose to run its auction first; indeed, we persisted in the policy of running the first auction even when others were advising us to delay (see Binmore and Klemperer, 2002). Yet in more than one country auction theorists advising on 3G auction design seemed either unaware of(!), or at least unaffected in their thinking by, the fact that there was to be a sequence of auctions across Europe. Clearly, these designers had far too narrow a view of the problem.[59]

Of course, other auctions are only the most obvious aspects of the wider context that auction designers need to consider. There are many other ways in which designers showed themselves very poor at thinking about the wider game. For example, many of the 3G auction designers had a very limited understanding of how the auction process affected, and was affected by, the series of telecom mergers and alliances that the advent of 3G engendered—in the UK alone, there were no fewer than *five* mergers involving the four incumbent 2G operators, in less than a year around the auction.[60]

USING ECONOMIC THEORY

I have argued that while a good understanding of elementary undergraduate economic theory is essential to successful auction design, advanced graduate auction theory is often less important. It is important to emphasize, therefore, the crucially important role that advanced formal theory plays in developing our economic understanding. In particular, advanced theory often develops deeper connections between superficially distinct economic questions.

For example, Klemperer (2003b) demonstrates that auction-theoretic tools provide useful arguments in a broad range of mainstream economic contexts. As a further illustration, I will discuss how a part of the received auction theory—the effect of affiliation—that was, I have argued, not central to the auctions of 3G licenses, can develop useful insights about the economics of the "M-commerce" industry ("mobile commerce," in which people purchase through their mobile phones, and which is predicted to expand rapidly as a result of 3G technology). [61]

---

[58] Klemperer (2002a) develops the arguments in this paragraph in much more detail.

[59] Some of the incumbent bidders, by contrast, may possibly have had a clearer understanding. In an interesting example of the importance of political pressures, the Dutch operators successfully lobbied to delay the Netherlands auction and the clear gap that was thereby created between the British and Dutch auctions may have been a contributory factor to the Dutch fiasco.

[60] Klemperer (2002d) gives another illustration of how real-world context that was non-obvious to outsiders was important to the UK 3G auction.

[61] Klemperer (2003b) uses the other main piece of the received auction theory—the revenue equivalence theorem—to solve a war of attrition between several technologies competing to become an industry standard in, for example, 3G (see also Bulow and Klemperer, 1999), and to compute the value

Do e-commerce and M-commerce raise consumer prices?

Some commentators and regulators have expressed concern that e-commerce and M-commerce allow firms to easily identify and collect information about their customers which they can use to "rip them off."[62]

A simple analysis realizes that each consumer is analogous to an auctioneer, while firms are bidders competing to sell to that consumer. As we discussed in the section "The received auction theory," bidders' expected profits derive from their private information, and the auctioneer generally gains by reducing the amount of bidders' private information. So if all firms learn the same piece of information about a given consumer, this (weakly) reduces the private information that any bidder has relative to the other bidders, and so often benefits the auctioneer, that is, lowers the consumer's expected transaction price.

Although this result is a good start, it is not very novel,[63] nor does it address the bigger concern that e-commerce and M-commerce allow different firms to learn different information about any given consumer. However, Bulow and Klemperer (forthcoming) show how to use the mathematics of affiliation to address this issue too; in our model, even if firms learn different information about the consumers, this makes the market more competitive. In other words, a quick application of Milgrom and Weber's (1982) analysis suggests that the "loss of privacy" caused by 3G and the Internet is actually *good* for consumers.

Of course, having been cautious about the practical significance of affiliation in auction design, we should also be cautious about asserting that Bulow and Klemperer's argument shows that 3G is not as valuable to firms as some people once thought.[64] However, our model suggests a possibility which needs further study—including considering any empirical evidence and the plausibility of the intuitions—to confirm or disconfirm. Moreover, it certainly demonstrates that just because firms learn more about consumers, it does not follow that they can exploit them better—just as the RET refutes any simple presumption that one form of auction is always the most profitable. Our analysis therefore shows that firms' learning has other effects in addition to the very obvious one that firms can price-discriminate more effectively, and it helps us to see what these effects are[65]—we can then consider further whether these effects are

of new customers to firms when consumers have switching costs as they do for, for example, 3G phones (see also Bulow and Klemperer, 1998). Klemperer (2003b) also uses auction theory to address how e-commerce (and likewise M-commerce) affects pricing.

[62] The US Federal Trade Commission has held hearings on this issue, and the European Commission is currently studying it. Amazon has admitted charging different prices to different consumers.

[63] Thisse and Vives (1988), Ulph and Vulkan (2001), and Esteves (2005), for example, have developed similar results.

[64] Of course, there are more important reasons why 3G is no longer thought as valuable as it once was (see Klemperer, 2002a).

[65] In this case, while a firm may raise prices against consumers who particularly value its product, in a competitive environment it will also lower prices to other consumers who like it less—and other firms will then have to respond.

plausibly significant. It also provides a structure which suggests what other factors not in the simplest model might in fact be important, and might perhaps yield the originally hypothesized result.[66] And it very quickly and efficiently yields results that provide a good starting point for such further analysis.

Bulow and Klemperer pursue these issues in the context of this specific application. Klemperer (2003b) considers a range of other applications, including some that at first glance seem quite distant from auctions. The moral is that the "received auction theory" *is* of great value in developing our understanding of practical issues. But it needs to be used in conjunction with developing intuition and gathering empirical evidence to check its applicability to specific situations.

Conclusion

This chapter is *not* attacking the value of economic theory. I have argued that elementary economic theory is essential to successful economic policy. Furthermore, the methods of thinking that undergraduate economics teaches are very valuable, for example in understanding the important distinction between Hong Kong's two superficially similar auction designs (the one proposed and the one actually implemented). I have focused on examples from auctions, but the more I have been involved in public policy (for example, as a UK Competition Commissioner), the more I have been impressed by the importance of elementary undergraduate economics.

Nor is this chapter intended as an attack on modern, or sophisticated, or graduate economics. True, the emphasis of some graduate courses is misleading, and the relative importance of different parts of the theory is not always well understood, but almost all of it is useful when appropriately applied; it is *not* true that all economic problems can be tackled using undergraduate economics alone.[67]

Policy errors are also less likely when expertise is not too narrowly focused in one subdiscipline—for example, auction designers should remember their industrial economics and political economy (at least) in addition to pure auction theory.

While advanced theory can be misapplied, the correct answer is not to shy away from it, but rather to develop it further to bring in the important issues that have been

---

[66] For example, the analysis shows that even though it may be no bad thing for consumers if different firms learn different pieces of information about them, the result depends on firms learning the same amount of information about any given consumer. It probably is costly for a consumer to "lose his privacy" to only one firm, just as having asymmetrically informed bidders may be a bad thing for an auctioneer. Furthermore, even when firms learn the same amount of information about consumers' tastes, this information may sometimes lead to inefficient price discrimination, which reduces total welfare, in which case consumers may be made worse off, even though firms' profits are lowered, just as inefficient auctions may be bad for both auctioneers and bidders. Learning information may also affect firms' abilities to collude, and the ease of new entry.

[67] Furthermore, it is often only the process of thinking through the sophisticated graduate theory that puts the elementary undergraduate theory in proper perspective.

omitted. It may sometimes be true that "a little bit too much economics is a dangerous thing," but it is surely also true that a great deal of economic knowledge is best of all. Moreover, auction theory also illustrates that when a subdiscipline of economics becomes more widely used in practical policy-making, its development becomes more heavily influenced by the practical problems that really matter. Like a rapidly growing bush, theory may sometimes sprout and develop in unhelpful directions, but when pruned with the shears of practical experience it will quickly bear fruit!

Furthermore, advanced economic theory is of practical importance in developing our economic understanding of the world, even when it cannot be directly applied to an immediate practical problem. To recapitulate only the incomplete list of its merits that was illustrated by our example in the section "Using economic theory," it refutes oversimple arguments, makes precise and quantifies other arguments, allows us to see the relationship between superficially unconnected problems, organizes our ideas, brings out the important features of problems, shows possibilities, and quickly develops general results which, even when they are not final answers, provide good starting points for further analysis.

Nevertheless, the main lesson of this chapter is that the blinkered use of economic theory can be dangerous. Policy advisors need to learn from Marshall's example to be aware of the wider context, anticipate political pressures, and, above all, remember that the most sophisticated theory may not be the most relevant.

References

Abbink, K., Irlenbusch, B., Rockenbach, B., Sadrieh, A. and Selten, R. (2002) "The behavioural approach to the strategic analysis of spectrum auctions: the case of the German DCS-1800 auction," *Ifo Studien*, 48: 457–80.

Ausubel, L. M. and Schwartz, J. A. (1999) "The ascending auction paradox," Working Paper, University of Maryland.

Bikhchandani, S. (1988) "Reputation in repeated second-price auctions," *Journal of Economic Theory*, 46: 97–119.

Binmore, K. and Klemperer, P. (2002) "The biggest auction ever: the sale of the British 3G Telecom licences," *Economic Journal*, 112(478): C74–C96. Also published as Chapter 6 in Klemperer, P. (ed.) (2004) *Auctions: Theory and Practice*, Princeton University Press.

Brusco, S. and Lopomo, G. (2002a) "Collusion via signalling in simultaneous ascending bid auctions with heterogeneous objects, with and without complementarities," *Review of Economic Studies*, 69: 407–36.

—— —— (2002b) "Simultaneous ascending auctions with budget constraints," Working Paper, Stern School of Business, New York University.

Bulow, J. and Klemperer, P. (1996) "Auctions vs. negotiations," *American Economic Review*, 86(1): 180–94.

—— —— (1998) "The tobacco deal," *Brookings Papers on Economic Activity: Microeconomics*: 323–94.

—— (1999) "The generalized war of attrition," *American Economic Review*, 89: 175–89.

—— —— —— (2002) "Prices and the winner's curse," *Rand Journal of Economics*, 33(1): 1–21.

Bulow, J. and Klemperer, P. (forthcoming) "Privacy and pricing," Discussion Paper, Nuffield College, Oxford University.

____ Huang, M. and Klemperer, P. (1999) "Toeholds and takeovers," *Journal of Political Economy*, 107: 427–54. Reprinted in B. Biais and M. Pagano (eds) (2002) *New Research in Corporate Finance and Banking*, Oxford University Press, pp. 91–116.

Clark, C. (1940) *The Conditions of Economic Progress*, Macmillan.

Cramton, P. and Schwartz, J. A. (2000) "Collusive bidding: lessons from the FCC spectrum auctions," *Journal of Regulatory Economics*, 17(3): 229–52.

____ Gibbons, R. and Klemperer, P. (1987) "Dissolving a partnership efficiently," *Econometrica*, 55: 615–32.

Crossland, D. (2000) "Austrian UMTS auction unlikely to scale peaks," Reuters, October 31. Available at <http://www.totaltele.com>.

Engelbrecht-Wiggans, R. (1987) "Optimal reservation prices in auctions," *Management Science*, 33: 763–70.

____ (1993) "Optimal auctions revisited," *Games and Economic Behaviour*, 5: 227–39.

____ and Kahn, C. M. (2005) "Low revenue equilibria in simultaneous ascending bid auctions," *Management Science*, 51(3): 508–15.

Esteves, R. (2005) "Targeted advertising and price discrimination in the new media," DPhil Thesis, Oxford University.

Ewerhart, C. and Moldovanu, B. (2002) "The German UMTS design: insights from multi-object auction theory," *Ifo Studien*, 48(1): 158–74.

Friedman, M. (1953) *Essays in Positive Economics*, University of Chicago Press.

Gilbert, R. and Klemperer, P. D. (2000) "An equilibrium theory of rationing," *Rand Journal of Economics*, 31(1): 1–21.

Graham, D. A. and Marshall, R. C. (1987) "Collusive bidder behavior at single-object second-price and English auctions," *Journal of Political Economy*, 95: 1217–39.

____ ____ and Richard, J-F. (1990) "Differential payments within a bidder coalition and the Shapley value," *American Economic Review*, 80: 493–510.

Grimm, V., Riedel, F. and Wolfstetter, E. (2002) "The third generation (UMTS) spectrum auction in Germany," *Ifo Studien*, 48(1): 123–43.

____ ____ ____ (2003) "Low price equilibrium in multi-unit auctions: the GSM spectrum auction in Germany," *International Journal of Industrial Organisation*, 21: 1557–69.

Haile, P. (1996) "Auctions with resale markets," PhD Dissertation, Northwestern University.

Hansen, R. G. (1986) "Sealed bids versus open auctions: the evidence," *Economic Inquiry*, 24: 125–42.

Harstad, R. M. (1990) "Alternative common values auction procedures: revenue comparisons with free entry," *Journal of Political Economy*, 98: 421–9.

Hendricks, K. and Porter, R. H. (1989) "Collusion in auctions," *Annales D'Économie et de Statistique*, 15/16: 217–30.

____ ____ and Tan, G. (1999) "Joint bidding in federal offshore oil and gas lease auctions," Working Paper, University of British Columbia.

Jehiel, P. and Moldovanu, B. (2001) "The UMTS/IMT-2000 license auctions," Working Paper, University College London and University of Mannheim.

____ ____ (2003) "An economic perspective on auctions," *Economic Policy*, 36: 271–308.

Kagel, J. H. and Roth, A. E. (eds) (1995) *The Handbook of Experimental Economics*, Princeton University Press.

Keynes, J. M. (1933) *Essays in Biography*, Macmillan.

Klemperer, P. (1998) "Auctions with almost common values," *European Economic Review*, 42(3/5): 757–69.

_____ (1999a) "Auction theory: a guide to the literature," *Journal of Economic Surveys*, 13(3): 227–286. Reprinted in S. Dahiya (ed.) (1999) *The Current State of Economic Science,*. vol. 2, pp. 711–66, and as Chapter 1 in Klemperer, P. (ed.) (2004) *Auctions: Theory and Practice*, Princeton University Press.

_____ (1999b) "Applying auction theory to economics," Invited Lecture to Eighth World Congress of the Econometric Society, at <http://www.paulklemperer.org>.

_____ (ed.) (2000a) *The Economic Theory of Auctions*, Edward Elgar.

_____ (2000b) "What really matters in auction design," May 2000 version, at <http://www.paulklemperer.org>.

_____ (2002a) "How (not) to run auctions: the European 3G Telecom auctions," *European Economic Review*, 46(4/5): 829–45. Also published as Chapter 5 in Klemperer, P. (ed.) (2004) *Auctions: Theory and Practice*, Princeton University Press.

_____ (2002b) "What really matters in auction design," *Journal of Economic Perspectives*, 16(1): 169–89. Also published as Chapter 3 in Klemperer, P. (ed.) (2004) *Auctions: Theory and Practice*, Princeton University Press.

_____ (2002c) "Some observations on the German 3G Telecom auction," *Ifo Studien*, 48(1): 145–56. Also published as Chapter 7B in Klemperer, P. (ed.) (2004) *Auctions: Theory and Practice*, Princeton University Press.

_____ (2002d) "Some observations on the British 3G Telecom auction," *Ifo Studien*, 48(1): 115–20. Also published as Chapter 7A in Klemperer, P. (2004) *Auctions: Theory and Practice*, Princeton University Press.

_____ (2003a) "Using and Abusing Economic Theory—Lessons from Auction Design," *Journal of the European Economic Association* (2003), 1(2–3): 272–300. February 2003 version, at <http:/www.paulklemperer.org>.

_____ (2003b) "Why every economist should learn some auction theory", in M. Dewatripont, L. Hansen, S. Turnovsky (eds), *Advances in Economics and Econometrics: Invited Lectures to Eighth World Congress of the Econometric Society*, Cambridge University Press. Also published as Chapter 2 in Klemperer, P. (ed.) (2004) *Auctions: Theory and Practice*, Princeton University Press.

_____ (2004a) *Auctions: Theory and Practice (the Toulouse Lectures in Economics)*, Princeton University Press.

_____ (2004b) "Competition: its power, its fragility, and where we need more of it" (presentation to No.11 Downing St, March 2004), *HM Treasury Microeconomics Lecture Series* 2004–05.

_____ (2005) "Bidding markets," Occasional Paper No. 1, UK Competition Commission. Also published in *Journal of Competition Law and Economics* (2007), 3: 1–47.

_____ (2008) "Competition policy in auctions and 'bidding markets'," in P. Buccirossi (ed.), *Handbook of Antitrust Economics*, MIT Press, pp. 583–624.

Levin, D. and Smith, J. L. (1994) "Equilibrium in auctions with entry," *American Economic Review*, 84: 585–99.

_____ Kagel, J. H. and Richard, J-F. (1996) "Revenue effects and information processing in English common value actions," *American Economic Review*, 86(3): 442–60.

Maasland, E. (2000) "Veilingmiljarden Zijn een Fictie (billions from auctions: wishful thinking)," *Economisch Statistische Berichten*, 85: 479. Translation available at <http://www.paulklemperer.org>.

Machlup, F. (1946) "Marginal analysis and empirical research," *American Economic Review*, 36: 519–54.

Mailath, G. J. and Zemsky, P. (1991) "Collusion in second price auctions with heterogeneous bidders," *Games and Economic Behavior*, 3: 467–86.

Marshall, A. (1890) *Principles of Economics*, Macmillan.

——— (1906) Letter to A. L. Bowley, February 27, 1906, in A. C. Pigou (ed.) (1925) *Memorials of Alfred Marshall*, Macmillan, pp. 427–8.

Maskin, E. S. and Riley, J. G. (1984) "Optimal auctions with risk averse buyers," *Econometrica*, 52: 1473–518.

——— ——— (2000) "Asymmetric auctions," *Review of Economic Studies*, 67(3): 413–38.

Matthews, S. A. (1983) "Selling to risk averse buyers with unobservable tastes," *Journal of Economic Theory*, 3: 370–400.

——— (1984) "Information acquisition in discriminatory auctions", in M. Boyer and R. E. Kihlstrom (eds), *Bayesian Models in Economic Theory* (pp. 181–207), North Holland, pp. 181–207.

McAfee, R. P. and McMillan, J. (1987) "Auctions with entry," *Economics Letters*, 23: 343–47.

——— ——— (1988) "Search mechanisms," *Journal of Economic Theory*, 44: 99–123.

——— ——— (1992) "Bidding rings," *American Economic Review*, 82: 579–99.

McMillan, J. (1994) "Selling spectrum rights," *Journal of Economic Perspectives*, 8: 145–62.

Mead, W. J. and Schneipp, M. (1989) "Competitive bidding for federal timber in region 6, an update: 1983–1988," Community and Organization Research Institute, University of California, Santa Barbara.

Meeks, R. (2001) "An event study of the Swiss UMTS auction," Research Note, Nuffield College, Oxford University.

Menezes, F. (1996) "Multiple-unit English auctions," *European Journal of Political Economy*, 12: 671–84.

——— and Monteiro, P. K. (2000) "Auctions with endogenous participation," *Review of Economic Design*, 5: 71–89.

Milgrom, P. R. (1987) "Auction theory," in T. F. Bewley (ed.), *Advances in Economic Theory: Fifth World Congress*, Cambridge University Press.

——— and Weber. R. J. (1982) "A theory of auctions and competitive bidding, II," *Econometrica*, 50: 1089–122.

——— ——— (2000) "A theory of auctions and competitive bidding," in P. Klemperer (ed.), *The Economic Theory of Auctions*, Edward Elgar, vol. 2, pp. 179–94.

Myerson, R. B. (1981) "Optimal auction design," *Mathematics of Operations Research*, 6: 58–73.

Paarsch, H. J. (1991) "Empirical models of auctions and an application to British Columbian timber sales," Discussion Paper, University of British Columbia.

Perry, M. and Reny, P. J. (1999) "On the failure of the linkage principle in multi-unit auctions," *Econometrica*, 67(4): 895–900.

Persico, N. (2000) "Information acquisition in auctions," *Econometrica*, 68: 135–48.

Riley, J. G. and Li, H. (1997) "Auction choice: a numerical analysis," Mimeo, University of California at Los Angeles.

——— and Samuelson, W. F. (1981) "Optimal auctions," *American Economic Review*, 71: 381–92.

Robinson, M. S. (1985) "Collusion and the choice of auction," *Rand Journal of Economics*, 16: 141–5.

Roth, A. E. (2002) "The economist as engineer: game theory, experimentations, and computation as tools for design economics," *Econometrica*, 70(4): 1341–78.

Rothkopf, M. H. and Engelbrecht-Wiggans, R. (1993) "Misapplications reviews: getting the
model right—the case of competitive bidding," *Interfaces*, 23: 99–106.

Sills, David L. (ed.) (1968) *International Encyclopedia of the Social Sciences*. Macmillan and
Free Press, p. 10.

Stuewe, H. (1999) "Auktion von Telefonfrequenzen: Spannung bis zur letzten Minute," Frank-
furter Allgemeine Zeitung, October 29.

Thisse, J. and Vives, X. (1988) "On the strategic choice of spatial price policy," *American
Economic Review*, 78: 122–37.

Ulph, D. and Vulkan, N. (2001) "E-commerce, mass customisation and price discrimination,"
Mimeo, UCL and University of Bristol.

van Damme, E. (1999) "The Dutch DCS-1800 auction," in F. Patrone, I. Garcia-Jurado and
S. Tijs (eds), *Game Practice: Contributions From Applied Game Theory*, Kluwer Academic,
pp. 53–73.

——— (2002) "The European UMTS auctions," *European Economic Review*, 45(4/5): 846–58.

Vickrey, W. (1961) "Counterspeculation, auctions, and competitive sealed tenders," *Journal of
Finance*, 16: 8–37.

Waterson, M. (1984) *Economic Theory of the Industry*, Cambridge University Press.

Weber, R. J. (1997) "Making more from less: strategic demand reduction in the FCC spectrum
auctions," *Journal of Economics and Management Strategy*, 6(3): 529–48.

Wilson, R. (2002) "Architecture of power markets," *Econometrica*, 70(4): 1299–340.

MATCHING MARKETS

CHAPTER 4

MARKET DESIGN FOR KIDNEY EXCHANGE

TAYFUN SÖNMEZ AND M. UTKU ÜNVER

INTRODUCTION

THE National Organ Transplant Act of 1984 makes it illegal to buy or sell a kidney in the US, thus making donation the only viable option for kidney transplantation. A transplanted kidney from a live donor survives significantly longer than one from a deceased donor (see e.g. Mandal et al., 2003). Hence, live donation is always the first choice for a patient. Moreover, there is a significant shortage of deceased donor kidneys.[1] There are two kidneys in the human body, but just one healthy kidney is more than enough for everyday life. Since the risks associated with donation surgery and follow-up have decreased with the advancement of medical and surgical techniques, live donation has increased as a proportion. Usually, a live donor is a relative or friend of the recipient, and is willing to donate only if that particular recipient is going to receive a transplant. That is, she is a *directed live donor*. However, a recipient is often unable to receive a willing live donor's kidney because of blood-type incompatibility or antibodies to one of the donor's proteins (a 'positive crossmatch'). Medical doctor F. T. Rapaport (1986) proposed *live-donor paired kidney exchanges* between two such incompatible recipient–donor pairs: the donor in each pair gives a kidney to the other pair's compatible recipient.[2]

In the 1990s, Korea and the Netherlands started to build databases to organize such swaps. Both programs recently reported that live-donor kidney exchanges make up

---

[1] About 79,000 patients were waiting for a deceased donor kidney transplant in the United States as of March 2009. In 2008, about 16,500 transplants were conducted, 10,500 from deceased donors and 6,000 from living donors, while about 32,500 new patients joined the deceased donor waiting list and 4,200 patients died while waiting for a kidney (according to SRTR/OPTN national data retrieved at <http://www.optn.org> on March 17, 2009).

[2] Recently medical literature started to use the term *kidney paired donation* instead of *kidney exchange*.

more than 10% of the live-donor transplants in both countries (Park et al., 2004; de Klerk et al., 2005). Once the medical community in the US deemed the practice ethical (Abecassis et al., 2000), New England,[3] Ohio,[4] and Johns Hopkins transplant programs started conducting live-donor kidney exchange operations. The potential number of such exchanges has been estimated to be 2,000 additional transplants per year in the US; however, it has yet to live up to expectations. The initial hurdle in organizing kidney exchanges was the lack of mechanisms to clear the market in an efficient and incentive-compatible manner. Roth et al. (2004) proposed the first such mechanism. It was based on the core mechanism for the housing markets of Shapley and Scarf (1974), namely Gale's top trading cycles algorithm,[5] and a mechanism designed for the house allocation problem with existing tenants of Abdulkadiroğlu and Sönmez (1999), namely the "you-request-my-house-I-get-your-turn" algorithm.[6] This new mechanism, called the **top trading cycles and chains** (TTCC), is strategy proof, that is, it makes it the dominant strategy for recipients to reveal their preferences over compatible kidneys and all of their paired donors to the system. Moreover, it is Pareto efficient. As the two coauthors of that study (Roth, Sönmez, and Ünver, 2004) we showed through simulations that the potential benefits of switching to such a system would be huge.

However, one important aspect of kidney exchanges is that, regardless of the number of pairs participating in an exchange, all transplants in the exchange must be conducted simultaneously. Otherwise, one or more of the live donors whose recipients receive a kidney in the previously conducted part of an exchange may back out from future donations of the same exchange. Since kidney donations are gifts, the donor can change her mind at any moment prior to the actual transplant, and it is not legal to contractually bind a donor to make future donations. This may put some recipient, whose paired donor previously donated a kidney in the exchange, at harm.

Naturally, these is an upper limit on the number of kidney transplants that can be conducted simultaneously. The simulations showed that the TTCC mechanism may lead to large exchanges, with many recipient–donor pairs.

Another controversial issue in the market design for kidney exchange concerns the preferences of recipients over kidneys. A respected assumption in the field is that all compatible live-donor kidneys have the same likelihood of survival, following Gjertson and Cecka (2000), who statistically show this in their data-set (see also Delmonico, 2004).

Medical doctors also point out that if the paired donor is compatible with the recipient, the latter will directly receive a kidney from her paired donor and will not participate in the exchange.[7]

---

[3] New England Program for Kidney Exchange, <http://www.nepke.org>.

[4] Ohio Solid Organ Consortium, <http://www.paireddonationnetwork.org>.

[5] See also Roth and Postlewaite (1977), Roth (1982), and Ma (1994).

[6] See also Pápai (2000), Sönmez and Ünver (2005, 2010a), Pycia and Ünver (2009), and a literature survey of discrete resource allocation by Sönmez and Ünver (2011).

[7] This is a controversial point. Using European data, Opelz (1997) shows that, indeed, tissue-type matching matters even in living donations. Thus, there is no consensus in the medical community that

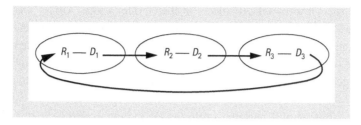

FIGURE 4.1. A three-way kidney exchange. R_i denotes the recipient and D_i denotes the donor in each pair of the exchange.

These institutional restrictions limit the applicability of the TTCC mechanism, which uses strict preferences information, opts in compatible pairs to the system, and results in possibly arbitrary lengths of exchange cycles. Thus, it is not immediately practical to implement this mechanism in the field.

Based on these restrictions, Roth et al. (2005a) focused on exchanges consisting of two pairs, assuming recipients are indifferent among all compatible donors. They proposed two mechanisms, a priority mechanism and an egalitarian mechanism, for strategy-proof and Pareto-efficient exchanges when recipients are indifferent among compatible donors.

The New England Program for Kidney Exchange (NEPKE) is the first US kidney exchange program that started to implement mechanisms for kidney exchange, and was established in 2004 as a collaboration between surgeon Francis Delmonico, tissue-typing expert Susan Saidman, Alvin Roth, and the authors. NEPKE started to implement a version of the priority mechanism proposed by Roth et al. (2005a) in 2004 (see also Roth et al., 2005b). It was followed by the Johns Hopkins Kidney Exchange Program (Segev et al., 2005), which adopted a similar algorithm due to Edmonds (1965) as proposed by Roth et al. (2005a).

However, there was a significant gap between theory and implementation. Two-way exchanges were clearly the cornerstone of the kidney exchange paradigm. However, it was not clear what society at large was losing by restricting exchanges to two-way ones. Roth et al. (2007) showed that in a large population, all the gains from exchange can be obtained by using two-, three-, and four-way exchanges. Especially, two- and three-way exchanges capture almost all the gains from exchange, and the marginal contribution of three-way exchanges is significantly large. Thus, going from two-way to two- and three-way exchanges nearly captures all the gains from exchange. The arrangement for a three-way exchange is shown in Figure 4.1.

Based on these observations, NEPKE started to implement a priority mechanism that could induce up to four-way exchanges.

tissue-type matching matters for long-term survival of live-donor kidneys (other than immediate rejection). Of course, there are certain properties of donors that all authors agree to be important, such as the age and health of the donor. Following the field practice of live donation, the models and field applications surveyed here do not directly take these points into consideration, other than the ability of a recipient to report her willingness to receive or not to receive a compatible kidney.

In 2005, the Ohio-based Alliance for Paired Donation (APD)[8] was established through the collaboration of surgeon Michael Rees, computer programmer Jon Kopke, Alvin Roth, and the authors. This program immediately started to implement a mechanism based on maximizing the number of patients to be matched through up to four-way exchanges. It uses a priority-based solution in case there is more than one maximal matching.

The establishment of a national program for kidney exchange is in progress. The United Network for Organ Sharing (UNOS), the contractor for the national organization that maintains the deceased-donor waiting list, the Organ Procurement and Transplantation Network (OPTN), is developing this program with the consultation of economists, computer scientists, medical doctors, and administrators who have worked on the development and in the executive body of the exchange programs mentioned here and some other independent organizations. In late 2010, they launched a pilot program and two match runs have already been concluded.

In this survey, we will summarize the works of Roth et al. (2005a, 2007), which we mentioned above, and Ünver (2010). The last extends the agenda of the first two papers, and analyzes the kidney exchange problem as a dynamic problem in which patients arrive over time under a stochastic distribution. Then it proposes efficient mechanisms that maximize the total discounted number of patients matched under different institutional restrictions.

We will also discuss computational issues involved in solving the optimization problems with the mechanism design approach. Finally, we will talk about other paradigms in kidney exchange that are in implementation, such as *list exchange, altruistic donor exchange*, and *altruistic donor chains*, and how these are incorporated in the market design paradigm.

MECHANICS OF DONATION

In this section, we summarize the mechanics governing kidney donations. There are two sources of donation: *deceased donors* and *living donors.*

In the US and Europe, a centralized priority mechanism is used for the allocation of deceased-donor kidneys, which are considered social endowments. There have been studies regarding the effect of the choice of priority mechanism on efficiency, equity, and incentives, starting with Zenios (1996) (see also Zenios et al., 2000; Votruba, 2002; Su and Zenios, 2006). In the US, a *soft opt-in* system is used to recruit such donors. On their drivers' licenses, candidates can opt in to be deceased donors; that is, they give consent to have their organs be used for transplantation upon their death. However, upon their death their relatives can override this decision. There are also other regimes in practice around the world, such as hard opt-in, hard opt-out, and soft opt-out.

---

[8] See <http://www.paireddonation.org>.

As mentioned, live donations have been an increasing source of donations in the last decade. Live donors are generally significant others, family members, or friends of recipients. There are also some altruistic live donors who are kind enough to donate a kidney to a stranger. There is no single regulation governing live donations in the US. The only rule of thumb used is that live donors should not be coerced into donation through economic, psychological, or social pressure. In some countries, live donors are required to be blood related or emotionally related (i.e., romantically related) to the recipient.

In this survey, we will deal with directed living donations, more specifically, the cases in which a living donor is willing to donate a kidney to a specific recipient but is incompatible with her intended recipient. We will also briefly comment on non-directed, i.e., altruistic, donations.

There are two tests that a donor must pass before she is deemed compatible with the recipient, *blood compatibility* and *tissue compatibility (or crossmatch)* tests:

- *Blood compatibility test.* There are four human blood types, O, A, B, and AB. Blood type is determined by the existence or absence of one or two of the blood-type proteins called A and B. As a rule of thumb, a donor can donate a kidney to a recipient who has all the blood-type proteins that the donor possesses.[9] Thus:
  - O blood-type kidneys are blood-type compatible with all recipients;
  - A blood-type kidneys are blood-type compatible with A and AB blood-type recipients;
  - B blood-type kidneys are blood-type compatible with B and AB blood-type recipients;
  - AB blood-type kidneys are blood-type compatible with AB blood-type recipients.
- *Tissue compatibility (or crossmatch) test.* Six human leukocyte antigen (HLA) proteins on DNA determine tissue type. There does not need to be a 100% match of the HLA proteins between the donor and the recipient for tissue compatibility. If antibodies form in the blood of the recipient against the donor's tissue types, then there is *tissue rejection* (or *positive crossmatch*), and the donor is tissue-type incompatible with the recipient. The reported chance of positive crossmatch in the literature is around 11% between a random blood-type compatible donor and a random recipient (Zenios et al., 2001).

If either test fails, the donation cannot go forward. We refer to such a pair as *incompatible*. This pair then becomes available for paired kidney exchange, which is the topic of the rest of the survey.

---

[9] O type is referred to as 0 (zero) in many languages, and it refers to the non-existence of any blood-type proteins.

A MODEL OF KIDNEY EXCHANGES

Let N be the set of groups of incompatible donors and their recipients, that is: each $i = \left(R_i, \{D_i^1, \ldots, D_i^{n_i}\}\right) \in N$ is a *group* (if $n_i = 1$, a *pair*) and is represented by a *recipient*, R_i and her paired incompatible *donors*, $D_i^1, \ldots, D_i^{n_i}$. We permit each recipient to have more than one incompatible donor. However, only one of these donors will donate a kidney if, and only if, the recipient receives one. We will sometimes refer to i simply as a *recipient*, since we treat the donors through their kidneys, which are objects, and consider the recipients as the decision makers, i.e. agents.

For each $i \in N$, let $\succsim_i$ be a preference relation on N with three indifference classes. Option $j \in N \setminus \{i\}$ refers to the recipient i receiving a kidney from the best donor of j for i. Option i refers to remaining unmatched. Let $\succ_i$ be the acyclic (i.e. strict preference) portion of $\succsim_i$ and $\sim_i$ be the cyclic (i.e. indifference) portion of $\succsim_i$. For any $j, k \in N \setminus \{i\}$, we have

- $j \succ_i i$ if at least one donor of j is compatible with i;
- $j \sim_i k$ if at least one donor of each of j and k is compatible with i;
- $i \succ_i j$ if all donors of j are incompatible with i; and
- $j \sim_i k$ if all donors of j and k are incompatible with i.

That is, a recipient with a compatible donor is preferred by i to remaining unmatched, which is, in turn, preferred to a recipient with incompatible donors. All recipients with only incompatible donors are indifferent for i. Similarly, all recipients each with at least one compatible donor are indifferent for i.

A *problem* is denoted by the recipients, their donors, and preferences. An outcome of a problem is a *matching*. A *matching* $\mu : N \to N$ is a one-to-one and onto mapping. For each $i \in N$, recipient i_1 receives a kidney from some donor of recipient $\mu(i)$. We do not specify which donor in our notation, since at most one donor of a recipient is going to make a donation in any matching. Thus, for our purposes i can be matched with any compatible donor of $\mu(i)$. A matching μ is *individually rational* if for all recipients $i \in N$, $\mu(i) \succsim_i i$. We will focus on only individually rational matchings. Thus, when we say a matching it will be individually rational from now on. Let $\mathcal{M}$ be the set of matchings. A $k-way$ *exchange* for some $k \geq 1$ is a list $(i_1, i_2, \ldots, i_k)$ such that i_1 receives a kidney from a compatible donor of i_k, i_2 receives a kidney from a compatible donor of $i_1, \ldots$, and i_k receives a kidney from a compatible donor of i_{k-1}. Similarly, all exchanges we consider will be individually rational. A degenerate exchange (i) denotes the case in which recipient i is unmatched. Alternately, we represent a matching μ as a set of exchanges such that each recipient participates in one and only one exchange.

Besides deterministic outcomes, we will also define stochastic outcomes. A stochastic outcome is a *lottery*, $\lambda = \left(\lambda_\mu\right)_{\mu \in \mathcal{M}}$, that is a probability distribution on all matchings. Although in many matching problems, there is no natural definition of von Neumann–Morgenstern utility functions, there is one for this problem: It takes value 1 if the recipient is matched and 0 otherwise. We can define the (*expected*) *utility* of the recipient of a pair i under a lottery λ as the probability of the recipient getting a transplant and we denote it by $u_i(\lambda)$. The *utility profile* of lottery λ is denoted by $u(\lambda) = (u_i(\lambda))_{i \in N}$.

A matching is *Pareto efficient* if there is no other matching that makes every recipient weakly better off and some recipients strictly better off. A lottery is *ex post efficient* if it gives positive weight to only Pareto-efficient matchings. A lottery is *ex ante efficient* if there is no other lottery that makes every recipient weakly better off and some recipient strictly better off.

A *mechanism* is a systematic procedure that assigns a lottery for each problem.

A mechanism is *strategy-proof* if, for each problem $(N, \succsim)$, it is a dominant strategy for each pair i

- to report its true preference $\succsim_i$ in a preference profile set $\mathcal{P}(\succsim_i)$ where for all $\succsim_i' \in \mathcal{P}(\succsim_i)$, $j \succ_i' i \Longrightarrow j \succ_i i$, i.e. one pair can never report a group with only incompatible donors as compatible; and
- to report full set of incompatible donors to the problem.

The first bullet point above underlines the fact that it is possible to detect incompatible donors through blood tests; thus, we will assume that no recipient can reveal an incompatible donor to be compatible. On the other hand, some idiosyncratic factors can lead a recipient to reveal compatible donors to be incompatible.

We will survey different Pareto-efficient and strategy-proof mechanisms for different institutional constraints.

Two-way kidney exchanges

First, we restrict our attention in this section to individually rational two-way exchanges. This section follows Roth et al. (2005a). Formally, for given any problem $(N, \succsim)$, we are interested in matchings $\mu \in \mathcal{M}$ such that for all $i \in N$, $\mu(\mu(i)) = i$. To make our notation simpler, we define the following concept: Recipients i, j are *mutually compatible* if j has a compatible donor for i, and i has a compatible donor for j. We can focus on a *mutual compatibility matrix* that summarizes the feasible exchanges and preferences. A *mutual compatibility matrix*, $C = [c_{i,j}]_{i \in N, j \in N}$, is defined as for any $i, j \in N$,

$$c_{i,j} = \begin{cases} 1 & \text{if } i \text{ and } j \text{ are mutually compatible} \\ 0 & \text{otherwise} \end{cases}.$$

The induced *two-way kidney exchange problem* from problem $(N, \succsim)$ is denoted by (N, C). A *subproblem* of (N, C) is denoted as (I, C_I) where $I \subseteq N$ and C_I is the restriction of C to the pairs in I. Thus, all relevant information regarding preferences is summarized by the mutual compatibility matrix C.

Observe that a problem (N, C) can be represented by an undirected *graph* in which each recipient is a *node*, and there is an *edge* between two nodes if and only if these two recipients are mutually compatible. Hence, we define the following graph-theoretic concepts for two-way kidney exchange problems:

A problem is *connected* if the corresponding graph of the problem is connected, i.e., one can traverse between any two nodes of the graph using the edges of the graph. A *component* is a largest connected subproblem. We refer to a component as *odd* if it has an odd number of recipients, and as *even* if it has an even number of recipients.

Although in many matching domains ex ante and ex post efficiency are not equivalent (see e.g. Bogomolnaia and Moulin, 2001), they are equivalent for two-way kidney exchanges with 0–1 preferences because of the following lemma:

Lemma 1 (*Roth et al., 2005a*). *The same number of recipients is matched at each Pareto-efficient matching, which is the maximum number of recipients that can be matched.*

Thus, finding a Pareto-efficient matching is equivalent to finding a matching that matches the maximum number of recipients. In graph theory, such a problem is known as a *cardinality matching problem* (see e.g. Lóvasz and Plummer, 1986, for an excellent survey of this and other matching problems regarding graphs). Various intuitive polynomial time algorithms are known to find one Pareto-efficient matching, starting with Edmonds' (1965) algorithm.

The above lemma would not hold if exchange were possible among three or more recipients. Moreover, we can state the following lemma regarding efficient lotteries:

Lemma 2 (*Roth et al., 2005a*). *A lottery is ex ante efficient if and only it is ex post efficient.*

There are many Pareto-efficient matchings, and finding all of them is not computationally feasible (i.e. it is NP complete). Therefore, we will focus on two selections of Pareto-efficient matchings and lotteries that have nice fairness features.

Priority mechanism

In many situations, recipients may be ordered by natural priority. For example, the sensitivity of a recipient to the tissue types of others, known as panel reactive antibody (PRA), is a criterion also accepted by medical doctors. Some recipients may be sensitive to almost all tissue types other than their own and have a PRA=99%, meaning that they will reject based solely on tissue incompatibility 99% of donors from a random sample. So, one can order the recipients from high to low with respect to their PRAs and use the following *priority mechanism*:

Given a priority ordering of recipients, a *priority mechanism*
matches *Priority 1* recipient if she is mutually compatible with a recipient, and skips her
otherwise.

$$\vdots$$

matches *Priority k* recipient in addition to all the previously matched recipients if possible, and skips her otherwise.

Thus, the mechanism determines which recipients are to be matched first, and then
one can select a Pareto-efficient matching that matches those recipients. Thus, the mechanism is only uniquely valued for the utility profile induced. Any matching inducing this
utility profile can be the final outcome. The following result makes a priority mechanism
very appealing:

Theorem 1. *A two-way priority mechanism is Pareto efficient and strategy proof.*

The structure of Pareto-efficient matchings

We can determine additional properties of Pareto-efficient matchings (even though
finding all such matchings is exhaustive and, hence, NP complete) thanks to the results
of Gallai (1963, 1964) and Edmonds (1965) in graph theory. We can partition the
recipients into three sets as N^U, N^O, N^P. The members of these sets are defined as
follows:

An *underdemanded recipient* is one for whom there exists a Pareto-efficient matching
that leaves her unmatched. Set N^U is formed by underdemanded recipients, and we will
refer to this set as the set of underdemanded recipients. An *overdemanded recipient* is
one who is not underdemanded, yet is mutually compatible with an underdemanded
recipient. Set N^O is formed by overdemanded recipients. A *perfectly matched recipient* is
one who is neither underdemanded nor mutually compatible with any underdemanded
recipient. Set N^P is formed by perfectly matched recipients.

The following result, due to Gallai and Edmonds, is the key to understanding the
structure of Pareto-efficient matchings:

Lemma 3 (*The Gallai (1963, 1964) and Edmonds (1965) decomposition (GED)*). *Let μ
be any Pareto-efficient matching for the original problem (N, C) and (I, C_I) be the **subproblem** for $I = N \setminus N^O$. Then we have:*

1. *Each overdemanded recipient is matched with an underdemanded recipient under μ.*
2. *$J \subseteq N^P$ for any even component J of the subproblem (I, C_I) and all recipients in J
   are matched with each other under μ.*
3. *$J \subseteq N^U$ for any odd component J of the subproblem (I, C_I) and for any recipient
   $i \in J$, it is possible to match all remaining recipients with each other under μ. Moreover, under μ*

- *either one recipient in J is matched with an overdemanded recipient and all others are matched with each other,*

  or

- *one recipient in J remains unmatched while the others are matched with each other.*

We can interpret this lemma as follows: There exists a competition among odd components of the subproblem (I, C_I) for overdemanded recipients. Let $\mathcal{O} = \{O_1, \ldots, O_p\}$ be the set of odd components remaining in the problem when overdemanded recipients are removed. By the GED lemma, all recipients in each odd component are matched but at most one, and all of the other recipients are matched under each Pareto-efficient matching. Thus, such a matching leaves $|\mathcal{O}| - |N^O|$ unmatched recipients, each of whom is in a distinct odd component.

First, suppose that we determine the set of overdemanded recipients, N^O. After removing those from the problem, we mark the recipients in odd components as *underdemanded*, and recipients in even components as *perfectly matched*. Moreover, we can think of each odd component as a single entity, which is competing to get one overdemanded recipient for its recipients under a Pareto-efficient matching.

It turns out that the sets N^U, N^O, N^P and the GED decomposition can also be found in polynomial time thanks to Edmonds' algorithm and related results in the literature.

Egalitarian mechanism

Recall that the utility for a recipient under a lottery is the probability of receiving a transplant. Equalizing utilities as much as possible may be considered very desirable from an equity perspective, which is also in line with the Rawlsian notion of fairness (Rawls, 1971). We define a central notion in Rawlsian egalitarianism:

A feasible utility profile is *Lorenz dominant* if

- the least fortunate recipient receives the highest utility among all feasible utility profiles, and

  ⋮

- the sum of utilities of the k least fortunate recipients is the highest among all feasible utility profiles. [10]

Is there a feasible Lorenz-dominant utility profile? Roth et al. (2005a) answer this question affirmatively. This utility profile is constructed with the help of the GED of the problem. Let

---

[10] By *k least fortunate recipients* under a utility profile, we refer to the k recipients whose utilities are lowest in this utility profile.

- $\mathcal{J} \subseteq \mathcal{O}$ be an arbitrary set of odd components of the subproblem obtained by removing the overdemanded recipients;
- $I \subseteq N^O$ be an arbitrary set of overdemanded recipients; and
- $N(\mathcal{J}, I) \subseteq I$ denote the *neighbors* of $\mathcal{J}$ among I, that is, each overdemanded recipient in $N(\mathcal{J}, I)$ is in I and is mutually compatible with a recipient in an odd component of the collection $\mathcal{J}$.

Suppose only overdemanded recipients in I are available to be matched with underdemanded recipients in $\bigcup_{J \in \mathcal{J}} J$. Then, what is the upper bound of the utility that can be received by the *least fortunate* recipient in $\bigcup_{J \in \mathcal{J}} J$? The answer is

$$f(\mathcal{J}, I) = \frac{\left| \bigcup_{J \in \mathcal{J}} J \right| - (|\mathcal{J}| - |N(\mathcal{J}, I)|)}{\left| \bigcup_{J \in \mathcal{J}} J \right|}$$

and it can be received only if

1. all underdemanded recipients in $\bigcup_{J \in \mathcal{J}} J$ receive the same utility; and
2. all overdemanded recipients in $N(\mathcal{J}, I)$ are committed for recipients in $\bigcup_{J \in \mathcal{J}} J$.

The function f is the key in constructing an *egalitarian utility profile* u^E. The following procedure can be used to construct it:

Partition $\mathcal{O}$ as $\mathcal{O}_1, \mathcal{O}_2, \ldots$ and N^O as $N_1^O, N_2^O, \ldots$ as follows:

Step 1.

$$\mathcal{O}_1 = \arg \min_{\mathcal{J} \subseteq \mathcal{O}} f(\mathcal{J}, N^O) \text{ and}$$

$$N_1^O = N(\mathcal{O}_1, N^O)$$

$$\vdots$$

Step k.

$$\mathcal{O}_k = \arg \min_{\mathcal{J} \subseteq \mathcal{O} \setminus \bigcup_{\ell=1}^{k-1} \mathcal{O}_\ell} f\left(\mathcal{J}, N^O \setminus \bigcup_{\ell=1}^{k-1} N_\ell^O\right) \text{ and}$$

$$N_k^O = N\left(\mathcal{O}_k, N^O \setminus \bigcup_{\ell=1}^{k-1} N_\ell^O\right)$$

Construct the vector $u^E = (u_i^E)_{i \in N}$ as follows:

1. For any overdemanded recipient and perfectly matched recipient $i \in N \setminus N^U$,

$$u_i^E = 1.$$

2. For any underdemanded recipient i whose odd component left the above procedure at step $k(i)$,

$$u_i^E = f(\mathcal{O}_{k(i)}, N_{k(i)}^O).$$

We provide an example explaining this construction:

Example 1. *Let $N = \{1, \ldots, 16\}$ be the set of recipients and let the reduced problem be given by the system in Figure 4.2. $N^U = \{3, \ldots, 16\}$ is the set of underdemanded recipients. Since both recipients 1 and 2 have edges with recipients in N^U, $N^O = \{1, 2\}$ is the set of overdemanded recipients.*

$$\mathcal{O} = \{O_1, \ldots, O_6\}$$

where

$$O_1 = \{3\}, O_2 = \{4\}, O_3 = \{5\}, O_4 = \{6, 7, 8\}$$

$$O_5 = \{9, 10, 11\}, O_6 = \{12, 13, 14, 15, 16\}$$

Consider $J_1 = \{O_1, O_2\} = \{\{3\}, \{4\}\}$. Note that, by the GED lemma, an odd component that has k recipients guarantees $\frac{k-1}{k}$ utility for each of its recipients. Since $f(J_1, N^O) = \frac{1}{2} < \frac{2}{3} < \frac{4}{5}$, none of the multi-recipient odd components is an element of O_1. Moreover, recipient 5 has two overdemanded neighbors and $f(J, N^O) > f(J_1, N^O)$ for any $J \subseteq \{\{3\}, \{4\}, \{5\}\}$ with $\{5\} \in J$. Therefore

$$\mathcal{O}_1 = \mathcal{J}_1 = \{\{3\}, \{4\}\}, \quad N_1^O = \{1\},$$

$$u_3^E = u_4^E = \frac{1}{2}.$$

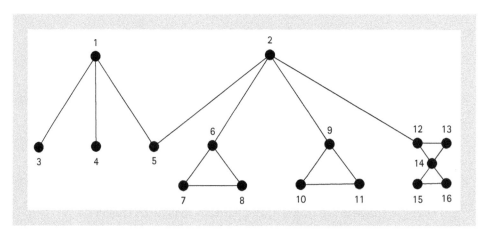

FIGURE 4.2. Graphical representation for the set of recipients in example 1.

Next, consider $J_2 = \{O_3, O_4, O_5\} = \{\{5\}, \{6, 7, 8\}, \{9, 10, 11\}\}$. Note that $f(J_2, N^O \setminus N_1^O) = \frac{7-(3-1)}{7} = \frac{5}{7}$. Since $f(J_2, N^O \setminus N_1^O) = \frac{5}{7} < \frac{4}{5}$, the five-recipient odd component O_6 is not an element of O_2. Moreover,

$$f(\{O_3\}, N^O \setminus N_1^O) = f(\{O_4\}, N^O \setminus N_1^O)$$
$$= f(\{O_5\}, N^O \setminus N_1^O) = 1,$$
$$f(\{O_3, O_4\}, N^O \setminus N_1^O) = f(\{O_3, O_5\}, N^O \setminus N_1^O) = \frac{3}{4},$$
$$f(\{O_4, O_5\}, N^O \setminus N_1^O) = \frac{5}{6}.$$

Therefore,

$$\mathcal{O}_2 = \mathcal{J}_2 = \{\{5\}, \{6, 7, 8\}, \{9, 10, 11\}\},$$
$$N_2^O = \{2\},$$
$$\text{and} \quad u_5^E = \cdots = u_{11}^E = \frac{5}{7}.$$

Finally since $N^O \setminus (N_1^O \cup N_2^O) = \emptyset$,

$$\mathcal{O}_3 = \{\{12, 13, 14, 15, 16\}\},$$
$$N_3^O = \emptyset,$$
$$\text{and} \quad u_{12}^E = \cdots = u_{16}^E = \frac{4}{5}.$$

Hence the egalitarian utility profile is

$$u^E = (1, 1, \frac{1}{2}, \frac{1}{2}, \frac{5}{7}, \frac{5}{7}, \frac{5}{7}, \frac{5}{7}, \frac{5}{7}, \frac{5}{7}, \frac{5}{7}, \frac{4}{5}, \frac{4}{5}, \frac{4}{5}, \frac{4}{5}, \frac{4}{5}).$$

Roth et al. (2005a) proved the following results:

Theorem 2 (*Roth et al., 2005a*). *The vector u^E is a feasible utility profile.*

In particular, the proof of theorem 2 shows how a lottery that implements u^E can be constructed.

Theorem 3 (*Roth et al., 2005a*). *The utility profile u^E Lorenz dominates any other feasible utility profile (efficient or not).*

The egalitarian mechanism is a lottery mechanism that selects a lottery whose utility profile is u^E. It is only uniquely valued for the utility profile induced. As a mechanism, the egalitarian approach also has appealing properties:

Theorem 4 (*Roth et al., 2005a*). *The egalitarian mechanism is ex ante efficient and strategy proof.*

The egalitarian mechanism can be used for cases in which there is no exogenous way to distinguish among recipients.

The related literature for this section includes four other papers, two of which are by Bogomolnaia and Moulin (2004), who inspect a two-sided matching problem with the same setup as the model above, and by Dutta and Ray (1989), who introduce the egalitarian approach for convex TU-cooperative games. Morrill (2008) inspects a model similar to the one surveyed here for two-way exchanges, with the exception that preferences are strict. He considers Pareto-efficient matchings and proposes a polynomial time algorithm for finding one starting from a status quo matching (see the section on dynamic kidney exchange later in this chapter). Yilmaz (2011) considers an egalitarian kidney exchange mechanism when multi-way list exchanges are possible. He considers a hybrid model between Roth et al. (2004) and (2005a).

MULTI-WAY KIDNEY EXCHANGES

Roth et al. (2007) explored what is lost when the central authority conducts only two-way kidney exchanges rather than multi-way exchanges. More specifically, they examined the upper bound of marginal gains from conducting two- and three-way exchanges instead of only two-way exchanges, two-, three, and four-way exchanges instead of only two- and three-way exchanges, and unrestricted multi-way exchanges instead of only two-, three-, and four-way exchanges. The setup is very similar to that given in the previous section, with only one difference: a matching does not necessarily consist of two-way exchanges. All results in this section are due to Roth et al. (2007) unless otherwise noted.

In this section, a recipient will be assumed to have a single incompatible donor, and thus, the recipient and her incompatible donor will be referred to as a pair. The blood types of the recipient R_i and donor D_i are denoted as X–Y for pair i, where the recipient is of blood type X and donor is of blood type Y.

An example helps illustrate why the possibility of a three-way exchange is important:

Example 2. *Consider a sample of fourteen incompatible recipient–donor pairs. There are nine pairs who are blood-type incompatible, of types A–AB, B–AB, O–A, O–A, O–B, A–B, A–B, A–B, and B–A; and five pairs who are incompatible because of tissue rejection, of types A–A, A–A, A–A, B–O, and AB–O. For simplicity in this example there is no tissue rejection between recipients and other recipients' donors.*

- If only two-way exchanges are possible:
  (A–B,B–A); (A–A,A–A); (B–O,O–B); (AB–O,A–AB) is a possible Pareto-efficient matching.
- If three-way exchanges are also feasible:
  (A–B,B–A); (A–A,A–A,A–A); (B–O,O–A,A–B); (AB–O, O–A, A–AB) is a possible maximal Pareto-efficient matching.

The three-way exchanges allow:

1. an odd number of A–A pairs to be transplanted (instead of only an even number with two-way exchanges), and
2. a pair with a donor who has a blood type more desirable than her recipient's to facilitate three transplants rather than only two. Here, the AB–O type pair helps two pairs with recipients having less desirable blood type than their donors (O–A and A–AB), while the B–O type pair helps one pair with a recipient having a less desirable blood type than her donor (O–A) and a pair of type A–B. Here, note that another A–B type pair is already matched with a B–A type, and this second A–B type pair is in excess.

First, we introduce two upper-bound assumptions and find the size of Pareto-efficient exchanges with only two-way exchanges:

Assumption 1 (*upper-bound assumption*). *No recipient is tissue-type incompatible with another recipient's donor.*

Assumption 2 (*large population of incompatible recipient–donor pairs*). *Regardless of the maximum number of pairs allowed in each exchange, pairs of types O–A, O–B, O–AB, A–AB, and B–AB are on the "long side" of the exchange, in the sense that at least one pair of each type remains unmatched in each feasible set of exchanges. We simply assume there is an arbitrarily many number of O–A, O–B, O–AB, A–AB, and B–AB type pairs.*

The following observations concern the feasibility of exchanges:

Observation 1. *A pair of type X-Y ∈ {O–A, O–B, O–AB, A–AB, B–AB} can participate in a two-way exchange only with a pair of its reciprocal type Y-X or type AB–O.*

Observation 2. *A pair of O–O, A–A, B–B, AB–AB, A–B, or B–A can participate in a two-way exchange only with its reciprocal type pair or a pair belonging to some of the types among A–O, B–O, AB–O, AB–A, AB–B.*

Observation 3. *A pair of type X-Y ∈ {A–O, B–O, AB–O, AB–A, AB–B} can participate in a two-way exchange with a pair of not only its own type (and possibly some other types in the same set), but also some types among O–A, O–B, O–AB, A–AB, B–AB, O–O, A–A, B–B, AB–AB, A–B, B–A, as well.*

Based on the above observations and the intuition given in example 2, we formally classify the types of pairs into four (Ünver, 2010):

- *overdemanded* types: $\mathcal{T}^O = \{$A–O, B–O, AB–O, AB–A, AB–B$\}$
- *underdemanded* types: $\mathcal{T}^U = \{$O–A, O–B, O–AB, A–AB, B–AB$\}$
- *self-demanded* types: $\mathcal{T}^S = \{$O–O, A–A, B–B, AB–AB$\}$
- *reciprocally demanded* types: $\mathcal{T}^R = \{$A–B, B–A$\}$

Observe that the definitions of overdemanded and underdemanded types in this chapter are different from their definitions used earlier for the GED lemma. We will use these definitions in the next two sections as well. Both definitions are in the same flavor, yet they are not equivalent.

The first result is about the greatest lower bound of the size of two-way Pareto-efficient matchings:

Proposition 1 (*Roth et al., 2007*). *The **maximal size of two-way matchings:** For any recipient population obeying assumptions 1 and 2, the maximum number of recipients who can be matched with only two-way exchanges is:*

$$2 \left(\#(A\text{–}O) + \#(B\text{–}O) + \#(AB\text{–}O) + \#(AB\text{–}A) + \#(AB\text{–}B) \right)$$

$$+ \left(\#(A\text{–}B) + \#(B\text{–}A) - |\#(A\text{–}B) - \#(B\text{–}A)| \right)$$

$$+ 2 \left(\left\lfloor \frac{\#(A\text{–}A)}{2} \right\rfloor + \left\lfloor \frac{\#(B\text{–}B)}{2} \right\rfloor + \left\lfloor \frac{\#(O\text{–}O)}{2} \right\rfloor + \left\lfloor \frac{\#(AB\text{–}AB)}{2} \right\rfloor \right)$$

where $\lfloor a \rfloor$ refers to the largest integer smaller than or equal to a and $\#(x\text{–}y)$ refers to the number of x–y type pairs.

We can generalize example 2 in a proposition for three-way exchanges. We introduce an additional assumption for ease of notation. The symmetric case implies replacing types "A" with "B" and " B" with "A" in all of the following results.

Assumption 3. $\#(A\text{–}B) > \#(B\text{–}A)$.

The following is a simplifying assumption.

Assumption 4. *There is either no type A–A pair or there are at least two of them. The same is also true for each of the types B–B, AB–AB, and O–O.*

When three-way exchanges are also feasible, as we noted earlier, lemma 1 no longer holds. Thus, we consider the largest of the Pareto-efficient matchings under two- and three-way matching technology.

On the other hand, an overdemanded AB–O type pair can potentially save two underdemanded type pairs of types O–A and A–AB, or O–B and B–AB, under a three-way exchange (see Figure 4.3).

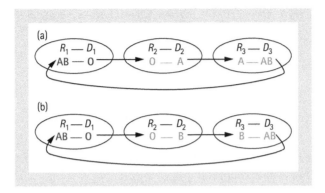

FIGURE 4.3. AB–O type pair saving two underdemanded pairs in a three-way exchange.

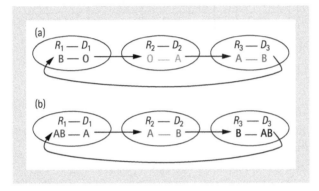

FIGURE 4.4. Overdemanded pairs B–O / AB–A each saving one underdemanded pair and an A–B type pair in a three-way exchange.

When the number of A–B type pairs is larger than the number of B–A type pairs in a static pool (assumption 3):

- All B–A type pairs can be matched with A–B type pairs in two-way exchanges.
- Each B–O type pair can potentially save one O–A type pair and one *excess* A–B type pair in a three-way exchange.
- Each AB–A type pair can potentially save one *excess* A–B type and one B–AB type pair in a three-way exchange (see Figure 4.4).

The above intuition can be stated as a formal result:

Proposition 2 (*Roth et al., 2007*). *The maximal size of two- and three-way matchings: For any recipient population for which assumptions 1–4 hold, the maximum number of recipients who can be matched with two-way and three-way exchanges is:*

$$2 \left(\#(A{-}O) + \#(B{-}O) + \#(AB{-}O) + \#(AB{-}A) + \#(AB{-}B)\right)$$

$$+ \left(\#(A{-}B) + \#(B{-}A) - |\#(A{-}B) - \#(B{-}A)|\right)$$

$$+ \left(\#(A{-}A) + \#(B{-}B) + \#(O{-}O) + \#(AB{-}AB)\right)$$

$$+ \#(AB{-}O)$$

$$+ \min\{(\#(A{-}B) - \#(B{-}A)), (\#(B{-}O) + \#(AB{-}A))\}$$

And to summarize, the marginal effect of availability of two- and three-way kidney exchanges over two-way exchanges is:

$$\#(A{-}A) + \#(B{-}B) + \#(O{-}O) + \#(AB{-}AB)$$

$$- 2 \left(\left[\frac{\#(A{-}A)}{2}\right] + \left[\frac{\#(B{-}B)}{2}\right] + \left[\frac{\#(O{-}O)}{2}\right] + \left[\frac{\#(AB{-}AB)}{2}\right] \right)$$

$$+ \#(AB{-}O)$$

$$+ \min\{(\#(A{-}B) - \#(B{-}A)), (\#(B{-}O) + \#(AB{-}A))\}$$

What about the marginal effect of two-, three-, and four-way exchanges over two- and three-way exchanges? It turns out that there is only a slight improvement in the maximal matching size with the possibility of four-way exchanges.

We illustrate this using the above example:

Example 3 (*example 2 continued*). *If four-way exchanges are also feasible, instead of the exchange (AB–O, O–A, A–AB) we can now conduct a four-way exchange (AB–O, O–A, A–B, B–AB). Here, the valuable AB–O type pair helps an additional A–B type pair in excess in addition to two pairs with less desirable blood-type donors than their recipients.*

Thus, each AB–O type pair can potentially save one O–A type pair, one *excess* A–B type pair, and one B–AB type pair in a four-way exchange (see Figure 4.5).

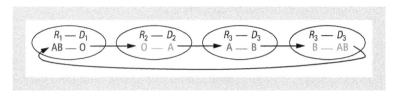

FIGURE 4.5. An overdemanded AB–O type pair can save three underdemanded pairs in a four-way kidney exchange.

We formalize this intuition as the following result:

Proposition 3 (*Roth et al., 2007*). *The maximal size of two-, three-, and four-way matchings: For any recipient population in which assumptions 1–4 hold, the maximum number of recipients who can be matched with two-way, three-way, and four-way exchanges is:*

$$2\,(\#(A\text{–}O) + \#(B\text{–}O) + \#(AB\text{–}O) + \#(AB\text{–}A) + \#(AB\text{–}B))$$

$$+ \,(\#(A\text{–}B) + \#(B\text{–}A) - |\#(A\text{–}B) - \#(B\text{–}A)|)$$

$$+ \,(\#(A\text{–}A) + \#(B\text{–}B) + \#(O\text{–}O) + \#(AB\text{–}AB))$$

$$+ \,\#(AB\text{–}O)$$

$$+ \,\min\{(\#(A\text{–}B) - \#(B\text{–}A)),$$

$$(\#(B\text{–}O) + \#(AB\text{–}A) + \#(AB\text{–}O))\}$$

Therefore, in the absence of tissue-type incompatibilities between recipients and other recipients' donors, the marginal effect of four-way kidney exchanges is bounded from above by the rate of the very rare AB–O type.

It turns out that under the above assumptions, larger exchanges do not help to match more recipients. This is stated as follows:

Theorem 5 (*Roth et al., 2007*). *Availability of four-way exchange suffices: Consider a recipient population for which assumptions 1, 2, and 4 hold and let μ be any maximal matching (when there is no restriction on the size of the exchanges). Then there exists a maximal matching v that consists only of two-way, three-way, and four-way exchanges, under which the same set of recipients benefits from exchange as in matching μ.*

What about incentives, when these maximal solution concepts are adopted in a kidney exchange mechanism? The strategic properties of multi-way kidney exchange mechanisms are inspected by Hatfield (2005) in the 0–1 preference domain. This result is a generalization of theorem 1.

A deterministic kidney exchange mechanism is *consistent* if whenever it only selects a multi-way matching in set $\mathcal{X} \subseteq \mathcal{M}$ as its outcome, where all matchings in $\mathcal{X}$ generate the same utility profile when the set of feasible individually rational matchings is $\mathcal{M}$, then for any other problem for the same set of pairs such that the set of feasible individually rational matchings is $\mathcal{N} \subset \mathcal{M}$ with $\mathcal{X} \cap \mathcal{N} \neq \varnothing$, it selects a multi-way matching in set $\mathcal{X} \cap \mathcal{N}$.[11]

---

[11] Recall that a kidney exchange mechanism may select many matchings that are utility-wise equivalent in the 0–1 preference domain. A two-way priority mechanism is an example.

A deterministic mechanism is *non-bossy* if whenever one recipient manipulates her preferences/number of donors and cannot change her outcome, defined as either being matched to a compatible donor or remaining unmatched, then she cannot change other recipients' outcome under this mechanism with the same manipulation.

The last result of this section is as follows:

Theorem 6 (*Hatfield, 2005*). *If a deterministic mechanism is non-bossy and strategy proof then it is consistent. Moreover, a consistent mechanism is strategy proof.*

Thus, it is straightforward to create strategy-proof mechanisms using *maximal-priority* or *priority* multi-way exchange rules. By maximal-priority mechanisms, we mean mechanisms that maximize the number of patients matched (under an exchange restriction such as two, three, four, etc., or no exchange size restriction), and then use a priority criterion to select among such matchings.

SIMULATIONS USING NATIONAL RECIPIENT CHARACTERISTICS

In this section we dispense with the simplifying assumptions made so far, and turn to simulated data reflecting national recipient characteristics. Specifically, we now look at populations in which a recipient may have tissue type incompatibilities with many donors. This will allow us to assess the accuracy of the approximations derived under the above assumption that exchange is limited only by blood-type incompatibilities.

The simulations reported here follow those of Saidman et al. (2006), and Roth et al. (2007). We will see that the formulas predict the actual number of exchanges surprisingly well. That is, the upper bounds on the maximal number of exchanges when exchange is limited only by blood-type incompatibility are not far above the numbers of exchanges that can actually be realized. In addition, only a small number of exchanges involving more than four pairs are needed to achieve efficiency in the simulated data.

Recipient–donor population construction

We consider samples of non-blood-related recipient–donor pairs, to avoid complications due to the impact of genetics on immunological incompatibilities. The characteristics such as the blood types of recipients and donors, the PRA distribution of the recipients, donor relation of recipients, and the gender of the recipients are generated using the empirical distributions of the data from an OPTN subsidiary in the US, the Scientific Registry of Transplant Recipients (SRTR) (see Table 4.1). We consider all ethnicity in the data.

Table 4.1. Patient and living–donor distributions used in simulations

	Frequency (percent)
A. Patient ABO blood type	
0	48.14
A	33.73
B	14.28
AB	3.85
B. Patient gender	
Female	40.90
Male	59.10
C. Unrelated living donors	
Spouse	48.97
Other	51.03
E. PRA distribution	
Low PRA	70.19
Medium PRA	20.00
High PRA	9.81

Based on OPTN/SRTR Annual Report in 2003, for the period 1993–2002, retrieved from <http://www.optn.org> on November 22, 2004. Patient characteristics are obtained using the new waiting list registrations data, and living-donor relational type distribution is obtained from living-donor transplants data.

In our simulations, we randomly simulate a series of recipient–donor pairs using the population characteristics explained above. Whenever a pair is compatible (both blood-type compatible and tissue-type compatible), the donor can directly donate to the intended recipient and therefore we do not include them in our sample. Only when they are either blood-type or tissue-type incompatible do we keep them, until we reach a sample size of n incompatible pairs. We use a Monte-Carlo simulation size of 500 random population constructions for three population sizes of 25, 50, and 100.

Tissue-type incompatibility

Tissue-type incompatibility (a *positive crossmatch*) is independent of blood-type incompatibility, and arises when a recipient has preformed antibodies against a donor tissue type.

Recipients in the OPTN/SRTR database are divided into the following three groups based on the odds that they have a crossmatch with a random donor:

1. Low-PRA (percent reactive antibody) recipients: Recipients who have a positive crossmatch with less than 10% of the population.

2. Medium-PRA recipients: Recipients who have a positive crossmatch with 10–80% of the population.
3. High-PRA recipients: Recipients who have a positive crossmatch with more than 80% of the population.

Frequencies of low-, medium-, and high-PRA recipients reported in the OPTN/SRTR database are given in Table 4.1. Since a more detailed PRA distribution is unavailable in the medical literature, we will simply assume that:

- each low-PRA recipient has a positive crossmatch probability of 5% with a random donor;
- each medium-PRA recipient has a positive crossmatch probability of 45% with a random donor; and
- each high-PRA recipient has a positive crossmatch probability of 90% with a random donor.

We have already indicated that when the recipient is female and the potential donor is her husband, it is more likely that they have a positive crossmatch due to pregnancies. Zenios et al. (2001) indicate that while positive crossmatch probability is 11.1% between random pairs, it is 33.3% between female recipients and their donor husbands. Equivalently, female recipients' *negative crossmatch* probability (i.e. the odds that there is no tissue-type incompatibility) with their husbands is approximately 75% of the negative crossmatch probability with a random donor. Therefore, we accordingly adjust the positive crossmatch probability between a female recipient and her donor husband using the formula

$$PRA^* = 100 - 0.75(100 - PRA)$$

and assume that

- each low-PRA female recipient has a positive crossmatch probability of 28.75% with her husband;
- each medium-PRA female recipient has a positive crossmatch probability of 58.75% with her husband; and
- each high-PRA female recipient has a positive crossmatch probability of 92.25% with her husband.

Outline of the simulations

For each sample of n incompatible recipient–donor pairs, we find the maximum number of recipients who can benefit from an exchange when both blood-type and tissue-type incompatibilities are considered, and

- only two-way exchanges are allowed;
- two-way and three-way exchanges are allowed;
- two-way, three-way, and four-way exchanges are allowed; and
- any size exchange is allowed.

In our simulations, to find the maximal number of recipients who can benefit from an exchange when only two-way exchanges are allowed, we use a version of Edmonds' (1965) algorithm (see Roth et al., 2005a), and to find the maximal number of recipients who can benefit from an exchange when larger exchanges are allowed, we use various integer programming techniques.

We compare these numbers with those implied by the analytical expressions in the above propositions, to see whether those formulas are close approximations or merely crude upper bounds. Since many high-PRA recipients cannot be part of any exchange due to tissue-type incompatibilities, we report two sets of upper bounds induced by the formulas we developed:

1. For each sample we use the formulas with the raw data.
2. For each sample we restrict our attention to recipients each of whom can participate in at least one feasible exchange.

That is, in Table 4.2, "upper bound 1" for each maximal allowable size exchange is the formula developed above for that size exchange (i.e. propositions 1, 2, and 3 for maximal exchange sizes two, three, or four pairs) with the population size of $n = 25$, 50, or 100. However, in a given sample of size $n = 25$, for example, there may be some recipients who have no compatible donor because of tissue-type incompatibilities, and hence cannot possibly participate in an exchange. In this population there is therefore a smaller number $n' < n$ of pairs actually available for exchange, and "upper bound 2" in Table 4.2 reports the average over all populations for the formulas using this smaller population of incompatible recipient–donor pairs. Clearly upper bound 2 provides a more precise (i.e. lower) upper bound to the number of exchanges that can be found. The fact that the difference between the two upper bounds diminishes as the population size increases reflects that, in larger populations, even highly sensitized recipients are likely to find a compatible donor.

Discussion of the simulation results

The static simulation results (which include tissue-type incompatibilities) are very similar to the theoretical upper bounds we develop for the case with only blood-type incompatibilities. While two-way exchanges account for most of the potential gains from exchange, the number of recipients who benefit from exchange significantly increases when three-way or more exchanges are allowed, and, consistent with the theory, three-way exchanges account for a large share of the remaining potential gains. For example, for a population size of 25 pairs, an average of:

Table 4.2. Simulation results for the average number of patients actually matched and predicted by the formulas to be matched.

Pop. size	Method	Two-way	Two-way, three-way	Two-way, three-way, four-way	No constraint
				Type of exchange	
$n=25$	Simulation	8.86	11.272	11.824	11.992
		(3.4866)	(4.0003)	(3.9886)	(3.9536)
	Upper bound 1	12.5	14.634	14.702	
		(3.6847)	(3.9552)	(3.9896)	
	Upper bound 2	9.812	12.66	12.892	
		(3.8599)	(4.3144)	(4.3417)	
$n=50$	Simulation	21.792	27.266	27.986	28.09
		(5.0063)	(5.5133)	(5.4296)	(5.3658)
	Upper bound 1	27.1	30.47	30.574	
		(5.205)	(5.424)	(5.4073)	
	Upper bound 2	23.932	29.136	29.458	
		(5.5093)	(5.734)	(5.6724)	
$n=100$	Simulation	49.708	59.714	60.354	60.39
		(7.3353)	(7.432)	(7.3078)	(7.29)
	Upper bound 1	56.816	62.048	62.194	
		(7.2972)	(7.3508)	(7.3127)	
	Upper bound 2	53.496	61.418	61.648	
		(7.6214)	(7.5523)	(7.4897)	

The standard errors of the population are reported in parentheses. The standard errors of the averages are obtained by dividing population standard errors by the square root of the simulation number, 22.36.

- 11.99 pairs can be matched when any size exchange is feasible;
- 11.27 pairs can be matched when only two-way and three-way exchange are feasible; and
- 8.86 pairs can be matched when only two-way exchange is feasible.

Hence for $n = 25$, two-way exchanges account for 74% (i.e. $\frac{8.86}{11.99}$) of the potential gains from exchange, whereas three-way exchanges account for 77% (i.e. $\frac{11.27-8.86}{11.99-8.86}$) of the remaining potential gains. These rates are 78% and 87% for a population size of 50 pairs, and 82% and 94% for a population size of 100 pairs. The theory developed in the absence of crossmatches is still predictive when there are crossmatches: virtually all possible gains from trade are achieved with two-way, three-way, and four-way exchanges, especially when the population size is large (see Table 4.2).[12]

[12] When the population size is 100 incompatible pairs, in 485 of the 500 simulated populations the maximum possible gains from trade are achieved when no more than four pairs are allowed to participate in an exchange.

DYNAMIC KIDNEY EXCHANGE

The above two models consider a static situation: a pool of recipients with their directed incompatible donors. These models answer how we can organize kidney exchanges in an efficient and incentive-compatible way. However, in real life, the recipient pool is not static but evolves over time. Ünver (2010) considered a model in which the exchange pool evolves over time by pairs of a recipient and her directed donor arriving with a Poisson distribution in continuous time with an expected arrival rate of λ. The question answered by this paper is that if there is a constant unit cost of waiting in the pool for each recipient, what is the mechanism that should be run to conduct the exchanges so that the expected discounted exchange surplus is maximized? (It turns out that this is equivalent to maximizing the expected discounted number of recipients to be matched.)

There are also operation research and computer science articles answering different aspects of the dynamic problem. Zenios (2002) considers a continuous-arrival model with pairs of recipients and their directed donors. The model is stylistic in the sense that all blood types are not modeled, and all exchanges are two way. However, the preferences are not 0–1 and the outside option is list exchange. Awasthi and Sandholm (2009) consider an online mechanism design approach to find optimal dynamic mechanisms for kidney exchange when there are no waiting costs but pairs can exit the pool randomly. They look at mechanisms that are obtained heuristically by sampling future possibilities depending on the current and past matches. Their model has a very large state space; thus, online sampling is used to simplify the optimization problem.

Exchange pool evolution

We continue with Ünver's (2010) model. For any pair type X–Y $\in \mathcal{T}$, let q_{X-Y} be the probability of a random pair being of type X–Y. We refer to q_{X-Y} as the *arrival probability* of pair type X–Y $\in \mathcal{T}$. We have $\sum_{X-Y \in \mathcal{T}} q_{X-Y} = 1$.

Once a pair arrives, if it is not compatible, it becomes available for exchange. If it is compatible, the donor immediately donates a kidney to the recipient of the pair, and the pair does not participate in exchanges. The *exchange pool* is the set of the pairs which have arrived over time and whose recipient has not yet received a transplant.

Let p_c be the *positive crossmatch* probability that determines the probability that a donor and a recipient will be *tissue-type incompatible*. Let p_{X-Y} denote the *pool entry probability* of any arriving pair type X–Y. Since blood-type incompatible pairs always join the exchange pool, we have $p_{X-Y} = 1$ for any blood-type incompatible X–Y.

Since blood-type compatible pairs join the pool if and only if they are not tissue-type compatible, we have $p_{X-Y} = p_c$ for any blood-type compatible X–Y. Let $\lambda^P = \lambda \sum_{X-Y \in \mathcal{T}} p_{X-Y} q_{X-Y}$ be the expected number of pairs that enter the pool for exchange per unit time interval.

Time- and compatibility-based preferences

Each recipient has preferences over donors and time of waiting in the pool. For any incompatible pair i, recipient R_i's preferences are denoted by $\succsim_i$ and defined over donor–time interval pairs. Recipient R_i's preferences over donors fall into three indifference classes (as in earlier sections): compatible donors are preferred to being unmatched—an option denoted by being matched with her paired *incompatible* donor D_i—and, in turn, being unmatched is preferred to being matched with incompatible donors. Moreover, time spent in the exchange pool is another dimension in the preferences of recipients: waiting is costly. Formally, preferences of R_i over donors and time spent in the pool are defined as follows: [13]

1. for any two compatible donors D and D' with R_i, and time period t, $(D, t) \sim_i (D', t)$
    (indifference over compatible donors if both transplants occur at the same time);
2. for any compatible donor D with R_i and time periods t and t' such that $t < t'$,
    $(D, t) \succ_i (D, t')$ (waiting for a compatible donor is costly);
3. for any compatible donor D with R_i and time periods t and t', $(D, t) \succ_i (D_i, t')$
    (compatible donors are preferred to remaining unmatched);
4. for any incompatible donor $D \neq D_i$ and time periods t and t', $(D_i, t) \succ_i (D, t')$
    (remaining unmatched is preferred to being matched with incompatible donors).

For each pair, we associate waiting in the pool with a monetary cost and we assume that the *unit time cost* of waiting for a transplant by undergoing continuous dialysis is equal to $c > 0$ for each recipient. The alternative to a transplant is dialysis. A recipient can undergo dialysis continuously. It is well known that receiving a transplant causes the recipient to resume a better life (Overbeck et al., 2005). Also, health-care costs for dialysis are higher than those for transplantation in the long term (Schweitzer et al., 1998). We model all the costs associated with undergoing continuous dialysis by the unit time cost c.

Dynamically efficient mechanisms

A *(dynamic) matching mechanism* is a *dynamic* procedure such that at each time $t \geq 0$ it selects a (possibly empty) matching of the pairs available in the pool. Once a pair is

---

[13] Let $\sim_i$ denote the indifference relation and $\succ_i$ denote the strict preference relation associated with the preference relation $\succsim_i$.

matched at time t by a matching mechanism, it leaves the pool and its recipient receives the assigned transplant.

Let $\#^A(t)$ be the total number of pairs that have arrived until time t. If mechanism ϕ is executed (starting time 0), $\#^\phi(t)$ is the total number of pairs matched by mechanism ϕ. There are $\#^A(t) - \#^\phi(t)$ pairs available at the pool at time t.

There is a health authority that oversees the exchanges.

Suppose that the health authority implements a matching mechanism, ϕ. For any time t, the current value of expected cost at time t under matching mechanism ϕ is given as:[14]

$$E_t\left[C^\phi(t)\right] = \int_t^\infty cE_t\left[\#^A(\tau) - \#^\phi(\tau)\right]e^{-\rho(\tau-t)}d\tau,$$

where ρ is the *discount rate*.

For any time τ, t such that $\tau > t$, we have $E_t\left[\#^A(\tau)\right] = \lambda^p(\tau - t) + \#^A(t)$, where the first term is the expected number of recipients to arrive at the exchange pool in the interval $[t, \tau]$ and the second term is the number of recipients that arrived at the pool until time t. Therefore, we can rewrite $E_t\left[C^\phi(t)\right]$ as:

$$E_t\left[C^\phi(t)\right] = \int_t^\infty c(\lambda^p(\tau - t) + \#^A(t) - E_t\left[\#^\phi(\tau)\right])e^{-\rho(\tau-t)}d\tau.$$

Since $\int_t^\infty e^{-\rho(\tau-t)}d\tau = \frac{1}{\rho}$ and $\int_t^\infty (\tau - t)e^{-\rho(\tau-t)}d\tau = \frac{1}{\rho^2}$, we can rewrite $E_t\left[C^\phi(t)\right]$ as:

$$E_t\left[C^\phi(t)\right] = \frac{c\lambda^p}{\rho^2} + \frac{\#^A(t)}{\rho} - \int_t^\infty cE_t\left[\#^\phi(\tau)\right]e^{-\rho(\tau-t)}d\tau. \tag{1}$$

Only the last term in equation 1 depends on the choice of mechanism ϕ. The previous terms cannot be controlled by the health authority, since they are the costs associated with the number of recipients arriving at the pool. We refer to this last term as the *exchange surplus at time t* **for** mechanism ϕ and denote it by:

$$\mathcal{ES}^\phi(t) = \int_t^\infty cE_t\left[\#^\phi(\tau)\right]e^{-\rho(\tau-t)}d\tau.$$

We can rewrite it as:

$$\mathcal{ES}^\phi(t) = \int_t^\infty c\left(E_t\left[\#^\phi(\tau) - \#^\phi(t)\right] + \#^\phi(t)\right)e^{-\rho(\tau-t)}d\tau$$

$$= \frac{c\#^\phi(t)}{\rho} + \int_t^\infty c\left(E_t\left[\#^\phi(\tau) - \#^\phi(t)\right]\right)e^{-\rho(\tau-t)}d\tau.$$

The first term above is the exchange surplus attributable to all exchanges that have been done until time t and at time t, and the second term is the *future exchange surplus* attributable to the exchanges to be done in the future. The central health authority

---

[14] E_t refers to the expected value at time t.

cannot control the number of past exchanges at time t either. Let $n^\phi(\tau)$ be the *number of matched recipients at time τ by mechanism ϕ*, and we have:

$$\#^\phi(t) = \left(\sum_{\tau < t} n^\phi(\tau)\right) + n^\phi(t).$$

We focus on the *present* and *future exchange surplus*, which is given as:

$$\widetilde{\mathcal{ES}}^\phi(t) = \frac{cn^\phi(t)}{\rho} + \int_t^\infty c\left(E_t\left[\#^\phi(\tau) - \#^\phi(t)\right]\right) e^{-\rho(\tau - t)} d\tau. \tag{2}$$

A dynamic matching mechanism v is (*dynamically*) **efficient** if, for any t, it maximizes the *present and future exchange surplus* at time t given in equation 2. We look for solutions of the problem independent of initial conditions and time t. We will define a steady-state formally. If such solutions exist, they depend only on the "current state of the pool" (defined appropriately) but not on time t or the initial conditions.

Dynamically efficient *two-way* exchange

In this subsection, we derive the dynamically optimal *two-way* matching mechanism. Throughout this subsection we will maintain two assumptions, assumptions 1 and 2, introduced earlier.

We are ready to state theorem 7.

Theorem 7 (Ünver, 2010). *Let dynamic matching mechanism v be defined as a mechanism that matches only X–Y type pairs with their reciprocal Y–X type pairs, immediately when such an exchange is feasible. Then, under assumptions 1 and 2, mechanism v is a dynamically optimal two-way matching mechanism.*

Moreover, a dynamically optimal two-way matching mechanism conducts a two-way exchange whenever one becomes feasible.

Next we show that assumption 2 will hold in the long run under the most reasonable pair-type arrival distributions; thus, it is not a restrictive assumption.

Proposition 4 (Ünver, 2010). *Suppose that $p_c\left(q_{AB-O} + q_{X-O}\right) < q_{O-X}$ for all $X \in \{A, B\}$, $p_c\left(q_{AB-O} + q_{AB-X}\right) < q_{X-AB}$ for all $X \in \{A, B\}$ and $p_c q_{AB-O} < q_{O-AB}$. Then, assumption 2 holds in the long run regardless of the two-way matching mechanism used.*

The hypothesis of the above proposition is very mild and will hold for sufficiently small crossmatch probability. Moreover, it holds for real-life blood frequencies. For example, assuming that the recipient and the paired donor are blood unrelated, the arrival rates reported in the earlier simulations satisfy these assumptions, when the crossmatch probability is $p_c = 0.11$, as reported by Zenios et al. (2001).

Dynamically efficient *multi-way* exchanges

In this subsection, we consider matching mechanisms that allow for not only two-way exchanges, but larger exchanges as well. Roth et al. (2010) have studied the importance of three-way and larger exchanges in a static environment, and we summarized these results earlier. The results in this subsection follow this intuition, and are due to Ünver (2010). We can state the following observation motivated by the results reported earlier:

Observation 4. *In an exchange that matches an underdemanded pair, there should be at least one overdemanded pair. In an exchange that matches a reciprocally demanded pair, there should at least be one reciprocal type pair or an overdemanded pair.*

Using the above illustration, under realistic blood-type distribution assumptions, we will prove that assumption 2 still holds, when the applied matching mechanism is unrestricted. Recall that through assumption 2, we assumed these were arbitrarily many underdemanded type pairs available in the long-run states of the exchange pool, regardless of the dynamic matching mechanism used in the long run.

Proposition 5 (*Ünver, 2010*). *Suppose that $p_c \left(q_{AB-O} + q_{X-O} \right) + \min \left\{ p_c q_{Y-O}, q_{X-Y} \right\} < q_{O-X}$ for all $\{X, Y\} = \{A, B\}$, $p_c \left(q_{AB-O} + q_{AB-X} \right) + \min \left\{ p_c q_{AB-Y}, q_{Y-X} \right\} < q_{X-AB}$ for all $\{X, Y\} = \{A, B\}$ and $p_c q_{AB-O} < q_{O-AB}$. Then, assumption 2 holds in the long run regardless of the unrestricted matching mechanism used.*

The hypothesis of the above proposition is also very mild and will hold for sufficiently small crossmatch probability p_c. Moreover, it holds for real-life blood frequencies and crossmatch probability. For example, assuming that the recipient and the paired donor are blood unrelated, the arrival rates reported in the simulations section of the paper satisfy these assumptions. Thus, we can safely use assumption 2 in this section, as well.

Next, we characterize the dynamically efficient mechanism.

In a dynamic setting, the structure of three-way and four-way exchanges discussed earlier may cause the second part of theorem 7 not to hold when these larger exchanges are feasible. More specifically, we can benefit from not conducting all feasible exchanges currently available, and holding on to some of the pairs that can currently participate in an exchange in expectation of saving more pairs in the near future.

We maintain assumption 1 as well as assumption 2 in this subsection. We state one other assumption. First, we state that as long as the difference between A–B and B–A type arrival frequencies is not large, overdemanded type pairs will be matched immediately.

Proposition 6 (*Ünver, 2010*). *Suppose assumptions 1 and 2 hold. If q_{A-B} and q_{B-A} are sufficiently close, then under the dynamically efficient multi-way matching mechanism, overdemanded type pairs are matched as soon as they arrive at the exchange pool.*

Assumption 5 (*assumption on generic arrival rates of reciprocally demanded types*). *A–B and B–A type pairs arrive at relatively close frequency to each other so that proposition 6 holds.*

Under assumptions 1, 2, and 5, we will only need to make decisions in situations in which multiple exchanges of different sizes are feasible: For example, consider a situation in which an A–O type pair arrives at the pool, while a B–A type pair is also available. Since, by assumption 2, there is an excess number of O–A and O–B type pairs in the long run, there are two sizes of feasible exchanges, a three-way exchange (for example, involving A–O, O–B, and B–A type pairs) or a two-way exchange (for example, involving A–O and O–A type pairs). Which exchange should the health authority choose?

To answer this question, we analyze the dynamic optimization problem. Since the pairs arrive according to a Poisson process, we can convert the problem to an embedded *Markov decision process*. We need to define a state space for our analysis. Since the pairs in each type are symmetric by assumption 1, the natural candidate for a *state* is a sixteen-dimensional vector, which shows the number of pairs in each type available. In our exchange problem, there is additional structure to eliminate some of these state variables. We look at overdemanded, underdemanded, self-demanded, and reciprocally demanded types separately:

- *Overdemanded types.* If an overdemanded pair i of type X–Y $\in \mathcal{T}^O$ arrives, by proposition 6, pair i will be matched immediately in some exchange. Hence, the number of overdemanded pairs remaining in the pool is always 0.
- *Underdemanded types.* By assumption 2 as well as assumption 1, there will be an arbitrarily large number of underdemanded pairs. Hence, the number of underdemanded pairs is always ∞.
- *Self-demanded types.* Whenever a self-demanded pair i of type X–X $\in \mathcal{T}^S$ is available in the exchange pool, it can be matched through two ways under a multi-way matching mechanism:
  1. If another X–X type pair j arrives, by assumption 1, i and j will be mutually compatible, and a two-way exchange (i, j) can be conducted.
  2. If an exchange $E = (i_1, i_2, \ldots, i_k)$, with Y blood-type donor D_{i_k} and Z blood-type recipient R_{i_1}, becomes feasible, and blood-type Y donors are blood-type compatible with blood-type X recipients, while blood-type X donors are blood-type compatible with blood-type Z recipients, then pair i can be inserted in exchange E just after i_k, and by assumption 1, the new exchange $E' = (i_1, i_2, \ldots, i_k, i)$ will be feasible.

By observation 4, a self-demanded type can never save an overdemanded or reciprocally demanded pair without the help of an overdemanded or another reciprocally demanded pair. Suppose that there are n X–X type pairs. Then, they should be matched in two-way exchanges to save $2 \lfloor \frac{n}{2} \rfloor$ of them (which is possible by assumption 1). This and the above observations imply that under a dynamically efficient matching mechanism, for any X–X $\in \mathcal{T}^S$, at steady-state there will be either 0 or 1 X–X type pair.

Therefore, in our analysis, the existence of self-demanded types will be reflected by four additional state variables, each of which gets values either 0 or 1. We will derive the efficient dynamic matching mechanism by ignoring the self-demanded type pairs:

Assumption 6 (*no self-demanded types assumption*). *There are no self-demanded types available for exchange and $q_{X\text{-}X} = 0$ for all X–X $\in \mathcal{T}$.*

- *Reciprocally demanded types*: By the above analysis, there are no overdemanded or self-demanded type pairs available and there are infinitely many underdemanded type pairs. Therefore, the state of the exchange pool can simply be denoted by the number of A–B type pairs and B–A type pairs. By assumption 1, an A–B type pair and B–A type pair are mutually compatible with each other, and they can be matched in a two-way exchange. Moreover, by observation 4, an A–B or B–A type pair cannot save an underdemanded pair in an exchange without the help of an overdemanded pair. Hence, the most optimal use of A–B and B–A type pairs is being matched with each other in a two-way exchange. Therefore, under the optimal matching mechanism, an A–B and B–A type pair will never remain in the pool together but will be matched via a two-way exchange. By this observation, we can simply denote the state of the exchange pool by an integer s, such that if $s > 0$, then s refers to the number of A–B type pairs in the exchange pool, and if $s < 0$, then $|s|$ refers to the number of B–A type pairs in the exchange pool. Formally s is the difference between the number of A–B type pairs and B–A type pairs in the pool, and *only one of these two numbers can be non-zero*. Let $S = \mathbb{Z}$ be the state space (i.e., the set of integers).

Markov chain representation

In this subsection, we characterize the transition from one state to another under a dynamically optimal matching mechanism by a Markov chain given assumptions 1, 2, 5, and 6:

First, suppose $s > 0$, i.e. there are some A–B type pairs and no B–A type pairs. Suppose a pair of type X–Y $\in \mathcal{T}$ becomes available. In this case, three subcases are possible for pair i:

1. X–Y $\in \mathcal{T}^U = \{$O–A, O–B, O–AB, A–AB, B–AB$\}$. By observation 4, in any exchange involving an underdemanded pair, there should be an overdemanded

pair. Since there are no overdemanded pairs available under the optimal mechanism, no new exchanges are feasible. Moreover, the state of the exchange pool remains as s.

2. $X–Y \in \mathcal{T}^O = \{A–O, B–O, AB–O, AB–A, AB–B\}$: If pair i is compatible (which occurs with probability $1 - p_c$), donor D_i donates a kidney to recipient R_i, and pair i does not arrive at the exchange pool. If pair i is incompatible (which occurs with probability p_c), pair i becomes available for exchange. Three cases are possible:

   • $X–Y \in \{A–O, AB–B\}$. Since $s > 0$, there are no B–A type pairs available. In this case, there is one type of exchange feasible: a two-way exchange including pair i, and a mutually compatible pair j of type Y–X. By assumption 2, such a Y–X type pair exists. By proposition 6, this exchange is conducted, resulting in two matched pairs, and the state of the pool remains as s. There is no decision problem in this state.

   • $X–Y \in \{B–O, AB–A\}$. Since $s > 0$, there are A–B type pairs available. There are two types of exchanges that can be conducted: a two-way exchange and a three-way exchange:

      • By assumption 2, there is a mutually compatible pair j of type Y–X, and (i, j) is a feasible two-way exchange.
      • If X–Y = B–O, then, by assumption 2, there is an arbitrary number of O–A type pairs. Let pair j be an O–A type pair. Let k be an A–B type pair in the pool. By assumption 2, (i, j, k) is a feasible three-way exchange (see Figure 4.4).
      If X–Y = AB–A, then, by assumption 2, there is an arbitrary number of B–AB type pairs. Let k be a B–AB type pair. Let j be an A–B type pair in the pool. By assumption 1, (i, j, k) is a feasible three-way exchange.

   Let action a_1 refer to *conducting a smaller exchange* (i.e. two-way), and action a_2 refer to *conducting a larger exchange* (i.e. three-way). If action a_1 is chosen, two pairs are matched, and the state of the pool does not change. If action a_2 is chosen, then three pairs are matched, and the state of the pool decreases to $s - 1$.

   • X–Y = AB–O. Since $s > 0$, there are three types of exchanges that can be conducted: a two-way exchange, a three-way exchange, or a four-way exchange:

      • By assumption 2 and observation 1, for any $W–Z \in \mathcal{T}^U$, there is a mutually compatible pair j of type W–Z for pair i. Hence, (i, j) is a feasible two-way exchange.
      • By assumption 2, there are a pair j of type O–B and pair k of type B–AB such that (i, j, k) is a feasible three-way exchange. Also by assumption 2, there are a pair g of type O–A and a pair h of type A–AB such that (g, h, i) is a feasible three-way exchange (see Figure 4.4). By assumption 2, there is an arbitrarily large number of underdemanded pairs independent of the matching mechanism, therefore, conducting either of these two three-way

exchanges has the same effect on the future states of the pool. Hence, we will not distinguish these two types of exchanges.

- By assumptions 1 and 2, a pair h of type B–AB, a pair j of type O–A, and a pair k of type A–B form the four-way exchange (h, i, j, k) with pair i (see Figure 4.5).

Two-way and three-way exchanges do not change the state of the pool. Therefore, conducting a three-way exchange dominates conducting a two-way exchange. Hence, under the optimal mechanism, we rule out conducting a two-way exchange, when an AB–O type pair arrives. Let action a_1 refer to *conducting a smaller* (i.e. three-way) *exchange*, and let action a_2 refer to *conducting a larger* (i.e. four-way) *exchange*. If action a_1 is chosen, three pairs are matched, and the state of the pool remains as s. If action a_2 is chosen, four pairs are matched, and the state of the pool decreases to $s - 1$.

3. $\text{X–Y} \in \mathcal{T}^R = \{\text{A–B, B–A}\}$. Two cases are possible:

    (a) X–Y = A–B. By observation 4, an A–B type pair can only be matched using a B–A type pair or an overdemanded pair. Since there are no overdemanded and B–A type pairs, there is no possible exchange. The state of the pool increases to $s + 1$.

    (b) X–Y = B–A. By assumption 1, a feasible two-way exchange can be conducted using an A–B type pair j in the pool and pair i. This is the only feasible type of exchange. Since matching a B–A type pair with an A–B type pair is the most optimal use of these types of pairs, we need to conduct such a two-way exchange and the state of the pool decreases to $s - 1$.

Note that we do not need to distinguish decisions regarding two-way versus three-way exchanges, and three-way versus four-way exchanges. We denote all actions regarding smaller exchanges by a_1, and all actions regarding larger exchanges by a_2. Since the difference between a smaller exchange and a larger exchange is always one pair, i.e. an A–B type pair, whenever the state of the pool dictates that a three-way exchange is chosen instead of a two-way exchange when a B–O or AB–A type pair arrives, then it will also dictate that a four-way exchange will be chosen instead of a three-way exchange when an AB–O type pair arrives.

For $s < 0$, that is, when $|s|$ B–A type pairs are available in the exchange pool, we observe the symmetric version of the above evolution. For $s = 0$, that is, when there are no A–B or B–A type pairs available in the exchange pool, the evolution is somewhat simpler. At state 0, the only state transition occurs, when an A–B type pair arrives (to state 1), or when a B–A type pair arrives (to state −1). Actions involving largest exchanges for the case $s > 0$, referred to as *action a_2*, are infeasible at state 0, implying that there is no decision problem. Moreover, there are no exchanges involving A–B or B–A type pairs. In this state, a maximum size exchange is conducted when it becomes feasible.

The dynamically efficient multi-way matching mechanism

A *(deterministic) Markov matching mechanism*, ϕ, is a matching mechanism that chooses the same action whenever the Markov chain is in the same state. In our reduced state and action problem, a Markov matching mechanism chooses either action a_1, conducting the smaller exchange, or action a_2, conducting the largest exchange, at each state, except state o. The remaining choices of the Markov mechanism are straightforward: It chooses a maximal exchange when such an exchange becomes feasible (for negative states by interchanging the roles of A and B blood types as outlined in the previous subsection). Formally, $\phi : S \rightarrow \{a_1, a_2\}$ is a Markov matching mechanism.

Next we define a class of Markov matching mechanisms. A Markov matching mechanism $\phi^{\bar{s},\underline{s}} : S \rightarrow \{a_1, a_2\}$ is a *threshold matching mechanism*, **with thresholds** $\bar{s} \geq 0$ and $\underline{s} \leq 0$, if

$$\phi^{\bar{s},\underline{s}}(s) = \begin{cases} a_1 & \text{if } \underline{s} \leq s \leq \bar{s} \\ a_2 & \text{if } s < \underline{s} \text{ or } s > \bar{s} \end{cases}.$$

A threshold matching mechanism conducts the largest exchanges that do not use existing A–B or B–A type pairs ("the smaller exchanges") as long as the numbers of A–B or B–A type pairs are not greater than the threshold numbers, $\bar{s}$ and $|\underline{s}|$ respectively; otherwise, it conducts the largest possible exchanges including the existing A–B or B–A type pairs ("the larger exchanges").

Our next theorem is as follows:

Theorem 8 (Ünver, 2010). *Suppose assumptions 1, 2, 5, and 6 hold. There exist $\bar{s}^* = 0$ and $\underline{s}^* \leq 0$, or $\bar{s}^* \geq 0$ and $\underline{s}^* = 0$ such that $\phi^{\bar{s}^*,\underline{s}^*}$ is a dynamically efficient multi-way matching mechanism.*

The dynamically optimal matching mechanism uses a threshold mechanism. It stocks A–B or B–A type pairs, and does not use them in larger exchanges as long as the stock of the control group is less than or equal to $\bar{s}^*$ or $|\underline{s}^*|$ respectively. Under the optimal matching mechanism, either the number of A–B type pairs or B–A type pairs is the state variable, but not both. Under the first type of solution, the number of B–A type pairs is the state variable. As long as the number of B–A type pairs in the pool is zero, regardless of the number of A–B type pairs, when the next arrival occurs, the first type of optimal mechanism conducts the maximal size exchanges possible. If there are B–A type pairs and their number does not exceed the threshold number $|\underline{s}^*|$, then these pairs are exclusively used to match incoming A–B type pairs in two-way exchanges. On the other hand, if the number of B–A type pairs exceeds the threshold number $|\underline{s}^*|$, they should be used in maximal exchanges, which can be (1) a two-way exchange involving an A–B type pair if the incoming pair type is A–B, (2) a three-way exchange involving A–O and O–B type pairs or A–AB and AB–B type pairs if the incoming pair type is A–O or AB–B, respectively, and (3) a four-way exchange involving A–AB, AB–O, and O–B type pairs if the incoming pair type is AB–O. The other types of maximal exchanges are conducted by the optimal mechanism as soon as they become feasible. The second possible solution is the symmetric version of the above mechanism taking the number of A–B type pairs as a state variable.

Next, we specify the optimal mechanism more precisely.

Theorem 9 (Ünver, 2010). *Suppose assumptions 1, 2, 5, and 6 hold. Then,*

- *If $q_{A-B} \geq q_{B-A}$, that is, A–B type arrives at least as frequently as B–A type, and $q_{B-O} + q_{AB-A} < q_{A-O} + q_{AB-B}$, that is, the types that can match A–B type pairs in larger exchanges arrive less frequently than those for the B–A type, then $\phi^{0,\underline{s}^*}$ is the dynamically efficient multi-way matching mechanism for some $\underline{s}^* \leq 0$.*
- *If $q_{A-B} = q_{B-A}$ and $q_{B-O} + q_{AB-A} = q_{A-O} + q_{AB-B}$, then $\phi^{0,0}$ is the dynamically efficient multi-way matching mechanism. That is, maximal size exchanges are conducted whenever they become feasible.*
- *If $q_{A-B} \leq q_{B-A}$ and $q_{B-O} + q_{AB-A} > q_{A-O} + q_{AB-B}$, then $\phi^{\bar{s}^*,0}$ is the dynamically efficient multi-way matching mechanism for some $\bar{s}^* \geq 0$.*

According to the arrival frequencies reported in Table 4.1, for pairs forming between random donors and recipients, we expect the mechanism reported in the first bullet point to be the efficient mechanism.

Concluding remarks

We conclude our survey by surveying other topics that have attracted the attention of researchers and practitioners alike.

Computational issues

Following Roth et al. (2007), one can write an integer program to solve the maximal kidney exchange problem.

We give the explicit formulation of finding the maximal number of patients who can benefit from two-way and up to k-way exchanges for any number k such that $|N| \geq k \geq 2$.

Suppose $E = \left(R_{i_1} - D_{i_1}, \ldots, R_{i_k} - D_{i_k} \right)$ denotes a k-way exchange in which pairs $i_1, \ldots, i_k$ participate. Let $|E|$ be the number of transplants possible under E; hence we have $|E| = k$.

Let $\mathcal{E}^k$ be the set of feasible two-way through k-way exchanges possible among the pairs in N. For any pair i, let $\mathcal{E}^k(i)$ denote the set of exchanges in $\mathcal{E}^k$ such that pair i can participate. Let $x = (x_E)_{E \in \mathcal{E}^k}$ be a vector of 0s and 1s such that $x_E = 1$ denotes that exchange E is going to be conducted, and $x_E = 0$ denotes that exchange E is not going to be conducted. Our problem of finding a maximal set of patients who will benefit from two-way, ..., and k-way exchanges is given by the following integer program:

$$\max_{x} \sum_{E \in \mathcal{E}^k} |E| \, x_E$$

subject to

$$x_E \in \{0, 1\} \quad \forall E \in \mathcal{E}^k,$$

$$\sum_{E \in \mathcal{E}^k(i)} x_E \leq 1 \quad \forall i \in N.$$

This problem is solved using Edmonds' (1965) algorithm for $k = 2$ (i.e. only for two-way exchanges) in polynomial time. However, for $k \geq 3$ this problem is NP complete [15] (see also Abraham et al., 2007.)

We also formulate the following version of the integer programming problem, which does not require ex ante construction of the sets $\mathcal{E}^k$:

Let $C^* = \left[c^*_{i,j} \right]_{i \in N, j \in N}$ be a matrix of 0s and 1s such that if recipient R_i is compatible with donor D_j we have $c^*_{i,j} = 1$ and if R_i is not compatible with donor D_j we have $c^*_{i,j} = 0$. Let $X = \left[x_{i,j} \right]_{i \in N, j \in N}$ be the *assignment matrix* of 0s and 1s such that $x_{i,j} = 1$ denotes that recipient R_i receives a kidney from donor D_j, and $x_{i,j} = 0$ denotes that recipient R_i does not receive a kidney from donor D_j under the proposed assignment X. We solve the following integer program to find a maximal set of two-way, ..., and k-way exchanges:

$$\max_{X} \sum_{i \in N, j \in N} x_{i,j}$$

subject to

$$x_{i,j} \in \{0, 1\} \quad \forall i, j \in N, \tag{3}$$

$$x_{i,j} \leq c^*_{i,j} \quad \forall i, j \in N, \tag{4}$$

$$\sum_{j \in N} x_{i,j} \leq 1 \quad \forall i \in N, \tag{5}$$

$$\sum_{j \in N} x_{i,j} = \sum_{j \in N} x_{j,i} \quad \forall i \in N, \tag{6}$$

$$x_{i_1, i_2} + x_{i_2, i_3} + \ldots + x_{i_k, i_{k+1}} \leq k - 1 \quad \forall \{i_1, i_2, \ldots, i_k, i_{k+1}\} \subseteq N. \tag{7}$$

A solution of this problem determines a maximal set of patients who can benefit from two-way, ..., and k-way exchanges for any $k < |N|$. A maximal set of patients who can benefit from unrestricted exchanges is found by setting $k = |N|$. In this case constraint

---

[15] The observation that the mixed two- and three-way problem is NP complete was made by Kevin Cheung and Michel Goemans (personal communication).

7 becomes redundant. This formulation is used to find the maximal set of unrestricted multi-way exchanges.

Since the problems are NP complete for $k > 2$, there is no known algorithm that runs in worst-case time that is polynomial in the size of the input. Simulations have shown that for more than a certain number of pairs in the exchange pool, commercial integer programming software programs have difficulty solving these optimization problems. Abraham et al. (2007) proposed a tailored integer programming algorithm designed specifically to solve kidney large exchange problems.[16] This algorithm increases the scalability of a computable problem size considerably more than commercial integer programming software capabilities, and can solve the problem optimally in less than two hours at the full projected scale of the nationwide kidney exchange (10,000 pairs). The US national kidney exchange program, whose pilot runs started to be conducted in late 2010, uses this tailored algorithm, while some regional programs continue to use commercial integer programming software versions of the computational implementation.

List exchange chains

Another concept that is being implemented in NEPKE is that of *list exchange chains* (Roth et al., 2004; see also Roth et al., 2007). A k-way list exchange chain is similar to a k-way paired kidney exchange, with the exception that one of the pairs in the exchange is a *virtual pair* with the property that

- the donor of this pair is a priority on the deceased-donor waiting list; that is, whomever is assigned this donor gets priority to receive the next incoming deceased-donor kidney; and
- the recipient of this pair is the highest-priority recipient who is waiting for a kidney on the deceased-donor waiting list.

Thus, in a list exchange chain, one recipient of a pair receives a priority to receive the next incoming compatible deceased-donor kidney (by trading her own paired live-donor's kidney); and one donor of a pair in the exchange does not donate to anybody in the exchange pool but donates to the top-priority recipient waiting for a deceased-donor kidney (Figure 4.6).

There are two ethical concerns regarding list exchanges in the medical community; therefore, not all regions implement it (Ross et al., 1997; Ross and Woodle, 2000).

The first concern regards the imbalance between the blood type of the recipient at the top of the waiting list who receives a kidney and the recipient in the exchange pool who receives top priority on the waiting list. Because of blood-type compatibility

---

[16] There is also a recent strand of literature that deals with different computability issues under various solution concepts for the kidney exchange problem. See e.g. Cechlárová et al. (2005), Biró and Cechlárová (2007), Irving (2007), Biró and McDermid (2008).

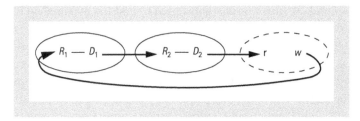

FIGURE 4.6. A three-way list exchange chain. Here *r* refers to the recipient on the deceased-donor waiting list and *w* refers to priority on the deceased-donor waiting list.

requirements, most of the time the recipient who gets a live-donor kidney will be of an inferior type, such as AB, A, or B, while the recipient who is sent to the top of the waiting list will be of O blood type. Thus, this will increase the waiting time for O blood-type patients on the waiting list. The second concern regards the inferior quality of deceased-donor kidneys compared with live-donor kidneys. Many medical doctors are not willing to leave such a decision to patients, i.e., whether to exchange a live-donor kidney for a deceased-donor kidney.

Altruistic donor chains

A new form of exchange is finding many applications in the field. In a year, there are about 100 *altruistic donors*, live donors who are willing to donate one of their kidneys to a stranger, in the US. Such donations are not regulated and traditionally have been treated like deceased-donor donations. However, a recent paradigm suggests that an altruistic donor can donate to a pair in the exchange pool, and in return this pair can donate to another pair, ..., and finally the last pair donates to the top-priority recipient on the waiting list. This is referred to as a *simultaneous altruistic donor chain* (Montgomery et al., 2006; see also Roth et al., 2007). Thus, instead of an altruistic donor helping a single recipient on the waiting list, he helps k recipients in a k-way closed altruistic donor chain. Figure 4.7 shows the example of a three-way chain.

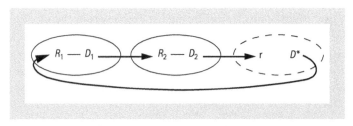

FIGURE 4.7. A simultaneous three-way altruistic donor chain. Here D^* refers to the altruistic donor and *r* refers to a recipient on the top of the deceased-donor waiting list.

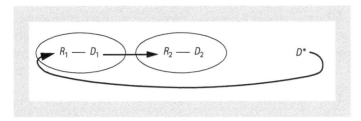

FIGURE 4.8. A non-simultaneous two-way altruistic donor chain. Here, D^* refers to the altruistic donor, and D_2 is the bridge donor who will act as an altruistic donor in a future altruistic donor chain.

A newer paradigm takes this idea one step forward. Instead of the last donor immediately donating a kidney to a recipient on the waiting list, he becomes a *bridge donor*, that is, he acts as an altruistic donor and may help a future incoming pair to the exchange. The problem with this approach is that the bridge donor can opt out from future donation after his paired recipient receives a kidney. However, field experimentation suggests that in APD no bridge donor has backed out yet in any of the six operational chains. Such an exchange is referred to as a *non-simultaneous altruistic donor chain* (Roth et al., 2007; Rees et al., 2009). Figure 4.8 shows the example of a two-way chain.

The potential impact of altruistic donor chains is quite large. For example, in APD, twenty-two transplantations were conducted through six non-simultaneous altruistic donor chains in ten states, all with active bridge donors (at the time this chapter was drafted).

Exchange with compatible pairs

Currently, compatible pairs are not part of the kidney exchange paradigm, since the recipient of the pair receives directly a kidney from her paired donor. Woodle and Ross (1998) proposed compatible pairs to be included in kidney exchanges, since they will contribute to a substantial increase in the number of transplants from exchanges. Indeed, the simulations by Roth et al. (2005b) show that when compatible pairs are used in exchanges, since the pairs will likely be of overdemanded types, they will increase the gains from exchange tremendously (also see Roth et al., 2004). Table 4.3 shows the results of this simulation for efficient two-way exchange mechanisms. This table shows the dramatic potential impact of including compatible pairs in exchange. When list exchange is not possible for $n = 100$, about 70% of the pairs receive a kidney when only incompatible pairs participate in exchange. This number increases to 91% when compatible pairs also participate in exchange.

Sönmez and Ünver (2010b), the authors of this survey, model the two-way kidney exchange problem with compatible pairs. We obtain favorable graph-theoretical results analogous to the problem without compatible pairs (see Roth et al., 2005a).

Table 4.3. A Pareto–efficient two-way exchange mechanism outcome for *n* pairs randomly generated using national population characteristics (including compatible and incompatible pairs) when compatible pairs are in/out of exchange, when *n=25/100*, when list exchanges are impossible/possible and *40%* of the pairs are willing to use this option.

Compatible pairs	Population size	% wait-list option	Total no. of transplants[a]		
			Own	Exchange	w-List
Out of the exchange	*n* = 25	0		15.52	
		%	11.56	3.96	0
		40		21.03	
		%	11.56	5.76	3.71
	n = 100	0		70.53	
		%	47.49	23.04	0
		40		87.76	
		%	47.49	28.79	11.48
In the exchange	*n* = 25	0		20.33	
		%	1.33	19.00	0
		40		23.08	
		%	1.33	19.63	2.12
	n = 100	0		91.15	
		%	1.01	90.14	0
		40		97.06	
		%	1.01	91.35	4.70

[a] *Own* refers to the patients receiving their own-donor kidneys (i.e., when compatible pairs are out, this is the number of compatible pairs generated in the population). *Exchange* refers to the number of patients who receive a kidney through exchange. *w-List* refers to the number of patients who get priority on the waiting list when list exchange is possible.

We show that the latter is a special case of the former general model and extend the Gallai–Edmonds decomposition to this domain. We introduce an algorithm that finds a Pareto-efficient matching with polynomial time and space requirements. We generalize the most economically relevant results and the *priority* mechanisms to this domain. Moreover, our results generalize to a domain that includes altruistic donors that are incorporated through simultaneous two-way chains.

False-negative crossmatches

Detection of tissue-type incompatibility without a *crossmatch test* is not a perfect science. Since this test, which involves mixing blood samples from the donor and the recipient, is expensive to conduct between all donors and recipients, exchange programs usually rely on a different method to determine whether a donor is tissue-type compatible with a recipient. Using a simple *antibody test*, doctors determine the HLA proteins that trigger antibodies in a recipient. Also taking into account the previous rejection and sensitivity history of the recipient, they determine the HLA proteins that are compatible (or incompatible) with her. Hence, the donors who have the compatible (or incompatible) HLAs are deemed tissue-type compatible (or incompatible) with the recipient. However, this test has a flaw: the false-negative crossmatch (false tissue-type compatibility) rate is sometimes high. As a result, some exchanges found by the matching mechanism do not go through. Such cases affect the whole match, since different outcomes could have been found if these incompatibilities had been taken into account. Kidney large exchange programs with an extended history can partially avoid this problem, since many actual crossmatch tests have already been conducted between many donors and recipients over the years. They can simply use the data in matching instead of the simple test results. Morrill (2008) introduces a mechanism for the two-way matching problem (the *roommates* problem) to find a Pareto-efficient matching starting from a Pareto-inefficient matching. His model's preference domain is strict preferences. An application of this mechanism is as follows: after a set of kidney exchanges are fixed, if some of these fail to go through for some reason, we can use Morrill's mechanism to find a matching that Pareto dominates the initial one. This mechanism has a novel polynomial time algorithm that synthesizes the intuition from Gale's top trading cycles algorithm (used to find the core for strict preferences with unrestricted multi-way exchanges) with Edmonds' algorithm (used to find a Pareto-efficient matching for 0–1 preferences with two-way exchanges).

Transplant center incentives

Transplant centers decide voluntarily whether to participate in a larger exchange program, such as the APD or the national program. Moreover, if they do, they are free to determine which recipients of their center will be matched through the larger program. Thus, centers can strategically decide which of their patients will be matched through the larger program. If centers care about maximizing the number of recipients to be matched through exchanges, the following result shows that no efficient mechanism is immune to manipulation:

Theorem 10 (*Roth et al., 2005c*). *Even if there is no tissue-type incompatibility between recipients and donors of different pairs, there exists no Pareto-efficient mechanism where full participation is always a dominant strategy for each transplant center.*

The proof is through an example: There are two transplant centers, A, B, three pairs, $a_1, a_2, a_3 \in I_A$, in center A, and four pairs, $b_1, b_2, b_3, b_4 \in I_B$, in center B. Suppose that the list of feasible exchanges are as follows: (a_1, a_2), (a_1, b_1), (a_2, b_2), (a_3, b_4), (b_2, b_3), (b_3, b_4). Figure 4.9 shows all feasible exchanges among the pairs.

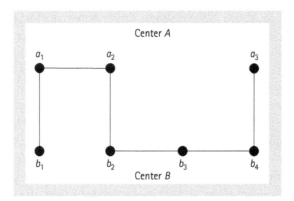

FIGURE 4.9. All feasible exchanges between three pairs at two centers.

In all Pareto efficient matchings, six pairs receive transplants (an example is $\{(a_1, b_1), (a_2, b_2), (b_3, b_4)\}$). Since there are seven pairs, one of the pairs remains unmatched under any Pareto-efficient matching. Let ϕ be a Pareto-efficient mechanism. Since ϕ chooses a Pareto-efficient matching, there is a single pair that does not receive a transplant. This pair is either in center A or in center B.

- The pair that does not receive a transplant is in center A. In this case, if center A does not submit pairs a_1 and a_2 to the centralized match, and instead matches them internally to each other, then there is a single multi-center Pareto-efficient matching $\{(a_3, b_4), (b_2, b_4)\}$, and ϕ chooses this matching. As a result, center A succeeds in matching all three of its pairs.
- The pair that does not receive a transplant is in center B. In this case, if center B does not submit pairs b_3 and b_4 to the centralized match, and instead matches them internally to each other, then there is a single multi-center Pareto-efficient matching $\{(a_1, b_1), (a_2, b_2)\}$, and ϕ chooses this matching. As a result, center B succeeds in matching all four of its pairs.

In either case, we showed that there is a center that can successfully manipulate the Pareto-efficient multi-center matching mechanism ϕ.

Future research in this area involves finding mechanisms that have good incentive and efficiency properties for centers, using different solution and modeling concepts. A recent example of this line of research is by Ashlagi and Roth (2011), who investigate the participation problem using computer science techniques for large populations.

REFERENCES

Abdulkadiroğlu, A. and Sönmez, T. (1999) "House allocation with existing tenants," *Journal of Economic Theory*, 88: 233–60.

Abecassis, M., Adams, M., Adams, P., Arnold, R. M., Atkins, C. R., Barr, M. L., Bennett, W. M., Bia, M., Briscoe, D. M., Burdick, J., Corry, R. J., Davis, J., Delmonico, F. L., Gaston, R. S., Harmon, W., Jacobs, C. L., Kahn, J., Leichtman, A., Miller, C., Moss, D., Newmann, J. M., Rosen, L. S., Siminoff, L., Spital, A, Starnes, V. A., Thomas, C., Tyler, L. S., Williams, L., Wright, F. H., and Youngner, S. (2000) "Consensus statement on the live organ donor," *Journal of the American Medical Association*, 284: 2919–926.

Abraham, D. J., Blum, A., and Sandholm, T. (2007) "Clearing algorithms for barter exchange markets: enabling nationwide kidney exchanges," in *Proceedings of ACM–EC 2007: the Eighth ACM Conference on Electronic Commerce*.

Ashlagi, I. and Roth, A. E. (2011) "Individual rationality and participation in large scale, multi-hospital kidney exchange," Working Paper.

Awasthi, P. and Sandholm, T. (2009) "Online stochastic optimization in the large: application to kidney exchange," in *Proceedings of the International Joint Conference on Artificial Intelligence (IJCAI)*.

Biró, P., and Cechlárová, K. (2007) "Inapproximability of the kidney exchange problem," *Information Processing Letters*, 101: 199–202.

_____ and McDermid, E. (2008) "Three-sided stable matchings with cyclic preferences and the kidney exchange," U. Endriss and P. W. Goldberg (eds), *COMSOC-2008: Proceedings of the 2nd International Workshop on Computational Social Choice*, pp. 97–108.

Bogomolnaia, A. and Moulin, H. (2001) "A new solution to the random assignment problem." *Journal of Economic Theory*, 100: 295–328.

_____ _____ (2004) "Random matching under dichotomous preferences." *Econometrica*, 72: 257–79.

Cechlárová, K., Fleiner, T. and Manlove, D. F. (2005) "The kidney exchange game," in *Proceedings of SOR'05: the 8th International Symposium on Operations Research in Slovenia*, pp. 77–83.

Delmonico, F. L. (2004) "Exchanging kidneys—advances in living-donor transplantation," *New England Journal of Medicine*, 350: 1812–14.

Dutta, B. and Ray, D. (1989) "A concept of egalitarianism under participation constraints." *Econometrica*, 57: 615–35.

Edmonds, J. (1965) "Paths, trees, and flowers," *Canadian Journal of Mathematics*, 17: 449–67.

Gallai, T. (1963) "Kritische Graphen II," *Magyar Tudumdnyos Akademia—Matematikai Kutató Intezenek Közlemengei*, 8: 373–95.

Gallai, Tibor (1964) "Maximale Systeme unabhängiger kanten," *Magyar Tudumdnyos Akademia—Matematikai Kutató Intezenek Közlemengei*, 9: 401–13.

Gjertson, D. W. and Cecka, J. M. (2000) "Living unrelated donor kidney transplantation," *Kidney International*, 58: 491–9.

Hatfield, J. W. (2005) "Pairwise kidney exchange: comment," *Journal of Economic Theory*, 125: 189–93.

Irving, R. W. (2007) "The cycle roommates problem: a hard case of kidney exchange," *Information Processing Letters*, 103: 1–4.

Klerk, M. de, Keizer, K. M., Claas, F. H. J., Witvliet, M., Haase-Kromwijk, B. J. J. M., and Weimar, W. (2005) "The Dutch national living donor kidney exchange program," *American Journal of Transplantation* 5: 2302–5.

Lovász, L. and Plummer, M. D. (1986) *Matching Theory*, North-Holland.

Ma, J. (1994) "Strategy-proofness and the strict core in a market with indivisibilities," *International Journal of Game Theory*, 23: 75–83.

Mandal, A. K., Snyder, J. J., Gilbertson, D. T., Collins, A. J. and Silkensen, J. R. (2003) "Does cadavaric donor renal transplantation ever provide better outcomes than live-donor renal transplantation?" *Transplantation*, 75: 494–500.

Montgomery, R. A., Gentry, S. E., Marks, W. H., Warren, D. S., Hiller, J., Houp, J., Zachary, A. A., Melancon, J. K., Maley, W. R., Rabb, H., Simpkins, C. E., and Segev, D. L. (2006) "Domino paired kidney donation: a strategy to make best use of live non-directed donation," *Lancet*, 368: 419–21.

Morrill, T. (2008) "The roommates problem revisited," *Working Paper*.

Opelz, G. (1997) "Impact of HLA compatibility on survival of kidney transplants from unrelated live donors," *Transplantation*, 64: 1473–5.

Overbeck, I., Bartels, M., Decker, O., Harms, J., Hauss, J. and Fangmann, J. (2005) "Changes in quality of life after renal transplantation," *Transplantation Proceedings* 37: 1618–21.

Pápai, S. (2000) "Strategyproof assignment by hierarchical exchange," *Econometrica*, 68: 1403–33.

Park, K., Lee, J. H., Huh, K. H., Kim, S. I. and Kim, Y. S. (2004) "Exchange living donor kidney transplantation: diminution of donor organ shortage," *Transplantation Proceedings*, 36: 2949—51.

Pycia, M., and Ünver, M. U. (2009) "A theory of house allocation and exchange mechanisms," *Working Paper*.

Rapaport, F. T. (1986) "The case for a living emotionally related international kidney donor exchange registry," *Transplantation Proceedings*, 18: 5–9.

Rawls, J. (1971) *A Theory of Justice*, Harvard University Press.

Rees, M. A., Kopke, J. E., Pelletier, R. P., Segev, D. L., Rutter, M. E., Fabrega, A. J., Rogers, J., Pankewycz, O. G., Hiller, J., Roth, A. E., Sandholm, T., Ünver, M. U. and Montgomery, R. A. (2009) "A non-simultaneous extended altruistic-donor chain." *New England Journal of Medicine*, 360: 1096–101.

Ross, L. F., Rubin, D. T., Siegler, M., Josephson, M. A., Thistlethwaite, J. R., Jr, and Woodle, E. S. (1997) "Ethics of a paired-kidney-exchange program," *New England Journal of Medicine*, 336: 1752–5.

—— and Woodle, E. S. (2000) "Ethical issues in increasing living kidney donations by expanding kidney paired exchange programs," *Transplantation*, 69: 1539–43.

Roth, A. E. (1982) "Incentive compatibility in a market with indivisibilities," *Economics Letters*, 9: 127–32.

—— and Postlewaite, A. (1977) "Weak versus strong domination in a market with indivisible goods," *Journal of Mathematical Economics*, 4: 131–7.

—— Sönmez, T. and Ünver, M. U. (2004) "Kidney exchange," *Quarterly Journal of Economics*, 119: 457–88.

—— —— —— (2005a) "Pairwise kidney exchange," *Journal of Economic Theory*, 125: 151–88.

—— —— —— (2005b) "A kidney exchange clearinghouse in New England," *American Economic Review Papers and Proceedings*, 95(2): 376–80.

—— —— —— (2005c) "Transplant center incentives in kidney exchange," *Unpublished*.

—— —— —— (2007) "Efficient kidney exchange: coincidence of wants in markets with compatibility-based preferences," *American Economic Review*, 97(3): 828–51.

—— —— —— Delmonico, F. L., and Saidman, S. L. (2006) "Utilizing list exchange and nondirected donation through 'chain' paired kidney donations," *American Journal of Transportation*, 6: 2694–705.

Saidman, S. L., Roth, A. E. Sönmez, T., Ünver, M. U. and Delmonico, F. L. (2006) "Increasing the opportunity of live kidney donation by matching for two and three way exchanges," *Transplantation*, 81: 773—82.

Segev, D., Gentry, S., Warren, D. S., Reeb, B. and Montgomery, R. A. (2005) "Kidney paired donation: Optimizing the use of live donor organs," *Journal of the American Medical Association*, 293: 1883–90.

Shapley, L. and Scarf, H. (1974) "On cores and indivisibility," *Journal of Mathematical Economics*, 1: 23–8.

Sönmez, T. and Ünver, M. U. (2005) "House allocation with existing tenants: an equivalence," *Games and Economic Behavior*, 52: 153–85.

—— —— (2010a) "House allocation with existing tenants: a Characterization," *Games and Economic Behavior*, 69(2): 425–45.

—— —— (2010b) "Altruistic kidney exchange," *Working Paper.*

—— —— (2011) "Matching, allocation, and exchange of discrete resources," in *J. Benhabib, A. Bisin, and M. Jackson (eds), Handbook of Social Economics,* North-Holland, Vol. 1A, pp. 781–52.

Su, X. and Zenios, S. A. (2006) "Recipient choice can address the efficiency-equity trade-off in kidney transplantation: a mechanism design model," *Management Science*, 52: 1647–60.

Schweitzer, E. J., Wiland, A., Evans, D., Novak, M., Connerny, I., Norris, L., Colonna, J. O., Philosophe, B., Farney, A. C., Jarrell, B. E., and Bartlett, S. T. (1998) "The shrinking renal replacement therapy break-even point." *Transplantation*, 107: 1702—8.

Ünver, M. U. (2010) "Dynamic kidney exchange," *Review of Economic Studies*, 77 (1): 372–414.

Votruba, M. (2002) "Efficiency–equity tradeoffs in the allocation of cadaveric kidneys," Working Paper.

Woodle, E. S. and Ross, L. F. (1998) "Paired exchanges should be part of the Solution to ABO incompatibility in living donor kidney transplantation." *Transplantation*, 66(3): 406–7.

Yilmaz, Ö (2011) "Kidney Exchange: an egalitarian mechanism," *Journal of Economic Theory*, 146(2): 592–618.

Zenios, S. (1996) "Health care applications of optimal control theory," PhD Thesis, Massachusetts Institute of Technology.

Zenios, S. A. (2002) "Optimal control of a paired-kidney exchange program." *Management Science*, 48: 328–42.

—— Chertow, G. M. and Wein, L. M. (2000) "Dynamic Allocation of Kidneys to Candidates on the Transplant Waiting List." *Operations Research*, 48, 549–569.

—— Woodle, E. S. and Ross, L. F. (2001) "Primum non nocere: avoiding increased waiting times for individual racial and blood-type subsets of kidney wait list candidates in a living donor/cadaveric donor exchange program," *Transplantation*, 72: 648–54.

CHAPTER 5

···

SCHOOL CHOICE

···

ATİLA ABDULKADİROĞLU

INTRODUCTION

GOOD public schools are scarce, and admissions to those will always matter. Public schools are free of charge and admissions in many districts have been defined by location of schools and the home addresses of pupils. As traditional neighborhood-based assignment has led to the segregation of neighborhoods along socioeconomic lines, recent decades have witnessed a surge in programs that offer parental choice over public schools, expanding families' access to schools beyond their residential area. In fact the origins of school choice in the United States can be traced back to *Brown* v. *Board of Education*, 1954. Boston's renowned controlled choice program evolved out of a 1974 ruling that enforced desegregation of Boston public schools. Today, there are other reasons for public-school choice; school districts have been increasingly leaving the one-size-fits-all model of schooling and developing alternative curricula to better meet educational needs of a highly heterogenous student population. As districts offer more options for parents and students, choice and therefore *student assignment* become an integral part of enrollment planning.

Since the introduction of this problem by Abdulkadiroğlu and Sönmez (2003), economists have found great opportunity to study and design student assignment systems around the US. Most notable of these are the redesign of the student assignment systems in Boston and New York City. The former was initiated by a *Boston Globe* article on Abdulkadiroğlu and Sönmez (2003),[1] which described flaws with the student assignment in Boston at the time. The latter was initiated independently when, being aware of his pioneering work on market design in the entry-level labor markets (Roth, 1984; Roth and Peranson, 1999), the New York City Department of Education (NYC DOE) contacted Alvin E. Roth to inquire about the possibility of adopting a system like

---

[1] See "School assignment flaws detailed" by Cook (2003).

the National Residency Matching Program (NRMP)[2] for their high-school admissions. The school choice problem and its market design applications have fostered a new line of research in mechanism design theory. The goal of this chapter is to summarize recent developments in the field and in mechanism design theory.

We divide the chapter into two parts. The next section discusses the school choice problem, and the issues in the canonical model of Abdulkadiroğlu and Sönmez (2003). It also gives a brief discussion to various student assignment mechanisms. The section is intended for the general audience and practitioners in the field. The remainder of the chapter expands on the developments.[3]

THE SCHOOL CHOICE PROBLEM

A *school choice problem* (Abdulkadiroğlu and Sönmez, 2003) consists of a finite set of students and a finite set of schools with finitely many seats available for enrollment. In this section, we will refer to students by a, b, c, d, and schools by s_1, s_2, s_3 and s. Students have preferences over schools. We represent a student's preferences as a linear order of schools to which she prefers to be assigned rather than accept her outside option. For example, $a : s_1 - s_2 - s_3$ means that student a prefers school s_1 to school s_2 and school s_2 to school s_3; she prefers her outside option to being assigned any other school. Her outside option is not specified in the model; it may be a private school or home schooling, or some other option. We assume that students form their preferences based on exogenous school characteristics, such as curricula, extra-curricular activities, distance to home, average test scores, and graduation rates in the past years. This rules out, for example, conditioning one's preferences on the composition of the incoming class.

Admissions to schools are usually regulated via assignment priorities. For instance, for most schools in Boston, for half of the seats at the school, the students are priority ordered as follows:

1. students who are guaranteed a space at the school by virtue of already attending that school or a feeder school (guaranteed priority);
2. students who have a sibling at the school and live in the walk zone of the school (sibling–walk priority);
3. students who have a sibling at the school (but who do not live in the walk zone of the school) (sibling priority);

---

[2] The National Resident Matching Program is a United States-based non-profit non-governmental organization created in 1952 to help match medical school students with residency programs in the US.

[3] The focus of this chapter is limited to school choice. Therefore it may miss many important references in matching theory. Naturally, it may also be biased toward my own work on the topic and my experience in the field. For another recent survey on school choice, see Pathak (2011).

4. students who live in the walk zone of the school (but who do not have a sibling at the school) (walk zone priority); and
5. other students in the zone.

A random lottery number for each student breaks ties in each category (random tie-breaker). For the other half of the seats, walk zone priorities do not apply, and students are priority ordered based on guaranteed and sibling priority, and the random tie-breaker (Abdulkadiroğlu and Sönmez, 2003; Abdulkadiroğlu et al., 2006).

Such priority structure may reflect a district's policy choice. Neighborhood priority may be granted to promote involvement of neighborhood parents in school activities; sibling priority may be adopted to reduce transportation and organizational costs for parents, and to promote spillover benefits of siblings attending the same school. Priorities may be determined differently at different schools. In fact, the priority list of a school may even reflect preferences of the school staff over students. For instance, some high schools in New York City can access students' academic records and rank students in a preference order. When priorities are determined by some exogenous rules, such as in Boston, we say that the market is *one-sided*. When priorities at some schools reflect preferences of the school staff, as in New York City, we say that the market is *two-sided*.

Regardless of its resource, we represent the priority list at a school as a linear order of all the students that are eligible for enrollment at that school. For example, $s_1 : b - a - c$ means that student b has the highest priority at school s_1, a has the next highest priority and c has the lowest priority; student d is not eligible for enrollment at that school. The number of available seats at schools completes the model.

A matching of students and schools determines the assignment of each student. Hereafter, we use *matching*, *assignment*, and *enrollment* interchangeably. Each student is matched with at most one school or remains unmatched. A school can be matched with students up to its capacity. We will utilize examples of the following type in our discussions:

Example 1. *There are three students $\{a, b, c\}$, and three schools $\{s_1, s_2, s_3\}$, each with one seat. Student preferences and school priorities are given as follows:*

$$
\begin{array}{ll}
a : s_2 - s_1 - s_3 & s_1 : a - c - b \\
b : s_1 - s_2 - s_3 \quad and \quad & s_2 : b - a - c \\
c : s_1 - s_2 - s_3 & s_3 : b - a - c
\end{array}
$$

We will denote a matching that assigns a to s_1, b to s_2 and leaves c unmatched as

$$
m_1 = \begin{pmatrix} a & b & c \\ s_1 & s_2 & - \end{pmatrix}
$$

Issues and policy goals

What are the goals of a successful choice plan? Are these policy goals compatible with each other? What are the trade-offs and how should one compromise? These questions are closely related to the design of student assignment mechanisms. The education literature provides guidance for the design of assignment mechanisms but does not offer a specific one. Also, flaws in the existing school choice plans result in difficult and stressful decision making for parents, gaming and behind-closed-doors strategies by savvy parents, as well as appeals in the US courts by unsatisfied parents (Abdulkadiroğlu and Sönmez, 2003).

Economists' approach to such allocation problems is to translate the relevant policy goals into normative theoretical criteria, and look for solutions that meet these criteria, and if no such solution exists, then find one with optimal compromise. The following notions emerge naturally in the context of school choice.

Feasibility

Overcrowding at schools is controlled by school capacities. A matching is deemed feasible in our model if enrollment at each school does not exceed the school capacity and only eligible students are enrolled at every school.

Individual rationality

If a student is assigned a school that is not in her choice list, one may expect her family to opt out for its outside option, which may be a private school, home schooling, or some other option. A matching is individually rational if it matches every student only with schools in her choice list, and leaves her unassigned otherwise. Hereafter we consider only feasible and individually rational matchings.

Efficiency

Perhaps the most obvious desideratum that guides a design is that the match process should promote student welfare to the greatest extent possible; that is, it should be efficient for students. We say that a matching *wastes* a seat at school s if there remains an empty seat at s and an eligible student prefers s to her match. In example 1, m_1 wastes a seat at s_3 because student c is unassigned, a seat at s_3 remains available and c prefers s_3 to being unassigned. The matching

$$m_2 = \begin{pmatrix} a & b & c \\ s_1 & s_2 & s_3 \end{pmatrix}$$

improves c's welfare without harming other students. Identifying and remedying such wastefulness is relatively easy. A more subtle wastefulness occurs in the assignment of a and b. Notice that both a and b are assigned their second choices. They become better off if they swap their assignments. In other words, in comparison to m_2, the matching

$$m_3 = \begin{pmatrix} a & b & c \\ s_2 & s_1 & s_3 \end{pmatrix}$$

improves a and b's welfare without harming c. We say that a matching *Pareto dominates* another matching if the former improves some student's welfare without harming others in comparison to the latter. In our example, m_3 Pareto dominates m_2, which Pareto dominates m_1. We say that a matching is *Pareto efficient* or simply *efficient* if it is not Pareto dominated by another matching. In particular, m_3 is efficient in our example; both a and b are assigned their first choices, and c cannot be assigned a better choice without harming a or b's assignments.

Note that the following matchings are also efficient:

$$m_4 = \begin{pmatrix} a & b & c \\ s_2 & s_3 & s_1 \end{pmatrix}, \; m_5 = \begin{pmatrix} a & b & c \\ s_3 & s_2 & s_1 \end{pmatrix}$$

Respecting or violating priorities in assignment

An integral input to our model is school priorities. Districts utilize priorities to ration seats when schools are oversubscribed. How priorities restrict assignment is a matter of policy choice.

In their weakest form, priorities simply determine eligibility. If a student is eligible for an empty seat at a school and she prefers it to her match, one might expect her parents to file an appeal to the district. Therefore, a wasteful matching is not desirable from a policy standpoint. However, if determining eligibility were the only role priorities are supposed to play, an unordered list of eligible students would be sufficient. To give priorities a broader role in rationing seats in assignment, we say that a matching *violates a student's priority* at school s if the student ranks s higher than her assigned school and has higher priority at s than some other student who is assigned s. We say that a matching is *stable* if it does not violate priorities and does not waste any seat.

In the elaboration of example 1, m_3 violates c's priority at s_1, because c prefers s_1 to her assigned school s_3 and she has higher priority at s_1 than b, who is assigned s_1. Therefore it is not stable. In fact m_2 is the only stable matching in this example. Note that students a and b get their second choices, at m_2, and would have been better off had they swapped their matchings. In that case c's priority at s_1 would have been violated. This is the first trade-off we encounter: stability comes at the cost of student welfare. A stable matching need not be efficient, and an efficient matching need not be stable.

As example 2 demonstrates, there may be multiple stable matchings:

Example 2. *There are three students $\{a, b, c\}$ and three schools $\{s_1, s_2, s_3\}$, each with one seat. Student preferences and school priorities are given as follows:*

$$
\begin{aligned}
a &: s_2 - s_1 - s_3 & s_1 &: a - b - c \\
b &: s_1 - s_2 - s_3 \text{ and } s_2 &: b - a - c \\
c &: s_1 - s_2 - s_3 & s_3 &: b - a - c
\end{aligned}
$$

We have only changed the priorities at s_1 from example 1. Now there are two stable matchings:

$$n_1 = \begin{pmatrix} a & b & c \\ s_1 & s_2 & s_3 \end{pmatrix}, \; n_2 = \begin{pmatrix} a & b & c \\ s_2 & s_1 & s_3 \end{pmatrix}$$

n_2 would not be stable for example 1, because c's priority at s_1 would be violated. In this example, c has the lowest priority at every school, so her priority is not violated by n_2.

We say that a stable matching is *student-optimal stable* if it is not Pareto dominated by any other stable matchings. In example 2, n_1 is Pareto dominated by n_2 since it assigns a and b to their higher choices without changing c's assignment. n_2 is not Pareto dominated by any stable matching, so it is student-optimal stable.

So far we have talked only about student welfare. The preferences of schools in a two-sided market may also matter. For example, if the priorities reflect school preferences in example 2, then n_2 no longer Pareto dominates n_1, because while n_2 assigns a and b better, it matches both s_1 and s_2 with their less preferred students. We cannot improve any student's assignment in n_1 without harming the assignment of another student or *school*. In other words, n_1 is efficient when priorities reflect school preferences. In general, stability implies efficiency in such two-sided markets.

School preferences may stem from different comparative advantages. For example, different EdOpt schools in New York City seem to have different preferences even for students with low reading scores, with some schools preferring higher scores, and others preferring students who had good attendance. Even when student welfare is the primary concern in such two-sided markets, allowing scope for school preferences via stability may be desirable to utilize such comparative advantages (Abdulkadiroğlu et al., 2009).

Whether or not it is acceptable for priorities to be violated is determined by the circumstances of the specific problem. For instance, during the redesign of student assignment in Boston, violating priorities was initially favored in order to promote student welfare. Boston public schools (BPS) decided to respect priorities in the final design. We will discuss these issues in more detail later.

Incentives to game the system

If student preferences were known a priori, it would be easy for a district to meet the goal of efficiency or student-optimal stability. However, preference data is unknown to the admissions office. Eliciting that information truthfully during application is not a trivial task. Indeed, student assignment systems in most school choice programs force parents to submit a choice list that is different than their true preference list. We will see a prevalent example later.

A student assignment system, or simply a mechanism, determines the matching of students with schools for every profile of preferences, priorities, and school capacities. Since an assignment mechanism responds to student preferences, a student can presumably affect her assignment by changing the list of schools she submits in her application form. We say that an assignment mechanism is *strategy-proof (for students)* if listing schools in true preference order in the application form is optimal for every student, regardless of the priority structure and other students' applications. In other words, a strategy-proof assignment system ensures that a student gets her best assignment—not necessarily her first choice—under every circumstance by filling in her true preference list. We can define strategy proofness for schools in a similar manner in two-sided markets, in which schools also rank students in preference order.

Strategy proofness has at least three policy advantages. First, it simplifies the decision-making process for parents by making truthful listing of preferences a best strategy. Under a strategy-proof mechanism, parents may focus solely on determining schools that would best fit their children's educational needs; they do not need to navigate the system via preference manipulation in their application form; indeed, doing so may even harm them. This also allows school districts to give straightforward advice on filling in application forms. Second, some parents may lack the information or the ability required to navigate a system that is prone to gaming. By removing the need for gaming, a strategy-proof mechanism levels the playing field among parents. Finally, a strategy-proof mechanism provides reliable demand data for districts, which can play a crucial role in enrollment planning.

When priorities reflect school preferences, if there is a student–school pair that prefer each other to their match, the school has an incentive to circumvent the match to enroll the students it prefers. Stability eliminates such circumstances. Therefore, stability also offers scope for eliminating gaming of the system by schools.

Armed with these notions, next we will discuss and compare three prominent student assignment mechanisms.

Three student assignment mechanisms

One way to think about these design concerns is that Pareto efficiency for the students is the primary welfare goal, and strategy proofness in the elicitation of student preferences is an incentive constraint that has to be met. Moreover, stability of the matching may enter as a policy choice when priorities reflect district policies, or as an incentive constraint in two-sided markets in which priorities reflect school preferences. Mechanisms can be evaluated and formulated from this "mechanism design" perspective.

The Boston mechanism

Probably the most prevalent student assignment mechanism is the so-called Boston mechanism, developed in Cambridge in the 1980s. The Boston mechanism tries to assign as many students as possible to their first choices, assigning higher-priority students to overdemanded schools; and only after first choice assignments are made, it considers unassigned students at their second choices in the same fashion, and so on. That is, given student preferences and school priorities, the matching is determined by the following algorithm:

- Step 1. For each school, consider the students who have listed it as their first choice in the application form. Assign seats of the school to these students one at a time in the order of priority at that school until either there are no seats left or there is no student left who has listed it as her first choice.

In general, in step kth: Consider only the kth choices of the students who are not assigned in an earlier step. For each school with seats still available, assign the remaining seats to the students who have listed it as their kth choice in the order of priority until either there are no seats left or there is no student left who has listed it as her kth choice.

The algorithm terminates when no more students are assigned. Let us apply this in example 1. In the first step, student a is considered for and assigned s_2; b and c are considered for s_1; since there is only one seat and c has higher priority, c is assigned s_1. b remains unassigned. Since there is no seat available at s_2, b is not considered for s_2 in the second step. She is considered for and assigned s_3 in the third step, and the algorithm terminates. The Boston matching is:

$$m_{Boston} = \begin{pmatrix} a & b & c \\ s_2 & s_3 & s_1 \end{pmatrix}$$

Notice that b is assigned her third choice even though she has the highest priority at her second choice, s_2. Therefore the Boston mechanism is not stable. Moreover, by ranking s_2 as second choice, b loses her priority to a, who ranks s_2 as first choice. If she instead ranked s_2 as her first choice, she would have been assigned s_2, which she prefers to s_3. That is, the Boston mechanism is not strategy proof, and a student can improve her odds of getting into a school by ranking it higher in her application. Indeed, the BPS school guide (2004, p. 3) explicitly advised parents to follow that strategy when submitting their preferences (quotes in original):

> For a better chance of your "first choice" school... consider choosing less popular schools. Ask Family Resource Center staff for information on "underchosen" schools.

The feature that one may gain from manipulating her choice list in the Boston mechanism is also recognized by parents in Boston and elsewhere. Indeed the West Zone Parent Group (WZPG), a parent group in Boston, recommends strategies to take advantage of the mechanism:[4]

> One school choice strategy is to find a school you like that is undersubscribed and put it as a top choice, OR, find a school that you like that is popular and put it as a first choice and find a school that is less popular for a "safe" second choice.

Efficient transfer mechanism

The efficient transfer mechanism (ETM), proposed by Abdulkadiroğlu and Sönmez (2003),[5] lines up students at schools with respect to their priorities. It tentatively assigns

---

[4] For more references to anecdotal evidence see Abdulkadiroğlu and Sönmez (2003), Ergin and Sönmez (2006), and Abdulkadiroğlu et al. (2006).

[5] ETM is known as the top trading cycles mechanism (TTC) in the literature. "Efficient transfers" reflect the nature of the algorithm equally well, if not better than "top trading cycles." In our experience

one empty seat at a time to the highest-priority student. If a student is happy with her assignment, she keeps it. Otherwise, ETM looks for welfare-enhancing transfers among those students. Once such transfers are exhausted, it continues in the same fashion by assigning seats to the next-highest-priority student. In slightly different but more formal language, given student preferences and school priorities, the matching is determined by the following algorithm:

- Step 1. Every school points to its highest-priority student; every student points to her most preferred school. A transfer cycle is an ordered list of schools and students (school 1–student 1–school 2–...–school k–student k), with school 1 pointing to student 1, student 1 to school 2, ..., school k to student k, and student k pointing to school 1. All the cycles are found. Every student in a cycle is assigned a seat at the school she points to and is removed; the number of seats at that school is decreased by one.

  In general, in step k. Every school with seats still available points to its highest-priority student; every student points to her most preferred school with seats still available. All the cycles are found. Every student in a cycle is assigned a seat at the school she points to and is removed; the number of seats at that school is decreased by one.

The algorithm terminates when no more students are assigned. Applying this in example 1, s_1 points to a, both s_2, and s_3 point to b; a points to s_2, and b and c both point to s_1. (s_1, a, s_2, b) form a cycle, a is assigned s_2, b is assigned s_1, they are removed, there are no more available seats at s_1 and s_2. In the second step, only s_3 has an available seat; s_3 points to c, the highest-priority student among remaining students, and c points back to s_3, her most preferred school among all with seats still available; (s_3, c) forms a cycle, c is assigned s_3. Note that the ETM matching

$$m_{ETM} = \begin{pmatrix} a & b & c \\ s_2 & s_1 & s_3 \end{pmatrix}$$

is efficient. In fact, ETM is a strategy-proof and efficient mechanism (Abdulkadiroğlu and Sönmez, 2003). However, m_3 violates c's priority at s_1, so ETM does not guarantee stability.

The student-optimal stable matching mechanism

Gale–Shapley's student-optimal stable matching mechanism (SOSM) operates like the Boston mechanism (Gale and Shapley, 1962). However, a student does not lose her priority at a school to students who rank it higher in their choice lists. To achieve this, SOSM makes *tentative* assignments and reconsiders them at every step. Formally, given

---

in the field, parents tend to have a dislike for the word "trade," complicating an objective discussion of the mechanisms for policy makers. Therefore we will refer to the mechanism as the efficient transfer mechanism.

student preferences and school priorities, the matching is determined by the following algorithm:

- Step 1. Each students applies to her first choice. Each school tentatively assigns its seats to its applicants one at a time in their priority order until capacity is reached. Any remaining applicants are rejected.

  In general, in step k. Each student who was rejected in the previous step applies to her next best choice, if one remains. Each school considers the set consisting of the students it has been holding from previous steps and its new applicants, and tentatively assigns its seats to these students one at a time in priority order. Any students in the set remaining after all the seats are filled are rejected.

The algorithm terminates when no more students are assigned, then tentative assignments are finalized. Let us find the SOSM matching in example 1. In the first step, student a applies to and is tentatively assigned s_2; b and c apply to s_1; since there is only one seat and c has higher priority, c is tentatively assigned s_1. b is rejected. Then b applies to s_2, which considers b along with a. Since b has a higher priority, b is tentatively assigned s_2, and a is rejected. Then a applies to s_1, which considers a along with c. a is tentatively assigned, and c is rejected. Then c applies to and is rejected by s_2, and finally she applies to and is tentatively assigned s_3. Since no more students are assigned, the tentative assignments are finalized, and the SOSM produces

$$m_{SOSM} = \begin{pmatrix} a & b & c \\ s_1 & s_2 & s_3 \end{pmatrix}$$

In contrast with the Boston algorithm, SOSM assigns seats only tentatively at each step, and students with higher priorities may be considered in subsequent steps. That feature guarantees that SOSM is stable in the sense that there is no student who loses a seat to a lower-priority student and receives a less-preferred assignment. More importantly, all students prefer their SOSM outcome to any other stable matching (Gale and Shapley, 1962), and SOSM is strategy proof (Dubins and Freedman, 1981; Roth, 1982b). When priorities reflect school preferences, stability eliminates the need for schools to circumvent the match to enroll the students they would prefer. However, in general, there is no stable matching mechanism, student-optimal or not, that is strategy proof for schools in two-sided markets (Roth, 1985).

Comparison of the mechanisms

The Boston mechanism is not stable. Notice that b's priority at s_2 is violated at m_{Boston}. On the other hand, it is not possible to improve the assignment of a student who gets her first choice at the Boston matching, since she is already getting her first choice. Consider a student who gets his second choice. His first choice is filled with students who rank it as first choice. Therefore, it is not possible to assign him his first choice

Table 5.1. Properties of the mechanisms

	Boston	ETM	SOSM
Strategy proof	No	Yes	Yes
Efficient	No	Yes	No
Stable	No	No	Yes
Student-optimal stable	No	No	Yes

without assigning another student at that school lower in her choice list. In general, a student cannot be assigned better than her Boston matching without harming another student's assignment. That is, the Boston mechanism is efficient with respect to the *submitted* preferences.

However, the Boston mechanism is not strategy proof. As a result, parents are forced to play a complicated game of navigating the system through preference manipulation during applications. Therefore a more important question from a policy point of view is whether the outcome resulting from this strategic interaction will be efficient or stable with respect to *true* preferences. When every parent has access to full information, and therefore knows the true preferences of other parents, and the priority orderings at schools, and this is common knowledge among parents, the outcome of the Boston mechanism that emerges from parents' strategic interaction[6] is stable with respect to the true preference profile, even though some parents manipulate their preferences (Ergin and Sönmez, 2006). This implies that SOSM is preferred to the Boston mechanism by students in such full-information environments, since SOSM produces the stable matching that students prefer to any other stable matching. However it is easy to find examples of the failure of stability and efficiency with the Boston mechanism when the full information assumption is violated.[7]

Both ETM and SOSM are strategy proof. ETM is efficient but not stable; SOSM is not efficient but it is student-optimal stable. We summarize these results in Table 5.1.

Note that the ETM outcome Pareto dominates the SOSM outcome in example 1. However, despite its superior efficiency property, the ETM outcome is not always better for every student:

Example 3. *There are three students $\{a, b, c\}$ and three schools $\{s_1, s_2, s_3\}$, each with one seat. Student preferences and school priorities are given as follows:*

$$a : s_2 - s_1 - s_3 \qquad\qquad s_1 : a - c - b$$
$$b : s_1 - s_3 - s_2 \quad and \quad s_2 : b - a - c$$
$$c : s_1 - s_2 - s_3 \qquad\qquad s_3 : b - a - c$$

---

[6] Formally, we are referring to a Nash equilibrium outcome of the complete information game induced by the Boston mechanism.

[7] Ergin and Sönmez (2006) provide an example with informational asymmetry among parents, in which the resulting outcome of the Boston mechanism fails to be stable with respect to the true preferences. Failure of efficiency is apparent even in the full-information game, since a full-information equilibrium is stable, and stability does not imply efficiency.

The SOSM outcome is

$$n_{SOSM} = \begin{pmatrix} a & b & c \\ s_2 & s_3 & s_1 \end{pmatrix}$$

and the ETM outcome is

$$n_{ETM} = \begin{pmatrix} a & b & c \\ s_2 & s_1 & s_3 \end{pmatrix}$$

Student c prefers n_{SOSM} and b prefers n_{ETM}.

As noted, SOSM and ETM simplify the task of advising parents in filing applications. All an official needs to recommend to parents is that they identify the best-fit schools for their child, and rank them in the order of their preferences.

A second concern for school districts is to explain the match to parents whose children are not assigned one of their higher choices. The outcome of SOSM is easily justified. If a student does not get into, say, her first choice under SOSM, it is because every student that is enrolled in her first choice has higher priority than she does. The outcome of the ETM can be justified in a similar fashion. Whereas SOSM tentatively assigns seats to applicants in the order of their preferences, ETM tentatively assigns seats to students in the order of school priorities. Therefore, each seat is associated with the priority of the student that it is initially assigned. If a student does not get into her first choice under ETM, it is because every seat at her first choice was initially assigned to a student with higher priority than hers. Furthermore, she could not be transferred to her first choice because she did not have high enough priority at other schools to qualify for such a transfer.

We discuss these mechanisms in further detail later. For now, a brief discussion of mechanism choices in Boston and New York City will illuminate the interplay between theory and the design.

Market design at work

The differences in the initiation of the redesign efforts and the decision-making processes in Boston and New York City (NYC) illuminate the contrasting features and challenges in both markets.

School choice in Boston has been partly shaped by desegregation. In 1974, Judge W. Arthur Garrity ordered busing for racial balance. In 1987, the US Court of Appeals freed BPS to adopt a new, choice-based assignment plan with racial preferences. In 1999, BPS eliminated racial preferences in assignment. Despite its poor incentive properties, the Boston mechanism continued to clear the market for public-school choice until 2003. Although the gaming aspect of the mechanism had apparently been known in certain Boston parent circles, it was brought to light by Abdulkadiroğlu and Sönmez (2003).

A public debate initiated by a *Boston Globe* piece on the article led to the redesign of the system. In December 2003, the Boston School Committee initiated an evaluation of all aspects of student assignment, which yielded a task-force report with a recommendation of adopting ETM. After intensive discussions, public meetings organized by BPS, and analysis of the existing school choice system and the behavior it elicited, in July 2005, the Boston School Committee voted to replace the existing school choice mechanism with SOSM. It is the first time that "strategy-proofness," a central concept in the game theory literature on mechanism design, has been adopted as a public policy concern related to transparency, fairness, and equal access to public facilities (Abdulkadiroğlu et al., 2005b).

In contrast, NYC was failing to assign more than 30,000 of the approximately 100,000 incoming high-school students to a school of their choice, yielding public outcry during the assignment period every March.[8] The NYC DOE was aware of the matching process for American physicians, the National Resident Matching Program (Roth, 1984). They contacted Alvin E. Roth in the fall of 2003 to inquire if it could be appropriately adapted to the city's schools. After an intense sequence of meetings with economists, the NYC DOE adopted a new system by January 2004 (Abdulkadiroğlu et al., 2005a). In this respect, "Boston was like a patient with high blood pressure, a potentially deadly disease that has no easily visible symptoms; the NYC high-school admission process was like a patient with a heart attack, where the best treatment might not be obvious, but there was little dispute that treatment was needed" (Abdulkadiroğlu et al., 2006).

Two features of the NYC high-school choice favored SOSM over ETM. The first was that schools withheld capacity to match with students they preferred. The fact that school administrators gamed the system indicated they were strategic players. Stable assignments eliminate part of the incentives for gaming the system. Furthermore, empirical observations suggest that centralized matching mechanisms in two-sided markets are most often successful if they produce stable matchings (Roth, 1991). Second, principals of EdOpt schools can express preferences over students. Discussions indicated that principals of different EdOpt schools had different preferences even for students with low reading scores, with some schools preferring higher scores, and others preferring students who had good attendance. If schools have different comparative advantages, allowing scope for their preferences seemed sensible.

The performance of the mechanisms also differs across markets. SOSM generates greater efficiency loss in NYC, whereas it is almost efficient in Boston. We will compare the two mechanisms in more detail later.

A special form of ETM is utilized in the supplementary round of the NYC high-school match.[9] The Louisiana Recovery School District adopted ETM in 2012. Also,

---

[8] See Goodnough (2003).

[9] The supplementary round is designed to match students who have not been matched in the main round of the process. Those students fill out a new application form on which they rank from the list of schools that still have seats available at the end of the main round. Due to the time constraint, priority information is no longer collected from schools in that round. Instead, students are ordered randomly,

after consulting with economists, [10] the San Francisco Board of Education unanimously approved a new system based on ETM in March 2010. [11]

As in any market design exercise, choice programs offered by school districts may involve distinctive features that are not captured by the basic model. Next we discuss some of those features brought to light by applications and the developments in the literature led by them.

EXTENSIONS

For the sake of completeness, we provide the formal definitions in mathematical notation in this section. A (school choice) *problem* consists of

- a finite set of students I,
- a finite set of schools S,
- school capacities $q = (q_s)_{s \in S}$, where q_s is the number of available seats at school $s \in S$,
- a profile of student preferences $P = (P_i)_{i \in I}$,
- and a profile of school priorities $\underset{\sim}{\succeq} = (\underset{\sim}{\succeq}_s)_{s \in S}$.

Each student $i \in I$ has a strict preference relation P_i over schools and her outside option o. [12] sP_is' means i prefers s to s'. Let R_i denote the weak preference relation induced by P_i, that is, sR_is' if and only if sP_is' or $s = s'$. A school s is *acceptable* for i if i prefers s to her outside option.

Each school $s \in S$ has a weak priority relation $\underset{\sim}{\succeq}_s$ over $I \cup \{\varnothing\}$, where $\varnothing$ represents leaving a seat empty. [13] A student i is *eligible* for school s if $i \succ_s \varnothing$. A student i is either eligible for school s or not, that is, either $i \succ_s \varnothing$ or $\varnothing \succ_s i$ for all i, s.

A *matching* of students to schools is a set valued function $\mu : I \cup S \rightrightarrows 2^{I \cup S}$ such that

- $\mu(i) \subset S \cup \{o\}$, $|\mu(i)| = 1$ for all $i \in I$,
- $\mu(s) \subset I$, $|\mu(s)| \le q_s$ for all $s \in S$, and
- $s \in \mu(i)$ if and only if $i \in \mu(s)$ for all $i \in I$ and $s \in S$.

and are matched one by one in that order with their most preferred school that still has available seats. This mechanism is a special form of ETM; therefore it is strategy proof and efficient.

[10] Clayton Featherstone, Muriel Niederle, Parag Pathak, Alvin Roth and I teamed up to assist the San Francisco Unified School District (SFUSD) in the redesign. Featherstone and Niederle led the discussions with SFUSD.

[11] The SFUSD decided to develop the matching software on their own, without consulting us any further. Their decision was due to concerns about sharing confidential data for monitoring the effects of the new system.

[12] Formally, P_i is a complete, irreflexive, and transitive binary relation over $S \cup \{o\}$.

[13] When $\underset{\sim}{\succeq}_s$ represents the preferences of s over students, we extend $\underset{\sim}{\succeq}_s$ over subsets of I as follows: each $\underset{\sim}{\succeq}_s$ is responsive (to its restriction on $I \cup \{\varnothing\}$). That is, for every $I' \subset I$ and $i, j \in I \backslash I'$, (i) $I' \cup \{i\} \underset{\sim}{\succeq}_s I'$ if and only if $\{i\} \underset{\sim}{\succeq}_s \varnothing$, and (ii) $I' \cup \{i\} \underset{\sim}{\succeq}_s I' \cup \{j\}$ if and only if $\{i\} \underset{\sim}{\succeq}_s \{j\}$ (Roth, 1985).

That is, a student is matched with a school or her outside option, the number of students matched with a school cannot exceed its capacity, and a student is matched with a school if and only if the school is also matched with the student. We will equivalently use $\mu(i) = s$ for $s \in \mu(i)$.

Given $(\succsim_s, P_I)$, μ *violates i's priority at s* if i prefers s to her match and another student with lower priority is matched with s, that is, $sP_i\mu(i)$ and there is a student $j \in \mu(s)$ such that $i \succ_s j$.[14]

In the *one-sided* matching models of school choice, priorities can be violated to promote student welfare. In contrast, the *two-sided* matching models do not allow priority violations at *any* school. To provide a unified treatment (Abdulkadiroğlu, 2011), in addition to the standard model, we say that a school has *a strict priority policy* if priorities may not be violated at the school, and has *a flexible priority policy* otherwise. If the priority list of a school reflects its preferences, one may assume the school has a strict priority policy. We assume that S is partitioned into S_{strict}, the set of schools with a strict priority policy, and $S_{flexible}$, the set of schools with a flexible priority policy. Formally, $S = S_{strict} \cup S_{flexible}$, and $S_{strict} \cap S_{flexible} = \varnothing$.

Next we define the policy-relevant mathematical properties, or axioms, that a matching may possess.

A matching μ is *feasible* if every student that is matched with a school is eligible for that school. We restrict our attention to feasible matchings only. A matching μ is *individually rational* if every student weakly prefers her match to her outside option. To simplify the exposition, we assume that a student can rank a school only if she is eligible for that school, that is, if $s \succ_i o$ then $i \succ_s \varnothing$. Then individual rationality implies feasibility.

In our unified model, a priority violation at school s is a cause of concern only if s has a strict priority policy. Accordingly, a matching μ is *pseudo-stable* if it is individually rational and it does not violate priorities at any school with a strict priority policy, that is, there is no $i \in I, s \in S_{strict}$ and $j \in \mu(s)$ such that $sP_i\mu(i)$ and $i \succ_s j$. The null matching that matches every student to her outside option is trivially pseudo-stable.

A matching μ *wastes a seat at s* if $|\mu(s)| < q_s$ and there is a student who is eligible for s and prefers it to her match (Balinski and Sönmez, 1999); that is, there exists $i \in I$ such that $i \succ_s \varnothing$ and $sP_i\mu(i)$.[15] A matching μ is *stable* if it is pseudo-stable and it does not waste any seat. Although the null matching is trivially pseudo-stable, it wastes all the seats so it is not stable.

A matching μ *Pareto dominates* another matching ν if every student weakly prefers her μ-match to her ν-match and some strictly, i.e. $\mu(i)R_i\nu(i)$ for all $i \in I$ and $\mu(i)P_i\nu(i)$ for some $i \in I$. A matching is *Pareto efficient* if it is not Pareto dominated by another

---

[14] In the standard two-sided matching literature, such an (i, s) pair is said to block μ, and it is referred to as a blocking pair. Alternately, Balinski and Sönmez (1999) refer to it as envy, by i at s. The naming of violating priorities is due to Ergin (2002).

[15] Such a pair is also refered to as a blocking pair in the two-sided matching literature. The renaming of it as wastefulness is due to Balinski and Sönmez (1999).

matching. A matching is *student-optimal stable* if it is stable and not Pareto dominated by another pseudo-stable matching.

Our unified model reduces to the standard two-sided matching model when $S = S_{strict}$, that is, every school has a strict priority policy. It reduces to the standard one-sided matching model of school choice when $S = S_{flexible}$, that is, every school has a flexible priority policy. In that case, every matching is pseudo-stable and every student-optimal stable matching is Pareto efficient. In other words, Pareto efficiency becomes a special case of our notion of student-optimal stable matching when priority violations are allowed at all schools.

A student admissions procedure is defined as a mechanism. A (deterministic) *mechanism* selects a matching for every problem. The definitions for matching trivially extend to a mechanism. For example, a mechanism is stable if it selects a stable matching for every problem. Suppressing school priorities, let $\varphi(P)$ denote the matching selected by a mechanism φ. A mechanism φ is *strategy proof for students* if reporting true preferences is a dominant strategy for every student in the preference revelation game induced by φ. That is:

$$\varphi(P)(i)R_i\varphi(P'_i, P_{-i})(i)$$

for all P, $i \in I$ and P'_i, where $P_{-i} = (P_j)_{j \in I\setminus\{i\}}$. Strategy proofness for schools is defined similarly. A mechanism φ *Pareto dominates* another mechanism φ' if for every problem $< I, S, q, P, \succsim >$, every student prefers her φ-match to her φ'-match and some strictly, that is $\varphi(P)(i)R_i\varphi'(P)(i)$ for all i and $\varphi(P)(i)P_i\varphi'(P)(i)$ for some i.

Further discussion of the mechanisms

When all schools have a strict priority policy, the problem turns into a two-sided matching problem. In that case, SOSM is the unique stable mechanism that is strategy-proof for students (Alcalde and Barberà, 1994). When priorities do not reflect school preferences, the notion of respecting priorities can be interpreted as the elimination of justified envy (Balinski and Sönmez, 1999). When a student's standing in the priority list of school improves, the student is assigned a weakly better school by SOSM. In fact, SOSM is the only stable mechanism with that property (Balinski and Sönmez, 1999).

SOSM is not efficient from the students' perspective. Ergin (2002) shows that the outcome of SOSM is efficient if and only if school priorities satisfy a certain acyclicity condition. Ehlers and Erdil (2010) generalize that result when school priorities are coarse. Although the ETM outcome may Pareto dominate the SOSM outcome for some problems, no Pareto-efficient and strategy-proof mechanism Pareto dominates SOSM when school priorities do not involve ties (Kesten, 2010). Kesten (2010) proposes a new algorithm that eliminates the efficiency loss associated with SOSM by allowing students to give up certain priorities whenever it does not hurt them to do so.

When all schools have a flexible priority policy, the problem turns into a one-sided matching problem. Starting with Shapley and Scarf (1974), ETM has mostly been studied in exchange markets for indivisible objects. That model corresponds to a special case of our model in which each school has a single seat, and a student is ranked highest by at most one school. In that environment, ETM is strategy proof (Roth, 1982a), and it is the only mechanism that is Pareto efficient, strategy proof, and that guarantees every student that is top ranked at a school an assignment that she weakly prefers to that school (Ma, 1994). When students are allowed to be ranked highest by more than one school, ETM is a special subclass of Pápai's (2000) hierarchical exchange rules. In that case, Pápai characterizes hierarchical exchange rules by Pareto efficiency, group strategy proofness (which rules out beneficial preference manipulation by groups of individuals), and reallocation proofness (which rules out manipulation by two individuals via misrepresenting preferences and swapping objects ex post). ETM is a hierarchical exchange rule defined by the priority lists of schools. In a similar vein, Pycia and Ünver (2010) introduce and characterize trading cycles with brokers and owners by Pareto efficiency and group strategy proofness. Bogomolnaia et al. (2005) provide a characterization for a general class of Pareto-efficient and strategy-proof mechanisms for the case in which schools have multiple seats and no priorities.

Despite the lack of a Pareto ranking between SOSM and ETM, there exists a clear-cut comparison between SOSM and Boston when market participants have full information about others' preferences and priorities, and that is common knowledge. In particular, given strict school priorities, every Nash equilibrium outcome of the Boston mechanism is stable under true preferences. Therefore the dominant strategy equilibrium of SOSM weakly Pareto dominates every Nash equilibrium outcome of the Boston mechanism (Ergin and Sönmez, 2006).[16]

Further characterizations of SOSM and the Boston mechanism are provided via monotonicity conditions on preferences by Kojima and Manea (2010) and Kojima and Unver (2010) respectively. Roth (2008) provides a survey of the history, theory, and practice of SOSM.

Ties in school priorities

Much of the earlier theory of two-sided matching focuses on the case where all parties have strict preferences, mainly because indifferences in preferences were viewed as a "knife-edge" phenomenon in applications like labor markets (Roth and Sotomayor, 1990). In contrast, a primary feature of school choice is that there are indifferences—"ties"—in how students are ordered by at least some schools. How to break these ties raises some significant design decisions, which bring in new trade-offs between

---

[16] Kojima (2008) generalizes this finding to more complicated priority structures that, for instance, can favor specific student populations via quotas.

efficiency, stability, and strategy proofness (Erdil and Ergin, 2008; Abdulkadiroğlu et al., 2009).

The mechanism of choice must specify how to order equal-priority students from the point of view of schools with limited space. For instance, one can assign each student a distinct number, breaking ties in school priorities according to those assigned numbers—single tie breaker—or one can assign each student a distinct number at each school—multiple tie breakers—breaking ties according to school specific numbers. Since any non-random assignment of such numbers can be incorporated into the priority structure at the outset, we will consider randomly generated tie breakers.

Ex post efficiency

ETM remains Pareto efficient and strategy proof with single and multiple tie breakers. Furthermore, when there are no priorities at schools, i.e. all students tie in priority at every school, ETM produces the same probability distribution over matchings when a single or a multiple tie breaker is drawn uniformly randomly (Pathak and Sethuraman, 2011).

If one applies SOSM to the strict priorities that result from tie breaking, the stability and strategy proofness of SOSM is preserved. However, tie breaking introduces artificial stability constraints (since, after tie breaking, schools appear to have strict rankings between equal priority students), and these constraints can harm student welfare. In other words, when SOSM is applied to the strict priorities that result from tie breaking, the outcome it produces may not in fact be a student-optimal stable matching in terms of the original priorities.

When school priorities are weak, there may be multiple student-optimal stable matchings that are not Pareto ranked with each other. Every student-optimal stable matching can be obtained by SOSM with some tie breakers (Ehlers, 2006). However, some forms of tie breaking may be preferable to others. For instance, during the course of designing the NYC high-school match, policy makers from the Department of Education were concerned with the fairness of tie breaking; they believed that each student should receive a different random number at each program they applied to, and this number should be used to construct strict preferences of schools for students. Their rationale was that if a student draws a bad number in a single tie breaker, her bad luck would apply to every school of her choice, whereas multiple tie breakers would give a new life line at her lower-ranked schools if that student is rejected by a school. However, we show via simulations with NYC high-school match data that significantly more students get their first choices when ties are broken by a single lottery (Abdulkadiroğlu et al., 2009). Table 5.2 summarizes our simulation results for 250 random draws of tie breakers for grade 8 applicants in 2006–07. In particular, on average SOSM with single breakers matches about 2,255 more students to their first choices. Note also that SOSM with single breakers leaves about 186 more students unassigned, which implies that there is no comparison between SOSM with single breakers, and SOSM with multiple tie breakers, in terms of first-order stochastic dominance.

Table 5.2. Welfare consequences of tie breaking and strategy proofness for grade 8 applicants in NYC in 2006–07

Choice	Single tie breakers	Multiple tie breakers	SIC	Efficient
1	29,849.9 (67.7)	32,105.3 (62.2)	32,701.5 (58.4)	34,707.8 (50.5)
2	14,562.3 (59.0)	14,296.0 (53.2)	14,382.6 (50.9)	14,511.4 (51.1)
3	9,859.7 (52.5)	9,279.4 (47.4)	9,208.6 (46.0)	8,894.4 (41.2)
4	6,653.3 (47.5)	6,112.8 (43.5)	5,999.8 (41.4)	5,582.1 (40.3)
5	4,386.8 (39.4)	3,988.2 (34.4)	3,883.4 (33.8)	3,492.7 (31.4)
6	2,910.1 (33.5)	2,628.8 (29.6)	2,519.5 (28.4)	2,222.9 (24.3)
7	1,919.1 (28.0)	1,732.7 (26.0)	1,654.6 (24.1)	1,430.3 (22.4)
8	1,212.2 (26.8)	1,099.1 (23.3)	1,034.8 (22.1)	860.5 (20.0)
9	817.1 (21.7)	761.9 (17.8)	716.7 (17.4)	592.6 (16.0)
10	548.4 (19.4)	526.4 (15.4)	485.6 (15.1)	395.6 (13.7)
11	353.2 (12.8)	348.0 (13.2)	316.3 (12.3)	255.0 (10.8)
12	229.3 (10.5)	236.0 (10.9)	211.2 (10.4)	169.2 (9.3)
Unassigned	5,426.7 (21.4)	5,613.4 (26.5)	5,613.4 (26.5)	5,613.4 (26.5)

Data from the main round of the New York City high-school admissions process in 2006–07 for students requesting an assignment for grade 9 (high school). The table reports the average choice received distribution of applicants from SOSM with single tie breakers, SOSM with multiple tie breakers, stable improvement cycles (SIC) algorithm, and efficient matchings which are produced by TTC by using the SIC assignment as endowment. The averages are based on 250 random draws. Simulation standard errors are reported in parentheses. Reproduced from Abdulkadiroğlu et al. (2009).

Some theoretical insight for that observation comes from the fact that, when school priorities are weak, all student-optimal stable matchings can be found by SOSM with single breakers (Abdulkadiroğlu et al., 2009; Erdil, 2006). In other words, if there is a matching produced by SOSM with multiple breakers that cannot be produced by any SOSM with single breakers, then it is not a student-optimal stable matching.

However, a single lottery is not sufficient for student optimality (Erdil and Ergin, 2008; Abdulkadiroğlu et al., 2009). Given a matching, a stable improvement cycle of students $\{a_1, \ldots, a_{n+1} \equiv a_1\}$ is such that every student in the cycle is matched with a school, every a_k, $k = 1, \ldots, n$, prefers a_{k+1}'s match to her match, and she has the highest priority among all students who prefer a_{k+1}'s match to their match (Erdil and Ergin, 2008). If the cycle is implemented by transferring a_k to a_{k+1}'s matched school, the resulting matching is stable and Pareto dominates the original matching. Based on this novel observation, Erdil and Ergin (2008) show that a stable matching μ is student optimal if and only if it does not admit a stable improvement cycle. They also introduce a stable improvement cycles (SIC) algorithm, which starts with an arbitrary stable matching and finds and implements a cycle until no cycle is found. SIC is student-optimal stable. Employing SIC on top of SOSM with single breakers, Table 5.2 shows that about 596 more students can be matched with their first choices.

Incentives and ex post efficiency

More interestingly, ties in school priorities introduce a trade-off between efficiency and strategy proofness. In particular, there is no strategy-proof mechanism that always selects a student-optimal stable matching (Erdil and Ergin 2008). Therefore SOSM with any breakers may yield inefficient outcomes and removal of such inefficiency harms students' incentives. Furthermore, given a set of tie breakers, the associated SOSM is not Pareto dominated by any strategy-proof mechanism (Abdulkadiroğlu et al., 2009). This observation generalizes two earlier results: SIC is not strategy-proof (Erdil and Ergin, 2008), and no Pareto-efficient and strategy-proof mechanism Pareto-dominates SOSM when school priorities are strict (Kesten, 2010).

In other words, SOSM with a tie breaker lies on the Pareto frontier of strategy-proof mechanisms. This theoretical observation gives us an empirical strategy to assess the cost of strategy proofness. In particular, the additional 596 students who get their first choices under SIC in Table 5.2 can be interpreted as the efficiency cost of strategy proofness for students in SOSM with single breakers.

In Table 5.2, when students start with their SIC matches and welfare-improving transfers are exhausted among students via ETM, [17] on average an additional 2,006 students can be matched with their first choice. Similarly, this number can be interpreted as the welfare cost of limiting the scope of manipulation for schools in NYC.

Ex ante efficiency

The earlier literature, in particular all the results stated so far, relies on a notion of efficiency from an ex post point of view, that is, after the resolution of all potential uncertainties. When too many students demand a seat at a school, admissions to the school are regulated by priorities. When priorities are strict, both ETM and SOSM uniquely determine the outcome. In contrast, with weak priorities, there remains a great deal of freedom in placing students according to their preferences. Furthermore, a new scope of efficiency from an ex ante point of view emerges. These points are illustrated in the following example by Abdulkadiroğlu et al. (2011).

Example 4. *There are three students, $\{1, 2, 3\}$ and three schools, $\{s_1, s_2, s_3\}$, each with one seat. Schools have no intrinsic priorities over students, and student i has a von-Neumann Morgenstern (henceforth, vNM) utility value of v_j^i when she is assigned to school j:*

	v_s^1	v_s^2	v_s^3
$s = s_1$	0.8	0.8	0.6
$s = s_2$	0.2	0.2	0.4
$s = s_3$	0	0	0

[17] That is, start with the SIC matching. Run the following version of ETM: Every student points to her most preferred school among those remaining. Every school points to remaining students that it currently enrolls. Cycles are found. Every student in a cycle is transferred to the school she points to and she is removed. Continue in the same fashion until no more students are transferred.

Every feasible matching is stable due to schools' indifferences. More importantly, any such assignment is ex post Pareto efficient, and hence student-optimal stable, since students have the same ordinal preferences.

Since SOSM with any tie breaker is strategy proof, all three students submit true (ordinal) preferences of $s_1 - s_2 - s_3$. SOSM with a single tie-breaker that is drawn uniformly randomly matches every student to each school with equal probability of $\frac{1}{3}$, which yields an expected payoff of $\frac{1}{3}$ for each student. This random matching is ex ante Pareto dominated by the following random matching: Assign student 3 to s_2, and students 1 and 2 randomly between s_1 and s_3, which yields expected payoff of $0.4 > \frac{1}{3}$ for every student. This Pareto-dominating random matching arises as the unique equilibrium outcome of the Boston mechanism. In fact, this observation holds more generally. Suppose that all students tie in priorities at every school, students have the same ordinal ranking of schools, and their cardinal utilities are private information that are drawn from a commonly known distribution. Consider the Boston mechanism and SOSM with a single tie breaker that is drawn uniformly randomly. Then each student's expected utility in every symmetric Bayesian equilibrium of the Boston mechanism is weakly greater than her expected utility in the dominant-strategy equilibrium of SOSM (Abdulkadiroğlu et al., 2011). This finding contrasts with but does not contradict Ergin and Sönmez (2006), who analyze a complete information setup with strict school priorities and heterogenous ordinal preferences for students.

SOSM is strategy proof and therefore in the dominant strategy equilibrium of SOSM, every student submits her true preference list to the mechanism regardless of her cardinal utilities. In contrast, a student takes her cardinal utilities into account while submitting her equilibrium strategy under the Boston mechanism. That allows the Boston mechanism to break ties based on cardinal information, as opposed to the fully random tie breaking under SOSM.

Independently, Featherstone and Niederle (2008) show that truth telling becomes a Bayesian Nash equilibrium of the Boston mechanism when informational asymmetry on student preferences are introduced in a symmetric environment, in which all schools have the same capacity, all students tie in priorities at every school, and preferences of each student are drawn uniformly randomly on the set of all possible rank orderings of the set of schools. Then more students are matched with their first choices in the truth-telling equilibrium of the Boston mechanism than in the dominant-strategy truth-telling equilibrium of SOSM.

Troyan (2011) takes a more ex ante approach, and examines welfare before students know their cardinal utilities and priorities. He shows that, from this perspective, the Boston mechanism ex ante Pareto dominates any strategy-proof and anonymous mechanism, including SOSM and ETM, even with arbitrary priority structures.

These complementary works draw a picture of the Boston mechanism that has been overlooked by the earlier literature, which relies on the complete information assumption.

Motivated by their observation for the Boston mechanism, Abdulkadiroğlu et al. (2008) propose an SOSM with "preferential" tie breaking. Every student submits her ordinal preference list, and picks one school as a target, at which she will be favored in tie breaking. When two students tie at a school, the one who picks it as a target is favored in tie breaking; otherwise, the ties are broken randomly. It is still a dominant strategy to submit true preferences to their mechanism, and gaming is limited to the choice of the target school. They show that their modified mechanism results in ex ante efficiency gains in large economies. In a similar vein, Miralles (2008) shows that a variant of the Boston mechanism that utilizes a new lottery in every round of the assignment algorithm obtains similar efficiency gains over SOSM in a continuum economy.

Budish et al. (2013), on the other hand, generalize the theory of randomized assignment to accommodate multi-unit allocations and various real-world constraints, including group-specific quotas in school choice. They also provide new mechanisms that are ex ante efficient and fair.

Ex ante stability

When school priorities are weak, random tie breaking with SOSM yields randomization over stable matchings. In that setup, Kesten and Ünver (2010) introduce two notions of stability from an ex ante point of view: A random matching is ex ante stable if there are no students a, b, and a school s such that a has a higher priority at s than b, b, is matched with s with positive probability, and a is matched with positive probability with a school that she prefers less than s. An ex ante stable random matching is strongly ex ante stable if it avoids the following case among equal priority students, which they refer to as ex ante discrimination: a and b have equal priority at s, b enjoys a higher probability of being assigned to s than a, and a is matched with positive probability with a school that she prefers less than s. Kesten and Ünver (2010) propose an algorithm to select the strongly ex ante stable random matching that is ordinally Pareto dominant among all strongly ex ante stable random matchings.

Leveling the playing field

Strategy proofness has emerged as a major public policy concern related to transparency, fairness, and equal access to public facilities in the redesign of the Boston school assignment system (Abdulkadiroğlu et al., 2006). In July 2005, the Boston School Committee voted to adopt SOSM, which removes the incentives to "game the system" that handicapped the Boston mechanism. In his memo to the School Committee on May 25, 2005, Superintendent Payzant wrote:

> The most compelling argument for moving to a new algorithm is to enable families to list their true choices of schools without jeopardizing their chances of being assigned to any school by doing so.... A strategy-proof algorithm levels the playing field by diminishing the harm done to parents who do not strategize or do not strategize well.

Pathak and Sönmez (2008a) investigate this issue by studying a complete-information model with strict school priorities, and with both sincere students, who always submit their true preference rankings, and sophisticated students, who respond strategically. They find that the Nash equilibrium outcomes of the Boston mechanism are equivalent to the set of stable matchings of a modified economy where sincere students lose their priorities to sophisticated students at all but their first-choice schools; furthermore, every sophisticated student weakly prefers her assignment under the Pareto-dominant Nash equilibrium outcome of the Boston mechanism to the dominant-strategy outcome of SOSM.

A second issue raised by Abdulkadiroğlu et al. (2010) is related to neighborhood priorities, a common feature of many school choice programs. For instance, BPS gives priority to students who live within 1 mile from an elementary school, within 1.5 miles from a middle school, and within 2 miles from a high school in attending those schools. At the same time, one of the major goals of public school choice is to provide equal access to good schools for every student, especially for those in poor neighborhoods with failing schools. This goal is compromised by neighborhood priority. The extent to which the neighborhood priority inhibits access to good schools by students in failing schools districts differs across mechanisms. Under the SOSM, a student does not need to give up her neighborhood priority when applying for other (better) schools. This is in sharp contrast to what happens under the Boston mechanism. When a student does not rank her neighborhood school as first choice under the Boston mechanism, she loses her neighborhood priority at that school to those who rank it higher in their choice list. Similarly, if she ranks her neighborhood school as first choice, then she gives up priority at the other schools. In either case, another student would be able to improve her odds at that school or some other school. Abdulkadiroğlu, et al. (2011) provide examples in which this feature of the Boston mechanism provides greater access to good schools for students without neighborhood priority at those schools.

Controlled choice

Controlled school choice in the United States attempts to provide parental choice over public schools while maintaining racial, ethnic, and socioeconomic balance at schools. Boston's renowned controlled choice program emerged out of concerns for economically and racially segregated neighborhoods that were a consequence of traditional neighborhood-based assignment to public schools. Today, many school districts adopt desegregation guidelines either voluntarily or because of a court order. Other forms of control exist in choice programs in the US. Miami-Dade County Public Schools control for the socioeconomic status of students in order to diminish concentrations of low-income students at schools. In New York City, Educational Option (EdOpt) schools have to accept students of wide-ranging abilities. In particular, 16% of students that attend an EdOpt school must score above grade level on the standardized English Language Arts test, 68% must score at grade level, and the remaining 16% must score below grade level (Abdulkadiroglu et al., 2005).

It is easy to modify the mechanisms when each student can be of one type from a finite set, such as {Asian, Black, Hispanic, White, Other}, and the number of students of a type matched with a school cannot exceed a type specific quota at that school. In ETM, a student points to her most preferred school among all schools at which there is an available seat and the quota for her type is not met yet. ETM with quotas is Pareto efficient and strategy proof (Abdulkadiroğlu and Sönmez, 2003).

In SOSM, a school tentatively admits students in the order of priority up to its capacity among those students for whom the type-specific capacity has not yet been met. Given strict school priorities and quotas, SOSM with quotas produces a stable matching that respects quotas and is weakly preferred by every student to any other stable matching that respects quotas (Roth, 1984). Under the same assumptions, it is also strategy proof (Abdulkadiroğlu, 2005). These properties extend to a more general setting with substitutable preferences (Hatfield and Milgrom, 2005).[18]

Ehlers (2009) introduces quotas for the *minimum* number of students of each type who have to be assigned to schools. He shows that minimum quotas are incompatible with stability, relaxes the stability requirement, and studies student-optimal stable matchings.

Kojima (2010) shows that affirmative-action quotas can make majority students as well as every minority student worse off under both SOSM and ETM. Hafalir, Yenmez and Yildirim (2013) offer an alternative policy that gives preferential treatment to minorities for a number of reserved seats at each school. They also provide a group strategy-proof mechanism, which gives priority to minority students for reserved seats at schools. Their mechanism also Pareto dominates SOSM with quotas. Westcamp (2010) offers a strategy-proof SOSM for the allocation of German public universities for medicine and related fields, in which floating quotas are employed to prioritize students according to their grades or waiting time.

The generalized theory of randomized assignment with minimum as well as maximum type-specific quotas by Budish et al. (2013) applies to the controlled school choice problem when student assignment involves randomization.

Short preference lists

Some school districts impose a limit on the number of schools that can be listed in an application. For instance, students could list at most five schools in Boston before 2005; and the NYC high school admissions process allows students to rank at most twelve schools in their applications.

---

[18] Let the choice of school s from a set of students X be defined as $Ch(X; \succ_s) \subset X$ such that $Ch(X; \succ_s) \succ_s Z$ for all $Z \subset X, Z \neq Ch(X; \succ_s)$. Then a preference relation $\succ_s$ has the property of substitutability if $i \in Ch(X \backslash \{j\}; \succ_s)$ for every $X \subset I, i \in Ch(X; \succsim_s), j \in Ch(X; \succsim_s) \backslash \{i\}$ (Kelso and Crawford, 1982; Roth, 1984). That is, whenever i is chosen from a set, i will be chosen even if some other student is removed from the set.

Haeringer and Klijn (2009) study the preference revelation game induced by different mechanisms when students can only list up to a fixed number of schools. They focus on the stability and efficiency of the Nash equilibrium outcomes in a model with strict school priorities. They find that, when students can list a limited number of schools, (1) SOSM may have a Nash equilibrium in undominated strategies that produce a matching that is not stable under true preferences; (2) ETM may have a Nash equilibrium in undominated strategies that produce a matching that is not Pareto efficient under true preferences.

Pathak and Sönmez (2013) show that an SOSM with a cap of maximum k choices in more manipulable than an SOSM with a cap of maximum $l > k$ choices, in the sense that the former mechanism can be manipulated at a larger set of preference profiles.

Large markets

Size matters. Some of the trade-offs vanish as the number of participants increases. Whereas the number of stable matchings can be arbitrarily large in finite economies, Roth and Peranson (1999) observe that the set of stable matchings has been small in the NRMP, which they explain via simulations by the short preference lists submitted by the applicants in relatively large markets. [19] In contrast, Azevedo and Leshno (2012) give general conditions under which a model with finitely many schools and a continuum of students admits a unique stable matching.

There is no stable mechanism that is strategy proof for students as well as schools (Roth, 1982b). Also, when schools have more than one seat, there is no stable mechanism that is strategy proof for schools (Roth, 1985). These results can be proved via examples with a few students and schools. However, in a model with one seat at every school, Immorlica and Mahdian (2005) show that as the size of the market becomes large, the set of stable matchings shrinks. Kojima and Pathak (2009) generalize this finding to the model with multiple seats at schools and strict school priorities which reflect school preferences. They show that when schools are also strategic, reporting true preferences becomes an approximate Bayesian equilibrium for schools as the market power of schools vanishes in large markets.

Several applications, including the school choice programs in Korea and the second round of the NYC high-school match, involve no priorities on the school side. In that a case, the random priority mechanism (RP)[20] which assigns every student her most preferred school among the remaining schools one at a time in the order of a randomly drawn order of students, is strategy proof and ex post Pareto efficient. Bogomolnaia and

---

[19] We discuss the large market findings within the context of school choice, although some of them have been formulated outside the school choice context.

[20] This mechanism is also known as random serial dictatorship, and can be implemented as SOSM with a uniformly randomly drawn single tie breaker.

Moulin (2001) observe that RP allocation can be improved for some students in the sense of first order stochastic dominance without harming other students' allocations. An allocation which cannot be improved that way is ordinally efficient. Bogomolnaia and Moulin (2001) provide an ordinally efficient probabilistic serial mechanism (PS). However they also show that no ordinally efficient mechanism is strategy proof for students. Che and Kojima (2010) show that, as the number of students and school capacities grow, the RP becomes equivalent to the PS mechanism, of which the former is strategy proof and the latter is ordinally efficient. Therefore, the trade-off between strategy proofness and ordinal efficiency vanishes in such large markets.

Azevedo and Budish (2012) introduce a new notion to study incentives in large markets. Accordingly, a mechanism is strategy proof in the large if all of its profitable manipulations vanish with market size. They show that the outcomes of a large class of mechanisms can be implemented approximately by mechanisms that are strategy proof in the large. Budish (2011) studies an assignment problem in which all the known mechanisms are either unfair ex post or manipulable even in large markets. He introduces a slightly different strategy proofness in the large notion, and proposes a combinatorial assignment mechanism that is strategy proof in the large, approximately efficient, and fair.

Hybrid matching problems

A close look at the real-life cases reveals that the school choice problem exhibits features of one-sided matching and two-sided matching, simultaneously. For instance, many school districts offer admissions to some selective *exam* schools via an entrance exam. Violating priorities induced by an entrance exam proves to be a political and legal challenge even when such violations are justified by court-ordered desegregation guidelines (Abdulkadiroğlu, 2011). On the other hand, as in the case of Boston, violating priorities at regular schools may be considered in order to promote student welfare (Abdulkadiroğlu et al., 2005a). A similar issue may arise when some schools are strategic and have preferences over students while others are not, as in the case of the NYC high-school match (Abdulkadiroğlu et al., 2005a, 2010). In that case, violating the preferences of a strategic school would create an instance at which the school would prefer to circumvent the assignment to match with a more preferred student who also prefers the school to her match.

Ehlers and Westcamp (2010) study a school choice problem with exam schools and regular schools. They assume that exam schools rank students in strict priority order, and regular schools are indifferent among all students. Their model is a special case of Erdil and Ergin (2008), and Abdulkadiroğlu et al. (2009); however, their scope is quite different. In particular, they identify conditions on priorities of exam schools under which strategy proofness is preserved.

Abdulkadiroğlu (2011) studies a generalized matching model that encompasses one-sided and two-sided matching as well as their hybrid. In his model, every school is endowed with a priority list that may involve ties. However, a school may have a strict or flexible priority policy, and a stable matching may violate priorities at schools with a flexible priority policy. He characterizes student-optimal stable matchings via stable transfer cycles. A stable transfer cycle is an application of SIC (Erdil and Ergin, 2008). It operates like ETM but puts restrictions on schools with strict priority policies as in SIC. In particular, in a stable transfer cycle, a student can point to any school that she prefers to her current match as long as the school has a flexible priority policy. Otherwise, in order to be able to point to it, she has to be ranked highest among all students who prefer that school to their current match. Schools, on the other hand, point to the highest-priority students among those remaining.

Experiments

Starting with Chen and Sönmez (2006), there is a growing experimental literature with a focus on school choice. Consistent with theory, Chen and Sönmez (2006) observe a high preference manipulation rate under the Boston mechanism. They also find that efficiency under Boston is significantly lower than that of ETM and SOSM. However, contrary to theory, they find that SOSM outperforms ETM in terms of efficiency in their experimental environment.

Pais and Pinter (2007), on the other hand, show that, when the experiment is conducted in an incomplete information setup, ETM outperforms both SOSM and Boston in terms of efficiency. Moreover, it is slightly more successful than SOSM regarding the proportion of truthful preference revelation and manipulation is stronger under the Boston mechanism; even though agents are much more likely to revert to truth telling in lack of information about the others' payoffs, ETM results are less sensitive to the amount of information that participants hold.

Calsamiglia et al. (2010) analyze the impact of imposing limit on the number of schools in choice lists. They show that manipulation is drastically increased, which is consistent with Pathak and Sönmez's (2013) theoretical argument; including a safety school in the constrained list explains most manipulations; both efficiency and stability of the final allocations are also negatively affected.

Featherstone and Niederle (2008) observe that, when school priorities involve ties and are broken randomly, and preferences are private information, the Boston mechanism obtains better efficiency than SOSM.

Klijn et al. (2010) study how individual behavior is influenced by risk aversion and preference intensities. They find that SOSM is more robust to changes in cardinal preferences than the Boston mechanism, independently of whether individuals are allowed to submit a complete or a restricted ranking over the set of schools, and subjects with a higher degree of risk aversion are more likely to play "safer" strategies under the SOSM, but not under the Boston mechanism.

Conclusion

School choice has provided economists with new opportunities to study and design student assignment systems, which in turn have helped push forward the frontiers of mechanism design theory. This chapter aims at demonstrating this point. Many interesting questions remain open. To what extent is the stable improvement cycles mechanism manipulable in the field? How restrictive are the minimum quotas for minorities in controlled school choice programs? To what extent do they preclude stability, and foster gaming in the field? Can we design and implement mechanisms with better efficiency properties? Are there simple mechanisms that elicit not only ordinal preferences, but also some information on the underlying cardinal preferences? In fact, how do we define a simple mechanism; are they robust (Milgrom, 2009)? Theory gives impossibilities for some of these questions, and it is silent on others. Designing better market mechanisms will require not only further new theory, but also new engineering approaches that rely on careful synthesis of the theory, empirical analysis, and experiments (Roth, 2002).

In addition, in contrast to other market design applications, school choice has a direct public policy appeal. For example, how does information impact choice patterns and academic achievement for disadvantaged students (Hastings and Weinstein, 2008); does school choice foster competition among schools; does it help eliminate achievement gap (Hastings et al., 2008)? Second, school choice programs in the US present economists with unprecedented data with randomized assignments. Such data allow researchers to study the impact of different schooling options on student outcomes without suffering from selection bias issues, such as charter schools and their public-school alternatives (Abdulkadiroğlu et al., 2010; Angrist et al., 2011; Hoxby et al., 2009), and small schools (Bloom et al., 2010). While designing student assignment systems as market designers, we can also think about and address such broader questions as social scientists. Can we also incorporate sound econometric tools into our designs that would help districts evaluate their schooling alternatives beyond simple descriptive statistics and free of selection bias?

References

Abdulkadiroğlu, A. (2005) "College admissions with affirmative action," *International Journal of Game Theory*, 33: 535–549.
_____ (2011) "Generalized matching for school choice", Working Paper, Duke University.

Abdulkadiroğlu, A. and Sönmez, T. (2003) "School choice: a mechanism design approach," *American Economic Review*, 93: 729–47.

_____ Pathak, P. A. and Roth, A. E. (2005a) "The New York City high school match", *American Economic Review, Papers and Proceedings*, 95, 364–367.

_____ _____ _____ and Sönmez, T. (2005b) "The Boston public school match," *American Economic Review, Papers and Proceedings*, 95: 368–71.

_____ _____ _____ _____ (2006) "Changing the Boston school choice mechanism: strategy-proofness as equal access", Mimeo.

_____ Che, Y-K. and Yasuda, Y. (2008) "Expanding 'Choice' in school choice", Mimeo.

_____ Pathak, P. A. and Roth, A. E. (2009) "Strategy-proofness versus efficiency in matching with indifferences: redesigning the NYC high school match," *American Economic Review*, 99(5): 1954–78.

Abdulkadiroglu, A. J. D., Angrist, S. M., Dynarski, Kane, T. J. and Pathak, P. (2010) "Accountability and flexibility in public schools: evidence from Boston's charters and pilots," *Quarterly Journal of Economics*, 126(2): 699–748.

_____ Che, Y-K. and Yasuda, Y. (2011) "Resolving conflicting preferences in school choice: the "Boston" mechanism reconsidered," *American Economic Review*, 101(1): 399–410.

Alcalde, J. and Barberà, S. (1994) "Top dominance and the possibility of strategy-proof stable solutions to matching problems," *Economic Theory*, 4: 417–35.

Angrist, J., Cohodes, S. R., Dynarski, S., Fullerton, J. B., Kane, T. J., Pathak, P., and Walters, C. R. (2011) "Student achievement in Massachusetts charter schools," Report.

Azevedo, E. M. and Budish, E. (2012) "Strategyproofness in the large as a desideratum for market design," Mimeo.

_____ and Leshno, J. D. (2012) "A supply and demand framework for two-sided matching markets," Mimeo.

Balinski, M. and Sönmez, T. (1999) "A tale of two mechanisms: student placement," *Journal of Economic Theory*, 84: 73–94.

Bloom, H. S., Thompson, S. L. and Unterman, R. (2010) "Transforming the high school experience: how New York City's new small schools are boosting student achievement and graduation rates," MDRC Report.

Bogomolnaia, A. and Moulin, H. (2001) "A new solution to the random assignment problem," *Journal of Economic Theory*, 100: 295–328.

_____ Deb, R. and Ehlers, L. (2005) "Strategy-proof assignment on the full preference domain," *Journal of Economic Theory*, 123: 161–86.

Budish, E. (2011) "The combinatorial assignment problem: approximate competitive equilibrium from equal incomes," *Journal of Political Economy*, 119(6): 1061–103.

_____ Che, Y.-K., Kojima, F. and Milgrom, P. (2013) "Designing Random Allocation Mechanisms: Theory and Applications," *American Economic Review*, 103(2): 585–623.

Calsamiglia, C., Haeringer, G. and Klijn, F. (2010) "Constrained school choice: an experimental study," Mimeo.

Che, Y.-K, and Kojima, F. (2010) "Asymptotic equivalence of probabilistic serial and random priority mechanisms," *Econometrica*, 78(5): 1625–72.

Chen, Y. and Sönmez, T. (2006) "School choice: an experimental study," *Journal of Economic Theory*, 127: 2002–31.

Cook, G. (2003) "School assignment; faws detailed," *Boston Globe*, Metro Desk, September 12.

Dubins, L. E. and Freedman, D. A. (1981) "Machiavelli and the Gale–Shapley algorithm," *American Mathematical Monthly*, 88: 485–94.

Ehlers, L. (2006) "Respecting priorities when assigning students to schools," Mimeo.

_____ (2009) "School choice with control," Working Paper.

_____ and Erdil, A. (2010) "Efficient assignment respecting priorities," *Journal of Economic Theory*, 145: 1269–82.

_____ and Westcamp, A. (2010) "Breaking ties in school choice: (non-)specialized schools," Mimeo.

Erdil, A. (2006) "Two sided matching under weak preferences," PhD Thesis, University of Chicago.

Ergin, H. (2002) "Efficient resource allocation on the basis of priorities," *Econometrica*, 70: 2489–97.

Erdil, A. and Ergin, H. (2006) "Two-sided matching with indifferences," Mimeo.

_____ _____ (2008) "What's the matter with tie breaking? Improving efficiency in school choice," *American Economic Review*, 98: 669–89.

Ergin, H. and Sönmez, T. (2006) "Games of school choice under the Boston mechanism," *Journal of Public Economics*, 90: 215–37.

Featherstone, C. and Niederle, M. (2008) "Manipulation in school choice mechanisms," Mimeo.

Gale, D. and Shapley, L. (1962) "College admissions and the stability of marriage," *American Mathematical Monthly*, 69: 9–15.

Goodnough, A. (2003) "Many are shut out in high school choice," *New York Times*, March 11, 2003, section B; column 4; Metropolitan Desk; p. 3.

Haeringer, G. and Klijn, F. (2009) "Constrained school choice," *Journal of Economic Theory*, 144: 1921–47.

Hafalir, I., Yenmez, B. and Yildirim, M. A. (2013) "Effective affirmative action with school choice", *Theoretical Economics*, 8(2): 325–363.

Hastings, J. and Weinstein, J. (2008) "Information, school choice, and academic achievement: evidence from two experiments," *Quarterly Journal of Economics*, November.

_____ Kane, T. and Staiger, D. (2008) "Heterogeneous preferences and the efficacy of public school choice", Mimeo.

Hatfield, J. W. and Milgrom, P. R. (2005) "Matching with contracts," *American Economic Review*, 95: 913–35.

Hoxby, C. M., Murarka, S., and Kang, J. (2009) "How New York City's charter schools affect achievement," August report., Second report in Series, New York City Charter Schools Evaluation Project.

Immorlica, N. and Mahdian, M. (2005) "Marriage, honesty, and stability," in *Proceedings of the Sixteenth Annual ACM -SIAM Symposium on Discrete Algorithms*, Society for Industrial and Applied Mathematics, pp. 53–62.

Kelso, A. S. and Crawford, V. P. (1982) "Job matchings, coalition formation, and gross substitutes," *Econometrica*, 50: 1483–504.

Kesten, O. (2010) "School choice with consent," *Quarterly Journal of Economics,* 125(3): 1297–348.

_____ and Ünver, M. U. (2010) "A theory of school choice lotteries: why ties should not be broken randomly," Mimeo.

Klijn, F., Pais, J. and Vorsatz, M. (2010) "Preference intensities and risk aversion in School Choice: a laboratory experiment," Working Paper.

Kojima, F. (2008) "Games of school choice under the Boston mechanism with general priority structures," *Social Choice Welfare*, 31: 357–65.

_____ (2010) "School choice: impossibilities for a rmative action," Working Paper.

_____ and Manea, M. (2010) "Axioms for deferred acceptance (2007)," *Econometrica*, 78: 633–53.

_____ and Pathak, P. A. (2009) "Incentives and stability in large two-sided matching markets," *American Economic Review*, 99:3: 608–27.

_____ and Ünver, U. (2010) "The 'Boston' school choice mechanism", Working Paper.

Ma, J. (1994) "Strategy-proofness and the strict core in a market with indivisibilities," *International Journal of Game Theory*, 23: 75–83.

Milgrom, P. (2009) "The promise and problems of (auction) market design," Nemmers Prize Lecture, Northwestern University.

Miralles, A. (2008) "School choice: the case for the Boston mechanism," Working Paper.

Pais, J. and Pinter, Á. (2007) "School choice and information: an experimental study on matching mechanisms," *Games and Economic Behavior*, 64(1): 303–328, forthcoming.

Pápai, S. (2000) "Strategyproof assignment by hierarchical exchange," *Econometrica*, 68: 1403–33.

Pathak, P. A. (2011) "The mechanism design approach to student assignment", *Annual Reviews of Economics*, 3: 513–536.

_____ and Sethuraman, J. (2011) "Lotteries in student assignment: an equivalence result," *Theoretical Economics*, 6: 1–17.

_____ and Sönmez, T. (2008) "Leveling the playing field: sincere and sophisticated players in the Boston mechanism," *American Economic Review*, 98: 1636–52.

_____ _____ (2013) "School admissions reform in Chicago and England: comparing mechanisms by their vulnerability to manipulation," *American Economic Review*, 103(1): 80–106

Pycia, M. and Ünver, U. (2010) "Incentive compatible allocation and exchange of discrete resources," Working Paper.

Roth, A. E. (1982a) "Incentive compatibility in a market with indivisibilities," *Economics Letters*, 9: 127–32.

_____ (1982b) "The economics of matching: stability and incentives," *Mathematics of Operations Research*, 7: 617–28.

_____ (1984) "The evolution of the labor market for medical interns and residents: a case study in game theory", *Journal of Political Economy*, 92: 991–1016.

_____ (1985) "The college admissions problem is not equivalent to the marriage problem," *Journal of Economic Theory*, 36(2): 277–88.

_____ (1991) "A natural experiment in the organization of entry-level labor markets: regional markets for new physicians and surgeons in the United Kingdom," *American Economic Review*, 81(3): 415–40.

_____ (2002) "The economist as engineer: game theory, experimentation, and computation as tools for design economics. Fisher-Schultz Lecture," *Econometrica*, 70: 1341–78.

_____ (2008) "Deferred acceptance algorithms: history, theory, practice, and open questions," *International Journal of Game Theory*, 36: 537–69.

_____ and Peranson, E. (1999) "The redesign of the matching market for American physicians: some engineering aspects of economic design," *American Economic Review*, 89(4): 748–80.

_____ and Sotomayor, M. (1990) *"Two-sided matching: a study in game-theoretic modeling and analysis,* econometric society Monograph Series, Cambridge University Press.

Shapley, L. and Scarf, H. (1974) "On cores and indivisibility," *Journal of Mathematical Economics*, 1: 23–8.

Sönmez, T. and Ünver, M. U. (2010) "Matching, allocation, and exchange of discrete resources," in J. Benhabib, A. Bisin, and M. Jackson (eds) *Handbook of Social Economics*, San Diego: North Holland Elsevier, 781–852.

Troyan, P. (2011) "Comparing school choice mechanisms by interim and ex-ante welfare," SIEPR Discussion Paper No. 10-021.

Westcamp, A. (2010) "An analysis of the German university admissions system," Working Paper.

CHAPTER 6

....

IMPROVING EFFICIENCY
IN SCHOOL CHOICE

....

AYTEK ERDIL AND HALUK ERGIN

INTRODUCTION

....

EDUCATIONAL authorities which assign children to schools automatically by the district they live in often fail to take into account the preferences of their families. Such systems overlook reallocations of seats which could Pareto improve welfare. Motivated by such concerns, several cities[1] started centralized school choice programs. Typically in these programs, each family submits a preference list of schools, including those outside of their district, and then a centralized mechanism assigns students to schools based on the preferences. The mechanisms initially adopted by school choice programs were ad hoc, and did not perform well in terms of efficiency, incentives, and/or stability. Abdulkadiroğlu and Sönmez (2003) brought these to light, which triggered an interest in the matching literature about further analysis and design of school choice mechanisms.

The most common practice in assigning scarce (i.e. popular) school seats is to use some exogenously fixed priority ranking of students. *Respecting priorities*, formalized by the familiar *stability* concept from two-sided matching, constrains which assignments are deemed acceptable, and therefore can have welfare consequences. While priorities put constraints on which outcomes are considered feasible, the need to have straightforward incentives for truthful revelation of preferences constrains the mechanisms available to the designer. This interplay between efficiency, stability, and strategy proofness is the subject of this chapter.[2]

---

[1] Including New York City, Boston, Cambridge, Charlotte, Columbus, Denver, Minneapolis, Seattle, and St. Petersburg-Tampa, in the US, and most major cities in the UK.

[2] The specific choice of the material included in this chapter is influenced heavily by our own work. The related literature goes well beyond the scope of this review. As a starting point to explore further, see chapter 5.

Background

A school choice problem consists of a set of students and a set of schools, where each school, x, has a quota, q_x, of seats. Each student has a preference ranking of schools and an "outside" option, which corresponds to remaining unassigned or going to a private school, and each school has a priority ranking of students. The school choice model is closely related to the college admissions model of Gale and Shapley (1962). The important difference between the two models is that in school choice, the priority rankings are determined by local (state or city) laws and education policies, and do not reflect the school preferences, whereas in the college admissions model these rankings correspond to college preferences.[3] As a consequence in the college admissions model, students' as well as colleges' preferences are taken into account in welfare considerations. On the other hand, in the school choice model, schools are treated as indivisible objects to be consumed by the students, and only student preferences constitute the welfare criteria.

Given a priority ranking for each school and a preference profile of the students, a matching *violates the priority of student i*, if there are a student j and a school x such that i prefers x to her current assignment, and j is assigned to x while he has less priority for school x than i. A matching is *stable* if (1) it does not violate any priorities, (2) every student weakly prefers his assigned seat to remaining unassigned, and (3) no student would rather be matched to a school which has empty seats. Stability has been a property of central interest in two-sided matching models. In addition to the theoretical plausibility of the notion, Roth (2002) draws from both empirical and experimental evidence to show how stability has been an important criterion for a successful clearinghouse in matching markets ranging from the entry-level labor market for new physicians in the US to college sorority rush. In the context of school choice, legal and political concerns appear to strongly favor stable mechanisms. For instance, if the priority of student i for school x is violated, then the family of student i has incentives to seek legal action against the school district for not assigning her a seat at school x, and the district authorities seem to be extremely averse to such violations of priorities.[4]

Gale and Shapley (1962) gave a constructive proof of the existence of a stable matching by describing a simple algorithm. This is known as the student-proposing *deferred acceptance (DA) algorithm*:

- At the first step, every student applies to her favorite acceptable school. For each school x, q_x applicants who have highest priority for x (all applicants if there are fewer than q_x) are placed on the hold list of x, and the others are rejected.

---

[3] There are certain exceptions like New York City, where a number of schools determine their own priority orders. See Abdulkadiroğlu and Sönmez (2003), Balinski and Sönmez (1999), and Ergin (2002) for a more detailed discussion of the relationship between the two models.

[4] For example, along these concerns, Boston officials decided to adopt a mechanism that always produces stable matchings at the expense of efficiency, rather than the top trading cycles mechanism, which would ensure efficiency, yet not stability.

- At step $t > 1$, those applicants who were rejected at step $t - 1$ apply to their next best acceptable schools. For each school x, the highest-priority q_x students among the new applicants and those in the hold list are placed on the new hold list, and the rest are rejected.

The algorithm terminates when every student is either on a hold list or has been rejected by every school that is acceptable to her. After this procedure ends, schools admit students on their hold lists.

Gale and Shapley (1962) show that, when preferences and priorities are strict, the DA algorithm yields a unique stable matching that is Pareto superior to any other stable matching from the viewpoint of the students. Hence the outcome of the student-proposing DA algorithm is also called the *student-optimal stable matching*, and the mechanism that associates the student-optimal stable matching to any school choice problem is known as the *student-optimal stable mechanism (SOSM)*.[5] Besides the fact that it gives the most efficient stable matching, another appealing feature of the SOSM when priorities are strict is that it is strategy proof; that is, no student has an incentive to misstate her true preference ranking over schools (Dubins and Freedman, 1981; Roth, 1982). Due to these desirable features, the DA algorithm has been adopted by the school choice programs of New York City (in 2003) and Boston (in 2005), in consultation with economists Abdulkadiroğlu, Pathak, Roth, and Sönmez.

Inefficiency of the deferred acceptance

Respecting priorities is not completely costless, as it imposes constraints on which assignments are allowed. Stability might rule out all Pareto-efficient assignments, and hence lead to an inefficient outcome. Example 1 illustrates the nature of this inefficiency.

Example 1. *Consider a school choice problem with three students 1, 2, 3, three schools x, y, z, each having one seat, and the following priority orders:*

$\succsim_x$	$\succsim_y$	$\succsim_z$
1	2	3
2	1	1
3	3	2

Now, suppose that the preferences of the students are:

R_1	R_2	R_3
y	z	y
x	y	z
z	x	x

---

[5] The SOSM played a key role in the redesign of the US hospital-intern market in 1998. See Roth and Peranson (1999), and Roth (2003).

The student-optimal stable matching for this instance of the problem is

$$\mu = \begin{pmatrix} 1\ 2\ 3 \\ x\ y\ z \end{pmatrix}$$

However, if students 2 and 3 could swap their seats, they would both be better off, and we would get the matching

$$\nu = \begin{pmatrix} 1\ 2\ 3 \\ x\ z\ y \end{pmatrix}$$

which Pareto dominates μ. The fact that student 1 prefers school y to her assigned school and that she has higher priority for school y than student 3 means the latter cannot be assigned this school. Thus, we end up with Pareto inefficiency due to having to respect priorities.

Whether we will actually observe this tension between stability and efficiency depends on the particular realization of preferences. Abdulkadiroğlu et al. (2009) find empirical evidence in the data from the main round of the New York City high-school admissions process in 2006–07. If stability constraints were ignored to let students "exchange" their seats after the match is announced, one could find a Pareto improvement which makes about 5,800 students (around 7.4% of the eight-graders requesting a high-school seat) better off. A complete characterization of priority structures for which the student-optimal stable matchings would never suffer from Pareto inefficiency is given by Ergin (2002) in the case of strict priorities, and by Erdil and Ehlers (2010) in general.

Inconsistency

A second issue that comes up in school choice programs has to do with participants *appealing* after the match is announced. For example, in 2006/07, some 80,000 appeals were lodged in the UK.[6] A standard notion of *consistency* would require that when the appealing individuals, and the school seats they have received, are considered a smaller assignment problem with the preferences and priorities inherited from the bigger problem, the assignment rule applied to this smaller problem should yield the same assignments as in the bigger problem.

Let us turn to example 1 again to see whether the assignment mechanism is consistent. The outcome of the DA algorithm is $\mu = (1x, 2y, 3z)$. Now, take the subproblem in which we consider only students 2 and 3, and their assigned schools, y and z. The priorities and preferences are inherited from the original problem, so when we apply the DA algorithm to the reduced problem we get

---

[6] See Rooney (2009). In addition to several guidebooks on appeals, there are dozens of professional consultancy firms and websites advising, in exchange for fees as high as £2,000, on how to appeal.

$$\begin{array}{cc|cc} \succsim_y & \succsim_z \\ \hline 2 & 3 \\ 3 & 2 \end{array} \qquad \begin{array}{c|c} R_2 & R_3 \\ \hline z & y \\ y & z \end{array} \quad \Rightarrow \quad \begin{pmatrix} 2 & 3 \\ z & y \end{pmatrix}$$

which is different from the outcome $(2y, 3z)$ inherited from the larger problem. Hence, the DA mechanism is inconsistent.

Constrained inefficiency when there are ties

The DA algorithm, as described above, requires that both the preference orders and priority orders be strict for it to be deterministic and single valued. This is because whenever a student proposes, she chooses her next best school, and a school rejects the lowest-priority students among those who applied. Obviously, indifference classes would create ambiguities in those choices. In the context of school choice, it might be reasonable to assume that the students have strict preferences, but school priority orders are typically determined according to criteria that do not provide a strict ordering of all the students. Instead, school priorities are weak orderings with quite large indifference classes. For instance, in Boston there are mainly four indifference classes for each school in the following order: (1) the students who have siblings at that school (sibling) and are in the reference area of the school (walk zone), (2) sibling, (3) walk zone, and (4) all other students.[7] Common practice in these cases is to exogenously fix an ordering of the students, chosen randomly, and break all the indifference classes according to this fixed strict ordering. Then one can apply the DA algorithm to obtain the student-optimal stable matching with respect to the strict priority profile derived from the original one. Tie-breaking the enlarges the set of stability constraints that need to be satisfied, so the outcome would be stable with respect to the original priority structure too. However, these extra constraints may be costly (example 2).

Example 2. *Consider a school choice problem with three students 1, 2, 3, three schools* x, y, z, *each having one seat, and the following priority orders*

$$\begin{array}{c|c|c} \succsim_x & \succsim_y & \succsim_z \\ \hline 1 & 2 & 3 \\ 2,3 & 1,3 & 1,2 \end{array}$$

If the ties in the priority orders are broken, favoring 1 over 2 over 3, to obtain the strict priority structure $\succsim'$, we find ourselves back in example 1:

---

[7] There are also students who have a guaranteed priority to a given school. For a complete description, see Abdulkadiroğlu et al. (2006) or "Introducing the Boston public schools 2007: a guide for parents and students," available at <http://www.boston.k12.ma.us/schools/assign.asp> (accessed September 12, 2007).

R_1	R_2	R_3		$\succsim'_x$	$\succsim'_y$	$\succsim'_z$
y	z	y		1	2	3
x	y	z		2	1	1
z	x	x		3	3	2

We already observed in example 1 that the student-optimal stable matching for the preference profile R and the strict priority structure $\succsim'$ is $\mu = (1x, 2y, 3z)$, which is Pareto dominated by $\nu = (1x, 2z, 3y)$. However, note that while ν violates the derived priorities $\succsim'$, it actually respects the original priorities. Hence, under the original priority structure with ties, μ is not constrained efficient, and the arbitrariness of the tie breaking can lead to even constrained inefficiency.

The *stable improvement cycles* procedure introduced in Erdil and Ergin (2008) is an effective way to identify the inefficiency that is due to the arbitrariness of the tie breaking. By taking us from the outcome of DA with arbitrary tie breaking to a student-optimal stable matching, this algorithm allows one to measure the extent of the illustrated welfare loss. Thus, Abdulkadiroğlu et al. (2009) find in the data from the New York high-school match that the stable improvement cycles could make about 1,500 students (around 1.9% of the applicants) better off without hurting others.

Strategy-proof improvement

It is well known that when the priorities are strict, the deferred acceptance mechanism is strategy proof (Dubins and Freedman, 1981; Roth, 1982). On the other hand, we have already seen that it may not be efficient. Secondly, if there are ties in priorities, arbitrariness of a tie-breaking rule can add further inefficiency, i.e., can lead to even constrained inefficient outcomes. Alternative mechanisms can Pareto improve these mechanisms, either by relaxing stability (Kesten, 2010), or by finding stability-preserving improvement in the case of constrained inefficiency (Erdil and Ergin, 2008). However, the additional stage of Pareto improvement may introduce incentives for misreporting preferences (Abdulkadiroğlu et al., 2009). Thus, strategy proofness might limit the extent of Pareto improvement over the inefficient mechanisms. A strategy-proof mechanism is on the "efficient frontier of strategy-proof mechanisms" if it is not dominated by another strategy-proof mechanism. While a randomization over such mechanisms preserves strategy proofness, the random mechanism might fail to be on that efficient frontier, i.e., might admit strategy-proof improvement (Erdil, 2011).

After introducing the model below, we revisit each issue, and present formally the aforementioned results on the extent of these issues, potential solutions, and their limitations. We refer the reader to the cited papers for the proofs.

THE MODEL

Let N denote a finite set of students and X a finite set of schools. Let $q_x \geq 1$ denote the number of available seats in school x. Throughout we will maintain the assumption that student preferences are strict: A *preference profile* is a vector of linear orders (complete, transitive, and antisymmetric relations) $R = (R_i)_{i \in N}$ where R_i denotes the preference of student i over $X \cup \{i\}$. Being assigned to oneself is interpreted as not being assigned to any school. Let P_i denote the asymmetric part of R_i. A *matching* is a function $\mu: N \to X \cup N$ satisfying: (1) $\forall i \in N : \mu(i) \in X \cup \{i\}$, and (2) $\forall x \in X : |\mu^{-1}(x)| \leq q_x$. A *rule* is a function that associates a non-empty set of matchings with every preference profile, whereas a *mechanism* f, is a singleton-valued rule. A *random mechanism* F, associates a probability distribution over matchings with every preference profile R.

A *priority structure* is a profile of weak orders (complete and transitive relations) $\succsim = (\succsim_x)_{x \in X}$ where for each $x \in X$, $\succsim_x$ ranks students with respect to their priority for x. Let $\succ_x$ denote the asymmetric part of $\succsim_x$. We say that $\succsim$ is *strict* if, for any $x \in X$, $\succsim_x$ is antisymmetric. Let $\mathcal{T}(\succsim)$ denote the set of strict priority profiles $\succsim'$ obtained by breaking the ties in $\succsim$.[8] Given $\succsim$ and R, the matching μ *violates the priority of i for x* if there is a student j such that j is assigned to x whereas i both desires x and has strictly higher priority for it, i.e., $\mu(j) = x$, $xP_i\mu(i)$, and $i \succ_x j$. The matching μ is *stable* if (1) it does not violate any priorities, (2) $\mu(i)R_i i$ for any i, and (3) there do not exist i and x such that $xP_i\mu(i)$ and $q_x > |\mu^{-1}(x)|$. Let $S^{\succsim}$ denote the *stable rule*, i.e., the rule that associates to each R the set of stable matchings with respect to $\succsim$ and R.

Given R, the matching μ' *Pareto dominates* the matching μ if $\mu'(i)R_i\mu(i)$ for every $i \in N$, and $\mu'(j)P_j\mu(j)$ for some $j \in N$. Given $\succsim$ and R, the matching μ is *constrained efficient* (or *student-optimal stable*) if (1) $\mu \in S^{\succsim}(R)$, and (2) μ is not Pareto dominated by any other $\mu' \in S^{\succsim}(R)$. Let $f^{\succsim}$ denote the *student-optimal stable rule (SOSR)*, i.e., the rule that associates to each R the set of constrained efficient matchings with respect to $\succsim$ and R. Given $\succsim$, a rule f is *constrained efficient* if, for any R, $f(R) \subseteq f^{\succsim}(R)$.

Theorem 1 (*Gale and Shapley, 1962*). *For any strict $\succsim$ and R, $f^{\succsim}(R)$ consists exactly of the matching given by the DA algorithm.*

When the priorities have ties, the DA algorithm can still be run by arbitrarily breaking the ties. The following are well known facts about how tie breaking affects the stable and the student-optimal stable rules.

Observation 1. $S^{\succsim} = \bigcup_{\succsim' \in \mathcal{T}(\succsim)} S^{\succsim'}$.

Observation 2. $f^{\succsim} \subseteq \bigcup_{\succsim' \in \mathcal{T}(\succsim)} f^{\succsim'}$.

---

[8] Formally, $\mathcal{T}(\succsim)$ is the set of strict priority structures $\succsim'$ such that $i \succ_x j$ implies $i \succ'_x j$ for all $x \in X$ and $i, j \in N$.

In other words: (1) any matching stable with respect to $\succsim$ is stable with respect to some tie breaking, and a matching stable with respect to an arbitrary tie breaking is stable with respect to the original priorities; (2) any student-optimal stable matching is student-optimal stable with respect to some tie breaking. The fact that the second inclusion might be proper means arbitrary tie breaking may lead to constrained inefficiency.

A CONSTRAINED EFFICIENT SOLUTION

Example 1 showed that arbitrarily breaking the ties in priorities and running the DA algorithm does not necessarily lead to a constrained efficient outcome. Motivated by this welfare loss, Erdil and Ergin (2008) introduce a particular Pareto improvement over a given stable matching.

Stable improvement cycles

Let μ be a stable matching for some fixed $\succsim$ and R. We will say that a student i *desires* school x if she prefers x to her assignment at μ, that is $x P_i \mu(i)$. For each school x, let D_x denote the set of highest $\succsim_x$-priority students among those who desire x. We will suppress the dependence of D_x on μ.

Definition 1. A stable improvement cycle *consists of distinct students* $i_1, \ldots, i_n \equiv i_0$ ($n \geq 2$) *such that*

    (1) $\mu(i_\ell) \in X$ *(each student in the cycle is assigned to a school),*
    (2) i_ℓ *desires* $\mu(i_{\ell+1})$, *and*
    (3) $i_\ell \in D_{\mu(i_{\ell+1})}$,

for any $\ell = 0, \ldots, n - 1$.

Given a stable improvement cycle define a new matching μ' by:

$$\mu'(j) = \begin{cases} \mu(j) & \text{if } j \notin \{i_1, \ldots, i_n\} \\ \mu(i_{\ell+1}) & \text{if } j = i_\ell \end{cases}$$

Note that the matching μ' continues to be stable and it Pareto dominates μ. The following result sheds light on the nature of Pareto-comparable stable matchings:

Theorem 2 (*Erdil and Ergin, 2008*). *Fix* $\succsim$ *and* R, *and let* μ *be a stable matching. If* μ *is Pareto dominated by another stable matching* v, *then it admits a stable improvement cycle.*[9]

---

[9] We could actually "squeeze in" a stable improvement cycle between any two Pareto-ranked stable matchings. Formally, we could guarantee that the new stable matching μ' obtained from μ by applying the improvement cycle lies weakly below v in a Pareto sense.

If a stable matching is not constrained efficient, then there must exist a Pareto improvement which is still stable. Theorem 2 says in order to find such a Pareto improvement, it is enough to look for a stable improvement cycle. Successive application of this result gives what Erdil and Ergin (2008) call the *stable improvement cycles algorithm*.

- **Step 0.** *Select* a strict priority structure $\succsim'$ from $\mathcal{T}(\succsim)$. Run the DA algorithm and obtain a temporary matching μ^0.
- **Step t ≥ 1.**

  (*t.a*) Given μ^{t-1}, let the schools stand for the vertices of a directed graph, where for each pair of schools x and y, there is an edge $x \longrightarrow y$ if and only if there is a student i who is matched to x under μ^{t-1}, and $i \in D_y$.

  (*t.b*) If there are any cycles in this directed graph, *select* one. For each edge $x \longrightarrow y$ on this cycle *select* a student $i \in D_y$ with $\mu^{t-1}(i) = x$. Carry out this stable improvement cycle to obtain μ^t, and go to step $(t + 1.a)$. If there is no such cycle, then return μ^{t-1} as the outcome of the algorithm.

In the above description, it is left open how the procedure should select $\succsim'$ in step 0, and how it should select the cycle and the student in step (*t.b*). Therefore one can think of the above description as corresponding to a *class* of algorithms, where an algorithm is determined only after we fully specify how to act when confronted with multiplicity. One can imagine these selections to be random or dependent on the earlier selections. Let $F^{\succsim}$ denote the random mechanism induced by the above algorithm when the selections are made independently and randomly with equal probabilities each time the algorithm faces a multiplicity. Remember that, given $\succsim, R$, and $\mu \in f^{\succsim}(R)$, there is a tie-breaking $\succsim' \in \mathcal{T}(\succsim)$ such that the DA algorithm applied to $(R, \succsim')$ returns μ. Since each tie breaking has a positive probability of being selected at step 0 of the algorithm corresponding to $F^{\succsim}$, $F^{\succsim}(R)$ gives a positive probability to every constrained efficient matching.

Note that observation 2 also yields an algorithm to find a student-optimal stable matching. Namely, one could apply the DA algorithm to all possible tie breakings of the given priority structure, record the outcomes, and Pareto compare them to find a student-optimal stable matching. However, even with a single indifference class of only 100 students, this would amount to running the DA algorithm more than 10^{90} times, a computationally infeasible task. From a practical perspective, the value of the stable improvement cycles algorithm comes from its remarkably small computational complexity.[10]

Stable improvement cycles are closely related to Gale's top trading cycles, originally introduced in Shapley and Scarf (1974), and later studied in detail by Pápai (2000) and

---

[10] In addition to the DA algorithm used in practice, it involves a repetition of cycle search in a directed graph. The latter is known to be of complexity $O(|V| + |E|)$, where V is the set of vertices and E the set of edges (Cormen et al., 2003). This obviously is very fast; the question is then how many times one has to repeat the cycle search. Notice that with every cycle, at least two students improve, therefore each cycle brings at least two *moves up* with respect to the students' preferences. Since there are $|N|$ students and the student preferences involve $|X|$ schools, there could be at most $|N|(|X| - 1)$ moves up. Therefore cycle search has to be repeated at most $\frac{1}{2}|N|(|X| - 1)$ times.

Abdulkadiroğlu and Sönmez (2003). At a matching μ, a *top trading cycle* consists of students $i_1, \ldots, i_n \equiv i_0$ ($n \geq 2$) such that conditions (1) and (2) in our definition of a stable improvement cycle are satisfied, and additionally $\mu(i_{\ell+1})$ is student i_ℓ's top ranked school for $\ell = 0, \ldots, n-1$. Suppose that matching μ is stable to start with. There are two reasons for which we could not make use of top trading cycles in the above construction. First, since condition (3) is not required in a top trading cycle, there is no guarantee that the matching μ' obtained after executing the top trading cycle will continue to be stable. Secondly, top trading cycles are *too demanding* for our purposes, since even when there exist Pareto-improving trading cycles which preserve stability, there may not exist such a cycle where all participating students receive their top choices.

Strategic properties

A mechanism, f, is *strategy proof* if for any preference profile R, student i and R_i', we have $f_i(R_i, R_{-i}) R_i f_i(R_i', R_{-i})$. We know from Dubins and Freedman (1981) and Roth (1982) that in the case of strict priorities, the constrained efficient mechanism, $f^{\succeq}$, is strategy proof. When we allow the priority orders to be weak, the constrained efficient set is not necessarily a singleton. In this case, it is natural to ask whether there is a mechanism, $f \subseteq f^{\succeq}$, that is strategy proof. The following example gives a negative answer to this question.

Example 3. *Consider a school choice problem with three schools x, y, z, each having one seat, three students 1, 2, 3 who find all schools acceptable, and*

R_1	R_2	R_3		$\succeq_x$	$\succeq_y$	$\succeq_z$
y	y	x		1	3	3
z	z	y		2	1, 2	2
x	x	z		3		1

The constrained efficient set consists of only two matchings:

$$f^{\succeq}(R) = \left\{ \begin{pmatrix} 1\ 2\ 3 \\ y\ z\ x \end{pmatrix}, \begin{pmatrix} 1\ 2\ 3 \\ z\ y\ x \end{pmatrix} \right\}$$

Consider the following manipulations

R_1'	R_2'
y	y
x	x
z	z

If student 1 announces R_1' when the other students announce truthfully, then

$$f^{\succeq}(R_1', R_{-1}) = \left\{ \begin{pmatrix} 1\ 2\ 3 \\ y\ z\ x \end{pmatrix} \right\}$$

Similarly, if student 2 announces R'_2 when the other students announce truthfully, then

$$f^{\succsim}(R'_2, R_{-2}) = \left\{ \begin{pmatrix} 1\ 2\ 3 \\ z\ y\ x \end{pmatrix} \right\}$$

Consider any mechanism $f \subseteq f^{\succsim}$. For the preference profile R, f has to select one of the matchings $(1y, 2z, 3x)$ or $(1z, 2y, 3x)$. If it selects $(1y, 2z, 3x)$, then student 2 has an incentive to misrepresent her preference and submit R'_2. On the other hand, if it selects $(1z, 2y, 3x)$, then student 1 has an incentive to misrepresent her preference and submit R'_1. Therefore f is not strategy proof.

For each student i, our model specifies only an ordinal ranking R_i over $X \cup \{i\}$. Assuming that the student is an expected utility maximizer, we need to know her cardinal (vNM) utility function $u_i: X \cup \{i\} \to \mathbb{R}$ to fully specify her risk preferences. Given two probability distributions p and q over $X \cup \{i\}$, p [strictly] first-order stochastically dominates q with respect to R_i if

$$\sum_{y \in X \cup \{i\}\,:\,yR_iz} p(y) \geq \sum_{y \in X \cup \{i\}\,:\,yR_iz} q(y)$$

for all $z \in X \cup \{i\}$ [with strict inequality for some $z \in X \cup \{i\}$]. It is a standard fact that p [strictly] first-order stochastically dominates q with respect to R_i if and only if for any vNM utility function u_i that gives the same ordinal ranking as R_i, the expected utility of p is [strictly] weakly more than the expected utility of q. Given a random mechanism F, a preference profile R, and a student i, let $F_i(R)$ denote the random allocation of i with respect to $F(R)$. The argument in example 3 can be adapted to conclude that the above impossibility persists even for random mechanisms.

Theorem 3 (Erdil and Ergin, 2008). *Let F be any mechanism which gives a constrained efficient allocation with probability one in each preference profile. Then there exist R, i, and R'_i, such that $F_i(R'_i, R_{-i})$ strictly first-order stochastically dominates $F_i(R_i, R_{-i})$ with respect to R_i.*

Hence strategy proofness and constrained efficiency are incompatible. In the example above, the strategic manipulation was aimed at ruling out the less preferred constrained efficient allocation, and consequently singling out the preferred one. Could a student manipulate her submitted ranking to induce a new matching, where she is assigned to a school more preferable than every school she could possibly be assigned to under her truthful statement? It turns out that she cannot achieve a school better than her best possibility in the constrained efficient set.

Remember the random mechanism $F^{\succsim}$. Even when a student has perfect knowledge of the priority structure and the preferences of all students, since the algorithm involves random selections, there is uncertainty to what outcome will be returned. The computation of the likelihood of a particular constrained efficient solution being returned is highly involved, and when faced with such uncertainty, what would an "optimist" do? Someone who tends to base her actions on her best assignment possible among the student-optimal solutions would consider manipulating the system only if such strategic announcement brought her a school more preferable than any school she could be assigned under her truthful revelation. Moreover, if for a particular preference profile there is only one constrained efficient matching, then no student would have any incentive to unilaterally misstate her preferences.

We have seen in example 3 that every selection from the SOSR was manipulable, but student 1 needed significant information regarding the preferences of students 2 and 3 in order to be able to correctly evaluate the consequences of her switching schools x and z in her preference list.[11] One may ask if a student with low information about the preferences and priorities of others would find it profitable to employ such manipulation.

As a benchmark for a low-information environment, consider the framework of Roth and Rothblum (1999).[12] A student's beliefs about two schools x and y are *symmetric* if when one changes the roles of x and y in the random variable interpreted as her beliefs on $(\succsim, R_{-i})$, the distribution of the random variable does not change. When this is the case, under the random mechanism $F^{\succsim}$, it is *never* profitable for a student to misstate her preferences by switching those two schools in her preference. In the firms–workers model of Roth and Rothblum (1999) with strict preferences on both sides, it was found that under the firm-proposing DA algorithm it may be profitable for a worker to submit a truncated preference, where a *truncation* of a preference list R_i containing r acceptable firms is a list R_i' containing $r' \leq r$ acceptable firms such that the r' firms in R_i' are the top r' in R_i with the same order. Since we are analysing the SOSR, with strict priorities, the truthful statement of a student would be her dominant strategy, ruling out any manipulation including truncation strategies. It turns out that, in the case of weak priorities too, truncation strategies are *never* profitable for students, independently of their beliefs about the preferences and priorities of others. However, another set of strategies might emerge, even when the student has almost no information allowing her to distinguish between others' priorities and preferences. Formally, an *extension* of a preference list R_i containing r acceptable schools is a list R_i' containing $r' \geq r$ acceptable schools such that the r elements of R_i are the top r in R_i' with the same order. Under $F^{\succsim}$, manipulation by announcing an extension strategy may be profitable even under symmetric information, as illustrated in example 4.

Example 4. *Consider three students 1, 2, and 3, and two schools x and y each having one seat. Suppose that every student has equal priority for all schools. Student 1's vNM*

<hr/>

[11] It is possible that a student may have an incentive to manipulate mechanism $F^{\succsim}$ under an incomplete-information environment, that is, without having detailed information about the others' preferences. An example is when certain schools are commonly recognized as being popular, i.e., ex ante more likely to be highly ranked by the students. In that case a student i who has high priority at a popular school x may find it profitable to lift school x in her submitted ranking. The rationale is that she may gain if she is temporarily assigned to x at step 0 of the algorithm and if she is able to "trade" x at subsequent stages of the algorithm. Such a manipulation would be profitable only if student i does not rank x very highly but has sufficient confidence in the popularity of x. Hence one would expect the ex ante likelihood of this manipulation to be low.

[12] See Erdil and Ergin (2008) for a detailed analysis of strategic behavior under low information.

preference is given by $u_1(y) = 1$, $u_1(1) = 0$, and $u_1(x) = -\epsilon$ for some $\epsilon > 0$, hence her true ordinal preference R_1 is such that $yP_1 1 P_1 x$. Her beliefs over $\succsim_x$, $\succsim_y$, R_2, and R_3 are independent and uniform over the respective domains; in particular, they are symmetric for x and y. Suppose that the random mechanism $F^{\succsim}$ is being used and that student 1 is contemplating to manipulate her true ranking and announcing the extension R'_1 such that $yP'_1 x P'_1 1$.

Recall the algorithm corresponding to our random mechanism and fix a realization of $\succsim$, R_{-1}, and $\succsim' \in \mathcal{T}(\succsim)$. Conditional on $(R_{-1}, \succsim')$, if student 1 submits R_1, and the algorithm assigns her to y, then this assignment must have been reached in step 0 as a result of the DA algorithm being applied to $(R_1, R_{-1}, \succsim')$. In this case, if she submits R'_1, the algorithm would again assign her to y in step 0 as a result of the DA algorithm being applied to $(R'_1, R_{-1}, \succsim')$. Therefore student 1 can lose by announcing R'_1 instead of R_1, only if the realization $(R_{-1}, \succsim')$ is such that she is left unassigned if she announces R_1. Before the realization of $(R_{-1}, \succsim')$, this expected loss is bounded above by ϵ from the point of view of student 1.

On the other hand, if the realization of $(R_{-1}, \succsim')$ is such that

R_2	R_3
x	y
y	x
2	3

$\succsim'_x$	$\succsim'_y$
1	2
3	3
2	1

then student 1 is left unassigned if she submits R_1 and she is assigned to y if she submits R'_1. Let $p > 0$ denote the probability of the above realization. If the student's risk preferences are such that $\epsilon < p$, then she will prefer to announce R'_1 when her true ordinal ranking is R_1.

The only profitable strategic manipulation in a low-information environment is to lengthen one's list. If, in addition, it is common knowledge that all schools are acceptable for all students, then being truthful is a Bayesian Nash equilibrium of the preference revelation game under the random mechanism $F^{\succsim}$.

Unconstrained efficiency and consistency

The efficiency and consistency of the SOSR are intimately related. What it means for a rule to be efficient is more or less standard, and we already gave the formal definition earlier. So, let us now discuss *consistency* in a bit more detail.

Many school choice programs allow parents to appeal the outcome of the match. The appeals process, which can be considered as a second round, can be very costly, as mentioned in footnote 6. Ideally, an assignment rule would not lead the participants to challenge the outcome and go on to a second round of matching with hopes of getting a better match. In other words, it is desirable for a rule to be robust to non-simultaneous allocation of school seats. Example 1 above also points to a tension between respecting priorities and such robustness. Suppose SOSR is applied in two rounds, and 1's final allocation is determined in the first round. Since $f^{\succsim}(R) = (1x, 2y, 3z)$, student 1 must be assigned to x. If in the second round, the SOSR is applied to the reduced problem to which 1 no longer belongs, then the assignment $(2z, 3y)$ is selected. The latter is not only inconsistent with the assignment selected in the first round, but it also violates the priority of 1 for y. Which priority structures guarantee that the SOSR is robust to non-simultaneous assignment of schools?

This property is known as *consistency*. [13] For any non-empty subset of students $N' \subseteq N$, a preference profile R, a priority structure $\succsim$, and an assignment μ, let $R_{N'} = (R_i)_{i \in N'}$, $\succsim |_{N'} = (\succsim_x |_{N'})_{x \in X}$, and $\mu_{|N'} : N' \to X \cup N'$, respectively denote the restrictions of the preference profile, the priority structure, and the assignment to N'. Given a pair $(\succsim, q)$, a non-empty subset of students $N' \subseteq N$, a subset of $q' = (q'_x)_{x \in X}$ seats of the schools, where $q'_x \leq q_x$ for each school x, and a preference profile R, consider the set of constrained efficient assignments for the smaller assignment problem $\mathcal{E}' = (N', q', R_{N'})$ with respect to $\succsim_{|N'}$. Let us call the map that associates the set of constrained efficient assignments with any such smaller problem $\mathcal{E}' = (N', q', R_{N'})$, the *extended SOSR* associated with $\succsim$ and denote it by $\bar{f}^{\succsim}$.

Given an assignment problem $\mathcal{E} = (N, q, R)$, an assignment μ for $\mathcal{E}$ and a non-empty subset of students $N' \subseteq N$, the *reduced problem* $r^{\mu}_{N'}(\mathcal{E})$ *of* $\mathcal{E}$ *with respect to* μ *and* N' is the smaller problem consisting of students N' and the remaining school seats after students in $N \setminus N'$ have left with their school seats under μ, i.e., $r^{\mu}_{N'}(\mathcal{E}) = (N', q', R_{N'})$, where $q'_x = q_x - |\mu^{-1}(x) \setminus N'|$ for each $x \in X$. Consistency requires that once an assignment is determined and some students have been assigned to their seats before the others, the rule should not change the assignments of the remaining students in the reduced problem involving the remaining students and seats. Formally, $\bar{f}^{\succsim}$ is *consistent* if, for any problem $\mathcal{E} = (N, q, R)$, one has $\mu_{|N'} \in \bar{f}^{\succsim}\left(r^{\mu}_{N'}(\mathcal{E})\right)$ for all $\mu \in \bar{f}^{\succsim}(\mathcal{E})$. Consistent rules are coherent in their outcomes for problems involving different groups of students and robust to non-simultaneous assignment of seats.

When priorities are assumed to be strict, Ergin (2002) gives a concise characterization of priority structures for which $f^{\succsim}$ is efficient. In fact, he shows that the same "no-cycle property" also characterizes the priority structures $\succsim$ for which $f^{\succsim}$ is consistent. When $q = 1$, a *cycle* of $\succsim$ is defined as: $i \succ_x j \succ_x k \succ_y i$, where i, j, k are distinct students, and x, y are distinct schools. When there are multiple seats in some schools, we need a *scarcity condition*, which requires that there exist (possibly empty) disjoint sets of students $N_x, N_y \subseteq N \setminus \{i, j, k\}$ such that the students in N_x have strictly higher $\succsim_x$-priority than j, the students in N_y have strictly higher $\succsim_y$-priority than i, $|N_x| = q_x - 1$, and $|N_y| = q_y - 1$. If $\succsim$ has no cycles, it is called *acyclic*. Ergin (2002) shows that for any

---

[13] See Thomson (2006) for a survey of the consistency principle in allocation problems. In indivisible-object assignment, see Ergin (2000), and Ehlers and Klaus (2007) for a discussion of consistency principle for deterministic rules, and Chambers (2004) for when randomizations are allowed.

strict priority structure $\succsim$, the following are equivalent: (1) $f^{\succsim}$ is efficient, (2) $f^{\succsim}$ is group strategy proof, (3) $\bar{f}^{\succsim}$ is consistent, (4) $\succsim$ is acyclic.

When priorities are weak, acyclicity is still a necessary condition for efficiency and consistency separately. However, it is no longer sufficient. For instance, in example 2, each priority ranking has only two indifference classes, whereas the cycle condition requires $i \succ_x j \succ_x k$ for some school x. Hence $\succsim$ is acyclic, yet as we have seen, the (extended) SOSR is not efficient (consistent). Therefore one needs a stronger condition on $\succsim$ than acyclicity in order to ensure that $f^{\succsim}$ is efficient ($\bar{f}^{\succsim}$ is consistent). For every $\succsim, x \in X$ and $\ell \in N$, let $W_x(\ell) = \{m \in N \mid m \succsim_x \ell\}$ denote the set of students who have weakly higher $\succsim_x$-priority than ℓ.

Definition 2. *A weak cycle of $\succsim$ comprises distinct $x, y \in X$ and $i, j, k \in N$ such that the following are satisfied:*

- *Cycle condition: $i \succsim_x j \succ_x k \succsim_y i$,*
- *Scarcity condition: There exist (possibly empty) disjoint sets of students $N_x, N_y \subseteq N \setminus \{i, j, k\}$ such that $N_x \subseteq W_x(j)$, $N_y \subseteq W_y(i)$, $|N_x| = q_x - 1$ and $|N_y| = q_y - 1$.*

A priority structure is *strongly acyclic* if it has no weak cycle.

Now we are ready to express the characterization:

Theorem 4 (*Ehlers and Erdil, 2010*). *Given any priority structure $\succsim$, the following are equivalent:*

(1) *$f^{\succsim}$ is efficient,*
(2) *$\bar{f}^{\succsim}$ is consistent,*
(3) *$\succsim$ is strongly acyclic.*

Hence, efficiency and consistency of the (extended) SOSR go together, and can be determined simply by checking whether the priority structure has strong cycles or not. This easy-to-verify criterion can serve as a guide to the designer who sets the priorities in terms of ensuring ex post efficiency. One consequence of the above theorem is that strong acyclicity of the priority structure ensures efficiency of the stable improvement cycles algorithm.

Note that if $\succsim$ is strict, a weak cycle is a cycle, and acyclicity is equivalent to strong acyclicity, therefore theorem 4 implies Ergin's theorem. Applying Ergin's theorem to strict resolutions of $\succsim$, and checking whether they are acyclic or not is also not the correct criterion for efficiency of the SOSR. For instance, suppose that there are three students i, j, k, and three schools x, y, z with priorities as:

$\succsim_x$	$\succsim_y$	$\succsim_z$
i	j	k
j, k	i, k	i, j

It is straightforward to verify that $\succsim$ is strongly acyclic. Hence, by theorem 4, $f^{\succsim}$ is an efficient rule. However, any tie breaking would lead to a cyclic strict priority structure, and accordingly the SOSR associated with any fixed tie-breaking rule will necessarily be an inefficient rule. For example, say the tie breaking in $\succsim_x$ favors j over k, so the derived priorities imply $i \succ'_x j \succ'_x k$ and $k \succ'_z i$. The "$\succ'_x$ priority j has over k" in this new priority structure $\succsim'$ constrains further the mechanism's flexibility to assign x to k, and can lead to inefficiency. Thus, the artificial priorities generated in tie breaking created too many extra constraints. This is in contrast with the fact that, here, the stable improvement cycles algorithm would always return an efficient assignment.

Strategy-proof Pareto improvement

Theorem 4 points out the tension between stability and efficiency for a large class of priority structures, namely those which fail to be strongly acyclic. Theorem 3, on the other hand, highlights the tension between constrained efficiency and strategy proofness. One way to remedy the inefficiency associated with cyclical priority structures could be to coarsen them to remove the cycles. Ehlers and Erdil (2010) discuss one means of coarsening, but of course such redesigning of priorities may end up disregarding some of the original objectives of prioritizing some students over others. Another approach to Pareto improve the student-optimal stable matchings is to relax the stability notion in a way which does not hurt any student compared with the SOSR. For example, when the student-optimal stable matchings are not efficient, Kesten (2010) suggests allowing violations of priorities as long as everyone is guaranteed a school which is at least as good as what they would have under the original stable mechanism. Such Pareto improvement over the SOSM can be achieved in many ways. For instance, we can simply implement the *core* of the market in which every student is endowed with her optimal stable school seat.[14] In Kesten's model, in addition to expressing their preference rankings, students can declare whether they consent to the violation of their priorities as long as such consent does not harm themselves. Now, if the student-optimal stable matching is not stable, one can look for Pareto improvements which would violate the priorities of only those who had given consent. This attractive proposal has one drawback though. It skews the incentives associated with preference revelation. Kesten (2010) shows that there is no strategy-proof and Pareto-efficient mechanism which always returns a matching that Pareto dominates the student-optimal stable matching.

More generally, say a mechanism g *dominates* another mechanism f if for every preference profile R, the matching $g(R)$ weakly Pareto dominates $f(R)$, and the domination is strict for at least one preference profile. Abdulkadiroğlu et al. (2009) prove that if the priorities are strict, there is no strategy-proof mechanism which Pareto dominates the DA mechanism. In particular, it is impossible to take advantage of consent in a strategy-proof way. Another implication of this impossibility is regarding the constrained inefficiency of the deferred acceptance with *arbitrary* tie breaking. Once the ties in priorities

[14] See Kesten (2010) for a more attractive approach which has the attractive property of keeping the number of eventual violations of priorities to a minimum.

are broken in some exogenous way, what we are implementing is the DA mechanism with strict priorities. Therefore, if the stable improvement cycles algorithm begins with some fixed tie-breaking rule, it will not be strategy proof. On the the other hand, this does not imply the impossibility result given in theorem 3. This is because the random mechanism $F^{\succsim}$ does not necessarily dominate the DA with some particular tie-breaking rule.

It turns out that it is the *non-wastefulness* of the DA mechanism which makes it undominated within the class of strategy-proof mechanisms. A (deterministic) mechanism f is called *non-wasteful* if for all R and for each student i, if some school x has an empty seat under $f(R)$, then $f_i(R)R_i x$. If a mechanism is wasteful, then at some preference profile, the outcome of the mechanism would have a school with an empty seat, while some student prefers this school to her assignment. Erdil (2011) shows that a strategy-proof, non-wasteful mechanism is not dominated by a strategy-proof mechanism. While this fairly general result subsumes the earlier impossibilities for deterministic mechanisms, it is mute about the actual random mechanism used in various school choice programs.

For example, in New York City and Boston, a uniform lottery chooses a linear order of students. This linear order is then used to break the ties before running the DA algorithm. So if $\mathcal{T}$ is the set of tie-breaking rules, each of which follows a linear order on the set of agents N, then $|\mathcal{T}| = (n!)$. Denoting by DA^τ the deferred acceptance applied after the tie-breaking rule τ, the *random deferred acceptance (RDA)* mechanism is

$$RDA = \frac{1}{n!} \sum_{\tau \in \mathcal{T}} DA^\tau.$$

For random mechanisms, first-order stochastic domination (FOSD) provides an unambiguous way of telling when one mechanism dominates another. Formally speaking, a mechanism g *dominates* f if for every preference profile R, and every student i the lottery $g_i(R)$ FOSD $f_i(R)$; and the domination is strict for at least one student at some preference profile.

Erdil (2011) finds that the RDA is *not* on the efficient frontier of strategy-proof mechanisms. In other words, there is a strategy-proof mechanism which every student prefers to the RDA. The proof is constructive, and the idea behind the construction is generalized to explore the extent of *strategy-proof improvement* whenever a mechanism admits strategy-proof improvement.

References

Abdulkadiroğlu, Atila, Parag A. Pathak, and Alvin E. Roth. (2005) "The New York City High School Match." *American Economic Review, Papers and Proceedings*, 95: 364–367.

Abdulkadiroğlu, A., Pathak, P. A. and Roth, A. E. (2009) "Strategy-proofness versus efficiency in matching with indifferences: redesigning the NYC high school match," *American Economic Review*, 99: 1954–78.

_____ Parag A. Pathak, Alvin E. Roth, and Tayfun Sönmez. (2005) "The Boston Public School Match." *American Economic Review, Papers and Proceedings*, 95: 368–371.

_____ Pathak, P. A., Roth, A. E. and Sönmez, T. (2006) "Changing the Boston school mechanism: strategy-proofness as equal access." <http://kuznets.fas.harvard.edu/~aroth>.

_____ and Sönmez, T. (2003) "School choice: a mechanism design approach," *American Economic Review*, 93: 729–47.

Balinski, M. and Sönmez, T. (1999) "A tale of two mechanisms: student placement," *Journal of Economic Theory*, 84: 73–94.

Chambers, C. P. (2004) "Consistency in the probabilistic assignment model," *Journal of Mathematical Economics*, 40: 953–62.

Cormen, T. H., Leiserson, C. E., Rivest, R. L. and Stein, C. (2003) *Introduction to Algorithms*, MIT Press.

Dubins, L. E. and Freedman, D. A. (1981) "Machiavelli and the Gale–Shapley algorithm," *American Mathematical Monthly*, 88: 485–94.

Ehlers, L. and Erdil, A. (2010) "Efficient assignment respecting priorities," *Journal of Economic Theory*, 145: 1269–82.

_____ and Bettina Klaus. (2006) "Efficient priority rules," *Games and Economic Behavior*, 55: 372–384.

_____ and Klaus, B. (2007) "Consistent house allocation," *Games and Economic Behavior*, 30: 561–74.

Erdil, A. (2011) "Strategy-proof stochastic assignment," Working Paper.

_____ and Ergin, H. (2006) "Two-sided Matching with Indifferences." Working Paper.

_____ _____ (2008) "What's the matter with tie-breaking? Improving efficiency in school choice," *American Economic Review*, 98: 669–89.

Ergin, H. (2000) "Consistency in house allocation problems," *Journal of Mathematical Economics*, 34: 77–97.

_____ (2002) "Efficient resource allocation on the basis of priorities," *Econometrica*, 70: 2489–97.

_____ and Sönmez, T. (2006) "Games of school choice under the Boston mechanism," *Journal of Public Economics*, 90: 215–37.

Gale, D., and Shapley, L. S. (1962) "College admissions and the stability of marriage," *American Mathematical Monthly*, 69: 9–15.

Halldórsson, Magnús, Robert W. Irving, Kazuo Iwama, David F. Manlove, Shuichi Miyazaki, Yasufumi Morita, and Sandy Scott. (2003) "Approximability Results for Stable Marriage Problems with Ties." *Theoretical Computer Science*, 306: 431–447.

Kesten, Onur. (2006) "On two Competing Mechanisms for Priority Based Allocation Problems." *Journal of Economic Theory*, 127: 155–171.

Kesten, O. (2010) "School choice with consent," *Quarterly Journal of Economics*, 125: 1297–348.

Manlove, David F., Robert W. Irving, Kazuo Iwama, Shuichi Miyazaki, and Yasufimi Morita. (2002) "Hard Variants of the Stable Marriage." *Theoretical Computer Science*, 276(1–2): 261–279.

Pápai, S. (2000) "Strategy-proof assignment by hierarchical exchange." *Econometrica*, 68: 1403–33.

Rooney, R. (2009) *How to Win Your School Choice Appeal*, A. & C. Black Publishers Ltd.

Roth, A. E. (1982) "The economics of matching: stability and incentives," *Mathematics of Operations Research*, 7: 617–28.

Roth, A. E. (1984) "The Evolution of the Labor Market for Medical Interns and Residents: A Case Study in Game Theory." *Journal of Political Economy*, 92(6): 991–1016.

—— (2002) "The economist as engineer: game theory, experimentation, and computation as tools for design economics," *Econometrica*, 70: 1341–78.

—— (2003) "The origins, history, and design of the resident match," *Journal of the American Medical Association*, 289(7): 909–12.

—— and Peranson, E. (1999) "The redesign of the matching market for American physicians: some engineering aspects of economic design," *American Economic Review*, 89: 748–80.

—— and Rothblum, U. G. (1999) "Truncation strategies in matching markets—in search for advice for participants," *Econometrica*, 67: 21–43.

—— and Sotomayor, M. (1990) *Two-sided Matching*. New York: Cambridge University Press.

Shapley, L. S., and Scarf, H. (1974) "On cores and indivisibility," *Journal of Mathematical Economics*, 1: 23–8.

Thomson, W. (2006) "Consistent allocation rules," Mimeo.

Zhou, L. (1990) "On a Conjecture by Gale about One-Sided Matching Problems," *Journal of Economic Theory*, 52: 125–135.

CHAPTER 7

CAN THE JOB MARKET FOR ECONOMISTS BE IMPROVED?

SARBARTHA BANDYOPADHYAY,
FEDOR ISKHAKOV, TERENCE JOHNSON,
SOOHYUNG LEE, DAVID McARTHUR,
JOHN RUST, JOEL WATSON,
AND JOHN WATSON[1]

Well-functioning markets do not always spring up spontaneously. As economists, we are well-positioned to monitor and modify the market through which new members enter our profession. (concluding sentence, p. 205, from Coles et al., 2010)

INTRODUCTION

In this chapter we discuss attempts to improve the operation of the job market for academic economists via the creation of EconJobMarket.org (EJM), which was launched in the fall of 2007.[2] While we shall define more precisely what we mean by the "economics job market" shortly, it consists primarily of the annual market for jobs for young

[1] While all authors of this chapter have an affiliation with EconJobMarket.org (EJM), not all of them are officers or members of the board of directors, and none of the statements or views expressed herein should be taken as being endorsed or approved by Econ Job Market Inc.

[2] Econ Job Market Inc. was founded by Martin Osborne, John Rust, and Joel Watson. The views expressed herein are those of the authors and do not necessarily represent the positions or policies of Econ Job Market Inc. or those of Martin Osborne. The authors include some of those who have volunteered to help develop and manage EJM, and others who are interested in job matching and research on alternative job market mechanisms, but do not include all directors and officers of EJM.

economists who either recently completed or who are about to complete their PhD degrees. As stated on the EJM website <https://EconJobMarket.org>, this service "seeks to reduce the costs of information flow in the economics job market by providing a secure central repository for the files of job-market candidates (including papers, reference letters, and other materials) accessed on line." A secondary goal of EJM is to use some of the data in this central repository to support research that focuses on the operation of the economics job market (subject to all restrictions necessary to preserve confidentiality of participants and comply with all relevant privacy laws and human subjects protections). We feel that lack of adequate data has impeded research on the operation of many labor markets, and a comprehensive database could prove invaluable to better understanding our own.

The primary role for EJM is not research, however, but to serve as a *labor market intermediary* with the goal of reducing search and transactions costs to market participants. As such, we view EJM as a modest innovation that does not otherwise attempt to alter the basic *decentralized search and matching process* that has characterized the operation of the economics job market since its inception. Examples of more ambitious and radical market designs include computerized matching services, such as those used in the market for medical residents (see, for example, Roth 1984; and Niederle and Roth 2003).

Even though EJM serves the limited role of online transmission of applications to reduce search and transaction costs, previous studies have shown that similar institutions can have large effects on labor market outcomes for both sides of the market. One such service, called Alma Laurea, was established by Italian universities in 1994 to improve the labor market for graduates of a consortium of Italian universities. The effect of this intermediary on this job market was analyzed by Bagues and Labini (2009). Their main conclusion is that "the adoption of the online labor market intermediary under study improved graduates' labor market outcomes three years after graduation" and their analysis suggests that "online labor market intermediaries may have a positive effect on matching quality" (p. 153).

Economic theories of market design often presume the existence of a central planner with the authority to impose virtually any chosen system of market rules on market participants. In reality, in most markets, no single person or organization has the authority to impose such changes, because most markets are *commons* that are not owned by any single organization. Various practical and legal obstacles, as well as coordination problems, make it difficult for individuals to significantly alter many markets, because any change in the market invariably has adverse welfare consequences for at least some market participants who may have strong vested interests in the status quo. This is certainly true in the market for academic economists: the creation of EJM offers a case study in the challenges confronting even modest attempts to improve market outcomes.

Despite these challenges, we show that the use of EJM has grown exponentially since its introduction in 2007, to the point where it is now handling a significant share of all job applications in the economics job market. This suggests that even modest interventions with the limited objective of reducing transaction costs may be able to alter the operation and structure of the market, making the information available to market participants more *centralized*. Further, EJM provides a technological platform that may facilitate more ambitious and useful changes to the market in the future, changes that would likely be much more difficult to implement in a completely decentralized market without EJM. In particular, we discuss a promising alternative design—computerized matching systems—that has the potential to further improve job market performance.

In the next section, we describe the economics job market and some of the problems that motivated the creation of EJM in order to operate more efficiently. While the idea of using information technology, and particularly the power of the web to transmit the information necessary for this market to help it operate, is quite natural, we argue that uncoordinated, unrestricted entry of labor market intermediaries in a fundamentally decentralized market has the paradoxical effect of *increasing* search costs, and *worsening* market outcomes—an outcome we refer to as *market fragmentation*.

In the third section, we describe how EJM attempts to achieve the key benefits of reduced search and transaction costs that modern information technology can provide, while avoiding the harm that can be caused by excess entry of labor market interme-diaries and market fragmentation. EJM is a non-profit organization that provides a limited set of services to the economics market so inexpensively that long-term survival of for-profit competitors may be difficult in its presence. We argue that there is a natural monopoly aspect to the primary function that EJM provides, namely, its role as a *data repository* where most of the relevant data that market participants need can be accessed. If a single site such as EJM can emerge as a *market maker* that provides essentially all of the relevant data on jobs and job candidates, then the problem of market fragmentation can be solved and market efficiency can be significantly improved. However, to the extent that there is a natural monopoly aspect to this market-making function, we argue that EJM needs to operate as a non-profit whose operation is *regulated* so that it serves the interests of the profession as a whole.

We discuss how various forms of beneficial competition can be encouraged once a dominant non-profit market marker is in place. In particular, there can be competition among various intermediaries that provide various "front end" and "back end" data connection services to the central data repository. The key participants in the economics job market are recruiters, applicants, and recommenders. A "front end" is a software interface to EJM that serves applicants, and assists them in searching and applying for ads, or assists recommenders in uploading and transmitting reference letters to the central data repository. A "back end" is a software interface to EJM that transfers applica-tions received by a specific recruiter from the central data repository to a separate secure computer database to permit further confidential analysis of applicants. EJM encourages competition among firms that provide these sorts of additional front- and back-end services, and we argue that unrestricted competition among such intermediaries will be beneficial (resulting in better software at a lower price) without the negative side-effects of market fragmentation *provided they all have equal access to, and agree to be interoperable with this single central data repository.*

EJM's objectives may be compared to the role that the non-profit organiza-tion ICANN <http://www.icann.org> plays in managing private competition in the provision of registered domain names for the Internet. By centralizing the role of assigning domain names and allowing other intermediaries to compete on other service dimensions like price, ICANN has substantially centralized the market while fostering competition. ICANN has recently considered adopting auctions as a method of selling top-level domains, providing another example of how centralization can be the first step to institutions that incrementally improve their design over time.

In the fourth section, we present several models that illustrate how the entry of a non-profit intermediary similar to EJM can reduce market fragmentation, and the associated search and transaction costs, and thereby improve overall market efficiency. A secondary efficiency question is whether an intermediary such as EJM, by successfully reducing market fragmentation and search and transactions costs, would create incentives for candidates to make excessive numbers of job applications. Labor market intermediaries such as EJM operate primarily to reduce the cost of *transmitting* information but they may do relatively little to help recruiters reduce the cost of *evaluating* this information. One might wonder if an intermediary such as EJM could worsen market outcomes if recruiters, flooded with many more applications than they previously received, end up devoting less effort evaluating each application, thereby compromising their ability to identify the best candidates. One solution is for recruiters to set application fees, which EJM facilitates as contributions to support the service. But few recruiters choose to impose application fees, so, there still is the question of whether the number of applications is excessively high.

In the fifth section, we discuss some of these problems and the potential role for more radical *centralized mechanisms* for operating the economics job market such as computerized matching algorithms or position auctions. We discuss recent contributions to the analysis of matching mechanisms from a mechanism design perspective, and the feasibility of implementing efficient outcomes via methods such as auctions. While these mechanisms have the potential to overcome problems that the more decentralized mechanisms cannot solve, the main challenge is that market participants cannot be compelled to use them. As we noted earlier, since there is no single individual or organization that "owns" the economics job market, the success in establishing these more ambitious types of market mechanisms is limited by *voluntary participation constraints*. Niederle and Roth (2003) have noted the problem of *unravelling* (akin to the problem of adverse selection in insurance markets) that can make more centralized designs unviable if groups of recruiters and candidates choose not to participate in a proposed mechanism.

Note that a completely different meaning for the term "unravelling" was introduced by Neeman and Vulkan (2010). They proved that decentralized trade via matching and bilateral bargaining is also subject to unravelling in the sense that when traders can choose whether to engage in bilateral bargaining or to trade in a central marketplace, there are strong forces that ensure that "all trade takes place in the centralized market" (p. 1). We believe the key insight underlying Neeman and Vulkan's sense of "unravelling" is the key to the rapid growth in EJM, at least to the extent EJM constitutes the "centralized market." However, the Niederele–Roth sense of unravelling may also be the key explanation of why the adoption of more centralized designs such as computerized matching mechanisms may be a far more challenging objective.

In the sixth section we conclude with comments and ideas for future research as well as ideas for future market experiments that build on the EJM platform, assuming that it continues to remain a viable entity, with resources and support for undertaking more radical types of market experiments.

THE ECONOMICS JOB MARKET

In comparison with many other labor markets, the existing economics job market is actually quite organized. The American Economic Association (AEA) has facilitated the market for new PhD economists in the US by supporting job interviews in hotel rooms during the annual Allied Social Science Association (ASSA) meetings (currently held annually, in early January), and creating the Job Openings for Economists (JOE) advertising service in 1974. In 2002 the JOE became an exclusively online service and, according to Coles et al. (2010), in 2008 over 1,900 academic jobs and over 1,000 non-academic jobs for PhD-level economists (both senior and junior) were advertised on JOE.

Services such as JOE use the web only to publicly *advertise* the existence of jobs, and fail to provide additional *online application and reference letter transmittal services*. Since each recruiter must review each job candidate in a relatively short time span, efficient information processing becomes a crucial feature of the market. Each year, roughly from late October until early December, thousands of recruiters advertise positions they seek to fill, and thousands of job candidates submit applications for these job advertisements. Each application typically involves the transmission of the candidate's curriculum vitae (résumé), his or her job-market paper or other writing samples, and letters of recommendation from several references. Often, a candidate might specify three or more letters of recommendation in each application, and these must be transmitted to the recruiter separately, since they are intended to be confidential and not seen by the candidate.

Prior to the entry of intermediaries, such as EJM, most applications in the economics job market were submitted in paper by traditional mail. Applicants needed to copy their curriculum vitae and writing samples, and mail these by post to dozens of different prospective employers—in many cases 100 or more. Coles et al. (2010) report that in 2008, several thousand candidates were applying to nearly 3,000 job advertisements in the US and North America region alone, and that a typical candidate might make eighty applications. If there are at least three references per application, then the operation of the job market also involves transmission of more than 500,000 reference letters. The collective time and other resources necessary to copy and mail all of this information in each job market season is, by itself, a potential source of significant economic inefficiency. In addition, there is substantial additional secretarial effort necessary to maintain and to file paper-based applications, since many recruiters may receive 500 or more applications to each job advertisement they post.

Online labor market intermediaries

With the advent of the Internet and the web, many of the transaction costs associated with the simple transmission of the application materials and references can be greatly reduced by creating efficient web-based advertising and application services. EJM was not the first and is certainly not the only organization to provide such services, even within the economics job market. For example, one of the largest such companies, Monster.com, was founded in 1994 with the goal of facilitating digital recruiting in general labor markets.

In the narrower area of academic recruiting, several companies exist, such as AcademicKeys.com, which started taking online job applications in 2002, and HigherEdJobs.com and the Chronicle of Higher Education. Within economics, there are several other for-profit and non-profit services that offer or previously offered approximately the same set of online services that EJM provides, including jobmarket-economist.com (founded in 2005, merged with EJM in 2009), AcademicJobsOnline.org, (launched in 2006), Econ-Jobs.com, econjobs.com, www.thesupplycurve.com (founded in 2008) and Walras.org (founded in 2007 and which began providing online application services in 2010 and merged with EJM in 2012).

In addition to the systems and organizations named above, there are other for-profit companies that are starting to capture a significant share of the *human resource (HR) administration market* and that provide database tracking of all aspects of behavior and records for employees of large companies from the date of hire. One example is PeopleAdmin.com, founded in 2000 "to reduce the cost, risk, and time spent managing human resources for government, higher education, and non-profit organizations." PeopleAdmin's systems include online application components that are now used by many large universities, including Columbia, University of Pennsylvania, New York University, and the University of Maryland. These online application services can also collect letters of recommendation from references named by an applicant in their online application.

Excess entry and market fragmentation

Given all of the various organizations and new online systems providing online application and reference letter transmittal services, is there a need for yet one more entrant, such as EJM? Could additional intermediaries actually degrade the functioning of the market? When recruiters must choose among many intermediaries there is a danger of *market fragmentation*. The problem is that market participants—especially candidates and the recommenders who submit reference letters—generally have to duplicate their efforts for each online system that recruiters use to collect application materials. These duplicative tasks include establishing accounts, submitting applications, uploading documents, and uploading reference letters.

A casual analysis of the situation suggests that a single intermediary could integrate all the cost-reducing features that other intermediaries provide, and eliminate the inefficiencies associated with fragmentation, thereby making a step towards an efficient outcome. Due to the natural economies of scale of information centralization, a market where multiple intermediaries are operating can be said to suffer from *excess entry*. If there were a single online system then the market participants would need to visit only one site to make an application, then post an advertisement, or upload a recommendation letter, and tasks of establishing accounts, uploading documents, and creating biographical profiles would be done just once.

Such a casual analysis, however, ignores a number of issues. First, the services offered by different intermediaries have different advantages and disadvantages, and this process of experimentation and innovation is potentially valuable. Second, competition and the threat of entry discipline incumbent firms. For example, a monopolist may decide to restrict the focus of his service to increase profits, shutting some portions of the market out from access to more efficient intermediation. Finally, a market might pursue both competition and centralization by adopting policies that centralize the information, but encourage firms to compete on other service dimensions, such as their interface or algorithms that search for potential candidates.

The problem of excess entry of intermediaries is present to an extreme degree in a closely related market: online applications to graduate schools. Faculty members are now familiar with the various services such as *Embark.com, ApplyYourSelf.com, College-Net.com,* and dozens of other home-grown application systems designed by individual universities for taking applications by undergraduates for admission to graduate school, and corresponding websites that faculty must negotiate to upload letters of recommendation on the students who name them as references.

Because of poor software design and lack of standardization, many of these sites force faculty to hunt their email boxes for requests to provide letters of recommendation, to find or to request the requisite account and password, to go to the site to login to enter and re-enter contact information, to fill out extended questionnaires about the students they are recommending, and then finally to upload the letters of recommendation. All this must be done *per recommendation* and it can take between 15 and 30 minutes to negotiate a single form. A typical undergraduate student may apply to a dozen or many more graduate schools. Thus, the huge collective time burden on faculty of simply transmitting the reference information on their students who apply to graduate school becomes immediately apparent. Of course, students who are applying to graduate schools face these costs as well, even more so, since in addition to the time burden they may have to pay an application fee ranging from $50 to $100 per application.

There is increasing concern that the problems we see in the market for graduate school applications will start to spread to the next level up, to the job market for new PhDs. Indeed, we are starting to see the same sort of lack of coordination and excess entry of labor market intermediaries in the economics job market, and this is already creating an unnecessary burden on faculty who write recommendations letters for their graduating PhD students applying for jobs. In a private communication John Siegfried, Secretary-Treasurer of the American Economics Association and the Director of the JOE since 1997, noted that "By far the most annoying part of the process is the unique websites adopted by the Human Resource Departments of various employers, and especially those that can detect that it is our departmental assistant who is pretending to be us, and block her from entering the data." Also, in a private communication, Nancy Rose expressed similar frustration from her perspective as placement officer at MIT, particularly for recruiters that use "employer-specific URLs" which she feels have become "a complete nightmare." Rose concluded that "I think this system is inefficient and much, much too burdensome for PhD granting departments with any sizable number of students on the market in a given year. Financial pressures at many universities (including MIT) have led to staff reductions that exacerbate this cost for faculty."

EconJobMarket.org

In this section, we provide a brief description of the EJM software/site and and some of the services it offers, and provide some data on the level of usage and rate of adoption of EJM by the market. In particular, the descriptive analysis outlines the rapid growth of EJM, which has doubled in size each year since its introduction. In general, we see that candidates are making more applications using EJM, and that the number of applications received per post has grown very rapidly. These findings suggest a number of interpretations and market design issues, which we discuss.

Functionality of EJM

The EJM software is undergoing continual development and improvement, but in this subsection we describe the state of the EJM software as of March 2012. As noted earlier, there are three types of user accounts on EJM: (1) recruiters, (2) candidates, and (3) recommenders. All of these accounts are free, but there is provision in the EJM software for recruiters and candidates to make voluntary contributions. While virtually any organization wishing to recruit economists is allowed to have a free account on EJM, these accounts must be approved, and recruiters who attempt to post advertisements requiring skills that are not deemed to be sufficiently close to economics can be prohibited from using the site.

Recruiters typically receive a permanent account, allowing their personnel to post new job advertisements at any time. A posted advertisement can utilize the EJM functionality for the transmission of application materials, or simply explain the details of the job and give candidates further instructions on how to apply elsewhere.

As applications are submitted for job postings, recruiters can search the application files of individual candidates interactively by logging in and selecting a search/view applicants function. Also, recruiters are also allowed to download an Excel-compatible file listing the name, organization, degree, and other key information of the applicants, and a zip file that contains the material submitted by candidates, as well as any reference letters. Recruiters can also download individual PDF-formatted "virtual application folders" consisting of a cover page, the curriculum vitae reference letters, and all other files uploaded by the candidate as part of the application. This allows the authorized members of the recruiting organization to view the files at their convenience.

On the other side of the market, candidates obtain free accounts from which they can upload their vita, job-market papers, teaching or research materials, and other papers or writing samples. While logged into EJM, candidates can search or browse the available advertisements, and apply to any for which recruiters accept online applications via EJM. Typically, different recruiters request different materials, and candidates personalize their applications slightly to each job. One of the features of the EJM interface is that it provides recruiters great flexibility to design their application forms and required documents, and it gives candidates similar flexibility in applying to different job postings.

When candidates submit applications through EJM, they also specify their references. Recommenders can select whether to be notified every time they are named by candidates, and a new, free account is created if a person named as a reference does not already have a recommender account on EJM. As described below, EJM provides a great deal of flexibility and confidentiality to recommenders concerning how their reference letters are distributed through the EJM system. Also, EJM allows recommenders to specify other individuals to serve as their *proxies*, allowing authorized individuals such as administrative assistants or placement secretaries to manage the recommendation requests on their behalf. Since all of this information is centralized, the system notifies candidates when letters have been submitted, which provides a greater sense of assurance to candidates that their files will not be discarded for being incomplete.

EJM also conducts *identity verification* of all references to ensure that letters of reference on each applicant are really written and uploaded by the reference, minimizing the chance that EJM could be used to transmit fraudulent reference letters. To our knowledge, none of the other commercial intermediaries serving the economics job market provides this functionality: instead, the other services transmit applications, but not reference letters. Applicants may be able to name their references using the commercial sites, but make recruiters responsible for collecting the reference letters on their applicants separately, perhaps using a separate intermediary such as Interfolio.com. EJM provides a complete service: transmitting both the applications filed by applicants, and the reference letters provided by references, so that recruiters have all relevant information they need to evaluate the applicants to the positions they advertise on EJM.

Descriptive analysis of the EJM growth, users, and application decisions

The first year EJM became operational, in the 2007/08 job market season, it was running in "beta-test mode," and the number of job advertisements was deliberately restricted since the officers of EJM were reluctant to take the risk of fully scaling up the site until the software had been sufficiently tested. After the first year, and after a number of minor bugs were discovered and fixed, the EJM site was scaled up, and allowed to operate on an unrestricted basis. As a result, we restrict our analysis to the subsequent four full "job market seasons" that EJM has served, over the period of time August 1, 2008 until 2012. We define a *job market season* to be from August 1st in year t to July 31st in year $t + 1$, because job activity on the EJM site tends to be highest in the months of November and December, and lowest in the months of July and August. Note that we only have partial data for the most recent job market season, for the period August 1, 2011 to February 1, 2012.

In Figure 7.1 we plot the growth in various measures of EJM's size over the last four job market seasons. Overall, EJM grew exponentially, with annual growth rates for nearly all of the measures of EJM size and activity well in excess of 100% per year. The top left-hand panel of Figure 7.1 depicts the number of job advertisements placed on the EJM on a daily basis since the site went live in August 2007. Notice the dramatic peaking in the number of job advertisements during the period November to December in each year, the period of maximum activity in the primary economics market. As noted, the summer is the slow season for EJM, and the number of advertisements falls off considerably during these periods. At the peak there were over 220 advertisements posted on EJM in the 2011/12 job market season. By comparison, the December 2011 JOE had 315 job advertisements, and the November JOE had 581 job advertisements. Thus, EJM has grown very rapidly and already accounts for a significant share of all job advertisements posted in the primary market for economists (since JOE is widely known to be the main site where job advertisements for economists are placed, particularly for jobs in North America).

The top right-hand panel of Figure 7.1 plots the cumulative number of job advertisements posted on EJM as of the first of each month. By February 1, 2012, a cumulative total of 1,099 advertisements had been posted on EJM, and the annualized growth rate in the number of advertisements on the site was 105%. We note that this unusually rapid growth occurred during "recession years" when the demand for economists was weak, and the number of job advertisements significantly lower than what would be expected in normal economic times. The pronounced undulations in the number of cumulative job advertisements posted reflect the seasonality of the job market, where new advertisements posted increase most rapidly during the fall and then increase much more slowly during the slow season in the spring and summer of each year.

The top right-hand panel also plots the cumulative number of recruiter accounts on EJM. As of February 1, 2012, there were a total of 512 recruiter accounts. Further

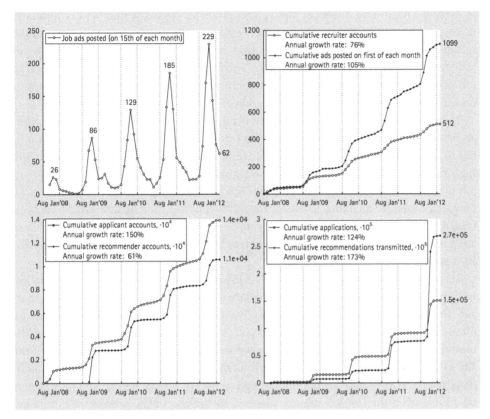

FIGURE 7.1. Growth in EJM ads, applicants, recruiters and recommenders.

information on the types of recruiters and their nationalities will be provided shortly. However, the main point is that the number of new recruiters registering to use EJM was growing at a rate of over 75% per year, and showed no obvious signs of slowing down. Of course, this growth must eventually slow if EJM continues to grow and capture a large share of all recruiters who are searching for economists with PhD degrees. Although it is difficult to estimate the number of all potential recruiters of PhD economists worldwide, we believe it to be at least several thousand organizations. Thus, the rapid rate of adoption of EJM by new recruiters could continue for several more years.

In addition to the numbers of recruiters, the overall "quality" of the various recruiters has been steadily increasing as well. During the 2011–12 job market season, job advertisements were posted by the highest-ranked economics and business schools worldwide, including MIT, Stanford, Harvard, Penn, Brown, Cambridge, Oxford, University College London, Columbia, Berkeley, and many others. Also, leading private companies such as the RAND Corporation, Yahoo! and Microsoft Research as well as leading government organizations such as the US Federal Reserve System, Banque de France, Sveriges Riksbank (National Bank of Sweden), and Congressional Budget Office have established accounts and posted advertisements on the site.

From the beginning, the most rapid growth in EJM was in the number of candidates using the service. The lower panels of Figure 7.1 plot the number of applicants and the number of applications made using the EJM website. These have grown at 150% per year with a particularly large jump in the number of applications during the 2011/12 job market season. By February 1, 2012, there were nearly 14,000 candidate accounts on EJM and over 150,000 applications had been processed by the EJM website.

The lower panels of Figure 7.1 also plot the growth in the number of recommenders and recommendation letters that have been transmitted by the EJM website. The number of recommenders with accounts on EJM is growing at a somewhat slower but still high rate of more than 60% per year. This growth slowed somewhat in the 2011–12 job market season since EJM adopted a policy of *mandatory identity verification* of all new recommender accounts. However, the number of recommendations that have been transmitted by the EJM system is increasing by over 170% per year, and by February 1, 2012, nearly 270,000 recommendation letters had been delivered to recruiters by the EJM system.

These rapid growth rates suggest that EJM is indeed serving a need that was not well met by other labor market intermediaries operating in the economics job market. The numbers also suggest strong positive self-reinforcing feedback effects that are often observed in other "two-sided markets" (see, e.g., Rysman 2004): The greater the number and quality of candidates with accounts on EJM, the greater the value of the site to recruiters, and vice versa. It is our impression that virtually all job market candidates from the top-ranked economics departments worldwide had candidate accounts on EJM during the last job market seasons, so the use of the service by candidates appears to be nearly universal already. There is still some distance to go in terms of recruiter accounts, and the number of recruiters and job ads placed on EJM could well double or triple before we start to see diminishing growth. Furthermore, this growth occurred entirely by word of mouth, since EJM cannot afford any significant amount of advertising. EJM's visibility has also been increasing following endorsements from the Econometric Society and the Canadian Economics Association, as well as an important collaboration with the European Economic Association.

We now turn to a descriptive analysis of the types of recruiters and candidates who have accounts on EJM, and an analysis of the application behavior by EJM candidates. Table 7.1 lists the number of candidates who used EJM in each academic year, and their characteristics. The number of candidates who registered for new EJM applicant accounts increased from 2,344 in 2008/09 to 3,466 in the 2011/12 job market season. Not all of these account holders actually made online applications via EJM: some may not have found appropriate positions on the site, and some advertisements on EJM are links that redirect applicants to apply on another site (such as the recruiter's own application system), and the statistics reported here refer only to the subset of applications that were actually processed on the EJM site. Thus, in 2008/09 only 1,613 out of the 2,344 new account holders actually submitted applications via the EJM system itself, but, by 2011/12, 2,439 of the 3,436 new account holders submitted applications using the EJM system. The higher fraction of candidates who actually submit applications via EJM no

Table 7.1. Location and employment status of EJM candidates

	2008/09	2009/10	2010/11	2011/12
Number of candidates	2,344	3,330	3,470	3,436
Panel A: Geographical location of candidates (%)				
US	65.7	55.1	55.0	57.8
Europe (excluding UK)	10.2	18.0	19.2	19.4
Canada	10.6	8.2	6.6	5.7
UK	4.9	7.5	8.3	6.5
Asia	2.5	2.3	2.5	2.5
Indian subcontinent	0.8	2.4	2.4	2.0
Australia and New Zealand	1.9	2.0	1.9	2.5
Middle East	1.3	1.4	1.2	0.7
Latin America	0.6	1.1	1.0	1.4
Africa	0.3	1.3	1.0	0.7
Russia	0.0	0.3	0.1	0.2
Others or N.A.	1.2	0.5	0.6	0.6
Panel B: Existing employment status of candidates (%)				
PhD student/dissertator	47.9	43.8	46.3	49.8
Postdoc/visiting scholar	9.0	10.6	9.8	8.3
Assistant professor	11.5	11.4	10.7	11.2
Associate professor	3.6	3.3	3.5	4.0
Full professor	2.8	2.6	2.9	2.2
Lecturer	5.0	4.9	5.1	4.6
Other academic	6.1	6.6	6.4	6.5
Non-academic	8.8	10.3	9.5	8.0

doubt reflects the larger number of advertisements that are posted on EJM, and the increasing fraction that process applications via EJM instead of redirecting applicants to apply elsewhere.

Table 7.1 reveals that over half of EJM applicants are based in the US although there is a clear trend toward "internationalization" of EJM over time, with a significant share of candidates based in Europe, the UK, and Canada. Panel B of Table 7.1 shows that nearly half of all EJM applicants are PhD students who are expecting to receive their degrees, and thus constitute what we refer to as the "primary market" for new PhDs. This was the primary market that EJM was created to serve, but we see that EJM is equally serving the "secondary market" for assistant, associate, and full professors who wish to change positions, and a significant component of non-academic economists looking for jobs.

In Table 7.2 we list the percentage distribution of new applicants signing up for applicant accounts by their self-designated primary field. These distributions are fairly stable across the four successive job market seasons that EJM has served, except that we see an increasing share of candidates in finance. We believe that this could be a

Table 7.2. Distribution (%) of candidates' primary fields

	2008/09	2009/10	2010/11	2011/12
Behavioral economics	1.7	2.3	2.2	2.2
Business economics	0.0	3.3	2.9	2.7
Computational economics	0.0	0.3	0.4	0.5
Development; growth	7.6	8.8	7.8	7.7
Econometrics	8.0	6.3	7.1	6.4
Economic history	0.9	1.3	1.0	1.0
Environmental; agricultural economics	4.8	5.7	7.2	6.0
Experimental economics	1.1	1.1	1.7	1.3
Finance	10.8	11.6	12.5	12.0
Health; education; welfare	0.7	3.9	4.5	4.6
Industrial organization	8.1	6.2	5.6	4.9
International finance/macro	6.1	4.7	4.3	4.2
International trade	4.9	5.1	4.8	4.7
Labor; demographic economics	7.6	6.4	6.8	7.0
Law and economics	0.6	0.9	0.6	0.4
Macroeconomics; monetary	12.0	11.6	10.1	10.2
Microeconomics	9.6	8.0	7.0	7.8
Political economy	0.1	0.2	2.4	2.3
Public economics	5.4	4.6	3.9	3.6
Theory	2.4	1.8	2.2	1.5
Urban; rural; regional economics	0.1	1.6	1.9	1.7
Other business/management	0.0	0.0	0.0	1.7
Other, any field, or N.A.	1.3	0.0	0.0	0.0

consequence of loss of "Wall Street jobs" in the aftermath of the Wall Street crash in fall 2008, and subsequent downsizing in large Wall Street firms and banks. As a result, many individuals who sought employment in the financial sector might diversify their job search to include government and academic positions. Later we will also see this reflected in an unusually large increase in applications submitted for a decreasing number of positions in finance.

In Table 7.3 we list the number of recruiters—institutions that posted their job openings on EJM—for each academic year and the composition of their characteristics. The number of job ads posted on EJM increased from 134 in 2008/09 to 328 in 2011/12. The most common type of position advertised on EJM was for assistant professors, accounting for 43% of all job advertisements on the site. However, we also see a significant number of higher-ranked tenured and untenured associate professor advertisements, full-professor advertisements, and advertisements for consultants and economists needed for non-academic positions.

Panel B of Table 7.3 shows the geographic breakdown of where the advertised positions are located. The majority of the advertised positions are located in the US, though we see that over a third of all advertisements are for positions based in Europe

Table 7.3. Characteristics of EJM job advertisements

	2008/09	2009/10	2010/11	2011/12
Number of advertisements placed on EJM	134	256	338	328
Panel A: Positions advertised (%)				
Assistant professors	51.5	42.2	36.1	43.0
Associate professors	3.7	2.0	1.2	1.8
Full professors	5.2	3.9	4.1	5.2
Assistant or associate	0.0	0.0	4.1	6.7
Professor, any level	14.8	15.3	17.2	15.9
Post-doctorate	2.2	12.5	10.9	7.3
Lecturers and other academic positions	7.4	10.5	15.1	8.9
Consultant	0.7	2.7	0.6	2.4
Non-academic	9.7	10.5	10.4	8.8
N.A.	4.5	0.4	0.3	0.0
Panel B: Geographical location of job (%)				
US	56.7	41.0	40.2	52.7
Canada	12.7	7.4	8.0	7.0
UK	7.5	6.6	8.6	6.4
Europe (excluding UK)	13.4	36.3	34.3	27.1
Australia and New Zealand	0.8	2.3	3.3	3.4
Asia	3.0	1.6	1.2	1.5
Latin America	1.5	1.6	2.7	0.0
Others or N.A.	4.5	3.0	1.8	1.9

and the UK. Similar to our findings related to candidates, recruiters from the UK and Europe are increasingly represented in EJM over time, and the large increase in the representation of European recruiters in 2009/10 may have reflected the endorsement of EJM by the European Economic Association in 2009.

In Table 7.4 we list the distribution of recruiter interest across research fields. The number of advertised fields in Table 7.4 is larger than the number of recruiters because one employer may list multiple research fields for its job advertisement. A noticeable pattern is that the fraction of advertisements in the areas of "Development and growth" and "Finance" decreased over this period. Although it is not conclusive, the increasing supply of candidates specialized in "Finance" shown in Table 7.2 and the relative decrease in the demand for finance PhDs among recruiters suggests that the market for "Finance" is becoming more competitive for candidates relative to other fields. We also observe a smaller number of job advertisements in the field "Macroeconomics; monetary" and "International finance/macro," which may be a bitter irony given that macroeconomic problems seem largely responsible for the weak job market for economists in recent years.

Table 7.4. Distributions of advertised research fields

	2008/09	2009/10	2010/11	2011/12
Number of advertised fields	326	667	734	854
Any field	13.8	11.2	13.1	13.6
Behavioral economics	2.1	2.8	3.0	2.7
Business economics	0.0	3.0	3.7	2.1
Computational economics	0.0	0.7	1.4	1.3
Development; growth	4.0	4.0	4.3	4.3
Econometrics	8.0	7.6	8.4	7.5
Economic history	2.1	1.5	1.3	0.8
Environmental; agriculture economics	6.7	3.6	4.0	3.5
Experimental economics	1.5	2.1	1.9	2.0
Finance	8.0	6.3	6.7	7.3
Health; education; welfare	0.0	3.4	4.7	3.6
Industrial organization	5.5	6.1	5.3	5.9
International finance/macro	5.8	4.8	2.9	3.7
International trade	5.5	4.0	3.9	4.1
Labor; demographic economics	4.9	5.7	4.7	4.6
Law and economics	3.1	2.2	2.5	2.7
Macroeconomics; monetary	8.3	8.2	6.0	5.6
Microeconomics	7.7	7.3	6.7	7.4
Political economy	0.0	0.0	1.9	1.8
Public economics	5.5	5.1	5.3	5.6
Theory	1.5	2.2	2.2	2.1
Urban; rural; regional economics	0.0	2.8	2.4	2.5
Others, N.A.	5.8	4.8	3.4	2.9

In Tables 7.5 and 7.6, we present the average number of applications that a job seeker sent via EJM, and that an employer received from EJM respectively. As we noted above, not all EJM applicant account holders use EJM to apply for jobs, though the fraction who do submit applications via EJM has been increasing, and exceeded 70% in the 2011/12 job market season. We see that the total number of applications processed per job season via EJM has quadrupled, from 12,869 in 2008/09 to 48,557 in 2011/12. This increase is a result of two main factors: (1) the increasing number of advertisements that are posted, and which take applications via the EJM site; and (2) the increase in the number of applications made by each applicant. As we noted, these effects are clearly interrelated, since the greater number of job advertisements on EJM increase the chance that applicants will find suitable attractive opportunities to apply to. Thus, the average number of applications submitted per applicant via EJM increased from eight in 2008/09 to twenty in 2011/12.

Panels B and C of Table 7.5 illustrate that, regardless of the current position or the geographical location or primary field of research of the applicant, all are making increasing use of EJM, and submitting a higher number of applications through it over time.

Table 7.5. Applications submitted by applicants

	2008/09	2009/10	2010/11	2011/12
Panel A: Statistics on applicants, job advertisements, and applications				
Number who submitted applications	1,613	1,982	2,254	2,439
Number of job advertisements posted	134	256	338	362
Total number of applications submitted	12,869	29,711	36,028	48,557
Average number of applications per candidate	8	15	16	20
Maximum applications by a candidate	49	305	201	128
Std dev in average applications per applicant	0.2	0.6	0.5	0.5
Panel B: Average number of applications by employment status of applicant at time of application				
PhD student/dissertator	9.3	16.4	16.5	22.2
Postdoc/visiting scholar	9.0	18.2	20.3	25.3
Assistant professor	8.4	18.2	22.3	22.3
Associate professor	8.0	20.9	17.1	23.0
Full professor	9.8	8.4	11.3	18.1
Lecturer	4.8	7.3	7.8	12.4
Other academic	5.8	9.1	12.0	14.1
Non-academic	9.0	40.3	20.2	19.8
Panel C: Average number of applications by geographical location of applicant at time of application				
US	8.5	17.0	18.0	21.3
Canada	7.0	15.8	13.6	20.2
UK	7.6	12.5	12.7	19.3
Europe (excluding UK)	5.2	13.0	14.0	16.2
Australia and New Zealand	3.9	3.6	4.8	5.9
Latin America	1.6	4.7	10.2	11.8
Asia	3.4	3.4	7.2	6.9
Middle East	6.7	6.8	18.2	6.0
Indian subcontinent	5.0	23.3	3.7	6.3
Africa	2.0	19.1	1.8	3.5
Russia	0.0	25.0	9.5	5.0
Others or N.A.	0.0	0.0	3.4	18.0

Table 7.6 illustrates the average number of applications submitted by the primary field of applicants. We see that there is generally increasing use of EJM by candidates in all fields, with particularly strong growth (and a tripling of applications submitted per applicant) in fields such as finance, development, macro, and industrial organization.

In Table 7.7 we list the average number of applications received by recruiters who placed advertisements on EJM. For example, the average number of applications that an employer received per advertisement posted on EJM nearly doubled, from 134 in 2008/09 to 242 in 2011/12. The increasing trend in the number of applications received per advertisement is clearest for advertisements for jobs in the US and Canada, but more variable for jobs located in various other regions of the world. This effect is likely

Table 7.6. Average number of applications submitted by primary field of applicant

	2008/09	2009/10	2010/11	2011/12
Behavioral economics	7.5	14.5	16.3	16.2
Business economics	6.0	8.7	4.8	4.0
Computational economics	11.0	1.4	1.9	6.3
Development; growth	8.0	20.9	17.1	23.1
Econometrics	9.3	16.4	16.6	22.2
Economic history	9.8	8.4	11.3	18.1
Environmental; agriculture economics	4.9	7.6	7.8	12.5
Experimental economics	9.0	40.3	20.2	19.8
Finance	3.4	5.7	8.4	9.7
Health; education; welfare	6.2	12.0	12.6	15.9
Industrial organization	8.7	15.5	19.8	25.2
International finance/macro	7.5	14.4	13.6	25.3
International trade	8.4	18.2	22.3	22.3
Labor; demographic economics	9.0	18.2	20.4	25.3
Law and economics	3.8	5.4	4.7	9.9
Macroeconomics; monetary	10.4	21.8	20.7	27.3
Microeconomics	9.4	17.9	24.2	25.4
Political economy	NA	10.0	17.5	21.2
Public economics	8.1	13.4	16.6	23.2
Theory	11.7	19.0	24.3	33.6
Urban; rural; regional economics	3.0	6.8	9.2	11.6
Other	5.2	6.9	7.0	3.2
Any field	1.9	6.6	7.0	10.9

representing greater "sampling variability" to the greater heterogeneity in the type of positions offered, and the smaller number of advertisements on EJM for jobs based outside North America.

Panel B of Table 7.7 illustrates the average number of applications received per advertisement, by primary field of the applicant. These are *conditional expectations* since they are not averages of applications received over *all* advertisements on EJM, but rather only averages over the subset of advertisements to which candidates in a given field apply. For example, in 2008/09 there were five advertisements on EJM to which candidates whose primary field was "experimental economics" applied, and the total number of applications submitted was sixteen, or an average of 3.2 applicants per advertisement. Thus, these numbers can be viewed as a measure of the "supply/demand" imbalance that we observe on EJM, with generally far more applications received for any advertisement than the total number of positions for which the recruiter can hire. Viewed from this perspective, we see that some of the most competitive fields include macro, micro, labor, development, econometrics, and finance. In general, all of the fields appear to have become more competitive over the period, which is in part a reflection of the effects

Table 7.7. Applications received by recruiters

	2008/09	2009/10	2010/11	2011/12
Advertisements receiving applications via EJM	101	189	198	240
Mean applications per advertisement	134	152	203	242
Max applications	620	690	2,758	775
Standard deviation	150	136	261	212
Panel A: Applications by geographical location of recruiter/position				
US	139	239	252	262
Canada	102	138	175	256
UK	184	125	238	393
Europe (excluding UK)	227	80	145	208
Australia and New Zealand	364	155	216	220
Asia	26	86	53	171
Latin America	44	39	154	124
Middle East and North Africa	1	47	41	62
Panel B: Average applications received per advertisement by primary field of research of applicant				
Behavioral economics	2.5	2.7	6.9	5.2
Business economics	0	1.8	1.8	1.4
Computational economics	0	1.0	0.0	1.3
Development; growth	5.6	7.5	14.9	23.7
Econometrics	7.0	8.0	11.9	19.6
Economic history	1.9	2.3	2.6	3.0
Environmental; agriculture economics	3.9	3.8	10.7	12.5
Experimental economics	3.2	2.9	6.2	5.0
Finance	4.0	4.9	11.2	17.6
Health; education; welfare	1.0	4.9	10.2	12.0
Industrial organization	7.6	8.2	13.8	18.7
International finance/macro	5.6	7.6	8.1	14.9
International trade	4.5	7.6	11.6	15.9
Labor; demographic economics	7.6	10.1	18.2	24.4
Law and economics	1.0	2.3	1.3	1.5
Macroeconomics; monetary	14.2	14.1	22.8	40.2
Microeconomics	9.2	11.2	19.0	27.8
Political economy	0.0	0.0	6.3	6.6
Public economics	5.3	4.6	8.7	13.1
Theory	4.3	3.5	10.0	9.7
Urban; rural; regional economics	0.0	1.9	2.1	4.0
Other	1.9	1.6	1.9	1.5
Any field	1.0	1.5	1.6	2.7

Table 7.8. Recommenders, recommendees and recommendations

	2008/09	2009/10	2010/11	2011/12
Number of recommenders who provided letters	1,638	2,443	3,322	5,023
Average number of recommendees per recommender	1.44	1.44	1.47	1.62
Number of recommendations sent per recommendee	2.30	3.73	6.08	21.38

of the recession. As a result, the economics job market appears to have been generally a "buyers' market," but some part of this effect might also be attributed to the relatively slower growth rate of advertisements placed on the EJM site relative to the number of applicants who are using EJM to apply for jobs.

Finally, Table 7.8 lists the number of recommenders who used EJM to transmit letters of recommendation over the four job market cycles. We see that the number of recommenders more than tripled, from 1,638 in 2008/09 to 5,023 in 2011/12. In addition, the number of recommendees per recommender has increased, though at a much slower rate: from 1.44 in 2008/09 to 1.62 in 2011/12. Besides the tripling of the number of recommenders using EJM, the reason for the explosive growth in the number of recommendation letters transmitted by EJM that we observed in the right-hand panel of Figure 7.1 is that the number of recommendation letters transmitted per recommendee increased nearly ten-fold, from 2.3 letters per recommendee in 2008/09 to 21.4 letters per recommendee in 2011/12. As we noted, average number of applications per applicant increased by a smaller amount, from eight applications per applicant in 2008/09 to twenty applications per applicant in 2011/12. We believe that, over time, an increasing number of recruiters who place advertisements on EJM are requiring letters of recommendation to be transmitted to them via EJM, and this explains why recommendations per recommendee has increased at a more rapid rate than the average number of applications per applicant.

Excess entry of intermediaries and market
fragmentation

There are several theories which at least partially capture the intuition that unrestricted entry of firms that supply intermediation services—*middlemen*—does not always lead to good outcomes, and can actually increase search and transactions costs. We have referred to this phenomenon as *market fragmentation*.

Ordinarily, the term *marketplace* connotes a single location where all relevant information and items to be traded are readily available to the individuals participating in the market. A fragmented marketplace is one in which there is no single location where all of the information and traders are located, but instead there are many separate "islands" or places where bargaining and trade can occur, and the information on prices and items for sale in these other markets are not readily available unless one visits them. As a result, traders need to incur significant costs to travel to other markets to search for and collect the information necessary to make good trading decisions. When the expected gains to searching in multiple marketplaces (or over multiple intermediaries) is sufficiently high, traders in these markets have to compare the payoff from arranging a potentially suboptimal transaction immediately with the discounted gains from continuing to search for better opportunities.

Neeman and Vulkan (2010) have argued that separate markets have a strong tendency toward consolidation into a single *central marketplace* where all trade occurs. They showed that consolidation not only reduces search and transaction costs, but improves the terms of trade for participants as the markets thicken. Neeman and Vulkan refer to their prediction that trade outside a single central marketplace should decline, and ultimately disappear as the *unravelling of the decentralized market*. Specifically, they considered a model of trade in a homogeneous commmodity, and considered the consequences of competition between two widely used exchange mechanisms: a "decentralized bargaining market," and a "centralized market." In their model, "in every period, members of a large heterogeneous group of privately-informed traders who each wish to buy or sell one unit of some homogeneous good may opt for trading through one exchange mechanism. Traders may also postpone their trade to a future period" (p. 1). Neeman and Vulkan's central result is that "*trade outside the centralized market completely unravels. In every perfect-like equilibrium, all trade takes place in the centralized market. No trade ever occurs through direct negotiations*" (p. 1).

Self-reinforcing mechanisms very similar to network externalities are at play in Neeman and Vulkan's unravelling result: the more valuable a central market is to buyers, the more valuable it is to sellers, and vice versa, and both will expect to achieve higher gains from trade from participating in the central market than in the decentralized bargaining market. We expect this intuition carries over to the economics job market as well: when a central market arises where employers can place job ads, this is also the place where job seekers will want to search, and when this happens there are strong self-reinforcing dynamics leading all buyers and sellers to participate exclusively in this central market.

While Neeman and Vulkan's argument is convincing in some cases, there are other markets where we fail to see the complete consolidation their model predicts, including the economics job market. Hall and Rust (2003) developed a different model that shows that a central market can coexist with a fringe of other intermediaries they call *middlemen*. Their model also captures the notion that market fragmentation drives up search and transaction costs, resulting in allocative inefficiencies.

Hall and Rust extended Spulber's (1996) model of search and matching where trade occurs via competing middlemen (intermediaries). Spulber's model can be viewed as a market that is completely fragmented: there are a continuum of buyers, sellers, and middlemen, and Spulber assumes that a buyer and seller can trade with each other only if they are matched by one of these middlemen. Buyers and sellers must engage in a costly search process to choose a middleman to buy or sell from. There is free entry of such middlemen, who have heterogeneous costs of intermediating trades. Spulber established the existence of a heterogeneous price search equilibrium in which buyers and sellers have heterogeneous reservation values (depending on their privately known valuation of the commodity). Most buyers and sellers will eventually trade when they find a middleman whose bid (ask) price is lower than (exceeds) their reservation value (for buyer and seller respectively).

We view Spulber's equilibrium as constituting a classic and extreme example of a fragmented market. There are no publicly posted prices at which individuals can trade at in this model. Instead, buyers and sellers are forced to engage in a costly search process to find a middleman that offers the most attractive price. Using this completely fragmented market as a point of departure, Hall and Rust showed how the equilibrium to Spulber's model changes when there is the possibility of entry by a monopolist market maker who posts *publicly observable* bid and ask prices. In that event, the majority of the trade occurs via the market maker, at the publicly posted bid and ask prices. Only a small fraction of residual traders choose to try to find prices that are better than the bid and ask prices posted by the market maker by searching in a much smaller residual market populated by the most efficient surviving middlemen.

Compared to Neeman and Vulkan's result, the entry of a monopolist market maker in Hall and Rust's model does not always cause the search and matching market to completely unravel, but it does succeed in driving out the majority of the least efficient middlemen. Thus, the entry of a market maker, i.e. an intermediary who posts publicly observable prices, *reduces, but may not eliminate market fragmentation.* However, if the market maker is not a profit maximizer but is rather a non-profit organization that only attempts to cover its operating costs, then in the event its marginal costs of intermediating trades is zero, complete unravelling in the Neeman and Vulkan sense will occur, and the entry of the non-profit market maker enables the market to achieve the fully Pareto-efficient Walrasian equilibrium solution.

We now consider a different model that illustrates how a natural contractual imperfection leads to market fragmentation, and how the entry of a non-profit charity (i.e. an organization similar to EJM) can help to alleviate the market fragmentation and improve market outcomes.

Suppose that there is a continuum of recruiters arranged on the unit circle, with a unit mass in total. Let $r \in [0, 1)$ denote an individual recruiter. For simplicity, let candidates and references be modeled collectively and assume there is a unit mass of candidates. Finally, suppose there are n intermediaries competing to serve recruiters to attract candidates. The intermediaries are equally spaced on the unit circle, at points $0, 1/n, 2/n, \ldots, (n-1)/n$.

Each recruiter wants to hire a single candidate, and makes a single job posting on one of the intermediaries' "websites." Every candidate wants to submit an application to every recruiter. Assume that, by law, recruiters must accept applications by regular mail even if they use web-based systems. Thus, an individual candidate has a choice of sending an application on paper by regular mail or submitting it via the electronic system of the intermediary that the recruiter has chosen. Suppose that a candidate pays a cost, c, for each intermediary that the candidate uses to submit applications online. A candidate also pays a cost, dm, per paper application, where m is the mass of recruiters to which he applies via regular mail. We assume $d > c > 0$ so that the cost of sending all applications by mail exceeds the cost of using a single electronic system to submit them all. Suppose the benefit to candidates of submitting applications exceeds these costs, so candidates will apply to every recruiter; thus, the issue is whether candidates use one of the web-based systems or submit paper applications.

To keep things simple, assume that if a recruiter has to deal with any paper applications then it pays a cost k. Also, a recruiter in location r that adopts the recruitment system of a firm in location x must pay a cost $\alpha(\min\{|x - r|, 1 - |x - r|\})^2$ due to the specifications of the recruitment system x being different than the recruiter's ideal r. (Note that $\alpha \min\{|x - r|, 1 - |x - r|\}$ is the distance between x and r on the unit circle.) Thus, recruiter r would be willing to adopt an electronic system from a firm at location i only if it is offered at a price that does not exceed $k - \alpha(\min\{|x - r|, 1 - |x - r|\})^2$ and will induce all of the candidates to apply electronically. Suppose the firms can provide recruitment systems at no cost. Payoffs are all measured in transferable monetary units.

This model exhibits two opposing efficiency concerns. First, note that recruiters like specialized software. Thus, to maximize their welfare without consideration of other market participants, it is optimal to have all of the intermediaries in the market supplying recruitment systems. In particular, if α is small so that $\alpha/4n^2 < k$, then to narrowly maximize recruiter welfare all n intermediaries should supply online application systems, and all recruiters should adopt such systems. If $\alpha/4n^2 > k$, then it is better to have a fraction of the recruiters use paper and regular mail. On the other hand, candidates (and the references they also represent in this model) benefit when recruiters use the same recruitment system.

Consider a three-stage game: first, the firms simultaneously select their contract offers; second, the recruiters observe the firms' pricing policies and simultaneously choose whether to accept contracts for recruitment systems; third, candidates observe the outcome of the first two stages, and simultaneously submit applications, by paper or electronically. We consider the coalition-proof subgame perfect equilibria of this game. Coalition proofness is applied to the recruiters' second-stage actions to deal with the fact that the recruiters are an atomless group (where an individual deviation would not directly affect the payoffs of the other parties).

We examine three cases.

Case 1: Full contracting

Suppose that the intermediaries are able to obtain fees from both recruiters and candidates but, for simplicity, assume that intermediaries cannot price discriminate. Thus, intermediary i's contract offer is a pair (p_i, q_i), where p_i is the price charged to recruiters for use of intermediary i's system, and q_i is the price per application charged to candidates. A candidate would then pay $q_i m$ to firm i to submit a mass m of applications using firm i's website.

Proposition 1. *If α is sufficiently close to zero then, with full contracting, there is a coalition-proof subgame perfect equilibrium of the game in which a single, centralized recruitment system prevails in the market.*

Proof sketch

Consider a strategy profile in which all of the intermediaries charge the same prices $p = c - d$ and $q = d - c$. In this case, the recruiters are supposed to coordinate by all selecting the recruitment system of firm 1, and then the candidates submit all of their applications via this system. It is clear that neither candidates nor any coalition of recruiters want to deviate from this specification. For instance, if a mass, m, of recruiters adopted one of the other intermediaries' systems, then no candidate would use it because the candidate would have to pay an additional lump sum, c, to use the second system. This would entail a cost $m(d - c) + c$, which exceeds the cost, dm, of submitting applications by regular mail to these recruiters.

Note that all of the firms get zero profits if the game plays out as just described. If an intermediary were to deviate by picking different prices (p', q'), then we prescribe a continuation of the game that is sensitive to whether $p' < c - d$ and/or $q' > d - c$. If $p' < c - d$ and $q' \leq d - c$, then we prescribe that the recruiters all adopt the system of the deviating firm, and the candidates apply using this website. If $p' < c - d$ and $q' > d - c$, then we prescribe that the recruiters all adopt the system of a single non-deviating firm, and the candidates apply using this website. In this second case, if the recruiters were to coordinate on the deviating firm, then the candidates would all opt for paper applications. If $p' > c - d$ then we prescribe that the recruiters coordinate by picking a single non-deviating firm. Thus, no intermediary can gain by deviating.

We argue that the setting just described is unrealistic because intermediaries typically cannot fully extract rents from candidates and references (the "candidates" in this model). In particular, we think that there are contractual imperfections that make it difficult to expropriate the benefit that references get from submitting letters through a centralized system. To understand the implications of this limitation, we look at the extreme case in which the intermediaries cannot exact payments from candidates.

Case 2: Partial contracting

Suppose that the intermediaries are able to obtain fees only from recruiters, so intermediary i's contract offer is a single price, p_i, that is charged to recruiters for use of firm i's system.

Proposition 2. *If c is sufficiently close to zero and there is partial contracting, in all coalition-proof subgame perfect equilibria of the game, all n firms have recruitment systems in use. Thus, the market for recruitment systems is fragmented.*

Proof sketch

Equilibrium prices must be non-negative, since firms cannot extract rents from candidates. Assume that, in equilibrium, intermediary i's recruitment system is not in use. It must be that, for some $\varepsilon > 0$, recruiters within ε of intermediary i's location

$(i-1)/n$ are obtaining a payoff no greater than $k - \frac{\alpha}{n^2} + \varepsilon$. But then intermediary i could offer a price close to zero so that the coalition of recruiters $[\frac{i-1}{n} - \varepsilon, \frac{i-1}{n} + \varepsilon]$ would prefer to purchase from firm i if they anticipate that the candidates would apply via intermediary i's system. A sufficient condition for candidates to behave in this way is that c is small. Thus, by offering such a price, firm i has positive sales, and earns positive profit, contradicting that this intermediary has no sales (and zero profit) in equilibrium.

So we conclude that realistic contractual imperfections not only lead to inefficiency as standard models predict, but they also lead to a particular form of inefficiency characterized by market fragmentation. An escape may come from the existence of an intermediary that internalizes the candidates' benefit of a centralized recruitment system.

Case 3: Partial contracting, non-profit

In our view, some non-profit charities play an important role of internalizing externalities through the preferences of the directors, managers, and financiers. In our model, for instance, suppose one of the n intermediaries is formed as a charitable organization, whose managers seek to increase the welfare of candidates (and references). In the extreme case, this firm obtains a value equal to its monetary profit plus the welfare of candidates. Assume partial contracting, as in case 2.

Proposition 3. *In the partial contracting setting with a charitable firm, and with α sufficiently small, if the charity's interests are enough aligned with that of the candidates then there is a coalition-proof subgame perfect equilibrium in which the charity runs a centralized recruitment system that all recruiters adopt.*

Proof sketch

Suppose that the charity offers the price $p = -\alpha\frac{1}{4}$. If all recruiters were to adopt the charity's system then all candidates would apply electronically, and the recruiters would all get payoffs of at least zero. No other firm could earn positive profits. If α is small, then the charity's loss is also small, and is dominated by the charity's satisfaction of serving the candidates.

This model is simplistic and merely suggestive; it does not capture the full richness and complexity of the economics job market, or the complicated dynamics of competition between intermediaries. However, it does succeed in illustrating circumstances where unrestricted entry of intermediaries can result in suboptimal outcomes, and even where competition among a fixed number of intermediaries (i.e. ignoring entry) results in market fragmentation. Further, the model suggests that these inefficiencies can be reduced by establishing a single central marketplace operated by a market maker whose role is to provide information to market participants, and match buyers and sellers. In the case where the market maker is a non-profit charity that can operate at nearly zero cost, the results indicate that nearly fully efficient outcomes can be achieved when all trade is conducted via this central market maker. Further, Neeman and Vulkan's unravelling results suggest that such an outcome should be stable: once a central market exists, there are no gains to individuals or even coalitions of buyers and sellers from trying to trade outside the central marketplace.

Our discussion considers how the presence of intermediaries in markets can affect welfare through fragmentation, but informational intermediaries can also have other, direct effects. Johnson and Rust (2012) considered a market where recruiters and candidates have publicly observable characteristics, but only learn their match value once a candidate has paid the cost of submitting an application, and the recruiter has incurred a cost of reviewing it. Due to these costs, recruiters strategically choose which received applications to review, and candidates strategically decide where to apply. Once the reviewers have moved, the allocation is decided through use of the Gale–Shapley algorithm, where candidates are considered unacceptable by any recruiter who did not review an application from them. Such a game has a large number of Nash equilibria, so Johnson and Rust focus on the perfect Bayesian equilibrium of a similar game in which the candidates and recruiters with better public signals are assumed to move first, creating a pattern of matching that is broadly assortative but incorporates some idiosyncratic tastes, similar to what is observed in the economics job market. This gives better candidates and recruiters a first-mover advantage, and selects a particular equilibrium to study.

Johnson and Rust found that candidates and recruiters tend to optimally use safety strategies, where they focus their search in a certain quality range, but include some lower-ranked options in case their preferred outcomes fall through. By lowering the costs of applying or reviewing applications, the agents tend to broaden their search downward, resulting in fairly dense competition in the middle of the pack. This benefits both sides of the market, since more of the "true preferences" are passed into the Gale–Shapley algorithm, leading to better matches. However, if the cost of reviewing applications is held fixed while the cost of applying is further reduced, the efficiency gains reach a threshold where further reductions in application cost fail to improve welfare. So although intermediaries like EJM can reduce costs dramatically on the applicant side, this translates into efficiency gains in terms of match quality only if the recruiting side is also optimally reviewing more applications.

OTHER POTENTIAL DESIGNS AND IMPROVEMENTS

EJM addresses many issues associated with the costs of applying, but other problems remain. For this reason, it is useful to consider how other markets and mechanisms overcome the transactional and informational challenges faced by the economics job market. In this section we will study several potential additional or alternative search mechanisms: job market signaling, guided search, centralized matching, and pricing mechanisms.

Signaling

The fact that the average number of applications per position advertised is large raises the concern that it may be costly for an employer to review all applications, and the employer may have multiple applications from job seekers who are indistinguishable in terms of observable characteristics, such as primary field of research, ranking of their degree program, and geographical location. In this environment, the employer may be able to reduce its search costs if it can focus its efforts on candidates who are more likely to accept the job offer than competitors who ex ante appear similar. The AEA signaling mechanism introduced in 2006/07 attempted to resolve some of this uncertainty, by allowing each job seeker to send a signal of particular interest to two employers via an AEA website. In theory, since these signals are scarce, they could be used to reveal information about the candidate's idiosyncratic preferences. Coles et al. (2010) provided the details of the AEA signaling mechanism and suggestive evidence that job seekers who used signals had a larger number of interviews. There is a growing number of studies that examine the role of signaling mechanisms in two-sided matching environments. In the context of college admission, Avery et al. (2004) compared the admission outcomes of students who used early application (thus sending their special interest in the college) with those who applied for regular admissions. In the context of online dating, Lee et al. (2009) analyzed a field experiment suggesting that signaling can improve search outcomes. Coles et al. (2013) examined the welfare implication of introducing a signaling mechanism in a model of a labor market.

Guided search

Rather than a simple central repository for information, an intermediary might provide tools for finding participants satisfying particular criteria, or even take an active role in making non-binding recommendations. This type of intermediation is often observed in dating service providers, such as eHarmony.com. Such "guided search" intermediaries could be useful in the economics job market as well. For example, suppose that intermediaries have better access to or lower costs of processing information about the pool of candidates, as well as a historical perspective on the search outcomes of recruiters. Then, by suggesting candidates who are especially suitable to a recruiter, the intermediary can assist the recruiters in focusing on candidates who are likely to meet their needs, instead of sifting through a large number of applications.

Second, applying in itself may be interpreted as a signal. A recruiter who receives an application from a candidate whom the recruiter perceives as overqualified may conclude the candidate must suffer some hidden deficiency, rather than infer that the candidate has an idiosyncratic interest in that recruiter. If an intermediary has better information about these idiosyncratic preferences, then it can make credible recommendations to the recruiters. Using data from an online matchmaking service, Lee (2009) found evidence supportive of this hypothesis. She found that the probability of a person's accepting a first date with another user is significantly higher if the online matchmaker introduces the two to each other, as compared with the case where the other user directly asks the person out.

Centralized matching

Many markets that share similar characteristics with the junior economics market have adopted some version of a centralized matching market. By centralized market, we mean that the participants report their preferences to a central authority whose function in the market is to aggregate this information, then use an algorithm to translate the preferences into a match. Notable examples include the matches between hospitals and gastroenterologists, and assignments of children to public schools; see, for example, Roth (1984), Roth (1991), Roth and Xing (1994), and Niederle and Roth (2003).

A growing number of empirical studies have compared market outcomes under decentralized matching with outcomes from centralized matching mechanisms. Niederle and Roth (2003) found that the likelihood of a medical student finding a residency in a hospital where he had no prior affiliation increased under centralized matching in the gastroenterology market. In the context of marriage markets, Hitsch et al. (2010), Banerjee et al. (2013), and Lee (2009) inferred mate preferences of individuals based on their dating history and used the estimated preferences to compute stable matchings using the Gale–Shapley algorithm. Hitsch et al. (2010) and Banerjee et al. (2013) found that, overall, the sorting pattern generated by the Gale–Shapley algorithm is comparable with that observed in their decentralized marriage markets, for example the US online dating market (Hitsch et al. 2010), and the Indian marriage market (Banerjee et al. 2013). In contrast, using a South Korean data-set, Lee (2009) found that marital sorting under the Gale–Shapley algorithm exhibits less sorting along geography and industry, compared with the sorting observed in actual marriages. These findings suggest that the extent to which the introduction of a centralized matching market will change outcomes may vary across the current market conditions.

Price-based mechanisms

By reducing application costs, there is a substantial risk that candidates will reach "corner solutions" where they apply to all opportunities, and the informational signal generated by submitting an application is wiped out. Consequently, recruiters will be unable to infer anything from the receipt of an application about the candidate's likelihood of accepting an offer, leading to an *increase* in inefficiency. Moreover, since the recruiters bear the burden of evaluating the candidates, the bottleneck on efficiency is likely to be a lack of attention paid to many of the applications received.

One way to address this issue is to introduce price-based mechanisms, like auctions or application fees, which can be used to reveal information about the participants. Studies such as Damiano and Li (2006), Hoppe et al. (2009), and Johnson (2010) examined how to design such mechanisms. Hoppe et al. (2009) and Johnson (2010) examined environments in which agents bid competitively for partners to signal their quality, leading to assortative matching based on the intensity of the signals. Johnson (2010) showed that profit-maximizing intermediaries, however, may be tempted to deviate from assortative matching, as well as refuse to arrange some socially valuable matches. Damiano and Li (2006) studied a mechanism where, instead of bidding, agents pay a fee for access to a smaller *pool* of agents for a match. By charging an increasing fee schedule for access to the pools on each side of the market, agents are incentivized to sort themselves by quality, resulting in more efficient matching.

While it is unlikely that such a "fine-tuned" mechanism would ever appear in the economics job market, the concept may be a useful one. A paper-based system imposes uniform costs across all candidates and recruiters for applying and reviewing. Since a centralized market would allow recruiters to decide on an application fee, a substantial number of "spurious" applications could be avoided. Moreover, the informational content of receiving an application will be restored, since candidates will once again be forced to think strategically about which opportunities to pursue. Rather than being wasted in the less informative signaling process of postal mail, this set-up could allow both sides of the market to better signal their intentions while still pursuing the goal of reduced inefficiency.

CONCLUSION

In this chapter we posed the question: "can the economics job market be improved?" Thanks to the efforts of the American Economic Association to promote the job interviews at the ASSA meetings and create the JOE website, the economics job market already operates much more efficiently than most other labor markets. Nevertheless, we have identified several key areas where further improvements can be made to improve the operation and efficiency of the economics job market.

An important precondition for any well functioning marketplace is that market participants have easy access to all the relevant information they need to make informed decisions. Prior to the advent of the web and online labor market intermediaries such as EJM and other services we have discussed in this chapter, assembling and transmitting this information to market participants was a major task that consumed substantial physical resources. The high cost of operation of paper-based systems caused market participants to operate on far less than the full set of available information.

While the adoption of information technology and the entry of intermediaries offering online advertisement posting, application, and reference letter delivery services has greatly reduced these costs, the proliferation of these competing labor market intermediaries has had offsetting negative effects. Each of these intermediaries offers only a subset of the full set of information that market participants would ideally like to have to make informed decisions. Since the competing labor market intermediaries do not generally share their information or attempt to be *interoperable,* we have argued that information technology has had a paradoxical negative effect on the operation of the economics job market, leading to an outcome we refer to as *market fragmentation.* When this happens, search and transaction costs can be driven up rather than driven down by the use of information technology, and this can worsen rather than improve market outcomes. We showed that the "market" for applications to graduate schools is already badly fragmented, and the inefficiencies this causes are a serious collective waste of scarce time of faculty members and students, even if these systems do benefit admissions committees of graduate schools.

The creation of EJM was motivated by the concern that the economics job market could eventually become as badly fragmented as the market for applications to graduate schools. The goal of EJM is to centralize the information to market participants, and reduce or eliminate market fragmentation, resulting in a far more efficient market that benefits *all* participants, rather than primarily benefiting recruiters through electronic delivery of application files to their recruiting committees.

To the extent that EJM is just another intermediary, however, it is fair to ask whether the entry of EJM is contributing to market fragmentation or ameliorating it. Although we have shown that EJM is growing at exponential rates and currently intermediates a significant fraction of the total number of job applications, it is too soon to know whether EJM will have a lasting, positive impact on the operation of the economics job market. We have shown that existing theoretical analyses, including the influential model of Neeman and Vulkan (2010), suggest that even in the absence of any explicit coordination, there are strong self-reinforcing dynamics at play that lead fragmented markets to "unravel" so that trade concentrates in a single central marketplace. Whether this will ultimately happen in the economics job market remains to be seen.

Although previous empirical studies that have shown that labor market intermediaries similar to EJM have resulted in significant improvements in other labor markets where the problem of market fragmentation can be managed (such as the Alma Laurea system operated by a consortium of Italian universities), it is unlikely that the current iteration of EJM will solve several other potential problems that we identified in the economics job market.

Perhaps the most significant problem is that even though EJM might drive down the cost of *transmitting* the critical information necessary at the first stages of the job market, it may have only a small effect on reducing the cost of *evaluating* this information. Although web-based candidate evaluation systems have a number of advantages over paper-based technology for recruiters, nevertheless the dominant bottleneck in market efficiency is the human time cost involved in reading applications and evaluating the information about the candidate to try to determine what the candidate's "true quality" is.

We have raised the possibility that technologies that reduce the cost of application may drive up the number of applications, and this could result in less "self-selection" by applicants, and cause recruiters to devote less time to evaluating each candidate. Indeed, we have documented a dramatic rise in the number of applications received by recruiters who use EJM. Once again, this could produce a paradoxical result that an improvement in information technology could worsen market outcomes.

These problems led us to consider several other strategies for improving the economics job market, including the use of computerized "match-making" services as part of a "guided search" strategy that Lee (2009) has shown to be effective in producing better matches in online dating contexts, to much more radical approaches, such as the use of computerized matching algorithms or price-based signaling mechanisms.

Computerized matching and auctions are highly centralized approaches because they require a high degree of coordination, and possibly even compulsory involvement on the part of market participants to be successful. While these mechanisms are potentially of the most interest from a market design perspective (and potentially could yield the greatest improvements in match quality), we do need to keep in mind the practical constraint that our power to design markets is quite limited in practice, given that our market is more akin to a commons that no single individual or organization owns or controls.

In particular, we have emphasized the critical *voluntary participation constraint* that can make it hard to implement centralized solutions, particularly when they result in improvements in payoffs to one group at the expense of another. Consequently, our focus has been more on attempting to *improve* the economics job market via an innovation that might be voluntarily adopted rather than attempting to design the economics job market which would presume a level of control and influence that none of us possesses.

The future evolution of the economics job market is likely to depend on how much improvement can be achieved by more modest interventions such as EJM that do not involve any compulsion or obligation in order to achieve wide scale use by market participants. If these sorts of systems can ameliorate the most severe inefficiencies, then there may be much less need for more radical interventions that do require some degree of compulsion in order to be successful. As we noted, Hitsch et al. (2010) and Lee (2009) come to different conclusions about the extent to which decentralized, privately determined matching outcomes from a dating service approximate the matches produced by a centralized approach—the Gale–Shapley matching algorithm. The extent to which decentralized outcomes in labor markets with intermediaries that provide guided search and matching services approximate outcomes produced by centralized matching algorithms is an interesting open question.

We conclude that more experience and further empirical and theoretical research are necessary to determine whether the decentralized search and matching process—perhaps intermediated by systems such as EJM and guided search—could result in an acceptably high level of efficiency in matching outcomes in the economics job market, or whether significant inefficiencies persist that would provide a strong case for adopting more ambitious mechanisms such as matching algorithms or price-based mechanisms to further improve the operation of the economics job market. However, the informational centralization of the economics job market provides a useful starting point, and suggests many avenues for future research.

References

Avery, C., Fairbanks, A. and Zeckhauser, R. (2004) *The Early Admissions Game: Joining the Elite*, Harvard University Press.

Bagues, M. and Labini, S. (2009) "Do online labor market intermediaries matter? The impact of *Alma Laurea* on the university-to-work transition," in *Studies in Labour Market Intermediation*, University of Chicago Press, pp. 127–154.

Banerjee, A., Duflo, E. Ghatak, M. and Lafortune, J. (2013) "Marry for what? Caste and Mate selection in Modern India," *American Economic Journal: Microeconomics*, 5(2): 33–72.

Coles, P., Kushnir, A. and Niederle, M. (2013) "Preference signaling in matching markets," *American Economic Journal: Microeconomics*, 5(2): 99–134.

——— Cawley, J., Levine, P., Niederle, M., Roth, A. and Siegfried, J. (2010) "The job market for new economists: a market design perspective," *Journal of Economic Perspectives*, 24(4): 187–206.

Damiano, E. and Li, H. (2006) "Price discrimination and efficient matching," *Economic Theory*, 30: 243–63.

Hall, G. and Rust, J. (2003) "Middlemen versus market makers: a theory of competitive exchange," *Journal of Political Economy*, 111: 353–403.

Hitsch, G., Hortaçsu, A. and Ariely, D. (2010) "Matching and sorting in online dating markets," *American Economic Review*, 100(1): 130–163.

Hoppe, H. Moldovanu, B. and Sela, A. (2009) "The theory of assortative matching based on costly signals," *Review of Economic Studies*, 76: 253–81.

Johnson, T. (2010) "Matching through position auctions", *Journal of Economic Theory*, 148: 1700–1713.

——— and Rust, J. (2012) "A two sided matching model of the economics job market," University of Notre Dame and Georgetown University.

Lee, S. (2009) "Marriage and online mate-search services: evidence from South Korea," Working Paper, University of Maryland.

——— Niederle, M., Kim, H. and Kim, W. (2009) "Propose with a rose? Signaling in Internet dating markets," University of Maryland and Stanford University.

Neeman, Z. and Vulkan, N. (2010) "Markets versus negotiations: The predominance of centralized markets," *BE Journal of Theoretical Economics*, 10(1): 6.

Niederle, M. and Roth, A. (2003) "Unraveling reduces mobility in a labor market: gastroenterology with and without a centralized match," *Journal of Political Economy*, 111: 1342–52.

Roth, A. (1984) "The evolution of the labor market for medical interns and residents: a case study in game theory," *Journal of Political Economy*, 92(6): 991–1016.

———— (1991) "A natural experiment in the organization of entry-level labor markets: regional markets for new physicians and surgeons in the United Kingdom," *American Economic Review*, 81(3): 415–40.

———— and Xing, X. (1994) "Jumping the gun: imperfections and institutions related to the timing of market transactions," *American Economic Review*, 84(4): 992–1044.

Rysman, M. (2004) "Competition between networks: a study of the market for yellow pages," *Review of Economic Studies*, 71: 483–512.

Spulber, D. (1996) "Market making by price-setting firms," *Review of Economic Studies*, 63: 559–80.

CHAPTER 8

....

DESIGNING MARKETS
FOR IDEAS

....

JOSHUA S. GANS AND SCOTT STERN[1]

INTRODUCTION

....

MARKETS have emerged and been designed for all manner of physical goods and services. In some cases, they have been designed for seemingly intangible assets (such
as spectrum). However, it is fair to say when it comes to ideas—which have a clear
economic value and also a value in exchange—that the emergence of markets has been
relatively sparse. Specifically, ideas may be valuable to many users and may be produc-
tively employed in applications or contexts far removed from the locus of the idea's
generation or invention. When the value of an idea can be realized only by "matching"
that idea with key complementary assets (Teece, 1986), markets that facilitate matching
and efficient distribution in a timely fashion will provide significant social returns.

Nonetheless, markets for ideas are relatively uncommon. While there are transactional exchanges in both ideas and technologies, and the rate of "ideas trading" seems
to be increasing over time (Arora, Fosfuri, and Gambardella, 2001), it is still very rare
for ideas or technologies to be traded in what economists would traditionally refer to
as an organized market. Instead, most exchanges of abstract knowledge or prototype
technology occur under conditions that are best described as a *bilateral monopoly*: the
buyer and seller engage in negotiations with limited outside options in terms of alternative exchanges. Buyers (sellers) are unable to play potential sellers (buyers) off against
one another, limiting the potential for competition to generate a stable equilibrium price
and evenly distribute gains from trade. Successful negotiations vary widely in terms of
the price and terms over which knowledge is transferred. Mark Lemley and Nathan

[1] Parts of this paper are drawn from Gans and Stern (2010). We thank the Australian Research
Council for financial assistance. Responsibility for all errors lies with the authors. The latest version of
this paper is available at <http://research.joshuagans.com>.

Myrvhold label the market for patents as "blind." "Want to know if you are getting a good deal on a patent license or technology acquisition? Too bad" (Lemley and Myrvhold, 2008; see also Troy and Werle, 2008). Not simply a matter of how the rents from a given idea are distributed between buyer and seller, the lack of transparent price signals results in distorted and inconsistent incentives to produce and commercialize new ideas.

The purpose of this chapter is to examine design issues associated with markets for ideas, with the aim of understanding what barriers might exist to their creation or emergence. Our analysis here is both qualitative and speculative. Our purpose is to identify potential areas for further study rather than to provide a theoretical and empirical treatment of the broad issue. In that sense, we aim here to provoke thought and promote further investigation into this largely untapped area of study in the design literature.

To this end, we employ the recent synthesis by Al Roth in characterizing the principles and challenges faced by market designers. Roth (2008; see also Chapter 1 of the present volume) draws on the emerging body of evidence from real-world market design applications to offer a framework and conditions upon which market designers can evaluate the effectiveness of their prescriptions. Specifically, Roth highlights three outcomes that are associated with efficient market operation: market thickness (a market is "thick" if both buyers and sellers have opportunities to trade with a wide range of potential transactors), lack of congestion (i.e. the speed of transactions is sufficiently rapid to ensure market clearing but slow enough so that individuals, when considering an offer, have the opportunity to seek alternatives), and market safety (a market is "safe" if agents do not have incentives for misrepresentation or strategic action that undermine the ability of others to evaluate potential trades). When these outcomes arise, market participants are able to consider trading with full access and knowledge of potential alternative transactions, yielding efficiency above and beyond bilateral exchange. Roth also identifies an important (and, to traditional economic theorists, surprising) feature of some real-world markets that he terms *repugnance*. In some markets, such as those for kidneys or sex, market designers are significantly constrained by social norms or legal restrictions that limit the use of the price system as an allocation mechanism. Importantly, while repugnance might impact on the uncoordinated evolution of market-based exchange, Roth argues that effective market design will proactively manage the constraints arising from repugnance.

To apply this framework to the case of ideas and technological innovation, we draw on insights from research on markets for technology (MfT). The MfT literature explores how technological innovation (as well as intangible knowledge goods) differs from more traditional goods and services, and considers the implications of these differences for business and public policy. In order to develop specific insights, we highlight three important characteristics of ideas that may impact the formation and efficient operation of a market. The salience of each may vary in different settings.[2] First, *idea*

---

[2] These three characteristics are synthesized from prior research in the MfT literature, and result from the potential for ideas and technology to be both non-rivalrous and non-excludable (Romer,

complementarity recognizes that ideas are rarely of value in isolation: to be of most value, ideas require matching with both complementary assets and complementary ideas (Teece, 1986; Bresnahan and Trajtenberg, 1995). Second, *user reproducibility* can mean that it is often difficult, as a seller, to appropriate an idea's full value (Arrow, 1962; Teece, 1986): specifically, in the absence of strongly delineated and easily enforceable intellectual property rights, disclosures or access may allow users to reproduce or expropriate ideas. Finally, even though ideas may be partially non-rivalrous in use—that is, a single idea may be able to be used by many individuals, and ideas may be replicated at low (even zero) marginal cost (Romer, 1990)—the economic exploitation of ideas may be subject to *value rivalry*. That is, users' willingness to pay for ideas may decline with the level of diffusion of that idea. The main contribution of this chapter is to use the market design evaluation scheme proposed by Roth to assess how these three economic properties of ideas impact the viability, operation, and structure of a multilateral market for ideas.

We highlight three main findings. First, *the nature of ideas undermines the spontaneous and uncoordinated evolution of a corresponding market for ideas.* Idea complementarity, user reproducibility, and value rivalry significantly undermine the ability to achieve certain types of contracts and engage in certain types of bargaining which are essential for an effective multilateral trading mechanism. For example, both the market thickness and market safety conditions identified by Roth suggest that buyers of ideas should be able to consider multiple offers from multiple potential sellers before contracting with a particular seller. However, when user reproducibility is high, the initial seller of an idea in an organized market faces the prospect that the *first* buyer is likely to subsequently become a seller (and competitor). In this case, the very existence of an organized exchange undermines the ability to conduct any trade at all. Our second central finding is a corollary of the first. *Specific institutions, most notably formal intellectual property rights such as patents, play a crucial role in addressing the challenges raised by market design.* For example, when patents are effective and enforceable, sellers are able to overcome both the disclosure problem and the potential for resale by buyers, which facilitates multilateral bargaining and raises the potential for efficient matching. Indeed, the rise of formalized patent exchanges and auctions such as Ocean Tomo demonstrates the potential for organized markets for abstract ideas that are protected through the patent system. At the same time, there are some environments where there may be a patent thicket—where overlapping and uncertain intellectual property rights make it difficult for a potential buyer to negotiate for access from multiple owners of an intellectual property. When the potential for patent thickets is particularly salient,

1990). In particular, the characteristics we emphasize are drawn from studies that examine the prevalence and rise of ideas and technology trading across different economic sectors (Arora et al, 2001; Lamoreaux and Sokoloff, 2001; Gans and Stern, 2003), the determinants of the innovative division of labor, particularly with respect to "general purpose" technologies (Arora and Gambardella, 1994; Bresnahan and Trajtenberg, 1995; Gambardella and Giarratana, 2008), and the special role played by formal intellectual property rights (such as patents) in facilitating knowledge transfer across firm boundaries (Arora, 1995; Gans, Hsu, and Stern, 2002, 2008; Troy and Werle, 2008).

enhancing the strength of intellectual property rights (e.g., by allowing for injunctive relief) can actually undermine the potential for a multilateral and coordinated market for ideas by enhancing individual incentives to engage in hold-up. Our final and perhaps most speculative observation is that *the most robust markets for ideas are those where ideas are free*. This is not only because, in many respects, those markets satisfy Roth's three conditions for effective market design, but also because those markets overcome some of the key constraints arising from repugnance.

The outline of this chapter is as follows. In the next section, we consider how the nature of ideas themselves impacts upon the effectiveness of markets for ideas. The subsequent section then examines the impact of repugnance and how specific real-world institutions and norms (such as those associated with open science) can be understood as attempts to facilitate multilateral idea exchange while managing the repugnance associated with idea trading. The final section concludes our analysis.

HOW DOES THE NATURE OF IDEAS IMPACT THE DESIGN OF MARKETS FOR IDEAS?

In this section, we explore some distinctive characteristics of ideas that pose challenges for designers of markets for the exchange of ideas. Our approach is to consider Roth's (2008) three essential criteria—thickness, lack of congestion, and safety—for efficient market design and to identify how particular aspects of ideas as economic commodities impact on each (see Chapter 1). The aspects of ideas we focus on were identified because they highlighted challenges in meeting Roth's criteria and are not aspects commonly salient across different ideas. We focus on three central characteristics of ideas that we believe offer insight into the feasibility and efficiency of the market for ideas: *ideas complementarity*, *value rivalry*, and *user reproducibility*. Each of these characteristics is a distinct aspect of innovation, and each may be more important for some types of ideas or technologies than others. Consequently, market designers will likely have different challenges depending upon the type of ideas being examined.

It is useful to review each of these aspects of ideas in turn prior to relating them to market design issues. First, ideas complementarity concerns the fact that the value of any given idea depends on its combination with others. For example, the value of a software algorithm depends crucially on the availability and properties of related pieces of software (and hardware, for that matter). Ideas complementarity arises from the interdependence among different ideas in particular applications and contexts (Rosenberg, 1998). The ability to trade a given idea (and the terms of that trade) may depend crucially on the availability and terms of access to other ideas for which such a strong interdependency exists. For instance, when ideas are of little value in isolation, downstream users may require access to multiple ideas in order to gain value from each idea.

Second, value rivalry is a subtle consequence of the non-rivalry of ideas (Romer, 1990). In the traditional study of innovation, ideas and knowledge are non-rivalrous in use but also in valuation: the ability to read a story does not depend on whether others have read that same story, and the enjoyment that one reader gets from a story is independent of whether others have also read the same story. However, in many applications and contexts, while ideas may be non-rivalrous in use (many people can have access to the same piece of information), they may be rivalrous in value (the value gained from having access to that information declines with an increase in the number of other individuals who have access to the same idea). To take but one extreme case, insider information about a financial asset is non-rival in use (many people could in principle have access to that information) but the advantage received from the information depends on it being maintained as a secret. A less extreme case of value rivalry arises in the context of drug development—while many firms can, in principle, take advantage of a cure to a disease, the private value of that scientific knowledge, to an individual firm, is higher if no other downstream firm is allowed to take advantage of this knowledge in the commercialization process of the drug. The *degree* of value rivalry, thus, depends on whether the value of an idea to a potential user/buyer declines when others have access to the same idea.

Finally, user reproducibility is a particular manifestation of the low cost of replication of information and ideas. While the low replication cost of information is well studied, little consideration has been given to the case when the *buyer* of an idea can also be in a position to replicate that idea for use by others—we consider this in our discussion. To take but one extreme example, the replication cost of music has been low since the development of recording technologies such as the phonograph and magnetic tapes; however, it was not until the development of both digital music formats such as CDs and MP3s and also the connectivity of the Internet that individual music consumers have been able to share (or even sell) recordings to a large number of other potential listeners (as indeed occurred with the rise of Napster and other music-sharing exchanges). The degree of user reproducibility is measured by the extent to which potential buyers of ideas are able to replicate that idea at low cost and share that idea with, or sell it to, other potential buyers.

These three distinctive properties of ideas—ideas complementarity, value rivalry, and user reproducibility—are likely to pose distinctive challenges for the feasibility and operation of a market for ideas. The remainder of this section focuses on how each of these factors impacts the Roth criteria.

Market thickness and ideas complementarity

While market thickness is a challenge in many settings, of particular note is the lack of thickness in the market for ideas and knowledge (Lemley and Myhrvold, 2008; Troy and Werle, 2008). Even when strong intellectual property rights exist (e.g., ideas are embedded in patents), market development has been of only limited scale and scope

(Lamoreaux and Sokoloff, 2001).[3] Notably, while patent auctions have long been dis-cussed (Barzel, 1968; Kitch, 1977; Kremer, 1998; Abramowicz, 2006), formal patent auctions have operated for only the past few years. As we discuss in more detail later in the chapter, most analyses of patent auctions, such as those organized by Ocean Tomo, suggest that they cover a relatively narrow range of innovation, and winning bids are at relatively modest prices (Kanellos, 2006). The lack of a thick market in patented ideas seems at first puzzling, given that there should (in principle) be little difference between a patent market and a secondary market for a more traditional capital good such as machinery, property, or collectibles.

While the lack of market thickness for knowledge—even patented knowledge—may be due to many reasons, the most significant is likely to be related to ideas com-plementarity, which can pose a central challenge to market design. If the value of a given (patented) idea depends on access to other (patented) ideas, then the returns to participation in a market depend on whether the market is likely to include all of the required knowledge inputs. In the absence of being able to aggregate across a "package" of intellectual property assets, potential buyers do not have incentives to offer a high price for any individual asset. From the perspective of a potential seller, it would indeed be preferable if all *other* sellers first engaged in trade with a particular buyer, thus offering a significant opportunity for hold-up as the last remaining intellectual property bottleneck. While the challenges of hold-up over intellectual property and the potential for patent thickets have been extensively discussed (Grindley and Teece, 1997; Shapiro, 2001; Heller, 2008), we are making the more nuanced claim that the potential for hold-up undermines the incentives for both buyers and sellers to participate in an organized exchange where many (but not all) relevant intellectual property assets may be offered.[4]

It is important to emphasize that the lack of market thickness is not simply due to the potential for hold-up (we discuss potential institutions to mitigate hold-up below). In particular, a key challenge in commercialization is that the value from a single innovation is only realized over time, during which the emergence of complementary ideas and technologies can be uncertain (Rosenberg, 1998). When ideas are developed over time, and ideas are complementary with one another, it is extremely difficult to develop a market mechanism in which each idea receives an appropriate market valu-ation (McDonald and Ryall, 2004).[5] The market design challenge is heightened when the precise form and timing of future ideas and technologies are difficult to anticipate,

[3] Levine (2009) finds that innovator returns to new drug development are related to the number of firms that market within a given physician specialty, with the share of returns less related to market size when such marketing functions are concentrated.

[4] The market design problem that arises from ideas complementarity is analogous to the more general problem in auction design when different items have interdependent valuations. Milgrom (2007) emphasizes that the problem of interdependency is among the most challenging issues in effective auction design, and proposes a framework for evaluating how to develop a mechanism that allows for such interdependencies to be taken into account.

[5] This is again analogous to the problems of combinatorial auction design emphasized by Milgrom (2007). Though we do not pursue it here, the market design challenge of aggregating ideas developed

and some of the most valuable "packages" are serendipitous combinations that emerge from disparate settings.[6]

When complementarity between ideas is important, and assuming that effective intellectual property rights are available (a topic we return to below), it is still possible to aggregate different ideas into a single package. Both patent pools and formal standard-setting processes reflect partial attempts to solve this aggregation problem. Patent pools combine different pieces of intellectual property owned by different property rights holders into a package which potential users can license in order to gain the freedom to use a set of interdependent technologies. These cooperative marketing agreements by the owners of intellectual property rights have the potential to overcome the coordination problem involved in selling overlapping ideas, and seem to serve as a mechanism in which a single (aggregate) seller encourages participation by potential buyers through the availability of "one-stop shopping" (Lerner and Tirole, 2004; Chiao et al., 2007). Standard-setting organizations also play a role in encouraging market thickness, and do so in an institutional context in which the values of both buyers and sellers are explicitly taken into account in the standard-setting process (Lerner and Tirole, 2006; Simcoe, 2008). Each of these institutional responses to ideas complementarity—patent pools and standard-setting—achieve market thickness by (1) limiting the range of technical alternatives that can be combined (i.e., by creating a 'standard' mode of operation, such as PAL codes for DVDs), and (2) leaving the status of *future* ideas and technologies ambiguous.

over time offers a potentially useful social function for so-called patent trolls or speculators, who acquire intellectual property rights during an embryonic phase.

[6] It is useful to note that, when ideas complementarity is relatively unimportant, it is possible to support thick markets for knowledge and ideas. For example, the recent rise of modular platforms for buying and selling applications software—such as Apple's iTunes Application Store—seem to provide concrete examples where an exchange mechanism can exist as long as the interdependency among different offerings is not too severe. While the management of technology literature has already emphasized the role of modularity in the creation of technical platforms that encourage third-party applications (Baldwin and Clark, 2000; Gawer and Cusumano, 2002), it is still useful to consider the market design role that such platforms play. Apple offers developers cheap (and easily accessible) product development and digital rights management tools to develop their ideas and applications. Then, while Apple assesses potential applications to ensure that they meet minimum quality thresholds and technical standards, Apple allows developers to offer their iTunes applications for sale on an integrated platform, choose their own pricing (including the option of free distribution), and has established a standard revenue-sharing plan (in which Apple retains 30% of all revenue). By designing a platform technology that minimizes the interdependency between individual innovations, the iTunes Application Store induced the development and exchange potential for more than 50,000 different application ideas and more than 1 billion application transactions within the first year of its founding. (<http://en.wikipedia.org/wiki/App_Store#cite_note-AppleFrontWeb1-12>, retrieved July 8, 2009).

A final institutional response is to simply design the market in a way that allows the entire "solution" of complementary ideas to be combined (and valued) in a single package. This is the essence of a prize system. While the concept of innovation prizes have been around for centuries (Mokyr, 2008), there has been a recent flurry of innovation prize offerings (mostly by philanthropic organizations) ranging from reusable spacecraft to energy efficient cars to the development of specific vaccines.[7] Of course, while a prize mechanism does encourage supply and provides a particular type of predetermined demand for an innovation, most ideas production is resistant to a prize mechanism because of the inability to completely and accurately specify the performance criteria and relevant parameters in advance (indeed, specifying the fundamental requirements of a design is often the most important "idea" regarding that design). More generally, it is useful to emphasize that each of the three institutional responses to ideas complementarity—patent pools, standard-setting, and prizes—achieve market thickness by (1) limiting the range of technical alternatives that can be combined (i.e., one may not be able to achieve operability outside the "standard" or one may ignore key design elements in the prize specification) and (2) leaving the status of *future* ideas and technologies ambiguous.

Congestion and value rivalry

We now turn to examine the impact of value rivalry on market congestion. When there is a high degree of value rivalry, the disclosure of the idea (even if not the sale) to one potential buyer reduces the value of that idea to other potential buyers. As emphasized by Anton and Yao (1994) and Gans and Stern (2000), the bargaining power of an idea's seller in a bilateral negotiation arises in part from their ability to agree to keep the idea a bilateral secret, conditional on a sale.[8] However, bilateral secrecy is at odds with the ability of an idea's seller to play multiple potential buyers off against one another *before* agreeing to an exclusive sale to the highest bidder. If the disclosure of the idea to all potential bidders undermines the valuation of the idea by each of those bidders, sellers of ideas may be very limited in their ability to consider multiple offers for a single idea.

---

[7] Prizes and forward contracts need not be large scale (Kremer & Williams, 2010). For example, InnoCentive allows established firms (which are vetted for credibility) to post problems they seek to have solved. One challenge set $100,000 for the delivery of a non-ion-sensitive super-absorbent polymer, while another by Kraft looks for bakeable cheese technology partners and many have been awarded (100 in all). Overall more than 140,000 people from most countries in the world have registered as potential solvers on the site (<http://www.innocentive.com/>).

[8] This is a refinement of Arrow's classical statement on disclosure (1962), and is emphasized in the literature on the impact of appropriability and the commercialization of new technology (Teece, 1986; Levin et al., 1987).

There are, of course, some sectors in which a limited market for ideas exists, and where it is possible to observe the consequences of value rivalry and limited appropriability. For example, in the market for movie scripts, a screenwriter will prepare a short treatment that, in some circumstances, can be marketed simultaneously to multiple potential movie production companies. While this facilitates effective matching (and, in the best of circumstances, allows the screenwriter to play different producers off against one another), the history of the movie industry is littered with stories in which a movie treatment is originally "rejected" by a producer who then develops the idea or a very similar variation. In some cases, this can lead to multiple studios producing very similar movies at the same time, limiting the box office success of each offering.[9]

It is perhaps not surprising that the main consequence of value rivalry is likely to be congestion. Rather than dilute the valuation of all potential buyers by disclosing (at least part of) the idea broadly, a buyer and seller may agree to engage in bilateral negotiations for a fixed period of time, with significant penalties for disclosure to third parties. That is, they retain value by limiting use. For example, in high-technology industries such as biotechnology and software, bargaining over the details of a license (including the detailed disclosures of the underlying technology) is often conducted on an exclusive basis, with both parties agreeing to limit contact with other potential buyers and sellers for a certain amount of time. These due-diligence periods imply that the detailed negotiations over the precise terms and conditions of a license take place in a bilateral rather than multilateral environment. This potentially leads to efficiency losses resulting from poor match quality and significant uncertainty regarding the "fair" price for an idea of a given quality. As emphasized by Lemley and Myhrvold (2008):

> Willing licensors and licensees can't find each other . . . no one can know whether they are getting a steal or being had. When parties do license patents, the prices are (to the extent we can tell) all over the map. And the rest of the world has no idea what those prices are. This, in turn, means that courts lack adequate benchmarks to determine a "reasonable royalty" when companies infringe patents. The lack of a real, rational market for patent licenses encourages companies to ignore patent rights altogether, because they cannot make any reasonable forecast of what it would cost them to obtain the licenses they need and because they fear that they will pay too much for a technology their competitors ignore or get on the cheap. At the same time, ignorance of prices permits unscrupulous patent owners to "hold up" companies that make products by demanding a high royalty from a jury that has no way of knowing what the patent is actually worth.

---

[9] See McAfee (2002, table 7.1, p. 155). Similar releases around the same time include movies whose main themes are Robin Hood (1991), volcanos (1997), animated ants (1998), asteroids (1998), Mars missions (2000), animated urban to wild animals (2005), animated penguins (2007), and Truman Capote (2007).

In other words, value rivalry poses a market design challenge that results by and large in a sharp tradeoff for buyers and sellers in the market for ideas: engage in either isolated bilateral transactions that involve inefficient allocation, or multilateral market-based bargaining that can reduce the productive value of completed trades.

It is useful to emphasize that when intellectual property rights are not costlessly enforceable (a topic we return to later), the use of bargaining protocols that induce congestion may be privately optimal to any particular buyer and seller while nonetheless being socially inefficient. Each potential buyer's value may depend on whether other buyers have had access to the technology or not (since rival access would allow competitors to expropriate some portion of the value by imitating technology, and raising the level of competition in the market). In this circumstance, a particular buyer–seller pair will seek to minimize informational leakages—by maintaining the idea as a bilateral secret—in order to retain the value created by their transaction. In such a circumstance, very few buyers will be able to evaluate and compete for access to the idea ex ante, lowering the probability that the ultimate buyer is a good match. Importantly, in the absence of an effective matching mechanism, the value of each sale in the market for ideas goes down, as the willingness to pay of a poorly matched buyer is lower than the willingness to pay of the "ideal" buyer.

Safety and the control of user reproducibility

Finally, we consider the challenges involved in ensuring market safety and transparency in the context of the buying and selling of ideas. While the unique properties of ideas may pose several additional limitations on market safety in ways not encountered in other markets, it is useful to focus our attention on the impact of user reproducibility on market safety. When users can reproduce an idea at a zero or very low marginal cost, there are often significant limitations on whether the seller can control how users exploit or distribute the idea. For example, it may be that the majority of potential customers for a digital song intend to use it themselves and value that song at $3 per user. However, there may exist another type of agent, indistinguishable from ordinary users, who has the capacity to resell or otherwise distribute that song. In other words, a small fraction of potential buyers may subsequently plan to also become sellers by taking advantage of the non-rivalry of digital information goods. When the original seller cannot distinguish between the two types of buyers, the sellers cannot simply charge "non-reproducing" users $3 per song, and "reproducing" users a much higher price. Instead, sellers need to develop a pricing scheme that takes into account the potential competition from resellers. Moreover, since the entry of reproducing users into the market will lower the price, the sellers need to take into account the non-reproducing users' expectations of the likelihood of entry. In the extreme—if buyers can replicate the idea at zero cost, and replication can be achieved instantaneously once the good has been acquired—it is possible that no positive price will be feasible, and the good may never be introduced into the market. [10,11]

The ability to expropriate ideas is particularly salient in the presence of an organized market mechanism. While most discussions of Arrow's disclosure problem tend to emphasize its impact on bilateral negotiations, the potential losses arising from disclosure may be more salient when a competing seller has an opportunity to offer a competing "version" of the same or a very similar idea to the same potential customers as the seller of an initial idea. The ability of buyers to also sell ideas (thus undercutting the sales of the original seller) is greater when there is a well functioning organized market that facilitates transactions. For example, in the case of digital music, the potential for some modest level of copyright infringement has always been present (e.g., through bootlegged tapes, etc.). However, the development of the Internet, and, more specifically, technology platforms such as Napster and BitTorrent dramatically increased the ability of users to share music with each other (by significantly lowering the cost of user reproduction). This dramatic increase in the share of music distributed through copyright-infringing behavior has further resulted in strategic behavior by record companies, who invest large sums of money in the development of ever-more-complicated digital rights management technologies (each of which has subsequently been neutralized by committed hackers) and aggressive rights enforcement against individual downloaders (Rob and Waldfogel, 2006). This has distortionary effects, as it requires large sums of money to be invested in activities that are not, in themselves, productive. Without a centralized exchange system, low user reproducibility may have little impact on market

[10] Boldrin and Levine (2008) offer an interesting analysis in which they consider a setting with (possibly small) frictions in ex post replication (either as the result of a small but non-zero replication cost or delays in the time required for replication) to argue that a positive price might indeed be feasible (and would therefore give producers of ideas incentives to develop innovations even in the absence of formal intellectual property protection). While a full discussion of the analysis of Boldrin and Levine is beyond the scope of this chapter (as the relationship between their assumptions and the MfT literature is a bit complex), it is worth noting that their focus on the role of alternative market institutions in shaping the welfare arising from ideas production offers an intriguing perspective, grounded in a market design approach, about the welfare consequences of formal intellectual property rights.

[11] The challenges arising from user reproducibility are in no way limited to digital information goods or ideas whose only form of intellectual property protection may be in the form of copyright. Consider the case of agricultural biotechnology. Over the past decade, Monsanto has commercialized a wide range of genetically modified seed crops. While many (though not all) of these crops could in principle have their seed be used over multiple generations (so-called seed-sharing), the Monsanto license permits only a single use from each seed (i.e., no seed-saving across generations). Monsanto enforces these agreements aggressively, including proactive monitoring of potential license violations, and maintaining a large capacity for litigation against potential infringers (see <http://www.wired.com/science/discoveries/news/2005/01/66282> retrieved July 2009). Monsanto claims that, in the absence of enforcement, farmer-competitors would be able to enter the market, undermining their property rights as granted in patent law, with the potential for significant distortions to their pricing and research and development incentives. In other words, in the absence of effective intellectual property rights enforcement, the potential ability to "replicate" Monsanto's seed technology (in this case, through natural reproduction) has the potential to undermine Monsanto's ability to sell its technology even to non-infringing farmers.

outcomes, as the ability of any individual user to compete with the original seller is limited.

Our analysis suggests that striking facets of the nature of innovation and ideas—ideas complementarity, value rivalry, and user reproducibility—each pose specific and fundamental challenges for the market design criteria proposed by Roth. In many contexts, the *lack* of organized markets is not simply a historical accident or a reflection of the fact that a market would have little value; instead, there are significant limitations on the feasibility of the market for ideas given the inherent challenges in market design. In other words, in the absence of specific institutional mechanisms to overcome these challenges, *the nature of ideas undermines the spontaneous and uncoordinated evolution of a corresponding market for ideas.*

The role of intellectual property on the design of markets for ideas

One of the central findings of the MfT literature is that formal intellectual property rights such as patents are closely associated with technological trade (Arora et al., 2001; Gans, Hsu, and Stern, 2002, 2008). This prior literature, however, does not distinguish between the role of intellectual property rights in facilitating bilateral transactions (the focus of nearly all of the prior literature) and in multilateral market mechanisms. [12] While this emerging body of empirical evidence offers support for the causal impact of the patent system on the feasibility of licensing, there is little empirical evidence as to whether such licensing is efficient, and whether intellectual property rights facilitate competition between multiple potential licensees and licensors. To evaluate the impact of intellectual property on the feasibility of an effective multilateral market for ideas, it is useful to consider the interplay between intellectual property and the three facets of ideas that we have highlighted throughout our analysis: value rivalry, user reproducibility, and ideas complementarity.

When the value of an idea to a potential buyer depends on their ability to have *exclusive* use of that idea (i.e., there is a high degree of value rivalry), formal intellectual property rights play a direct role in enhancing the potential for a market for ideas. In order to increase the expected sale price, a seller would like to disclose a nascent idea to multiple potential buyers, and then allocate the idea using an efficient and feasible mechanism such as an auction. As we discussed earlier, disclosing the idea to multiple buyers can limit the valuation of each buyer (since those who do not purchase will nonetheless benefit from the idea to a certain extent, and so limit the opportunities for monopolistic exploitation of the idea by the successful bidder). Effective and enforceable

---

[12] For example, in Gans et al. (2008), we find direct evidence that the date an innovation is licensed (by a technology entrepreneur) is increasing in whether a patent for that innovation has been *granted*; prior to patent grant, the property rights covering an innovation are more uncertain, reducing the ability to license to a downstream commercialization partner.

formal intellectual property rights directly overcome this constraint by offering an ex post mechanism to discipline those who expropriate the idea for their own use without payment or a formal agreement. Indeed, this role for formal intellectual property rights to facilitate organized exchange markets can be seen across numerous settings. Over the past decade, so-called innovation exchanges such as Ocean Tomo have emerged, and have evolved to focus almost exclusively on technologies covered by formal intellectual property protection (in Ocean Tomo's 2009 auction, the entire portfolio of auctioned items were covered under a US patent grant). Similarly, the overwhelming share of university licenses are linked to specific pieces of intellectual property, and a very high share of all university licensing activity involves inventions for which a patent has been applied.

A similar case can be made for the impact of intellectual property on ideas characterized by high user reproducibility. When the marginal cost for replicating an idea is extremely low (or even potentially zero beyond the first unit), sellers of ideas in organized markets face the possibility that the most aggressive early buyers of their ideas are precisely those who are planning to market that idea to others, thus undermining the ability of the originator of the idea to appropriate the value of their idea even when that idea achieves a high level of diffusion. Because effective intellectual property protection provides an ex post mechanism to punish such behavior, it is possible to limit such behavior ex ante. Of course, the mere existence of a property right is not enough; there must be effective and credible enforcement when buyers abridge the terms of their licensing agreements to distribute the idea more widely. In some sense, the often-criticized litigation behavior of the Recording Industry Association of America (RIAA) and individual record companies highlights the dilemma: the massive scope and scale of copyright-infringing file-sharing networks such as BitTorrent (and Napster in an earlier era) limit the credibility of the litigation threat for any particular buyer, while the punishments in the small number of "example" cases seem to many like an abuse of the intellectual property right itself. The broader point, though, is that intellectual property rights do not simply enhance bilateral exchange but, by enhancing market safety, enhance the potential for multilateral exchange.

Interestingly, intellectual property has a more ambiguous impact in environments characterized by a high degree of ideas complementarity. When the value of any one idea depends on its combination with other ideas, the ability to extract value from the market for ideas depends on bargaining position and strength of each idea holder and potential buyer. If intellectual property rights are extremely strong (e.g., a successful law suit allows for injunctive relief, including the cessation of commercial operations infringing the patent), the relative bargaining power of different holders of property rights need not be determined by the intrinsic value and marginal contribution of their idea, but instead may be determined by the credibility to threaten hold-up after specific investments have been made. In other words, when ideas complementarity is strong, there is a greater incentive on the part of each seller of ideas to forgo participation (undermining market thickness), and these incentives can potentially be exacerbated by formal intellectual property rights. At the same time, intellectual property can play a crucial role in helping to design institutional responses to mitigate the potential for hold-up. As mentioned earlier, standard-setting organizations in the information technology industry have evolved to serve as both a mechanism for coordination among multiple sellers of ideas and also as a clearinghouse to disclose and occasionally even pool intellectual property claims into a coherent bundle so that potential buyers can avoid a patent thicket.

More generally, this discussion highlights the fact that *formal intellectual property rights play a special but subtle role in facilitating the operation of a market for ideas.* Whereas Gans et al. (2002) emphasize that formal intellectual property rights such as patents encourage collaborative (but bilateral) commercialization, the analysis here suggests that patents play an arguably more central role in multilateral settings. The ability of an expropriator to exploit the market by taking advantage of the seller of idea's disclosure, can potentially lead to large *costs* of expropriation. While intellectual property straightforwardly overcomes the disclosure problem and so enhances the potential for multilateral bargaining over ideas, the enforcement of intellectual property rights— most notably the ability to assert a marginal claim and threaten injunctive relief in a probabilistic patents system—may enhance incentives for hold-up and so undermine market thickness. Which of these effects dominates is an empirical question, and is likely to differ in alternative environments. For example, while it is likely that intellectual property rights have facilitated more centralized bargaining in areas such as biotechnology where ideas complementarity tends to be relatively low, it is possible that the converse is true in areas such as software or business method patents.

REPUGNANCE IN THE MARKET FOR IDEAS

Our previous discussion has analyzed the challenges, from an economics perspective, in the design and operation of markets for ideas. Nonetheless, those economists who have engaged in practical market design have noted that other, non-economic factors, can play a role—even a decisive one—in driving what is possible. Roth (2007) classified a large number of such constraints under the rubric of *repugnance*. [13] In particular, repugnance refers to social constraints preventing exchange from taking place at positive prices. For example, there are legal restrictions on establishing markets in areas such as organ trading or child adoption; specifically, *on the use of money to facilitate such trade.* To Roth, these were reasons markets did not exist and also factors that market designers need to work around.

We have noted the paucity of idea exchange at a positive price. However, the exchange of ideas and knowledge does indeed take place throughout society and over time.

---

[13] While Roth considers repugnance issues in the study of markets only from an economics perspective, such constraints have been identified and explored in other contexts by sociologists. This includes the seminal work of Zelizer (2005) on the pricing of child care and Titmuss (1971) on the use of (or lack of) monetary incentives in blood donation.

Indeed, it is the unpriced flow of ideas and knowledge—knowledge spillovers—that have come to be taken as the crucial building block for modern theories of endogenous economic growth. In other words, while the inability to place a positive price on some types of transaction may be a puzzle within a particular setting, our understanding of the broader process by which economic growth occurs depends heavily on the historical fact that (at least some) producers of ideas have only limited appropriability over their ideas and are unable to earn their marginal product through an organized and competitive market for ideas.

The notion that repugnance might be an important constraint on the exchange of ideas and knowledge is perhaps best exemplified by the wide body of historical and contemporary evidence that, at least for some types of idea such as scientific knowledge, producers of ideas explicitly value the dissemination and future use of that knowledge over the monetization of the idea. Consider the famous words of Benjamin Franklin, a noted Enlightenment inventor and ideas producer:

> as we enjoy great advantages from the inventions of others, we should be glad of an opportunity to serve others by any invention of ours; and this we should do freely and generously. (Franklin, 2003, pp. 117–18)

Though expressions of the value of free exchange by suppliers of ideas and knowledge are pervasive—from scientists to journalists to advocates for diverse religious expression—there are very few analyses that take on the consequences of such sentiments for the incentives to produce knowledge or the impact on the design of efficient institutions for the exchange and dissemination of that knowledge.

Such norms go beyond a simple desire to "pay less" or offer a "discount." Instead, we observe a bimodal structure to transactions in the ideas market. On the one hand, some ideas are associated with either bilateral or multilateral exchanges, and there are significant premiums placed on successful innovations (potential drug candidates, promising software algorithms, etc.). At the other extreme, there is a wide body of knowledge that is distributed for free. Interestingly, there are few transactions that take place at a low but positive price (particularly for goods that are themselves considered pure "knowledge" or "ideas"). For the class of ideas where *both* buyers and sellers believe that trade is repugnant at any price, the equilibrium that emerges is that only a small number of (very valuable) ideas will have a high and positive price (and be criticized for that monopolistic pricing) while a larger number of ideas will effectively be sold at a price of zero.

In the remainder of this section, we raise the hypothesis that this is not simply a matter of market design but also the result of repugnance. We certainly acknowledge that this hypothesis requires careful empirical evaluation in future work. However, we also think it is useful, in the spirit of Roth, to consider the impact and role that repugnance might play in the market for ideas, and evaluate the potential impact of alternative policies and institutions designed to promote the exchange of ideas and knowledge in the presence of a repugnance constraint. We emphasize that this part of our analysis is quite speculative, as our main contribution is simply to highlight settings where repugnance may impact the efficiency of the exchange of pure ideas.

Sources of repugnance

The potential origins of repugnance over ideas trading are likely diverse and subtle, and our examination here is necessarily incomplete; we are highlighting what we think may be the most important drivers of repugnance while fully acknowledging that we are in no way completing a comprehensive survey.

First, as emphasized by Arrow, there appears to be a complicated set of essentially psychological *intrinsic drivers*:

> It seems to me that that there is a motive for action not taken account of in standard economic models. It is a motive that operates in a social context and cannot fully be discussed in the terms standard in "methodological individualism." I refer to what appears to me to be a tendency for individuals to exchange information, to engage in gossip at all levels. There is some form of satisfaction not only in receiving information but also in conveying it. Currently, this is exemplified by the curious phenomenon of *Wikepedia* [sic], where individuals spend time and effort to convey information without pay and without fame. Probably, there is even an evolutionary basis for this phenomenon, though explanations of social traits (other than those involving kin) on the basis of natural selection have proved to be difficult. (Arrow, 2008, p. 2)

In other words, disclosure is fundamental to human communication. The dividing line between social communication and the disclosure of knowledge is often blurry, particularly in the context of embryonic ideas. An important component of human creativity is the communication of that novelty to others, in the desire both to impress and to share (Amabile, 1983, 1996). Simply put, while economists have essentially abstracted away from the joy and excitement of discovery in the study of innovation, discovery and creativity are nonetheless important stimuli that are shared through communication (requiring disclosure that most economic theory suggests inventors will keep to a minimum).

A second potential driver is grounded in the sociology of collective sharing and gift exchange (Gouldner, 1960; Iannaccone, 1992).[14] While the conditions in which communities establish norms regarding free exchange are subtle (as we discuss later), it is possible that the willingness of suppliers to provide ideas and knowledge for free is grounded in their membership of a community in which they also receive free ideas and knowledge from their peers. Indeed, this form of communal sharing flips the challenge arising from the low costs of user reproducibility on its head; rather than serving as a deterrent to an organized market, an entire community acts as both suppliers and demanders, and enforces an equilibrium norm in which exchange takes place at a zero

---

[14] This is similar to the emotional commitments described by Frank (1988).

price. From a broad market design perspective, this collective (equilibrium) choice to exclude monetary exchange and other forms of profit can manifest itself in the form of repugnance for cash transactions.

Finally, it is possible that the origin of repugnance might be due to an aversion to complex contracting over the uses and applications of intangible goods. One of the distinctive properties of information is that potential buyers may not be able to anticipate precisely how they might use a particular idea or new technology once it is acquired. Consequently, buyers may be extremely averse to negotiating contracts (particularly contracts in which they have an informational disadvantage) about how they might use or exploit an idea once it is exchanged. In such an environment, potential buyers would have an extreme control-rights preference against paying for an idea in a way that involved significant ex post monitoring regarding the use of that idea. For example, there would be significant aversion to contract terms that involved metering of restrictions on the scope of application. From a market design perspective, an inability to charge a positive price for the use of an idea (even when that may be "efficient" from the perspective of traditional economic theory) can be interpreted as a repugnance-based constraint on certain types of licensing and intellectual exchange arrangements.

Transaction costs versus repugnance

Before turning to the impact of institutions that seem to account for repugnance in ideas markets, it is useful to consider whether the lack of exchange of ideas at a positive price is simply the result of transaction costs. While transaction costs certainly mitigate the viability of certain types of opportunistic transactions that might involve considerable negotiation (even in the absence of the types of challenges we described earlier), it is also worth considering the fact that the dynamics of markets for technology or ideas with positive prices versus zero prices are strikingly different:

> From the consumer's perspective, though, there is a huge difference between cheap and free. Give a product away and it can go viral. Charge a single cent for it and you're in an entirely different business, one of clawing and scratching for every customer. The psychology of "free" is powerful indeed, as any marketer will tell you. . . . People think demand is elastic and that volume falls in a straight line as price rises, but the truth is that zero is one market and any other price is another. In many cases, that's the difference between a great market and none at all.

> The huge psychological gap between "almost zero" and "zero" is why micropayments failed. It's why Google doesn't show up on your credit card. It's why modern Web companies don't charge their users anything. And it's why Yahoo gives away disk drive space. The question of infinite storage was not if but when. The winners made their stuff free first. (Anderson, 2008)

To an economist, what Anderson is implying is that not only is the cost of information replication low, but the demand curve for information goods becomes highly elastic at a zero price (and relatively inelastic at any positive price). In other words, even a very small monetary cost can engender a dramatic shift in realized demand. While certain types of "micro-payments" have emerged in certain contexts (e.g., iTunes' 99 cents pricing), participants in many ideas transactions seem willing to negotiate over whether and when knowledge will be exchanged (incurring significant transaction costs), but not price—there seems to be significant aversion to transactions at low but positive monetary prices. Thus, even where transaction costs have fallen dramatically (e.g. news delivery), this has not translated into the emergence of monetary payments.

The design of markets for free ideas

Roth emphasizes that repugnance need not be a fundamental constraint on efficient exchanges (though of course it does raise some difficult challenges). When Roth confronted repugnance in the market for kidney donation, he began to design markets that involved exchanges among voluntary donor pairs, essentially allowing for exchanges across families. Working within the repugnance constraint, Roth has organized an emerging set of markets for kidney exchange that operate without monetary payments but do indeed save lives through effective market design (see Chapter 1).

In the market for ideas, there are a striking number of real-world institutions that are premised on a price of zero. [15] Consider Wikipedia (Tapscott and Williams, 2008; Greenstein and Devereux, 2006). On the one hand, the traditional encyclopedias such as the *Encyclopedia Britannica* involved the solicitation of articles by leading scholars along with a modest monetary payment, and the encyclopedias themselves were sold at extremely high margins (e.g. the 1980s-era *Encyclopedia Britannica* sold for about $3,000 and was produced for a marginal cost of about $300) (Devereux and Greenstein, 2006). Wikipedia, on the other hand, is organized according to a very different principle. Both the provision of content and the use of the online encyclopedia are not only free but open to debate and interpretation by the user community. Rather than soliciting articles from leading "experts," Wikipedia allows any user to also serve as a contributor and has developed subtle protocols to adjudicate debates when different users/contributors hold different perspectives. Once an entry or contribution is submitted, individuals do not even have an absolute right of "control" over their own prior contributions; not only are there no prices, there are no property rights. Despite this quite idiosyncratic "design" for an encyclopedia, Wikipedia has quickly emerged as the single most utilized reference source in the world. In less than a decade, Wikipedia has essentially supplanted the positively priced expert-based system that had existed for nearly 200 years. Of course, the reliance on mere users and free contributions has raised concerns about quality and accuracy. Perhaps surprisingly, however, most independent tests suggest that the

---

[15] It is interesting to note that while Roth's examples usually involve a law or regulation that prohibits monetary transfers, institutions for free ideas tend to operate according to (strongly enforced) informal norms and practices.

overall rate of error is similar across free, user-based systems and expert-based systems with positive prices (and, along some dimensions, Wikipedia is in fact superior) (Giles, 2005). Intriguingly, given the complexity and need for debate and adjudication within the Wikipedia user and contributor community, the decisive issue for Wikipedia is not a lack of "transaction costs" (indeed, there are significant transaction costs to make a contribution and understand the information upon which individual entries are based); instead, the key issue seems to be the complete transparency of the process by which information is provided, the ability to debate alternative ways of organizing a particular set of facts, and the ability of the worldwide user community to access that information for free (Tapscott and Williams, 2008). Put simply, the "wiki" model (which now extends well beyond Wikipedia) has emerged as a market for free ideas that simultaneously relies on free exchange and requires significant investment on the part of the contributors of ideas.

Whereas Wikipedia is a quite recent phenomenon, the development of institutions involving the free exchange of ideas is, of course, much older, and realized most durably and strikingly in the context of "open science" (Merton, 1973; Dasgupta and David, 1994; Stern, 2004; David, 2008). Open science is a complex system in which researchers participate within a scientific community by drawing upon and advancing a specialized field of knowledge through pursuing research directions of their own interest. The hallmark of this system is the priority-based reward system: to receive credit for their discoveries, scientists publicize their findings as quickly as possible and retain no formal intellectual property rights over their ideas (Merton, 1957; Dasgupta and David, 1994). In turn, the institutions supporting scientific research—from universities to public funding agencies to non-profit foundations—offer status-based rewards such as tenure and prizes to recognize significant achievements; these awards are publicly announced. The priority-based reward system not only serves to provide incentives for scientists, but also enables a system of efficient disclosure that (at least in principle) minimizes the duplication of research efforts among scientists (assuming that scientists can access and replicate each other's work at relatively low cost) and enhances the growth in the stock of knowledge within the boundaries of particular scientific disciplines (Dasgupta and David, 1994). While the origins of open science are grounded in a complex set of motives and incentives facing researchers and funders (David, 2008), the norms of open science have evolved in a more evident manner. They ensure a high level of participation (allowing researchers to build on ideas in an unstructured way over time), allow for multiple researchers to both collaborate and compete with each other in a (relatively) transparent way, and, strikingly, provide status-based rewards to those who can credibly claim to have initially made a discovery (rather than those who simply learn about it and diffuse it to others). As a market design, open science overcomes the challenges arising from ideas complementarity, value rivalry, and user reproducibility.[16]

---

[16] Indeed, it is precisely the violation of these norms that are at the heart of contemporary policy debates about the limits of open science when knowledge traditionally maintained within the public domain is also protected by formal intellectual property rights. As emphasized by Murray (2009) and

It is, of course, feasible to consider a wide range of institutions that support markets for free ideas, and examine each from the perspective of market design. Without claiming to provide a comprehensive list, such institutions range from enduring arrangements such as the freedom of the press and religion, to more contemporary phenomena such as the open-source software movement, the blogosphere, and YouTube. In each case, ideas that are costly to develop are nonetheless offered at essentially a zero price. One dramatic consequence of a zero price is that, conditional on participation by suppliers of ideas, it is relatively easy to ensure market thickness and to take advantage of the non-rivalry of ideas among users. Market safety is likely to be more of an issue, particularly when ideas can be used or manipulated in ways that are adverse to the interests of the supplier of ideas. While each of these institutions supports both the production and diffusion of free ideas—ranging from political rhetoric to well defined technological innovation— it is striking to us that there has been little systematic analysis of the institutional requirements for such arrangements to exist, the role that repugnance plays in shaping these institutions, and the contribution of these eclectic institutions to economic and social well-being.

Market design and the limits of repugnance

One of the most striking aspects of repugnant markets is that the constraints on pricing are rarely comprehensive and often emerge in relatively subtle ways. For example, while there are sharp constraints on organ trading at a positive price, there is certainly no expectation that physicians involved in kidney exchange should operate for free, nor are there constraints on charging for other human parts such as hair. How do the limitations and nature of repugnance impact the pricing of ideas and knowledge?

Consider the emergence of online two-sided markets such as Internet search. From a theoretical perspective, it is possible that, for technologies such as Google web search, the equilibrium involves (1) consumers paying for web search and access to advertisers, (2) advertisers paying for access to consumers, who are able to search for free, and (3) a mixture of payments on both sides of this technology platform. However, if consumers have a deep aversion to paying for "information," it becomes much more likely that the equilibrium will involve free consumer search alongside paid advertising content. It is useful to compare this model with the pricing of physical newspapers. Even for a newspaper in which the marginal cost was positive, consumers have traditionally paid a nominal charge and the bulk of newspaper revenues have been through the advertising channel. In other words, the existence of repugnance did not necessitate public funding in order to achieve a positive level of supply; instead, media and advertising have evolved to complement each other in order to overcome some of the key

Murray and Stern (2008), patents in particular seem to have emerged as an alternative non-monetary "currency" that has been adapted by the scientific community to promote the underlying norms of the open science system.

constraints that would arise if newspapers or other media could be accessed only at a high price.

Examining markets for ideas that involve significant limitations on the use of those ideas highlights a second type of nuanced constraint on pricing. For example, while the market for prerecorded magnetic videotape was by and large served in the form of a rental market (placing significant time limitations on use, opening up users to the potential for late fees, etc.), the pricing of DVDs and CDs is in the form of a flat fee for unlimited private exploitation. [17] More generally, different technologies and types of knowledge are associated with very different pricing schedules, and there has been little detailed examination of the conditions under which different arrangements are effective, and, in particular, what role repugnance over certain types of monetary transactions plays in the emergence of different types of pricing structure.

This can be seen perhaps most dramatically in the case of fixed-fee versus subscription services. While some types of information products can be sold through a subscription service (from newspapers to cable television), attempts to establish subscription services have failed in a wide range of settings, including software. While most consumers (and particularly business consumers) are fully aware that upgrades are likely to occur on a regular schedule, and that they are likely to purchase such upgrades (either as the result of enhanced quality or to ensure interoperability), software companies such as Microsoft and Intuit have largely failed in their efforts to establish subscription services for their products. In the absence of repugnance, this is surprising, since the availability of a subscription service likely reduces the riskiness of expenditures of a potential buyer and most subscription services have been offered in a way that made them an attractive option for those who were likely to upgrade anyway (which turns out to be most consumers). However, if buyers have a preference for control over the decision (even one that likely involves paying a premium ex post), the repugnance associated with subscription pricing likely undermines the market viability of what would otherwise be an efficient pricing mechanism.

Taken together, these examples suggest that understanding the form in which repugnance takes in particular circumstances, and considering how that particular form of repugnance impacts the broader challenge of designing an effective market for ideas, can deepen our analysis of repugnance.

---

[17] As well, except for media that have been protected by digital rights management software, it is also possible to share these materials with others in violation of the license agreement imposed on buyers. Indeed, Boldrin and Levine (2008) suggest that fixed-fee pricing with no limitations on use (including resale and replication) can be optimal. Their analysis captures the idea that if you allow idea buyers to resell the idea, you are able to charge a premium to early buyers and so avoid the costs imposed by the restrictions. When imitation is not immediate, first-mover advantages may allow ideas sellers to appropriate rents even in the absence of intellectual property protection. See also Gans and King (2007).

CONCLUSION

Our aim, in this chapter has been to develop an agenda and framework for understanding the apparent lack of formal markets for ideas. In so doing, we have combined insights from the economic literature on market design and the literature on markets for technology. We have noted that the latter has mostly studied bilateral exchange of ideas rather than "markets" as characterized by large numbers of buyers and sellers engaging in large numbers of transactions. Such markets enable participants to better evaluate options for combining ideas with each other and with other assets in a timely and stable manner. Consequently, markets for ideas can both enhance the useful application of ideas and also harness the force of competition to ensure that creators of ideas earn an appropriate return.

Several conclusions emerge from this exercise. First, ideas possess particular characteristics that make the efficient design of markets challenging and impede the unplanned emergence of markets. The fact that many ideas require access and perhaps ownership of other, complementary ideas in order to be of value makes it difficult to coordinate transactions so that participants can evaluate their choices over different bundles of ideas. In addition, the fact that ideas might be easily reproduced by users or expropriated by them through pre-contractual disclosures can make sales of an idea to many buyers unsafe, resulting in bilateral exchange.

To this end, Lemley and Myhrvold (2008) argue that changes in the rules regarding licensing can have a dramatic impact on the effectiveness of the market for ideas:

> The solution is straightforward—require publication of patent assignment and license terms. Doing so will not magically make the market for patents work like a stock exchange; there will still be significant uncertainty about whether a patent is valid and what it covers, particularly since patents tend by their nature to be unique goods. But it will permit the aggregate record of what companies pay for rights to signal what particular patents are worth and how strong they are, just as derivative financial instruments allow markets to evaluate and price other forms of risk. It will help rationalize patent transactions, turning them from secret, one-off negotiations into a real, working market for patents. And by making it clear to courts and the world at large what the normal price is for patent rights, it will make it that much harder for a few unscrupulous patent owners to hold up legitimate innovators, and for established companies to systematically infringe the rights of others.

While this would certainly allow some benchmarking and make it easier to define prices, enforcement might be costly. However, Lemley and Myhrvold's contention does highlight the potential for alterations to patent right obligations to facilitate the establishment of markets. Importantly, it shows that in terms of market design there are options available to policy-makers that may facilitate the emergence of markets for ideas.

Following that theme, formal intellectual property protection can in many cases assist in alleviating the challenges to the design of an efficient market for ideas. It can make intangible ideas into assets that can be easily traded and understood. By protecting against reproduction and expropriation, intellectual property protection can make idea selling safe. At the same time, intellectual property can in some cases enhance incentives for hold-up and exacerbate the coordination challenges in bringing together multiple complementary ideas. Our analysis therefore gives policy-makers a new set of challenges to consider when evaluating the design of intellectual property instruments. For example, enhancing the strength of patent protection may play a crucial role in enabling effective technology transfer by preventing disclosure to multiple potential users of the technology; at the same time, however, when multiple (overlapping, complementary) producers of ideas can use the patent system to foreclose commercial activity, it is possible that strengthening intellectual property rights may only serve to further fragment the technology transfer process.

Finally, we have identified the exchange of ideas for money as an activity that can be understood as being constrained by repugnance. We noted that the resistance to selling certain ideas comes from sellers as much as buyers and that it also appears to generate a desire for extreme control rights in the use of ideas. Repugnance is something, we argue, that has constrained the development of markets for ideas (at least with positive prices). Because in so many situations and communities (especially those that are creative) the sellers of ideas also benefit from the ideas of others, and gain value from the use of their own ideas by others, the most market-like areas of the exchange of ideas have occurred precisely where norms or repugnance have constrained the price to be zero. In this situation, the lack of monetary flows can itself be seen as a means of generating market thickness, avoiding congestion, making exchange safe, and adhering to repugnance. Put simply, by finding areas where sellers and buyers value idea dissemination, it is possible to design effective markets even though no monetary exchange takes place.

We believe that the analysis we have provided and the issues we have identified are critical for the study of idea dissemination and ensuring returns to innovators; in particular, this study aids our understanding of the complexities faced by business and government in their attempt to facilitate these objectives. However, it also suggests substantive areas for future study.

First, the exploration in this chapter was qualitative and intuitive but far short of the sort of formal theoretical model that market designers now rely upon for predictions. Formal modeling can assist in more precisely defining the aspects of the nature of ideas that pose particular market design challenges and also the possibility that institutions—in particular, formal intellectual property protection—may alleviate some of these challenges. In addition, in relation to repugnance, formal theoretical modeling is required to properly distinguish alternative hypotheses regarding the source of that repugnance; for instance, are zero prices a norm or a symptom of market breakdown?

Second, there is considerable scope for empirical work—drawn from both real-world data and experimental evidence—to identify quantitatively the magnitude of challenges in designing markets for ideas as well as the rate of return in terms of efficient matching from overcoming those particular challenges. For example, we identified several areas where idea exchange proceeded freely and multilaterally—science and open source communities—that may shed light on how to unlock similar liquidity in other areas where idea exchange may be fruitful. Studying how institutional changes and the impact of commercial incentives have impacted on these domains where ideas exchange in market-like ways will surely be an important first step in understanding whether market design can be brought to bear in other areas.

References

Abramowicz, M. (2006) "Patent auctions," Mimeo, University of Chicago.

Amabile, T. M. (1983) *The Social Psychology of Creativity*, Springer-Verlag.

_____ (1996) *Creativity in Context*, Westview Press.

Anderson, C. (2008) "Free! Why $0.00 is the future of business," *Wired*, March 16.

Anton, J. J. and Yao, D. A. (1994) "Expropriation and inventions: appropriable rents in the absence of property rights," *American Economic Review*, 84(1): 190–209.

Arora, A. (1995) "Licensing tacit knowledge: intellectual property rights and the market for know-how," *Economics of Innovation and New Technology*, 4: 41–59.

_____ and Gambardella, A. (1994) "The changing technology of technological change: general and abstract knowledge and the division of innovative labour," *Research Policy*, 32: 523–32.

_____ Fosfuri, A. and Gambardella, A. (2001) *Markets for Technology: The Economics of Innovation and Corporate Strategy*, MIT Press.

Arrow, K. J. (1951) *Social Choice and Individual Values*, Yale University Press.

_____ (1962) "Economic welfare and the allocation of resources for invention," in *The Rate and Direction of Inventive Activity*, Princeton University Press, pp. 609–25.

_____ (2008) "Comment on "The historical origins of 'open science'" (by Paul David)," *Capitalism and Society*, 3(2): article 6.

Barzel, Y. (1968) "The optimal timing of innovations," *Review of Economics and Statistics*, 50: 348–55.

Boldrin, M. and Levine, D. (2008) *Against Intellectual Monopoly*, Cambridge University Press.

Bresnahan, T. and Trajtenberg, M. (1995) "General purpose technologies: 'engines of growth'?" *Journal of Econometrics*, special issue, 65(1): 83–108.

Chiao, B., Lerner, J. and Tirole, J. (2007) "The rules of standard setting organizations: an empirical analysis," *Rand Journal of Economics*, 38: 905–30.

Cohen, W. M., Nelson, R. R. and Walsh, J. P. (2000) "Protecting their intellectual assets: appropriability conditions and why U.S. manufacturing firms patent (or not)," NBER Working Paper No.7552.

Cramton, P. (2002) "Spectrum auctions," in M. Cave, S. Majumdar, and I. Vogelsang (eds), *Handbook of Telecommunications Economics*, Elsevier Science, pp. 605–39.

_____ (2008) "Innovation and market design," in J. Lerner and S. Stern (eds), *Innovation Policy and the Economy*, National Bureau of Economic Research.

Dasgupta, P. and David, P. (1994) "Towards a new economics of science," *Research Policy*, 23(5): 487–521.

David, P. A. (1998) "Common agency contracting and the emergence of open science institutions," *American Economic Review*, 88(2): 15–21.

_____ (2008) "The historical origins of 'open science,'" *Capitalism and Society*, 3(2): article 5.

Demsetz, H. (1967) "Towards a theory of property rights," *American Economic Review*, 57(2): 347–59.

Devereux, M. and Greenstein, S. (2006) *The Crisis at Encyclopedia Britannica*, Kellogg case, Northwestern University.

Edelman, B., Ostrovsky, M. and Schwarz, M. (2007) "Internet advertising and the generalized second price auction: selling billions of dollars worth of keywords," *American Economic Review*, 97(1): 242–59.

Frank, R. (1988) *Passions Within Reason*, Norton.

Franklin, B. (2003) *The Autobiography and Other Writings*, K. Silverman (ed.), Penguin Classics.

Gambardella, A. and Giarratana, M. S. (2008) "General technologies, product market fragmentation, and markets for technology: evidence from the software security industry," Mimeo, University of Bocconi.

Gans, J. S. and King, S. P. (2007) "Price discrimination with costless arbitrage," *International Journal of Industrial Organization*, 25: 431–40.

_____ and Stern, S. (2000) "Incumbency and R&D incentives: licensing the gale of creative destruction," *Journal of Economics and Management Strategy*, 9(4): 485–511.

_____ _____ (2003) "The product market and the market for 'ideas': commercialization strategies for technology entrepreneurs," *Research Policy*, 32: 333–50.

_____ _____ (2010) "Is there a market for ideas?" *Industrial and Corporate Change*, 19(3): 805–37.

_____ Hsu, D. H. and Stern, S. (2002) "When does start-up innovation spur the gale of creative destruction?" *RAND Journal of Economics*, 33: 571–86.

_____ _____ _____ (2008) "The impact of uncertain intellectual property rights on the market for ideas: evidence for patent grant delays," *Management Science*, 54(5): 982–97.

Giles, J. (2005) "Internet encyclopaedias go head to head," *Nature*, December 15: 900–1.

Gouldner, A. W. (1960) "The norm of reciprocity: a preliminary statement," *American Sociological Review*, 25: 161–78.

Greenstein, S. and Devereux, M. (2006) "Wikipedia in the spotlight," *Kellogg School of Management*, Case 5-306-507.

Grindley, P. C. and Teece, D. J. (1997) "Managing intellectual capital: licensing and cross-licensing in semiconductors and electronics," *California Management Review*, 39(2): 1–34.

Heller, M. (2008) *The Gridlock Economy*, Basic Books.

Hurwicz, L. (1972) "On informationally decentralized systems," in C. B. McGuire and R. Radner (eds), *Decision and Organization: A Volume in Honor of Jacob Marshak*, North-Holland, pp. 297–336.

_____ (1973) "The design of mechanisms for resource allocations," *American Economic Review* 63(2): 1–30.

Iannaccone, L. R. (1992) "Sacrifice and stigma: reducing free-riding in cults, communes, and other collectives," *Journal of Political Economy*, April.

Kanellos, M. (2006) "Few buyers at patent auction," CNET News, April 6.

Kitch, E. (1977) "The nature and function of the patent system," *Journal of Law and Economics*, 20: 265–90.

Klemperer, P. (2004) *Auctions: Theory and Practice*, Princeton University Press.

Kremer, M. (1998) "Patent buyouts: a mechanism for encouraging innovation," *Quarterly Journal of Economics*, 1137–67.

——— and Williams, H. (2010) Incentivizing Innovation: Adding to the Toolkit Innovation Policy and the Economy, Vol.10, NBER pp.1–17.

Lamoreaux, N. R. and Sokoloff, K. L. (2001) "Market trade in patents and the rise of a class of specialized inventors in the nineteenth-century United States," *American Economic Review: Papers and Proceedings*, 91(2): 39–44.

Lemley, M. and Myhrvold, N. (2008) "How to make a patent market," *Hofstra Law Review*, 102 (forthcoming).

Lerner, J. and Tirole, J. (2004) "Efficient patent pools," *American Economic Review*, 94(3): 691–711.

——— ——— (2006) "A model of forum shopping," *American Economic Review*, 96(4): 1091–113.

Levin, R., Klevorick, A., Nelson, R. R. and Winter, S. (1987) "Appropriating the returns from industrial research and development," *Brookings Papers on Economic Activity*: 783–820.

Levine, A. (2009) "Licensing and scale economies in the biotechnology pharmaceutical industry," Mimeo, Harvard.

McAfee, P. (2002) *Competitive Solutions*, Princeton University Press.

McDonald, G. and Ryall, M. A. (2004) "How do value creation and competition determine whether a firm appropriates value?" *Management Science*, 50(10): 1319–33.

Merton, R. (1957) "Priorities in scientific discovery: a chapter in the sociology of science," *American Sociological Review*, 22(6): 635–59.

——— (1973) *The Sociology of Science: Theoretical and Empirical Investigation*, University of Chicago Press.

Milgrom, P. (2004) *Putting Auction Theory to Work*, Cambridge University Press.

——— (2007) "Package auctions and package exchanges (2004 Fisher-Schultz lecture)," *Econometrica*, 75(4): 935–66.

Mokyr, J. (2008) "Intellectual property rights, the industrial revolution, and the beginnings of modern economic growth," Mimeo, Northwestern University.

Murray, F. (2009) "The oncomouse that roared: hybrid exchange strategies as a source of productive tension at the boundary of overlapping institutions," *American Journal of Sociology* (forthcoming).

——— and Stern, S. (2008) "Learning to live with patents: assessing the impact of legal institutional change on the life science community," MIT Sloan Working Paper.

Rob, R. and Waldfogel, J. (2006) "Piracy on the high C's: music downloading, sales displacement, and social welfare in a sample of college students," *Journal of Law and Economics*, 49(1): 29–62.

Romer, P. (1990) "Endogenous technological change," *Journal of Political Economy*, 98(5): S71–102.

Rosenberg, N. (1998) "Uncertainty and technological change," in D. Neef, G. A. Siesfeld, and J. Cefola (eds), *The Economic Impact of Knowledge*, Butterworth-Heinemann, Chapter 1.

Roth, A. E. Repugnance as a constraint on markets," *Journal of Economic Perspectives*, 21(3): 37–58.

Roth, A. E. (2008) "What have we learned from market design?" Hahn Lecture, *Economic Journal*, 118 (March): 285–310.

——— and Xing, X. (1994) "Jumping the gun: imperfections and institutions related to the timing of market transactions," *American Economic Review*, 84(4): 992–1044.

Seabrook, J. (1994) "The flash of genius," *New Yorker*, January 11: 38–52.

Shapiro, C. (2001) "Navigating the patent thicket: cross licenses, patent pools and standard setting," in A. Jaffe, J. Lerner, and S. Stern (eds), *Innovation Policy and the Economy*, National Bureau of Economic Research, vol. 1, pp. 1190–250.

Simcoe, T. (2008) "Standard setting committees," Mimeo, Toronto.

Stern, S. (2004) "Do scientists pay to be scientists?" *Management Science*, 50(6):835–53.

Tapscott, D. and Williams, A. D. (2008) *Wikinomics: How Mass Collaboration Changes Everything*, Penguin.

Teece, D. J. (1986) "Profiting from technological innovation: implications for integration, collaboration, licensing, and public policy," *Research Policy*, 15: 285–305.

Titmuss, R. (1971) "The gift of blood," *Society*, 8(3): 18–26.

Troy, I. and Werle, R. (2008) "Uncertainty and the market for patents," MPIfG Working Paper, Cologne.

von Hippel, E. (2005) *Democratizing Innovation*, MIT Press.

Zelizer, V. A. (2005) *The Purchase of Intimacy*, Princeton University Press.

CHAPTER 9

REDESIGNING
MICROCREDIT

ASHOK RAI AND TOMAS SJÖSTRÖM[1]

INTRODUCTION

ECONOMIC theory explains credit market imperfections in terms of informational and enforcement problems. Financial constraints arise if lenders are unsure about the borrower's riskiness, effort, or project choice (Stiglitz and Weiss, 1981), or about the borrower's actual realized return (Townsend, 1979). These financial constraints are aggravated by a lack of collateral. Accordingly, poor households may be unable to finance high-return investments in entrepreneurial activities, durable consumption goods, and human capital. The result is underdevelopment and poverty. There is increasing microevidence that such financial constraints are important. For example, McKenzie and Woodruff (2008) find that the average real return to capital for small entrepreneurs in a Mexican town is 20–33% per month, substantially higher than the prevailing market interest rates.

Microcredit, the practice of making small uncollateralized loans to the poor, has appeared as a possible solution to these credit market imperfections. The Grameen Bank in Bangladesh, the world's flagship microcredit program, was honored with the 2006 Nobel Peace Prize for its poverty-reduction efforts, and its lending model has been replicated worldwide. Many microcredit programs are subsidized (Cull et al., 2009). But in view of the informational and enforcement problems that afflict credit markets, the success of microcredit programs in achieving high rates of repayment on loans that are not secured by traditional collateral is remarkable (Armendariz de Aghion and Morduch, 2005).

[1] We thank Ethan Ligon and seminar participants at the Second European Microfinance conference in Groningen for their comments.

Here we will reconsider the design of uncollateralized lending programs in light of recent field evidence. Originally, theoretical interest was stimulated by the use of joint liability in the lending scheme referred to as Grameen I (Yunus, 1999). A group of five borrowers were given individual loans, but held jointly liable for repayment. If any member defaulted, future loans to all group members would be denied or delayed. However, Grameen I included other intriguing features as well, such as public repayment meetings, frequent weekly repayments of loans, regular savings deposits, and emergency loans in times of natural disasters (Armendariz de Aghion and Morduch, 2005). Unfortunately, there was very little variation in the microcredit programs that replicated Grameen I, so it was hard to know if joint liability or something else was key to Grameen I's success. We were at a bit of an academic impasse.[2]

Recent evidence from the field, discussed in the following section and surveyed by Banerjee and Duflo (2010), has jolted us out of this impasse. First, in a remarkable institutional change, the Grameen Bank's revised lending contract, dubbed Grameen II, no longer involves joint liability. This institutional change is part of growing dissatisfaction with joint liability lending (Armendariz de Aghion and Morduch, 2005). Secondly, Giné and Karlan (2009) conducted an innovative field experiment with the Green Bank, a Grameen replica in the Philippines, in which they compared randomly selected branches with joint liability to those with individual liability, and found no difference in repayment rates.[3] Thus, even though theoretical models inspired by Grameen I explained why joint liability might dominate individual liability (Ghatak and Guinnane, 1999), the field evidence did not provide much support for this.

A striking feature of both Grameen I and Grameen II, as well as both the joint liability and individual liability branches of the Green Bank, is the use of public repayment meetings.[4] One can imagine various reasons why public repayments may be preferable to private ones. For instance, the transaction costs of collecting payments from a large group of assembled people at a pre-specified time is low. Public repayments may also serve as a way to keep loan officers in check and to prevent fraud. Alternatively, the public meetings may allow the bank to tap into information borrowers have about each other (Rai and Sjöström, 2004). Or public meetings may be a venue for publicly shaming defaulters (Rahman, 1999).

---

[2] Testing the effect of joint liability would require variation in real-world mechanisms, i.e., experimentation. There would be social benefits from trying out different lending schemes but private first-mover disadvantages (Besley 1994). If donors had been willing to subsidize such experiments, contractual alternatives might have emerged. By and large, however, there was a policy push toward financial sufficiency and away from subsidies (Cull et al., 2009).

[3] More recently, Attanasio et al. (2011) compared repayment performance in individual and joint liability loans in an experiment in Mongolia, and again found no significant difference. However, they found that joint liability borrowers are more likely to own businesses and spend more on food consumption than individual liability borrowers, and less likely to make transfers to family and friends. The impact of joint liability microcredit has also been analyzed by Banerjee et al. (2010).

[4] Other aspects of uncollateralized lending have been investigated recently, such as dynamic incentives (Bond and Rai, 2009; Giné et al., 2010), and repayment frequency (Fischer and Ghatak, 2010; Feigenberg et al., 2009; Field and Pande, 2008).

Public repayment meetings may also have a more subtle benefit: they can help borrowers make mutually beneficial informal insurance arrangements. As observed by Armendariz de Aghion and Morduch (2005), when repayments are made in public, "the villagers know who among them is moving forward and who may be running into difficulties." This anticipated shared knowledge can be used by borrowers ex ante to expand the set of incentive-compatible informal agreements. The informal agreements among the borrowers, which are not regulated by the formal contract offered by the microcredit lender, are referred to as side-contracts.

We will consider the role of public repayments, but our intention is broader. Like Townsend (2003), our starting point is a mechanism design approach which emphasizes the interplay between formal and informal contractual arrangements. Formal credit arrangements are limited by insurance market imperfections (Besley, 1995). If it is not possible to insure against negative exogenous shocks, then entrepreneurial activities with high expected return might not occur even if financing could be obtained, because they might be considered too risky. Side-contracts may provide some mutual insurance, but can suffer from the same kind of informational and enforcement problems as formal contracts (Ligon, 1998; Ligon et al., 2002; Townsend, 1994; Udry, 1994). However, enforcement problems may be less severe in informal arrangements which are enforced by social sanctions, i.e., which rely on social capital instead of traditional collateral. Informational problems may also be less severe in informal arrangements among neighbors who know a lot about each other, and can observe each other's behavior.

Informal mutual insurance arrangements are ineffective when borrowers face hard times simultaneously, and therefore are unable to help each other out. The microcredit lender may provide better insurance by not insisting on repayment after a verifiable exogenous shock, such as a natural disaster. But the microlender is at an informational disadvantage, and some exogenous shocks may be hard to verify. If default is costless, then the borrower has a strategic incentive to default, claiming she cannot repay for some exogenous reason that the lender cannot verify. To prevent strategic default, default must be costly to the borrower. An efficient contract minimizes the expected cost of default, subject to the incentive-compatibility constraint that strategic default should not pay.

It is helpful to use the terminology *external frictions* for the outside lender's problem of observing what goes on inside a village, and enforcing repayment on loans that are not secured by traditional collateral. These external frictions impede formal contracting between the outside lender and the villagers. In contrast, *internal frictions* are caused by the incomplete information the villagers have about each other, and the difficulties they face in enforcing side-contracts. We will discuss how microcredit design is influenced by both external and internal frictions. In theory, public repayment meetings might help alleviate informational frictions, both external (Rai and Sjöström, 2004) and internal (as mentioned earlier). The field evidence suggests to us that the latter effect may be more significant.

In theory, internal and external frictions should be treated symmetrically: a side-contracting group of agents face the same type of mechanism design problem as the outside mechanism designer or principal (Laffont and Martimort, 1997, 2000; Baliga and Sjöström, 1998).[5] In either case, incentive compatibility and enforcement constraints must be respected. Since side-contracts interact with formal contracts, understanding the former is important for the optimal design of the latter. The principal must take into account that his mechanism will influence the agents' side-contracting ability, for example by determining how much information they have about each other (e.g., by making messages sent to the principal publicly available).

This is not a comprehensive survey of the large literature on credit, savings, and insurance in developing countries. We focus on microcredit, and do not discuss broader issues of microfinance (see Karlan and Morduch, 2009, for a wide-ranging survey). Moreover, we assume the external friction to contracting is due to the possibility of strategic default. Thus, we abstract from problems of adverse selection and moral hazard. See Laffont and N'Guessan (2000) for adverse selection, Laffont and Rey (2000) for moral hazard, and Ghatak and Guinnane (1999) for a broad survey of joint liability contracting.

Field evidence

In this section we discuss how microcredit is redesigned in the field.

Grameen II in Bangladesh

In 2002, after several years of experimentation and learning, the Grameen Bank radically transformed its lending mechanism (Dowla and Barua, 2006). Under Grameen I, a group of borrowers who failed to repay would typically have been "punished" by having future loans denied or delayed. But according to Grameen's founder, Muhammad Yunus, Grameen I had been too rigid about enforcing repayment:

> There is no reason for a credit institution dedicated to providing financial services to the poor to get uptight because a borrower could not pay back the entire amount of a loan on a date fixed at the beginning.... many things can go wrong for a poor person during the loan period. After all, the circumstances are beyond the control of the poor people. (Muhammad Yunus, quoted in Dowla and Barua, 2006, p. 5)

Dowla and Barua (2006) add that

> aggressive insistence by the bank on strict adherence to rigid rules may lead borrowers back to destitution. Borrowers forced into involuntary default because of a bad

---

[5] The idea of imperfectly side-contracting agents (bidders) is familiar from auction theory (Graham and Marshall, 1987; Mailath and Zemsky, 1991; McAfee and McMillan, 1992; Lopomo et al., 2005).

shock did not have a way back to rebuilding their credit relationships with the bank. (Dowla and Barua, 2006, p. 95).

Below are some of the main design changes associated with Grameen II:

1. Grameen II explicitly dropped the joint liability requirement that was a feature of Grameen I. In Grameen II, borrowers who do not repay are offered flexible renegotiated loan terms, but are threatened with credit denial if they fail to repay the renegotiated loan. The original Grameen loans also relied on credit denial as a repayment incentive; the difference is that in Grameen II the promise of future credit for an individual borrower is not conditional on the performance of others in the group. Instead, the loan ceiling for an individual borrower depends primarily on her own repayment performance, attendance at public meetings, and on her own savings.
2. Grameen I typically required all borrowers to make weekly repayments on loans, a feature Armendariz de Aghion and Morduch (2005) pointed to as a potential selection device. Grameen II allows for more flexible repayments which could be structured more in line with the borrower's cash flows.
3. Borrowers in a group were given staggered loans under Grameen I, with one borrower receiving a loan first, then the next two receiving a loan after the first had repaid a few installments, and so on. Such staggering has been justified by Chowdhury (2005) for incentive reasons. But Grameen II disburses loans at the same time to all borrowers who have repaid previous loans in full.
4. Under Grameen I borrowers were forced to put regular savings into a group account. Withdrawals from this group account required the consent of all the group members. Such a group account has been eliminated under Grameen II. Each borrower must make deposits into a special savings account that acts as a form of collateral, but also has access to a voluntary savings account that pays interest. Thus, while regular savings deposits are required under both Grameen I and II, the opportunity for demand deposits has been created in Grameen II.
5. One significant feature of Grameen I was preserved in Grameen II: repayments are made at public meetings in which all borrowers at a particular center (or branch) are present. Public meetings might simply make it easier for loan officers to collect repayments. Further, the transparency of a public meeting might serve to discipline the loan officers, prevent embezzlement, or deter collusion. As discussed later, the public meetings also allow the borrowers to learn things about each other.

Making inferences about efficient contractual design from this institutional redesign is difficult. We lack the appropriate counterfactual. Grameen II has flourished, and recorded high repayment rates after the 2002 reforms, but it is unclear whether Grameen I would not have had the same success.

Green Bank in the Philippines

Giné and Karlan (2009) provided the appropriate counterfactual through a field experiment in the Philippines. Randomization allowed a clean evaluation of changes in microfinance design. They conducted two experiments with the Green Bank, a Grameen-style lender in the Philippines, which conducted its redesign in stages and at centers chosen randomly. In the first experiment, some of the existing Green Bank centers, in which borrowers were receiving joint liability loans, were chosen at random to have their loans converted to individual liability loans. Giné and Karlan (2009) found no differences in the repayment rates between the treatment centers (with individual liability loans) and control centers (where joint liability loans continued), three years into the conversion.

Since borrowers formed groups expecting joint liability in both treatment and control centers, the first experiment did not rule out a potential role for joint liability in preventing adverse selection. In the second experiment, the Green Bank randomly offered one of three types of loan contracts to *newly* created centers: joint liability loans, individual liability loans, and phased-in individual liability. In the last, borrowers started with joint liability loans and then switched to individual liability. Again, Giné and Karlan found no differences in default rates between these three types of loan contracts.

In both of Giné and Karlan's (2009) experiments, loan sizes were smaller in individual liability loan centers, which could indicate some welfare loss. Still, the results seem to suggest that joint liability loans give no better repayment incentives than individual liability loans. Years of experimentation and learning also led the Grameen Bank to drop joint liability, suggesting it may not be as crucial as previously thought. The public meetings to collect repayments were preserved. As Giné and Karlan (2009) note, social influences on repayment might be important. We discuss related theoretical issues in the next section.

THE THEORY OF STRATEGIC DEFAULT

Rai and Sjöström (2004) adapted the model of Diamond (1984) in order to study mechanism design by an outside bank in villages subject to internal contractual frictions. In the simplest possible model, there are two villagers. Each villager $i \in \{1, 2\}$ has an investment opportunity, project i, that requires an investment of one dollar. The project succeeds with probability p. A successful project yields output $h > 0$, while a failed project yields output 0. Project returns are independently distributed across the villagers. If both villagers invest, then there are four possible outcomes or "states": $(0, h)$ is the state where project 1 fails and project 2 succeeds, $(0, 0)$ means both projects fail, etc. The villagers are risk neutral but have no assets, so self-financing is impossible.

The bank can be thought of as a benevolent not-for-profit microcredit organization, or as a for-profit bank operating in a competitive market. For simplicity, assume the risk-free interest rate is zero, so to break even the bank must expect to get one dollar back for every dollar it lends. An efficient contract maximizes the expected welfare of the borrowers, subject to the bank's break-even constraint.

To simplify the presentation, assume h is large enough to satisfy

$$(1 - (1 - p)^2)h > 2 \tag{1}$$

This inequality implies $ph > 1$, so the investment opportunities have positive net present value, and therefore should be funded. In a world with no frictions, each villager would get a one-dollar loan from the bank with required repayment of $1/p < h$ if the project succeeds (and nothing if the project fails). The expected repayment would be $p(1/p) = 1$, so the bank breaks even. Each villager's expected surplus would be $ph - 1 > 0$.

As discussed in the Introduction, external frictions impede contracting between the villagers and the bank. Here we shall assume the bank cannot observe whether a project succeeds or fails. In traditional banking relationships, a borrower who defaults loses her collateral, and this prevents her from defaulting strategically. But in our village economy, traditional collateral is lacking, and borrowers have nothing but the project returns with which to repay their loans. Grameen I punished default by denying or delaying future loans. But this cost of defaulting would sometimes be incurred by borrowers who did nothing wrong, since projects sometimes fail for exogenous reasons.

Rather than specifying the details of how default is punished, we will simply assume default is costly to the borrower. For example, future loans may be delayed or denied.[6] Let C denote the cost of default to the borrower. The cost is a net loss of social surplus; there is no corresponding gain to the bank.[7] It follows that an efficient contract minimizes the expected cost of default, subject to the bank's break-even constraint. Recall the concerns, discussed in the previous section, that prompted the redesign of the Grameen Bank: Grameen I was too inflexible toward unlucky borrowers who were unable to repay.

Coasean benchmark: perfect side-contracting

If default is costly, and project returns are not perfectly correlated, then the villagers can benefit from mutual insurance. If one of them fails while the other succeeds, the successful one should help the unlucky one repay, thereby avoiding the cost of default. But such insurance contracts may be impeded by internal frictions within the village (informational or enforcement problems). Empirical work suggests that these frictions are important (Townsend, 1994; Udry, 1994). However, as a benchmark, consider in

---

[6] In a dynamic model where, following a default, the borrower can save in order to self-finance future investment projects, denial of access to future loans may not be a sufficient punishment to ensure repayment (Bulow and Rogoff, 1989). But in reality, microfinance programs such as Grameen II provide better savings opportunities than would otherwise exist, and default implies a reduction in the ability to save (Bond and Krishnamurthy, 2004). This might contribute to low default rates under Grameen II.

[7] In contrast, seizure of traditional collateral is not socially wasteful if it is costlessly seized, and if it is no less valuable to the bank than to the borrower. But here we assume no traditional collateral exists.

this subsection a village with no internal frictions to contracting. In particular, the villagers can credibly promise to make side-payments as a function of the state (which they observe perfectly). The Coase theorem applies: whatever contract is offered by the bank, the villagers will agree on a joint surplus-maximizing side-contract. Since they can enforce mutually advantageous insurance arrangements, the village behaves as a composite agent that minimizes the expected cost of default.[8]

Suppose the bank offers each villager a one-dollar loan with individual liability. The required repayment is $1 + r$, where r is the interest rate on the loan. Let r^* be defined by

$$r^* \equiv \frac{1}{1 - (1-p)^2} - 1 \tag{2}$$

Notice that $1 + r^* < h/2$ by equation (1). To enforce repayment, the bank imposes a cost C on any borrower who defaults. Individual liability means that neither borrower is formally responsible for the repayment of the other. However, as long as $C \geq 1 + r^*$, the surplus-maximizing village will repay both loans whenever possible. By the Coase theorem, the villagers will agree ex ante to mutually insure each other against failure. Specifically, villager 1 promises to repay both loans (i.e. give the bank $2(1 + r^*)$) in state $(h, 0)$ where she has $h > 2(1 + r^*)$ and villager 2 has nothing.[9] In return, villager 2 promises to repay both loans in state $(0, h)$. In state $(0, 0)$ where both projects fail, no repayment is possible, so in this state each borrower suffers the cost C. In state (h, h) each repays her own loan. Accordingly, the bank collects $2(1 + r^*)$ in states (h, h), $(h, 0)$, and $(0, h)$. The bank will break even, because equation (2) implies

$$(1 - (1-p)^2) \times 2(1 + r^*) = 2.$$

Joint liability is sometimes justified as a way to encourage the group members to help each other in bad times. However, our Coasean village behaves like that anyway. A joint liability loan would formalize the mutual insurance, but it would not improve on individual liability loans, as long as there are no internal contractual frictions. To

---

[8] For the sake of clarity, and due to space constraints, we assume uncorrelated project returns. If project returns were correlated, the same kind of arguments would apply, but of course mutual insurance would be less valuable in this case (having no value at all in the limiting case of perfect correlation). In reality, returns might be highly correlated for two close neighbors working on similar projects, but the contractual frictions between these two neighbors might be relatively small. In contrast, two borrowers with projects that are uncorrelated, e.g. because they are located far away from each other, might find it difficult to contract with each other, because the informational and enforcement problems would be more serious in this case. This trade-off between correlation of returns and contractual frictions could be formalized in a spatial model, where close neighbors have more highly correlated projects but also better information about each other, and thus better contracting ability. Of course, good side-contracting ability has costs as well as benefits, because it can be used by the borrowers to collude against the bank. A spatial model might shed light on the optimal distance between group members.

[9] Notice that equation (1) guarantees that one successful project generates enough revenue to repay both loans. In the more general case, it may allow full repayment of one loan, and partial repayment of the other. The argument would then be similar, with partial repayment leading to a reduced punishment (see Rai and Sjöström, 2004).

see this, suppose the bank makes a one-dollar loan to each villager, but the villagers are jointly held responsible for a repayment of $2(1 + r^*)$. That is, if the sum of the repayments is less than $2(1 + r^*)$, each group member incurs the cost $C \geq 1 + r^*$. By the Coase theorem, this joint liability loan would result in an identical outcome as the individual liability loans, i.e. both loans would be repaid in the states (h, h), $(h, 0)$, and $(0, h)$. Including a formal joint liability clause in the loan contract would be redundant, because a Coasean village can replicate any such joint liability arrangement by side-contracting (cf. Ghatak and Guinnane, 1999).

The more general point is this: if there are no internal frictions to contracting within the village, then the design of the lending contract is relatively unimportant. In such a Coasean world, the main objective of a benevolent outsider should be to provide adequate resources to the village. The method by which they are provided would not matter much, because by the Coase theorem, the resources will be efficiently used by the villagers to maximize the joint welfare of the group. Of course, when side-contracting is not perfect, the Coase theorem no longer holds. We will now discuss the efficient lending contract under different assumptions about internal contracting frictions, and reconsider the optimality of joint liability.

No side-contracts

In the previous subsection we considered the extreme case of perfect side-contracting. In this subsection, we consider the opposite extreme: for whatever reason, the villagers are completely unable to enforce state-contingent side-contracts. They can observe the true state, but cannot credibly promise to make side-payments contingent on the state. The Coase theorem no longer applies, since promises to help each other in bad times are not enforceable.

Suppose each villager gets an individual liability loan of one dollar. Whenever a villager's project fails, she must default, so each loan is repaid with probability p. To satisfy the bank's break-even constraint, the interest rate must equal

$$\hat{r} \equiv \frac{1 - p}{p} \tag{3}$$

The expected repayment is $p(1 + \hat{r}) = 1$. Suppose a borrower who defaults on an individual liability loan suffers a cost C_I. To prevent strategic default when the project has succeeded, it must be more costly to default than to give the bank $1 + \hat{r}$. Thus, the following incentive compatibility constraint must hold:

$$C_I \geq C_I^{\min} \equiv 1 + \hat{r} = \frac{1}{p} \tag{4}$$

Each borrower's expected cost of default is $(1 - p)C_I$.

Suppose the bank instead offers a joint liability loan with interest rate r^*, defined by equation (2). With joint liability, the villagers must *jointly* repay a total amount of $2(1+r^*)$, or else *each* suffers a cost C_J. The incentive compatibility constraint is

$$C_J \geq C_J^{\min} \equiv 2(1+r^*) \tag{5}$$

If this incentive compatibility constraint holds, then villager 1 has an incentive to fully repay both loans (i.e. give the bank $2(1+r^*)$) in state $(h, 0)$, although no side-contract forces her to do so. By the same logic, villager 2 will repay both loans in state $(0, h)$. Defaults occur only in state $(0, 0)$, so each villager defaults with probability $(1-p)^2$. The expected cost of default for each villager is therefore $(1-p)^2 C_J$. The joint liability loan dominates the individual liability loan if it carries a lower expected cost, i.e. if

$$(1-p)^2 C_J < (1-p) C_I \tag{6}$$

This inequality certainly holds if we assume the cost of default is the same for both types of loans, $C_J = C_I$. This would be the case, for example, if the cost is due to a fixed action such as the complete denial of all future loans. However, for a joint liability loan to induce a successful borrower to repay both loans requires a very large cost of default. Indeed, it can be verified that $C_J^{\min} > C_I^{\min}$. If the cost of default is a continuous variable which can be minimized subject to incentive compatibility, then the bank will set the cost of default equal to $C_J^{\min}$ with joint liability and $C_I^{\min}$ with individual liability.[10] It turns out that joint liability loans still dominate individual liability loans, because

$$(1-p)^2 C_J^{\min} < (1-p) C_I^{\min}$$

The bank improves risk sharing by offering joint liability loans which induce the borrowers to help each other in bad times, something they would not do with individual liability. (Recall that we are ruling out side-contracts in this subsection.) As long as the incentive compatibility constraints are satisfied, switching from individual liability to joint liability reduces default rates and increases efficiency.

We have assumed so far that it is feasible to set the cost of default high enough to prevent strategic default. If this is not true, then individual liability loans may dominate joint liability loans. Specifically, suppose the cost of default has an upper bound, $\bar{C}$. Thus, we impose $C_I \leq \bar{C}$ and $C_J \leq \bar{C}$. Suppose $\bar{C}$ satisfies

$$C_I^{\min} < \bar{C} < C_J^{\min} \tag{7}$$

Then joint liability loans cannot satisfy the incentive compatibility constraint given by equation (5), because $C_J \leq \bar{C} < C_J^{\min}$. Encouraging successful individuals to help unsuccessful ones requires an impossibly large cost of default in this case. On the other hand, individual liability loans can satisfy the incentive compatibility constraint given by equation (4): just choose C_I so that $C_I^{\min} \leq C_I \leq \bar{C}$. Simply put, the inequalities in

---

[10] For example, new loans might be delayed for some time, which can be variable.

equation (7) imply individual liability dominates joint liability, as the former can be incentive compatible but the latter cannot (cf. Besley and Coate, 1995).

Returning to the case where there is no upper bound on C (or, equivalently, the upper bound is large enough so it is not constraining the contract), Rai and Sjöström (2004) found that joint liability loans can be improved upon by adding a message game. Suppose the bank offers a joint liability loan, but after project returns are realized, the bank organizes a village meeting. At this meeting, the bank asks each villager whether or not they are able to jointly repay $2(1 + r^*)$, and each villager i makes a repayment b_i to the bank. If both said "yes we can repay," and the loan is in fact repaid in full, $b_1 + b_2 = 2(1 + r^*)$, then—of course—neither incurs any cost of default. The key point is that if both said "no we cannot repay," and neither repays anything, then again no cost of default is incurred. As long as they agree, the bank trusts them. They suffer a cost C only if they disagree with each other, or if there is some other inconsistency (e.g. they claim they can repay but don't do it). This game has a truthful equilibrium such that whenever at least one project succeeds, the amount $2(1 + r^*)$ is repaid in full, but no agent ever incurs any cost of default, whether the loan is repaid or not! Intuitively, this is a (non-cooperative) equilibrium because any unilateral deviation from the truth leads to a disagreement, and hence to a punishment (a cost), so it doesn't pay. We can even choose the disagreement payoffs such that this is the unique equilibrium outcome. [11] Since there is never any costly default in equilibrium, the outcome is first best. Thus, a joint liability loan augmented with a message game strictly dominates the simple individual and joint liability loans discussed earlier, since these simple loans always had costly default in sufficiently bad states.

In the context of joint liability lending, Rai and Sjöström (2004) suggested that a message game played out during the public repayment meeting may allow the bank to extract information about repayment ability. However, the Grameen II reforms and

[11] Consider the following message game. The bank asks each villager whether they can repay $2(1 + r^*)$ in full, and each villager responds "yes" or "no." Simultaneously, each villager i makes a repayment b_i to the bank. (1) If both said "yes," and the loan is in fact repaid in full, $b_1 + b_2 = 2(1 + r^*)$, then neither is punished. (2) If there are no repayments ($b_1 = b_2 = 0$), then anyone who said "no" escapes punishment, but anyone who said "yes" is punished by a large amount (e.g. denied all future loans). (3) If villager i said "no" and made no repayment ($b_i = 0$), but villager j said "yes" and repaid $b_j = 2(1 + r^*)$, then villager i is punished by a large amount, while villager j receives a reward: the repayment b_j is returned to her, plus a small "bonus" $\varepsilon > 0$. (4) In all other cases: both villagers are punished by a large amount. It can be verified that if the villagers play a Nash equilibrium in each state of the world, then whenever at least one project succeeds, the amount $2(1 + r^*)$ is repaid in full. No agent is ever punished in equilibrium. There are no other, "bad," Nash equilibria. As it stands, the message game is vulnerable to collusion. Indeed, suppose both projects succeed. In Nash equilibrium, they are meant to repay in full whenever feasible. But suppose the villagers collude against the bank: they both claim that they cannot repay, and make no repayment. By definition of the mechanism, neither villager incurs any cost in this case, so the joint deviation makes both strictly better off (because they don't have to repay the loan). Rai and Sjöström (2004) show, however, that a modified message game can improve efficiency even in the presence of collusion, as long as the agents can only collude imperfectly. With perfect side-contracting (i.e. perfect collusion), the Coase theorem implies that message games are of no use whatsoever. As always, the internal contracting ability of the agents is a critical component of the design problem.

the Giné and Karlan (2009) experiments suggest that in fact joint liability is not a key component of successful microlending. Switching from joint liability to individual loans does not seem to reduce repayment rates. This would be consistent with frictionless side-contracting: as shown in the previous subsection, in such Coasean environments the form of the lending mechanism doesn't matter much. But in reality, empirical research suggests that risk sharing in village economies is far from perfect.

The Grameen II reforms and the Giné and Karlan (2009) experiments kept the public repayment meetings even with individual liability loans. In the next subsection, we consider the usefulness of public repayments in a world of imperfect risk sharing.

Imperfect side-contracts

Empirical evidence (e.g. Townsend, 1994; Udry, 1994) suggests that informal arrangements within poor villages are extensive but subject to significant contractual frictions. To capture this in a simple model, assume villagers can side-contract, but neither villager knows the other's project outcome. This internal friction impedes their side-contracting ability, so the Coase theorem does not apply.

Suppose the bank offers individual liability loans. Ideally, the villagers should agree that if one project fails and the other succeeds, the successful villager repays both loans. But if repayments are made in private, there will be no way for a villager to know if her neighbor's project succeeded, so mutual insurance is not incentive compatible. A successful villager can tell her unlucky neighbor that she, too, is facing hard times, and cannot even repay her own loan, much less help anyone else. Meanwhile, she privately makes all her repayments on time, thus avoiding the cost of default. Since mutual insurance is impossible with private repayments, each must repay with probability p. The bank's break-even constraint therefore requires that the interest rate is given by equation (3). To prevent strategic default, the cost of default, C_{priv}, must satisfy

$$C_{priv} \geq C_{priv}^{min} \equiv 1 + \hat{r} = \frac{1}{p} \tag{8}$$

Each borrower's expected cost of default is $(1 - p)C_{priv}$.

The Bangladeshi villages served by Grameen II, and the Philippine villages served by the Green Bank, collect repayments at public meetings. Thus, suppose the bank instead offers individual liability loans with public repayments. Suppose the interest rate on each loan is r^*, as defined in equation (2). Repayments are publicly observed, which gives the villagers something to contract on. The bank requires each villager to repay $1 + r^*$, and imposes a cost C_{pub} on any individual who defaults. Suppose during the meeting, the borrowers simultaneously announce whether their own projects failed or succeeded. If they have not made any mutual insurance arrangement, then the bank simply collects

$1 + r^*$ from any successful borrower, and imposes the cost C_{pub} on any unsuccessful borrower.

Now suppose the borrowers have side-contracted to help each other repay in full. We must verify that such mutual insurance is incentive compatible. Of course, if both announce that their projects succeeded, then each simply repays her own loan and avoids default. But a borrower who announces that her project succeeded is obligated by the mutual insurance agreement to repay both loans in full if the other villager announces that her project has failed. (A threat of social sanctions deters her from violating the agreement in public). Since projects succeed with probability p, a villager who truthfully announces that her project succeeded expects to pay

$$p(1 + r^*) + (1 - p)2(1 + r^*)$$

On the other hand, if she lies and reports that her project failed, she suffers the cost C_{pub} if the other borrower also reports failure. But if the other borrower reports success, the mutual insurance agreement kicks in, and there is no default. Thus, the expected cost of default is $(1 - p)C_{pub}$. Incentive compatibility requires that if her project succeeds, she prefers to be truthful:

$$(1 - p)C_{pub} \geq p(1 + r^*) + (1 - p)2(1 + r^*)$$

Thus, C_{pub} must satisfy

$$C_{pub} \geq C_{pub}^{min} \equiv \frac{2 - p}{1 - p}(1 + r^*)$$

If this inequality holds, with public repayments, it is incentive compatible for the borrowers to agree ex ante to help fully repay each other's (individual liability) loans whenever possible. Each borrower's expected cost of default is $(1 - p)^2 C_{pub}$. Public repayments are welfare enhancing if they reduce the expected cost, i.e., if

$$(1 - p)^2 C_{pub} < (1 - p)C_{priv}$$

This is certainly true if $C_{pub} = C_{priv}$. Thus, if the cost of default is fixed, then public repayments raise welfare. However, $C_{pub}^{min} > C_{priv}^{min}$. That is, the punishment required to encourage mutual insurance with public repayments is greater than the punishment required for incentive compatibility of individual loans with private repayment. The reasoning is similar to that in the previous subsection (see also Besley and Coate, 1995). Public repayment meetings can encourage a successful borrower to help repay her unsuccessful partner's loan. While this is welfare improving, incentive compatibility requires default to be very costly. In fact, it can be verified that these two effects of public repayments exactly cancel each other out:

$$(1 - p)^2 C_{pub}^{min} = (1 - p)C_{priv}^{min}$$

Therefore, if the cost of default is a continuous variable which can be minimized subject to incentive compatibility, then it is irrelevant whether repayments take place in private or in public. But if the cost cannot be fine-tuned like this, public repayments dominate—as long as incentive compatibility holds. Intuitively, public repayment meetings enhance side-contracting possibilities, by forcing the borrowers to reveal information to each other. In a non-Coasean environment, the bank can improve efficiency by helping the villagers insure each other against default. In particular, our highly stylized model suggests that public repayment meetings can make mutual insurance easier, because the villagers get more information about each other. [12]

We have assumed so far in this subsection that it is feasible to set the cost of default high enough to prevent strategic default. If this is not true, there is no longer a case for public repayment meetings. Specifically, suppose the cost of default has an upper bound $\bar{C}$ which satisfies

$$C_{priv}^{min} < \bar{C} < C_{pub}^{min} \qquad (9)$$

In this case, public repayment meetings cannot make mutual insurance incentive compatible, because the cost of default C_{pub} is constrained to satisfy $C_{pub} \leq \bar{C} < C_{pub}^{min}$. But individual liability loans will satisfy the incentive compatibility condition given by equation (8), as long as the cost of default C_{priv} satisfies $C_{priv}^{min} \leq C_{priv} \leq \bar{C}$.

To summarize, if the cost of default is fixed at some level large enough that the villagers prefer to help each other out rather than defaulting, then with public repayment meetings, they will mutually insure each other, whether liability is individual or joint. So, if public meetings are maintained, then a change from joint to individual liability (as in Grameen II or the Green Bank experiment in the Philippines) would not affect repayment rates. Eliminating public repayment meetings would, however, reduce repayment rates on individual liability loans, by making mutual insurance impossible. If the cost of default is constrained to be quite small, however, public repayment meetings are not useful, as mutual insurance cannot be incentive compatible. There is, of course, no inefficiency involved in having a public meeting anyway, which may generate other forms of social benefits.

CONCLUSION

Efficient design of microcredit is impossible without an understanding of informal side-contracting. If side-contracting is perfect, the design problem is not very interesting ("Coasean benchmark"). If side-contracting is impossible, the theoretical welfare comparison between joint and individual liability is ambiguous. Joint liability encourages

---

[12] Just as in previous subsections, the argument has to be modified for the case where a successful borrower does not have enough to fully repay both loans. The modified argument involves a graduated cost of default, where a partial repayment reduces the cost proportionally, but the logic will be the same. Further, the model can be extended to allow variable effort to influence the probability of success. In this case, the borrowers may not want to provide complete mutual insurance, because of moral hazard concerns, but public repayment meetings still expand the set of incentive-compatible side-contracts.

the borrowers to help each other in hard times, which mitigates insurance market imperfections and enhances efficiency, but a large (perhaps infeasible) cost of default is required for this help to be incentive compatible. Both types of loans are dominated by more complicated lending mechanisms ("No side-contracts").

In reality, side-contracting seems to be extensive but far from perfect, due to internal informational and enforcement problems. To evaluate a microlending mechanism, we need to consider how it will influence the set of incentive compatible side-contracts. In the subsection "Imperfect side-contracts", we argued that public repayment meetings can enhance mutual insurance possibilities. The general point is that an outside intervention will influence the set of incentive compatible side-contracts: there is no reason to believe side-contracting ability is exogenously fixed once and for all. This idea is well known in general, although not usually expressed in this way (e.g. Ostrom, 1999).

Field experiments have been extremely useful in sorting between the mechanisms underlying microfinance contracts (Banerjee and Duflo, 2010). The two field experiments most relevant to our discussion of microcredit contracts are Giné and Karlan (2009) and Attanasio et al. (2011). The former compares the repayment performance of individual and joint liability loans in randomly treated villages, and finds no difference in repayment rates. Repayments are made at public meetings in both the treated and control villages. The latter compares repayment performance and borrower expenditure under joint and individual liability loans, but the repayments are made in private. To identify the value of public repayments, an ideal experiment would vary not just the liability structure of the loan contract but also the public or private nature of repayment. Additional information about the cost of default, and mutual insurance arrangements, could help us evaluate the risk-sharing theory of public repayment discussed here.

In our model, a key role is played by the cost of default C, usually interpreted as future credit denial. Such future credit denial is effective only if the bank is established, and borrowers believe that it will be solvent and able to make loans in the future (Bond and Rai, 2009). The borrowers' ability to side-contract depends on C. The very well established Grameen Bank in Bangladesh and the Green Bank in Philippines may achieve a high C because the threat of credit denial is strong. Lenders without a track record might be constrained to set a smaller C. Thus, the optimal microcredit contract may vary depending on the history of the lender. In addition, the cost of default is low if the borrower can turn to other lenders (unless lenders share information about defaulting borrowers), so competition will force C to be small (de Janvry et al., 2005). Thus, it is unlikely that one design will fit all environments, and much research remains to be done.

REFERENCES

Armendariz de Aghion, B. and Morduch, J. (2005) *The Economics of Microfinance*, MIT Press.
Attanasio, O., Augsburg, B., De Haas, R., Fitzsimons, E. and Harmgart, H. (2011) "Group lending or individual lending? Evidence from a randomised field experiment in Mongolia," Working Paper, UCL.

Baliga, S. and Sjöström, T. (1998) "Decentralization and collusion," *Journal of Economic Theory*, 83: 196–232.

Banerjee, A. and Duflo, E. (2010) "Giving credit where it is due," *Journal of Economic Perspectives*, 24: 61–79.

‗‗‗‗ ‗‗‗‗ Glennerster, R. and Kinnan, C. (2010) "The miracle of microfinance: evidence from a randomized evaluation," Working Paper, Duke University.

Besley, T. (1994) "How do market failures justify interventions in rural credit markets?" *World Bank Research Observer*, 9: 22–47.

‗‗‗‗ (1995) "Nonmarket institutions for credit and risk sharing in low-income countries," *Journal of Economic Perspectives*, 9: 115–27.

‗‗‗‗ and Coate, S. (1995) "Group lending, repayment incentives and social collateral," *Journal of Development Economics*, 46: 1–18.

Bond, P. and Krishnamurthy, A. (2004) "Regulating exclusion from financial markets," *Review of Economic Studies*, 71: 681–707.

‗‗‗‗ and Rai, A. S. (2009) "Borrower runs," *Journal of Development Economics*, 88: 185–91.

Bulow, J. and Rogoff, K. (1989) "Sovereign debt: is to forgive to forget?" *American Economic Review*, 79: 43–50.

Chowdhury, P. R. (2005) "Group-lending: sequential financing, lender monitoring and joint liability," *Journal of Development Economics*, 77: 415–39.

Cull, R., Demirgüç-Kunt, A. and Morduch, J. (2009) "Microfinance meets the market," *Journal of Economic Perspectives*, 23: 167–92.

Diamond, D. (1984) "Financial intermediation and delegated monitoring," *Review of Economic Studies*, 51: 393–414.

Dowla, A. and Barua, D. (2006) *The Poor Always Pay Back: The Grameen II story*, Kumarian Press.

de Janvry, A., McIntosh, C. and Sadoulet, E. (2005) "How rising competition among microfinance institutions affects incumbent lenders," *Economic Journal*, 115: 987–1004.

Field, E. and Pande, R. (2008) "Repayment frequency and default in microfinance: evidence from India," *Journal of the European Economic Association*, 6: 501–9.

Feigenberg, B., Field, E. and Pande, R. (2009) "Building social capital through microfinance," Working Paper, Harvard University.

Fischer, G. and Ghatak, M. (2010) "Repayment frequency in microfinance contracts with present-biased borrowers," Working Paper, LSE.

Ghatak, M. and Guinnane, T. (1999) "The economics of lending with joint liability: A review of theory and practice," *Journal of Development Economics*, 60: 195–228.

Giné, X., Jakiela, P., Karlan, D. S. and Morduch, J. (2010) "Microfinance games," *American Economic Journal: Applied Economics*, 2: 60–95.

‗‗‗‗ and Karlan, D. S. (2009) Group versus individual liability: Long term evidence from Philippine microcredit lending groups. Working Paper 970, Economic Growth Center, Yale University.

Graham, D. and Marshall, R. (1987) "Collusive bidder behavior at single-object second price and English auctions," *Journal of Political Economy*, 95: 1217–39.

Karlan, D. and Morduch, J. (2009) "Access to finance: credit markets, insurance and savings," in D. Rodrik and M. Rosenzweig (eds), *Handbook of Development Economics*, North Holland, Vol. 5.

Laffont, J.-J. and Martimort, D. (1997) "Collusion under asymmetric information," *Econometrica*, 65: 875–911.

―――― ―――― (2000) "Mechanism design with collusion and correlation," *Econometrica*, 68: 309–42.

―――― and N'Guessan, T. (2000) "Group lending with adverse selection," *European Economic Review*, 44: 773–784.

―――― and Rey, P. (2000) "Collusion and group lending with moral hazard," Working Paper, IDEI.

Ligon, E., Thomas, J. P. and Worrall, T. (2002) "Informal insurance arrangements with limited commitment: Theory and evidence from village economies," *Review of Economic Studies*, 69: 209–44.

―――― (1998) "Risk-sharing and information in village economies," *Review of Economic Studies*, 65: 847–64.

―――― (2004) "Targeting and informal insurance," in S. Dercon, (ed), *Insurance Against Poverty*, Oxford University Press.

Lopomo, L., Marshall, R. and Marx, L. (2005) "Inefficiency of collusion at English auctions," *Contributions to Theoretical Economics*, 5(1): article 4.

Mailath, G. and Zemsky, P. (1991) "Collusion in second price auctions with heterogeneous bidders," *Games and Economic Behavior*, 3: 467–86.

McAfee, P. and McMillan, J. (1992) "Bidding rings," *American Economic Review*, 82: 579–99.

McKenzie, D. and Woodruff, C. (2008) "Experimental evidence on returns to capital and access to finance in Mexico," *World Bank Economic Review*, 22: 457–82.

Ostrom, E. (1999) "Social capital: a fad or a fundamental concept?," in P. Dasgupta and I. Serageldin, (eds), *Social Capital: A Multifaceted Perspective*, The World Bank.

Rai, A. S. and Sjöström, T. (2004) "Is Grameen lending efficient? Repayment incentives and insurance in village economies," *Review of Economic Studies*, 71: 217–34.

Rahman, A. (1999) *Women and Microcredit in Rural Bangladeshm*, Westview Press.

Stiglitz, J. E. and Weiss, A. (1981) "Credit rationing in markets with imperfect information," *American Economic Review*, 71: 393–410.

Townsend, R. (1979) "Optimal contracts and competitive markets with costly state verification," *Journal of Economic Theory*, 21: 265–93.

―――― (1994) "Risk and insurance in village India," *Econometrica*, 62: 539–91.

―――― (2003) "Microcredit and mechanism design," *Journal of the European Economic Association*, 1: 468–77.

Udry, C. (1994) "Risk and insurance in a rural credit market: an empirical investigation in northern Nigeria," *Review of Economic Studies*, 61: 495–526.

Yunus, M. (1999) "The Grameen Bank," *Scientific American*, November: 114–19.

PART II
SECTION B

AUCTIONS

CHAPTER 10

THE PRODUCT-MIX AUCTION

A New Auction Design for Differentiated Goods

PAUL KLEMPERER[1]

INTRODUCTION

How should goods that both seller(s) and buyers view as imperfect substitutes be sold, especially when multi-round auctions are impractical?

This was the Bank of England's problem in autumn 2007 as the credit crunch began.[2] The Bank urgently wanted to supply liquidity to banks, and was therefore willing to accept a wider-than-usual range of collateral, but it wanted a correspondingly higher interest rate against any weaker collateral it took. A similar problem was the US Treasury's autumn 2008 Troubled Asset Recovery Program (TARP) plan to spend up to $700

[1] This chapter was originally published in the *Journal of the European Economic Association* (2010) 8(2–3): 526–36, and is reproduced here with the kind permission of the European Economic Association and the MIT Press. Minor revisions have been made to the original paper. The Bank of England continues to use this auction design regularly and enthusiastically—the Governor of the Bank (Mervyn King) wrote that "[it] is a marvellous application of theoretical economics to a practical problem of vital importance to financial markets." I have been a *pro bono* adviser to the Bank of England since autumn 2007, and I have also given *pro bono* advice to the US Treasury, other central banks, government agencies, etc., about these issues. I thank the relevant officials for help, but the views here are my own and do not represent those of any organization. I am very grateful to Jeremy Bulow and Daniel Marszalec for their help in advising the Bank of England. I also particularly benefited from discussions with Elizabeth Baldwin and Marco Pagnozzi, and thank Olivier Armantier, Eric Budish, Vince Crawford, Aytek Erdil, Meg Meyer, Moritz Meyer-ter-Vehn, Rakesh Vohra, the editor, and anonymous referees, and many other friends and colleagues for helpful advice.

[2] The crisis began in early August 2007, and a bank run led to Northern Rock's collapse in mid-September. Immediately subsequently, the Bank of England first ran four very unsuccessful auctions to supply additional liquidity to banks and then consulted me. I got valuable assistance from Jeremy Bulow and Daniel Marszalec.

billion buying "toxic assets" from among 25,000 closely related but distinct subprime mortgage-backed securities.

Because financial markets move fast, in both cases it was highly desirable that any auction take place at a single instant. In a multi-stage auction, bidders who had entered the highest bids early on might change their minds about wanting to be winners before the auction closed,[3] and the financial markets might themselves be influenced by the evolution of the auction, which magnifies the difficulties of bidding and invites manipulation.[4]

An equivalent problem is that of a firm choosing its "product mix": it can supply multiple varieties of a product (at different costs), but with a total capacity constraint, to customers with different preferences between those product varieties, and where transaction costs or other time pressures make multiple-round auctions infeasible.[5] The different varieties of a product could include different points of delivery, different warranties, or different restrictive covenants on use.

This paper outlines a solution to all these problems—the product-mix auction. I first developed it for the Bank of England, which now uses it routinely.[6] Indications of its success are that the Governor of the Bank of England (Mervyn King) wrote, after the Bank had been using it regularly for over eighteen months and auctioned £80 billion worth of repos using it, that "The Bank of England's use of Klemperer auctions in our liquidity insurance operations is a marvellous application of theoretical economics to a practical problem of vital importance to financial markets"; he made a similar statement to the *Economist* a year later; and an Executive Director of the Bank described the auction as "A world first in central banking . . . potentially a major step forward in practical policies to support financial stability."[7]

I subsequently made a similar proposal to the US Treasury, which would probably have used a related design if it had not abandoned its plans to buy toxic assets.[8] At

---

[3] Some evidence for this is that most bids in standard Treasury auctions are made in the last few minutes, and a large fraction in the last few seconds. For a multi-round auction to have any merit, untopped bids cannot be withdrawn without incurring penalties.

[4] The Bank of England insisted on a single-stage auction. Ausubel and Cramton (2008) argued a multi-stage auction was feasible for the US Treasury.

[5] That is, the Bank of England can be thought of as a "firm" whose "product" is loans; the different "varieties" of loans correspond to the different collaterals they are made against, and their total supply may be constrained. The Bank's "customers" are its counterparties, and the "prices" they bid are interest rates.

[6] See note 2. I do *not* give full details of the Bank's objectives and constraints here, and not all the issues I discuss are relevant to it. Although the auction was designed in response to the crisis, the Bank wanted a solution that would be used in normal times too (in part, so that the use of a specific auction design would convey no information).

[7] See Bank of England (2010, 2011), Fisher (2011), Milnes (2010), Fisher et al. (2011), the *Economist* (2012), and the Bank of England's website. The Bank's current auctions correspond closely to the design described in the second section of this chapter; future auctions may use some of the enhancements described in the third section.

[8] After I proposed my solution to the Bank of England, I learned that Paul Milgrom was independently pursuing related ideas. He and I therefore made a joint proposal to the US Treasury, together with Jeremy Bulow and Jon Levin, in September–October 2008. Other consultants, too, proposed a static (sealed-bid) design, although of a simpler form, and the Treasury planned to run a

the time of writing, another central bank is exploring my design, and a regulator is considering a proposal to use my product-mix auction for selling two close-substitute "types" of contracts to supply gas.

My design is straightforward in concept—each bidder can make one or more bids, and *each* bid contains a *set* of mutually exclusive offers. Each offer specifies a price (or, in the Bank of England's auction, an interest rate) for a quantity of a specific "variety." The auctioneer looks at all the bids and then selects a price for each "variety." From each bid offered by each bidder, the auctioneer accepts (only) the offer that gives the bidder the greatest surplus at the selected prices, or no offer if all the offers would give the bidder negative surplus. All accepted offers for a variety pay the same (uniform) price for that variety.

The idea is that the menu of mutually exclusive sets of offers allows each bidder to approximate a demand function, so bidders can, in effect, decide how much of each variety to buy *after* seeing the prices chosen. Meanwhile, the auctioneer can look at demand *before* choosing the prices; allowing it to choose the prices ex post creates no problem here, because it allocates each bidder precisely what that bidder would have chosen for itself given those prices.[9] Importantly, offers for each variety provide a competitive discipline on the offers for the other varieties, because they are all being auctioned simultaneously.

Compare this with the "standard" approach of running a separate auction for each different "variety." In this case, outcomes are erratic and inefficient, because the auctioneer has to choose how much of each variety to offer before learning bidders' preferences, and bidders have to guess how much to bid for in each auction without knowing what the price differences between varieties will turn out to be; the wrong bidders may win, and those who do win may be inefficiently allocated across varieties. Furthermore, each individual auction is much more sensitive to market power, to manipulation, and to informational asymmetries than if all offers compete directly with each other in a single auction. The auctioneer's revenues are correspondingly generally lower.[10] All these problems also reduce the auctions' value as a source of information. They may

first set of simple sealed-bid auctions, each for a related group of assets, and then enhance the design using some of the Bulow–Klemperer–Levin–Milgrom ideas in later auctions. However, it then suddenly abandoned its plans to buy subprime assets (in November 2008). Note also, however, that Larry Ausubel and Peter Cramton—who played an important role in demonstrating the value of using auctions for TARP (see e.g. Ausubel et al., 2008)—had proposed running dynamic auctions, and the possibility of doing this at a later stage was also still being explored. Milgrom (2009) shows how to represent a wide range of bidders' preferences such that goods are substitutes, and shows that a linear-programming approach yields integer allocations when demands and constraints are integer, but my proposal seems more straightforward and transparent in a context such as the Bank of England's.

[9] That is, it chooses prices like a Walrasian auctioneer who is equating bidders' demand with the bid-taker's supply in a decentralized process (in which the privately held information needed to determine the allocation is directly revealed by the choices of those who hold it). The result assumes the conditions for "truthful" bidding are satisfied—see later.

[10] Thus, for example, if the US Treasury had simply predetermined the amount of each type of security to purchase, ignoring the information about demand for the large number of closely related securities, competition would have been inadequate. There were perhaps 300 likely sellers, but the largest ten held of the order of two-thirds of the total volume, and ownership of many individual securities was far more highly concentrated.

also reduce participation, which can create "second-round" feedback effects further magnifying the problems.[11]

Another common approach is to set fixed price supplements for "superior" varieties, and then auction all units as if they are otherwise homogenous. This can sometimes work well, but such an auction cannot take any account of the auctioneer's preferences about the proportions of different varieties transacted.[12] Furthermore, the auctioneer suffers from adverse selection.[13]

The question, of course, is whether my alternative approach can actually be implemented, and—crucially—whether it can be done in a way that is simple and robust, and easy for bidders to understand, so that they are happy to participate.

The following section shows how my product-mix auction does this. The third section discusses extensions. In particular, it is easy to include multiple buyers and multiple sellers, and "swappers" who may be on either, or both, sides of the market. The fourth section observes that the product-mix auction is essentially a "proxy" implementation of a "two-sided" simultaneous multiple-round auction (SMRA)—but because my design is static, it is simpler and cheaper and less susceptible to collusion and other abuses of market power than is a standard dynamic SMRA. The fifth section concludes.

A SIMPLE TWO-VARIETY EXAMPLE

The application this auction was originally designed for provides a simple illustration. A single seller, the Bank of England (henceforth "the Bank"), auctioned just two "goods," namely a loan of funds secured against strong collateral, and a loan of funds secured against weak collateral. For simplicity I refer to the two goods as "strong" and "weak."[14] In this context, a per-unit price is an interest rate. The rules of the auction are as follows:

1. Each bidder can make any number of bids. *Each bid* specifies a *single* quantity and an offer of a per-unit price for *each* variety. The offers in each bid are mutually exclusive.
2. The auctioneer looks at all the bids and chooses a minimum "cut-off" price for each variety—I will describe later in this section how it uses the construction illustrated

[11] The feedback effects by which low participation reduces liquidity, which further reduces participation and liquidity, etc., are much more important when there are multiple agents on both sides of the market—see Klemperer (2008).

[12] Moreover, a central bank might not want to signal its view of appropriate price differentials for different collaterals to the market in advance of the auction.

[13] If, for example, the US Treasury had simply developed a "reference price" for each asset, the bidders would have sold it large quantities of the assets whose reference prices were set too high—and mistakes would have been inevitable, since the government had so much less information than the sellers.

[14] We assume (as did the Bank) that there is no adverse selection problem regarding collateral. For the case in which bidders have private information regarding the value of the collateral they offer, see Manelli and Vincent (1995).

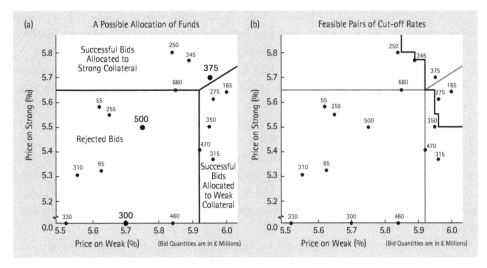

FIGURE 10.1. An example of bids in the Bank of England's auction.

in Figures 10.1 and 10.2 to determine these minimum prices uniquely, for any given set of bids, and given its own preferences.

3. The auctioneer accepts all offers that exceed the minimum price for the corresponding variety, *except* that it accepts at most one offer from each bid. If both price offers in any bid exceed the minimum price for the corresponding variety, the auctioneer accepts the offer that maximizes the bidder's surplus, as measured by the offer's distance above the minimum price. [15]

4. All accepted offers pay the minimum price for the corresponding variety—that is, there is "uniform pricing" for each variety. [16]

Thus, for example, one bidder might make three separate bids: a bid for £375 million at {5.95% for (funds secured against) weak *or* 5.7% for (funds secured against) strong}; a bid for an additional £500 million at {5.75% for weak *or* 5.5% for strong}; and a bid for a further £300 million at {5.7% for weak *or* 0% for strong}. Note that since offers at a price of zero are never selected, the last bid is equivalent to a traditional bid on only a single collateral. [17]

An example of the universe of all the bids submitted by all the bidders is illustrated in Figure 10.1a. The prices (i.e., interest rates) for weak and strong are plotted vertically and horizontally respectively; each dot in the chart represents an "either/or" bid. The number by each dot is the quantity of the bid (in £millions). The three bids made by the bidder described above are the enlarged dots highlighted in bold.

[15] See notes 18 and 21 for how to break ties, and ration offers that equal the minimum price.

[16] Klemperer (2008) discusses alternative rules.

[17] A bidder can, of course, restrict each of its bids to a single variety. Note also that a bidder who wants to guarantee winning a fixed total quantity can do so by making a bid at an arbitrarily large price for its preferred variety, and at an appropriate discount from this price for the other variety.

The cut-off prices and the winning bids are determined by the Bank's objectives. If, for example, the Bank wants to lend £2.5 billion, and there are a total of £5.5 billion in bids, then it must choose £3 billion in bids to reject.

Any possible set of rejected bids must lie in a rectangle with a vertex at the origin. Figure 10.1a shows one possible rectangle of rejected bids, bounded by the vertical line at 5.92% and the horizontal line at 5.65%. If the Bank were to reject this rectangle of bids, then all the accepted bids—those outside the rectangle—would pay the cut-off prices given by the boundaries: 5.92% for weak, and 5.65% for strong.

Bids to the north-east of the rectangle (i.e. those which could be accepted for either variety) are allocated to the variety for which the price is further below the offer. So bids that are both north of the rectangle, and north-west of the diagonal 45° line drawn up from the upper-right corner of the rectangle, receive strong, and the other accepted bids receive weak.

Of course, there are many possible rectangles that contain the correct volume of bids to reject. On any 45° line on the plane, there is generally exactly one point that is the upper-right corner of such a rectangle.[18] It is easy to see that the set of all these points forms the stepped downward-sloping line shown in Figure 10.1b.[19] This stepped line is therefore the set of feasible pairs of cut-off prices that accept exactly the correct volume of bids.

Every point on Figure 10.1b's stepped line (i.e., every possible price pair) implies both a price difference and (by summing the accepted bids below the corresponding 45° line) a proportion of sales that are weak. As the price difference is increased, the proportion of weak sales decreases. Using this information we can construct the downward-sloping "demand curve" in Figure 10.2.

If it wished, the auctioneer (the Bank) could give itself discretion to choose any point on the "demand curve" (equivalently, any feasible rectangle in Figures 10.1 and 10.1b) after seeing the bids. In fact, the Bank prefers to precommit to a rule that will determine its choice. That is, the Bank chooses a "supply curve" or "supply schedule" such as the upward-sloping line in Figure 10.2 so the proportion allocated to weak increases with the price difference.[20]

---

[18] Moving north-east along any 45° line represents increasing all prices while maintaining a constant difference between them. Because the marginal bid(s) is usually rationed, there is usually a single critical point that rejects the correct volume of bids. But if exactly £3 billion of bids can be rejected by rejecting entire bids, there will be an interval of points between the last rejected and the first accepted bid. As a tie-breaking rule, I choose the most south-westerly of these points.

[19] The initial vertical segment starts at the highest price for weak such that enough can be accepted on weak when none is accepted on strong (this price is the weak price of the bid for 680), and continues down as far as the highest price bid for strong (the strong price of the bid for 250). At this point some strong replaces some weak in the accepted set, and there is then a horizontal segment until we reach the next price bid for weak (the weak price of the bid for 345) where more strong replaces weak in the accepted set and another vertical segment begins, etc.

[20] The proposal for the US TARP to employ a "reference price" for each asset corresponds to choosing the multidimensional equivalent of a horizontal supply curve; buying a predetermined quantity of each asset corresponds to using a vertical supply curve. As I noted earlier, both these approaches are flawed. Choosing an upward-sloping supply curve maintains the advantage of the

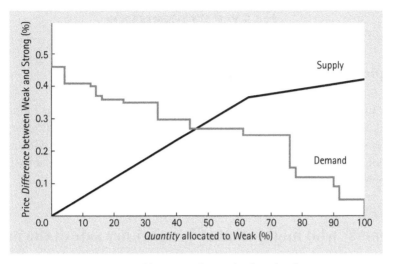

FIGURE 10.2. Equilibrium in the Bank of England's auction.

The point of intersection between the Bank's supply curve and the "demand curve" constructed from the bids determines the price differential and the percentage of weak sold in the auction. With the supply curve illustrated, the price difference is 0.27% and the proportion of weak is 45%—corresponding to the outcome shown in Figure 10.1a. [21]

This procedure ensures that bidders whose bids reflect their true preferences[22] receive precisely the quantities that they would have chosen for themselves if they had known the auction prices in advance. So unless a bidder thinks its own bids will affect the auction prices, its best strategy is to bid "truthfully;" if bidders all do this, and the Bank's supply curve also reflects its true preferences, the auction outcome is the competitive equilibrium. [23]

reference price approach, while limiting the costs of mispricing. (The optimal choice of supply-curve slope involves issues akin to those discussed in Poole (1970), Weitzman (1974), Klemperer and Meyer (1986), etc.; maintaining the reserve power to alter the supply curve after seeing the bids protects against collusion, etc.; see Klemperer and Meyer (1989), Kremer and Nyborg (2004), Back and Zender (2001), McAdams (2007), etc.)

[21] By determining the proportion of weak, Figure 10.2 also determines what fractions of any bids on the rectangle's borders are filled, and the allocation between goods of any bids on the 45° line.

[22] This does not require pure "private value" preferences, but does not allow bidders to change their bids in response to observing others' bids. We can extend our mechanism to allow bidders with "common values" to update their bids: the auctioneer takes bids as described earlier, and reports the "interim" auction prices that would result if its supply were scaled up by some predetermined multiple (e.g., 1.25). It then allows bidders to revise the prices of any bid that would win at the interim prices, except that the price on the variety that the bid would win cannot be reduced below that variety's interim price. Multiple such stages can be used, and/or more information can be reported at each stage, before final prices and allocations are determined—we offered such an option to the US Treasury, though it was not our main recommendation.

[23] Because on the order of forty commercial banks, building societies, etc., bid in the Bank of England's auctions, it is unlikely that any one of them can much affect the prices. I assume the Bank's

EASY EXTENSIONS

Multiple buyers and multiple sellers

It is easy to include additional potential sellers (i.e., additional lenders of funds, in our example). Simply add their maximum supply to the total that the auctioneer sells, but allow them to participate in the auction as usual. If a potential seller wins nothing in the auction, the auctioneer has sold the seller's supply for it. If a potential seller wins its total supply back, there is no change in its position.

"Swappers" who might want to be on either side of the market

Exactly the same approach permits a trader to be on either side, or both sides, of the market. If, for example, letting the auctioneer offer its current holdings of strong, a bidder in the auction wins the same amount of weak, it has simply swapped goods (paying the difference between the market-clearing prices).

Variable total quantity

Making the total quantity sold (as well as the proportions allocated to the different varieties) depend upon the prices is easy. The Bank might, for example, precommit to the total quantity being a particular increasing function of the price of strong. Using the procedure set out in the preceding section to solve for the strong price corresponding to every possible total quantity yields a weakly decreasing function, and the unique intersection of the two functions then determines the equilibrium.

Other easy extensions

Several other extensions are also easy. For example, bidders can be allowed to ask for different amounts of the different goods in a bid. Or a bidder can specify that a total quantity constraint applies across a group of bids. And there can, of course, be more than two goods, with a cut-off price for each, and a bid rejected only if *all* its offers are below the corresponding cut-off prices.

Bidders can express more complex preferences by using several bids in combination. For example, a bidder might be interested in £100 million weak at up to 7%, and £80 million strong at up to 5%. However, even if prices are high, the bidder wants an absolute

supply curve is upward sloping so, given our tie-breaking rule (see note 18), if there are multiple competitive equilibria the outcome is the unique one that is lowest in both prices.

minimum of £40 million. This can be implemented by making all of the following four bids, if negative bids are permitted:

1. £40 million of {weak at maximum permitted bid *or* strong at maximum permitted bid *less* 2%}.
2. £100 million of weak at 7%.
3. £80 million of strong at 5%.
4. *minus* £40 million of {weak at 7% *or* strong at 5%}.

The point is that the fourth (negative) bid kicks in exactly when one of the second and third bids is accepted, and then exactly cancels the first bid for £40 million "at any price" (since 2% = 7% – 5%).[24]

FURTHER EXTENSIONS, AND THE RELATIONSHIP TO THE SIMULTANEOUS MULTIPLE-ROUND AUCTION

My auction is equivalent to a static (sealed-bid) implementation of a simplified version of a "two-sided" simultaneous multiple-round auction (SMRA). (By "two-sided" I mean that sellers as well as buyers can make offers, as explained later.)

Begin by considering the special case in which the auctioneer has predetermined the quantity of each variety it wishes to offer, and the bids in my auction represent bidders' true preferences. Then the outcome will be exactly the same as the limit as bid increments tend to zero of a standard SMRA if each bidder bids at every step to maximize its profits at the current prices given those preferences,[25] since both mechanisms simply select the competitive-equilibrium price vector.[26]

[24] A bidder can perfectly represent any preferences across all allocations by using an appropriate pattern of positive and negative bids if the goods are imperfect substitutes such that the bidder's marginal value of a good is reduced at least as much by getting an additional unit of that good as by getting an additional unit of the other good (i.e., if $V(w,s)$ is the bidder's total value of £w of weak plus £s of strong, then $\partial^2 V/\partial w^2 \leq \partial^2 V/\partial w \partial s \leq 0$ and $\partial^2 V/\partial s^2 \leq \partial^2 V/\partial w \partial s \leq 0$). More general preferences than this require more complex representations—but the important point, of course, is that preferences can typically be well approximated by simple sets of bids. The geometric techniques used in the analysis of the product-mix auction also yield new results in the multidimensional analysis of demand: see Baldwin and Klemperer (2012).

[25] In a SMRA the bidders take turns to make bids in many ascending auctions that are run simultaneously (e.g., 55% of 2.5 billion = 1.375 billion auctions for a single £1 of strong, and 45% of 2.5 billion = 1.125 billion auctions for a single £1 of weak). When it is a bidder's turn, it can make any new bids it wishes that beat any existing winning bid by at least the bidding increment (though it cannot top up or withdraw any of its own existing bids). This continues until no one wants to submit any new bids. For more detail, including "activity rules" etc., see, e.g., Milgrom (2000), Binmore and Klemperer (2002), and Klemperer (2004).

[26] An exception is that an SMRA may not do this when bidders' preferences are such that they would ask for different amounts of the different goods in a single bid in my procedure. All the other types of

The general case in which the auctioneer offers a general supply curve relating the proportions of the different varieties sold to the price differences is not much harder. We now think of the auctioneer as acting *both* as the bid-taker selling the maximum possible quantity of both varieties, *and* as an additional buyer bidding to buy units back to achieve a point on its supply curve. That is, in our example in which the Bank auctions £2.5 billion, we consider an SMRA which supplies £2.5 billion weak *and* £2.5 billion strong, and we think of the Bank as an additional bidder that has an inelastic total demand for £2.5 billion and that bids in exactly the same way as any other bidder.[27, 28]

So my procedure is equivalent to a "proxy SMRA," that is, a procedure in which bidders submit their preferences, and the auctioneer (and other potential sellers) submit their supply curves, and a computer then calculates the equilibrium that the (two-sided) SMRA would yield.[29] However, my procedure restricts the preferences that the auction participants can express. Although I can permit more general forms of bidding than those discussed (see Klemperer, 2008),[30] some constraints are desirable. For example, I am cautious about allowing bids that express preferences under which varieties are complements.[31]

Importantly, exercising market power is much harder in my procedure than in a standard SMRA, precisely because my procedure does not allow bidders to express

bids discussed reflect preferences such that all individual units of all goods are substitutes for all bidders (so bidding as described in an SMRA is rational behavior if the number of bidders is large). I assume the auctioneer also has such preferences (i.e., the Bank's supply curve is upward sloping), so if there are multiple competitive equilibria, there is a unique one in which all prices are lowest and both mechanisms select it—see note 23 and Crawford and Knoer (1981), Kelso and Crawford (1982), Gul and Stacchetti (1999), and Milgrom (2000).

[27] That is, whenever it is the Bank's turn to bid, it makes the minimum bids both to restore its quantity of winning bids to £2.5 billion and to win the quantity of each variety that puts it back on its supply curve, given the current price difference. It can always do this to within one bid increment, since the weak-minus-strong price difference can only be more (less) than when it last bid if its weak (strong) bids have all been topped, so it can increase the quantity of strong (weak) it repurchases relative to its previous bids, as it will wish to do in this case.

[28] If there are other sellers (or "swappers"), add their potential sales (or "swaps") to those offered in the SMRA, and think of these participants as bidding for positive amounts like any other bidders.

[29] Although the description in the second section of the present chapter may have obscured this, our procedure is symmetric between buyers and sellers. (It is not quite symmetric if the auctioneer does not precommit to its supply curve, but if bidders behave competitively their bids are unaffected by this.)

[30] I could in principle allow any preferences subject to computational issues; these issues are not very challenging in the Bank of England's problem.

[31] The difficulty with complements is the standard one that there might be multiple unrankable competitive equilibria, or competitive equilibrium might not exist (see note 26), and an SMRA can yield different outcomes depending upon the order in which bidders take turns to bid. In independent work, Milgrom (2009) explores how to restrict bidders to expressing "substitutes preferences." Crawford's (2008) static mechanism for entry-level labor markets (e.g., the matching of new doctors to residency positions at hospitals) addresses related issues in a more restrictive environment. See also Budish (2004).

preferences that depend on others' bids. In particular, coordinated demand reduction (whether or not supported by explicit collusion) and predatory behavior may be almost impossible. In a standard dynamic SMRA, by contrast, bidders can learn from the bidding when such strategies are likely to be profitable, and how they can be implemented—in an SMRA, bidders can make bids that signal threats and offers to other bidders, and can easily punish those who fail to cooperate with them.[32, 33]

Finally, the parallel with standard sealed-bid auctions makes my mechanism more familiar and natural than the SMRA to counterparties. In contexts like the Bank of England's, my procedure is much simpler to understand.

Conclusion

The product-mix auction is a simple-to-use sealed-bid auction that allows bidders to bid on multiple differentiated assets simultaneously, and bid takers to choose supply functions across assets. It can be used in environments in which an SMRA is infeasible because of transaction costs, or the time required to run it. The design also seems more familiar and natural than the SMRA to bidders in many applications, and makes it harder for bidders to collude or exercise market power in other ways.

Relative to running separate auctions for separate goods, the product-mix auction yields better "matching" between suppliers and demanders, reduced market power, greater volume and liquidity, and therefore also improved efficiency, revenue, and quality of information. Its applications therefore extend well beyond the financial contexts for which I developed it.

References

Ausubel, L. and Cramton, P. (2008) "A troubled asset reverse auction," Mimeo, University of Maryland.

——— ——— Filiz-Ozbay, E., Higgins N., Ozbay, E. and Stocking, A. (2008) "Common-value auctions with liquidity needs: an experimental test of a troubled assets reverse auction," Working Paper, University of Maryland.

[32] In a standard SMRA, a bidder can follow "collusive" strategies such as "I will bid for (only) half the lots if my competitor does also, but I will bid for more lots if my competitor does" (see, e.g., Klemperer, 2002, 2004), but in our procedure the bidder has no way to respond to others' bids. Of course, a bidder who represents a significant fraction of total demand will bid less than its true demand in *any* procedure, including mine, which charges it constant per-unit prices. But it is much easier for a bidder to (ab)use its market power in this way in an SRMA.

[33] A multi-round procedure (either an SMRA, or an extension of our procedure—see note 22) may be desirable if bidders' valuations have important "common-value" components, but may discourage entry of bidders who feel less able than their rivals to use the information learned between rounds.

Back, K., and Zender, J. (2001) "Auctions of divisible goods with endogenous supply," *Economics Letters*, 73: 29–34.

Baldwin, E. and Klemperer, P. (2012) "Tropical Geometry to Analyse Demand," Mimeo, Oxford University.

Bank of England (2010) "The Bank's new indexed long-term repo operations," *Bank of England Quarterly Bulletin*, 50(2): 90–1.

_____ (2011) "The Bank's indexed long-term repo operations," *Bank of England Quarterly Bulletin*, 51/2: 93.

Binmore, K. and Klemperer, P. (2002) "The biggest auction ever: the sale of the British 3G telecom licenses," *Economic Journal*, 112: C74–96.

Budish, E. (2004) "Internet auctions for close substitutes," MPhil Thesis, University of Oxford.

Crawford, V. P. (2008) "The flexible-salary match: a proposal to increase the salary flexibility of the national resident matching program," *Journal of Economic Behavior and Organization*, 66(2): 149–60.

_____ and Knoer, E. M. (1981) "Job matching with heterogeneous firms and workers,". *Econometrica*, 49: 437–50.

The *Economist* (2012) "A golden age of micro." The *Economist*, Free Exchange, October 19, 2012. <http://www.economist.com/blogs/freeexchange/2012/10/microeconomics>.

Fisher, P. (2011) "Recent developments in the sterling monetary framework," at <http://www.bankofengland.co. uk/publications/speeches/2011/speech487.pdf>.

_____ Frost, T. and Weeken, O. (2011) "Pricing central bank liquidity through product-mix auctions—the Bank of England's indexed long-term repo operations," Working Paper, Bank of England.

Gul, F. and Stacchetti, E. (1999) "Walrasian equilibrium with gross substitutes," *Journal of Economic Theory*, 87: 95–124.

Kelso, A. S. Jr and Crawford, V. P. (1982) "Job matching, coalition formation, and gross substitutes," *Econometrica*, 50: 1483–504.

Klemperer, P. (1999) "Auction theory," *Journal of Economic Surveys*, 13(2): 227–86. Also reprinted in S. Dahiya (ed.), *The Current State of Economic Science* (1999), pp. 711–66.

_____ (2002) "What really matters in auction design," *Journal of Economic Perspectives*, 16: 169–89.

_____ (2004) *Auctions: Theory and Practice*, Princeton University Press.

_____ (2008) "A new auction for substitutes: central bank liquidity auctions, the U.S. TARP, and variable product-mix auctions," Mimeo, Oxford University.

_____ and Meyer, M. (1986) "Price competition vs. quantity competition: the role of uncertainty," *Rand Journal of Economics*, 17: 618–38.

_____ _____ (1989) "Supply function equilibria in oligopoly under uncertainty," *Econometrica*, 57: 1243–77.

Kremer, I. and Nyborg, K. (2004) "Underpricing and market power in uniform price auctions," *Review of Financial Studies*, 17: 849–77.

Manelli, A. M. and Vincent, D. (1995) "Optimal procurement mechanisms," *Econometrica*, 63: 591–620.

McAdams, D. (2007) "Uniform-price auctions with adjustable supply," *Economics Letters*, 95: 48–53.

Menezes, F. M. and Monteiro, P. K. (2005) *An Introduction to Auction Theory*, Oxford University Press.

Milgrom, P. R. (2000) "Putting auction theory to work: the simultaneous ascending auction," *Journal of Political Economy*, 108: 245–72.

____ (2004) *Putting Auction Theory to Work*, Cambridge University Press.

____ (2009) "Assignment messages and exchanges," *American Economic Journal: Micro economics*, 1: 95–113.

Milnes, A. (2010) "Creating confidence in cash," *Blueprint*, October.

Poole, W. (1970) "Optimal choice of monetary policy instruments in a simple stochastic macro model," *Quarterly Journal of Economics*, 84: 197–216.

Weitzman, M. (1974) "Prices vs. quantities," *Review of Economic Studies*, 41: 477–91.

CHAPTER 11

··

OPTIMAL INCENTIVES IN CORE-SELECTING AUCTIONS

··

ROBERT DAY AND PAUL MILGROM[1]

INTRODUCTION

IN early 2008, the UK's telecommunication authority, Ofcom, adopted a new pricing rule for its spectrum auction—a *minimum-revenue core-selecting* rule. The class of such rules had only recently been proposed and analyzed by Day and Milgrom (2007). Following the UK's lead, radio spectrum auctions with similar rules were planned in Austria, Denmark, Ireland, Portugal, and the Netherlands, and by the Federal Aviation Administration in the United States for the allocation of landing slot rights at New York City airports.[2]

The new pricing rule generalizes the familiar second-price auction rule for auctions of a single item. One way to characterize the outcome of a second-price auction is in terms of the *core*: the price is high enough that no bidder (or group of bidders) is willing to offer more to the seller to change the assignment and, among such prices, it is the lowest one. For multi-item auctions, a core price vector is one that is low enough to be individually rational and high enough that no group of bidders finds it profitable to offer a higher *total* price to the seller. Among core price vectors, the minimum-revenue core vectors are the ones with the smallest revenue for the seller.

Two general considerations inspired our development of the theory of core prices and core-selecting auctions. The first was discontent with the auction proposed by Vickrey (1961), whose weaknesses are reviewed by Ausubel and Milgrom (2006). Of particular

---

[1] This chapter updates and corrects work that we originally reported in Day and Milgrom (2007). Our text borrows liberally from our own earlier work.
[2] Most of these auctions also incorporated multiple rounds of bids following a suggestion of Ausubel et al. (2006).

concern is that Vickrey prices can be very low. The second was that similar core and stability concepts have been applied successfully in the design of real-world matching markets. The National Resident Matching Program is a famous example, but others include the mechanisms adopted by New York and Boston for assigning students to schools and the New England Kidney Exchange (Roth and Peranson, 1999; Roth et al. 2005; Abdulkadiroglu et al., 2005a,b).

There is both empirical and experimental evidence to suggest that the core is important, although most work in this area has focused on matching rather than on auctions. Stable matching mechanisms survive much longer in practical applications than related unstable mechanisms (Roth and Xing, 1994; Kagel and Roth, 2000). And there is a theoretical argument to explain this longevity: if a proposed match is stable, then no group would prefer to renege and make an alternative arrangement among themselves, because there is no feasible alternative that all group members would prefer. But if a proposed match is unstable, then some group would prefer to renege, and too much reneging would make the mechanism unreliable for its users.

Nothing limits this theoretical argument to the case matching. For an auction, if a mechanism produces a core allocation, then no group of bidders can profitably offer a higher total price to the seller.[3] And if the auction selects a point that is not in the core at least with respect to the submitted bids, then some group of bidders has already offered the seller a total price that is higher than the price prescribed by the auction. It is easy to see why sellers might want to renege and make a separate deal with that group of bidders.

Parts of these analyses assume that the recommended matching or auction mechanisms result in stable or core allocations, but whether that happens depends on the participants' strategies. Participant behavior in real mechanisms varies widely, from naïve to sophisticated, and the most sophisticated participants do not merely make truthful reports in the mechanism. Instead, they also make decisions about whether to make pre-emptive offers before the auction, to enter the auction as a single bidder or as several, to stay out of the auction and try to bargain with the winners afterwards, to buy extra units in the auction and resell some afterwards, to renege on deals, or to persuade the seller to make changes to the timing or rules of the mechanism. Each of these elements can be important in some auction settings.

Despite the variety of objectives and of important behavioral constraints in real auction settings, mechanism design researchers customarily impose truth-telling incentives first and then ask to what extent other objectives or constraints can be accommodated. Since optimization is at best an approximation to the correct behavioral theory for bidders, it is also interesting to reverse the exercise, asking: by how much do the incentives for truthful reporting fail when other design objectives are treated as constraints?

---

[3] The core is always non-empty in auction problems. Indeed, for any profile of reports, the allocation that assigns the items efficiently and charges each bidder the full amount of its bids selects a core allocation. This selection describes the "menu auction" analyzed by Bernheim and Whinston (1986). Other core-selecting auctions are described in Ausubel and Milgrom (2002), and Day and Raghavan (2007).

The modern literature does include some attempts to account for multiple performance criteria even when incentives are less than perfect. Consider, for example, the basic two-sided matching problem, commonly called the *marriage problem*, in which men have preferences regarding women and women have preferences regarding men. The early literature treats stability of the outcome as the primary objective, and only later turns its attention to the incentive properties of the mechanism. In the marriage problem, there always exists a unique *man-optimal* match and a unique *woman-optimal match*.[4] The direct mechanism that always selects the man-optimal match is strategy-proof for men but not for women,[5] and the reverse is true for the woman-optimal mechanism. Properties such as these are typically reported as advantages of the mechanism,[6] even though these incentives fall short of full strategy-proofness. Another argument is that even when strategy-proofness fails, finding profitable deviations may be so hard that most participants find it best just to report truthfully. A claim of this sort has been made for the pre-1998 algorithm used by National Resident Matching Program, which was not strategy-proof for doctors, but for which few doctors could have gained at all by misreporting and for which tactical misreporting was fraught with risks (Roth and Peranson, 1999).[7]

The analysis of multiple criteria is similarly important for the design of *package auctions* (also called "combinatorial auctions"), which are auctions for multiple items in which bidders can bid directly for non-trivial subsets ("packages") of the items being sold, rather than being restricted to submit bids on each item individually. In these auctions, revenues are an obvious criterion. Auctions are commonly run by an expert auctioneer on behalf of the actual seller and any failure to select a core allocation with respect to reported values implies that there is a group of bidders that have offered to pay more in total than the winning bidders, yet whose offer has been rejected. Imagine trying to explain such an outcome to the actual seller or, in a government-sponsored auction, to a skeptical public![8] Another possible design objective is that a bidder should not profit by entering and playing as multiple bidders, rather than as a single one.[9]

---

[4] As Gale and Shapley (1962) first showed, there is a stable match that is Pareto preferred by all men to any other stable match, which they called the "man-optimal" match.

[5] Hatfield and Milgrom (2005) identify the conditions under which strategy-proofness extends to cover the college admissions problem, in which one type of participant ("colleges") can accept multiple applicants, but the other kind ("students") can each be paired to only one college. Their analysis also covers problems in which wages and other contract terms are endogenous.

[6] For example, see Abdulkadiroglu et al. (2005a).

[7] There is quite a long tradition in economics of examining approximate incentives in markets, particularly when the number of participants is large. An early formal analysis is by Roberts and Postlewaite (1976).

[8] McMillan (1994) describes how heads rolled when second-price auctions were used to sell spectrum rights in New Zealand and the highest bid was sometimes orders of magnitude larger than the second-highest bid.

[9] Yokoo et al. (2004) were the first to emphasize the importance of "false name bidding" and how it could arise in the anonymous environment of Internet auctions. The problem they identified, however, is broader than just anonymous Internet auctions. For example, in the US radio spectrum auctions, several of the largest corporate bidders (including AT&T, Cingular, T-Mobile, Sprint, and Leap

We illustrate these conditions and how they fail in the Vickrey auction with an example of two identical items for sale. The first bidder wants both items and will pay up to 10 for the pair; it has zero value for acquiring a single item. The second and third bidders each have values of 10 for either one or two items, so their marginal values of the second item are zero. The Vickrey auction outcome assigns the items to the second and third bidders for prices of zero. Given that any of the three bidders would pay 10 for the pair of items, a zero price is surely too low: that is the low-revenue problem. Generally, the low-revenue problem for the Vickrey auction is that its payments to the seller may be less than those at any core allocation. [10] Also, suppose that the second and third bidders are both controlled by the same player, whose actual values are 10 for one item or 20 for two. If the bidder were to participate as a single entity, it would win the two items and pay a price of 10. By bidding as two entities, each of which demands a single item for a price of 10, the player reduces its total Vickrey price from 10 to 0: that is the shill bidding problem. These vulnerabilities are so severe that practical mechanism designers are compelled to investigate when and whether relaxing the incentive compatibility objective can alleviate these problems.

We have discussed matching and package auction mechanisms together not only because they are two of the currently mostly active areas of practical mechanism design but also because there are some remarkable parallels between their equilibrium theories. One parallel connects the cases where the doctors in the match are substitutes for the hospital and when the goods in the auction are substitutes for the bidders. In these cases, the mechanism that selects the doctor-optimal match is ex post incentive-compatible for doctors, and a mechanism, the ascending proxy auction of Ausubel and Milgrom (2002), which selects a bidder-optimal allocation (a core allocation that is Pareto optimal for bidders), is ex post incentive-compatible for bidders. [11]

A second important connection is the following one: for every stable match x and every stable matching mechanism, there exists an equilibrium in which each player adopts a certain *truncation strategy*, according to which it truthfully reports its ranking of all the outcomes at which it is not matched, but reports that it would prefer to be unmatched rather than to be assigned an outcome worse than x. What is remarkable about this theorem is that *one single profile of truncation strategies is a Nash equilibrium for* every *stable matching mechanism*. We will find that a similar property is true for core-selecting auctions, but with one difference. In matching mechanisms, it is usual to treat all the players as strategic, whereas in auctions it is not uncommon to treat the seller differently, with only a subset of the players—the *bidders*—treated as making decisions strategically. We are agnostic about whether to include the seller as a bidder

---

Wireless) have at times had contracts with, or financial interests in, multiple bidding entities in the same auction, enabling strategies that would not be possible for a single, unified bidder.

[10] In this example, the core outcomes are the outcomes in which 2 and 3 are the winning bidders, each pays a price between 0 and 10, and the total payments are at least 10. The seller's revenue in a core-selecting auction is thus at least 10.

[11] This is also related to results on wage auctions in labor markets as studied by Kelso and Crawford (1982), and Hatfield and Milgrom (2005), although those models do not employ package bidding.

or even whether to include all the buyers as strategic players. Regardless of how the set of strategic players is specified, we find that for every allocation on the Pareto frontier of the core for the players who report strategically, there is a single profile of truncation strategies that is an equilibrium profile for *every* core-selecting auction. [12]

The preceding results hinge on another similarity between package auctions and matching mechanisms. In any stable matching mechanism or core-selecting auction, and given any reports by the other players, a player's best reply achieves its maximum core payoff or best stable match given its actual preferences and the reported preferences of others. For auctions, there is an additional interesting connection: the maximum core payoff is exactly the Vickrey auction payoff.

Next are the interrelated results about incentives for *groups* of participants. Given a core-selecting auction, the incentives for misreporting are minimal for individuals in a particular group, S, if and only if the mechanism selects an S−best core allocation. If there is a unique S−best allocation, then truthful reporting by members of coalition S is an ex post equilibrium. This is related to the famous result from matching theory (for which there always exists a unique man-optimal match and a unique woman-optimal match) that it is an ex post equilibrium for men to report truthfully in the man-optimal mechanism and for women to report truthfully in the woman-optimal mechanism.

The remainder of this chapter is organized as follows. The following section formulates the package auction problem. The third section characterizes core-selecting mechanisms in terms of revenues that are never less than Vickrey revenues, even when bidders can use shills. The fourth section introduces definitions and notation, and introduces the theorems about best replies and full information equilibrium. The fifth section states and proves theorems about the core-selecting auctions with the smallest incentives to misreport. Various corresponding results for the marriage problem are developed in the sixth section. The seventh section notes an error regarding revenue monotonicity in an earlier version of this chapter (as it appeared in the *International Journal of Game Theory*), and makes connections to more recent research and applications. The eighth section concludes.

FORMULATION

We denote the seller as player 0, the bidders as players $j = 1, \ldots, J$, and the set of all players by N. Each bidder, j, has quasi-linear utility and a finite set of possible packages, X_j. Its value associated with any feasible package $x_j \in X_j$ is $u_j(x_j) \geq 0$. For convenience, we formulate our discussion mainly in terms of bidding applications, but the same mathematics accommodates much more, including some social-choice problems. In the central case of package bidding for predetermined items, x_j consists of a package of items

---

[12] These truncation strategies also coincide with what Bernheim and Whinston (1986) call "truthful strategies" in their analysis of a "menu auction," which is a kind of package auction.

that the bidder may buy. For procurement auctions, x_j could also usefully incorporate information about delivery dates, warranties, and various other product attributes or contract terms. Among the possible packages for each bidder is the null package, $\emptyset \in X_j$ and we normalize so that $u_j(\emptyset) = 0$.

For concreteness, we focus on the case where the auctioneer is a seller who has a feasible set $X_0 \subseteq X_1 \times \ldots \times X_J$ with $(\emptyset, \ldots, \emptyset) \in X_0$—so the no-sale package is feasible for the seller—and a valuation function $u_0 : X_0 \to \mathbb{R}$ is normalized so that $u_0(\emptyset, \ldots, \emptyset) = 0$. For example, if the seller must produce the goods to be sold, then u_0 may be the auctioneer-seller's variable cost function.

For any coalition S, a goods assignment $\hat{x}$ is *feasible* for coalition S, written $\hat{x} \in F(S)$, if (1) $\hat{x} \in X_0$ and (2) for all j, if $j \notin S$ or $0 \notin S$, then $\hat{x}_j = \emptyset$. That is, a bidder can have a non-null assignment when coalition S forms only if that bidder and the seller are both in the coalition.

The *coalition value function* or *characteristic function* is defined by:

$$w_u(S) = \max_{x \in F(S)} \sum_{j \in S} u_j(x_j) \tag{1}$$

In a *direct auction mechanism* (f, P), each bidder j reports a valuation function $\hat{u}_j$ and the profile of reports is $\hat{u} = \{\hat{u}_j\}_{j=1}^J$. The outcome of the mechanism, $(f(\hat{u}), (P(\hat{u}))) \in (X_0, \mathbb{R}_+^J)$, specifies the choice of $x = f(\hat{u}) \in X_0$ and the payments $p_j = P_j(\hat{u}) \in \mathbb{R}_+$ made to the seller by each bidder j. The associated payoffs are given by $\pi_0 = u_0(x) + \sum_{j \neq 0} p_j$ for the seller and $\pi_j = u_j(x) - p_j$ for each bidder j. The payoff profile is individually rational if $\pi \geq 0$.

A *cooperative game* (with transferable utility) is a pair (N, w) consisting of a set of players and a characteristic function. A payoff profile π is feasible if $\sum_{j \in N} \pi_j \leq w(N)$, and in that case it is associated with a feasible allocation. An *imputation* is a feasible, non-negative payoff profile. An imputation is in the *core* if it is efficient and unblocked:

$$Core(N, w) = \left\{ \pi \geq 0 \,\middle|\, \sum_{j \in N} \pi_j = w(N) \text{ and } (\forall S \subseteq N) \sum_{j \in S} \pi_j \geq w(S) \right\} \tag{2}$$

A direct auction mechanism (f, P) is *core-selecting* if for every report profile $\hat{u}$, $\pi_{\hat{u}} \in Core(N, w_{\hat{u}})$. Since the outcome of a core-selecting mechanism must be efficient with respect to the reported preferences, we have the following:

Lemma 1. *For every core-selecting mechanism (f, P) and every report profile $\hat{u}$,*

$$f(\hat{u}) \in \arg\max_{x \in X_0} \sum_{j \in N} \hat{u}_j(x_j) \tag{3}$$

The payoff of bidder j in a Vickrey auction is the bidder's marginal contribution to the coalition of the whole. In cooperative game notation, if the bidders' value profile is u, then bidder j's payoff is $\bar{\pi}_j = w_u(N) - w_u(N - j)$.[13]

---

[13] A detailed derivation can be found in Milgrom (2004).

REVENUES AND SHILLS: NECESSITY OF
CORE-SELECTING AUCTIONS

We have argued that the revenues from the Vickrey outcome are often too low to be acceptable to auctioneers. In order to avoid biasing the discussion too much, in this section we treat the Vickrey revenues as a just-acceptable lower bound and ask: what class of auctions have the properties that, for any set of reported values, they select the total-value maximizing outcome and lead always to bidder payoffs no higher than the Vickrey payoffs, even when bidders may be using shills? Our answer will be: exactly the class of core-selecting auctions.

In standard fashion, we call any mechanism with the first property, namely, that the auction selects the total-value-maximizing outcome, "efficient."

Theorem 1. *An efficient direct auction mechanism has the property that no bidder can ever earn more than its Vickrey payoff by disaggregating and bidding with shills if and only if it is a core-selecting auction mechanism.*

Proof. Fix a set of players (seller and bidders) N, let w be the coalitional value function implied by their reported values, and let π be the players' vector of reported payoffs. Efficiency means $\sum_{j \in N} \pi_j = w(N)$. Let $S \subseteq N$ be a coalition that excludes the seller. These bidders could be shills. Our condition requires that they earn no more than if they were to submit their merged valuation in a Vickrey auction, in which case the merged entity would acquire the same items and enjoy a total payoff equal to its marginal contribution to the coalition of the whole: $w(N) - w(N - S)$. Our restriction is therefore $\sum_{j \in S} \pi_j \leq w(N) - w(N - S)$. In view of efficiency, this holds if and only if $\sum_{j \in N-S} \pi_j \geq w(N - S)$. Since S was an arbitrary coalition of bidders, we have that for every coalition $T = N - S$ that includes the seller, $\sum_{j \in T} \pi_j \geq w(T)$. Since coalitions without the seller have value zero and can therefore never block, we have shown that there is no blocking coalition. Together with efficiency, this implies that $\pi \in Core(N, w)$.   □

TRUNCATION REPORTS AND EQUILIBRIUM

In the marriage problem, a *truncation report* refers to a reported ranking by person j that preserves the person's true ranking of possible partners, but which may falsely report that some partners are unacceptable. For an auction setting with transferable utility, a truncation report is similarly defined to correctly rank all pairs consisting of a non-null goods assignment and a payment, but which may falsely report that some of these are unacceptable. When valuations are quasi-linear, a reported valuation is a truncation report exactly when all reported values of non-null goods assignments are reduced by the same non-negative constant. We record that observation as a lemma.

Lemma 2. *A report $\hat{u}_j$ is a truncation report if and only if there exists some $\alpha \geq 0$ such that for all $x_j \in X_j$, $\hat{u}_j(x_j) = u_j(x_j) - \alpha$.*

Proof. Suppose that $\hat{u}_j$ is a truncation report. Let x_j and x_j' be two non-null packages and suppose that the reported value of x_j is $\hat{u}_j(x_j) = u_j(x_j) - \alpha$. Then, $(x_j, u_j(x_j) - \alpha)$ is reportedly indifferent to $(\emptyset, 0)$. Using the true preferences, $(x_j, u_j(x_j) - \alpha)$ is actually indifferent to $(x_j', u_j(x_j') - \alpha)$ and so must be reportedly indifferent as well: $\hat{u}_j(x_j) - u_j(x_j) - \alpha = \hat{u}_j(x_j') - u_j(x_j') - \alpha$. It follows that $u_j(x_j') - \hat{u}_j(x_j') = u_j(x_j) - \hat{u}_j(x_j) = \alpha$.

Conversely, suppose that there exists some $\alpha \geq 0$ such that for all $x_j \in X_j$, $\hat{u}_j(x_j) \equiv u_j(x_j) - \alpha$. Then for any two non-null packages, the reported ranking of (x_j, p) is higher than that of (x_j', p') if and only if $\hat{u}(x_j) - p \geq \hat{u}(x_j') - p'$, which holds if and only if $u(x_j) - p \geq u(x_j') - p'$. □

We refer to the truncation report in which the reported value of all non-null outcomes is $\hat{u}_j(x_j) = u_j(x_j) - \alpha_j$ as the "α_j truncation of u_j."

In full-information auction analyzes since that of Bertrand (1883), auction mechanisms have often been incompletely described by the payment rule and the rule that the unique highest bid, when that exists, determines the winner. Ties often occur at Nash equilibrium, however, and the way ties are broken is traditionally chosen in a way that depends on bidders' values and not just on their bids. For example, in a first-price auction with two bidders, both bidders make the same equilibrium bid, which is equal to the lower bidder's value. The analysis assumes that the bidder with the higher value is favored, that is, chosen to be the winner in the event of a tie. If the high-value bidder were not favored, then it would have no best reply. As Simon and Zame (1990) have explained, although breaking ties using value information prevents this from being a feasible mechanism, the practice of using this tie-breaking rule for analytical purposes is an innocent one, because, for any $\varepsilon > 0$, the selected outcome lies within ε of the equilibrium outcome of any related auction game in which the allowed bids are restricted to lie on a sufficiently fine discrete grid. [14]

In view of lemma 1, for almost all reports, assignments of goods differ among core-selecting auctions only when there is a tie; otherwise, the auction is described entirely by its payment rule. We henceforth denote the payment rule of an auction by $P(\hat{u}, x)$, to make explicit the idea that the payment may depend on the goods assignment in case of ties. For example, a first-price auction with only one good for sale is any mechanism which specifies that the winner is a bidder who has made the highest bid and the price is equal to that bid. The mechanism can have any tie-breaking rule to be used so long as equation 3 is satisfied. In traditional parlance, the payment rule, P, defines an *auction*, which comprises a set of mechanisms.

---

[14] See also Reny (1999).

Definition. $\hat{u}$ *is an equilibrium of the auction P if there is some core-selecting mechanism* (f, P) *such that $\hat{u}$ is a Nash equilibrium of the mechanism.*

For any auction, consider a tie-breaking rule in which bidder j is *favored*. This means that in the event that there are multiple goods assignments that maximize total reported value, if there is one at which bidder j is a winner, then the rule selects such a one. When a bidder is favored, that bidder always has some best reply.

Theorem 2. *Suppose that (f, P) is a core-selecting direct auction mechanism and bidder j is favored. Let $\hat{u}_{-j}$ be any profile of reports of bidders other than j. Denote j's actual value by u_j and let $\bar{\pi}_j = w_{\hat{u}_{-j}, u_j}(N) - w_{\hat{u}_{-j}, u_j}(N - j)$ be j's corresponding Vickrey payoff. Then, the $\bar{\pi}_j$ truncation of u_j is among bidder j's best replies in the mechanism and earns j the Vickrey payoff, $\bar{\pi}_j$. Moreover, this remains a best reply even in the expanded strategy space in which bidder j is free to use shills.*

Proof. Suppose j reports the $\bar{\pi}_j$ truncation of u_j. Since the mechanism is core selecting, it selects individually rational allocations with respect to reported values. Therefore, if bidder j is a winner, its payoff is at least zero with respect to the reported values and hence at least $\bar{\pi}_j$ with respect to its true values.

Suppose that some report $\hat{u}_j$ results in an allocation $\hat{x}$ and a payoff for j strictly exceeding $\bar{\pi}_j$. Then, the total payoff to the other bidders is less than $w_{\hat{u}_{-j}, u_j}(N) - \bar{\pi}_j \leq w_{\hat{u}_{-j}, u_j}(N - j)$, so $N - j$ is a blocking coalition for $\hat{x}$, contradicting the core-selection property. This argument applies also when bidder j uses shills. Hence, there is no report yielding a profit higher than $\bar{\pi}_j$, even on the expanded strategy space that incorporates shills.

Since reporting the $\bar{\pi}_j$ truncation of u_j results in a zero payoff for j if it loses and non-negative payoff otherwise, it is always a best reply when $\bar{\pi}_j = 0$.

Next, we show that the truncation report always wins for j, therefore yielding a profit of at least $\bar{\pi}_j$ so that it is a best reply. Regardless of j's reported valuation, the total reported payoff to any coalition excluding j is at most $w_{\hat{u}_{-j}, \hat{u}_j}(N - j) = \max_{x = (\emptyset, x_{-j}) \in X_0} \sum_{i \in N - j} \hat{u}_i(x)$. If j reports the $\bar{\pi}_j$ truncation of u_j, then the maximum value is at least $\max_{x \in X_0} \left(\sum_{i \in N - j} \hat{u}_i(x) + u_j(x) \right) - \bar{\pi}_j = w_{\hat{u}_{-j}, u_j}(N) - \bar{\pi}_j$, which is equal to the previous sum by the definition of $\bar{\pi}_j$. Applying lemma 1 and the hypothesis that j is favored establishes that j is a winner. $\qquad\square$

Definition. *An imputation π is bidder optimal if $\pi \in Core(N, u)$ and there is no $\hat{\pi} \in Core(N, u)$ such that for every bidder j, $\pi_j \leq \hat{\pi}_j$ with strict inequality for at least one bidder. (By extension, a feasible allocation is bidder optimal if the corresponding imputation is so.)*

Next is one of the main theorems, which establishes a kind of equilibrium equivalence among the various core-selecting auctions. We emphasize, however, that the strategies require each bidder j to know the equilibrium payoff π_j, so what is being described is a full-information equilibrium but not an equilibrium in the model where each bidder's own valuation is private information.

Theorem 3. *For every valuation profile u and corresponding bidder optimal imputation π, the profile of π_j truncations of u_j is a full-information equilibrium profile of every core-selecting auction. The equilibrium goods assignment x^* maximizes the true total value $\sum_{i \in N} u_i(x_i)$, and the equilibrium payoff vector is π (including π_0 for the seller).*[15]

Proof. For any given core-selecting auction, we study the equilibrium of the corresponding mechanism that, whenever possible, breaks ties in equation 3 in favor of the goods assignment that maximizes the total value according to valuations u. If there are many such goods assignments, any particular one can be fixed for the argument that follows.

First, we show that no goods assignment leads to a reported total value exceeding π_0. Indeed, let S be the smallest coalition for which the maximum total reported value exceeds π_0. By construction, the bidders in S must all be winners at the maximizing assignment, so $\pi_0 < \max_{x \in X_0, x_{-s} = \emptyset} u_0(x_0) + \sum_{i \in S-0} (u_i(x_i) - \pi_i) \leq w_u(S) - \sum_{i \in S-0} \pi_i$. This contradicts $\pi \in Core(N, w_u)$, so the winning assignment has a reported value of at most π_0: $w_{\hat{u}}(N) \leq \pi_0$. If j instead reports truthfully, it can increase the value of any goods allocation by at most π_j, so $w_{u_j, \hat{u}_{-j}}(N) \leq \pi_0 + \pi_j$.

Next, we show that for any bidder j, there is some coalition excluding j for which the maximum reported value is at least π_0. Since π is bidder optimal, for any $\varepsilon > 0$, $(\pi_0 - \varepsilon, \pi_j + \varepsilon, \pi_{-j}) \notin Core(N, w_u)$. So, there exists some coalition S_ε to block it: $\sum_{i \in S_\varepsilon} \pi_i - \varepsilon < w_u(S_\varepsilon)$. By inspection, this coalition includes the seller but not bidder j. Since this is true for every ε and there are only finitely many coalitions, there is some S such that $\sum_{i \in S} \pi_i \leq w_u(S)$. The reverse inequality is also implied because $\pi \in Core(N, w_u)$, so $\sum_{i \in S} \pi_i = w_u(S)$.

For the specified reports, $w_{\hat{u}}(S) = \max_{x \in X_0} \sum_{i \in S} \hat{u}_i(x_i) \geq \max_{x \in X_0} u_0(x_0) + \sum_{i \in S-0} (u_i(x_i) - \pi_i) \geq w_u(S) - \sum_{i \in S-0} \pi_i = \pi_0$. Since the coalition value cannot decrease as the coalition expands, $w_{\hat{u}}(N - j) \geq \pi_0$. By definition of the coalition value functions, $w_{\hat{u}}(N - j) = w_{u_j, \hat{u}_{-j}}(N - j)$.

Using theorem 2, j's maximum payoff if it responds optimally and is favored is $w_{u_j, \hat{u}_{-j}}(N) - w_{u_j, \hat{u}_{-j}}(N - j) \leq (\pi_0 + \pi_j) - \pi_0 = \pi_j$. So, to prove that the specified report profile is an equilibrium, it suffices to show that each player j earns π_j when these reports are made.

The reported value of the true efficient goods assignment is at least $\max_{x \in X_0} u_0(x_0) + \sum_{i \in N-0} (u_i(x_i) - \pi_i) = w(N) - \sum_{i \in N-0} \pi_i = \pi_0$. So, with the specified tie-breaking rule, if the bidders make the specified truncation reports, the selected goods assignment will maximize the true total value.

Since the auction is core selecting, each bidder j must have a reported profit of at least zero and hence a true profit of at least π_j, but we have already seen that these are also upper bounds on the payoff. Therefore, the reports form an equilibrium;

---

[15] Versions of this result were derived and reported independently by Day and Raghavan (2007) and by Milgrom (2006). The latter paper was folded into Day and Milgrom (2007).

each bidder j's equilibrium payoff is precisely π_j, and the seller's equilibrium payoff is $w_{\hat{u}}(N) - \sum_{i \in N-0} \pi_i = \pi_0$.   □

MINIMIZING INCENTIVES TO MISREPORT

Despite the similarities among the core-selecting mechanisms emphasized in the previous section, there are important differences among the mechanisms in terms of incentives to report valuations truthfully. For example, when there is only a single good for sale, both the first-price and the second-price auctions are core-selecting mechanisms, but only the latter is strategy-proof.

To evaluate simultaneously all bidders' incentives to deviate from truthful reporting, we introduce the following definition.

Definition. *The* incentive profile *for a core-selecting auction P at u is* $\varepsilon^P = \left\{ \varepsilon_j^P(u) \right\}_{j \in N-0}$, *where* $\varepsilon_j^P(u) \equiv \sup_{\hat{u}_j} u_j(f_j(u_{-j}, \hat{u}_j)) - P\left(u_{-j}, \hat{u}_j, f_j(u_{-j}, \hat{u}_j) \right)$ *is j's maximum gain from deviating from truthful reporting when j is favored.*

Our idea is to minimize these incentives to deviate from truthful reporting, subject to selecting a core allocation. Since the incentives are represented by a vector, we use a Pareto-like criterion.

Definition. *A core-selecting auction P provides* suboptimal incentives *at u if there is some core-selecting auction $\hat{P}$ such that for every bidder j, $\varepsilon_j^{\hat{P}}(u) \leq \varepsilon_j^P(u)$ with strict inequality for some bidder. A core-selecting auction provides* optimal incentives *if there is no u at which it provides suboptimal incentives.*

Theorem 4. *A core-selecting auction provides optimal incentives if and only if for every u it chooses a bidder-optimal allocation.*

Proof. Let P be a core-selecting auction, u a value profile, and π the corresponding auction payoff vector. From theorem 2, the maximum payoff to j upon a deviation is $\bar{\pi}_j$, so the maximum gain to deviation is $\bar{\pi}_j - \pi_j$. So, the auction is suboptimal exactly when there is another core-selecting auction with higher payoffs for all bidders, contradicting the assumption that π is bidder optimal.   □

Recall that when the Vickrey outcome is a core allocation, it is the unique bidder-optimal allocation. So, theorem 4 implies that any core-selecting auction that provides optimal incentives selects the Vickrey outcome whenever that outcome is in the core with respect to the reported preferences. Moreover, because truthful reporting then provides the bidders with their Vickrey payoffs, theorem 2 implies the following.

Corollary. *When the Vickrey outcome is a core allocation, then truthful reporting is an ex post equilibrium for any mechanism that always selects the bidder-optimal core.*

Among the bidder-optimal core-selecting auctions, one particularly interesting set is the class of minimum-revenue core-selecting auctions.

Definition. *A core-selecting auction $P(u,x)$ is a minimum-revenue core-selecting auction if there is no other core-selecting auction $\hat{P}(u, x)$ such that $\sum_{j \in J} \hat{P}_j < \sum_{i \in J} P_j$.*

Since the allocation x does not vary among core-selecting auctions, it is obvious from the defining inequality that no other core-selecting auction can lead to a higher payoff (and hence a lower price) for each bidder.

Lemma 3. *Every minimum-revenue core-selecting auction $P(u,x)$ is bidder optimal.*

The converse of lemma 3 is not true in general. As a counterexample, let suppose there are five bidders: $J = 5$.[16] Let each feasible X_j be a singleton; each bidder is interested in only one package, a condition often called *single-minded* bidding. Further, let $u_j(x_j) = 2$, for all j, and let x_1, x_2, x_3, be mutually disjoint, while $x_4 = x_1 \cup x_2$ and $x_5 = x_5 = x_2 \cup x_3$. For example, bidders could be interested in items from the set $\{A, B, C\}$ with bundles of interest $\{A\}$, $\{B\}$, $\{C\}$, $\{A, B\}$ and $\{B, C\}$, respectively. For these parameters, bidders 1, 2, and 3 win their bundles of interest in the unique efficient allocation. But a valid bidder-optimal rule may select payments $(1, 1, 1)$ with total revenue of 3, while the unique minimum-revenue solution is $(0, 2, 0)$, confirming that not all bidder-optimal payment rules minimize revenue within the core. To see that $(1, 1, 1)$ is indeed bidder optimal, note that any single or joint reduction in payment from that point will induce a blocking coalition involving one or other of the losing bidders.

Since minimum-revenue core-selecting auctions are bidder optimal, they inherit the properties of that larger class. The next theorem asserts that minimum-revenue core-selecting auctions have an additional optimality property.

Theorem 5. *If $\hat{P}$ is a minimum-revenue core-selecting auction, then for any fixed u and corresponding efficient allocation x:*

$$\hat{P}(u, x) \in \arg \min_P \sum_{i \in J} \varepsilon_j^P(u)$$

Proof. Again from theorem 2, we have a maximum possible gain from deviation given by $\varepsilon_j^P(u) = \bar{\pi}_j - \pi_j$ for each bidder, which, given any fixed value-maximizing x, is equal to $P_j - \bar{P}_j$. Thus, $\arg \min_P \sum_{j \in J} \varepsilon_j^P(u) = \arg \min_P \sum_{j \in J}(P_j - \bar{P}_j) = \arg \min_P \sum_{j \in J} P_j$, with the second equality following since $\bar{P}_j$ is a constant with respect to P, and the main result following by the revenue minimality of $\hat{P}$.   □

---

[16] Our counterexample has three winning bidders. There are no counterexamples with fewer than three winners.

CONNECTIONS TO THE MARRIAGE PROBLEM

Even though theorems 2–4 in this chapter are proved using transferable utility and do not extend to the case of budget-constrained bidders, they do all have analogs in the non-transferable utility marriage problem.

Consider theorem 2. Roth and Peranson (1999) have shown for a particular algorithm in the marriage problem that any fully informed player can guarantee its best stable match by a suitable truncation report. That report states that all mates less preferred than its best achievable mate are unacceptable. The proof in the original paper makes it clear that their result extends to any stable matching mechanism, that is, any mechanism that always selects a stable match.

Here, in correspondence to stable matching mechanisms, we study core-selecting auctions. For the auction problem, Ausubel and Milgrom (2002) showed that the best payoff for any bidder at any core allocation is its Vickrey payoff. So, the Vickrey payoff corresponds to the best mate assigned at any stable match. Thus, the auction and matching procedures are connected not just by the use of truncation strategies as best replies but by the point of the truncation, which is at the player's best core or stable outcome.

Theorem 3 concerns Nash equilibrium. Again, the known results of matching theory are similar. Suppose the participants in the match in some set S^C play non-strategically, like the seller in the auction model, while the participants in the complementary set S, whom we shall call bidders, play Nash equilibrium. Then, for a bidder-optimal stable match, [17] the profile at which each player in S reports that inferior matches are unacceptable is a full-information Nash equilibrium profile of *every* stable matching mechanism and it leads to that S-optimal stable match. This result is usually stated using only men or women as the set S, but extending to other sets of bidders using the notion of bidder optimality is entirely straightforward.

Finally, for theorem 4, suppose again that some players are non-strategic and that only the players in S report strategically. Then, if the stable matching mechanism selects an S-optimal stable match, there is no other stable matching mechanism that weakly improves the incentives of all players to report truthfully, with strict improvement for some. Again, this is usually stated only for the case where S is the set of men or the set of women, and the extension does require introducing the notion of a bidder-optimal match.

CORRECTIONS AND OTHER RELATED LITERATURE

The original paper on which this chapter was based (Day and Milgrom, 2007) claimed an additional theorem about revenue monotonicity of the minimum-revenue core-

---

[17] This is defined analogously to the bidder-optimal allocation.

selecting auction, namely, that the seller's revenue weakly increases as bid values increase or alternatively as additional bidders enter the auction. This claim later proved to be erroneous. This error was brought to our attention in independent contributions by Ott (2009) and Lamy (2009). Beck and Ott (2010) give necessary and sufficient conditions to characterize revenue-monotonic core-selecting auctions and find the ones with the best incentives in that set.

To illustrate the failure of revenue monotonicity in a revenue-minimizing core-selecting mechanism, consider the following simple example. Let bidders 1, 2, and 3 each bid $2 on a single item of interest (say A, B, and C respectively) and let bidder 4 bid $3 on $\{A, B\}$ while bidder 5 bids $3 on $\{B, C\}$. Bidders 1, 2, and 3 win in the efficient allocation, while the presence of losing bidders 4 and 5 dictates core constraints on the winning bidders' payments as follows: bidders 1 and 2 must pay at least $3 in total, and bidders 2 and 3 must pay at least $3 in total. The unique minimum-revenue solution is for bidders 1, 2, and 3 to pay $1, $2, and $1, respectively. But if bidder 2 were to increase her bid to $3, the unique set of payments becomes $0, $3, $0, and the seller's revenue has dropped from $5 to $3 following a $1 bid increase by bidder 2. Intuitively, though bidder 2's payments count only once from the perspective of the seller, they help to satisfy two core constraints at once, in contrast to the payments of bidders 1 and 3. If we consider further bid increases by bidder 2, we see that she need not pay any more than $3, illustrating *eventual* revenue invariance under increases in a truncation strategy—a property first described by Day and Cramton (2012).

Despite the non-monotonicity of some core-selecting auctions, this class continues to be studied and applied in practice. Goeree and Lien (2009) demonstrate revenue weaknesses of core-selecting auctions under Bayes–Nash equilibrium in a limited setting, while related work of Rastegari, Condon, and Leyton-Brown (2010) provide impossibility results for revenue monotonicity under a variety of assumptions. In a more positive stream, Erdil and Klemperer (2009) introduce refined rules for core-selecting auctions to mitigate incentives for small deviations (as opposed to maximal incentives to deviate, treated in theorems 4 and 5). Some of the strongest support for core-selecting auctions in the more recent literature is given by Othman and Sandholm (2010), who introduce envy-reduction auction protocols that result in core outcomes. Day and Cramton (2012) also demonstrate an envy-reduction result, that truncation strategies result in envy-free outcomes in core-selecting auctions.

CONCLUSION

Our study of core-selecting auctions was motivated both by their practical interest and by their relations to stable matching mechanisms. The evidence from case studies and from the Kagel–Roth laboratory experiments, which shows that participants are quick to stop using certain unstable matching mechanisms but that stable mechanisms persist, has usually been understood to be applicable in general to matching mechanisms. But there is no obvious reason to accept that as the relevant class. The usual theoretical arguments about the continued use of a mechanism distinguish core-selecting mechanisms from other mechanisms. That applies equally for auctions and matching problems, and the failure to reject the narrower theoretical hypothesis is also a failure to reject the broader one.

Despite the theoretical similarities between auction and matching mechanisms, stable matching mechanisms for multi-item applications have so far been more extensively used in practice. It is possible that this is about to change. The two complexity challenges that are posed by core-selecting auctions—computational complexity and communications complexity—are both being addressed in research and in practice.

The computations required by core-selecting auctions are, in general, much harder than those for matching, and computational tractability for problems of an interesting scale has only recently been achieved. Indeed, Day and Raghavan (2007) showed that the computational complexity of finding core outcomes is equivalent to the complexity of the corresponding efficient allocation problem, and is thus NP-hard in the most general case. The implementation of core-selecting auctions is limited primarily by our ability to solve larger and larger NP-hard problems, or to find reasonable application-specific restrictions on bidding that make the problem tractable. And efforts are being made to find just such restrictions. For example, the core-selecting European spectrum auctions to date have each described their sets of objects in ways that made for comfortably small optimization problems, which can be solved relatively quickly on a desktop computer.

The issue of communications complexity can be highlighted with some simple arithmetic. In an environment with N items for sale, the number of non-empty packages for which a bidder must report values is $2^N - 1$. That is unrealistically large for most applications if N is even a small two-digit number. For the general case, Segal (2003) has shown that communications cannot be much reduced without severely limiting the efficiency of the result.

But communication complexity need not definitively rule out core-selecting package auctions. In many real-world settings, the auctioneer can simplify the problem by limiting the packages that can be acquired or by engaging in *conflation*, according to which similar items are treated as if they were identical (Milgrom, 2010). An auctioneer may know that radio spectrum bands must be compatible with international standards, or that complementarities in electricity generation result from costs saved by operating continuously in time, minimizing time lost when the plant is ramped up or down, or that a collection of airport landing rights at 2:00–2:05 can be conflated without much loss with rights at 2:05–2:10 or 2:10–2:15. And for some classes of preferences, such as the case where goods are substitutes, substantial progress on compact expressions of values has already been made. [18] Practical designs that take advantage of such knowledge can still be core-selecting mechanisms and yet can entail compact reporting by bidders.

---

[18] Hatfield and Milgrom (2005) introduced the *endowed assignment valuations* for this purpose.

The class of core-selecting auctions includes the pay-as-bid "menu auction" design studied by Bernheim and Whinston (1986), the ascending proxy auction studied by Ausubel and Milgrom (2002) and Parkes and Ungar (2000), the assignment auction introduced in Milgrom (2009a,b), and any of the mechanisms resulting from the core computations in Day and Raghavan (2007), Day and Cramton (2012), or Erdil and Klemperer (2009). Several of these are the very minimum-revenue core-selecting auctions that continue to be proposed for high-stakes applications.

REFERENCES

Abdulkadiroglu, A., Pathak, P., Roth, A. and Sonmez, T. (2005a) "The Boston public school match," *AEA Papers and Proceedings*: 368–71.
————————— (2005b) "The New York city high school match," *AEA Papers and Proceedings*: 364–7.
Ausubel, L. and Milgrom, P. (2002) "Ascending auctions with package bidding," *Frontiers of Theoretical Economics*, 1(1): article 1.
———— (2006) "The lovely but lonely Vickrey auction," in P. Cramton, Y. Shoham, and R. Steinberg (eds), *Combinatorial Auctions*, MIT Press, pp. 1–40.
———— Cramton, P. and Milgrom, P. (2006) "The clock-proxy auction: a practical combinatorial auction design," in P. Cramton, Y. Shoham, and R. Steinberg (eds), *Combinatorial Auctions*, MIT Press, pp. 115–18.
Beck, M. and Ott, M. (2010) *Revenue Monotonicity in Core-Selecting Auctions*, Stanford University.
Bernheim, B. D. and Whinston, M. (1986) "Menu auctions, resource allocation and economic influence," *Quarterly Journal of Economics*, 101: 1–31.
Bertrand, J. (1883) "Théorie mathématique de la richesse sociale," *Journal des Savants*, 69: 499–508.
Day, R. W. and Cramton, P. (2012) "The Quadratic Core-Selecting Payment Rule for Combinatorial Auctions", *Operations Research*, 60(3): 588–603.
———— and Milgrom, P. (2007) "Core-selecting package auctions," *International Journal of Game Theory*, 36(3–4): 393–407.
———— and Raghavan, S. (2007) "Fair payments for efficient allocations in public sector combinatorial auctions," *Management Science*, 53(9): 1389–406.
Erdil, A. and Klemperer, P. (2009) "A new payment rule for core-selecting auctions," *Journal of the European Economic Association*, 8(2–3): 537–547.
Gale, D. and Shapley, L. (1962) "College admissions and the stability of marriage," *American Mathematical Monthly*, 69: 9–15.
Goeree, J. and Lien, Y. (2009) "On the Impossibility of Core-Selecting Auctions", *Institute for Empirical Research in Economics*, University of Zurich Working Paper (452).
Hatfield, J. and Milgrom, P. (2005) "Matching with contracts," *American Economic Review*, 95(4): 913–35.
Kagel, J. and Roth, A. (2000) "The dynamics of reorganization in matching markets: a laboratory experiment motivated by a natural experiment," *Quarterly Journal of Economics*, 115(1): 201–35.

Kelso, A. and Crawford, V. (1982) "Job matching, coalition formation, and gross substitutes," *Econometrica*, 50(6): 1483–504.

Lamy, L. (2009) "Core-selecting auctions: a comment on revenue monotonicity," *International Journal of Game Theory*, 39: 503–10.

McMillan, J. (1994) "Selling spectrum rights," *Journal of Economics Perspectives*, 8: 145–62.

Milgrom, P. (2004) *Putting Auction Theory to Work*, Cambridge University Press.

—— (2006) "Incentives in core-selecting auctions," Stanford University.

—— (2009a) "Assignment exchange and auction," Patent Application US 2009/0177555 A1.

—— (2009b) "Assignment messages and exchanges," *AEJ Micro*, 1(2): 95–113.

—— (2011) "Critical Issues in Market Design," *Economic Inquiry*, 48(2): 311–320.

Othman, A. and Sandholm, T. (2010) "Envy quotes and the iterated core-selecting combinatorial auction," *Proceedings of the National Conference on Artificial Intelligence.*

Ott, M. (2009) *Second-Price Proxy Auctions in Bidder–Seller Networks*, Thesis, (Universität Karlsruhe, 2009).

Parkes, D. and Ungar, L. (2000) "Iterative combinatorial auctions: theory and practice," *Proceedings of the 17th National Conference on Artificial Intelligence*: 74–81.

Rastegari, B., Condon, A. and Leyton-Brown, K. (2010) "Revenue monotonicity in deterministic, dominant-strategy combinatorial auctions," *Artificial Intelligence*, 175(2): 441–456.

Reny, P. (1999) "On the existence of pure and mixed strategy Nash equilibria in discontinuous games," *Econometrica*, 67(5): 1029–56.

Roberts, J. and Postlewaite, A. (1976) "The incentives for price-taking behavior in large exchange economies," *Econometrica*, 44(1): 115–29.

Roth, A. E. and Peranson, E. (1999) "The redesign of the matching market for American physicians: some engineering aspects of economic design," *American Economic Review*, 89: 748–80.

—— and Xing, X. (1994) "Jumping the gun: imperfections and institutions related to the timing of market transactions," *American Economic Review*, 84: 992–1044.

Roth, A., Sonmez, T. and Unver, U. (2005) "Kidney exchange," *AEA Papers and Proceedings*, 95(2): 376–80.

Segal, I. (2003) "The communication requirements of combinatorial auctions," in P. Cramton, Y. Shoham, and R. Steinberg (eds), *Combinatorial Auctions*, Princeton University Press.

Simon, L. K. and Zame, W. R. (1990) "Discontinuous games and endogenous sharing rules," *Econometrica*, 58: 861–72.

Vickrey, W. (1961) "Counterspeculation, auctions, and competitive sealed tenders," *Journal of Finance*, 16: 8–37.

Yokoo, M., Sakurai, Y. and Matsubara, S. (2004) "The effect of false-name bids in combinatorial auctions: new fraud in internet auctions," *Games and Economic Behavior*, 46(1): 174–88.

CHAPTER 12

....

AUCTIONING ROUGH DIAMONDS

A Competitive Sales Process for BHP Billiton's Ekati Diamonds

....

PETER CRAMTON, SAMUEL DINKIN, AND ROBERT WILSON[1]

INTRODUCTION

BHP Billiton produces approximately 6% of the world's diamonds from its Ekati mine in the Northwest Territory, Canada. These rough stones are then sold through various channels, primarily in Antwerp, Belgium. This chapter discusses the previous sales process and analyzes the transition to the current (auction) sales process. We address both the spot market and a longer-term market intended to capture a premium for supply regularity.

Three problems with the previous sales process (described in the third section of this chapter) were: (1) an excessive reliance on the price book for pricing, (2) the limited ability of customers to express preferences for quantities and types of stones, and (3) failure to capture a competitive premium for supply regularity. These shortcomings suggest that the allocation of stones may not have been best, and the pricing of the output may not have been competitive.

Beginning in January 2007, we worked with BHP Billiton to develop and implement a simple auction approach to improve the assignment and pricing of the mine's output. The auction follows the same sales cycle as before and a similar bundling of the

[1] The market design project discussed in this chapter grew from a long collaboration with BHP Billiton. We thank the many talented BHP Billiton staff who collaborated with us on every phase of this project. Special thanks to Alberto Calderon for initiating the collaboration, to Gordon R. Carlyle and Christopher J. Ryder for leading the design phase, and to Martin H. Leake for leading the successful implementation.

stones into a set of nineteen "deals" (products) grouped by size, color, and quality. The difference is that the auction lets the customers compete directly for quantity using either a uniform-price auction or an ascending-clock auction. Both auction formats are simple market mechanisms, commonly used to find the value-maximizing assignment and competitive prices of the goods. By putting the diamonds in the best hands, BHP Billiton better satisfies the needs of its customers and improves sales revenues from the Ekati mine. Customers focus on their business and being more competitive, rather than on efforts to please the producer to receive a more favorable allocation.

To provide supply regularity, a term auction is offered periodically in which customers bid a differential to the spot price for each deal for terms of one year or more. An ascending-clock auction was chosen to foster price and assignment discovery. This enables each customer to build a portfolio of quantity commitments across the deals. Each customer pays the same price premium or receives the same discount for locking in supply and demand long term for a particular deal.

Finally, two or three times a year, large stones are sold in a simultaneous ascending-clock auction, called a specials auction. Each lot is a single stone or a group of stones of like size, color, and quality. The ascending-clock format is used, since price discovery is especially important for these exceptionally rare and valuable stones.

Educating customers to the new approach was an important step in the transition. Some resistance was experienced from regular customers. Resellers, especially, felt they had the most to lose if the inefficiencies of the previous process were eliminated. BHP Billiton carefully managed customer relationships during the transition, and developed support for the approach. The main advantage is to customers with high values. These customers find it easier to acquire both the type and quantity of stones they desire.

The new approach combines many aspects of the previous sales process with well tested and understood auction methods. Most importantly, the new approach is a more transparent and market-responsive sales mechanism. Customers express their preferences directly and credibly through their bids in competitive auctions. The transition was carefully managed to gradually introduce new methods and build customer support. The transition entailed little risk because the demand side for rough stones is competitive. Individual customers do not benefit by boycott or exit, since there are many other manufacturers and resellers that desire to be BHP Billiton customers. Moreover, it will be BHP Billiton's best customers—those with high values—who benefit the most from the new approach, since these customers are able to win the quantities and types of stones they most desire.

EMPIRICAL EVIDENCE OF THE SUCCESS OF THE APPROACH

Vivid evidence of the success of transparent auctions is seen by comparing rough diamond price indices from several public sources. This is done in Figure 12.1 for the period

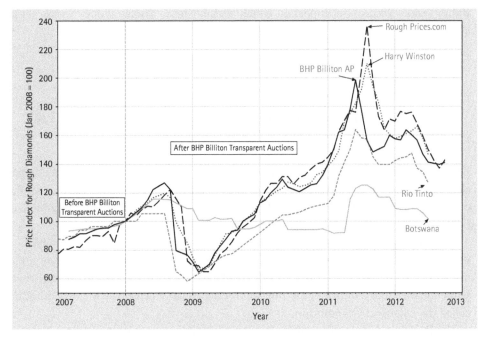

FIGURE 12.1. Price index for rough diamonds, 2007–13. Sources: BHP Billiton Customer Report 141; Harry Winston Investor Presentation September 2012; Rio Tinto Diamonds and Minerals 12 September 2012; Botswana Financial Statistics September 2012, tab 6.5: RoughPrices.com.

2007–13. To make the indices comparable, each index has been scaled so January 2008 = 100. Notice that all the indices are about the same in 2007, before the introduction of auctions. Then, in 2008, BHP Billiton introduced transparent auctions. After that, BHP Billiton becomes the price leader. The other indices either lag behind the BHP Billiton index by one month (Harry Winston, operating a modified sight system with market pricing, or RoughPrices.com) or are typically below BHP Billiton (Rio Tinto and Botswana, operating traditional sight systems). Importantly, the revenue advantage of our approach is even greater when one recognizes that most of BHP Billiton sales are at a price premium above the BHP Billiton API. The price premium is set at term auctions (about one per year) in which customers lock in quantity shares for particular categories of stones. Typically the price premium is 1–4% of the API.

Consistent with the traditional sales approach of De Beers, the Botswana price index is much flatter. For most of the five years it is significantly below the BHP Billiton index, suggesting a significant revenue loss by Botswana. The exception is the late 2008 to late 2009, during the global financial crisis, when the Botswana index is significantly above the others. However, in this period BHP Billiton was able to sell the entire output of Ekati and had revenues well above De Beers, which markets the Botswana diamonds, despite its much smaller size—a high price is of little value if it results in few sales. Overall, the transparent auction methodology has provided superior returns for BHP

Billiton and served as an important benchmark for price discovery for the market as a whole.

A BRIEF HISTORY OF THE DIAMOND INDUSTRY

BHP Billiton is the fourth largest source for rough diamonds. De Beers currently has about 45% of the market. ALROSA (Russia, 20%) and Rio Tinto (Australia, 8%) are the other two large producers of rough diamonds. Market demand is highly fragmented; there are over 1,500 potential customers for rough diamonds.

Until the 1990s, De Beers controlled the vast majority of the market and established its structure. In the 1880s, Cecil Rhodes started consolidating control of the major diamond mines which at that time were all located in South Africa. In the late 1920s, Ernest Oppenheimer took control of De Beers and established a central selling organization called the Diamond Corporation.

The Diamond Corporation offered rough diamonds to each customer in a box that would contain an assortment of rough diamonds picked by De Beers for the individual customer. The box had to be accepted or rejected as a package. If the box was rejected, De Beers might not invite the customer back for some years, if ever. De Beers priced these boxes at 25% below its estimate of market prices, but varied price and quantity to smooth price changes and to reward or penalize behavior. A customer found to be buying diamonds coming from outside the cartel might be penalized by being offered a box with poor-quality goods at high prices. The message, which was sometimes reinforced verbally, was stop cheating on the cartel or be excluded from the direct benefits of it.

De Beers organized the demand side of the market in this way to restrict the supply of polished diamonds. Manufacturers and resellers had strong incentives to continue to participate in the cartel. How the rough diamonds were allocated to customers was less important to De Beers than ensuring that overall supply was restricted and demand was growing.

Customers did their best to convince De Beers that they should get additional quantity and higher quality. This was challenging because all customers wanted more supply. Competition focused on gaming the opaque De Beers allocation process and staying in favor. One possible result of this gaming might have been the high number of customers.

Over the years, De Beers had to cope with discoveries in Russia, Zaire, and Angola, which it did by including these new players in its cartel. In the 1980s and 1990s, De Beers faced challenges as some mines in Zaire and Australia elected to sell directly to customers. This was the advent of open competition in the supply of rough diamonds. De Beers stopped restricting the supply of the stones that these mines specialized in, resulting in a steep price drop. This was effective in getting Zaire to rejoin the cartel, but others stayed independent.

In 1998, when the Ekati diamond mine in Canada was opened, BHP Billiton adopted many of the practices that were customary in the industry, but did not join the De Beers cartel. In 2004, BHP Billiton began offering portions of its supply by sealed tender. In 2008, BHP Billiton began selling more than half of its supply by ascending-clock auction and the rest in sealed-bid uniform-price auctions. This revolution in market pricing has benefited BHP Billiton and its best customers. The approach has performed well throughout the global financial crisis. The auction approach enabled BHP Billiton to quickly adjust to competitive market prices. This allowed it to keep sales volumes high when prices fell. In early 2009, BHP Billiton increased revenues while De Beers' revenue fell. Customers were allowed to bid for long-term supply contracts for the goods they wanted. Now, more producers are considering the BHP Billiton model—a model of pricing and assigning diamond supply in a transparent competitive process.

OUTLINE OF THE PREVIOUS BHP BILLITON SALES PROCESS

Like De Beers and other producers of rough diamonds, BHP Billiton had a proprietary price book that was used in setting prices. The output from the Ekati mine is sold on a five-week cycle, ten times per year. Each shipment is roughly $60 million, excluding large stones ("specials"), which are grouped for two or three sales per year. The rough diamonds are sorted by size, quality, and color into about 4,000 price points—each with a price per carat in the price book. The diamonds are then grouped into about nineteen deals. Each deal is an aggregation of closely related price points. There are about 200 price points in each deal.

About 15% of the total quantity, in value terms, was removed for Northwest Territories (10%) test polishing and direct sales to retailers (5%). The remaining stones were sold in regular (50%), elite (20%), and tender and window (15%) channels. Each of the deals was split into parcels, where each split was a representative sample of the deal.

There were eight regular customers. Each received about ten parcels per cycle and paid the total price for all parcels, based on the price book. This was the invoice price and was the only price that the regular customer saw.

There were between two and four splits of each deal for the regular customers. This was done to get comparable market feedback. Feedback was the actual or estimated sales price reported for each parcel by the regular customers. Reports were received after about ten days. Feedback impacted the price book and whether the regular customer was dropped. BHP Billiton targeted a long-run customer margin of a few per cent in setting the price points. Deal-level price information was hidden from customers to avoid cost-plus feedback, in which the customers simply reported, say, 4% more than cost.

Elite customers were like regular customers (indeed three of nine were regular customers), except they paid a premium over the price book. The premium was bid for a two-year period. Unlike regular customers, elite customers could reject the goods. On average, the elite customers paid significantly more than the price book.

About twenty parcels were tendered, each valued at about $200,000–$500,000. There was a secret reserve price based on the price book. The bidder examined some subset of the parcels, and submitted a sealed bid on each desired parcel within twenty-four hours of viewing. Viewing typically took about three hours. Parcels receiving bids above the reserve price were awarded to the highest bidder at the price bid. Tender sales were several per cent above the price book. Window sales, which were negotiated, also were about several per cent above the price book. Tender and window sales were by invitation only. Consistently poor performers were not asked back. Bidders learned the high bid on lots they bid on provided they won at least one parcel; otherwise, they learned nothing.

A final source of price information was from the sale prices of polished stones. BHP Billiton polished and sold some of the stones in the Canada Mark program. The rough-to-polished data provided valuable information for the pricing relationships in the price book. Sales to customers based in the Northwest Territory were priced at market prices. A premium was charged as the deals were tailored for polishing in the Northwest Territory.

PROBLEMS WITH THE PREVIOUS SALES PROCESS

There were four problems with the previous sales process.

First was the heavy reliance on the price book to set price. It was difficult for BHP Billiton to know if it was getting the best price. This problem was somewhat mitigated by using several methods to adjust the price book: (1) regular customer feedback, (2) elite bids, (3) tender and window sales, and (4) outcomes for polished stones. Still there was a potential gaming problem of the regular customer feedback. A customer might underreport in the hope that doing so would lead to better prices in the future. Alternatively, a customer might overreport in the hope of getting more goods in the future. Entry and exit from the regular channel provided a relatively weak and discontinuous incentive for truthful feedback. Regular customers were only rarely swapped out. Moreover, the criteria for becoming and remaining a regular customer were unclear.

The second problem was that customers, especially regular customers, had little means of expressing preferences for stones—in terms of either quantity or type. BHP Billiton fixed quantities for regular customers. There was little means to ensure that the goods were going to the right parties.

The third problem was that BHP Billiton failed to capture a premium for the supply regularity that its regular customers enjoyed.

A fourth problem was the complexity and non-transparency of the sales process. The incentives in each of the main channels were complex. Bidders wanting more quantity had to participate in more channels—or even demerge to become two customers—rather than directly expressing larger demands. The process lacked transparency, especially in the regular channel, where BHP Billiton set both prices and quantities.

A SPOT MARKET IN WHICH CUSTOMERS BID DIRECTLY FOR QUANTITY

We now consider a market in which bidders directly express preferences for various quantities. We begin with a spot market to be held at each cycle. This is the cornerstone of the newly introduced market. Under this approach, the diamonds are awarded to the highest bidders at market prices. The approach is simpler than the previous sales process. Most importantly, it creates value by seeing that the diamonds are put in the hands of customers with the highest values. In addition, customers can limit quantity risk—the possibility of winning more or less than desired—first, through a complementary long-term market and, second, through their bidding strategies in the spot market. In this way, BHP Billiton can maximize the value to its customers and thereby the revenues from the Ekati mine's output.

First consider a single deal. All customers for the deal compete together. This includes all the regular and elite customers, as well as many other customers with membership in the Responsible Jewellery Council. A representative split or sample of the deal, typically between a twelfth and a sixth, depending on the particular deal, is put in a parcel for viewing. Bidders know how many splits there will be for this deal (e.g. seven) and how they will be divided between the term and spot market (e.g. four splits in term and three in spot). The viewing parcel is selected carefully to be the most representative of the entire deal. Each customer views the parcel and then submits a bid schedule, indicating its demand for the deal with one or more price–quantity pairs. Price is the price per carat (e.g. $730/ct). Quantity is the number of splits desired (e.g. two splits) with price adjusted for quality based on the price book (e.g. a discount of 1.2% for a split of slightly lower quality than the viewing parcel). Each customer has a maximum quantity for the deal: three splits for deals with five or more splits in the term market, or two otherwise.

There are a number of possible auction formats within this structure. We describe three: the uniform-price auction, the pay-as-bid auction, and the ascending-clock auction.

Uniform-price auction

The auctioneer aggregates all the bid schedules to form the aggregate demand curve, as shown in Figure 12.2.

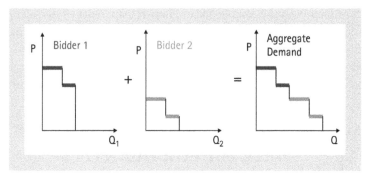

FIGURE 12.2. Forming aggregate demand from individual demands

The aggregate demand is then crossed with the supply curve. The intersection of supply and demand determines the market clearing price, as shown in Figure 12.3.

All bids above the clearing price win and pay the clearing price. Quantity for a bid at the clearing price may be subject to a random tie break, so the total sold equals 100%. In addition, bidders are aware that the quality and quantity won may vary by a few per cent due to the discrete nature of the product sold. Finally, the supply curve reflects the reserve price or, more generally, the desire of the seller to postpone sales if prices are too low. Goods not sold in the current auction, as a result of the supply curve, are sold at later auction prices once market prices exceed the reserve price. In the event that supply and demand intersect over a range of prices, the clearing price is the highest such price; in the event that supply and demand intersect over a range of quantities, the clearing quantity is the largest such quantity.

Figure 12.4 gives an example focusing on two bidders, blue and red. The table on the left gives the aggregate demand curve, as well as the bids of blue and red. On the right, we see that the demand curve intersects with supply at \$560. Both bidders' higher bids are accepted in full. Blue's lower bid at \$560 is "on the margin." It is partially accepted

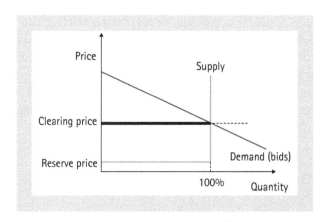

FIGURE 12.3. Price versus quantity in a uniform-price auction.

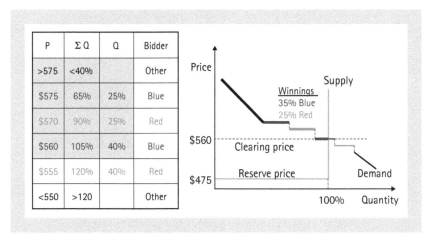

P	Σ Q	Q	Bidder
>575	<40%		Other
$575	65%	25%	Blue
$570	90%	25%	Red
$560	105%	40%	Blue
$555	120%	40%	Red
<550	>120		Other

FIGURE 12.4. Two bidders, blue and red, in a uniform-price auction.

(rationed), resulting in winnings of 35% for blue and 25% for red. Both pay $560/carat for their shares, appropriately adjusted for quality differences.

The uniform-price auction is the most common method for selling a divisible good. In this setting, the use of the price book to adjust splits for quality makes the deal a divisible good.

The frequent use of the uniform-price auction stems from its many desirable properties. Absent market power, each bidder has the incentive to bid its true demands, and the resulting assignment maximizes the total value of the goods. In the long-run, such an outcome should maximize BHP Billiton's revenue from the mine.

Bidders like the fact that they do not ever pay more than the market price for the quantity won. Moreover, uniform pricing lets the bidder better manage quantity risk. The bidder can bid its full value, knowing that it will be required to pay only the clearing price. In this way and through the long-term market, the bidder guarantees that it wins its desired minimum quantity. Both the bidders and BHP Billiton benefit from this reduction in quantity risk.

When bidders have market power, the uniform-price auction has incentives for demand reduction, causing each bidder to bid less than its true demand. The result is lower auction revenues and reduced auction efficiency. However, given the competitive market structure on the demand side, this is unlikely to be a problem, and in any event the reserve price provides substantial protection both from demand reduction and collusion.

Pay-as-bid auction

The most common alternative to the uniform-price auction is the pay-as-bid auction. The only difference between the two is the pricing rule. In a pay-as-bid auction, all bids

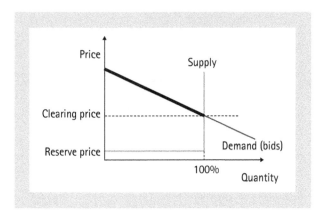

FIGURE 12.5. Price versus quantity in a pay-as-bid auction.

above the clearing price win, but the bidder pays its bid for any quantity it wins, as shown in Figure 12.5.

At first glance, it may appear that the pay-as-bid auction generates more revenue than the uniform-price auction, since the bidder pays its bid, which is at least as high and typically higher than the clearing price. This, however, is not the case. The pricing rule greatly impacts the bidding behavior. Figure 12.6 shows typical bid curves for a bidder, with the true demand shown as the thinner straight line to the right. Under pay-as-bid pricing (curved line), the bidder guesses the clearing price and tries not to bid much above it. Under uniform pricing, the bidder bids closer to its true demand, although the bidder does increasingly shade its bid for larger quantities, optimally taking account of its impact on price.

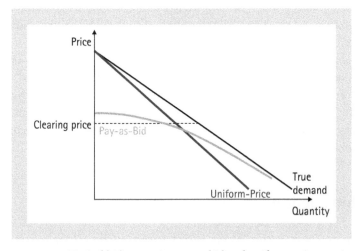

FIGURE 12.6. Typical bid curves in pay-as-bid and uniform-price auctions.

As a result, revenues may be either higher or lower with pay-as-bid pricing than with uniform pricing. Despite a vast theoretical, empirical, and experimental literature, results comparing revenues under these two pricing rules are decidedly ambiguous. What is known is that quantity risk is much greater under pay-as-bid pricing, whereas price risk is greater under uniform pricing. The reason is that the aggregate demand curve under pay-as-bid pricing is much flatter than under uniform pricing. As a result, with pay-as-bid pricing a modest change in a bidder's bid price can have a large impact on the quantity the bidder wins.

To reduce quantity risk, the pay-as-bid auction can be extended to include price-taker bids. These bids are awarded in full at the average sales price that is bid competitively. With this extension, customers can guarantee minimum quantities, as in a uniform-price auction.

There is some experimental evidence that, in repeated auction contexts, like this one, pay-as-bid pricing is more vulnerable to collusion, because the bidders have a stronger incentive to coordinate on a price, and thereby reduce the amount of money "left on the table"—the amount bid in excess of the clearing price.

In 1998, the US Treasury switched from pay-as-bid pricing to uniform pricing, after many years of study. The switch was motivated from the pay-as-bid auction's vulnerability to the "short squeeze," where one bidder attempts to corner the market of a particular product. The short squeeze is not an issue here, since short sales are not allowed, the BHP Billiton sales are only a fraction of the total market, and a cap was imposed on how much each customer can win of each deal (40–50%).

Finally, the uniform price rule has more resilient efficiency in the face of highly variable pricing such as that experienced during the recent financial crisis. The pay-as-bid auction provides a strong incentive to use ex ante expectations to try to guess the final price. If the final price is far from expectation, the goods will go disproportionately to the best guessers as opposed to the customers who value the goods the highest. The uniform-price auction continues to achieve high efficiency in this circumstance and there is little advantage to being a good guesser.

Ascending-clock auction

In recent years, thanks in part to the power of the Internet, it has been common to sell divisible goods using an ascending-clock auction. This is simply a dynamic version of the uniform-price auction. Rather than submitting a demand curve at a single time, the bidder submits demands over a sequence of rounds. The auctioneer announces a low starting price and the bidders respond with the quantity desired at that price. If there is excess demand, then the auctioneer raises the price and the bidders again respond with their demands at the higher price. The process continues until there is no excess demand. Each bidder then wins its bid quantities at the clearing price, just as in a uniform-price auction. The "clock" is simply the price, which ascends until supply and demand balance, as shown in Figure 12.7.

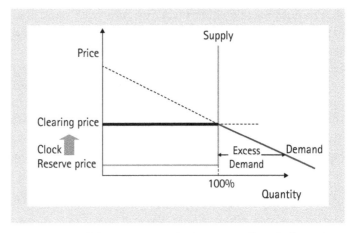

FIGURE 12.7. Price versus quantity in an ascending-clock auction.

The clock auction has all the advantages of the uniform-price format, but, in addition, allows for better discovery of the market price via iterative feedback. Price discovery is often important in contexts like this one in which there is common uncertainty about the value of the goods, and each bidder is estimating value.

To promote price discovery, there is an activity rule that prevents a bidder from increasing its demand as prices rise after an initial few rounds. Bidders can maintain or decrease their quantities only as prices rise. Thus, each bidder must bid in a manner consistent with a downward-sloping demand curve.

Clock auctions of this sort can be conducted in a matter of one to four hours over the Internet. A technique called intra-round bids typically is used to retain most of the advantages of a continuous price clock, and yet allow the auction to be conducted in, say, six to ten rounds.

A bidder, especially if it desires only a small quantity, may prefer to submit all its bids at one time. Such "proxy bidding" is accommodated easily, guaranteeing that bidders will not be discouraged from participating as a result of a lengthy (or overnight) bidding process. In particular, this allows a bidder to enter bids all at the start, if the bidder does not wish to take advantage of price discovery. A few bidders choose to bid in this simple way.

Collusion is mitigated by the reserve price and an information policy that limits the bidder's information to the aggregate demand at the end of each round. In particular, bidders do not learn the quantity bid by individual bidders and do not learn the identity of the other bidders for particular deals during the auction.

Handling multiple deals

It is straightforward to extend the single-deal format to nineteen deals.

With the sealed-bid methods (uniform-price and pay-as-bid), the bidder simply submits bid schedules for each of the deals. With multiple deals, quantity risk may be more of a factor, especially if all deals are auctioned simultaneously. This would favor uniform pricing, which lets the bidder better manage the quantity risk. For example, the bidder can guarantee winning a desired minimum quantity of each deal by bidding a high price—and being a price-taker—for this minimum quantity.

With multiple deals, the dynamic auction is conducted with multiple price clocks, one for each deal. The auctioneer announces a price for each deal, and the bidder responds with the quantity desired for each deal. Given the limited importance of complements in this setting, it makes sense to use a simple implementation. After an initial switching phase where customers can bid on any deal, each deal is treated as an independent, but simultaneous, sale. This means that monotonicity of bids is enforced deal by deal, and each deal closes independently. Independent closing limits substitution across deals, but a simultaneous ascending auction with limited switching still gives each bidder the ability to dynamically bid for a desirable portfolio of diamonds across all deals. This provides much greater flexibility than is allowed with the static methods. Bidder budget constraints are also much better managed.

The sealed-bid method has one important drawback, which is mitigated in the dynamic auction. Bidders are forced to decide which deals to bid on before seeing how many other bidders have decided to bid on the same deal. This can result in many bidders competing for some deals and few bidders competing for others. In the dynamic auction, the initial switching phase where customers can freely switch among deals resolves the coordination problem inherent in the sealed-bid method. The switching phase allows bidders to equalize competition across all deals, improving the efficiency of the auction. Both BHP Billiton and its customers benefit from the improved pricing of the dynamic auction with an initial switching phase.

What if demand curves are not downward sloping?

Some have argued that customer demand curves for diamonds are upward sloping, so that all or most customers will bid for the maximum quantity. We were suspicious that this perception of upward-sloping demand curves was an artifact of the previous system, in which regular customers were getting too small a quantity at too low a price. These regular customers were thus arguing for more quantity and providing reasons why they needed more quantity in the hope that they would get more. We suspected that as soon as customers could bid directly or the prices rose, we would observe the familiar downward-sloping demand curve. Typically, once a minimum sufficient scale is reached, dealers will have diminishing marginal value for additional quantity, for the simple reason that opportunities for using the stones will be ordered from best to worst, and the dealer will exploit the best opportunities first. The solution implemented forced customers to express either flat or downward-sloping demands. Very rarely did customers express flat demands. We infer that customers' previous willingness to pay more for higher quantity was a consequence of prices being below competitive market prices and the sales quantity being less than the quantity demanded.

Adjusting the price book

The price book is used in two ways.

First, it is used to account for quality differences in a split of a deal. For this use, only the relative price factors are relevant, such as the percentage price difference between an eight-carat and nine-carat stone, holding color and quality constant. It would be possible to periodically ask customers to give feedback on these relative price factors. Truthful responses should be expected, since bidder risk is reduced if the quality adjustments better reflect the bidder's true preferences. BHP Billiton could then update and potentially disclose the relative price factors based on the feedback. (These can also be estimated based on deal composition and prices bid.)

Second, the price book is used for setting reserve prices in the auction. For this purpose, the absolute price level is relevant. The price book can be adjusted with each cycle in a similar manner as it was adjusted using the previous sales process. However, greater weight can be placed on the recent auction outcomes.

Maintaining customer diversity and avoiding collusion

Collusion, both tacit and explicit, is sometimes a problem in situations where the same buyers compete on a repeated basis, as is true here. Explicit collusion is a concern given that the vast majority of customers are located within 1 km of the center of the diamond district in Antwerp and are likely to be personally known to each other.

One means of guarding against collusion is making sure there is a sufficient number of customers and that the customer pool is sufficiently diverse. Ideally, customers would come from several geographic locations, several ethnic groups, and several company structures. The upper limit on a customer's share in the deal (e.g. 40%) is the primary method of ensuring that there is a sufficient number and diversity of customers and manageable credit risk. If these means prove inadequate, then targeted marketing is used to attract desirable buyers.

The rough diamond industry is conducive to diversity, with many ethnicities and nationalities present in the diamond district. There are many customers based in Belgium, India, Israel, the Netherlands, and South Africa. The high number of customers, many of whom are fierce competitors in the polished market, makes it less likely that a collusive cartel will develop. Historically, there has been little need for a customer cartel. De Beers provided a below-market price to all of its customers. Rather than pushing De Beers further below market prices by forming a cartel, customers pushed to expand quantity with De Beers, perhaps by lobbying and gaming reports to De Beers. Some families have several companies, one per brother. This outcome may be an artifact of De Beers allocating quantity by customer qualifications.

The reserve price is an essential instrument to mitigate collusion. It does this by limiting the maximum gain from engaging in collusive activity. At the same time, it mitigates the harm from successful collusion.

There is a code of conduct that clearly spells out that any discussions about bidding strategy with other bidders is forbidden and is illegal under competition laws. Such discussions would be grounds for exclusion from the current and possibly any future auctions.

A final instrument to mitigate collusion is the information policy—what bidders learn during the bidding process. Limiting information often enhances competition in repeated auctions. Thus, rather than reporting prices and winning shares at the end of each auction, it is better to only report prices. Similarly, in the clock implementation it is better to report only prices and aggregate demand at the end of the round, rather than all the individual bids.

Physical facilities for securely viewing the parcels limited how much the customer pool could be expanded, which might have limited the effectiveness of recruiting to expand the customer pool and to target specific kinds of buyers. Noting that viewing rooms were a scarce resource led to tighter management of the resource to allow more customers to use it. Low-volume customers were asked to view during off-peak times. This allowed an expanded customer base to make collusion more difficult and for BHP Billiton to benefit from more robust competition and a broader range of private values.

The auctions provide valuable information for customer selection and the allocation of scarce viewing time. Customers are ranked based on the value they add. Poor performers are removed from the customer pool.

Auctioning large stones

Large stones (of seven carats or more), "specials," are auctioned separately, in independent auctions several times a year. The specials auctions attract the attention of the market participants even though these stones represent less than 10% of the revenue from the Ekati mine.

Depending on the size and quality, the stones are auctioned as individual stones or bundled with similar stones. A simultaneous ascending-clock auction is used in the specials auctions. Price discovery is particularly important for these stones, given their rarity and value. The ascending-clock process also lets the bidders better manage portfolio needs and budget constraints. Finally, by using the same approach as the term auction, the customers are able quickly to become familiar with the auction format.

Five specials auctions have been run to date, with extremely high demand at the start of each auction due to the low starting prices (below the reserve prices). In the first auction, demand averaged more than ten customers at the starting prices for each of forty parcels.

To enhance customers' ability to value large stones, a recent innovation is to include a DVD that includes a three-dimensional image of the stone and the data required by cut-optimization software. This allows the customer to see all imperfections and determine the optimal cut, and thereby get an excellent estimate of the polished outcome.

Further bundling or unbundling

For continuity, a similar deal and price point structure from the previous sales process is used today. Some changes were made in the deal structure in order to have critical mass for the spot and term markets. The bundling of about 4,000 price points into about nineteen spot and sixteen term deals is an effective structure in the auction market.

"Illusion" is sometimes mentioned as a reason for bundling. It is apparently effective because larger bundles tend to go unsold less often because reserve pricing errors sometimes offset. Furthermore, if two parcels are auctioned separately and one does not sell, it is common for the unsold parcel to subsequently sell at a reduced price in a negotiated sale following the tender. This is a rational market response. The fact that the parcel failed to sell in a tender is bad news about its value (those who inspected it were unwilling to bid more than the reserve price). Subsequent customers should value the parcel at less. This, however, does not mean that bundling the two parcels would raise seller revenue. Whether to bundle depends more on the heterogeneity of bidder demands. Less bundling can provide more transparency and better matching in situations of sufficient demand heterogeneity. Viewing times and costs may also be reduced with effective bundling.

Bundling does simplify the auction and reduce transaction costs. Technology can also lower some transaction costs. For example, the fourth specials auction had over fifty parcels. Customers interested in twenty-five of them might have to type quickly to enter all of their bids during the auction round. Labor-saving devices such as proxy bidding and displaying only those parcels a customer is still bidding on allow the auction to still be run in a few hours.

Our suspicion is that less bundling, not more, may be better.

SPOT MARKET COMPOSITION

The regular, elite, and tender/window were combined into one auction sale. A uniform-price auction was used in the initial years, although clock auctions are being used for more stones over time and are contemplated as an approach for all stones in the future. Both approaches are commonly used, are well understood, and are well suited to the setting.

Among the sealed-bid methods we prefer the uniform-price auction to the pay-as-bid auction and this was selected. Uniform pricing provides better incentives to bid true values, especially given the competitive market structure we anticipate. It also is easier for the customer to guarantee a desired minimum quantity. With this approach, customers get the benefits of quantity certainty on whatever quantity for which they wish to be price-takers. In addition, customers like the fact that they do not overpay the market clearing price on any quantity won. Uniform pricing has greater price variation across cycles than pay-as-bid pricing. However, given the fairly liquid market for rough stones and the use of a reserve price, we do not expect this greater price variation to be a problem.

The best alternative to the uniform-price auction is the ascending-clock auction. The clock auction is similar to the uniform-price auction, but has several additional benefits. In particular, the multiple-round approach of a clock auction provides valuable price discovery, and it allows bidders to better manage budget constraints and quantity risk. It is especially desirable when bidders care about the particular portfolio of stones won.

The clock auction is slightly more difficult to implement and entails slightly higher participation costs for the bidders. It takes about three hours to conduct the clock auction with fifty price clocks for a specials auction. In the spot market, the additional benefits of the clock auction may not exceed these extra costs when the market is less volatile. Work is ongoing to develop auction technology to allow a faster ascending auction for the spot market. A shorter clock auction will have somewhat less price and assignment discovery than the longer specials auction, but more than with the uniform-price auction. The clock auction does allow bidders to raise their bid if they are losing, so market feedback is immediate. This is important, especially when diamond prices are more volatile.

Both recommended approaches build on the previous approach, through the use of deals to bundle many related price points. This greatly facilitated the transition to the auction market.

A challenge for the clock auction was that it might be perceived by some customers as too dramatic a change. This was one of the reasons to begin with the uniform-price auction for the spot market, and then transition to a clock auction as needed, once customers are comfortable with the auction process. Switching from the uniform-price to the ascending-clock is a natural and modest step.

The critical assumption for the auction approach is that a bidder can examine a representative sample of the deal and bid on that, knowing that what it receives may be somewhat different than the sample it inspected, with a price adjustment based on the price book. This works fine provided the viewing parcels are representative of the deal, and care is taken in making sure that the parcels created for winning bidders are also representative of the viewed parcel to the extent possible. Thus far, the approach has worked well. The assumption seems modest, when compared to the De Beers approach of being presented with a take-it-or-leave-it offer for a parcel of stones selected by De Beers.

The big difference between the auction methods and the previous sales process is that with the auction approach the customers compete directly for quantity and pay the market price for any quantity won. With the previous process, competition for quantity is much less direct and much more complex. The auction approach does a much better job of putting the diamonds in the hands of those customers best able to use them. In addition, the pricing under the auction approach better reflects competitive market prices. The improved assignment and pricing of diamonds under the auction approach appears to translate into higher sales revenues for BHP Billiton and allows the best customers to expand.

Risk of collusion is another issue to address in the auction market. Our conclusion was that a well designed and implemented auction market would be less susceptible to collusion than the previous system, especially the reliance on the price book. Thus far, our conclusion appears sound. There have not been any instances of collusion observed.

A LONG-TERM MARKET TO FOSTER REGULARITY OF SUPPLY

Customers value regularity of supply. An important question is how one can create value by enhancing supply regularity.

After considering durations from six to thirty-six months, BHP Billiton decided to hold an auction to sell eighteen-month term supply contracts. The contracts are for a particular per cent of each deal in each of the fifteen cycles during the eighteen months. An ascending-clock auction was used, with a different clock (price) for sixteen deals, one for each deal. Bidders bid the quantity (number of splits) they desire for each of the deals, given the price, which is a percentage of the spot price. The auction is started at a discount to the spot price, such as 5%. Each clock price is raised until supply and demand balance. As described earlier, a uniform-price auction is used in the spot market to assign and price the residual portion of each deal that is not sold in the term auction.

The term auction was open to an expanded group of potential customers, rather than restricted to a set of regular customers. For each customer, there is an upper limit on quantity in each deal of two or three splits, representing 42–60% of available long-term supply for that deal or 25–35% of total supply for that deal. Deals that allow a higher percentage to be won by one bidder are deals that represent a smaller absolute amount of money. The number of splits for each deal is closely correlated to the expected total sales price for all splits in the deal.

The motivation for using an ascending-clock format for the term auction is that it allows the customers over the course of the auction to build a desirable portfolio of deal quantities, given the observed premiums over spot prices. The auction was conducted in four hours (an hour longer than expected, since prices exceeded expectation). An alternative design would use the uniform-price auction; however, we believe that the extra price and assignment discovery of the ascending clock was helpful to bidders in the term auction, given that each term auction allocated much more value than each spot or specials auction. Extra price and assignment discovery was especially important in early auctions, where there was more uncertainty.

To illustrate how a customer builds a portfolio of quantities that makes sense given the prices for each deal, imagine there are three deals (A, B, and C) up for auction. Suppose A and B are substitutes for the bidder, and that C complements the A or B purchase. Then, during the clock auction, the bidder can begin bidding for both A and B, and then reduce its purchase of the one with the larger premium. Similarly, as the premium for C increases, the bidder can reduce its demand for C as well as A and B.

Under this approach, the mine's output, excluding the portion set aside for the Northwest Territories and the large specials stones, is sold in two markets: a term market, which offers regular supply at a premium above spot, and a spot market. The division between these two markets depends on the customers' preferences for regular supply and the requirement to preserve critical mass for spot sales. A substantial premium for regular supply was observed. For each deal, as high an amount as possible was selected that would still preserve critical mass for the spot market price to be meaningful. Around 50–65% of supply of each deal was provided to the long-term market, except for three deals with insufficient supply, which went solely to the spot market.

Since the term contracts may be at a premium over the spot price, it was essential that the bidders have confidence in the spot market. This requires transparency in conducting the spot market. The spot market was run for a period of time, until the customers gained confidence that it was producing reasonable prices.

The term contracts are similar to the elite channel, except the contract is must-take—the customer does not have the option of rejecting its share of the deal unless the spot market fails to produce a price. Each bidder knows that it is committed to purchasing its particular per cent of the deal at the market-clearing price premium.

It is natural to ask why a customer would bid a premium over the spot price. Could the customer not achieve supply regularity in the spot market simply by bidding high for its desired quantity? Then it would get the quantity but not pay a premium above spot. The answer is subtle and has to do with commitment. The term supply contract commits the bidder to winning a particular fraction of the deal in each cycle, regardless of the spot price. This commitment puts the customer in a desirable position relative to others competing for supply, and thereby reduces quantity risk. However, the advantage is limited, and indeed may be negative if customers care more about price risk than quantity risk. Our sense, however, is that quantity risk is the primary concern, and, therefore, we expected and saw a clearing price premium for most of the deals. The premium was of the order of 3–5%. This is a large premium, about the same as BHP Billiton's estimate of its customer profit margin.

Even if the premium falls in the future, BHP Billiton should not be discouraged if the price premium is zero or negative for many deals. A zero price premium would result if a sufficient number of customers believed that they could successfully purchase stones in the spot market. In this case, BHP Billiton has successfully reduced its own quantity risk by selling a portion of its supply forward. We expect the premium to trend downward as customers become more expert on bidding in the spot and term markets.

Forward contracts often have the advantage of improving the performance of the spot market by reducing incentives for the exercise of spot market power. However, the term contracts discussed here, since they base the price on the spot price, do less on this score. Large winners of term contracts still have an incentive to reduce demands in the spot market, since the spot price is determining the price paid for the entire quantity, not just the spot quantity. Nonetheless, the contracts do limit how much a customer can reduce its demands. Hence, market power and collusion are somewhat improved by the term contracts, but both market power and collusion remain important problems to watch. BHP Billiton guarded against this by expanding the number of customers allowed to bid in the spot to encourage competition even if no term customers bid in the spot auction.

The term market provides supply regularity that is valuable not just to customers but to BHP Billiton as well. Customers with long-term commitments have a greater incentive to make investments that enhance the value of the mine's output. BHP Billiton shares in the value created from these investments. In turn, BHP can conduct long-term planning on the value of increasing mine production which the customers will benefit from.

TRANSITION

As anticipated, the regular customers reacted negatively to change, since they enjoyed purchasing at prices that were somewhat below competitive market prices. This reaction took the form of lobbying BHP Billiton to criticize the plan, talking down the plan—even predicting disaster. Many of these criticisms focused on a reduction in loyalty, price transparency reducing intermediary profit, and the effectiveness of auctions at achieving better prices, driving customers out of business.

Nonetheless, new customers and some regular customers were strongly in favor of the new approach. These customers were able to obtain more supply without lobbying or setting up new entities. Large expanding customers especially liked the ascending auction, as it allows tailoring of a supply portfolio during the auction as prices evolve.

Due to the fragmented nature of rough-diamond demand, it is likely in BHP Billiton's long-term interest to encourage industry consolidation. It will become more difficult for customers to profit from pricing inefficiency, which will put pressure on customers to innovate or merge. This will be especially true if De Beers' member countries turn to market methods to allocate a portion of their production among their customers.

BHP Billiton's careful attention to customer needs allowed it to maintain good relationships with its regular customers through the transition. Vigorous discussion with customers synthesized improvements in contract terms that helped both BHP Billiton and its customers. Some of these contract terms became viable only in the presence of a competitive market. For example, BHP Billiton provided a six-month contract with two six-month options to continue buying at the same price. This would have been a difficult option to price if BHP Billiton had to do so unilaterally. A competitive auction allows the market price to be discovered so that BHP Billiton need not be overly cautious in offering a favorable contract to customers.

Favorable contract terms help customers reframe their relationship with BHP Billiton. Customers no longer benefit from pursuing zero-sum bargaining over contract terms. Competition raises the market price to reflect the value of contract changes. Customers can focus on lobbying only for changes that create value, such as minimizing overall risk and figuring out which party is best suited to shoulder it.

To gradually get customers comfortable with the approach, the first term auction, in September 2008, was limited to a handful of deals. This was a full-scale test of the approach for the deals offered, since the entire deal was sold under the new approach. The gradual approach also allowed some fine-tuning based on experience. To avoid "gaming of the experiment," the subset of deals utilizing the new approach represented a significant fraction of the total value of mine production.

One issue requiring ongoing study is how best to set reserve prices to manage collusion, revenue risk, and other factors. This has been especially important during the global financial crisis.

The key to a successful transition was careful development and then education of customers. For regular customers, moving from the classic De Beers approach, in which both the price and the assignment are set by the seller, required some gearing up. One way to ease the transition was to start with the uniform-price auction for the spot market, and then switch to the clock auction only if needed and after the customers had gained experience with the auction approach. BHP Billiton instituted a comprehensive education campaign involving both large-group and small-group training sessions and practice auctions.

For the term market, we found that customers prefer and BHP Billiton benefits from the use of the ascending-clock auction. Given the higher stakes of the term market, we found that greater price and assignment discovery was well worth the slightly higher implementation cost.

In making these recommendations, we assumed that the demand side for rough stones was competitive. We have found no evidence to the contrary. This assumption is supported by the fact that BHP Billiton's initial steps away from the De Beers' model—the elite, tender, and window sales—were not met with customer revolt. A competitive demand side means that BHP Billiton cannot be harmed by the boycott or exit of an individual customer. There are many potential manufacturers and resellers that desire to become BHP Billiton customers.

RESULTS

BHP Billiton successfully ran spot sales every cycle for over a year and held two or three ascending auctions per year for large stones. BHP Billiton had a surprisingly good result

Table 12.1. BHP Billiton term auction, February 2009

Round	Average start price (per cent of SMCP)	Deals sold/deals	Aggregate demand*/supply (splits)
1	95.00%	0/16	222/81
2	95.56%	0/16	142/81
3	96.06%	0/16	137/81
4**	97.06%	0/16	232/81
5	98.06%	0/16	213/81
6	98.94%	2/16	196/81
7	99.94%	2/16	167/81
8	101.25%	6/16	131/81
9	102.26%	7/16	103/81
10	102.77%	12/16	90/81
11	103.02%	14/16	84/81
Final	103.03%	16/16	81/81

* Aggregate demand at the beginning of the round except for round 1, where it is at the end of the round.

** This reflects the final opportunity for customers to increase demand.

for its transition term auction in September 2008 for approximately 20% of annual mine output, with prices 5% higher than expected.

In February 2009, BHP Billiton held a term auction for 60% of Ekati mine production (Table 12.1), with the balance to be auctioned in the spot market. The auction concluded successfully. All eighty-one splits in sixteen deals were sold. The price was an average of 103% of the spot market clearing price (SMCP) for terms of six, twelve or eighteen months at the option of the winner. The auction result was consistent with competitive bidding. The 103% average price exceeded the expectations of BHP Billiton. The quantity result was also impressive, especially in the middle of a massive financial crisis. All splits of all deals selling is counter to an industry trend of lower volume sold and indicates a growing market share for BHP Billiton.

Actual aggregate demand going into round 4 was 232—nearly three times supply, which is consistent with a competitive auction. The 3% price premium above spot prices also suggests a competitive auction.

On the day after the auction, February 21, 2009, the headline of a business story in the *New York Times* was "Diamond Sales, and Prices, Plunge." This was a tough time for an auction, but the approach did well despite the challenges. Fortunately, the ascending-clock auction is excellent at establishing—or re-establishing—confidence in a difficult market.

Many factors contributed to the success. The value proposition of a term contract pegged to the spot price is clearly excellent, with the auction exceeding price expectations. The addition of options for the customers to extend a minimum six-month term to twelve or eighteen months improved the value of the contract to the customers further to offset the dismal market sentiment. Better utilization of client rooms allowed twice as many customers as in the previous term auction—and several times the number of regular customers under the prior approach. Customers were also targeted based on spot bidding profiles to enhance competition across all deals.

An excellent value proposition, targeting of new customers who have interest in specific deals, high visibility to potential customers, a simple auction design with a good implementation, excellent training, documentation, and outreach to prevent technical and conceptual issues getting in the way of bidding all helped facilitate this superb outcome.

Conclusion

In thinking about a new sales process, it is helpful to reflect on why De Beers established the rather peculiar institution where customers are given a sack of stones and told the price. De Beers needed this tight control of both price and assignment as it was developing the market for diamonds in the first hundred years of the industry. The approach was made possible by the near monopoly position of De Beers.

Today, the diamond market is well established. Large but non-dominant sellers like BHP Billiton do not benefit from the De Beers approach. Rather BHP Billiton benefits from getting the stones into the hands of those that value them the most. For this to happen, a more market-responsive sales method was needed.

We worked with BHP Billiton to develop and implement auction methods to replace several of the previous sales channels for the Ekati diamonds. The auction approach does a better job of assigning and pricing the mine's output. Customers compete directly in simple auctions. In this way, the diamonds are allocated to the customers with the highest values, and the prices paid reflect current market conditions. The auctions allow each customer to express preferences for various quantities and types of stones, and find the value-maximizing assignments. Prices are competitively determined, with much less reliance on the price book. The extra value created from the better assignment of the stones results in higher sales revenues for BHP Billiton.

Spot auctions are held ten times per year and currently use a uniform-price format.

To foster supply regularity, the approach includes an auction for term supply. A customer desiring a supply commitment of up to eighteen months bids a percentage differential to the spot price for the quantity of each deal it desires. An ascending-clock auction allows each customer to build a portfolio of supply commitments across deals that best meets its needs, and pays the market-clearing price premium. By satisfying demands for supply regularity, BHP Billiton further enhances the revenues it achieves from its Ekati mine, resulting in a premium of 3–5% above the spot market price in two successive term market sales.

Large stones also are sold two or three times per year in specials auctions. An ascending-clock auction is used to better facilitate the discovery of market prices, and allow bidders to manage portfolio and budget constraints.

The auction approach rewards BHP Billiton's best customers and keeps them focused on their business and being competitive.

A key benefit of the approach is transparent pricing consistent with market fundamentals. The approach has proven robust to the global financial crisis, which has rocked the diamond industry. Both prices and quantities have exceeded expectation.

PART II
SECTION C

E-COMMERCE

....

CHAPTER 13

ENDING RULES IN INTERNET AUCTIONS

Design and Behavior

....

AXEL OCKENFELS AND ALVIN E. ROTH[1]

INTRODUCTION: ENDING RULES
AND LAST-MINUTE BIDDING

....

THERE is no need for ending rules in simple textbook auctions. The reason is that there is no time dimension in sealed-bid auctions, and dynamic auctions are typically modeled as clock auctions, where "price clocks" determine the pace of the bidding. In practice, however, the pace of bidding is often determined by the bidders themselves, so rules that specify when bidding ends are needed. Simultaneous auctions for spectrum licenses, for instance, often end after there has been no bid on any license in a given bidding round. Internet auctions, however, are typically run in real time, not in rounds, and bidders do not continually monitor the auctions. The simplest rule for ending such auctions is a fixed end time (a "hard close"), as employed by eBay. Auctions run on other platforms such as those formerly run by Amazon, which operated under otherwise similar rules, were automatically extended if necessary past the scheduled end time until ten minutes passed without a bid (a "soft close"). Yahoo auctions let the seller decide whether the auction is hard or soft close. We note, however, that many of eBay's competitors such as Amazon and Yahoo do not offer auctions anymore. So, the studies reported in this

[1] We thank Dan Ariely, Gary Bolton, Ben Greiner, David Reiley, and Karim Sadrieh for having worked with us on Internet auctions. Ockenfels thanks the German Science Foundation (DFG) for financial support through the Leibniz program and through the research unit "Design & Behavior"; Roth thanks the NSF.

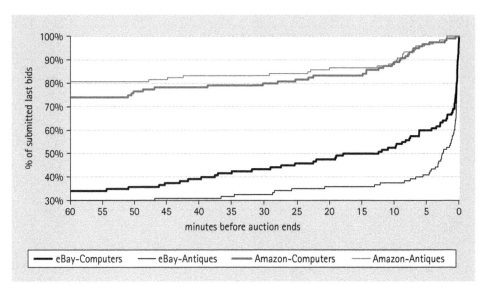

FIGURE 13.1. Cumulative distributions over time of auctions' last bids (Roth and Ockenfels, 2002).

chapter that compare behavior in eBay, Amazon and Yahoo auctions were lucky that there were such similar auctions that differed in their ending rules.[2]

A striking property of bidding on eBay is that a substantial fraction of bidders submit their bids in the closing seconds of an auction, just before the hard close, while there is almost no late bidding on Amazon-style auctions. Based on a study by Roth and Ockenfels (2002), Figure 13.1 shows the empirical cumulative probability distributions of the timing of the last bid in each auction for a sample of 480 eBay and Amazon auctions of antiques and computers with a total of 2,279 bidders. The timing of bids in Amazon is defined with respect to the initially scheduled deadline, which, with its soft close, can differ from the actual closing time.[3]

Figure 13.1 shows that there is significantly more late bidding on eBay than on Amazon. For instance, 40% of eBay computer auctions and 59% of eBay antiques auctions

[2] In 2005, Brown and Morgan (2009) conducted field experiments on eBay and Yahoo, and found evidence that behavior was inconsistent with equilibrium hypotheses for coexisting auction sites. They concluded that the eBay–Yahoo market was in the process of tipping. Yahoo shut down its North American auction website on June 16, 2007. In order to simplify our exposition, we will nevertheless use the present tense when we talk about Amazon and Yahoo auctions. We also remark that Google's and Yahoo's auctions of online ads are always accepting bids, and so do not need ending rules. For general surveys of online auction and in particular eBay research, see Bajari and Hortaçsu (2004), Lucking-Reiley (2000), Ockenfels et al. (2006), Hasker and Sickles (2010), and Greiner et al. (2012).

[3] This presentation may lead us to overestimate the extent to which Amazon bidders bid late, which would only strengthen our comparative results (see Roth and Ockenfels, 2002). We also note that in our 2002 study – one of the earliest on eBay – we collected the data by hand, which accounts for the small sample sizes; modern studies of eBay use millions of auctions as data. However, the results were clear and subsequently replicated by many other groups.

in our sample have last bids in the last five minutes, compared to about 3% of both Amazon computer and Amazon antiques auctions that have last bids in the last five minutes before the initially scheduled deadline or later. The pattern repeats in the last minute and even in the last ten seconds. In the 240 eBay auctions, 89 have bids in the last minute and 29 in the last ten seconds. In the Amazon auctions, on the other hand, only one bid arrived in the last minute. Figure 13.1 also indicates that within eBay, bidders bid later on antiques than on computers.

This chapter surveys the literature on how the rules for ending an auction can explain these different bidding dynamics, across online platforms and product categories, and how they impact price discovery and auction outcomes, both in theory and in practice.

LAST-MINUTE BIDDING IN THEORY AND PRACTICE

The basic setting: eBay's dynamic second-price auction and the risk of late bidding

One of the attractions of Internet auctions is that buyers do not all have to gather at the same place to participate, so that sellers can use Internet auctions to sell even relatively low-value items to a potentially wide audience. However, the size of the market would be limited if all potential bidders had to be online at the same time, and for this reason most auctions are conducted over a period of days, often a week.[4] To make it simple for bidders to participate in a week-long auction, without having to be constantly vigilant, or to be online at the close of the auction, most Internet auction houses make available a simple kind of software bidding agent; eBay calls it "proxy bidding."

On eBay bidders are asked to submit maximum bids (called "proxy bids") and explains that "eBay will bid incrementally on your behalf up to your maximum bid, which is kept secret from other eBay users." That is, once a bidder submits his "maximum bid," his resulting bid registers as the minimum increment above the previous high bid. As subsequent proxy bids by other bidders come in, the bid of the bidder in question automatically rises by the minimum increment until the second-highest submitted proxy bid is exceeded (or until his own maximum is exceeded by some other bidder). At the end of the auction, the bidder who submitted the highest proxy bid wins the object

---

[4] Unlike offline auctions, which typically last only a few minutes, Internet auctions such as those on eBay, Yahoo, and Amazon last many days. Lucking-Reiley et al. (2007) and Gonzales et al. (2009) observed that longer auction durations on eBay tend to attract more bidders and lead to higher prices. Lucking-Reiley et al. (2007) reported that while three-day and five-day auctions yield approximately the same prices on average, seven-day auction prices are approximately 24% higher and ten-day auction prices are 42% higher. Gonzales et al. (2009) observed that the change in the final sales price achieved by extending the auction from three to ten days is about 10.9%.

being auctioned and pays a price that is a small increment above the second-highest maximum (proxy) bid.[5]

To understand the bidding behavior that the proxy bidding system elicits, it will help to first consider how different the auction would be if, instead of informing all bidders about the bid history at each point of time during the auction, the auction were a second-price sealed-bid auction (in which nobody is informed about the proxy bids of other bidders until the auction is over). Then, the proxy bidding agent provided by eBay would make incremental or multiple bidding unnecessary. Suppose for instance that your maximum willingness to pay for an antique coin auctioned on eBay were $100. Then, bidding your maximum willingness to pay in a second-price sealed-bid auction is your dominant strategy, i.e., you can never do better than by bidding $100 (Vickrey, 1961).

The economics of second-price auctions are explained by eBay to its bidders along these lines, and it extends the conclusion to its own auctions, in which bids are processed as they come in:

> eBay always recommends bidding the absolute maximum that one is willing to pay for an item early in the auction. . . . If someone does outbid you toward the last minutes of an auction, it may feel unfair, but if you had bid your maximum amount up front and let the Proxy Bidding system work for you, the outcome would not be based on time.

The underlying idea is, of course, that eBay's bidding agent will bid up to the maximum bid only when some other bidder has bid as high or higher. If the bidder has submitted the highest proxy bid, he wins at the "lowest possible price" of one increment above the next highest bid. Thus, similar to the second-price sealed-bid auction described earlier, at the end of the auction a proxy bid wins only if it is the highest proxy bid, and the final price is the minimum increment above the second-highest submitted proxy bid, regardless of the timing of the bid. As we show later, however, proxy bidding does not necessarily remove the incentives for late or incremental bidding in these second-price auctions in which bids are processed as they come in, nor do bidders behave as if they thought it did.

An important institutional detail of eBay is that there are risks in last-minute bidding. As the time it takes to place a bid may vary considerably because of, for example, Internet congestion or connection times, last-minute bids have a positive probability of being lost. In a survey of seventy-three bidders who successfully bid at least once in the last minute of an eBay auction, 86% replied that it happened at least once to them that they started to make a bid, but the auction was closed before the bid was received (Roth and Ockenfels, 2002). Humans and artificial agents do not differ in this respect. The online sniping agent esnipe.com admits that it cannot make sure that all bids are actually placed:

---

[5] In case two bidders are tied for the highest bid, the one who submitted it first is the winner. In the following analyses we will assume for simplicity that the price increment is negligibly small. Ariely et al. (2005), for instance, provide a formal analysis that includes the minimum increment.

We certainly wish we could, but there are too many factors beyond our control to guarantee that bids always get placed. While we have a very good track record of placed bids, network traffic and eBay response time can sometimes prevent a bid from being completed successfully. This is the nature of sniping. (<http://www.esnipe.com/faq.asp>)

However, although this danger creates an incentive not to bid too late, there are also incentives not to bid early in the auction, when there is still time for other bidders to react, to avoid a bidding war that will raise the final transaction price. In particular, we identified three important and distinct kinds of bidding wars: bidding wars with like-minded late bidders; those with uninformed bidders who look to others' bids to determine the value of an item; and those with incremental bidders. Roth and Ockenfels (2002) and Ockenfels and Roth (2002, 2006) offer detailed game theoretic analyses of late and incremental bidding strategies, field evidence for strategic late bidding, and examples. The following examples and illustrations are taken from this work.

Bidding late to avoid bidding wars with like-minded bidders

Bidding late can be the best response to the late bidding strategies of like-minded bidders. As an example, suppose you are willing to pay up to $100 for an antique coin, and there is only one other potential bidder whom you believe also has a willingness to pay about $100. If both of you submit your value early, you will end up with a second-highest submitted proxy bid of about $100 implying a price of about $100. Thus, regardless of whether you win or not, your earnings (calculated as your value minus the final price if you are the winner, and zero if you are the loser) would be close to zero.

Now consider a strategy that calls for a bidder to bid $100 at the very last minute and not to bid earlier, unless the other bidder bids earlier. If the other bidder bids earlier, the strategy calls for a bidder to respond by promptly bidding his true value. If both bidders follow this strategy and mutually delay their bids until the last minute, both bidders have positive expected profits, because there is a positive probability that one of the last-minute bids will not be successfully transmitted, in which case the winner only has to pay the (small) minimum bid. However, if a bidder deviates from this strategy and bids early, his expected earnings are (approximately) zero because of the early price war triggered by the early bid. Thus, following the last-minute strategy, expected bidder profits will be higher and seller revenue lower than when everyone bids true values early.

Ockenfels and Roth (2006) develop a game theoretic model of eBay and prove, in an independent private-value environment, that mutual late bidding can constitute equilibrium behavior. Early bids are modeled for simplicity as taking place at times t on the half open interval $[0,1)$, while late bids happen at time $t = 1$. Thus there is always time to follow an early bid with another bid, but late bids happen simultaneously, when it is too late to submit a subsequent bid in response. Again for simplicity, early bids are transmitted with probability 1, while late bids are successfully transmitted with some probability p that may be smaller than 1. In this model, the above argument shows that it is not a dominant strategy to bid one's true value early. Indeed, the argument shows that it can be an equilibrium to bid late, even if $p < 1$.[6]

Bidding late to protect information in auctions with interdependent values

There are additional strategic reasons to bid late in auctions with interdependent values ("common-value auctions"). As an example, suppose you are a dealer of antique coins who can distinguish whether a coin is genuine or worthless. Suppose you identify an antique coin auctioned on eBay as genuine and that your maximum willingness to pay is $100. Another potential bidder, however, is not an expert and, thus, cannot tell whether the coin is genuine or worthless, but values a genuine coin higher than you, say at $110. What should you do?

When values are interdependent as in this example, the bids of others can carry valuable information about the item's value that can provoke a bidder to increase his willingness to pay. This creates incentives to bid late, because less informed bidders can incorporate into their bids the information they have gathered from the earlier bids of others, and experts can avoid giving information to others through their own early bids by bidding late. Specifically, in the scenario described earlier, if the minimum bid is positive and the probability that the coin is worthless is sufficiently high, the uninformed bidder should not bid unless the expert submitted a bid earlier and, thus, signaled that the coin is genuine. Bidding without such a signal from the expert would run the risk of losing money by paying the minimum price for a worthless coin. Such conditional bidding behavior of uninformed bidders creates, in turn, an incentive for experts to submit the bid for a genuine item very late in order to, as esnipe.com puts it, "prevent other bidders from cashing in on their expertise." Last-minute bids do not leave sufficient time for uninformed bidders to respond to and outbid experts' bids. See Ockenfels and Roth (2006) for a simple game theoretic model and Bajari and Hortaçsu (2003), who formalize this idea in an elegant symmetric common-value model.

As an illustration, Figure 13.2 displays the bid history of a completed auction that gives reason to speculate that we might be seeing an expert protecting information. The auction had only one bid, placed so late—five seconds before the deadline— that nobody could respond. This is an antiques auction, and antiques might reasonably be expected to have significant scope for asymmetric information among bidders as to the authenticity and value of items. The bidder's feedback number of 114 indicates that the bidder is familiar with the rules and bidding strategies in eBay auctions because the bidder must have completed at least 114 eBay auctions as a seller or a high bidder. Finally, the bidder's

---

[6] That it is also not a dominant strategy to bid one's true value late can be seen by supposing that any other potential bidders are planning not to bid at all; now the fact that late bids have a positive probability of not being transmitted makes it preferable to bid early.

eBay Bid History for
ANTIQUE GERMAN NIAGARA FALLS WIRE BASKET NR (Item #463085061)

Currently	$9.99	First bid	$9.99
Quantity	1	# of bids	1
Time left	Auction has ended.		
Started	Oct-08-00 21:10:39 PDT		
Ends	Oct-18-00 21:10:39 PDT		
Seller (Rating)			

Bidding History (Highest bids first)

User ID	Bid Amount	Date of Bid
llmuseum@city.niagarafalls.on.ca (114) ☆	$9.99	Oct-18-00 21:10:34 PDT

Remember that earlier bids of the same amount take precedence.

FIGURE 13.2. Late bidding to protect information.

ID is the email address of Lundy's Lane Historical Museum in the City of Niagara Falls, Canada, suggesting that the bidder is indeed likely to have special expertise on antiques related to Niagara Falls, such as the one in this auction.

A related idea is formalized and tested by Hossain (2008). He analyzes a dynamic second-price auction with an informed bidder and an uninformed bidder who, upon seeing a posted price, learns whether his valuation is above that price. In the essentially unique equilibrium, an informed bidder bids in the first period if her valuation is below some cutoff and bids only in the last period otherwise.

Bidding late to avoid bidding wars with incremental bidders

Last-minute bidding can also be a best reply to (naïve or strategic) incremental bidding. To see why, put yourself in the place of the bidder described earlier, who is willing to pay as much as $100 for an antique coin. Moreover, suppose that there is only one other potential bidder, and that you believe that this bidder is willing to pay more than you for the coin, say $110. This other bidder, however, bids incrementally, that is, he starts with a bid well below his maximum willingness to pay and is then prepared to raise his proxy bid whenever he is outbid, as long as the price is below his willingness to pay. Last-minute bids can be a best response to this kind of incremental bidding because bidding very near the deadline of the auction would not give the incremental bidder sufficient time to respond to being outbid. By bidding at the last moment, you might win the auction at the incremental bidder's initial, low bid, even though the incremental bidder's willingness to pay exceeds your willingness to pay. As esnipe.com puts it:

> A lot of people that bid on an item will actually bid again if they find they have been outbid, which can quickly lead to a bidding war. End result? Someone probably paid more than they had to for that item. By sniping, you can avoid bid wars. (<http://esnipe.com/faq.asp>)

eBay Bid History for
Antique style 10k SAPPHIRE DIAMOND Ring 22)

Currently	$65.51	.	$52.50
Quantity	1	;	6
Time left	Auction has ended.		
Started	Oct-11-00 14:16:38 PDT		
Ends	Oct-18-00 14:16:38 PDT		
Seller (Rating)			

Bidding History (Highest bids first)

User ID	Bid Amount	Date of Bid
rlvers1de (44) ☆ 👓	$65.51	Oct-18-00 14:14:22
aquette (24) ☆	$64.51	Oct-18-00 14:15:59
aquette (24) ☆	$62.51	Oct-18-00 14:15:36
aquette (24) ☆	$60.51	Oct-18-00 14:15:17
aquette (24) ☆	$57.51	Oct-18-00 14:15:03
aquette (24) ☆	$52.50	Oct-18-00 13:38:15

FIGURE 13.3. Late bidding as best response to incremental bidding.

Figure 13.3 shows the bid history of an auction that ended on October 11 2000 at 14:16:38 PDT. The history reveals that until 14:14:21 on the last day of the auction, just before the eventual high bidder rlvers1de submitted his bid, aquette was the high bidder. Then, rlvers1de became the high bidder. With about two minutes left, bidder aquette immediately responded and placed a new proxy bid, and, finding that this was not a winning bid, raised his bid three times in the last 95 seconds before the auction ended, without, however, becoming the high bidder. Thus, it appears likely that if rlvers1de had bid later, and too late for aquette to respond, he would have saved the $13 increase in price due to aquette's last-minute attempts to regain the high bidder status.

There are two types of reasons for incremental bidding: strategic and non-strategic. One non-strategic reason for incremental bidding is that bidders may not be aware of eBay's proxy system and thus behave as if they bid in an ascending (English) auction. Another explanation is an "endowment effect," as suggested by Roth and Ockenfels (2002) and Wolf et al. (2005), which posits that temporarily being the high bidder during an auction increases the bidder's value. Cotton (2009) incorporates the idea into a private-value, second-price auction model, and shows how it may drive both incremental and late bidding. Still other authors refer to "auction fever" as another potential explanation for incremental bidding (Heyman et al., 2004), or escalation of commitment and competitive arousal (Ku et al., 2005).[7]

---

[7] Late bidding is not only a good strategy to avoid incremental bidding wars with *other* emotional bidders, but may also serve as a *self*-commitment strategy to avoid one's own bids being affected by auction fever and endowment effects.

Incremental bidding can also have strategic reasons. One of these strategies is shill bidding by confederates of the seller in order to push up the price beyond the second-highest maximum bid. Engelberg and Williams (2009) demonstrate how shill bidders may use incremental bids and eBay's proxy-bid system to make bidders pay their full valuations. Barbaro and Bracht (2006), among others, argue that bidding late may protect a bidder from certain shill bidding strategies.

Also, according to a model by Rasmusen (2006), incremental bidding may be caused by uncertainty over one's own private valuation (see also Hossain, 2008; Cotton, 2009). He argues within a game-theoretic model that bidders are ignorant of their private values. Thus, rational bidders may refrain from incurring the cost of thinking hard about their values until the current price is high enough that such thinking becomes necessary. This, too, creates incentives for bidding late, because it prevents those incremental bidders from having time to acquire more precise information on their valuation of the object being auctioned.

Another well known, rational reason for incremental bidding is that bidders may be reluctant to report their values, fearing that the information they reveal will later be used against them (see Rothkopf et al., 1990). While the highest maximum bid is kept secret on eBay, it sometimes happens that the winner defaults and that then the seller contacts the bidder who submitted the second-highest bid. If this bidder revealed his value during the auction, the seller can make a take-it-or-leave-it offer squeezing the whole surplus from the trade. By bidding incrementally, private information can be protected—but only at the risk that a sniper will win at a price below one's value.

Finally, another direction for explaining late and multiple bidding is based on the multiplicity of listings of identical objects, which may create incentives to wait until the end of an auction in order to see how prices develop across auctions (see Budish, 2012, and our concluding section). Peters and Severinov (2006) propose a model with simultaneously competing auctions and argue that late bidding is consistent with this model. Stryszowska (2005a; see also 2005b,c) models online auctions as dynamic, private-value, multi-unit auctions. By submitting multiple bids, bidders coordinate between auctions and thus avoid bidding wars. In one class of Bayesian equilibria, multiple bidding also results in late bidding, even when late bids are accepted with a probability smaller than 1. Wang (2006) shows theoretically that in a twice repeated eBay auction model, last-minute bidding is in equilibrium and offers some field evidence for this. The models support the idea that the incentives to bid late are amplified when there are multiple listings of the same item.[8]

---

[8] Anwar et al. (2006) provide evidence suggesting that eBay bidders tend to bid across competing auctions and bid on the auction with the lowest standing bid. Regarding substitution across platforms, Brown and Morgan (2009) provide evidence indicating that revenues on eBay are consistently 20–70% higher than those on Yahoo, and that eBay auctions attract approximately two additional buyers per seller than equivalent Yahoo auctions, suggesting that cross-platform substitution is out of equilibrium. Two other studies (Zeithammer, 2009; Arora et al., 2003) of bidding behavior in sequential online auctions do not address the issue of bid timing within a given auction. Vadovic (2005) studies dynamic auctions in which bidders "coordinate" who searches for outside prices and shows that bidders with low search costs tend to bid late.

Field evidence for late bidding

The preceding sections show that there are a variety of reasons for bidding very near the scheduled end of an eBay auction, despite the risk that late bids may not be transmitted successfully. It is a best response to naïve or strategically motivated incremental bidding strategies, and can arise at equilibrium in both private-value and common-value auctions. In fact, there is also plenty of field evidence for late bidding on eBay. The first evidence comes from Roth and Ockenfels (2002) and is illustrated in Figure 13.1, and largely confirmed by other studies. Bajari and Hortaçsu (2003), for instance, found that 32% of the bids in their sample are submitted after 97% of the auction has passed. Anwar et al. (2006) noted that more than 40% of the bids in their eBay sample are submitted during the final 10% of the remaining auction time. Simonsohn (2010) reported that in his sample almost 20% of all winning bids are placed with just one minute left in the auction, and Hayne et al. (2003a,b) reported that bidding in the last minute occurs on average in 25% of their sample of 16,000 auctions. Regarding the whole distribution of the timing of bids, Roth and Ockenfels (2000) and Namazi (2005) observed that bid submission times on eBay follow a power-law distribution with most bids concentrated at the closing time.[9]

However, the field evidence regarding the profitability of sniping is less robust. Using eBay field data, Bajari and Hortaçsu (2003) could not statistically confirm whether early bids lead to higher final prices. Gonzales et al. (2009) as well as Wintr (2008) could not find evidence that the distribution of final prices is different for winning snipes and winning early bids on eBay. In a controlled field experiment, Ely and Hossain (2009) found a small and significant surplus-increasing effect of their sniping in DVD auctions as compared to early bidding. Gray and Reiley (2007) also found somewhat lower prices when the experimenter submitted the bid just ten seconds before the end of the auction compared to when the bid was submitted several days before the end, although the difference was not statistically significant here.

THE DESIGN OF THE ENDING RULE, AND WHY IT MATTERS

In this section, we show that the bidding dynamics are strongly affected by the ending rule, which thus may influence revenues and efficiency. We also show that sniping in hard-close auctions is likely to arise in part as a response to incremental bidding.

---

[9] There appear to be differences with respect to sniping frequencies across countries. Hayne et al. (2003a) reported that in their sample bidding occurs in the last minute of an auction with, for instance, 12% probability in the UK and 36.5% probability in Sweden. Shmueli et al. (2004) observed that the start of an auction also sees an unusual amount of bidding activity (see also Shmueli et al., 2007).

Last-minute bidding in hard-close vs. soft-close Internet auctions: field evidence

Amazon auctions are automatically extended if necessary past the scheduled end time until ten minutes have passed without a bid. Although the risks of last-minute bidding remain, the strategic advantages of last-minute bidding are eliminated or severely attenuated in Amazon-style auctions. That is, a bidder who waits to bid until the last seconds of the auction still runs the risk that his bid will not be transmitted in time. However, if his bid is successfully transmitted, the auction will be extended for ten minutes, so that, no matter how late the bid was placed, other bidders will have time to respond. Thus on Amazon, an attentive incremental bidder, for example, can respond whenever a bid is placed. As a result, the advantage that sniping confers in an auction with a fixed deadline is eliminated or greatly attenuated in an Amazon-style auction with an automatic extension (see Ockenfels and Roth, 2006, for formal results along these lines).

The difference in late bidding between eBay and Amazon auctions is illustrated in Figure 13.1. It suggests that late bidding arises in large part from the rational response of the bidders to the strategic environment. Non-strategic reasons for late bidding, including procrastination, use of search engines that make it easy to find auctions about to end, endowment effects, or management of bidding in multiple auctions in which similar objects may be offered, should be relatively unaffected by the difference in closing rules between eBay and Amazon. Moreover, Roth and Ockenfels (2002) observe an interesting correlation between bidders' feedback numbers and late bidding. The impact of the feedback number on late bidding is highly significantly positive in eBay and (weakly significantly) negative in Amazon. Similarly, Wilcox (2000), Ariely et al. (2005) and Borle et al. (2006) also observed in both laboratory and field studies that more experienced bidders snipe more often in an eBay environment than less experienced bidders.[10] This shows that more experienced bidders on eBay bid later than less experienced bidders, while experience in Amazon has the opposite effect, as suggested by the strategic hypotheses. It seems therefore safe to conclude that last-minute bidding is not simply due to naïve time-dependent bidding. Rather, it responds to the strategic structure of the auction rules in a predictable way. In addition, since significantly more late bidding is found in antiques auctions than in computer auctions on eBay, but not on Amazon, behavior responds to the strategic incentives created by the possession of information, in a way that interacts with the rules of the auction.[11]

---

[10] Borle et al. (2006) found that more experienced bidders are more active toward both the start and the end of the auction. Simonsohn (2010) investigated the consequences of such lateness on the strategic behavior of sellers. The idea is that because many bidders snipe, an auction's end time is likely to influence the number of bidders it receives. In fact, he found that a disproportionate fraction of sellers set the end time of their auctions to hours of peak demand.

[11] Borle et al. (2006) also found that the extent of late bidding observed on eBay varies significantly across product categories. However, while they suggest that this variation can be an important step toward constructing empirical measures of the extent of common/private values in online auctions, they do not find evidence that the measures are correlated.

Interpretation of such field data is complicated by the fact that there are differences between eBay and Amazon other than their ending rules. For instance, eBay has many more items for sale than Amazon, and many more bidders. Furthermore, buyers and sellers themselves decide in which auctions to participate, so there may be differences between the characteristics of sellers and buyers and among the objects that are offered for sale on eBay and Amazon. Some combination of these uncontrolled differences between eBay and Amazon might in fact cause the observed difference in bidding behavior, instead of the differences in rules. Laboratory experiments can control for such complexities. Moreover, experiments can better control of the effect of experience, [12] induce buyer and seller valuations and so easily allow observations of revenues and efficiency, and can separate the multiple reasons for late bidding that may contribute to the observed differences in bidding behavior on eBay and Amazon.

Ariely et al. (2005) conducted experiments in a controlled laboratory private-value setting, in which the only difference between auctions was the ending rule, to address these issues. One of the major design decisions in the experiment was to run all auctions in discrete time, so that "bidding late" could be precisely defined without running into problems of continuous-time decision-making such as individual differences in typing speed, which might differentially influence how late some bidders can bid. [13] Specifically, in all auctions, bidding was in two stages. Stage 1 was divided into discrete periods, and in each period, each trader had an opportunity to make a bid (simultaneously). At the end of each period, the high bidder and current price (typically the minimum increment over the second-highest bid) were displayed to all. Stage 1 ended only after a period during which no player made a bid. This design feature ensured that there was always time to respond to a bid submitted "early" in the auction, as is the case on eBay and in the theoretical models outlined in Ockenfels and Roth (2006). Stage 2 consisted of a single period. The bidders had the opportunity to submit one last bid with a probability $p = 0.8$ (in treatment eBay.8 and Amazon) or $p = 1$ (in treatment eBay1). The eBay auctions ended after stage 2. A successfully submitted stage-2 bid on Amazon, however, started stage-1 bidding again (and was followed by stage 2 again, etc.). Thus, in the Amazon condition, the risk of bidding late was the same as in the eBay.8 condition, but a successful stage-2 bid caused the auction to be extended.

---

[12] The proxies for experience in the field data (feedback ratings) are imperfect, because feedback ratings reflect only completed transactions, but not auctions in which the bidder was not the high bidder. In addition, more experienced buyers on eBay not only may have more experience with the strategic aspects of the auction, but may have other differences from new bidders; for example, they may also have more expertise concerning the goods for sale, they may have lower opportunity cost of time and thus can spend the time to bid late, or they may be more willing to pay the fixed cost of purchasing and learning to use a sniping program.

[13] Because eBay and Amazon are online auctions, it would have been possible to conduct the auction using precisely the eBay and Amazon interfaces, had that been desirable, by conducting an experiment in which the auctions were on the Internet auction sites; for a classroom demonstration experiment of this sort, in a common-value environment, see Asker et al. (2004). This would not have served the present purpose as well as the discrete version described. In this respect, it is worth noting that what makes an experimental design desirable is often what makes it different from some field environment, as well as what makes it similar.

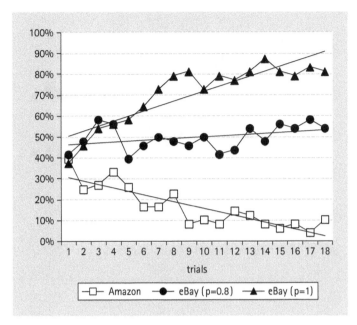

FIGURE 13.4. Percentage of bidders who snipe (bids in stage 2), and linear trends.

The experiment replicated the major field findings in a controlled laboratory private-value setting in which the only difference between auctions was the ending rule. Figure 13.4 illustrates that there was more late bidding in the hard-close (eBay) conditions than in the automatic-extension (Amazon) condition, and, as bidders gained experience, they were more likely to bid late in the eBay conditions, and less likely to bid late in the Amazon condition. Each of the three multi-period auction conditions started with about 40% of bidders submitting stage-2 bids, but by trial 18, Amazon had only about 10%, eBay.8 had 50%, and eBay1 had 80% late bidders.

The experiment also demonstrates that, *ceteris paribus*, "early" prices on Amazon are an increasingly good predictor of final prices, whereas price discovery on eBay became increasingly delayed (and frenzied). Figure 13.5 shows that, on Amazon, after bidders gained experience, the stage-1 price reached more than 90% of the final price, whereas the opposite is true on eBay.8 (about 70%) and eBay1 (less than 50%).

Regarding allocations, our data support the view that, in our eBay conditions, early bidding does not pay: a bidder's payoff is significantly negatively correlated with his own number of stage-1 bids, while the corresponding coefficient for the Amazon condition is not significant. Moreover, the Amazon condition is slightly more efficient and yields higher revenues than the other conditions. This seems to reflect the fact that Amazon is the only treatment in which low bidders always had time to respond to being outbid at prices below values, while eBay-bidders could only respond to stage-1 bids but not to stage-2 bids.

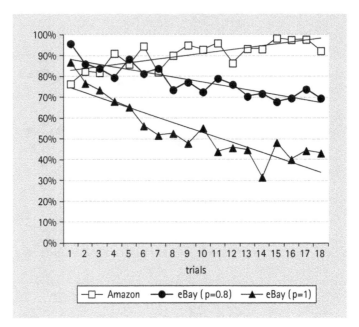

FIGURE 13.5. "Early" (final stage-1) prices as percentage of final price and linear trends.

As before, however, the field evidence is less clear. Brown and Morgan (2009) and Houser and Wooders (2005) took advantage of the fact that Yahoo sellers are allowed to choose whether to end the auction with a hard or a soft close. In both studies, identical items were sold using both ending rules. However, none of these studies found a significant effect of the ending rule on the amount of late bidding.[14] However, Houser and Wooders (2005) observed—as Ariely et al. (2005) did in the laboratory—that, *ceteris paribus*, hard-close auctions tend to raise less revenue than soft-close auctions.

Simulation experiments by Duffy and Ünver (2008) with artificial adaptive agents who can update their strategies via a genetic algorithm, replicate these findings and thus provide another robustness check.

Bidding wars and incremental bidding

As suggested in the section "Last-minute bidding in theory and practice" there can be equilibria where all bidders submit only one bid late in the auction, even in purely private-value auctions and even though this risks failing to bid at all. This kind of

---

[14] In a laboratory experiment, in which three sudden termination variants of hard-close auction (a.k.a. *candle auction*) were examined, Füllbrunn and Sadrieh (forthcoming) find that the extent of late bidding crucially depends on the first stage in which the probability of sudden termination is greater than zero.

equilibrium can be interpreted as collusion against the seller, because it has the effect of probabilistically suppressing some bids, and hence giving higher profits to the successful bidders. However, the model is generally rejected in favor of late bidding as a response to incremental bidding.

In fact, there is robust evidence in both the laboratory and the field that incremental bidding is common. Wilcox (2000) indicates that the average bidder submits 1.5–2 bids. Ockenfels and Roth (2006) report that 38% of bidders submit at least two bids. Among these bidders, the large majority submit a new bid after being outbid. In particular, 53% of the last bids of incremental bidders are placed after the previous bid was automatically outbid by eBay's proxy bidding agent (i.e. by another bidder's proxy that was submitted earlier in the auction), 34% are placed after the previous bid was outbid by a newly submitted proxy bid of another (human or artificial) bidder, and only 13% are placed by the current high bidder (so that the current price is not changed). Bids per bidder increase with the number of other bidders who bid multiple times in an auction, which suggests that incremental bidding may induce bidding wars with like-minded incremental bidders. In a regression study using eBay field data, Wintr (2008) found that the presence of incremental bidders leads to substantially later bids, supporting the view that sniping is reinforced by incremental bidding. Ely and Hossain (2009) conducted a field experiment on eBay to also test the benefit from late bidding. They show that the small gain from sniping together with some other patterns can be explained by a model in which multiple auctions are run concurrently and a fraction of the bidders are bidding incrementally.

Ockenfels and Roth (2006) note that naïve English-auction bidders may also have an incentive to come back to the auction near to the deadline in order to check whether they are outbid. However, the data indicate that among those bidders who submit a bid in the last ten minutes of an eBay auction, one-bid bidders submit their bid significantly later than incremental bidders. The data also reveal that bidders with a larger feedback score tend to submit fewer bids per auction, suggesting that incremental bidding is reduced with experience. This is in line with a study by Borle et al. (2006) who investigated more than 10,000 eBay auctions and found that more experienced bidders tend to indulge less in multiple bidding. However, in a study by Hayne et al. (2003b) the bidders who submitted multiple bids had a higher average feedback score than the average for all bidders.

Finally, Ariely et al. (2005) investigated the timing of bids in their pure private-value laboratory setting. They observed that early bids are mostly made in incremental bidding wars, when the low bidder raises his bid in an apparent attempt to gain the high-bidder status, while late bids are made almost equally often by the current high bidder and the current low bidder. That is, late bids appear to be planned by bidders regardless of their status at the end of the auction. Moreover, the amount of late bidding goes up significantly when the risk of sniping is removed (in treatment eBay1; see Figure 13.4). This indicates that the "implicit collusion" effect that results from the possibility of late bids not being transmitted is not the driving force here. Overall, there is substantial evidence from different sources showing that sniping arises, in part, as a best response to incremental bidding.

DISCUSSION AND CONCLUSION

With the advent of online and spectrum auctions, controlling the pace of an auction became an important topic in market design research. However, late bidding is a much older and more general phenomenon. Sniping was probably first observed in candle auctions, which were started about 1490 (see Cassady, 1967). The auctioneer lights a candle and accept bids only as long as the candle is burning. Here too, there is a risk to sniping, because the exact moment when no more bids will be accepted is not predictable. In his diary of his London life Samuel Pepys (1633–1703) records a hint from a highly successful bidder, who had observed that, just before expiring, a candle wick always flares up slightly: on seeing this, he would shout his final—and winning—bid.

Sniping is also a concern in other markets. Auctions in the German stock exchange, for instance, randomly select the exact moment when the auction will end. The idea is to prevent bidders from waiting until the very last second before submitting their final bids. In fact, the theoretical and laboratory work by Füllbrunn and Sadrieh (forthcoming) and Füllbrunn (2009) suggests that such auctions may perform better than a hard-close auction without any risk that late bids get lost.

Similarly, online negotiation sites that promise dispute resolution (such as e-commerce disputes and traditional litigation) via electronic and standardized communication also suffer from late bidding. One of the more prominent online negotiation sites, clicknsettle.com, experimented in 1999 with round-by-round demands and offers. But this format did not prove to be effective, because a deadline effect similar to what has been observed on eBay and in experimental bargaining games (Roth et al., 1988) hindered efficient negotiations. As clicknsettle.com put it: "After reviewing the early results with our clients, we discovered that in most negotiations, the first two rounds were being 'wasted' and the disputing parties really only had one opportunity to settle the case, the final round" (see Ockenfels, 2003).

We have seen that soft-close auctions can eliminate or severely attenuate the incentives to bid late. One alternative way to avoid late bidding and to control the pace of auctions is to create pressure on bidders to bid actively from the start. Milgrom and Wilson designed an activity rule that was applied to the US spectrum auctions (McAfee and McMillan, 1996; see also Milgrom 2004). The activity rule requires a bidder to be "active" (that is to be the current high bidder or to submit new bids) on a predetermined number of spectrum licenses. If a bidder falls short of the required activity level, the number of licenses it is eligible to buy shrinks. Thus, bidders are prevented from holding back.[15]

Another important feature of spectrum auctions is the fact that, most often, auctions for all licenses end simultaneously, that is, no auction is over until they are all over. The fact that eBay auctions a lot of items without this simultaneous closing rule, despite

---

[15] However, activity rules of this sort are incompatible with the flexibility needed on global Internet auction platforms.

the fact that many are close substitutes or complements, however, turns out not to be too problematic given the information they show on auctions that have not ended yet. Budish's (2012) work suggests that the provision of information about both current and near-future objects for sale substantially increases the social surplus generated by single-unit second-price auctions when the goods traded are imperfect substitutes, and that the remaining inefficiency from not using a multi-object auction is surprisingly small.

The research surveyed in this chapter shows that sniping is a robust strategy, robust in a game-theoretic sense (it is a best response to naïve and other incremental bidding strategies, and can even arise at equilibrium in both private-value and common-value auctions), but also against bounded rationality (such as various motivational and cognitive limits to behavior). In fact, much of the late-bidding phenomenon can be explained as a strategic response to naïve, incremental bidding. Obviously, the rule for ending an auction must take such irrationalities into account. While the traditional theoretical mechanism design literature compares equilibria of different mechanisms, market designers have to think about how beginners will play, and how experts will play against beginners, and so on. So, looking only at equilibria may not be enough to derive robust recommendations about the design of markets.[16]

Even the fact that we need ending rules at all (and do not just implement sealed-bid auctions) may be partly related to the fact that people do not behave like idealized perfectly rational agents Even in a purely private-value setting where, theoretically, fully efficient sealed-bid auctions can be devised, bidders sometimes perform better in open, dynamic auctions. Ariely et al. (2005) have shown, for instance, that the feedback delivered in open second-price auctions such as eBay substantially accelerates the speed of learning compared to second-price sealed-bid auctions. This improves the price discovery process and increases competition among bidders so that efficiency and revenues can be enhanced, even in purely private-value environments. In line with this finding, Ivanova-Stenzel and Salmon (2004) report that, when having the choice between sealed-bid and open, ascending-bid auctions, laboratory subjects in a private-value environment have a strong preference for the open format (for similar points see also Kagel and Levin, 2009; Cramton, 2006).

Summing up, one of the core challenges of market design is not only to take institutional but also behavioral complexities into account. The work on ending rules in online auctions demonstrates how theory, field, laboratory, simulation, and survey studies can work together to get a robust picture of how institutions and behavior interact (see also Roth, 2002, 2008; Bolton et al., 2012; Ockenfels 2009; Bolton and Ockenfels, 2012). This can be used to devise better and more robust systems.

---

[16] This is not to say that equilibrium analyses are not useful. For example, some observers of eBay believe that the amount of sniping will decrease over time because it is mainly due to inexperience and unfamiliarity with eBay's proxy bidding system. This is unlikely, however, because sniping is also an equilibrium phenomenon among rational bidders. Moreover, there is plenty of evidence that experienced bidders snipe more than inexperienced bidders. Thus, as long as the auction rules remain unchanged, it seems likely that late bidding will persist on eBay.

References

Anwar, S., McMillan, R. and Zheng, M. (2006) "Bidding behavior in competing auctions: evidence from eBay," *European Economic Review*, 50(2): 307–22.

Ariely, D., Ockenfels, A. and Roth, A. E. (2005) "An experimental analysis of ending rules in Internet auctions," *RAND Journal of Economics*, 36(4): 890–907.

Arora, A., Xu, H., Padman, R. and Vogt, W. (2003) "Optimal bidding in sequential online auctions." Carnegie Mellon University Working Paper No. 2003–4. <http://www.heinz.cmu.edu/research/121full.pdf>.

Asker, J., Grosskopf, B., McKinney, C. N., Niederle, M., Roth, A. E. and Weizsäcker, G. (2004) "Teaching auction strategy using experiments administered via the Internet," *Journal of Economic Education*, 35(4): 330–42.

Bajari, P. and Hortaçsu, A. (2003) "Winner's curse, reserve prices and endogenous entry: empirical insights from eBay auctions," *RAND Journal of Economics*, 34(2): 329–55.

—— —— (2004) "Economic insights from Internet auctions," *Journal of Economic Literature*, 42(2): 457–86.

Barbaro, S. and Bracht, B. (2006) "Shilling, squeezing, sniping: explaining late bidding in online second-price auctions." University of Mainz Working Paper.

Bolton, G. E., Greiner, B. and Ockenfels, A. (2012) "Engineering trust—reciprocity in the production of reputation information." *Management Science*, 58(12): 2225–2233.

—— and Ockenfels, A. (2012) "Behavioral economic engineering," *Journal of Economic Psychology* Vol. 33 (3): 665–676.

Borle, S., Boatwright, P. and Kadane, J. B. (2006) "The timing of bid placement and extent of multiple bidding: an empirical investigation using eBay online auctions," *Statistical Science*, 21(2): 194–205.

Brown, J. and Morgan, J. (2009) "How much is a dollar worth? Tipping versus equilibrium coexistence on competing online auction sites," *Journal of Political Economy*, 117(4): 668–700.

Budish, E. (2012) "Sequencing and information revelation in auctions for imperfect market design: understanding eBay's market design." Harvard University Working Paper.

Cassady Jr., R. (1967) *Auctions and Auctioneering*. California University Press.

Cotton, C. (2009) "Multiple bidding in auctions as bidders become confident of their private valuations," *Economics Letters*, 104(3): 148–50.

Cramton, P. (2006) "Simultaneous ascending auctions," n P. Cramton, Y. Shoham and R. Steinberg (eds), *Combinatorial Auctions*, MIT Press, pp. 99–114.

Duffy, J. and Ünver, M. U. (2008) "Internet auctions with artificial adaptive agents: a study on market design," *Journal of Economic Behavior and Organization*, 67(2): 394–417.

Ely, J. C. and Hossain, T. (2009) "Sniping and squatting in auction markets," *American Economic Journal: Microeconomics*, 1(2): 68–94.

Engelberg, J. and Williams, J. (2009) "Ebay's proxy bidding: a license to shill," *Journal of Economic Behavior and Organization*, 72(1): 509–26.

Füllbrunn, S. (2009) "A Comparison of Candle Auctions and Hard Close Auctions with Common Values." Otto-von-Guericke-University Magdeburg FEMM Working Paper 09019.

—— and Sadrieh, A. (forthcoming) "Sudden termination auctions—an experimental study." *Journal of Economics and Management Strategy*.

Gonzales, R., Hasker, K. and Sickles, R. C. (2009) "An analysis of strategic behavior in eBay auctions," *Singapore Economic Review*, 54(3): 441–72.

Gray, S. and Reiley, D. H. (2007) "Measuring the benefits to sniping on eBay: evidence from a field experiment," University of Arizona Working Paper.

Greiner, B., Ockenfels, A. and Sadrieh, A. (2012) "Internet auctions," in M. Peitz and J. Waldfogel (eds), *The Oxford Handbook of the Digital Economy*. New York: Oxford University Press, 306–342.

Hasker, K. and Sickles, R. C. (2010) "EBay in the economic literature: analysis of an auction marketplace," *Review of Industrial Organization*, 37: 3–42.

Hayne, S. C., Smith, C. A. P. and Vijayasarathy, L. R. (2003a) "Sniping in eBay: a cultural analysis." University of Colorado Working Paper.

―― ―― ―― (2003b) "Who wins on eBay: an analysis of bidders and their bid behaviours," *Electronic Markets*, 13(4): 282–93.

Heyman, J. E., Orhun, Y. and Ariely, D. (2004) "Auction fever: the effect of opponents and quasi-endowment on product valuations," *Journal of Interactive Marketing*, 18(4): 7–21.

Hossain, T. (2008) "Learning by bidding," *RAND Journal of Economics*, 39(2): 509–29.

Houser, D. and Wooders, J. (2005) "Hard and soft closes: a field experiment on auction closing rules," in R. Zwick and A. Rapoport (eds), *Experimental Business Research*, Springer, vol. 2, pp. 123–31.

Ivanova-Stenzel, R. and Salmon, T. C. (2004) "Bidder preferences among auction institutions," *Economic Inquiry*, 42(2): 223–36.

Kagel, J. H. and Levin, D. (2009) "Implementing efficient multi-object auction institutions: an experimental study of the performance of boundedly rational agents," *Games and Economic Behavior*, 66(1): 221–37.

Ku, G., Malhotra, D. and Murnighan, J. K. (2005) "Towards a competitive arousal model of decision-making: a study of auction fever in live and Internet auctions," *Organizational Behavior and Human Decision Processes*, 96(2): 89–103.

Lucking-Reiley, D. (2000) "Auctions on the Internet: what's being auctioned, and how?" *Journal of Industrial Economics*, 48(3): 227–52.

―― Bryan, D., Prasad, N. and Reeves, D. (2007) "Pennies from eBay: the determinants of price in online auctions," *Journal of Industrial Economics*, 55(2): 223–33.

McAfee, R. P. and McMillan, J. (1996) "Analyzing the airwaves auction," *Journal of Economic Perspectives*, 10(1): 159–75.

Milgrom, P. R. (2004) *Putting Auction Theory to Work*, Cambridge University Press.

Namazi, A. (2005) "Emergent behavior and criticality in online auctions," PhD dissertation, University of Cologne.

Ockenfels, A. (2003) "New institutional structures on the Internet: the economic design of online auctions," in M. J. Holler, H. Kliemt, D. Schmidtchen and M. Streit (eds), *Jahrbuch für Neue Politische Ökonomie*, Mohr Siebeck, vol. 20, pp. 57–78.

―― (2009) "Marktdesign und Experimentelle Wirtschaftsforschung," *Perspektiven der Wirtschaftspolitik*, 10 (supplement 1): 31–5.

―― and Roth, A. E. (2002) "The timing of bids in Internet auctions: market design, bidder behavior, and artificial agents," *Artificial Intelligence Magazine*, 23(3): 79–87.

―― ―― (2006) "Late and multiple bidding in second price Internet auctions: theory and evidence concerning different rules for ending an auction," *Games and Economic Behavior*, 55(2): 297–320.

―― Reiley, D. and Sadrieh, A. (2006) "Online auctions," in T. J. Hendershott (ed.), *Handbooks in Information Systems I: Handbook on Economics and Information Systems*, pp. 571–628.

Peters, M. and Severinov, S. (2006) "Internet auctions with many traders," *Journal of Economic Theory*, 130(1): 220–45.

Rasmusen, E. B. (2006) "Strategic implications of uncertainty over one's own private value in auctions." *B.E. Journal of Theoretical Economics: Advances in Theoretical Economics*, 6(1): article 7, available at <http://www.bepress.com/bejte/advances/vol6/iss1/art7>.

Roth, A. E. (2002) "The economist as engineer: game theory, experimental economics and computation as tools of design economics," *Econometrica*, 70(4): 1341–78.

———— (2008) "What have we learned from market design?" *Economic Journal*, 118(527): 285–310.

———— and Ockenfels, A. (2000) "Last minute bidding and the rules for ending second-price auctions: theory and evidence from a natural experiment on the Internet," NBER Working Paper 7729.

———— ———— (2002) "Last-minute bidding and the rules for ending second-price auctions: evidence from eBay and Amazon auctions on the Internet," *American Economic Review*, 92(4): 1093–103.

———— Murnighan, J. K. and Schoumaker, F. (1988) "The deadline effect in bargaining: some experimental evidence," *American Economic Review*, 78(4): 806–23.

Rothkopf, M. H., Teisberg, T. J. and Kahn, E. P. (1990) "Why are Vickrey auctions rare?" *Journal of Political Economy*, 98(1): 94–109.

Shmueli, G., Russo, R. P. and Jank, W. (2004) "Modeling bid arrivals in online auctions," Working paper. Available at <http://www.devsmith.umd.edu/ceme/pdfs_docs/papers/bidarrivals_jan04.pdf>.

———— ———— ———— (2007) "The barista: a model for bid arrivals in online auctions," *Annals of Applied Statistics*, 1(2): 412–41.

Simonsohn, U. (2010) "eBay's crowded evenings: competition neglect in market entry decisions," *Management Science*, 56(7): 1060–73.

Stryszowska, M. (2005a) "Last-minute and multiple bidding in simultaneous and overlapping second price Internet auctions," CentER Discussion Paper.

———— (2005b) "On the ending rule in sequential Internet auctions," CentER Discussion Paper.

———— (2005c) "Coordination failure in Internet auctions with costly bidding," CentER Discussion Paper.

Vadovic, R. (2005) "Bidding and searching for the best deal: strategic behavior in Internet auctions," Working Paper. Available at <http://www.cerge-ei.cz/pdf/events/papers/060224_t.pdf>.

Vickrey, W. (1961) "Counterspeculation, auctions, and competitive sealed tenders," *Journal of Finance*, 16(1): 8–37.

Wang, J. T. (2006) "Is last minute bidding bad?" UCLA Working Paper.

Wilcox, R. T. (2000) "Experts and amateurs: the role of experience in Internet auctions," *Marketing Letters*, 11(4): 363–74.

Wintr, L. (2008) "Some evidence on late bidding in eBay auctions," *Economic Inquiry*, 46(3): 369–79.

Wolf Jr., J. R., Arkes, H. R. and Muhanna, W. A. (2005) "Is overbidding in online auctions the result of a pseudo-endowment effect?" Ohio State University Working Paper.

Zeithammer, R. (2009) "Sequential auctions with information about future goods" UCLA Working Paper.

CHAPTER 14

DESIGNING MARKETS FOR MIXED USE OF HUMANS AND AUTOMATED AGENTS

ANDREW BYDE AND NIR VULKAN

INTRODUCTION

OVER the last three decades, rapid progress in market design has brought the subject to an engineering-like state, where a large number of well understood mechanisms can be prescribed for a given situation. Electronic markets—whether direct negotiations, auctions or exchanges—are a particularly good place to apply this theory, because the interactions between participants are regulated by the communication protocol. Since there are already rules in place, it seems wise to ensure that they are the right rules—the ones that will lead to efficient outcomes. In Chapter 17 of this *Handbook* (Vulkan and Preist) we look at market design issues for markets where all participants are automated agents. Elsewhere in the volume there are plenty of examples for market design for human agents, be it individuals or firms. In this chapter we look at the hybrid case where some of the organizations involved are using a software agent to directly aid them in their negotiations with other organizations where agents are not used. The context for this is procurement negotiations: Very large firms can force their suppliers to participate in an auction, but small and middle-size firms cannot. To these firms, the technology described in this chapter can be used to ensure almost the same efficiency of outcome without changing the way they negotiate. The system uses multiple one-to-one negotiations to try to mimic the process and hopefully outcome of an auction (in this case a reverse, price-lowering auction).

For any organization, saving on procurement costs has an impact on profitability that is multiplied by gross margin. Although much research focus has been placed on achieving the lowest possible cost for the goods/services purchased, we concern ourselves here with the operational procurement costs of the organization.

The AutONA (Automated One-to-one Negotiation Agent) system was conceived as a means of reducing these operational procurement costs, enabling procurement departments to automate as much price negotiation as possible, thus creating the option of reducing direct costs and/or redeployment of operational effort into strategic procurement requiring high human involvement (Byde et al., 2003). The problem domain has been limited to the automation of multiple 1:1 negotiations over price for quantities of a substitutable good subject to the organization's procurement constraints of target quantity, price ceiling, and deadline.

We present the design of the core reasoning system and preliminary results obtained from a number of experiments conducted in HP's Experimental Economics Lab. The architecture of AutONA is that there is a central reasoner that sets goals, targets, and price caps for each seller–quantity pair, and for each seller there is a *bidding agent* that interprets these control parameters, and handles negotiation with a given seller, maintaining a record of the history of interaction, and acting accordingly. Thus the reasoner is responsible for assessing the merit of each seller's position relative to the others, while the bidding agent chooses good local policy accordingly.

Our main conclusion is that AutONA could reasonably be deployed for automated negotiation, having shown no evidence for being identified as an automated system by suppliers, and having demonstrated comparable gains from trade.

In the next section we review previous work in the automated negotiation domain. In the third section we specify the reasoner. In the fourth section, we specify the bidding agents. In the fifth section we describe the experimental setup for the human-based evaluation of AutONA, and in the seventh section we summarize the results of these experiments. In the eighth section we conclude.

Background

The problem of bargaining is an old one. When studied from a game theoretic perspective (e.g. Muthoo, 1999), the problem of what offers to make (in a single 1–1 negotiation context) almost always reduces to a calculation of the *first* offer to make, which is immediately accepted by one's opponent—so that no actual negotiation occurs! This reduction is due to assumptions about the nature of rewards, information, and rationality that simply do not hold in the real world.

Faratin et al. (1998) presented a pragmatic approach to the replication of reasonable human 1–1 negotiation strategies in machines. The usage of the term *tactic* here mirrors the use there. They consider three main types of negotiation tactic: time-based, resource-based, and imitative. An important class of behavior missing from this taxonomy is that of competition-based behavior, which is only feasible in a context where there are several parallel streams of negotiation being conducted. As far as we are aware, there are no attempts in the literature to address this problem. Instead, researchers have investigated methods for conducting multi-variable 1–1 negotiation (e.g. Faratin et al., 2000; Sycara, 1989), or have focused on the application of negotiation technology to various distributed computing problems, such as resource allocation (e.g. Sathi and Fox, 1990).

Reasoner specification

Assumptions

The beginning of any procurement process is a *purchase request*, specifying the quantity desired, Q, and the maximum price acceptable for the full quantity, P. This quantity Q can be bought from one or more sellers, each of which has a minimum quantity they will consider selling, a maximum quantity they will consider selling, and whose potential sale quantities jump in some specified minimum increments. These parameters are specified (for a seller S) as q_{min}^S, q_{max}^S, and q_{step}^S respectively.

The reasoning about how much to offer for each quantity centers around *options*, where an option, o, is defined by a seller and a quantity. For each option, the system forms a series of estimates regarding the likely price of purchasing the specified quantity from the specified seller. These estimates are parametrized by a *risk parameter*. The possible values for the option risk parameter are *best*, *expected*, and, giving rise to three prices for each option o, $p_-(o)$, $p_e(o)$, and $p_+(o)$ respectively.

Price estimates

The price estimates for an option are calculated using estimates of the distribution of the lowest price a seller will accept for the options quantity. A *belief* is a probability distribution over prices *per unit*, parametrized by the properties that an option may have.

There are many possible ways to represent beliefs; observation of frequencies in historical data can be used to build non-parametric models, but when the amount of data is small, these methods are not suitable. We choose to assume a log-normal distribution on prices, and select the mean and standard deviation to minimize squared error with respect to observed closing prices in prior negotiations. The observed closing prices in previous negotiations are normalized with respect to a benchmark price that carries information on the market price on the date that the negotiation was concluded. By doing so, we reduce the impact of the variation of market prices over time. To this effect we also introduce a customizable scale factor that gives exponentially less weight to older data.

To each seller, S, we associate a belief function, $b_S(p, q)$, with the interpretation that the probability of the price for the option $o \in option(S)$ closing between prices p_1 and p_2 (per unit) is believed, prior to the start of negotiations, to be

$$Prob(p(o) \in [p_1, p_2]) = \int_{p_1}^{p_2} b_S(x, q_o)dx \tag{1}$$

The price estimates for a given option are generated as follows:

1. The *best* price $p_-(o)$ of an option, o, is defined to be the current highest offer that AutONA has made for the specified option, or some fixed minimum, $p_{min}(o)$ otherwise (i.e. if no offer has yet been made).
2. The *no-risk* price $p_+(o)$ of an option, o, is the larger of the best price and the largest number p such that for all $p' < p$, $Prob(p(o) \in [p', \infty]) > 0$. Informally, it is the highest price to which AutONA should attatch non-zero probability (via the belief).
3. Given an option, o, with quantity q_o and seller S, associated belief $b_S(p, q)$, best price $p_-(o)$, and no-risk price $p_+(o)$, the *expected* price of the option o is given as

$$p_e(o) = \frac{\int_{p_-(o)}^{p_+(o)} x\, b_S(x/q_o, q_o)dx}{Prob(p(o) \in [p_-(o), p_+(o)])} \tag{2}$$

Spreads

Although it may be that negotiations will be for the full quantity with each seller, it is also quite possible that, due to quantity constraints, it will be necessary to divide the full purchase quantity, Q, between several sellers; the trade-offs that AutONA then makes will be between alternative ways of dividing the quantity up between the available sellers. We call such a "dividing-up" a *spread*. Formally, a spread is a set of options.

Just like an option, associated to a spread are a quantity, and a range of estimates of the price at which it can be obtained, where the prices are parameterized by risk. The quantity of a spread $\sigma = \{o_1, \ldots, o_k\}$ is just the sum of the quantities of its corresponding options, and the prices are defined likewise:

$$quantity(\sigma) = \sum_{i=1}^{k} quantity(o_i)$$

$$p_*(\sigma) = \sum_{i=1}^{k} quantity(o_i)p_*(o_i) \tag{3}$$

where, as for options, $*$ can be $-$, e, or $+$. (Recall that option prices are *per-unit* spread prices are for the full quantity $quantity(\sigma)$).

Targets

In order to determine how hard to bargain for each option under consideration, the reasoner sets *targets* for each option, which are calculated with reference to the other sellers and the options they offer. The target of option o belonging to seller S is intuitively understood to be the maximum price per unit likely to be acceptable for o, and is calculated via a sort of "credible threat" reasoning: It is worth considering o at price p only if there is a completion of o to a spread no more expensive than the best spread available not including options belonging to S. This understanding is modified by risk parameters "$-$", "e", "$+$", that capture best-case, average-case, and worst-case qualifications of the above clauses.

Formally, for each option o, and some set of potential purchase spreads $\mathcal{M}$, we make the following definitions:

1. The set of *alternatives* to o in $\mathcal{M}$ is the set of those purchase spreads in $\mathcal{M}$ which do not contain any options belonging to the seller S of o:

$$alt_{\mathcal{M}}(o) = \{\sigma \in \mathcal{M} \mid options(seller(o)) \pitchfork \sigma\}$$

2. The set of *completions* to o in $\mathcal{M}$ is the set of spreads that, with o added, become an acceptable purchase spread:

$$comp_{\mathcal{M}}(o) = \{\sigma \mid \sigma \cup \{o\} \in \mathcal{M}, options(seller(o)) \pitchfork \sigma\}$$

3. The target of o relative to the set of purchase spreads $\mathcal{M}$ is defined for any pair of spread-risk preferences, $r_1, r_2 \in \{-, e, a, +\}$, as

$$t_{r_1, r_2}^{\mathcal{M}}(o) = \min(p_{r_1}(\sigma) \mid \sigma \in alt_{\mathcal{M}}(o)) \\ - \min(p_{r_2}(\sigma) \mid \sigma \in comp_{\mathcal{M}}(o)) \qquad (4)$$

Example 1. *Suppose that the options are $\{a, b, c, d, e, f\}$ with associated quantities 1,3,2,4,6,2, and suppose that each is associated to a unique seller. If the purchase request quantity Q is 5, then the set of all acceptable purchase bundles $\mathcal{M}$ is*

$$\mathcal{M} = \{\{a, d\}, \{b, c\}, \{b, f\}, \{a, c, f\}\}.$$

Selecting option a to calculate targets for, we have

$$alt_{\mathcal{M}}(a) = \{\{b, c\}, \{b, f\}\},$$

$$comp_{\mathcal{M}}(a) = \{\{d\}, \{c, f\}\}.$$

The target for a relative to $\mathcal{M}$, with expected prices for alternatives and no-risk prices for completions is therefore

$$t_{e,+}^{\mathcal{M}}(a) = \min(p_e(\sigma)|\sigma \in \{\{b,c\},\{b,f\}\})$$
$$- \min(p_+(\sigma)|\sigma \in \{\{d\},\{c,f\}\}),$$

$$= \min\left(3p_e(b) + 2p_e(c), 3p_e(b) + 2p_e(f)\right)$$
$$- \min\left(4p_+(d), 2p_+(c) + 2p_+(f)\right).$$

Acceptable purchase spreads

The set of acceptable purchase bundles $\mathcal{M}$ in equation (4) should ideally be the set of *all* possible spreads consistent with the purchase request, i.e.

$$\mathcal{M}_{Q,P} := \{\sigma|p_-(\sigma) \le P, quantity(\sigma) = Q\} \tag{5}$$

Notice that we require the spread's total quantity to be exactly Q, so that $\mathcal{M}_{Q,P}$ may be empty. Future implementations may allow flexibility in the purchase request, and hence the set of all acceptable purchase spreads.

When there are several sellers with small feasible quantity steps q_{step}, the set $\mathcal{M}_{Q,P}$ may be too large to reason over, in which case it is necessary to restrict attention to some subcollection of spreads.

It can be shown that if the price per additional unit is non-increasing with quantity for each seller, then the set of spreads that can minimize total price is given by the extreme points of the convex hull (in quantity space) of the set of all acceptable purchase spreads, $\mathcal{M}_{Q,P}$. This fact, and the intuition that at any given time there will be some seller that is "favorite," and from whom we should like to buy as much of the quantity Q as possible subject to quantity constraints, informed our choice of algorithm for restricting the set of spreads under consideration.

Starting and ending negotiations

Starting

We assumed that the procurement process begins with the buyer sending out a request for quotes to each seller, in response to which they will each quote an ask for the requested quantity (which is, of course, not always Q, depending on seller constraints). AutONA then has to make a counter offer; the seller counter-offers again, and from then on the tactics selected by the reasoner will specify counter offers. This process requires us to specify how AutONA's first bid is generated.

We chose the first bid on an option o to be $0.94p_-(o) + 0.06p_+(o)$, i.e. close to the best one could expect. This choice was made on the assumption that our first bid would almost certainly not succeed, but that a successful transaction would be concluded only after negotiations. If the initial bid were set too high, it would almost certainly be accepted, which could lead (via the construction of beliefs on the basis of historical trade information) to inflation in the price that AutONA would consider reasonable.

Ending

The reasoner controls completion of individual negotiations: AutONA continues trading until the difference between the worst-case and expected-case prices is less than a predefined (small) proportion, ϵ of the worst-case price:

$$best_+(\mathcal{M}) - best_e(\mathcal{M}) < \epsilon \cdot best_+(\mathcal{M}) \tag{6}$$

where

$$best_r(\mathcal{M}) := \min(p_r(\sigma)|\sigma \in \mathcal{M}) \tag{7}$$

BIDDING AGENT SPECIFICATION

Option choice

When negotiating with a direct seller S, there may be many options with respect to which negotiations could proceed. We choose to order the options according to the best expected price amongst acceptable purchase spreads containing them.

1. The *best spread* with respect to risk option r, $\mathbf{B}_r$ is any spread in the maximization set $\mathcal{M}$ such that $p_r(\mathbf{B}_r) = best_r(\mathcal{M})$. We assume that there is an implicit total ordering on spreads which allows us to select $\mathbf{B}_x$ consistently and unambiguously.
2. If $\mathbf{B}_e \cap options(S) \neq \emptyset$, then the option o which forms the intersection is the most favored option for seller S.
3. Otherwise, o is the smallest-quantity option which minimizes the expected price function over spreads containing an option from the given seller:

$$q_o p_e(o) + best_e(\mathcal{M}_o^c) = best_e(\mathcal{M} \setminus \mathcal{M}_o^a) \tag{8}$$

Tactics

A *tactic* is a rule specifying a new value to offer in response to the thread of negotiation that has so far taken place with a given seller. The tactics used by AutONA are all *alpha–beta* tactics, which are specified by two numbers, α and β. A new bid is given with respect to the preceding one, the last ask, and the most recent change to the ask, as

$$new\ bid = \min(old\ ask,\ old\ bid +$$
$$+ \alpha \times (change\ in\ ask) \tag{9}$$
$$+ \beta \times (ask - bid))$$

More specifically, we use two sub-families: pure alpha and pure beta tactics:

- the *fixed alpha tactics* A_j, $j = 0, 1, 2, 3, 4$ are the five alpha–beta tactics with $\beta = 0$, $\alpha = \{\alpha_0, \frac{1}{2}(1 + \alpha_0), \frac{1}{2}, 1, 0\}$ respectively; and
- the *fixed beta tactics* B_j, $j = 0, 1, 2$ are the three alpha–beta tactics with $\alpha = 0$, $\beta = \{0, \beta_{small}, \beta_{big}\}$ respectively;

here $\alpha_0 > 1$, $0 < \beta_{small} < \beta_{big} < 1$ are constants for which the values chosen were 2, $\frac{1}{5}$ and $\frac{1}{2}$ respectively. Note that $A_4 = B_0$.

Tactic selection

The choice of which tactic to use with each option o depends on the relative standing of that seller (for that quantity) with respect to the others.

The intuition behind tactic selection is that the value of the expected price relative to the expected-price alternatives governs the use of the α parameter; the β parameter is determined by "how far the seller has to go": the normalized difference between the current ask and the expected price.

If the change between the previous and current ask is non-zero, i.e. if the seller has conceded at all since his previous offer, we choose the tactic for option o to be the fixed alpha tactic A_j, with j selected according to the following algorithm:

1. Define[1]

$$
\begin{aligned}
t_0 &= t_{-,e}(o) \\
t_1 &= \tfrac{1}{2}(t_{-,e}(o) + t_{e,e}(o)) \\
t_2 &= t_{e,e}(o) \\
t_3 &= \tfrac{1}{2}(t_{e,e}(o) + t_{+,e}(o)) \\
t_4 &= t_{+,e}(o)
\end{aligned}
$$

2. Choose j such that $|t_j - p_e(o)|$ is minimized.

The intuition is that if the expected price of the option $p_e(o)$ is close to t_0, for example, then it is expected to be comparable to the *best case* for its best possible alternative, and hence is valuable, so that we should concede in order to keep the seller happy; if $p_e(o)$ is close to t_4 then we expect o to be comparable (when completed) to the *worst-case* alternative: hence it is the seller's responsibility to concede toward us if he wants to be considered seriously.

If the change between the previous and current ask is zero, the current tactic for option o is chosen to be the fixed beta tactic B_j according to the following algorithm:

---

[1] Recall that the most suitable option o to negotiate over is chosen using equation (8).

1. Let

$$
s = \begin{cases} \dfrac{p_+(o) - p_e(o)}{p_+(o) - p_-(o)} & \text{if } p_+(o) > p_-(o) \\[2mm] 0 & \text{otherwise} \end{cases}
$$

2. If $s < \frac{1}{4}$, choose $j = 0$;
3. If $\frac{1}{4} \le s < \frac{3}{4}$, choose $j = 1$; and
4. If $\frac{3}{4} \le s$, choose $j = 2$.

EXPERIMENTS

Overview

Since AutONA is designed for real procurement applications, it is essential to understand its performance before any deployment in real business environments. More specifically, there are three key questions that we seek to answer:

1. Are the negotiation algorithms on which AutONA is based exploitable by clever sellers? Is it possible for sellers to detect that they are bidding against a "machine" when negotiating with AutONA?
2. How well does AutONA perform in different trading environments? The goal here is to identify, as much as possible, a relationship between specific features of the purchasing environment and the performance of AutONA.
3. How well does AutONA perform compared to human traders in similar circumstances?

A sequence of laboratory experiments was conducted to perform the tests, following standard experimental economics methodology. The subjects were given accurate information about the game, in particular, how their actual monetary rewards depended on their aggregate performance over the course of the session. Experimental anonymity with respect to roles and payment was preserved, and no deception was used. Experiments with all human subjects were conducted to serve as benchmarks to measurements of AutONA's effectiveness. The same experiments were then run again with AutONA replacing one of the human buyers.

The experimental model

The goal of the experimental design phase was to capture important aspects of the true procurement environments in which AutONA is intended to participate. To remove

any conscious or unconscious biases in the experimental design, very little information about how AutONA works was provided to the experimenter who designed the
experiments. The primary information used to construct the experiments came from
the HP procurement organization, which provided detailed descriptions of, and data
from, their procurement operations.

Due to business and scientific considerations, we chose to examine a scenario similar
to that of DRAM procurement. Important aspects of this scenario, such as the small
numbers of buyers and sellers, their relative market power, the inflexibility of short-
term capability, and the possibilities of shortages were included in the design of the
experiments. Some complications, such as inventory carry-over and timing of delivery,
were ignored.

The experimental model has three central components: the buyers, the sellers, and
the negotiation process.

The buyers

Each buyer's objective is to procure a certain amount Q, which will be referred to as the
target quantity, of a single homogeneous commodity. Buyers are rewarded according to
the following formula: has a linear download sloping demand function with a cut-off
point at Q, and an additional bonus if he procures an amount not less than Q. Thus his
demand function is

$$Demand(q) = \begin{cases} a - b \times q & \text{if } q < Q, \\ a - b \times Q + bonus & \text{if } q = Q, \\ 0 & \text{if } q > Q, \end{cases} \tag{10}$$

where a and b are positive constants obeying the constraint $a - b.Q > 0$, so that buyers
are always incented to buy no less than Q goods. This demand function gives rise to the
reward function,

$$Reward(q) = \begin{cases} a.q - b.q^2 & \text{if } q < Q, \\ a.Q - b.Q^2 + bonus & \text{if } q \geq Q. \end{cases} \tag{11}$$

A player's total payoff for purchasing quantity q is given by $Reward(q) - C(q)$, where
$C(q)$ is what the buyers pay for the goods. This payoff function provides no incentive to
procure any amount more than Q, which is similar to the situation in which a buyer is
trying to procure enough DRAM to manufacture computers for a specific fixed quantity
contract with a downstream reseller.

The sellers

Each seller has a cost function $K(q)$ where q is the quantity they sell. Their payoff
function is $C(q) - K(q)$, where $C(q)$ is what the buyer(s) pay him. The cost function
$K(q)$ is assumed to have a fixed cost (F), a variable cost (c) and a capacity (k):

$$K(q) = \begin{cases} F + c \times q & \text{if } q \leq k \\ F + c \times k + 10c \times (q - k) & \text{if } q > k \end{cases} \qquad (12)$$

It is assumed that when a seller tries to sell above capacity, he has to incur ten times the normal costs. This is probably more realistic than assuming that it is impossible to sell more than capacity, since sellers can, if they wish, always procure goods on the spot market to cover short-falls in supply. The net result of the extra factor of ten is to make production beyond capacity expensive but not impossible, which is realistic in the DRAM environment. Sellers were always played by human subjects.

Supply and demand calibration

There are only a few major players in the DRAM market: Four major suppliers cover roughly 70–80% of the market. The market is a bit more fragmented on the buyer side, but there are only a few players (such as HP, IBM, and Dell) that have the market power to negotiate substantial deals with the major sellers.

The experiment was set up with four homogeneous buyers and four heterogeneous sellers. The sellers capabilities reflected true market share in the DRAM market. The total market capacity was normalized to 1,000. Both capacities and cost functions were fixed throughout the experiment, so that the only uncertainty existed on the demand side. The demand parameters were set up so that the market equilibrium quantity was the smaller of either the total capacity or the totals of all the buyer's target quantities, Q. This allowed us to measure the effectiveness of a buyer by simply looking at the amount he had procured.

Buyers target quantities were generated by a random process consistent with actual demand fluctuations. The HP Procurement Risk Organization has been analyzing the distribution of DRAM demand over the years. A normalized form of this distribution was used in the experiment.

Two supply and demand scenarios were considered. In the first scenario, the average total target quantity was slightly higher than the total capacity. However, demand was generated according to a log-normal distribution, so the chance of a shortage (total target more than total capacity) was roughly 50%. In the second scenario, the total target quantity was *always* greater than the total capacity. Thus, every trading period is in shortage, although it is uncertain of the degree of the shortage.

The negotiation process

The negotiation process was modeled as a round-based multiple 1–1 negotiation game. In each round, buyers and sellers take turns to make offers consisting of a price and a quantity, with no requirement to improve on previous offers. Each offer is directed at only *one* player on the other side of the market, and is private information between the buyer–seller pair. In each round, a player can make a new offer, accept the offer on the table, or stop the negotiation.

A limited form of cheap talk was allowed: A player could send a message consisting of a price and a quantity to anyone on the other side of the market, with no commitments: There were no consequences of this communication other than information exchanges.

A time cost was introduced to provide incentives for timely negotiation. The first eight rounds of negotiation were free, but after that each round cost a fixed amount to any player who had an active offer on the table. The trading period terminated if either side of the market (buyers or sellers) had no active offers. This process does not guarantee termination, but in practice negotiations usually terminated in about ten to fourteen rounds.

Customization

AutONA was designed before the experiments were. The design criteria behind AutONA were for it to be applicable to a wide range of procurement situations, to exhibit flexibility through customization. To play the game, AutONA needed to be customized; this section covers some of the customization choices that we made, and discusses the impact they had on the experimental results. Customization can be seen as consisting of two components: a set of parameter values for certain control parameters; and heuristics and rules relating to the way in which data are fed to and from the system by an operator.

Customization parameters

Termination condition

The parameter ϵ (see earlier) sets the point at which AutONA will recommend to the buyer that a price is accepted and that negotiation with the seller over a particular quantity should be concluded. We decided to set ϵ to 5%, meaning that AutONA will recommend to close a deal when the price that the seller offers is within 5% of the price it expects for that seller at that quantity.

History scale factor

AutONA was pre-loaded with a history of previous negotiations with the various sellers, and with the market price for previous rounds. Because of the accelerated time in the experiments, we had the freedom to place the actual periods in time at our will. We decided that the history scale factor would be set so that all the data of the previous experiments would count for about a half of the data of the current experiment. Between periods in the same experiment the time difference was considered to be negligible.

Heuristics and rules

Deal definition

During the deal definition phase, the user operating AutONA sets values for parameters such as quantity required and price ceiling. The obvious choice to make was to define the quantity to procure as the target quantity of the game. For the price ceiling, we use a value that is equivalent to the reward that AutONA would receive for procuring the target quantity, as defined in equation (12). With these settings, we ensure that AutONA will not form deals that will incur greater costs than the maximum reward.

Seller selection

In the seller selection phase, the quantities q^S_{min}, q^S_{max} and q^S_{step} are defined for each of the sellers. q^S_{max} is one of the most important parameters of the game, as it represents the capacity that sellers have available. But that piece of information was not available to AutONA (nor to any other buyer-side player). Nor had AutONA been designed to elicit that knowledge as the game progressed. The values of q^S_{max} for the sellers determine how AutONA builds its spreads. For the first experiment we used a heuristic that would have AutONA build spreads that divide the required quantity nearly equally among the four sellers. The rule was to set q^S_{max} for each seller at 27% of the target quantity. Having observed that AutONA was not so successful in procuring the target quantity (see discussion of the second result), later on we decided to have AutONA build spreads where one of the sellers was getting the biggest share of its target quantity. We did that by setting the maximum quantity available from each of the seller to be 75% of the target quantity.

Negotiation

The protocol used in the experiments prescribed that the buyer put in the first offer, whereas the protocol that AutONA used had been designed to play a game where the seller would submit the first offer, for a quantity requested by AutONA. To comply with the rules of the game, we had to define a heuristic for the first offer that was not suggested through the AutONA user interface. To play fairly, we needed to bind the heuristic to information that was available to AutONA. Our decision was that the first offer would be submitted as a percentage of the price that AutONA expected for negotiation from a given seller ($p_e(S)$). In the first experiment we guessed that 90% might be a fair value. Having observed that in the second experiment AutONA procured prices with a spread of 93.8% to 105.8% on the mean, we set it to be 94% for the fourth experiment. In both cases AutONA exhibited a less than brilliant performance in procuring the target quantity (see discussion on the second result). To improve things in the last experiment, we decided that the first bid was to be submitted at exactly $p_e(S)$, resulting in a better performance of AutONA quantity-wise.

Recomputing spreads

AutONA was designed to attempt to impose quantity on the suppliers, through the RFQ process. The game would go smoothly if suppliers did accept the quantities by responding with a counter-offer on the same quantity. We observed that this was not the case during the experiments. Whenever the seller proposes a different negotiation quantity, the AutONA operator faces a decision on whether to proceed negotiating over quantities appearing as options in AutONA spreads or restarting AutONA to recompute the spreads. In the spirit of making the experiments as repeatable as possible, we needed to put the operator in a condition to use deliberation as little as possible. So we defined a rule that if none of the sellers responded to the quantity suggested, AutONA should be restarted by the seventh round. Likewise, AutONA needed to be restarted if sellers would not respond even after a deal had just been struck. In that case, the operator shoud restart AutONA, subtracting the deal quantities achieved so far from the game target. Restarting AutONA is not ideal, but in both cases gives us the advantage that q^S_{max} can be set using information taken from offers that sellers have made. This tactic is useful in reducing the number of rounds required to achieve deals, thus avoiding round costs. More importantly, it is useful to actually secure the quantity that was needed, especially in cases of supply shortage. To respond to the problem that AutONA was having in procuring target quantity, in the fifth experiment we modified the rules so as to restart and recompute the spreads after the fifth round, using seller information on quantities to set q^S_{max}. A further rule was that after bundle recomputation, the AutONA operator would accept standing seller offers that would fall within the percentage of $p_e(o)$ that was set to determine the first offer.

RESULTS

A total of five experiments were conducted (Table 14.1), each with eight players (four buyers and four sellers). Two of the five experiments were all human experiment. In the rest, AutONA played the role of one buyer. In the fifth experiment, a modified version of AutONA was used, to counteract behavioral traits discovered in the first four experiments, discussed later. In each case, AutONA was provided with data from previous experiments as simulations of market inputs.

Our first result addresses the first experimental question.

Result 1. *AutONA passed a limited version of the Turing test. There is no obvious method for the human subjects to exploit AutONA.*

In the beginning of each experiment that involved AutONA we announced that one of the players would be played by a robot. At the end of the experiment, we informally quizzed all the subjects as to the identity of the robot. The answers we obtained were random. There is no evidence that human subjects could identify which player was

Table 14.1. Summary of the experiments

Experiment	Supply/Demand treatment	Players
1	Random shortage	All Human
2	Random shortage	AutONA
3	All shortage	All Human
4	All shortage	AutONA
5	All shortage	Modified AutONA

played by AutONA. Furthermore, we can conclude that no subjects found and used any logical loop-holes in AutONAs algorithms.

The other two experimental questions are concerned with performance. There are two primary measures we use to benchmark the performance of a buyer: price, and quantity with respect to target. Payoff is not relevant, because AutONA is not designed to optimize the experimental payoff, and indeed is not even aware of the existence of a payoff function.

All things being equal, quantity with respect to the target is the most important measure. Table 14.2 summarizes buyers' performance as measured by the quantity they procured as a percentage of their targets.

Since the buyers are homogeneous, their performance should be roughly the same if all of them are playing rationally. In all the experiments with two exceptions, human subjects procured roughly a similar amount (compared within experiment) with respect to their targets. The two exceptions are buyer 4 in experiment 1, and buyer 3 in experiment 5. Some variations are expected since humans do not negotiate equally. Experiment 1 seems to show a larger variation, which can probably be explained by the presence of inexperienced subjects.

The "Market" column in Table 14.3 lists the total quantities procured in the market (by the four buyers) as a percentage of the total capacity. Experiments 1 and 2 have the same supply and demand parameters, while experiments 3, 4, and 5 have another set of parameters. It is clear that aggregate results are consistent across experiments. The percentage bought with respect to target quantity is within 2 percentage points across buyers for each experiment. This is strong evidence that experimental results were repeatable, and that the human subjects understood their instructions and responded well to monetary incentives.

This brings us to our second major result.

Result 2. *The original AutONA was procuring substantially less, relative to its target quantity, than human buyers. This is particularly significant when there is a shortage.*

As can be seen from Table 14.3, the quantity procured by AutONA is substantially lower than that of human players in experiment 2 and 4. Table 14.3 also reports a summary of experiment 2 with only the periods in shortage. In those periods, AutONA was

Table 14.2. Summary of buyers' performance as measured by the average price of transactions

Experiment	Buyer 1	Buyer 2	Buyer 3	Buyer 4 / AutONA
1	$167	$161	$183	$167
2	$183	$172	$172	**$163**
2 (shortage)	$192	$174	$170	**$163**
3	$191	$191	$181	**$189**
4	$191	$210	$182	**$182**
5	$184	$185	$184	**$185**

procuring even less, at 53% of target, which is consistent with the results in the "all shortage" experiment (experiment 4).

On the basis of experiments 1 through 4, it is clear that AutONA has a severe behavioral bias. Roughly speaking, it is not aggressive enough in completing negotiations with successful transactions: it spends too long negotiating, and sellers go elsewhere. This problem is exacerbated by a shortage. Human buyers seem to be able to recognize the importance of grabbing supplies as fast and as aggressively as they can, while AutONA does not.

Result 3. *AutONA received lower prices than the human players.*

From Table 14.3, we see in experiments 2 and 4 that AutONA has the lowest average price.

It seems that AutONA was trading off prices with quantities: one reason why AutONA procured significantly less than expected was its strong stance on price. This bias toward aggressive price negotiation is due to a design assumption: that there was no competition

Table 14.3. Summary of buyers' performance as measured by the percentage of the target quantity purchased (results in bold are for AutONA)

Experiment	Buyer 1	Buyer 2	Buyer 3	Buyer 4 / AutONA	Market
1	100%	93%	87%	**72%**	87%
2	89%	92%	94%	**67%**	85%
2 (shortage)	96%	91%	91%	**53%**	82%
3	71%	85%	71%	**75%**	76%
4	89%	82%	87%	**52%**	77%
5	83%	80%	65%	**80%**	77%

against other buyers, and hence that time is much less of an issue. When time is not an important issue, there is no reason to negotiate speedily, except to meet purchasing deadlines, and so it is advisable to bargain hard. The DRAM procurement game, especially when there was a shortage, definitely involved competition between buyers, and although AutONA does well on price, it does poorly on quantity.

Bearing in mind the relative importance of the two performance measures, we decided to modify the behaviour of AutONA. To begin with, instead of opening on option o with a bid of $0.94p_-(o) + 0.06p_+(o)$ (we reconfigured AutONA to open with $p_e(o)$. This (unsurprisingly) led to many negotiations concluding immediately, and on average reduced the duration of negotiations considerably, at the expense of leading to more expensive trades. In addition, an adjustment of the seller constraints (encoded in q_{min}^S and q_{max}^S), such that the largest component in each spread took up about 75% of Q, seemed to result in superior performance. Both of these modifications were at the configuration level. We anticipate that each specific negotiating environment will place different requirements on AutONA, and hence will lead to different configurations. The fifth experiment was run with this modified version.

Result 4. *A modified version of AutONA performed significantly better on quantity, and not as well on price. Its payoff of price and quantity was similar to that of humans.*

See experiment 5 in Tables 14.2 and 14.3.

Conclusions

This chapter describes how markets using a mixture of automated and human agents could work. In it we present a system AutONA, for conducting multiple simultaneous 1–1 negotiations over price and quantity. The use of competition between sellers to guide negotiation tactics is key.

We have implemented this system, and conducted human trials to evaluate it on the basis of its ability to negotiate "reasonably," and on its performance with respect to a trading game that was designed independently of the system itself. We find that AutONA passes a limited version of the Turing test: The experiments did not reveal any obvious exploitation that a human trader can use against AutONA. On the other hand, AutONA in its original configuration exhibited significantly different aggregate behavior from human traders; it was less aggressive on quantity, and more aggressive on price—a behavioral bias that is non-desirable in the HP DRAM procurement context in which it was evaluated. Subsequently, AutONA was modified, and the modified version behaved more in line with human traders in the experiments. For market designers to successfully be able to come up with designs that can be used by hybrids of human and automated agents, much more research is needed. We hope the findings in this chapter can be seen as a step in that direction.

References

Byde, A., Yearworth, M., Chen, K. Y., and Bartolini, C. (2003) "Autona: a system for automated multiple 1–1 negotiation," *IEEE International Conference on ECommerce*, pp. 59–67.

Faratin, P., Sierra, C., and Jennings, N. (1998) "Negotiation decision functions for autonomous agents," *Robotics and Autonomous Systems*, 3–4(24): 159–82.

―――― ―――― ―――― (2000) "Using similarity criteria to make trade-offs," in *Proceedings of the Fourth International Conference on Multi-Agent Systems*, pp. 119–26.

Abhinay Muthoo, A. (1999) *Bargaining Theory with Applications*, Cambridge University Press.

Arvind Sathi, A. and Fox, M. S. (1990) "Constraint-directed negotiation of resource reallo-cations," in L. Gasser and M. N. Huhns (eds), *Distributed Artificial Intelligence*, Morgan Kaufmann Publishers Inc., vol. 2, pp. 163–93.

Sycara, K. P. (1989) *Multiagent Compromise Via Negotiation*, Morgan Kaufmann Publishers Inc., pp. 119–37.

CHAPTER 15

··

THE DESIGN OF ONLINE
ADVERTISING MARKETS

··

BENJAMIN EDELMAN

INTRODUCTION

··

ONLINE advertising is big business, reaching some $40 billion per year by 2012. For advertisers, online advertising offers the triple promises of reaching just the right consumers, at fair prices, with robust measurement of the effects of online campaigns. For website publishers, advertising offers an opportunity to make money from their sites— an important consideration, since few consumers appear willing to provide money payments for the sites and services they use. For users, in principle online ads can be useful in finding new products or suppliers. However, online ads are often easily overlooked (compared with, say, the temporal and auditory interruption of television advertisements).

Because the market for online advertising is both new and fast-changing, participants experiment with all manner of variations. Should an advertiser's payment reflect the number of times an ad was shown, the number of times it was clicked, the number of sales that resulted, or the dollar value of those sales? Should ads be text, images, video, or something else entirely? Should measurement be performed by an ad network, an advertiser, or some intermediary? Market participants have chosen all these options at various points, and prevailing views have changed repeatedly. Online advertising therefore presents a natural environment in which to evaluate alternatives for these and other design choices.

In this chapter, I review the basics of online advertising, then turn to design decisions as to ad pricing, measurement, incentives, and fraud.

Defining the Product: Payment Structure and Purchasing Incentives

The fundamental product in online advertising markets is a lead—a customer who might make a purchase from a given advertiser, or otherwise respond to an advertiser's offer. An advertiser typically prefers to reach customers especially likely to buy its product or service, and observable customer characteristics indicate varying degrees of interest in an advertiser's offer. For example, consider an advertiser selling motorcycles. The advertiser could attempt to reach consumers in particular demographic groups (say, males age eighteen to twenty-five), site browsing (reading a motorcycle enthusiast website), or search terms (searching for "motorcycle deals"). The advertiser's forecast of the likelihood of the user making a purchase would inform the advertiser's willingness to pay to present its offer to that consumer.

Meanwhile, from the perspective of an online publisher operating a website or other online resource, advertising is typically an ancillary component, to be integrated with, or at least juxtaposed against, a larger offering. If a publisher offers a search function, the publisher could show text ads related to users' search requests. Alternatively, a publisher could place "banner ads" (typically graphical images in industry-standard sizes—see Figure 15.1) adjacent to articles on its site. In principle, a publisher could even make individual words on its site into ads, by making them links to advertisers' sites—though with questions about who selects which words link where, and whether and how consumers know they are clicking on ads. A publisher's resource is typically space on its site or service. If the publisher's site presents too many ads, consumers may reach an unfavorable view of the site.

Online advertising can be measured and sold along any of several metrics. An advertiser could pay a fee each time its ad is shown—a "cost per impression" placement, often known as CPM ("cost per mille," being the price for 1,000 impressions). Alternatively, an advertiser could pay when its ad is clicked—"cost per click" (CPC). Or an advertiser could pay only when a user clicks and subsequently makes a purchase—"cost per action" (CPA) (Table 15.1). An advertiser could even offer payment proportional to the *amount* of the user's purchase, *ad valorem*, or differing payment scales could apply to the advertiser's various products. In expectation, advertisers and publishers might be indifferent among these payment metrics; with a known click rate, conversion rate, or order size, an advertiser and publisher could agree to use any of these metrics, and fees would be equal in expectation. That said, the metrics have importantly different implications for parties' incentives, moral hazard, and fraud, as discussed in subsequent sections.

Industry norms associate certain payment metrics with certain advertising formats. Historically, display advertisements were typically priced per impression—a natural approach from the perspective of a publisher who does not know which ads will attract many clicks, and who wants to be able to predict site revenues. That said, selling ads per impression influences participants' behavior: a CPM advertiser wants to attract as many

Table 15.1. Payment rules for different formats of online advertising

	Display ads	Search ads	Affiliate/links
Pay per impression	Standard	Unusual	
Pay per click	Also used	Standard	Unusual, though implemented at eBay beginning in 2009
Pay per action	Used for some campaigns	Brief experiment at Google	Standard

clicks as possible, even from customers who may ultimately be minimally interested in the advertiser's offer; perhaps some of those marginal customers can be convinced to buy the advertiser's product. CPM advertisers thus have a clear incentive to present banners with overstated claims of relevance of urgency, like those shown in Figure 15.1. Facing this onslaught of low-value ads, consumers seem to develop "banner blindness." As of 2009, practitioners at iMedia Connection report that for every 1,000 display ads shown to consumers, just 0.2–0.3 are clicked (Stern, 2010). Meanwhile, some display ad services have begun to price ads differently—selling ad placements on a per-click basis, encouraging advertisers to design offers that consumers choose to activate.

Ads on search engines typically follow a CPC model—not charging advertisers for their ads to be shown, but charging substantial fees when a user clicks an ad (for some keywords, as much as $20 or more per click). With CPC pricing, an advertiser

FIGURE 15.1. Deceptive banner ads overstate the urgency of clicking through.

seeks to attract only customers reasonably likely to purchase its product or otherwise offer the advertiser some benefit; attracting clicks from uninterested customers means unnecessary marketing expense. On the most favorable view, CPC pricing also invites users to click ads: Knowing that an advertiser was willing to pay to reach users searching for a given keyword, a user may expect that the advertiser's offer will match the user's request. Indeed, as Overture (later acquired by Yahoo) began offering pay-per-click ads, founder Bill Gross specifically boasted of the benefits of "us[ing] money as a filter" of which sites to show in search listings (Hansell, 2001).

Affiliate link systems typically follow a conversion-contingent CPA payment model—either paying a publisher only when a user signs up (e.g. a \$15 commission for referring a customer to Netflix), or in proportion to the dollar value of the user's purchase (e.g. a 6% commission on the user's purchase from Amazon). To date, few affiliate marketing programs have been willing to pay affiliates for impressions or clicks—seemingly on the view that little-known affiliates, without meaningful vetting or supervision, would have an overwhelming tendency to fake impressions and/or clicks, whereas actual sales are viewed as harder to fake. That said, as detailed in the section on advertising fraud later in the chapter, even conversion-based payment methods suffer strategic behavior that inflates advertisers' costs.

SEARCH ADS

Auctions and pricing

Historically, online ads were typically sold through posted prices, rate sheets, and person-to-person negotiations—much like ads in print, television, and radio. But auctions and auction-like mechanisms have proven particularly well suited to online advertising, for at least three reasons. For one, there are a multiplicity of items to be sold, including a large number of sites showing ads, as well as multiple ad placements on each such site. With so many items to sell, it would be difficult to announce a price for each or to negotiate the particulars of a placement. Furthermore, values change as market conditions change—making efforts to post or negotiate prices all the more difficult. Finally, the automated online delivery of advertisements seems to complement an automated online sales process; interconnected systems and servers can accept offers for a given placement, select an ad to be shown, show the ad to the corresponding users, and charge the advertiser accordingly.

The use of auctions and auction-like mechanisms presents a variety of questions of auction design. Should an advertiser be charged its own bid ("first price") or something less ("second price" or otherwise)? How often may bids be updated, and should an advertiser pay a fee for adjusting its bid? Should advertisers be able to see the bids of competitors seeking the same placements, see how many competitors are interested, or see something less? Should an auction impose a reserve price, below which ads are rejected, or is any payment better than nothing? Ad platforms have reached differing conclusions on all these questions.

The world of sponsored search advertising began in 1998 with pay-per-click text ads developed by Goto.com, later renamed Overture and purchased by Yahoo. Advertisers were suspicious of Overture's novel approach to pricing: With early fees often reaching $1 per click or even more, advertisers were concerned that competitors might click their ads, or Overture might charge for clicks that did not actually occur. To attempt to address these concerns, Overture showed advertisers the ads and bids of all competitors—confirming that an advertiser was not alone in its use of Overture's offering, and that others were onboard too. Showing all bids also helped advertisers adjust to the unfamiliar auction format: With competitors' bids visible for inspection, an advertiser could better assess the tradeoff between bidding higher (getting more clicks) and bidding lower (reduced price, but lesser exposure).

When a user clicked an advertiser's ad, Overture charged each advertiser the amount it had bid—a first-price auction. This system was intuitive: If an advertiser reported being willing to pay $0.70 for a click, why would Overture charge the advertiser anything less? But the game was infinitely repeated, with bid updates allowed frequently. (Initially, it seems, update frequency was limited only by the effort required to log into Overture's systems and make adjustments. Later, a rule limited updates to one every fifteen minutes, and a widespread automatic bidding agent adjusted bids every fifteen minutes.) In the Overture first-price auction, each advertiser had an incentive to lower its bid to the minimum increment ($0.01) above the next-highest advertiser—letting the advertiser retain the same position but pay a reduced price. Edelman and Ostrovsky (2007) show that the resulting instability led to an inefficient allocation of placements—often misordering advertisers, putting a lower-value advertiser above one that valued clicks more highly, and thereby destroying surplus. The resulting instability also reduced total revenue of the mechanism by at least 7% (a conservative bound reflecting the difficulty of estimating advertisers' valuations from historic bid data).

In 2002, Google began to use a mechanism with some characteristics of a second-price auction. Rather than paying its own bid, an advertiser would pay an amount linked to the bid of the next-highest advertiser—reducing the incentive to adjust bids continuously. Moreover, Google adjusted each bid by the estimated likelihood of a user clicking the corresponding ad, thereby selecting the ad with largest expected revenue to Google.

Edelman, Ostrovsky, and Schwarz (2007) (EOS) study this multi-unit second-price mechanism, calling it "generalized second price" or "GSP." EOS show that GSP has no dominant-strategy equilibrium, and truth-telling is not an equilibrium. However, the corresponding generalized English auction has a unique equilibrium, and that equilibrium is an ex post equilibrium with bidders' strategies independent of their beliefs about others' valuations. Moreover, Cary et al. (2007) show that a reasonable myopic bidding strategy converges to the equilibrium identified by EOS. Further overviews of sponsored search appear in Feldman et al. (2008), Lahaie et al. (2007), Liu et al. (2008), and Yao and Mela (2009).

Ad platforms continue to use reserve prices to rule out bids they view as undesirably low. In simulations, Edelman and Schwarz (2010) assess the revenue consequences of an optimally chosen reserve price. Which bidders face the largest cost increases from a rising reserve price? Edelman and Schwarz show that, for all advertisers that do not drop out as the reserve price increases, the increased reserve price yields an *identical* dollar-for-dollar increase in total payment.

Most ad platforms offer additional targeting of their ads based on at least the user's geographic region ("geotargeting") and day/time ("dayparting"). These targeting functions are typically operated on a binary basis: Either a user request matches the restrictions, and hence is eligible to see the advertiser's ad, or the advertiser specifies that its ad may not be shown. Microsoft adCenter offers further supplemental targeting based on user self-reports of age and gender at other Microsoft properties (such as Hotmail, MSN, and Windows Live). If a user matches the demographic characteristics an advertiser specifies, the advertiser may opt to increase its bid, potentially increasing its ranking relative to competitors. Thus, in adCenter, an advertiser's bid is not just a price, keyword, and vector of match conditions, but also additional price adjustments paired with demographic conditions. Despite the additional targeting possible under demographic bid adjustment, uptake of demographic targeting seems to be limited so far.

Transparency of pricing and ranking

Ad platforms limit the information available to advertisers, unlike the early listing of all advertisers and bids that Overture initially provided. For example, Google has never shown advertisers the bids or identities of competing bidders. Instead, Google provides advertisers a *traffic estimator* tool: An advertiser enters a possible bid, and Google reports the estimated number of clicks it would provide per day, as well as the advertiser's estimated average position in ad listings.

Ranking of advertisers sometimes raises concerns about favoritism or penalties—concerns that tend to focus on Google, given that company's large market share (discussed further in the section "Multihoming, competition, and barriers"). Google states that it ranks advertisers according to both their bids and Google's various assessments of site characteristics (Varian, 2009). If one site enjoys a more favorable assessment, it can obtain more prominent placements at considerably lower expense. On one view, a search engine is a private party entitled to show whatever links it sees fit, in whatever order and prominence it chooses. But some advertisers allege that Google singles out up-and-coming competitors for particularly unfavorable treatment, typically by demanding unreasonably high prices for ads from those would-be competitors.

For example, TradeComet styles itself as a vertical search engine, specifically a potential way for businesses to find the suppliers they require, and a potential competitor to Google, to the extent that companies use TradeComet, not Google, to find desired resources. In ongoing antitrust litigation in the United States, TradeComet claims

Google violated the Sherman Act by increasing TradeComet's prices from $0.05–$0.10 per click to $5–$10 per click, overnight. TradeComet says Google attributed the price increases to "landing page quality." But TradeComet claims Google itself had recently awarded TradeComet "site of the week," and says recognition from others was similarly positive—countering any suggestion that TradeComet was undesirable or low quality.

Foundem (of Bracknell, UK) made similar allegations. Foundem says Google dramatically reduced the prominence of organic (ordinary, unpaid) links to Foundem's site, which dropped overnight from top 10 to number 100 or lower for certain terms in Google, while remaining highly ranked (as high as number 1) in Yahoo and Bing searches for the same terms. Foundem also bought advertising placements from Google, but found it faced dramatically increased prices: Foundem says prices spiked from around 5p to £5, a 100-fold increase, overnight (Foundem, 2009).

Foundem attributes its penalties to Google manually cutting Foundem's "quality score" rating (Foundem, 2009). But quality scores are not available to the public, so it is difficult to confirm these allegations except through litigation and discovery. That said, Google policies indicate penalties for sites with "little or no original content" (Google, 2009). On one view, many such sites are traps that seek to ensnare users within mazes of advertisements. Yet the Google search service itself offers little or no original content; instead, Google links to content hosted elsewhere. Indeed, a lack of original content is distinctively characteristic of vertical search sites, like TradeComet and Foundem, that seek to compete with Google. Would-be competitors therefore take these Google exclusions and penalties to be an improper barrier to competition. This aspect of ranking remains a subject of dispute.

As Google develops offerings in new sectors, additional sites have expressed concern at the competitive implications of Google's ranking practices. For example, searches for hotels and restaurants now often yield prominent Google places. Meanwhile, links to other travel and dining sites, such as TripAdvisor and Yelp, have become somewhat less prominent. Expedia (corporate parent of TripAdvisor) recently criticized these changes (Catan, 2010), as did Yelp (Barnett, 2011)—each alleging that Google's prominent placement of its own links and apparent demotion of competitors' listings constitute an improper leveraging of Google's dominance in algorithmic search.

MATCHING DISPLAY ADS TO USERS AND SITES

In the realm of search ads, a user's search request provides most of the information required to select suitable advertisements. But in the area of display ads, a user's requests provide significantly less context. Knowing what webpage a user is viewing often does not suggest what commercial offers the user would be likely to accept.

Matching is made more difficult by the preferences of both advertisers and consumers. From an advertiser's perspective, sites are importantly different. Users at some sites may be significantly more likely to accept an advertiser's solicitation. Furthermore, some sites may be viewed as inappropriate for an advertiser's offer, for example due to inclusion of offensive, adult, or copyright-infringing material.

Meanwhile, from a user's perspective, ads are also importantly different. Some ads offer products or services users actually want or need. But other ads resort to trickery or deception to attract consumers' attention (see e.g. Edelman, 2009a).

Most display ad platforms offer relatively limited methods of matching advertisers with sites and users. Typically, platforms begin by excluding placements where the advertiser or site has rejected a counterpart specifically or through various characteristics viewed as undesirable. For example, Edelman (2009a) explores the various characteristics by which Yahoo Right Media allows sites to exclude ads that are deceptive, distracting, or otherwise undesirable. Then, platforms sort ads from highest expected revenue to lowest, conditioning on the advertiser and/or ad, the site, and sometimes an interaction between advertiser/ad and site. As a user browses a site, the site's chosen ad platform typically begins by showing the ad with highest expected revenue, then onwards to ads expected to yield lower revenue. If the site uses multiple ad platforms, the site typically attempts to pass each ad placement to the platform expected to pay the most for that placement, and some third-party services aim to assist sites in this effort.

To date, matching rules have been binary, without any notion of pricing or compensating differentials. For example, a publisher typically must either allow or deny a category of ads (e.g. ads that play sounds, deceptive ads), but the publisher ordinarily cannot demand an increased fee for showing disfavored ads. Similarly, an advertiser must either allow or reject placement of its ads on a given site, but ad platforms typically give the advertiser no clear mechanism to demand a lower price for ads placed on a site viewed as less desirable. In this context, it may seem natural to introduce prices for disfavored placements: Prices would increase complexity, but would also reduce deadweight loss by facilitating placements that current rules discard. That said, added payment for placement of unethical or otherwise undesirable ads may be viewed as repugnant (Roth, 2007). Moreover, such payments might have legal consequences. So far, sites have not faced legal liability for showing deceptive ads.[1] However, if sites could

---

[1] Google was sued for deceptive advertisements, namely ads for "free" ringtones that actually carried substantial monthly charges. However, Google presented a successful defense grounded in the Communications Decency Act §230, which prohibits treating the provider of an interactive computer service (here, Google) as the publisher of information provided by an independent entity (here, the advertiser who submitted the deceptive ad). See *Goddard v. Google*. N.D.Ca. 2008, Case No. 5:2008cv02738.

be shown to charge extra for deceptive ads, they would reveal that the publisher is both aware of the problem and, in an important sense, culpable.[2]

AD NETWORKS AND SYNDICATION

Advertisers typically prefer to buy online advertising in large blocks from known partners, so intermediaries organize multiple sites into *networks*. By helping advertisers buy placements on small to mid-sized sites, networks help fund such sites—fueling the diversity of web content. Furthermore, networks reduce transaction costs by aggregating many small sites into a single item that an advertiser can buy with a single contract and a single payment.

Information disclosure in ad networks

Ad networks present a clear question of disclosure of lists of participating sites. When buying online ad placements, advertisers naturally want to know where their ads appear. Some ad networks provide lists of their member sites. But most networks see a strategic downside in providing advertisers with site lists: With a site list, an advertiser could bypass the network—contacting member sites and negotiating direct placements that deny the network compensation for its effort in suggesting the placement. Citing this concern, many networks use a "blind" information structure—selling placements on a bundle of sites without telling advertisers which sites are included.

It is unclear whether the risk of bypass merits keeping network site lists confidential. For large advertisers running ads on just a few sites, bypassing a network might offer financial benefits sufficient to justify the effort. But such bypasses would require sacrificing networks' serving, tracking, contracting, and payment functions, which would require considerable effort to replace. Moreover, if networks' sole concern is bypass, they have other tools at their disposal. For example, affiliate network LinkShare requires that an advertiser commit not to run any affiliate marketing activities through competing networks, while affiliate network Commission Junction prohibits an advertiser from bypassing the network for any relationship initially brokered by the network.

An alternative explanation for blind networks comes from member sites that advertisers would not approve, if an advertiser's approval were requested. By keeping its member

---

[2] In *Gucci America, Inc. v. Frontline Processing Corp.*, 09 Civ. 6925 (HB), credit-card processing companies were held liable for contributory trademark infringement when they charged extra fees to "high risk" sellers selling counterfeit merchandise.

list confidential, a network can avoid advertiser scrutiny of its sites—thereby letting the network include sites of mixed desirability.

Pricing in ad networks

When a network bundles placements on multiple websites, billed to advertisers without itemization of included sites, a network must allocate payments within the network. If some sites will be paid more than others, what measure will allocate value among sites? Will each impression or each click yield an equal payment? Or are some impressions or clicks more valuable than others?

If a network pays the same price for each impression or each click, it risk under-paying sites where traffic is particularly valuable, that is, particularly likely to lead to purchases or other desired outcomes. If these top sites then leave, the network would retain only average to below-average sites—an unraveling that would reduce advertisers' valuation of the network's traffic. Indeed, there is some evidence for such unraveling: The web's top publishers often sell much of their advertising space directly to advertisers; they report that networks offer lower revenue than direct rela-tionships. At the same time, a few premium networks (e.g. Quigo) promise spe-cial care in selecting member sites, yielding higher revenues to sites that make the cut.

In response, networks recognize a need to offer different payments to different pub-lishers. For example, Google describes its "smart pricing" as follows: "If our data shows that a click is less likely to turn into business results (e.g. [an] online sale ...), we may reduce the price [an advertiser] pay[s] for that click" (Google, 2004). That said, it is difficult for networks to condition payments on user behavior at advertisers' sites. For one, such conditioning requires combining multiple data sources, including outcomes of many advertisers' ads on many publishers' sites. Furthermore, advertisers often view post-click outcomes as confidential, lest networks know advertisers' results and raise prices when results are favorable.

Intermediary counts and the prospect of disintermediation

Early intuition on online markets anticipated disintermediation—that online markets would let contracting parties eliminate brokers and middle men (Bambury, 1998). But disintermediation has not been the dominant outcome in online advertising, especially not in display advertising. Rather, a drop in transaction costs makes it easier and more common to build lengthy relationships not often seen in other contexts. For example, an advertiser's ad might pass through half a dozen brokers en route to a publisher's site—each taking a cut as small as a few per cent, such that even these complex relationships may leave adequate surplus to the ultimate buyer and seller. On the other hand, lengthy relationships reduce accountability when an ad ends up misplaced (e.g. Edelman, 2007), while also slowing ad loading times and sometimes yielding lost impressions or error messages.

MEASUREMENT, MISMEASUREMENT, AND FRAUD

Measuring the value of an ad placement

To optimize their spending, advertisers typically seek to assess the value of an advertisement placement—then buy more of the placements that seem to offer the largest value relative to cost. Simple as it sounds, such measurement often proves difficult. In principle, advertisers can measure the ratio of impressions or clicks to sales, including the gross profit from such sales, thereby calculating the benefit attributable to a given placement. But this measurement calls for an online sales process—a poor fit for those selling through offline channels. Offline sellers can attempt to collect data on ad effectiveness by collecting leads online, for example by asking would-be car-buyers to submit their contact information for referral to a local dealer. But customers often decline to submit such leads, adding bias or requiring ad hoc manual adjustments.

Most measurement assumes that, without an advertising expenditure, subsequent sales would not have occurred. For example, if a user clicks an ad and then makes a purchase, a typical measurement concludes that the ad "caused" the purchase—asserting that, without the ad, the purchase would not have occurred, and asserting that other advertising efforts did nothing to cause the sale. This assumption tends to reduce the apparent value of display ads, which often offer delayed benefits to advertisers. For example, a user might see an ad on a news site, then begin to consider a possible future purchase of the advertised product (Fowler, 2007). This assumption similarly discounts the value of offline advertising (TV, print, billboard, etc.), which is also hard to tie to specific purchases. Conversely, this assumption tends to increase the apparent value of search ads, which often immediately precede a purchase. For example, a user looking to buy a laptop might search for "laptop" or even "Thinkpad x300 laptop" right before completing the purchase. Yet the user running such a search might well buy the specified laptop even if no ad were presented. Thus, from the perspective of the advertiser, the relevant comparison may be "pay for the ad and sell the product" versus "don't pay for the ad, yet still sell the product." In that context, the ad may be poor value for money. Yet most measurement systems nonetheless assume that online advertising directly and solely causes subsequent purchases.

Moreover, all manner of spyware, adware, typosquatting sites, and other interlopers can claim to have referred customers who actually requested a merchant specifically and by name, as detailed in the subsequent section.

Advertising fraud

Delivered purely electronically, through computer systems without in-person checks or well developed verifications, online advertising can suffer from a variety of frauds, unjustified charges, and other complications. For example, a site can load many banner ads in invisible windows—then charge advertisers for the resulting "impressions" even though users could not see the resulting ads (Edelman, 2006b). Through spyware or adware installed on users' computers, or through certain JavaScripts within ordinary webpages, sites can fake or simulate pay-per-click ad clicks—imposing costs on advertisers that pay by the click (Edelman, 2006a). Rogue affiliate marketers can invisibly invoke affiliate links so that they receive commission on subsequent purchases from the corresponding affiliate advertisers—managing to overcharge even advertisers that chose what was believed to be a fraud-proof or low-fraud channel (Edelman, 2007).

For most advertisers, measurement efforts are the best defense against improper charges. But sophisticated fraudsters can manipulate the figures most advertisers measure. For example, if a display advertiser is wary of placements with a high ratio of impressions to clicks (too few clicks relative to the number of impressions), the fraudster can fake both clicks and impressions. If a pay-per-click advertiser is measuring the ratio of sales to clicks, the fraudster can design its systems to target users already likely to make a purchase from a given advertiser—for example, by faking clicks when the user is already at the advertiser's site (Edelman, 2006a). The resulting costs can be substantial. For example, June 2010 indictments allege that Brian Dunning and Shawn Hogan stole some $5 million and $15 million from eBay through eBay's Commission Junction affiliate program; the indictments allege that these affiliates actually sent eBay worthless traffic, yet eBay's measurement systems deemed them eBay's two largest and most productive affiliates.

Incentives, both between firms and within firms, sometimes dull efforts to uncover advertising fraud. Most large advertisers buy online ads through agencies which are paid on a commission basis. Catching fraud would reduce the measured spending and hence reduce the agency's commission—requiring an investment of time and effort yielding *lower* payment to the agency. Networks' incentives are also attenuated: In the long run, advertisers will distrust networks with a reputation for fraud. But in the short run, networks can increase revenue by retaining unsavory placements that increase volume.

Furthermore, within-firm incentives invite advertisers' staff to ignore or tolerate fraud. For many buyers of online advertising, the prestige of a position comes in part from the size of the budget under management—limiting the incentive to exclude fraudulent spending which would reduce budgets. Furthermore, some buyers face leveraged incentives that sharply discourage clean-up. For example, some companies pay their affiliate managers based on year-over-year growth of the programs they operate. Ejecting fraud would cut spending and yield a disproportionate drop in compensation. Finally, where a buyer has been defrauded, that person may hesitate to come forward, on the view that admitting the problem would reveal a personal failure. In a forthcoming draft, Edelman attempts to measure some of these effects based on variation in staff and network compensation schemes.

Ad placement arbitrage

Industry participants often use the term "arbitrage" to describe buying ad placements from a low-cost source, then showing ads through a network that offers higher payments. If both placements are equally desirable, such arbitrage might equalize prices across markets, improving efficiency and increasing surplus. But if a seller offers lower prices because its placements are of lower quality, resale of these resources to a high-paying buyer does not constitute "arbitrage" as economists use the term. Rather, such resale is more likely to constitute misrepresentation of a low-quality resource as a high-quality resource (Edelman, 2005).

MULTIHOMING, COMPETITION, AND BARRIERS

Ashlagi et al. (2010) reported that 58% of search advertisers use only Google, not Yahoo or Microsoft adCenter. This is arguably puzzling because, from an advertiser's perspective, competing search ad services seem to be at least orthogonal if not complementary: Some users favor one search engine, while others use another, and an advertiser who forgoes a top ad platform fails to reach those users who rely on the corresponding search engine. Prices cannot explain this puzzle because Google has both the most advertisers and the highest prices (Edelman, 2009b).

Instead, it seems advertisers distinctly favor Google because, despite Google's higher prices, Google offers access to more users and to a larger volume of searches. Ashlagi et al. (2010) show that the advertisers that use all of Google, Yahoo, and Microsoft are significantly larger than the advertisers that use just one or two of these platforms. Ashlagi attributes this difference to transaction costs: advertisers using multiple platforms face extra costs, including signup costs, copying and updating ads, monitoring performance, and adjusting bids.

In principle, advertisers could use automated software systems to copy their campaigns from one ad platform to another—avoiding most costs of transferring and updating ads. Each ad platform provides an application programming interface (API) to let advertisers and tool-makers update and check ads and bids. That said, Google's API contract historically limited how advertisers may use this API—prohibiting tools that copy ads from one platform to another. Edelman (2008) argues that these restrictions are an improper barrier to advertisers seeking to use smaller ad platforms. The restrictions remained in place until Google's January 2013 settlement with the US Federal Trade Commission (FTC), in which Google agreed to eliminate these provisions.

CONSUMER PROTECTION: DISCLOSURES AND DECEPTION

Online advertising raises all manner of consumer protection issues. For one, must advertisements be labeled as such? The FTC has called for "clear and conspicuous disclosures" that listings are advertisements, particularly in contexts such as search advertising, where users may reasonably fail to recognize advertisements as such. Through late 2010, most search engines used terms like "sponsored links" to label their advertisements. In an online experiment, Edelman and Gilchrist (2010) show that the more detailed label "paid advertisement" reduces users' clicks on ads by 23–26%, with drops particularly pronounced for users with low income, low education, and little online experience. Meanwhile, Edelman (2010) critiques Google's advertisement label, "Ads," pointing out that the new label is so tiny that it substantially fits within an "o" of Google, among other shortfalls.

Some pay-per-click advertisements seek to deceive or defraud users—for example, promising "free ringtones" when in fact the service carries a substantial charge. Edelman (2006c) documents all manner of such schemes. However, in *Goddard v. Google*, 640 F. Supp 2d 1193 (N.D.Cal. 2009), Google was found to be not responsible for deceptive ads it sold space for and presented to users—even when Google charged for each advertisement, was aware of the untrue statements, and even encouraged the deception through, for example, a "keyword suggestion tool" that suggested describing ringtones as "free." This decision reflects an interpretation of the Communications Decency Act §230, which instructs that a website must not "be treated as the publisher or speaker of any information provided by" anyone else.

OPEN QUESTIONS

The contracts, institutions, and norms of online advertising continue to evolve. Innovation continues even on questions as fundamental as when an advertiser pays—with new payment metrics based on "view-throughs" (a CPM–CPA hybrid requiring an impression followed by a conversion) and "impressions per connection" (a CPM–CPC hybrid charging advertisers for impressions, but providing bonus impressions if click-throughs are sufficiently frequent). These metrics alter incentives for advertisers and publishers, addressing some of the problems with standard approaches but simultaneously creating new concerns. With so much in flux, there remains ample opportunity to identify new metrics that better satisfy participants' requirements.

Meanwhile, Google's market share continues to grow—exceeding 90% of search volume in scores of countries. Does Google's auction mechanism fully determine prices? Or can Google use its increasing popularity to increase prices to advertisers and otherwise enjoy its market power?

The structure of online advertising markets is closely linked to issues of general public concern. For example, despite the rise of online advertising, newspapers receive significantly less revenue for readers reached online rather than in print. But newspapers serve important public functions, so online advertising shortfalls prompt a need to revisit the future of journalism. Funding newspapers through online ads is particularly challenging because it is often unclear what ads are most suitable: What advertiser seeks a placement adjacent to news of war, election, or natural disaster? Some ads could be selected based on a user's prior activities rather than current browsing, but this approach calls for collecting and retaining ever more information about users' activities. Balancing these concerns—while satisfying users, advertisers, publishers, and various intermediaries— presents challenging questions at the intersection of economics, computer science, law, and public policy.

References

Ashlagi, I., Edelman, B. and Lee, H. S. (2010) "Competing ad auctions: multi-homing and participation costs," Harvard Business School Working Paper No. 10-055.

Bambury, P. (1998) "A taxonomy of Internet commerce," *First Monday*, 3(10), October 5.

Barnett, E. (2011) "Google issues ultimatum to yelp: free content or no search indexing," *Telegraph*, March 1.

Cary, M., Das, A., Edelman, B., Giotis, I., Heimerl, K., Karlin, A., Mathieu, C. and Schwarz, M. (2007) "Greedy bidding strategies for keyword auctions," *Proceedings of the 8th ACM Conference on Electronic Commerce*.

Catan, T. (2010) "Travel sites ally to block Google deal," *Wall Street Journal*, pp. 262–271 October 25.

Edelman, B. (2005) "How Yahoo funds spyware," August 31, at <http://www.benedelman. org/news/083105-1.html>.

—— (2006a) "The spyware–click–fraud connection—and Yahoo's role revisited," April 4, at <http://www.benedelman.org/news/040406-1.html>.

—— (2006b) "Banner farms in the crosshairs," June 12, at <http://www.benedelman.org/ news/061206-1.html>.

—— (2006c) "False and deceptive pay-per-click ads," October 9, at <http://www. benedelman.org/ppc-scams>.

—— (2007) "Spyware still cheating merchants and legitimate affiliates," May 21, at <http://www.benedelman.org/news/052107-1.html>.

—— (2008) "PPC platform competition and Google's 'May not copy' restriction," June 27, at <http://www.benedelman.org/news/062708-1.html>.

—— (2009a) "False and deceptive display ads at Yahoo's right media," January 14, at <http://www.benedelman.org/rightmedia-deception>.

—— (2009b) "Towards a bill of rights for online advertisers," September 21, at <http://www. benedelman.org/advertisersrights>.

—— (2010) "A closer look at Google's advertisement labels," November 10, at <http://www. benedelman.org/adlabeling/google-nov2010.html>.

_____ and Gilchrist, D. (2010) "'Sponsored links' or 'advertisements'? Measuring labeling alternatives in internet search engines," HBS Working Paper No. 11-048.

_____ and Ostrovsky, M. (2007) "Strategic bidder behavior in sponsored search auctions. With Michael Ostrovsky," *Decision Support Systems*, 43(1): 192–8.

_____ and Schwarz, M. (2010) "Optimal auction design and equilibrium selection in sponsored search auctions," *American Economic Review*, 100(2): 597–602.

_____ Ostrovsky, M. and Schwarz, M. (2007) "Internet advertising and the generalized second price auction: selling billions of dollars worth of keywords," *American Economic Review*, 97(1): 242–59.

Feldman, J. and Muthukrishnan, S. (2008) "Algorithmic methods for sponsored search advertising," in Z. Liu and C. H. Xia (eds), *Performance Modeling and Engineering*, Springer, Chapter 4.

Foundem (2009) "Search neutrality—Foundem's Google story," August 18, at <http://www.searchneutrality.org/foundem-google-story>.

Fowler, J. (2007) "Overlap's impact on reach, frequency and conversions," Mimeo, June 5, Atlas Institute.

Google (2004) "Google AdWords news archive—April 2004," April, at <https://www.google.com/intl/en_us/adwords/select/news/sa_mar04.html>.

_____ (2009) "Little or no original content," Google Webmaster Central, June 10, at <http://www.google.com/support/webmasters/bin/answer.py?hl=en&answer=66361>.

Hansell, S. (2001) "Paid placement is catching on in web searches," *New York Times*, June 4.

Lahaie, S., Pennock, D. M., Saberi, A. and Vohra, R. V. (2007) "Sponsored search auctions," in N. Nisan, T. Roughgarden, E. Tardos, and V. Vazirani (eds), *Algorithmic Game Theory*, Cambridge University Press, Chapter 28.

Liu, D., Chen, J. and Whinston, A. B. (2008) "Current issues in keyword auctions," SSRN Working Paper No. 1008496.

Roth, A. (2007) "Repugnance as a constraint on markets," *Journal of Economic Perspectives*, 21(3): 37–58.

Stern, A. (2010) "8 ways to improve your click-through rate," iMedia Connection, February 1, at <http://www.imediaconnection.com/content/25781.asp>.

Varian, H. (2009) "Introduction to the Google ad auction," March 11, at <http://www.youtube.com/watch?v=K7loa2PVhPQ>.

Yao, S. and Mela, C. F. (2009) "A dynamic model of sponsored search advertising," SSRN Working Paper No. 1285775.

CHAPTER 16

....

VERY-LARGE-SCALE GENERALIZED COMBINATORIAL MULTI-ATTRIBUTE AUCTIONS

Lessons from Conducting $60 Billion of Sourcing

....

TUOMAS SANDHOLM[1]

INTRODUCTION

....

DRAWING from our experiences of designing and fielding over 800 sourcing auctions totaling over $60 billion, I will discuss issues that arise in very-large-scale generalized combinatorial auctions, as well as solutions that work (and ones that do not). These are by far the largest (in terms of the number of items as well as the number of side constraints) and most complex combinatorial markets ever developed.

I will discuss how combinatorial and multi-attribute auctions can be soundly hybridized. I will address preference and constraint expression languages for the bidders and the bid taker, as well as techniques for effectively using them. I will discuss scalable optimization techniques for the market clearing (a.k.a. winner determination) problem. I will also address a host of other issues that this work uncovered, and I will study the

[1] I thank the CombineNet employees for helping make the vision presented in this chapter a reality. Special thanks go to Subhash Suri, David Levine, Paul Martyn, Andrew Gilpin, Rob Shields, Bryan Bailey, Craig Boutilier, David Parkes, George Nemhauser, Egon Balas, Yuri Smirnov, and Andrew Fuqua. Early versions of parts of this chapter appeared in my 2007 article in *AI Magazine* (Sandholm, 2007), and the section on automated supply network configuration is largely based on our 2006 paper in *Interfaces* (Sandholm et al., 2006).

significant efficiency gains and other benefits that followed. While the experiences are mainly from sourcing, I believe that the lessons learned apply to many other combinatorial reverse auctions, combinatorial auctions, and combinatorial exchanges.

HISTORICAL BACKDROP ON SOURCING

Sourcing, the process by which companies acquire goods and services for their operations, entails a complex interaction of prices, preferences, and constraints. The buyer's problem is to decide how to allocate the business across the suppliers. Sourcing professionals buy several trillion dollars' worth of goods and services yearly.

Traditionally, sourcing decisions have been made via manual in-person negotiations. The advantage is that there is a very expressive language for finding, and agreeing to, win–win solutions between the supplier and the buyer. The solutions are implementable because operational constraints can be expressed and taken into account. On the downside, the process is slow, unstructured, and non-transparent. Furthermore, sequentially negotiating with the suppliers is difficult and leads to suboptimal decisions. (This is because what the buyer should agree to with a supplier depends on what other suppliers would have been willing to agree to in later negotiations.) The one-to-one nature of the process also curtails competition.

These problems have been exacerbated by a dramatic shift from plant-based sourcing to global corporate-wide (category-based rather than plant-based) sourcing since the mid-1990s. This transition is motivated by the desire of corporations to leverage their spend across plants in order to get better pricing and better understanding and control of the supply network while at the same time improving supplier relationships (see, e.g. Smock 2004). This transition has yielded significantly larger sourcing events that are inherently more complex.

During this transition, there has also been a shift to electronic sourcing where prospective suppliers submit offers electronically to the buyer. The buyer then decides, using software, how to allocate its business across the prospective suppliers. Advantages of this approach include speed of the process, structure and transparency, global competition, and simultaneous negotiation with all suppliers (which removes the difficulties associated with the speculation about later stages of the negotiation process, discussed earlier).

The most famous class of electronic sourcing systems—which became popular in the mid-1990s through vendors such as FreeMarkets (now part of Ariba), Frictionless Commerce (now part of SAP), and Procuri (now part of Ariba)—is the *reverse auction*. The buyer groups the items into lots in advance, and conducts an electronic descending-price auction for each lot. The lowest bidder wins. (In some cases "lowness" is not measured in terms of price, but in terms of an ad hoc score which is a weighted function that takes into account the price and some non-price attributes such as delivery time and reputation.)

Reverse auctions are not economically efficient, that is, they do not generally yield good allocation decisions. This is because the optimal bundling of the items depends on the suppliers' preferences (which arise, among other considerations, from the set, type, and time-varying state of their production resources), which the buyer does not know at the time of lotting. Lotting by the buyer also hinders the ability of small suppliers to compete. Furthermore, reverse auctions do not support side constraints, yielding two drastic deficiencies: (1) the buyer cannot express her business rules, thus the allocation of the auction is unimplementable and the "screen savings" of the auction do not materialize in reality; and (2) the suppliers cannot express their production efficiencies (or differentiation), and are exposed to bidding risks. In short, reverse auctions assume away the complexity that is inherent in the problem, and dumb down the events rather than embracing the complexity and viewing it as a driver of opportunity. It is therefore not surprising that there are strong broad-based signs that reverse auctions have fallen into disfavor.

THE NEW PARADIGM: *EXPRESSIVE COMMERCE*

In 1997 it dawned on me that it is possible to achieve the advantages of both manual negotiation and electronic auctions while avoiding the disadvantages. The idea is to allow supply and demand to be expressed in drastically more detail (as in manual negotiation) while conducting the events in a structured electronic marketplace where the supply and demand are algorithmically matched (as in reverse auctions). The new paradigm, which we called *expressive commerce* (or *expressive competition*), was so promising that I decided to found a company, CombineNet, Inc. to commercialize it.

I began technology development in 1997 and founded the company in 2000. I then served as its Chairman and Chief Technology Officer/Chief Scientist. I left the company after its acquisition in 2010. It continues to operate under the same name.

In expressive commerce, the finer-grained matching of supply and demand yields higher allocative efficiency (i.e. a win–win between the buyer and the suppliers in aggregate). However, matching the drastically more detailed supply and demand is an extremely complex combinatorial optimization problem. We developed the fastest algorithms for optimally solving this problem (discussed later). These algorithms are incorporated into the market-clearing engine at the core of our flagship product, the Advanced Sourcing Application Platform.

Expressive commerce has two sides: *expressive bidding* and *expressive allocation evaluation* (also called *expressive bid taking*) (Sandholm and Suri 2001), which I now describe.

EXPRESSIVE BIDDING

With *expressive bidding*, the suppliers can express their offers creatively, precisely, and conveniently using expressive and compact statements that naturally correspond to the suppliers' business rules, production constraints, efficiencies, etc. Our expressive bidding takes several forms. Our software supports the following forms of expressive bidding, among others, all in the same event.

- Bidding on an arbitrary number of self-constructed packages of items (rather than being restricted to bidding on predetermined lots, as in basic reverse auctions). The packages can be expressed in more flexible and more usable forms than that supported in vanilla combinatorial auctions. For example, the bidder can specify different prices on the items if the items are accepted in given proportions, and the bidder can specify ranges for these proportions, thus allowing an exponential number of packages to be captured by one compact expression.
- Conditional discount offers. Both the trigger conditions and the effects can be specified in highly flexible ways. For example, the trigger conditions can specify whether they should be evaluated before or after the effects of the current discount and other discounts are taken into account.
- Rich forms of discount schedules. Simpler forms of discount schedules have already been addressed in the literature (Sandholm and Suri, 2001a, 2002; Hohner et al., 2003). Figure 16.1 shows a fairly basic example. Richer forms allow the bidder to submit multiple discount offers and to control whether and how they can be combined. Also, discount triggers can be expressed as dollars or units, and as a percentage or an absolute.
- A broad variety of side constraints—such as capacity constraints (Sandholm and Suri, 2001a).
- Multi-attribute bidding (Sandholm and Suri, 2001a). This allows the buyer to leave the item specification partially open, so the suppliers can pick values for the item

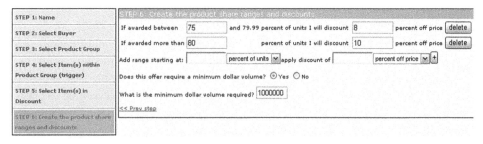

FIGURE 16.1. A relatively simple discount schedule. This screenshot is from an actual sourcing event of ours. The figure shows STEP 6 of the process. The scope of the trigger of the discount (STEP 4) can be different than the scope of the items to which the discount is to be applied (STEP 5).

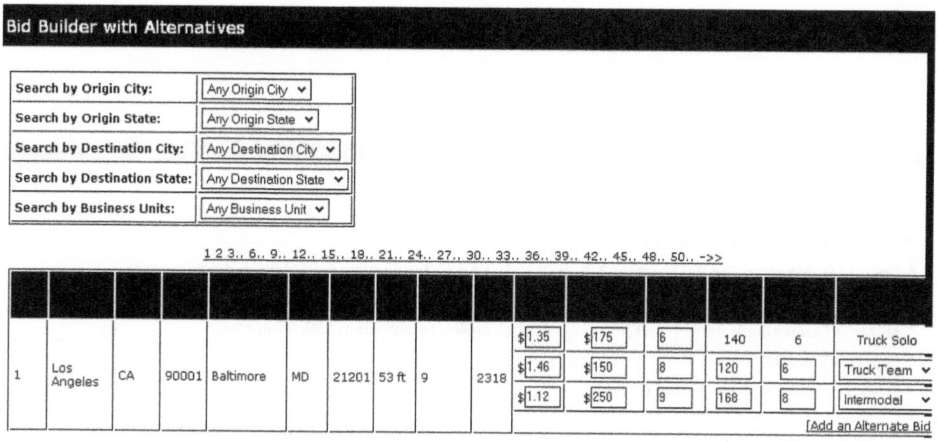

Bid Builder with Alternatives

Search by Origin City:	Any Origin City ▾
Search by Origin State:	Any Origin State ▾
Search by Destination City:	Any Destination City ▾
Search by Destination State:	Any Destination State ▾
Search by Business Units:	Any Business Unit ▾

1 2 3.. 6.. 9.. 12.. 15.. 18.. 21.. 24.. 27.. 30.. 33.. 36.. 39.. 42.. 45.. 48.. 50.. ->>

1	Los Angeles	CA	90001	Baltimore	MD	21201	53 ft	9	2318	$1.35	$175	6	140	6	Truck Solo
										$1.46	$150	8	120	6	Truck Team ▾
										$1.12	$250	9	168	8	Intermodal ▾

[Add an Alternate Bid

FIGURE 16.2. A simple example of bidding with alternates, cost drivers, attributes, and constraints. This piece of screen is from an actual event for where we sourced truckload transportation. The figure shows part of the bid by one bidder on one item.

attributes—such as material, color, and delivery date—in a way that matches their production efficiencies. This is one way in which the suppliers can also express alternate items.

- Free-form expression of alternates. This fosters unconfined creativity by the suppliers.
- Expression of cost drivers. In many of our events, the buyer collects tens or hundreds of cost drivers (sometimes per item) from the suppliers. By expressing cost drivers, the bidder can concisely and implicitly price huge numbers of items and alternates. Figures 16.2 and 16.3 illustrate bidding with attributes and cost drivers.

All of these expressive bidding features of our software have been extensively used by our customers. The software supports bidding through both web-based interfaces and spreadsheets. In some cases, catalog prices from databases have also been used.

Our expressive bidding approach is flexible in the sense that different suppliers can bid in different ways, using different offer constructs. In fact, some suppliers may not be sophisticated enough to bid expressively at all, yet they can participate in the same sourcing events using traditional bidding constructs in the same system. This paves a smooth road for adoption, which does not assume sudden process changes at the participating organizations.

Benefits of expressive bidding

The main benefit of expressive bidding is that it leads to greater efficiency of the allocation. In business terms, it creates a win–win between the buyer and the suppliers in aggregate. There are several reasons for this.

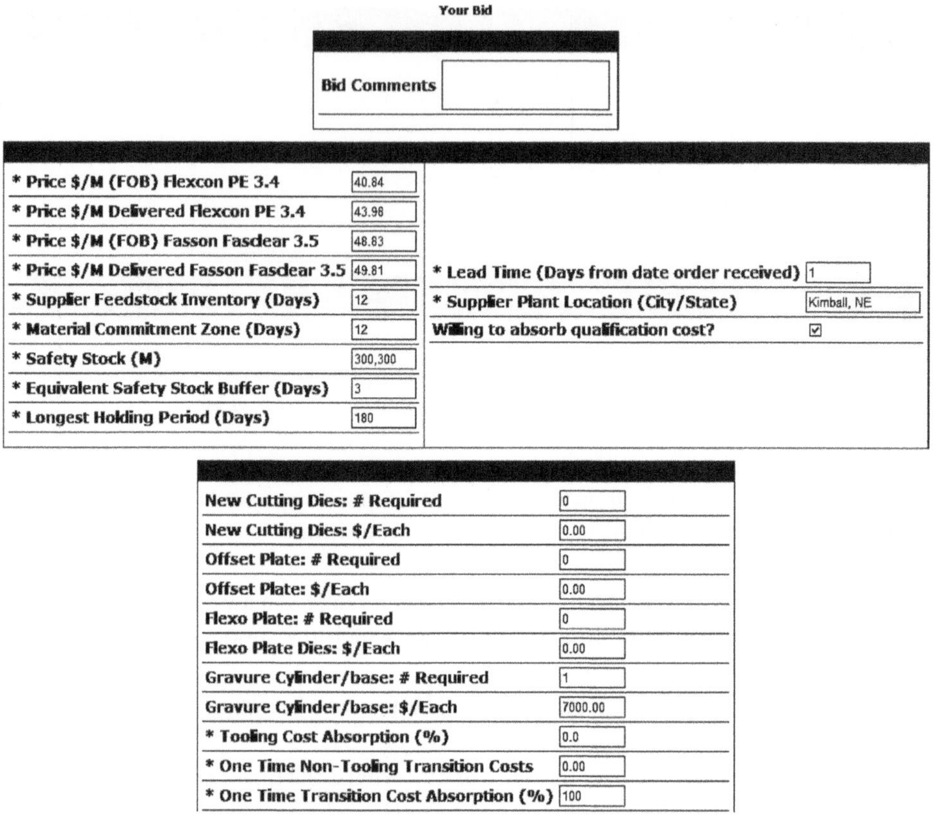

Your Bid

Bid Comments	

* Price $/M (FOB) Flexcon PE 3.4	40.84
* Price $/M Delivered Flexcon PE 3.4	43.98
* Price $/M (FOB) Fasson Fasclear 3.5	48.83
* Price $/M Delivered Fasson Fasclear 3.5	49.81
* Supplier Feedstock Inventory (Days)	12
* Material Commitment Zone (Days)	12
* Safety Stock (M)	300,300
* Equivalent Safety Stock Buffer (Days)	3
* Longest Holding Period (Days)	180

* Lead Time (Days from date order received) 1
* Supplier Plant Location (City/State) Kimball, NE
Willing to absorb qualification cost? ☑

New Cutting Dies: # Required	0
New Cutting Dies: $/Each	0.00
Offset Plate: # Required	0
Offset Plate: $/Each	0.00
Flexo Plate: # Required	0
Flexo Plate Dies: $/Each	0.00
Gravure Cylinder/base: # Required	1
Gravure Cylinder/base: $/Each	7000.00
* Tooling Cost Absorption (%)	0.0
* One Time Non-Tooling Transition Costs	0.00
* One Time Transition Cost Absorption (%)	100

FIGURE 16.3. An example of bidding with cost structures and attributes. This is part of an actual event where we sourced printed labels.

First, because the suppliers and the buyer can express their preferences completely (and easily), the market mechanism can make better allocation decisions (economically more efficient and less wasteful), which translates to higher societal welfare. In other words, the method yields better matching of supply and demand because they are expressed in more detail. The savings do not come from lowering supplier margins, but from reducing economic inefficiency. With expressive bidding, the suppliers can offer specifically what they are good at, and at lower prices because they end up supplying in a way that is economical for them. (They can consider factors such as production costs and capacities, raw-material inventories, market conditions, competitive pressures, and strategic initiatives.) This creates a win–win solution between the suppliers and the buyer. For example, in the sourcing of transportation services, a substantial increase in economic efficiency comes from bundling multiple deliveries in one route (back-haul deliveries and multi-leg routes). This reduces empty driving, leading to lower transportation costs and yielding environmental benefits as well: lower fuel consumption, less driver time, less frequent need to replace equipment, and less pollution.

Second, suppliers avoid exposure risks. In traditional inexpressive markets, the suppliers face exposure problems when bidding. That makes bidding difficult. To illustrate this point, consider a simple auction of two trucking tasks: the first from Pittsburgh to Los Angeles, and the second from Los Angeles to Pittsburgh. If a carrier wins one of the tasks, he has to factor in the cost of driving the other direction empty. Say that his cost for the task then is $1.60 per mile. On the other hand, if he gets both tasks, he does not have to drive empty, and his cost is $1.20 per mile. When bidding for the first task in an inexpressive auction, it is impossible to say where in the $1.20–$1.60 range he should bid, because his cost for the first task depends on whether he gets the second task, which in turns depends on how other carriers will bid. Any bid below $1.60 exposes the carrier to a loss in case he cannot profitably win the second task. Similarly, bidding above $1.20 may cause him to lose the deal on the first task, although it would be profitable to take on that task if he wins the second task. In an expressive auction, the buyer can price each of the tasks separately, and price the package of them together, so there is no exposure problem. (For example, he can bid $1.60 per mile for the first task, $1.60 per mile for the second task, and $1.20 per mile for the package of both tasks. Of course, he can also include a profit margin.) Therefore bidding is easier: the bidder does not have to speculate what other suppliers will bid in the later auctions. Also, the tasks get allocated optimally because no bidder gets stuck with an undesirable bundle, or misses the opportunity to win when he is the most efficient supplier. Furthermore, when there is an exposure problem, the suppliers hedge against it by higher prices. Removal of the bidders' exposure problems thus also lowers the buyer's sourcing cost.

Third, by expressive bidding with side constraints (such as capacity constraints), each supplier can bid on all bundles of interest without being exposed to winning so much that handling the business will be unprofitable or even infeasible. This again makes bidding easier because—unlike in inexpressive markets—the supplier does not have to guess which packages to commit his capacity to. (In an inexpressive market, making that guess requires counterspeculating what the other suppliers are going to bid, because that determines the prices at which this supplier can win different alternative packages.) This also leads to more efficient allocations compared to those in inexpressive markets because in those markets each bidder needs to make guesses as to what parts of the business he should bid on, and those might not be the parts for which he really is the most efficient supplier.

Fourth, expressive bidding allows more straightforward participation in markets because the strategic counterspeculation issues that are prevalent in non-combinatorial markets can be mitigated, as discussed earlier.[2] This leads to wider access to the benefits of ecommerce because less experienced market participants are raised to an equal playing field with experts. This yields an increase in the number of market participants, which itself leads to further economic efficiency and savings in sourcing costs. Broader

---

[2] In fact, with full expressiveness, in principle, truthful bidding can be made a dominant strategy by using Vickrey–Clarke–Groves pricing. However, as I will discuss later, that does not seem practical here.

access also stems from the buyer not lotting the items and thus facilitating competition from small suppliers as well.

Fifth, in basic reverse auctions, the buyer has to pre-bundle items into lots, but he cannot construct the optimal lotting because it depends on the suppliers' preferences. With expressive commerce, items do not have to be pre-bundled. Instead, the market determines the optimal lotting (specifically, the optimizer determines the optimal allocation based on the expressive bids and the expressions from the buyer). This way, the economically most efficient bundling is reached, weeks are not wasted on pre-bundling, and suppliers that are interested in different bundles compete. As a side-effect, small suppliers' bids taken together end up competing with large suppliers.

Sixth, expressive bidding fosters creativity and innovation by the suppliers. This aspect is highly prized by both the suppliers and buyers. It can also yield drastic savings for the buyer due to creative construction of lower-cost alternates.

Overall, expressive bidding yields both lower prices and better supplier relationships. In addition to the buyers (our customers), suppliers are also providing very positive feedback on the approach. They especially like that they (1) also benefit from expressive bidding (unlike in traditional reverse auctions, where their profit margins get squeezed), (2) can express their production efficiencies, and (3) can express differentiation and creative offers. In fact, suppliers like expressive commerce so much that they agree to participate in expressive commerce even in events that they boycotted when basic reverse auctions had been attempted. Furthermore, perhaps the best indication of supplier satisfaction is the fact the suppliers are recommending the use of our approach and software to buyers.

The benefits of expressiveness can be further enhanced by multiple buyers conducting their sourcing in the same event. This provides an opportunity for the bidders to bundle across the demands of the buyers, and also mitigates the exposure risks inherent in participating in separate events. As an example, in spring 2005 we conducted an event where Procter & Gamble and its two largest customers, Walmart and Target, jointly sourced their North America-wide truckload transportation services for the following year (Sandholm et al., 2006). This enabled the carriers to construct beneficial backhaul deliveries and multi-leg routes by packaging trucking lanes across the demand of the three buyers. This was a large event. Procter & Gamble's volume alone exceeded $885 million.

EXPRESSIVE ALLOCATION EVALUATION BY THE BID TAKER

The second half of expressive commerce is *expressive allocation evaluation*, where the bid taker (i.e. buyer in the case of sourcing) expresses preferences over allocations using a rich, precise, and compact language that is also natural in the buyer's business. It can be used to express legal constraints, business rules, prior contractual obligations, and strategic considerations.

In our experience, different types of side constraints offer a powerful form of expressiveness for this purpose. For example, the buyer can state: "I don't want more than 200 winners (in order to avoid overhead costs)," "I don't want any one supplier to win more than 15% (in order to keep the supply network competitive for the long term)," "I want minority suppliers to win at least 10% (because that is the law)," "Carrier X has to win at least \$3 million (because I have already agreed to that)," etc. Our system supports hundreds of types of side constraints.

Our system also has a rich language for the buyer to express how item attributes (such as delivery date or trans-shipment specifications) and supplier attributes (such as reputation) are to be taken into account when determining the allocation of business (Sandholm and Suri, 2001b).

A professional buyer—with typically no background in optimization—can set up a *scenario* in our system by adding constraints and preferences through an easy-to-use web-based interface. A simple example is shown in Figure 16.4. To set up each such expression, the buyer first chooses the template expression (e.g. "I don't want more than a certain number of winners," or "I want to favor incumbent suppliers by some amount") from a set of expressions that have been deemed potentially important for the sourcing event in question by the person who configured the event within our software tool. He then selects the *scope* to which that expression should apply: everywhere, or to a limited set of items, bid rounds, product groups, products, sites, and business groups. Finally, he selects the exact parameter(s) of the constraint, for example exactly how many winners are allowed. Constraints and preferences can also be uploaded from business rule databases. Once the buyer has defined the scenario consisting of side constraints and preferences, he calls the optimizer in our system to find the best allocation of business to suppliers under that scenario.

Our software takes these high-level supply and demand expressions, *automatically* converts them into an optimization model, and uses sophisticated tree search algorithms to solve the model. We have faced scenarios with over 2.6 million bids (on 160,000 items, multiple units of each) and over 300,000 side constraints, and solved them to optimality.

Scenario navigation

Once (at least some of) the bids have been collected, the buyer can engage in *scenario navigation* with the system. At each step of that process, the buyer specifies a set of side constraints and preferences (these define the scenario), and runs the optimizer to find an optimal allocation for that scenario. This way the buyer obtains a quantitative understanding of how different side constraints and preferences affect the sourcing cost and all other aspects of the allocation. In our system, the buyer can change/add/delete any number of side constraints and preferences in between optimizations.

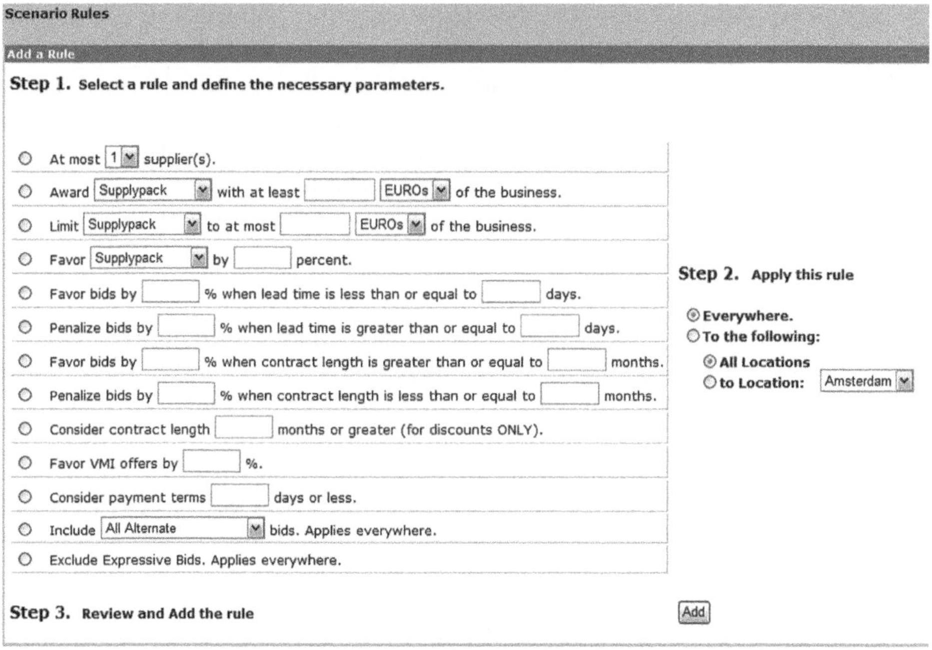

Scenario Rules

Add a Rule

Step 1. Select a rule and define the necessary parameters.

○ At most [1 ▼] supplier(s).

○ Award [Supplypack ▼] with at least [   ] [EUROs ▼] of the business.

○ Limit [Supplypack ▼] to at most [   ] [EUROs ▼] of the business.

○ Favor [Supplypack ▼] by [   ] percent.

○ Favor bids by [   ] % when lead time is less than or equal to [   ] days.

○ Penalize bids by [   ] % when lead time is greater than or equal to [   ] days.

○ Favor bids by [   ] % when contract length is greater than or equal to [   ] months.

○ Penalize bids by [   ] % when contract length is less than or equal to [   ] months.

○ Consider contract length [   ] months or greater (for discounts ONLY).

○ Favor VMI offers by [   ] %.

○ Consider payment terms [   ] days or less.

○ Include [All Alternate ▼] bids. Applies everywhere.

○ Exclude Expressive Bids. Applies everywhere.

Step 2. Apply this rule

⦿ Everywhere.
○ To the following:
  ⦿ All Locations
  ○ to Location: [Amsterdam ▼]

Step 3. Review and Add the rule   [Add]

FIGURE 16.4. A user interface for expressive allocation evaluation by the bid taker. Every sourcing event has different expressiveness forms (selected from a library or preconfigured in a template). This particular sourcing event was one of the simpler ones in that relatively few expressions
of constraints and preferences were included in the user interface. The buyer uses the interface
as follows. First, on the left (STEP 1), he selects which one of the expressiveness forms he wants
to work on, and sets the parameter(s) for that form. Then, on the right (STEP 2) he selects the
scope to which this rule is to be applied. In STEP 3, he presses the "Add" button to add the rule,
and a restatement of the rule appears in natural language for verification (not shown). The buyer
can then add more rules to the same scenario by repeating this process. Finally, the buyer presses
the "Optimize" button (not shown) to find the optimal allocation for the scenario. This triggers
the automated formulation of all these constraints and preferences—together with all the bid
information that the bidders have submitted—into an optimization problem, and the solving of
the problem via advanced tree search.

Studying our sourcing events around 2006, we found that a buying organization using
our system will navigate an average of about 100 scenarios per sourcing event. The
maximum we saw by then was 1,107. To navigate such large numbers of scenarios, fast
clearing is paramount.

Rapid clearing enables scenario navigation to be driven by the actual data (offers). In
contrast, most prior approaches required the scenario (side constraints and preferences,
if any) to be defined prior to analysis; there were insufficient time and expert modeling
resources to try even a small number of alternative scenarios. Data-driven approaches
are clearly superior because the actual offers provide accurate costs for the various
alternative scenarios.

Benefits of expressive allocation evaluation

Through side constraints and preference expressions, the buyer can include business rules, legal constraints, logistical constraints, and other operational considerations to be taken into account when determining the allocation. This makes the auction's allocation *implementable* in the real world: the plan and execution are aligned because the execution considerations are captured in the planning.

Second, the buyer can include prior (e.g. manually negotiated) contractual commitments in the optimization. This begets a sound hybrid between manual and electronic negotiation. For example, he may have the obligation that a certain supplier has to be allocated at least eighty truckloads. He can specify this as a side constraint in our system, and the system will decide which eighty truckloads (or more) are the best ones to allocate to that supplier in light of all other offers, side constraints, and preferences. This again makes the allocation implementable. (A poor man's way of accomplishing that would be to manually earmark a particular part of the business to the prior contract(s). Naturally, allowing the system to do that earmarking with all the pertinent information in hand yields better allocations.)

Third, the buyer obtains a quantitative understanding of the tradeoffs in his supply network by scenario navigation; that is, by changing side constraints and preferences and reoptimizing, the buyer can explore the tradeoffs in an objective manner. For example, he may add the side constraint that the supply base be rationalized from 200 to 190 suppliers. The resulting increase in procurement cost then gives the buyer an understanding of the tradeoff between cost and practical implementability. As another example, the buyer might ask: If I wanted my average supplier delivery-on-time rating to increase to 99%, how much would that cost? As a third example, the buyer might want to see what would happen if he allowed a supplier to win up to 20% of the business instead of only 15%. The system will tell the buyer how much the sourcing cost would decrease. The buyer can then decide whether the savings outweigh the added long-term strategic risks such as vulnerability to that supplier's default and the long-term financial downside of allowing one supplier to become dominant.

Fourth, quantitative understanding of the tradeoffs also fosters stakeholder alignment on the procurement team, because the team members with different preferences can base their discussion on facts rather than opinions, philosophies, and guesswork. The buyer is typically not an individual but an organization of several individuals with different preferences over allocations. Finance people want low sourcing cost, plant managers want small numbers of suppliers, marketing people want a high average carrier-delivery-on-time rating, etc. Scenario navigation enables the organization to better understand the available tradeoffs.

Feedback to bidders, and its interaction with scenario navigation

Allowing the bid taker to conduct scenario navigation introduces interesting issues. If scenario navigation is allowed, the sourcing mechanism is not uniquely defined from the perspective of the bidders. This is because scenario navigation by the bid taker changes the rules of the game for the bidders. Therefore, one could argue that a sourcing mechanism with scenario navigation is not an auction at all. Beyond semantics, this also affects the bidders' incentives, and this is one reason why we (and to my knowledge all other expressive sourcing vendors nowadays) use the first-price (i.e. pay your winning bids) pricing rule instead of mechanisms that are incentive compatible without scenario navigation, notably the Vickrey–Clarke–Groves (VCG) mechanism. While our market-clearing engine supports both first-price and VCG pricing, none of our customers has wanted to use the latter. Therefore, in the sourcing platform we offer the former.

Scenario navigation also has an interesting interaction with quotes:[3] if the bid taker changes the rules later on, then how should the price quote on a bundle (or any other kind of quote) be computed? The quote depends on the scenario, but when the scenario is not (yet) fixed, quotes are ill-defined.

One practical solution that we used is to provide quotes (or other feedback) based on the current scenario with the understanding that the feedback may not be accurate in light of the final scenario. This is in line with the standard understanding that quotes are not accurate anyway because they may change as other bidders change their bids.

Another, very practical, but limited, approach that our software offers is to provide quotes of the kind that do not depend on the scenario, such as giving bidders feedback on their bids on individual items only (e.g. "you are currently the seventh-lowest bidder on item x" or "your bid on item x is $3.54 higher than the lowest bid"). One can also offer feedback to bidders on what items are sparsely covered by bids; that helps bidders focus their bidding on "opportunities".

A third possibility in our system, which is sometimes used by our customers for fast auctions with only tens of items, is to force the bid taker to lock in the scenario before bidding starts. That enables accurate quotes but precludes the bid taker from conducting scenario navigation in a data-driven way based on the bids.

Another possible approach is to constrain how the bid taker can change the scenario. For instance, he may be allowed to tighten, but not loosen, constraints during the sourcing process. Depending on the constraints, non-trivial upper or lower bounds on quotes may then be well defined, and even bounds can give useful guidance to bidders on which bundles they should focus their valuation determination and bidding effort.

Our system supports sealed-bid events (winners are determined at the end), events that have multiple (usually two or three) rounds (winners are determined and feedback provided at the end of each round), and "live" events (winners are determined and feedback provided every time any participant expresses anything new). All three formats have been used extensively by our customers. In the live and multi-round formats, we require that a bidder's change in his offer not make the allocation worse (for one, otherwise early bids would be meaningless because they could be pulled out).[4] An idea

---

[3] Sandholm (2002a) discusses how one can compute quotes for bundles in a combinatorial auction.

[4] For example, a bidder can lower his price(s) or increase his capacity(ies).

for the future is to give detailed feedback only at the end of each bidding round but coarser feedback all the time. One could even use "mini-rounds" within each round for medium-granularity feedback, and so on.

Our customers have typically not been very concerned about collusion because the number of suppliers tends to be large, the bid taker often has the most negotiation power, and he knows what prices should be expected quite well. Live events run the risk of supporting collusion, while sealed-bid events tend to deter collusion because the parties of a potential coalition have an opportunity to "stab each other in the back" without the other parties observing their actions until the auction is over—when it is too late to respond. The multi-round format tries to get the best of both worlds: giving bidders feedback between rounds in order to help focus their valuation determination and bidding effort, while at the same time having a back-stabbing opportunity at the end.

It is not known what kind of feedback is best in order to minimize sourcing cost. Under stylized assumptions in the single-item auction setting, the revenue equivalence theorem states that the open-cry format and sealed-bid format yield equal expected revenue. However, if bidders' valuation distributions are asymmetric, either can have a higher expected revenue than the other (Maskin and Riley, 2000). In one metal-sourcing event we knew that one of the suppliers had a cost significantly below the rest. We therefore chose a sealed-bid format so that supplier could not know exactly how low he needed to bid to win; rather, he would ensure winning by going lower than that. I will discuss research related to revenue-maximizing (cost-minimizing) combinatorial auctions in the last section of this chapter.

Additional tools for the bid taker

Sometimes the buying organization can get carried away with controlling the allocation with side constraints to the extent that no feasible allocation exists. To help the bid taker in that situation, we developed a *feasibility obtainer* as an extension of our market-clearing (a.k.a. winner determination) technology where the optimizer finds a minimum-cost relaxation of the constraints if the scenario is overconstrained. Each constraint can have a different percentage relaxation cost. The feasibility obtainer can yield allocations that have high sourcing cost. To address this, we developed an *optimal constraint relaxer*. It finds a solution that minimizes the sum of sourcing cost and constraint relaxation cost.

Sometimes the bid taker can contact bidders to encourage certain bids, but this takes time and uses up favors with the bidders. Using optimization we developed a methodology and algorithms for deciding what favors to ask. In addition to the usual inputs to the market-clearing optimization, the bid taker can state how he expects he could improve the suppliers' offers by negotiating with them, and then asks the optimizer how he should negotiate in light of all of the inputs so as to minimize his total sourcing cost. For example, "If I can improve five discount schedules by 2% each, which five should I negotiate?"

Our system also provides coverage feedback to the bid taker so he can encourage bidders to submit bids on items that are thinly covered by bids.

The bid taker typically has the option to source items by means other than the current auction. He can buy from existing catalog prices, negotiate manually, or hold another auction. Which items should be bought in the current auction and which should be sourced in these other ways? Our system enables the bid taker to optimize this decision. This is accomplished by simply inserting *phantom bids* into our system to represent the cost at which items (or bundles) can be sourced by means other than the current auction. The clearing algorithm then optimizes as usual; whatever items are won by phantom bids are sourced by the other means. This approach is important because it allows some items not to be sourced in the auction if their bid prices are too high. This is a way of accomplishing automated demand reduction, and is used in most of our auctions. In some of our sourcing events—mainly single-item multi-unit ones—we also used another form of automated demand reduction, where the bid taker specifies a price-quantity curve and if the current point is above the curve, demand is automatically reduced.

Time to contract in expressive commerce

The time to contract is reduced from several months to weeks because no manual lotting is required, all suppliers can submit their offers in parallel, what-if scenarios can be rapidly generated and analyzed, and the allocation is implementable as is. This causes the cost savings to start to accrue earlier, and decreases the human hours invested.

Automated item configuration

An additional interesting aspect of bidding with cost drivers and alternates (e.g. using attributes) is that the market-clearing (a.k.a. winner determination) algorithm not only decides who wins, but also ends up optimizing the configuration (setting of attributes) for each item. In deciding this, the optimizer, of course, also takes into account the buyer's constraints and preferences.

Automated supply network configuration

In many of the sourcing events we conducted, we did not commit to a particular supply network up front, but rather collected offers for pieces of all possible supply networks. We then let winner determination—as a side effect—optimize the supply network multiple levels upstream from the buyer. This is in sharp contrast to the traditional approach, where the supply network is designed first, and then one sources to the given network.

As an example of this new paradigm, in a sourcing event where Procter & Gamble (P&G) sourced in-store displays using our hosting service and technology, we sourced items from different levels of the supply network in one event: buying colorants and cardboard of different types, the service of printing, the transportation, the installation services, etc. (Sandholm et al., 2006). Some suppliers made offers for some of those individual items while others offered complete ready-made displays (which are, in effect, packages of the lower-level items), and some bid for partial combinations. The market clearing determined the lowest-cost solution (adjusted for P&G's constraints and preferences) and thus, in effect, configured the supply network multiple levels upstream.

I will now discuss this event in more detail. P&G uses pre-packed displays to help retailers merchandise its products. A display can contain different sizes of one product or contain multiple products, for example Crest® toothpaste and Scope® mouthwash. Retail stores place displays in the aisles or in promotional areas when there is some special activity, such as a sale and/or a coupon for the brand. P&G spends $140 million annually in North America on these displays.

Based on individual product-promotion schedules and display requirements, managers typically used incumbent suppliers to design, produce, and assemble turnkey displays for easy setup in the stores. While these solutions were of high quality, there was little visibility into the costs and quality of alternate methods. P&G's corporate sourcing team thought that there could be a more efficient way to source displays, and wanted to understand the cost tradeoffs between buying the traditional turnkey displays and buying components, leveraging the size of P&G's entire operations.

Process

The P&G–CombineNet project team developed and executed a sourcing implementation designed to allocate P&G's annual spending on displays across a more efficiently utilized supplier base, while also improving the reliability and quality of display production and services. The plan contained three key elements:

- A bidding structure designed to capture component-specific information.
- A simple way for suppliers to understand and participate in the bidding process.
- Advantages for P&G's product managers that encouraged them to embrace the new process.

P&G's purchasing department invited all of the incumbents and some new suppliers to bid on the company's annual volume of displays. P&G's new capability to collect detailed cost information and solicit expressive or creative offers from suppliers allowed the purchasing organization to put up for bid each of the supply network cost drivers that contributed to the final cost of the display, such as display components as well as assembly and shipping costs that increase the base cost of the display materials. The purchasing department collected detailed information on the costs of materials, such as corrugated paper, film, and trays that hold the product, the costs of holding inventory, of freight, and of printing. It invited suppliers to bid on specification and then to make alternate off-specification bids that would allow suppliers to suggest ways to reduce the cost of the display. (For example, using three-color printing instead of four-color printing for the header card, which advertises the product, would reduce its cost.)

Of the forty suppliers that participated in the sourcing event, some were manufacturers only, some were assemblers only, and some could manufacture and assemble. There were four display categories (pallets, special packs, pigment/dye/quick trays, and wings and floor stands) covering fourteen benchmark and unique displays. For roughly 500 display components, suppliers offered piece prices, substrate fluctuations, other fixed and variable costs, assembly rates, packaging, and freight. There were two online rounds of bidding followed by one round of offline negotiation.

For suppliers, the flexibility of component-based bidding and the unique expressive bidding format allowed them to bid on their own terms, including volume discounts, bundled pricing, and alternate products or services. P&G encouraged the suppliers to submit two sets of bids, one identifying prices for full turnkey displays (including the aspects of production handled by others in their alliance networks) and a second bid for only those display components and services they could supply directly.

For P&G, the larger, more complex set of data generated greater business insight when analyzed using our scenario navigation tool that enabled P&G to quickly and easily consider a large number of what-if scenarios by changing side constraints and preferences.

Results

The unconstrained savings were nearly 60% compared to the previous year's prices. The implementable savings (that is, the savings P&G could achieve after applying its side constraints and preferences) were nearly 48% ($67 million annually). The collaborative planning produced insights into costs and strengthened P&G's relationships with its suppliers. P&G's annual procurement cycle dropped from twenty to eight weeks, with the time for finding allocations to scenarios reduced from days to seconds.

P&G used our scenario navigation to assess the cost impact of constraints and preferences, such as favoring incumbent suppliers, and the cost of different mixes of display components. P&G gained the ability to separate the true cost of must-have components and services from nice-to-haves. This let P&G compare the cost of a supplier's turnkey display to the total cost of sourcing the display as its components and then managing the process. P&G realized it could allocate much of its spending more efficiently.

The bidding and award process also improved P&G's relationships with its suppliers by promoting collaboration and allowing suppliers to leverage their strengths. Our expressive bidding format gave suppliers an opportunity to bid on their own terms and did not commoditize their offerings. Both P&G and its suppliers benefited from a consolidated and easy-to-manage sourcing cycle.

EXPRESSIVE COMMERCE AS A GENERALIZATION
OF COMBINATORIAL AUCTIONS

A relatively simple early form of expressive commerce was a *combinatorial reverse auction* (Sandholm et al., 2002), where the only form of expressiveness that the suppliers have is package bidding, and the buyer has no expressiveness. A predecessor of that was a *combinatorial auction* where the bidders are the buyers (and there is only one unit of each item and no side constraints). Combinatorial auctions (Rassenti et al., 1982; Sandholm 1993, 2002b; Ledyard et al., 1997; Rothkopf et al., 1998; Kwasnica et al., 2005; Sandholm et al., 2005; Sandholm and Suri, 2003; Hoos and Boutilier, 2000; Boutilier 2002; deVries and Vohra, 2003) enable bidders to express complementarity among items (the value of a package being more than the sum of its parts) via package bids. Substitutability (the value of a package being less than the sum of its parts) can also be expressed in some combinatorial auctions, usually using different languages for specifying mutual exclusivity between bids (Sandholm, 2002a,b; Fujishima et al., 1999; Nisan, 2000; Hoos and Boutilier, 2001).

Expressiveness leads to more economical allocations of the items because bidders do not get stuck with partial bundles that are of low value to them. This has been demonstrated, for example, in auctions for bandwidth (McMillan, 1994; McAfee and McMillan, 1996), transportation services (Sandholm, 1991, 1993, 1996; Caplice and Sheffi, 2003), pollution rights, airport landing slots (Rassenti et al., 1982), and carrier-of-last-resort responsibilities for universal services (Kelly and Steinberg, 2000).

However, package bids and exclusivity constraints are too impoverished a language for real-world sourcing. While any mapping from bundles to real numbers can be expressed in that language in principle, the real-world preferences in sourcing cannot be easily, naturally, and concisely expressed in it. Starting in 1997, we tackled this challenge and generalized the approach to expressive commerce, with the language constructs discussed earlier. Similar approaches have recently been adopted by others, but for drastically less complex events (orders of magnitude smaller and less expressive) (Hohner et al., 2003; Metty et al., 2005).

The use of our richer expressiveness forms (rather than mere canonical package bids with exclusivity constraints) is of key importance for several reasons:

- Bidders can express their preferences in the language that is natural in their domain.
- Bidders can express their preferences concisely. To illustrate this point, consider the following simple example. A bidder has no production efficiencies and thus has a fixed price for each item regardless of what other items he produces. However, he has a capacity constraint. In our bidding language, he can simply express a price for each item and a capacity constraint. In contrast, in the classical combinatorial

auction bidding languages, the supplier would have to submit bids for an exponential number of packages.

- Due to this conciseness, the bids are easy to communicate to the bid taker.
- Our bidding constructs maintain the natural structure of the problem (rather than mapping this structure into a format that allows only package bids with exclusivity constraints). The clearing algorithms take advantage of that structure in many ways, for example in generating cutting planes, deciding what variables to branch on, and so on.

OPTIMIZATION TECHNIQUES TO ENABLE
EXPRESSIVE COMMERCE

A significant challenge in making expressive commerce a reality is that the expressiveness makes the problem of allocating the business across the suppliers an extremely complex combinatorial optimization problem. Specifically, the *clearing problem* (a.k.a. *winner determination problem*) is that of deciding which bids to accept and reject (and to what extent in the case of partially acceptable bids) so as to minimize sourcing cost (adjusted for preferences) subject to satisfying all the demand and all side constraints. Even in the vanilla combinatorial reverse auction where the only form of bidding is package bidding, and no side constraints or preferences are allowed, the clearing problem is NP-complete (Rothkopf et al. 1998) and inapproximable in the worst case in polynomial time (Sandholm et al., 2002). Expressive commerce is a much richer problem; thus the NP-hardness and inapproximability carry over. (Müller et al. (2006) review the worst-case complexity of the clearing problem of different variants of combinatorial auctions). Thus sophisticated techniques are required.

Prior to our system, no technology was capable of solving clearing problems of the scale and expressiveness that our customers required; for example, Hohner et al. (2003) found integer programming techniques to be effective for problems only as large as 500 items and 5,000 bids. In 2001, P&G gave us a trial instance of trucking services sourcing that took a competing optimization product thirty minutes to solve. Our system solved it optimally in nine seconds. While that was already a decisive speed difference, since that time our technology development has yielded a further speed improvement of two to three orders of magnitude.

There is significant structure in the expressive commerce problem instances, and it is paramount that the optimizer be able to take advantage of the structure. Mixed integer programming (MIP) techniques, which use tree search, are quite good at this, and our software takes advantage of them. However, the techniques embodied in the leading general-purpose MIP solvers alone are not sufficient for the clearing problem.

Our system uses sophisticated tree search to find the optimal allocation. Given that the problem is NP-complete, in the worst case the run-time is super-polynomial in the size of the input (unless P=NP). However, in real-world sourcing optimization the algorithms run extremely fast: the median run time is less than a second and the average is twenty seconds, with some instances taking days. The algorithms are also anytime algorithms: they provide better and better solutions during the search process.

I began algorithm development in 1997, and over ten years CombineNet grew to have sixteen people on my team working on the algorithms, half of them full time. The team has tested hundreds of techniques (some from the artificial intelligence and operations research literature and some invented by us) to see which ones enhance speed on expressive commerce clearing problems. Some of the techniques are specific to market clearing, while others apply to combinatorial optimization more broadly. We published the first generations of our search algorithms (Sandholm, 2002a; Sandholm and Suri, 2003; Sandholm et al., 2005). The new ideas in these algorithms included:

- different formulations of the basic combinatorial auction clearing problem— branching on items (Sandholm, 2002a), branching on bids (Sandholm and Suri, 2003; Sandholm et al., 2005), and multi-variable branching (Gilpin and Sandholm, 2011)
- upper and lower bounding across components in dynamically detected decompositions (Sandholm and Suri, 2003; Sandholm et al., 2005)
- sophisticated strategies for branch question selection (Sandholm, 2002a, 2006; Sandholm and Suri, 2003; Sandholm et al., 2005)
- dynamically selecting the branch selection strategy at each search node (Sandholm, 2006; Sandholm et al., 2005)
- the information-theoretic approach to branching in search (Gilpin and Sandholm, 2011)
- sophisticated lookahead techniques (Sandholm, 2006; Gilpin and Sandholm, 2011)
- solution seeding (Sandholm, 2006)
- primal heuristics (Sandholm, 2006; Sandholm et al., 2005)
- identifying and solving tractable cases at nodes (Sandholm and Suri, 2003; Sandholm et al., 2005; Sandholm, 2006; Conitzer et al., 2004)
- techniques for exploiting *part* of the remaining problem falling into a tractable class (Sandholm, 2006; Sandholm and Suri, 2003)
- domain-specific preprocessing techniques (Sandholm, 2002a)
- fast data structures (Sandholm, 2002a; Sandholm and Suri, 2003; Sandholm et al., 2005)
- methods for handling reserve prices (Sandholm, 2002a; Sandholm and Suri, 2003)
- incremental winner determination and quote computation techniques (Sandholm, 2002a).

Sandholm (2006) provides an overview of the techniques.

We also invented a host of techniques for the search algorithms that we have decided to keep proprietary for now. They include different formulations of the clearing problem, new branching strategies, custom cutting plane families, cutting plane generation and selection techniques, etc.

Since around 2002, we have been using machine learning methods to predict how well different techniques will perform on specific instances at hand. (For this purpose, an instance is represented by about fifty hand-selected numeric features.) This information can be used to select the technique for the instance at hand, to give time estimates to the user, and so on. Our solver has several dozen important parameters and each of them can take on several values. Therefore, our machine learning approach of setting the parameters well on an instance-by-instance basis is significantly more challenging and more powerful than using machine learning to select among a handful of hard-wired solvers, an approach that has been pursued in academia in parallel (e.g. Leyton-Brown et al., 2006). More recently, machine learning-based parameter tuning for optimization algorithms has become a popular research topic in academia (see, e.g., Xu et al., 2011).

While the literature on combinatorial auctions has mainly focused on a variant where the only form of expressiveness is package bidding (sometimes supplemented with mutual exclusion constraints between bids), in our experience with sourcing problems the complexity is dominated by rich side constraints. Thus we have invested significant effort into developing techniques that deal with side constraints efficiently. We have faced several hundred different types of real-world side constraints. Our system supports all of them. We abstracted them into eight classes from an optimization perspective so that the speed improvements we build into the solver for one type of side constraint get leveraged across all side constraint types within the class.

The resulting optimal search algorithms are often 10,000 times faster than others. The main reason is that we specialize on a subclass of MIP problems and have over 100,000 real-world instances on which to improve the algorithms. This speed has allowed our customers to handle drastically larger and more expressive sourcing events. The events have sometimes had over 2.6 million bids (on 160,000 items, multiple units of each) and over 300,000 side constraints.

State-of-the-art general-purpose MIP solvers are inadequate also due to numeric instability. They err on feasibility, optimality, or both, on about 4% of the sourcing instances. We have invested significant effort on stability, yielding techniques that are significantly more robust.

HOSTED OPTIMIZATION FOR SOURCING PROFESSIONALS

Our backend clearing engine, ClearBox, is industry independent, and the interface to it is through our Combinatorial Exchange Description Language (CEDL), an XML-based language that allows ClearBox to be applied to a wide variety of applications by the company and its partners (see Figure 16.5).

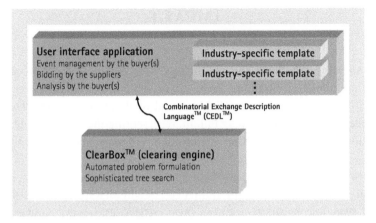

FIGURE 16.5. Advanced sourcing application platform. The platform is hosted on a server farm with multiple instantiations of each component. The system also includes modules for clearing management, server farm management, secure databases, etc. (not shown).

Intuitive web-based interfaces designed for the buyer and for the suppliers bring the power of optimization to users with expertise in sourcing, not in optimization. The users express their preferences through interfaces that use sourcing terminology. The interfaces support simple click-through interaction rather than requiring the user to know special syntax. The approach allows sourcing professionals to do what they are best at (incorporating sourcing knowledge such as strategic and operational considerations) and the optimizer to do what it is best at (sifting through huge numbers of allocations to pick the best one).

For every event, separate front ends are instantiated that support only those bidding and allocation evaluation features that are appropriate for that event. This makes the user interfaces easier and more natural to use by sourcing professionals. User training typically takes a few hours. New front ends typically take a few days or weeks to go from project specification to deployment. Instantiation of a front end can start from a clean slate with the entire configuration space of the system available. As an alternate, we also provide templates that are specific to industry and sourcing category, in order to accelerate the configuration process.

The user interfaces feed CEDL into ClearBox, and ClearBox then *automatically* formulates the optimization problem for our search algorithms. This contrasts with the traditional mode of using optimization, where a consultant with optimization expertise builds the model. The automated approach is drastically faster (seconds rather than months) and avoids errors.

Our web-based products and software-as-a-service (SaaS) business model make optimization available on demand. No client-side software installation is necessary. This also avoids hardware investments by customers. We buy the hardware and leverage it across customers, each with temporary load. The SaaS model allows us to quickly and transparently tune our algorithms, and to provide enhancements to all customers simultaneously. The solution is also offered through consulting firms.

IMPACT

The new sourcing paradigm and technology have already had significant impact. Between December 2001 and June 2010,[5] we used our system to host over 800 highly combinatorial sourcing events, totaling a spend of about $60 billion. The individual events ranged from $2 million to $7 billion, representing the most complex combinatorial auctions ever conducted. They spanned a broad range of categories such as:

- *transportation:* truckload, less-than-truckload, ocean freight, dray, bulk, intermodal, small parcel, air freight, train, fleet, freight forwarding, and other;
- *direct materials:* sugars/sweeteners, meat, vegetables, honey, starches, colorants, fibers/non-wovens, steel, fasteners, solvents, chemicals, casings, resins, and polymers;
- *packaging:* cans/ends, corrugates, corrugated displays, flexible film, folding cartons, labels, foam trays/pads, caps/closures, shrink and stretch films, bags, pulp, pallets, and printed instructions;
- *indirect materials:* management, repair, and operations (a.k.a. MRO) (electrical supplies, filters, pipes/valves/fittings, power transmissions, pumps, safety supplies, office supplies, lab supplies, file folders, solvents, and furnishings), chemicals (cylinder gasses, fuels, and other), technology (laptops/desktops and cameras), leased equipment, fleet vehicles, and promotional items;
- *services:* security, janitorial, legal, patent/trademark, consulting, equipment maintenance, temporary labor, marketing, customization, insurance, shuttling/towing, warehousing, pre-press, and advertising;
- *healthcare:* pharmaceuticals as well as medical/surgical equipment and supplies;
- *telecommunication:* sourcing wireless plans for employees of companies.

From this we conclude that the market design that we created for combinatorial multi-attribute markets indeed has an appropriate level of generality. We believe that it can serve as a model for the structuring of such markets in most applications.

At the point where we had cleared $35 billion of sourcing spend through our system, we conducted a study of those sourcing events. Over sixty buyer companies used our system and they were mostly among the Global 1,000. A total of over 12,000 supplier companies bid in our system. We had delivered hard-dollar savings of $4.4 billion (12.6% of spend) to our customers (i.e. the buyers) in lowered sourcing costs. The savings were measured compared to the prices that the buyer paid for the same items the previous time the buyer sourced them (usually twelve months earlier). The $4.4 billion is the *implementable savings* that the system yielded after the buyer had applied

---

[5] CombineNet was acquired in 2010. The company continues to operate, but my most recent data are as of the acquisition time.

side constraints and preferences; the *unconstrained savings* (which can be viewed as the savings arising from expressive bidding before the buyer expresses constraints and preferences) was $5.4 billion (15.4%). The savings figure is noteworthy especially taking into account that, during the same time period, the prices in the largest segment, transportation, increased by 6–9% in the market overall.

The savings number does not include the savings obtained by suppliers, which are harder to measure because the suppliers' true cost structures are proprietary. However, there is strong evidence that the suppliers also benefited, so a win–win was indeed achieved: (1) suppliers that have participated in expressive commerce events are recommending the use of that approach to other *buyers*, (2) on numerous occasions, suppliers that boycotted reverse auctions (offered by other sourcing system vendors) came back to the "negotiation table" once we introduced expressive commerce for the sourcing event, and (3) suppliers are giving very positive feedback about their ability to express differentiation and provide creative alternatives.

The savings number also does not include savings that stem from reduced effort and compression of the event timeline from months to weeks or even days.

The cost savings were achieved while at the same time attaining the other advantages of expressive commerce discussed earlier, such as better supplier relationships (and better participation in the events), redesign of the supply network, implementable solutions that satisfy operational considerations, and solutions that strike tradeoffs in a data-driven way and align the stakeholders in the buying organization. See also Sandholm et al. (2006).

CombineNet grew to 130 full-time employees (about half of them in engineering) and a dozen academics as advisors. The company has operations on four continents, with headquarters in Pittsburgh, Pennsylvania.

Summary of key lessons learned

In this section I will distill some of the key lessons learned from this experience.

- (Generalization of) combinatorial auctions can be practical even in settings with tens of thousands of items, large numbers of units of each item, hundreds of thousands of side constraints, and millions of bids.
- Combinatorial and multi-attribute auctions can be soundly hybridized into a general domain-independent market design. It can be embodied in easy-to-use, application-independent, hosted software.
- On larger scales, practicality requires a bidding language that is compact and natural. The canonical bidding language of combinatorial auctions—namely package bidding (with forms of exclusivity constraints between packages to express substitutability)—does not scale (if used alone).

- Bidders, as well as the bid taker, almost always have preferences that are so rich that they cannot be modeled in a linear program. Rather, both continuous and discrete variables are necessary for their modeling.
- For scalability of communicating the optimization problem to the market-clearing engine, and for scalability of that optimization, one should retain the structure inherent in the bidding language within the optimization model (rather than expanding that input into a canonical language of package bids and exclusivity constraints).
- By retaining this structure, the clearing problem can typically be solved to provable optimality quickly, so the fact that the problem is NP-complete and worst-case inapproximable in polynomial time is not a prohibitive obstacle. Even on instances where the optimization does not solve to optimality quickly, typically solutions that are provably near optimal are found quickly—ones with much better quality guarantees than those provided by approximation algorithms that run in worst-case polynomial time.
- Typically the bid taker has additional preferences and constraints beyond cost minimization, but these tend to be hard to articulate up front. Rather, bid takers like to conduct scenario navigation with (at least some of) the bids in hand.
- Incentive-compatible mechanisms seem to be undesirable, at least in sourcing. One reason is that such mechanisms are not actually incentive compatible in reality, since related auctions are conducted repeatedly over years with roughly the same bidders and items. Another reason is that scenario navigation compromises incentive compatibility even within a single sourcing event. Additional undesirable aspects of incentive-compatible auctions are discussed in Rothkopf et al. (1990), Sandholm (1996), Conitzer and Sandholm (2006), Ausubel and Milgrom (2006), and Rothkopf (2007). First-price (pay-your-winning-bids) mechanisms are natural and seem to work well.

CHALLENGES AND FUTURE WORK

Our experiences in developing and fielding large-scale combinatorial markets have uncovered many new issues that require further attention. In this section I will discuss some of the main ones.

Is more expressiveness always better?

The experiences covered in the chapter show that increasing the expressiveness offered to the market participants tends to increase the efficiency of the allocation. But is that always so?

We recently developed a theory that ties the expressiveness of mechanisms to their efficiency in a domain-independent manner (Benisch and Sandholm, 2010). We introduced two expressiveness measures, *maximum impact dimension*, which captures the number of ways that an agent can impact the outcome, and *shatterable outcome dimension*, which is based on the concept of *shattering* from computational learning theory. We derived an upper bound on the expected efficiency of any mechanism under its most efficient Nash equilibrium. Remarkably, it depends only on the mechanism's expressiveness. We proved that the bound increases strictly as we allow more expressiveness. We also showed that in some cases a small increase in expressiveness yields an arbitrarily large increase in the bound.

We then showed that in any private-values setting, the bound can always be reached in pure strategy Bayes–Nash equilibrium (while achieving budget balance in expectation). In contrast, without full expressiveness, dominant-strategy implementation is not always possible.

Finally, we studied *channel-based* mechanisms. They restrict the expressions of value through channels from agents to outcomes, and select the outcome with the largest sum. Channel-based mechanisms subsume most combinatorial and multi-attribute auctions, the VCG mechanism, etc. In this class, a natural measure of expressiveness is the number of channels allowed; this generalizes the k-wise dependence measure of expressiveness used earlier in the combinatorial auction literature (Conitzer et al., 2005). We showed that our domain-independent measures of expressiveness increase strictly with the number of channels allowed. Using this bridge, our general results yield interesting implications, and a better understanding of problems such as the exposure problem. For example, even a slight lack of expressiveness can result in arbitrarily large inefficiency—unless agents have no private information.

The results mentioned in this section assume that the mechanism designer can create the mechanism de novo for the allowed expressiveness level. Future research should study the relationship between expressiveness and efficiency (and other objectives such as sourcing cost) when the mechanism designer is forced to stay within some standard allocation and pricing rules.

Revenue-maximizing (cost-minimizing) mechanism design

As an example of how more expressiveness is not always better when operating within a given mechanism family, if one is using VCG, the bid taker can increase expected revenue (analogously, reduce sourcing cost) by reducing expressiveness via careful bundling. To see this, consider a simple auction for an apple and an orange with two bidders. One bidder wants the apple and the other wants the orange. Fully expressive VCG would allocate each item to the bidder who wants it, but there would be no competition and zero revenue. In contrast, if the bid taker bundles the items so they must sell together, VCG revenue will equal the lower of the two bidders' valuations for the bundle.[6,7]

More generally, it is open, even for just two items, how to design a revenue-maximizing combinatorial auction. Designing a revenue-maximizing deterministic combinatorial auction is NP-complete (Conitzer and Sandholm, 2004), so it is unlikely that a short characterization can exist. Instead, *automated mechanism design* holds promise for this (Sandholm et al., forthcoming). Designing the mechanism to make use of the bid taker's prior probability distribution over the suppliers' costs seems to make sense in these settings where roughly the same event is conducted year after year and the bid taker has the opportunity to learn. Of course, the bidders knowing that the bid taker is learning for the purpose of cost minimization in the future may affect their bidding.

Automated scenario navigation

In basic scenario navigation, discussed earlier in this chapter, the bid taker changes her preferences (hard and soft) and reoptimizes. She repeats this over and over (often hundreds of times) until she finds a solution that is satisfactory to her in terms of the tradeoff between her non-monetary preferences and economic value (sourcing cost in the case of sourcing). However, the space of scenarios to navigate is infinite, so there is no guarantee that even a reasonably good solution has been found. Furthermore, a good solution could have been found with much less scenario navigation than the user actually went through.

To address these issues, we pioneered *automated scenario navigation* (Boutilier et al., 2004), and built a prototype of it on top of our sourcing system. Compared to basic scenario navigation, it enables a more systematic and less wasteful navigation of the scenario space. The system queries the sourcing team about their preferences, using, for example, tradeoff queries ("how much hassle would an extra supplier be in dollars?— give me an upper or lower bound in terms of dollars") and comparison queries ("which of these two allocations do you prefer?"). These two kinds of query are shown in Figures 16.6 and 16.7, respectively. The system decides the queries to pose in a *data-directed way* so as to only ask the team to refine its preferences *on an as-needed basis*. (This is desirable because internal negotiation in the team is costly in terms of time and goodwill.) Specifically, based on all the offers that the suppliers have submitted, and all answers to previous queries, the system strives to minimize maximum regret. At each iteration of automated scenario navigation, the system finds a *robust solution* that minimizes maximum regret (the regret is due to the fact that the sourcing team has not fully specified its preferences, so for some preferences that are still consistent with the

---

[6] Vendors of inexpressive reverse auctions often justify bundling as a way to avoid cherry picking, that is, suppliers bidding on just the "desirable" items, which would lead to poor—or no—bid coverage of the "undesirable" items. This argument is incorrect, however, because items' desirability is determined by prices, which are set endogenously by the auction.

[7] Kroer and Sandholm (2013) study optimal bundling in a VCG context.

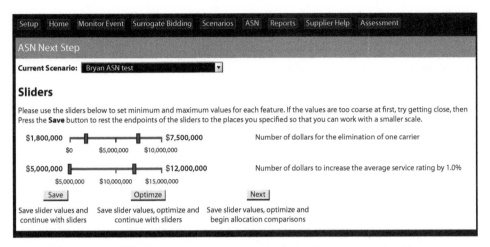

FIGURE 16.6. A tradeoff query in our system. The user gets to adjust the sliders to represent upper and lower bounds (in dollars) on the value of different aspects of allocations. That gives information to the system about her preferences.

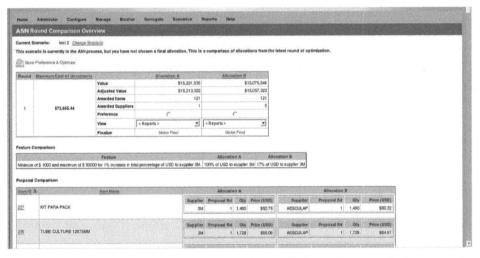

FIGURE 16.7. A comparison query in our system in a medical sourcing event. In the top matrix, the user gets to click on which allocation she prefers. That gives information to the system about her preferences.

answers so far, the system's recommended allocation is not optimal). As the other step of each iteration, the system poses a query to refine the team's preferences in order to be able to reduce the maximum regret further.

The maximum regret also provides a quantitative measure of when further negotiation within the team is no longer worth it, and the team should implement the current robust allocation.

With automated scenario navigation, provably good solutions (i.e. ones with low maximum regret) tend to be found while asking the user relatively little regarding her preferences. On the downside, the optimization problem of finding a regret-minimizing allocation is harder than the clearing problem discussed in most of this chapter. We discuss automated scenario navigation in detail, including its different design dimensions and algorithms, in Boutilier et al. (2004).

Eliciting the bidders' preferences

The section above discussed how the bid taker's preference information can be frugally collected on an as-needed basis to make good allocation decisions. Perhaps even more important is how the bidders' preference information gets collected. Most, but not all, game-theoretic mechanisms are so called *direct-revelation* mechanisms, where each agent (bidder) reveals its type completely up front. The revelation principle states that in the absence of computation/communication limitations, this restriction comes at no cost. However, in practice such mechanisms are problematic because the agents may need to determine their own preferences via costly deliberation, for example computing to generate plans such as routings and schedules (Sandholm, 1993, 1996), or information gathering, and communicating complete preferences may be undesirable from the perspective of privacy or conserving bandwidth. It would be highly desirable to be able to reach the right allocation without requiring all the preference information from all the bidders.

The bidders have to decide which combinations they want to evaluate (and how precisely) and communicate bids on. If they choose bundles which they do not win, evaluation effort and communication are wasted, and unnecessary private information is revealed. Also, economic efficiency will suffer if the bidders do not evaluate bundles on which they would win. However, it is difficult to anticipate which combinations one will win before knowing the others' bids!

As a general way of tackling this, in our academic research we introduced the idea of explicit *preference elicitation in combinatorial auctions* (Conen and Sandholm, 2001). The idea is to supplement the clearing algorithm with an *elicitor* (software) that incrementally queries the agents about their preferences, and builds a model of them. The elicitor decides the next query to ask (and whom to ask it from) based on the answers it has received from the agents so far. Once enough information has been elicited to determine the right (in any given sense) outcome, no more information is elicited.

Preference elicitation in multi-agent systems is fundamentally different from traditional preference elicitation, where there is only one party whose preferences are to be elicited, because what information is needed from an agent depends on what information other agents reveal. We showed that allowing the elicitor to condition the queries it asks of an agent on the answers of another agent (rather than eliciting the agents' preferences separately) can yield exponential savings in communication and preference determination effort.

Furthermore, if the elicitor conditions the query it poses to an agent on the answers given by other agents, the elicitor leaks information about the others' answers to the agent. This introduces the potential for new forms of strategic manipulations not present in single-shot mechanisms. Conen and Sandholm (2001) proposed a method that nevertheless motivates each bidder to answer the elicitor's questions truthfully in ex post equilibrium (so the method does not require any knowledge of prior probability distributions). This is accomplished by eliciting enough information to determine the optimal allocation and the VCG payments. This maintains the truth-promoting property even while allowing bidders to skip questions and to answer questions that were never asked. This yields a *push–pull mechanism* where the revelation of information is guided by where the bidders think they are competitive, and by where the elicitor knows that it does (and does not) need further information.

Preference elicitation in combinatorial auctions has become an entire research area. Sandholm and Boutilier (2006) provide a review, discussing a range of results for various query types and query policies. They also discuss the surprising power of bundle pricing. Ascending combinatorial auctions are a special case of preference elicitation that preceded the general approach. Parkes (2006) provides a review of those mechanisms.

To my knowledge, the general preference elicitation approach is not yet used in real combinatorial auctions. Rather, in our industrial work, we and others have focused on designing bidding languages via which bidders can express their preferences compactly and naturally in a push-only way. Future work should integrate explicit preference elicitation into combinatorial auctions in practice.

Bidding support tools versus extreme expressiveness

There has been quite a bit of discussion about tools for bidding in combinatorial auctions, especially in transportation domains. Bidders spend significant effort constructing and submitting bids, and there is demand for tools to make this easier. There has been some research on developing such tools for bidding in trucking (Sandholm, 1993; Song and Regan, 2005; Ergun et al., undated). A key challenge, though, is that what bids a bidder should submit depends not only on the bidders' local information but also on what others have bid. Therefore, any purely bidder-side bidding support tool is inherently limited, and tools should consider what information other bidders have submitted. This would bring bidding tools close to the preference elicitation approach discussed earlier.

Also, as discussed throughout this chapter, my view is that the market design should allow highly expressive, compact, natural bids rather than mere package bids. Taken to its extreme, under that view a bidder should be able to submit all the information he would submit to the bidding support tool to the auction directly! The auction clearing algorithm would then take the role of the bidding support tool as well, but would be able to make better, optimized decisions because it has the information from all bidders, not just one.

That said, one argument in favor of the current level of expressiveness—which does not require all the details about each bidders' local optimization problem to be submitted to the auction—is that the bidding language serves as an abstraction layer between the bidders and the auction. This allows different ways of generating bids (automated or manual) to be used by different bidders. This open architecture may serve to foster innovation in tools by the bidders. The bidding language that I described in this paper seems to be an appropriate interface layer since it has been successfully used across a wide range of sourcing categories and settings.

Planning versus execution

There are at least two key challenges in combinatorial sourcing auctions related to the discrepancy between planning (i.e. sourcing) and execution (i.e. procurement):

- In (e.g. year-ahead) combinatorial sourcing auctions, aspects of complementarity, substitutability, and feasibility are not known at bidding time. Rather, they become apparent only during the execution of the long-term contracts that the auction is used to construct. For example, if a supplier (carrier) wins ten truckloads per week on a lane from Pittsburgh to Chicago, and ten truckloads per week on a lane from Chicago to Pittsburgh, it is not clear that those can be executed as backhaul deliveries due to execution-time constraints—such as pickup and delivery time windows—that arise dynamically. As another example, once a carrier agrees to a long-term contract via a sourcing event, it does not mean that the carrier always has the capacity available at the moment to take on every load that the buyer calls on him to carry throughout the year based on the contract.
- Current downstream execution systems, *contract management systems* (*transportation management systems* in the case of logistics), do not understand the expressive contracts that a combinatorial auction generates. Rather, such systems assume that item prices are determined by sourcing and then, during execution, procurement personnel can simply order through the contract management systems using those prices.[8] One cannot fix this by simply splitting accepted package bids into item prices because an item's price should depend on what other items it gets procured with. This actually raises a deep issue of what a package bid means: does it mean that the items in the package have to always be procured together, or that the agreed-upon quantities of those items are ordered over the entire procurement period (year), or something in between? Similarly, does a discount trigger based on what is awarded at sourcing time or what is actually procured (in the entire year, per

---

[8] Similar issues occur in upstream *spend analysis systems* because they do not support expressive contracts either.

week, or per month)? Both practices have been used, for example, in truckload transportation sourcing.

One could try to address the problem by allowing bids to have their pricing conditioned on execution-time variables, for example whether lanes execute as a bundle.[9] However, in that setting the market-clearing engine would have to take into account what values those variables might end up taking. This would involve risk management issues. (One cannot postpone the market clearing until execution time; that would undermine the predictability of long-term contracts, which also allow the parties to make situation-specific investments in capacity, etc.) Furthermore, current downstream systems do not support that.

The status quo is that the buyer might not procure the volumes that are sourced. This can occur due to his demand deviating from projections. It can also occur due to certain spot market opportunities. For example, in truckload transportation, certain spot bids are considered acceptable to take in order to reduce deadheading.[10] Similarly, on the supply side, bidders cannot always fulfill the procurement needs that arise related to a contract that they have won. In summary, the contracts that sourcing yields are soft on both sides.

In the future, it would be important to formalize execution issues into expressive contracts themselves, and to monitor them, in order to make the system function more efficiently. That, in turn, begets the need to optimize one's procurement against one's contracts. This can even involve reoptimization in light of the execution state and new projections.[11]

REFERENCES

Ausubel, L. and Milgrom, P. (2006) "The lovely but lonely Vickrey auction," in P. Cramton, Y. Shoham, and R. Steinberg (eds), *Combinatorial Auctions*, MIT Press, ch. 1.

Benisch, M. and Sandholm, T. (2010) "A theory of expressiveness in mechanisms," Mimeo. Significantly extends an AAAI-08 paper.

Boutilier, C. (2002) "Solving concisely expressed combinatorial auction problems," in *Proceedings of the National Conference on Artificial Intelligence (AAAI-02)*.

---

[9] What values the execution-time variables take may be affected by other sourcing events (auctions or manual) that the bidders may be involved in. Thus the terms of a contract in one such auction could depend on the outcomes of other such auctions. However, this would be difficult to monitor and has the potential to lead to agency problems.

[10] The spot market opportunities should then be taken into account in the buyer's sourcing optimization. During sourcing time, they could be modeled stochastically. For example, in other work we have employed stochastic spot market modeling in the optimization used to decide which proposed Internet display advertising campaigns to accept versus reject (Boutilier et al., 2008).

[11] If there are multiple expressive contracts covering some of the same items with the same supplier, a procurement sequence can be interpreted in multiple ways: different (combinations of) contracts could have been used to accomplish that procurement sequence. Some of those alternatives can be better interpretations than others, for example, in terms of which discounts get triggered. Curiously, this begets the opportunity to optimize the past!

Boutilier, C., Sandholm, T. and Shields, R. (2004) "Eliciting bid taker non-price preferences in (combinatorial) auctions," in *Proceedings of the National Conference on Artificial Intelligence (AAAI-04)*.

Boutilier, C., Parkes, D., Sandholm, T. and Walsh, W. (2008) "Expressive banner ad auctions and model-based online optimization for clearing," in *Proceedings of the National Conference on Artificial Intelligence (AAAI-08)*.

Caplice, C. and Sheffi, Y. (2003) "Optimization-based procurement for transportation services," *Journal of Business Logistics*, 24(2): 109–28.

Conen, W. and Sandholm, T. (2001) "Preference elicitation in combinatorial auctions: extended abstract," in *Proceedings of the ACM Conference on Electronic Commerce (EC)*.

Conitzer, V. and Sandholm, T. (2004) "Self-interested automated mechanism design and implications for optimal combinatorial auctions," in *Proceedings of the ACM Conference on Electronic Commerce (EC)*.

———— (2006) "Failures of the VCG mechanism in combinatorial auctions and exchanges," in *Proceedings of the International Conference on Autonomous Agents and Multi-Agent Systems (AAMAS)*. Early version in the Agent-Mediated Electronic Commerce Workshop, 2004.

———— Derryberry, J. and Sandholm, T. (2004) "Combinatorial auctions with structured item graphs," in *Proceedings of the National Conference on Artificial Intelligence (AAAI-04)*.

———— Sandholm, T. and Santi, P. (2005) "Combinatorial auctions with k-wise dependent valuations," in *Proceedings of the National Conference on Artificial Intelligence (AAAI-05)*.

de Vries, S. and Vohra, R. (2003) "Combinatorial auctions: a survey," *INFORMS Journal on Computing*, 15(3): 284–309.

Ergun, Ö., Kuyzu, G. and Savelsbergh, M. (undated) "Bid price optimization for simultaneous truckload transportation procurement auctions," Mimeo.

Fujishima, Y., Leyton-Brown, K. and Shoham, Y. (1999) "Taming the computational complexity of combinatorial auctions: optimal and approximate approaches," in *Proceedings of the International Joint Conference on Artificial Intelligence (IJCAI)*.

Gilpin, A. and Sandholm, T. (2011) "Information-theoretic approaches to branching in search," *Discrete Optimization*, 8: 147–59.

Hohner, G., Rich, J., Ng, E., Reid, G., Davenport, A., Kalagnanam, J., Lee, H and An, C. (2003) "Combinatorial and quantity-discount procurement auctions benefit Mars, Incorporated and its suppliers," *Interfaces*, 33: 23–5.

Hoos, H. and Boutilier, C. (2000) "Solving combinatorial auctions using stochastic local search," in *Proceedings of the National Conference on Artificial Intelligence (AAAI-00)*.

———— (2001) "Bidding languages for combinatorial auctions," in *Proceedings of the International Joint Conference on Artificial Intelligence (IJCAI)*.

Kelly, F. and Steinberg, R. (2000) "A combinatorial auction with multiple winners for universal service," *Management Science*, 46(4): 586–96.

Kroer, C. and Sandholm, T. (2013) "Computational Bundling for Auctions," *Carnegie Mellon University Computer Science Department Technical Report* CMU-CS-13-111.

Kwasnica, A., Ledyard, J., Porter, D. and DeMartini, C. (200) "A new and improved design for multiobject iterative auctions," *Management Science*, 51(3): 419–34.

Ledyard, J., Porter, D. and Rangel, A. (1997) "Experiments testing multiobject allocation mechanisms," *Journal of Economics and Management Strategy*, 6(3): 639–75.

Leyton-Brown, K., Nudelman, E. and Shoham Y. (2006) "Empirical hardness models for combinatorial auctions," in P. Cramton, Y. Shoham, and R. Steinberg (eds), *Combinatorial Auctions*, MIT Press, ch. 19.

Maskin, E. and Riley, J. (2000) "Asymmetric auctions," *Review of Economic Studies*, 67: 413–38.

McAfee, P. and McMillan, J. (1996) "Analyzing the airwaves auction," *Journal of Economic Perspectives*, 10(1): 159–75.

McMillan, J. (1994) Selling spectrum rights," *Journal of Economic Perspectives*, 8(3): 145–62.

Metty, T., Harlan, R., Samelson, Q., Moore, T., Morris, T., Sorensen, R., Schneur, A., Raskina, O., Schneur, R., Kanner, J., Potts, K. and Robbins, J. (2005) "Reinventing the supplier negotiation process at Motorola," *Interfaces*, 35(1): 7–23.

Müller, R., Lehmann, D. and Sandholm, T. (2006) "The winner determination problem," in P. Cramton, Y. Shoham, and R. Steinberg (eds), *Combinatorial Auctions*, MIT Press, ch. 12.

Nisan, N. (2000) "Bidding and allocation in combinatorial auctions," in *Proceedings of the ACM Conference on Electronic Commerce (EC)*, 1–12.

Parkes, D. (2006) "Iterative combinatorial auctions," in P. Cramton, Y. Shoham, and R. Steinberg (eds), *Combinatorial Auctions*, MIT Press, ch. 2.

Rassenti, S., Smith, V. and Bulfin, R. (1982) "A combinatorial auction mechanism for airport time slot allocation," *Bell Journal of Economics*, 13(2): 402–17.

Rothkopf, M. (2007) "Thirteen reasons why the Vickrey–Clarke–Groves process is not practical," *Operations Research*, 55(2): 191–7.

—— Teisberg, T. and Kahn, E. (1990) "Why are Vickrey auctions rare?" *Journal of Political Economy*, 98(1): 94–109.

—— Pekec, A. and Harstad, R. (1998) "Computationally manageable combinatorial auctions," *Management Science*, 44(8): 131–1147.

Sandholm, T. (1991) "A strategy for decreasing the total transportation costs among area-distributed transportation centers," in *Nordic Operations Analysis in Cooperation: OR in Business Conference*, Turku School of Economics, Finland.

—— (1993) "An implementation of the contract net protocol based on marginal cost calculations," in *Proceedings of the National Conference on Artificial Intelligence (AAAI-93)*.

—— (1996) "Limitations of the Vickrey auction in computational multiagent systems," in *Proceedings of the International Conference on Multiagent Systems*. Extended version (2000) "Issues in computational Vickrey auctions," in *International Journal of Electronic Commerce*, 4: 107–29.

—— (2002a) "Algorithm for optimal winner determination in combinatorial auctions," *Artificial Intelligence*, 135: 1–54. Early versions: ICE-98, IJCAI-99.

—— (2002b) "eMediator: a next generation electronic commerce server," *Computational Intelligence*, 18(4): 656–676. Early version: AAMAS-00.

—— (2006) "Optimal winner determination algorithms," in P. Cramton, Y. Shoham, and R. Steinberg (eds), *Combinatorial Auctions*, MIT Press, ch. 14.

—— (2007) "Expressive commerce and its application to sourcing: how we conducted $35 billion of generalized combinatorial auctions," *AI Magazine*, 28(3): 45–58.

—— and Boutilier, C. (2006) "Preference elicitation in combinatorial auctions," in P. Cramton, Y. Shoham, and R. Steinberg (eds), *Combinatorial Auctions*, MIT Press, ch. 10.

—— and Suri, S. (2001a) "Market clearability," in *Proceedings of the International Joint Conference on Artificial Intelligence (IJCAI)*.

—— —— (2001b) "Side constraints and nonprice attributes in markets," in *Proceedings of the IJCAI Workshop on Distributed Constraint Reasoning*, Seattle, WA. Journal version (2006) *Games and Economic Behavior*, 55: 321–30.

—— —— (2002) "Optimal clearing of supply/demand curves," in *Proceedings of the International Symposium on Algorithms and Computation (ISAAC)*.

Sandholm, T. and Suri, S. (2003) "BOB: improved winner determination in combinatorial auctions and generalizations," *Artificial Intelligence*, 145: 3–58. Early version: AAAI-00.

_____ _____ Gilpin, A. and Levine, D. (2002) "Winner determination in combinatorial auction generalizations," in *Proceedings of the International Conference on Autonomous Agents and Multiagent Systems (AAMAS)*.

_____ _____ _____ _____ (2005) "CABOB: a fast optimal algorithm for winner determination in combinatorial auctions," *Management Science*, 51(3): 374–90 Early version: IJCAI-01.

_____ Levine, D., Concordia, M., Martyn, P., Hughes, R., Jacobs, J. and Begg, D. (2006) "Changing the game in strategic sourcing at Procter & Gamble: expressive competition enabled by optimization," *Interfaces*, 36(1), 55–68.

_____ Likhodedov, A. and Gilpin, A. (forthcoming) "Automated design of revenue-maximizing combinatorial auctions," *Operations Research*, special issue on computational economics,. Subsumes and extends over a AAAI-04 and AAAI-05 paper.

Smock, D. (2004) "How P&G buys plastics," *Modern Plastics*. Copyright © 2005 Canon Communications LLC.

Song, J. and Regan, A. (2005) "Approximation algorithms for the bid construction problem in combinatorial auctions for the procurement of freight transportation contracts," *Transportation Research Part B: Methodological*, 39(10): 914–33.

Xu, L., Hutter, F., Hoos, H. and Leyton-Brown, K. (2011) "Hydra-MIP: automated algorithm configuration and selection for mixed integer programming," RCRA Workshop on Experimental Evaluation of Algorithms for Solving Problems with Combinatorial Explosion at the International Joint Conference on Artificial Intelligence (IJCAI).

....

CHAPTER 17

DESIGNING AUTOMATED MARKETS FOR COMMUNICATION BANDWIDTH

....

NIR VULKAN AND CHRIS PREIST

INTRODUCTION

....

MUCH of market design these days takes place on the Internet. Market design on the Internet clearly is attractive in that it lowers set-up and participation costs and it allows for twenty-four-hour trading across time zones. However, it also presents market designers with unique challenges. One of these challenges, and the focus of this chapter, is the automation of user–market interaction. Agent technology is particularly instrumental in furthering the commercial advantage of the Internet because it has the potential of smoothing friction, for example by allowing asynchronous trading over different time zones; see Vulkan (1999) for more general discussion of the economic value of agent technology. In most markets a degree of automation is required to control the format and content of the messages users send to the market. Other sites go one further and automate some of the strategic content of the way users interact with the market—the eBay bidding agent, which automatically increases users' bids up to the limit they set if such an increase is needed, is a good example of that. But some markets require more. The market for communication bandwidth is an example of that.

This market builds on technological developments in electronic commerce which allow the buying and selling electronically of the right to transmit data over the net and the telecommunications infrastructure. Organizations like Band-X, RateXchange and Min-X provide bulletin boards and double auctions to buy and sell bandwidth and connection time. Currently, these transactions are wholesale, between large operators. However, we could imagine such transactions taking place far more frequently. If negotiation were cheap and easy, and appropriate billing infrastructure was in place, renegotiation of connection contracts could take place every few seconds. This would allow sources of traffic (such as local Internet service providers) to dynamically switch between long-distance carriers in response to price fluctuations. In this way, a spot market for bandwidth could develop; see Lehr and McKnight (1998) for an overview.

By shortening contract lengths (e.g. reducing the granularity of the bandwidth being traded) and standardizing contracts, Internet service providers (ISPs), telecommunication companies (which own the backbone), and large users (e.g. universities or corporations with many employees) can achieve greater efficiency and reduce transaction costs. Reduced costs and increased efficiency increase participation of smaller buyers and sellers, which in turn increase efficiency and reduce costs even further.

One of the keys to making this happen is the ability to negotiate automatically. If negotiations were to take place constantly it would be expensive for people to perform the task. Furthermore, they will not always be able to react fast enough to changes in market conditions. For this reason, communications bandwidth has inspired work on automated negotiation from its inception (Rosenschein and Zlotkin, 1994). In this chapter we focus on designing adaptive agents: agents capable of choosing their bids *online* based on current information from the user (e.g. maximum willingness to pay) and on past observations.

This chapter presents a high-level design algorithm for automated trading in online spot markets. The approach pursued here is particularly suitable for random environments, for example when demand (and possibly also supply) is "bursty," that is, when it fluctuates considerably and quickly. While our work is motivated by our interest in designing markets for communication bandwidth, our results are relevant to any market with similar characteristics.

We begin in the next two sections by examining bandwidth trading as a learning environment: at any given stage the spot market can be described as a game with a relatively small number of players. However, this number can change from one trading period to the next. The game-theoretical approach to learning, where each player holds beliefs about the behavior of all the other players (and typically optimizes their action given these beliefs), is therefore not suitable for our setting. We describe game-theoretical learning models in more detail in the seventh section.

Instead, we know that the "world" changes, in the sense that the payoffs (or profits) associated with each bid change from one period to the next. For example, a selling agent with a reservation price p submits a bid b. What can be said about the expected payoff of this agent in this case? At any given trading period we can describe the possible payoffs and the probabilities associated with these payoffs (e.g. "with probability 0.2 the clearing price will be p' and the payoff $p'-p$" and so on). But are these probabilities *fixed across consecutive trading periods*? If they are, then we say that the trading environment is *stochastically stable*, and the agent decision problem can be described as the *multi-arm bandit decision problem* as it is known in statistics and economics. (In our case each bid value corresponds to an arm.) If the environment is stochastically stable then there is one strategy (i.e. a function mapping from reservation prices to bids), which statistically dominates all other strategies (see Gittins, 1989). If the environment is not stochastically stable, then such a strategy does not exist.

The first contribution of this chapter, in the third section, is to normatively express the requirements of a "good" learning algorithm in an online spot market setting. An important characteristic of bandwidth markets is that demand is "bursty" (see e.g. Mindel and Sibru, 2001; Hwang et al., forthcoming). Since the pattern of bids changes dramatically between periods, it is not practical for the agent to treat all data as arising from a single distribution. Instead, the trading environment is better described as a number of changing stable distributions (plus periods of noise). Such environments are sometimes known as "semi-stochastic."

If the environment is semi-stochastic then a "good" learning algorithm should be able to identify when a new stable distribution begins and—once it has learned—best respond to it. Our first normative requirement is therefore that if the trading environment is stochastically stable (i.e. in a period of stochastic stability) then the learning algorithm will converge (quickly) to playing the stochastically dominant strategy. At the same time we require that if the environment is not stochastically stable (e.g. the distribution changes or we are in a noisy period), then the learning algorithm will be responsive to changes in the underlying structure. A "good" learning algorithm will quickly adapt to such changes.

Once we have characterized what constitutes good learning, we provide, in the fourth section, a high-level specification for an online algorithm for trading in bursty online spot markets. This algorithm is effectively the blueprint for designing agents that automatically trade in such a market. The algorithm uses the tools of statistical learning theory to test online whether the trading environment is consistent with the agent's model of the world. If the environment is stable and is consistent with the agent's model, then the agent best responds to its beliefs (i.e. chooses the bid which maximizes the user's expected utility). If the trading environment is not stable then the algorithm switches to its transitory mode, where the next period's bid is selected. Finally, the algorithm allows for new models of the environment to replace old ones. One of the key issues is the switching mechanism used by the agent. We do this by allowing the agent to test—in real time—its own success in predicting what happens next at any given stage. The agent then switches between its learning modes when sufficient evidence exists to suggest its learning is not consistent with recent outcomes.

In the fifth section we analyze the algorithm and show that it satisfies our definitions of convergence and responsiveness. We show also that it combines rapid convergence with maximum responsiveness to changes. The sixth section provides a set of simulations which demonstrate the performance of the agent in a number of trading environments. The seventh section discusses related work, and the eighth section concludes.

DOUBLE AUCTION MARKETS FOR AUTOMATED
AGENTS

In this section, we introduce some economic terminology, and describe the k-double auction marketplace that is used subsequently in this chapter.

A double auction is a fast and effective method of trading in markets with many buyers and sellers. The alternatives to a double auction are one-sided auctions or the use of fixed prices (e.g. tariffs). One-sided auctions (buyer or seller) are more suitable when there is a single or dominant buyer (alt. seller) and many small sellers (alt. buyers), because the auction favors the person running it (i.e. it puts him in the strongest possible negotiation position). A fixed price is not a suitable option if demand and/or supply change quickly, as they do in bandwidth markets (and many others). In a double auction the current price truly reflects the current state of supply and demand, because it is determined by aggregating the bids and asks of buyers and sellers in any given moment. In fact, the shorter the connection time (i.e. the traded contract size), the more efficient the double auction is in allocating bandwidth to those who are willing to pay for it.

More specifically, in a k-double auction, buyers announce their maximum willingness to pay, and sellers the minimum price they are willing to accept. Sorting out the buyers' bids in increasing order of price, and the sellers' in decreasing order of price, we let $[a,b]$ be the interval where b is the buyer's bid with the highest index (in the sorted list), which is still larger than the corresponding seller's bid, a, with the same index.[1] A k-double auction mechanism selects the clearing price $p = ka + (1 - k)b$, where $k \in (0,1)$. All buyers who bid at least p trade with all sellers who bid no more than p. All those who trade, trade at the clearing price, and all others do not trade. Note that, since the number of bidders is finite, there is a positive probability that a given buyer or seller enters the marginal bid (i.e. a or b). Hence, bidders can gain from setting their bids strategically (i.e. for a seller this is the real value under which it is not worth trading, and for a buyer a price above which trade is no longer profitable). For the case where the number of buyers and sellers and the distribution of reserve prices are commonly known, the exact formulae for optimal bidding strategies can be found in Rustachini et al. (1994).

In this chapter, we consider a repeated k-double auction for trading instantly perishable goods. A good is instantly perishable if it ceases to be useable soon after it becomes available for sale. The right to transmit on a network at a given time is an example of such a good; a network provider can offer connection time now, but if it is not used now, it ceases to be available. You cannot store up connection time for sale later. (Though, of course, new connection time becomes available.)

---

[1] Alternatively, bids can be aggregated into supply and demand curves. The interval $[a,b]$ now corresponds to the intercepting section of the two graphs.

Such a good suggests a marketplace where trades are taking place continuously, and buyers and sellers are continuously adjusting their price in response to the activities of others. Each buyer and seller has a certain good (or need for a good) at any given trading round, and announces the price they are willing to trade at. In the next round, any goods/needs from the previous round expire, but they receive new goods/needs. Hence goods are constantly flowing into the market, and traders are modifying their bids/offers in an attempt to trade successfully.

At each trading period organizations automatically communicate their reservation price privately to their trading agent, i.e. the maximum (alt. minimum) price it is willing to pay (alt. accept) without making a loss. These reservation prices may originate from fixed marginal costs (for sellers), or from contracts with end users (for ISP providers, who are buyers in this setting). This will be discussed in more detail in the following section.

Communicating priorities

Buyers of bandwidth may wish to communicate to their agents the degree of priority of trading in the next period. For example, continuous bandwidth may be required when transmitting voice or video, whereas data transmission could wait until the ISP provider gets a better deal on bandwidth rates.

There are a number of ways to transmit these priorities to the agent. Firstly, the priority of trade can be taken into account by the program that sets the reservation price: a seller who desperately needs to trade will accept lower prices, and similarly, the buyer will pay more. Secondly, the agent can be designed to maximize a function of two variables, reservation price and priority. More specifically, when the user installs the agents she is asked to express her utility in terms of combinations of price and priority. The second approach is of course more general than the first, but it may prove difficult for the user to express her requirements in terms of the relationship between price and trading priority.

A third possibility, which can be seen as a compromise between the first two, is for the user to indicate a high priority using the weight attached to the *probability* of trading. To see the idea behind this model, note that the agent is effectively participating in a lottery. Unless it has the highest possible valuation (if valuations are bound), or zero cost, then it will trade with probability less than 1. We therefore propose that we model a user with a high priority of trade as being more *risk averse*. We propose a method where this risk aversion is explicitly entered into the agent's decision choice.

To illustrate how this can be done, consider a seller with a reservation price (excluding any priority considerations) r. The monetary payoff for this seller from trading at price p is $p-r$ (similarly, the payoff for a buyer is $r-p$). At each stage the agent must decide how much to bid. Denote this bid $b(r)$. Denote by $P(b(r))$ the ex ante probability of trading in the next period as a function of the posted bid, b. A reasonable objective function for a risk-neutral agent with no special priority is $ArgMax_b P(b) \cdot (p(b)-r) + (1-P(b(r)) \cdot 0 =$

$P(b) \cdot (p(b)-r)$. A higher bid can increase the payoff from trading (because there is a positive probability that this agent will happen to be one of the two agents whose bids determine the clearance price), but decreases the probability of trading. A lower bid does the converse, meaning the agent is more likely to trade but will make less profit.

We can generalize the risk-neutral objective function to include a measure of risk, α. Given α, the agent chooses the bid that maximizes $P^{\alpha}(b) \cdot (p(b)-r)$. If $\alpha = 1$, the agent is risk neutral. If $\alpha > 1$, the agent is risk averse, and prioritises getting a trade over making a large profit. If $\alpha < 1$, the agent is risk seeking. This method of representing priorities has the following advantages. On the one hand, agents still maximize a single-dimension function, hence they can compute the (typically unique) optimal bid quickly. On the other hand, this method allows for a separate online representation of reservation prices and priorities. For an ISP purchasing bandwidth to service its customers, reservation prices are associated with variable costs in delivering a service, and expected payback from the customers. The payback depends on the nature of the customer's contract with the ISP—whether they pay a subscription charge, or pay per use. The priority with which the trade needs to be made depends on the nature of the service being supported. Is it a real-time service, such as a video feed, where delay makes a real difference to service quality, or is it a service where some delay can be accommodated (such as email)? If delay is acceptable, the associated purchase can be given a low priority. If it is not acceptable, the priority will depend on what quality-of-service guarantees the ISP has made to the end user, and the penalty it will suffer for failing to deliver on these. These tradeoffs can be made automatically by monitoring current end-user demand and using domain-specific financial forecasting tools. The results can then be passed directly to the agent.

LEARNING IN A DOUBLE AUCTION
FOR BANDWIDTH

In this section we consider the normative aspects of trading in double auction markets.

A history-independent strategy (HIS)[2] is a function from the set of reservation prices to the set of possible bids. Normalizing bids to $[0,1]$, an HIS for the buyer is therefore a function $B : [0,1] \rightarrow [0,1]$, and $S : [0,1] \rightarrow [0,1]$ for the seller, where $B(r)$ returns the bid corresponding to a reserve price of r. We restrict attention only to strategies that do not use dominated (loss-making) bids, that is, for sellers $S(r) \geq r$, and for buyers $B(r) \leq r$. An additional reasonable assumption is that $B'(r) \geq 0$ that is, bids are not a decreasing function of the reserve price (and similarly, $S'(r) \leq 0$).

---

[2] That is, a strategy in the one-shot double auction game.

We now define $Opt(r^t)$ to be the function that returns the (ex ante) optimal trading strategy ($B(r)$ or $S(r)$) at any time t (assuming a utility-maximizing strategy exists). We can divide trading environments into the following three classes:

Class 1: $B(r^t)$ and $S(r^t)$ are independent of t.
Class 2: $B(r^t)$ and $S(r^t)$ depend on t, but there exists a pair of deterministic functions, f and g, such that $B(r^t) = f(B(r^{t-1}))$ and $S(r^t) = g(S(r^{t-1}))$.
Class 3: All other trading environments.

It should be clear that learning cannot occur in trading environments of class 3 because the past is not relevant to the future. We are therefore interested in characterizing "good" learning algorithms for the first two classes.

We now illustrate what an optimal strategy might look like (in a trading environment belonging to class 1 or 2). Suppose that the number of bidders is first drawn according to a distribution $N^O Bids$, with finite or infinite support, and that all these bids are then drawn independently according to a distribution $Bids$, with support on the unit interval. Suppose both distributions were known. Then it is possible to compute the optimal bid, given the reservation price. For example, when the number of bidders becomes large, we know from Rustachini et al. (1994) that $B(r)$ and $S(r)$ converge to r (i.e. for every $\varepsilon > 0$ there exist $N(\varepsilon)$ such that if the number of bidders is greater than or equal to $N(\varepsilon)$ then $|B(r)-r| < \varepsilon$). In other words, truthful bidding becomes (almost) optimal (because the probability of being the marginal bidder goes to zero). Rustachini et al. (1994) also provide the general formula determining the optimal bidding strategy in k-double auction markets.

As we explained before, the game-theoretical approach does require that players know the number of other players, and that they can somehow reason their way to equilibrium. This approach is still useful as a benchmark to what *can* be achieved by a learning process, although it neglects any effect that the learning process might have. A weaker approach to computing the optimal bid, which still works well in environments where the success probabilities associated with each bid are fixed, is the *multi-armed bandit* approach (a bandit is a nickname for a slot machine). In general, the solution to these problems involves a period of experimenting followed by convergence to one solution; these are known in statistics and economics as the *Gittins indices* (see Gittins, 1989).

So we conclude that in trading environments belonging to classes 1 and 2 there could exist optimal trading strategies. Let L be a learning algorithm, where $L(r^t, h^{t-1})$ returns period t's bid given a history, h^{t-1}. Following the terminology of Friedman and Shenker (1998), we introduce the following definitions:

Optimality. We say that a learning algorithm $L(r^t, h^{t-1})$ is optimal if it converges (statistically) to $Opt(r^t)$.

Responsiveness. Suppose that the trading environment from period t' onwards belongs to class 1 or 2. Then L is responsive if $lim_t |L(r^t, h^{t-1}) - L(r^t, h^{t-1/t'})| = 0$, where $h^{t-1/t'}$ denotes the history from t' to $t - 1$.

Degree of convergence. We say that $L1$ converges faster than $L2$ if $|L1(r^t, h^{t-1}) - Opt(r^t)| < |L2(r^t, h^{t-1}) - Opt(r^t)|$ almost always.

Degree of responsiveness. We say that $L1$ is more responsive than $L2$ if $|L1(r^t, h^{t-1}) - L1(r^t, h^{t-1/t'})| < |L2(r^t, h^{t-1}) - L2(r^t, h^{t-1/t'})|$ almost always.

HIGH-LEVEL SPECIFICATION OF THE ONLINE TRADING ALGORITHM

The procedures used by the main algorithm are explained in sufficient detail for a programmer to create a working code, although "fine tuning" of some of the parameters (e.g. the degree of statistical confidence with which the null hypothesis is accepted) is deliberately left unspecified. A trial-and-error process is required to set the actual values of these parameters. We come back to this point towards the end of this chapter.

The top-level algorithm is defined in pseudocode as follows.

An agent with reservation price r at time t executes the following procedure:

```
IF modelStatus = 'stable'
THEN {
        bid := a prioriOptimalBid(r,worldModel);
        send(bid);
        receive(results);
        }
ELSE {
        bid := transitoryBid(r, reinforcementRules);
        send(bid);
        receive(results);
        reinforcementUpdate(reinforcementRules,results);
        }
update(worldModel,results);
updateModelStatus(worldModel,modelStatus);
```

Components of the model

The design consists of the following four components:

- The model of the world, *worldModel*, and updating procedures for this model, *update(worldModel,results)*.
- Consistency test for observed data, *consistent(WorldModel)*.
- Test for whether the model has become stable or unstable, and resetting the model if necessary, *updateModelStatus(worldModel,modelStatus)*.
- Transitory bidding rules, *reinforcementRules*, and the associated updating procedure, *reinforcementUpdate()*.

The model of the world and updating

The model of the world, *WorldModel*, aims to provide a probabilistic estimate of current behavior based on past statistics. If the world is stochastically stable, in the sense that the relevant underlying probability distributions are fixed over time, then the agent can estimate these probabilities using available data.

Depending on the information available to the agent, the world model can take different forms. Recall from the section "Learning in a double auction for bandwidth" the definitions of the distributions $N^O Bids$ and *Bids*. If the agent can observe the actual bids, then it can estimate these distributions directly. If, however, bids cannot be observed directly, then the agent can estimate the probability that a given bid is smaller than the market clearing price, which we denote $Trd(\cdot)$. The empirical distribution $Trd^t(s)$ is used to measure the proportion of time, in the previous t trading periods since Trd^t was last initialized, where the bid s would have been accepted. Denote by p^t the equilibrium price at period t. This implies that a seller bidding any price $s \le p^t$ would have succeeded in trading at period t. (Similarly, any buyer bidding $b \ge p^t$ would trade.) Agents start off with $Trd^t(s)$ initiated to zero everywhere.

After each round of trading, the world model is updated using the routine *update(worldModel, results(t))*. It does this by selecting the equilibrium price at time t from the results, and updating Trd^t as follows:

$$Trd^{t+1}(s) = \frac{t \cdot Trd^t(s) + 1}{t + 1} \quad \text{if } s \le p^t$$

$$Trd^{t+1}(s) = \frac{t \cdot Trd^t(s)}{t + 1} \quad \text{if } s > p^t$$

One can imagine more sophisticated forms of beliefs, for example beliefs about possible patterns in the market (e.g. "markets are bearish in January"). That is, the distribution of trading prices is stable, but time dependent. In any case, estimating and updating of the model will take a similar form (though the agent will be updating more sophisticated data structures).

The agent then solves its maximization problem using its empirical estimates in place of the real distributions. Since, in a stochastically stable world, these empirical distributions converge to the real distributions (according to the law of large numbers), the outcome of this maximization problem will also converge to $Opt(r)$.

Consistency tests

At any stage of trading, except for the first few rounds, the agent is able to compare its current model of the world (e.g. the empirical distribution $Trd^t(s)$), with recent data, using the procedure *consistent(WorldModel)*. This is done by defining and testing the appropriate null hypothesis. That is, the agent should be able to conclude that "There is, or is not, sufficient evidence with which to reject the hypothesis that the data are consistent with the model." This test is then carried out to a pre-specified error measure.

The agent does not have a preconceived idea about the nature of the distribution it is estimating. In practice, these distributions can take any form. Therefore a non-parametric test should be carried out. There are a number of such tests. For example, the Kolmogorov–Smirnov test considers the absolute value of the difference between the sample and the model. It turns out that this difference is distribution-free (see, for example, Silvey, 1975). The null hypothesis is then accepted if the difference is below a pre-specified threshold, and rejected otherwise.

There are many other non-parametric tests. In our simulations, we use convergence of means. Here, the null hypothesis is expressed in terms of the difference in means between the sample and the model. We come back to this issue in more detail, when we describe our simulations. Whichever test is used, the user must specify (1) the size of the buffer which defines the sample size, and (2) the acceptance threshold.

Determining if the model is stable/unstable, and resetting the model

The procedure *consistent(worldModel)* is not used directly by the top-level algorithm, but rather is used by *updateModelStatus()* to determine whether the model should be treated as stable or as unstable.

Initially, we assume that the model is unstable. If, after a number of model updates, we find that the model is consistent with the observations over k rounds (i.e. the null hypothesis is accepted consecutively a number of times), then we assume the model is stable, and set the *modelStatus* flag appropriately. When the model is considered stable, we continue to test the consistency of the model against the observed results. If the null hypothesis is rejected once, then this could be because of a "bad" sample. Of course, the larger the sample size, the less likely this is to happen. Still, our simulations suggest that even for finely tuned learning parameters, bad data will occur. We therefore take the approach that the algorithm should tolerate a small number of rejections before past history is thrown away.

More specifically, we check whether the null hypothesis is rejected consecutively a fixed number of times, j. Only if this happens do we conclude that a structural break-down is likely, and therefore re-initialize beliefs (effectively throwing away old data). In general, the user may specify a different test to determine when to throw old data away.

When data are discarded in this way, the model of the world is considered unstable, and the algorithm moves to its transitory mode. It is considered to return to stability when the null hypothesis is accepted a fixed number of times, and the algorithm becomes belief-based once more. The pseudocode for the procedure is as follows:

```
updateModelStatus(WorldModel,ModelStatus) {
IF modelStatus = 'stable' THEN {
    IF consistent(worldModel) = false THEN {
            count := count + 1;
            IF count = j THEN {
                    worldModel := initialWorldModel;
                    modelStatus = 'unstable';
                    count := 0;}
            }
    ELSE
            count := 0;}
ELSE
    IF consistent(worldModel) THEN {
            count := count + 1;
            IF count = k THEN {
                    modelStatus = 'stable';
                    count := 0;}
            }}
```

Transitory bidding rules and updating

The purpose of this learning mode is to choose bids in those cases where the agent's current model of the world is not consistent with recent data. Using our classification from the section "Learning in a double auction for bandwidth" we say that—according to its current information—the agent is in a trading environment of case 3. In other words, the probabilities linking payoffs to actions of one period are not informative with respect to the other periods. There is no theory of optimal trading in these cases (and, of course, there cannot be).

We specify a small number of rules of thumb and use a reinforcement mechanism to select among these rules. In principle these rules should mimic rules of thumb which human traders use in similar circumstances (like technical rules in markets). We now provide a few examples of such rules (for simplicity we describe rules only for a seller agent):

- *Maximal likelihood rule.* Ask for the reservation price, $s(r) = r$. This rule maximizes the likelihood of trade, but at the expense of abandoning profits.

- *Greedy rule.* A rule which asks for the highest possible price, $s(r) = 1$. Profits are maximized, but the likelihood of trade is minimized (the equivalent rule for the buyer is $b(r) = 0$).
- *Linear rules.* Any combination of the above two rules: Asks for the average between the maximum price and the reservation price, $s(r) = qr + (1 - q)$ for $q \in (0, 1)$ (equivalent rule for buyer agent is $b(r) = qr$).
- *Constant bid rules.* Always use the same bid, except in those cases where the bid is lower than the reservation price, for example $s(r) = max(0.75, r)$.
- *Decreasing surplus rules.* A rule whereby the demanded surplus (i.e. the ask price minus the reservation price) decreases with r. It can be any rule which satisfies the following three constraints: $(1)\, s(1) = 1\, (2)\, s(r) \geq r$ for all $r \in (0, 1)$, and $(3)\, s' < 1$.[3] For example, $s(r) = \sqrt{r}$. (Or $b(r) = 1 - \sqrt{(1 - r)}$) for the buyer agent).
- *History dependent rules.* For example $s = f(eq^{t-1})$, or $s = f(eq^{t-1}, eq^{t-2})$ and so on.[4]

If the set of rules used is relatively small we recommend adding a rule that returns a random number between r and 1 (so that any bid in the interval $[r, 1]$ is made with positive probability).

ANALYSIS OF THE LEARNING ALGORITHM

For a certain class of environments, we can prove useful properties of our algorithm. These environments are consistent with classes 1 and 2 as defined earlier, in the section "Learning in a double auction for bandwidth."

Stochastic stability. We say that a trading environment is *stochastically stable in the time interval* $[t_1, t_2]$ (where $t_2 >> t_1$), if, for every $s \in [0, 1]$, the probability that s^t is smaller than p^t, and the conditional (on trading) payoff from bidding s^t, are fixed for all $t \in [t_1, t_2]$.

Semi-stochastic stability. We say that a trading environment is *semi-stochastically stable* if there exists a set $\{t_1, t_2 \ldots\}$ such that the trading environment is stochastically stable at the intervals $[t_i, t_{i+1}]$, for $i = 0, 1 \ldots$

Denote by L^* any algorithm which fits the above high-level spec. Denote by α the probability that the model becomes unstable when the trading environment is stochastically

---

[3] That is, $s(r)-r$ is a decreasing function of r.

[4] An underlying assumption of the belief model is that the events eq^t are independent and identically distributed (i.i.d.) (although they are allowed to be time dependent). If, however, eq^t is dependent on previous outcomes, say on eq^{t-1} then this assumption is clearly violated. The purpose of this class of rules is therefore to detect and capitalise on such dependencies.

stable (i.e. that the null hypothesis is continuously rejected). Denote by β the probability that the algorithm fails to reset beliefs when the underlying structure changes (i.e. the null hypothesis is not continuously rejected immediately after a change in the underlying probabilities). One of the main contributions of our learning algorithm is that parameters α and β are determined by two different and independent procedures, and can therefore *both* be minimized *simultaneously*.

We make the following observations:

Observation 1. L^* is optimal.

Proof. Suppose that the trading environment is stochastically stable over $[t_1, t_2]$. Then the null hypothesis is accepted with probability $1-\alpha$ at any stage in that interval. Hence, with probability $1-\alpha$, L^* will not reset its empirical distributions. Because of the law of large numbers the empirical distributions will converge to the real distributions, and $L^*(r^t, h^{t-1})$ will converge to $Opt(r^t)$. □

Remark. *Notice that the above algorithm also has a fast rate of convergence. First, note that the tests are chosen to minimize α, hence the probability of counting valid data is maximized. Second, L^* does not keep old history (as opposed to most learning algorithms, like Q learning or game-theory fictitious play), which is likely to be irrelevant to the current underlying structure. Hence, with a large probability, the estimates are based on all, and nothing but, relevant observations. Hence, the rate of convergence is maximized.*

Let c denote the lag variable, that is, the number of periods before the algorithm throws away old observations. Clearly, c depends on the threshold parameter being used in the testing of the null hypothesis, which in turn depends on the distributions and the noise. The optimization of c needs to be carried out based on real trading data—in essence it is the "fine tuning" of the learning algorithm.

Given c is optimized, a further two observations can be made:

Observation 2. L^* is responsive.

Proof. Let the environment be stochastically stable from period t'. With probability $1 - \beta$, L^* will throw away observations made before period $t' + c$. Hence, with probability $1-\beta$, $L^*(r^t, h^{t-1}) = L^*(r^t, h^{t-1/t'})$, for $t > t' + c$. Hence, L^* is responsive. □

Observation 3. In a semi-stochastically stable trading environment, there is no other learning algorithm that is more responsive than the above algorithm.

Proof. Note from the proof of observation 2 that with probability $1 - \beta$, $L^*(r^t, h^{t-1}) = L^*(r^t, h^{t-1/t'})$, for $t > t' + c$, or $|L^*(r^t, h^{t-1}) - L^*(r^t, h^{t-1/t'})| = 0$. Since the difference is already zero, and since β is minimized, L^* maximizes responsiveness. □

The responsiveness of L^* is clearly demonstrated in our simulations and can be seen most clearly in the simulations section of Vulkan and Priest (2003: figs 1–3).

OVERVIEW OF SIMULATION RESULTS
FROM VULKAN AND PREIST (2003)

Using this framework, Vulkan and Preist (2003) carried out a number of simulations that we summarize here. They argued that markets for computational bandwidth can be approximated by a *semi-stochastically stable* environment. Such an environment can have dramatic changes in demand from moment to moment, but this demand is drawn from an underlying statistical distribution that can be estimated from past history. However, this distribution can change at different times—for example, the underlying distribution for 4pm EST would be expected to be different from that for 4am EST.

To simplify matters, we assume a one-dimensional distribution from which the maximum acceptable trading price is taken. If our selling agent bids anything above this price, then it does not trade. However, if the bid is below this price, the agent trades *at its bid*, which is somewhat similar to a 0-double auction (Rustachini et al., 1994). This provides sufficient structure on the agent's decision problem (so that choosing a higher bid increases expected payoff, but decreases the chances of trading), while allowing us to simulate a relatively simple process of generating a single price (and not all the other bids). We assume the agent is risk neutral. Hence it computes:

$$Arg_{s\in[r,1]} \ \max P(s)\cdot(s-r).$$

The agent's model of the world then takes the form of a function $P(s)$, which returns the probability of the bid s being accepted. Of course, if the world is stochastically stable then $P(s)$ is fixed over time for any given value of s. We start with, $P^0(s) \equiv 0$ and the agent updates its beliefs using:

$$P^{t+1}(s) = \frac{t\cdot P^t(s) + 1}{t+1} \quad \text{if s} \leq p^t$$

$$P^{t+1}(s) = \frac{t\cdot P^t(s)}{t+1} \quad \text{if } s > p^t$$

At each trading round we test whether P is consistent with the last eight observations. Specifically, we compute, at time t, the values x_1 and x_2, such that $P(x_1) \approx 0.4$ and $P(x_2) \approx 0.6$. We compute the average equilibrium price over the last eight periods of trade, A_t. P is considered consistent if $x_1 < A_t < x_2$. Otherwise P is inconsistent.

If P has been inconsistent for five consecutive rounds, then P is re-initialized. In other words, the agent detects a structural breakdown of the underlying probability distribution of equilibrium prices. Old information may be harmful in that it will outweigh new, relevant information. Hence, the agent starts learning P from scratch.

P will not be re-initialized again until it has regained stability; in other words, until the agent detects five consecutive periods where P is consistent.

We compared the effectiveness of our algorithm with that of a similar learning algorithm that does not re-initialize its belief function after detecting structural breakdowns. To do this, we performed a simple Monte Carlo simulation of the performance of each algorithm in several different environments, and compared their performance. Each simulation consists of 100 runs of 1,000 trading periods. In each trading period, the agent determines its bid and submits it to the simulated marketplace. The simulated marketplace determines the maximum acceptable trade price by a random sample from a probability distribution function representing the distribution of the maximum acceptable trade price. The agent is informed of this price and receives a profit of $s-r$ if s is below it, and no profit otherwise. In each trading period, we simultaneously calculate the payoff for the agent using the belief algorithm with reset and for the agent using the algorithm without reset; hence both are subject to the same sets of random trade prices.

Analysis of the resulting simulations determined the following:

- In environments where the mean of the underlying stochastic distribution changed every 200 trade rounds, the re-initializing algorithm considerably outperformed the continuous learning algorithm, resulting in 16–25% higher profits. It would identify the changed environment, reset, and relearn, to be placing optimal bids within twenty rounds of the change.
- In environments where the standard deviation of the underlying distribution changed every 200 trade rounds without a change to the mean, the re-initializing algorithm performed marginally better than the continuous learning algorithm, but the performance improvement was not statistically significant. The algorithm struggled to spot the changes in the distribution, and many of its resets were false positives.
- As the environment became more unstable, with changes to the mean of the underlying distribution occurring more frequently, the advantage gained by the re-initializing algorithm is reduced. This is because of the increasing number of switches that are required, and the resulting loss of income during the learning period immediately after a switch. This factor becomes increasingly dominant when a period lasts twenty-five rounds, when the two algorithms perform roughly equivalently, and ten rounds, when the belief algorithm marginally outperforms the switching algorithm.

RELATED WORK

Research into automated negotiation has long been an important part of distributed AI and multi-agent systems. Initially it focused primarily on negotiation in collaborative problem solving, as a means to improve the coordination of multiple agents working together on a common task. Laasri et al. (1992) provide an overview of the pioneering work in this area. As electronic commerce became increasingly important, the work expanded to encompass situations with agents representing individuals or businesses with potentially conflicting interests. The contract net (Smith, 1981) provides an early architecture for the distribution of contracts and subcontracts to suppliers. It uses a form of distributed request-for-proposals. However, it does not discuss algorithms for determining what price to ask in a proposal. Jennings et al. (1996) use a more sophisticated negotiation protocol to allow the subcontracting of aspects of a business process to third parties. This is primarily treated as a one-to-one negotiation problem, and various heuristic algorithms for negotiation in this context are discussed by Faratin et al. (1998). Vulkan and Jennings (2000) recast the problem as a one-to-many negotiation, and provide an appropriate negotiation protocol to handle this. Other relevant work in one-to-one negotiation includes the game-theoretic approach of Rosenschein and Zlotkin (1994) and the logic-based argumentation approach of Parsons et al. (1998).

As much electronic commerce involves one-to-many or many-to-many negotiation, the work in the agent community has broadened to explore these cases too. The Michigan AuctionBot (Wurman et al., 1998a) provides an automated auction house for experimentation with bidding algorithms, and provided impetus for theoretical analysis of appropriate mechanisms for auctions for electronic commerce (Wurman et al. 1998b). The Spanish fish market (Rodriquez-Aguilar et al., 1997) provides a sophisticated platform and problem specifications for comparison of different bidding strategies in a Dutch auction, where a variety of lots are offered sequentially. The Kasbah system (Chavez et al., 1997) featured agents involved in many-to-many negotiations to make purchases on behalf of their users. However, the algorithm used by the agents—a simple version of those by Faratin et al. (1998)—was more appropriate in one-to-one negotiation, and so gave rise to some counterintuitive behavior by the agents. Gjerstad and Dickhaut (1998) use a belief-based modeling approach to generating appropriate bids in a double auction, based on observation of previous bids. Cliff and Bruten (1997) and Preist and van Tol (1997) use an adaptive approach in a similar environment. Preist (1999) demonstrates how these can be used to produce a market mechanism with desirable properties. Park et al. (1998, 1999) present a stochastic-based algorithm for use in the University of Michigan Digital Library, another many-to-many market. Tesauro and Das (2001) have modified the Gjerstad–Dickhaut algorithm, resulting in performance improvements over other strategies. Tesauro and Bredin (2002) use dynamic programming resulting in further improvements. Walsh et al. (2002) analyze strategic interactions in the choice of strategies. Easley and Ledyard (1993) developed automated trading strategies that simulate human behavior in double auctions. Unlike our work, all these assume that the trading environment is relatively stable and not subject to the dramatic fluctuations in the environment we use. He et al. (2003) have developed a bidding agent based on fuzzy heuristics, and deployed it in an environment which is not stable, but is stochastically stable. They show that, in this environment, the agent outperforms other agents designed for stable and near-stable environments.

The use of economically inspired computational techniques for controlling the allocation of resources is discussed in Clearwater (1997). In some cases, an artificial economy is created. For example, Clearwater (1997) presents an artificial economy for regulating air conditioning. Mullen and Wellman (1995) present an agent-based economy where the agents spend a network resource allocation to access information services in the University of Michigan Digital Library. Yamaki et al. (1998) present an auction-based mechanism for allocation of services in a multimedia environment, including a market for future access rights. Others have proposed using a real economy to regulate bandwidth usage in a similar spirit. Miller et al. (1997) present an automated auction to regulate the use of ATM network bandwidth. Our work is in the spirit of the latter. We assume the existence of spot markets for bandwidth, as proposed by Lehr and McKnight (1998), and focus on appropriate agent algorithms for use within these.

There is now a significant body of literature on Internet traffic pricing. Notably, MacKie-Mason and Varian (1995) focus on the current inefficiencies of Internet traffic pricing, and Shenker et al. (1996) focus on the complexity issues of pricing implementation in the Internet environment. Furthermore, Gupta et al. (1999) discuss the (efficiency) implications of dynamic pricing in bursty environments. All these papers stress the importance of dynamic, decentralized pricing strategies for the overall efficiency of bandwidth traffic. The k-double auction model studied in the present chapter is a straightforward example of the type of mechanisms discussed by these authors. Moreover, pricing each short-term contract at the time significantly reduces the complexity of the decision-making process of agents and therefore addresses the concerns raised, for example, by Shenker et al. (1996). Once again, we *assume* the existence of an efficient pricing mechanism and focus on what constitutes optimal behavior for agents trading within these mechanisms.

Exchanges for telecommunications bandwidth are now regularly used in place of fixed-term contracts. Two recent studies, by Mindel and Sirbu (2001) and Weiss and Shin (2002), survey the current state of bandwidth exchanges, focusing on the granularity of contracts and payment schemes used. Hwang et al. (forthcoming), Fulp and Reeves (2001), and Vishal and Prashant (2001) investigate the pricing regimes that can be used by bandwidth exchanges. These papers assume naive behavior on the part of the agents (truth telling, no learning or updating). Based on this assumption they can then study the optimal mechanism design for the exchange. Our chapter can therefore be seen as complementary to these papers. Finally, Shin et al. (2002) and Semret et al. (2000) take a game-theoretical approach to the design of agents in bandwidth markets. This approach may be suitable if players are relatively "big," that is if their actions are likely to affect the market, and if the game is stable. When these two assumptions (and especially the second) are satisfied it is possible to study the Nash equilibrium of the game and use it as the blueprint for designing the trading agents. To further explain the difference between our approach and the standard game-theory approach, we now discuss in more detail the assumptions made and questions addressed by the literature on learning in games.

For a survey of the literature on learning in games see Fudenberg and Levine (1998). Fudenberg and Levine also suggest a number of new theories based on ideas from reinforcement learning theory. Other examples of reinforcement learning in games include McKelvey and Palfrey (1995), Erev and Roth (1998), and Roth and Erev (1995). These papers differ considerably from our chapter in two important respects. First, the literature on learning in games is motivated by the following three questions:

- Under what conditions will players converge to equilibrium?
- What degree of rationality should players possess in order to reach an optimum?
- Which rules best resemble how real people behave?

Instead, the focus of our research is to find useful conclusions for the engineers who design artificial traders in similar online markets.

Second, the literature on learning in games considers the actions of players who repeatedly play the *same* game. More specifically, these models compare the steady-state outcome of the learning rule to the equilibrium of the stage game. But this is meaningful only if there *is* a fixed-stage game, or at least if the equilibrium is fixed. As mentioned earlier, this is not the case in online spot markets with bursty demand, where the equilibrium of the underlying game changes considerably from one period to another.

For example, in Fudenberg and Levine's setting (1998), pairs of players are drawn from a large population to play the same game. In the reinforcement learning models studied by Erev and Roth (1998), and Roth and Erev (1995), and many more, the probability with which players will play a certain strategy increases between periods if this strategy proves successful (where success is measured by the relative payoff). For a survey of the literature on reinforcement learning see Vulkan (2000).

CONCLUSIONS

Market design is one of the most successful and active branches of economics in recent years. This form of economic engineering is particularly powerful for Internet-based markets where formal rules govern the interactions of users with the market. The automation of user–market interactions poses interesting challenges to market designers. Users should be able to trust their agents and be able to easily communicate to them their priorities. This kind of trust will develop only if users reasonably believe the agents they use are at least as good at achieving their goals as they themselves would be.

In this chapter, we have presented a general approach to the design of an automated trading algorithm for environments that are semi-stochastically stable. The algorithm combines belief-based learning with reinforcement learning, using the former during periods of stochastic stability and the latter in circumstances when its beliefs are no longer reliable. It identifies such structural breakdowns through the use of a statistical test. We have argued that real-time bandwidth trading is one example of a market that is likely to exhibit semi-stochastic stability, and so is a promising application area for such algorithms.

Good agent design is a crucial first step toward the automation of markets. Such markets, for example the market for communication bandwidth, can facilitate new and efficient kinds of economic interactions. While the problem of designing fully functioning, trustworthy trading agents is hard and complicated, we hope our efforts described in this chapter can be seen as a step in that direction.

REFERENCES

Chavez, A., Dreilinger, D., Guttman, R. and Maes, P. (1997) "A real-life experiment in creating an agent marketplace," in H. S. Nwana and N. Azarmi (eds), *Software Agents and Soft Computing: Lecture Notes in Computer Science 1198*, Springer, pp. 160–79.

Clearwater, S. H. (ed.) (1997) *Market Based Control: A Paradigm for Distributed Resource Allocation*, World Scientific.

Cliff, D. and Bruten, J. (1997) "Less than human: simple adaptive trading agents for CDA markets," technical report Hewlett-Packard Labs-97-155.

Easley, D. and Ledyard, J. (1993) "Theories of price formation and exchange in double aural auctions," in D. Friedman and J. Rust (eds), *The Double Auction Market: Institutions, Theories and Evidence*, Addison-Wesley, pp. 63–98.

Erev, I. and Roth, A. E. (1998) "Predicting how people play games: reinforcement learning in experimental games with unique, mixed strategy equilibria," *American Economic Review*, 88(4): 848–81.

Faratin, P., Sierra, C. and Jennings, N. (1998) "Negotiation decision functions for autonomous agents," *Robotics and Autonomous Systems*, 24(3–4): 159–82.

Friedman, E. J. and Shenker, S. (1998) "Learning and implementation on the Internet," Mimeo, Rutgers University.

Fudenberg, D. and Levine, D. (1998) *Theory of Learning in Games*, MIT Press.

Fulp, E. W. and Reeves, D. S. (2001) "Optimal provisioning and pricing of Internet differentiated services in hierarchical markets," *Proceedings of the IEEE International Conference on Networking*, pp. 409–18.

Gittins, J. (1989) *Allocation Indices for Multi-Armed Bandits*, Wiley.

Gjerstad, S. and Dickhaut, J. (1998) "Price formation in double auctions," *Games and Economic Behaviour*, 22(1): 1–29.

Gupta, A., Stahl, D. O. and Whinston, A. B. (1999) "The economics of network management," *Communications of the ACM*, 42(9): 57–63.

He, M., Leung, H. and Jennings, N. R. (2003) "A fuzzy logic based bidding strategy for autonomous agents in continuous double auctions," *IEEE Transactions on Knowledge and Data Engineering*, 15(6), 1345–63.

Hwang, J., Kim, H. J. and Weiss, M. B. (forthcoming) "Interprovider differentiated service interconnection management models in the Internet bandwidth commodity markets," *Telematics and Informatics, Special Issues on Electronic Commerce*.

Jennings, N. R., Faratin, P., Johnson, M. J., O'Brien, P. and Wiegand, M. E. (1996) "Using intelligent agents to manage business processes," in *Proceedings of the First International Conference on the Practical Application of Intelligent Agents and Multi-Agent Technology*, Practical Applications Ltd, pp. 345–60.

Laasri, B., Laasri, H., Lander, S. and Lesser, V. (1992) "A generic model for intelligent negotiating agents," *International Journal of Intelligent and Cooperative Information Systems*, 1(2): 291–317.

Lehr, W. and McKnight, L. (1998) "Next generation bandwidth markets," *Communications and Strategies*, 32: 91–106.

MacKie-Mason, J. and Varian, H. (1995) "Pricing the Internet," in B. Kahin and J. Keller (eds), *Public Access to the Internet*, Prentice-Hall.

McKelvey, R. and Palfrey, T. (1995) "Quantal response equilibria for normal form games," *Game and Economic Behaviour*, 10(1): 6–38.

Miller, M., Krieger, D., Hardy, N., Hibbert, C. and Dean Tribble, E. (1997) "An automated auction in ATM network bandwidth," in S. H. Clearwater (ed.), *Market Based Control: A Paradigm for Distributed Resource Allocation*, World Scientific, pp.94–125.

Mindel, J. and Sirbu, M. A. (2001) "Taxonomy of traded bandwidth," unpublished manuscript, Engineering and Public Policy Department, Carnegie Mellon University.

Mullen, T. and Wellman, M. (1995) "A simple computational market for network information services," in *Proceedings of the First International Conference on Multi-Agent Systems*, AAAI Press, pp. 283–9.

Park, S., Durfee, E. and Birmingham, W. (1998) "Emergent properties of a market-based digital library with strategic agents," in *Proceedings of the Third International Conference on Multi-Agent System*, IEEE Press, pp. 230–7.

―― ―― ―― (1999) "An adaptive agent bidding strategy based on stochastic modeling," in *Proceedings of the Third Conference on Autonomous Agents*, ACM Press, pp. 147–53.

Parsons, S., Sierra, C. and Jennings, N. (1998) "Agents that reason and negotiate by arguing," *Journal of Logic and Computation*, 8(3): 261–92.

Preist, C. (1999) "Commodity trading using an agent-based iterated double auction," in *Proceedings of the Third Conference on Autonomous Agents*, ACM Press, pp. 131–8.

―― and van Tol, M. (1998) "Adaptive agents in a persistent shout double auction," in *Proceedings of the First International Conference on the Information and Computation Economies*, ACM Press, pp. 11–18.

Rodriquez-Aguilar, J., Noriega, P., Sierra, C. and Padget, J. (1997) "Fm96.5: a Java-based electronic auction house," in *Proceedings of the Second International Conference on the Practical Application of Intelligent Agents and Multi-Agent Systems*, Practical Applications Ltd, pp. 207–24.

Rosenschein, J. and Zlotkin, G. (1994) *Rules of Encounter*, MIT Press.

Roth, A. E. and Erev, I. (1995) "Learning in extensive-form games: experimental data and simple dynamic models in intermediate term," *Games and Economic Behaviour*, 8: 164–212.

Rustachini A., Satterthwaite, M. A. and Williams, S. R. (1994) "Convergence to efficiency in a simple market with incomplete information," *Econometrica*, 62(5): 1041–63.

Semret, N., Liao, R., Campbell, A. T. and Lazar, A. A. (2000) "Pricing, provisioning and peering: dynamic markets for differentiated Internet services and implication for network interconnections," *IEEE Journal on Selected Areas in Communications*, 18(12): 2499–513.

Shenker, S., Clark, D., Estrin, D. and Herzog, S. (1996) "Pricing in computer networks: reshaping the research agenda," *Journal of Telecommunications Policy*, 20(3): 183–201.

Shin, S., Correa, H. and Weiss, M. B. (2002) "A game theoretic modeling and analysis for Internet access market," in *ITS-14 Conference Papers, Regulation and Public Policy*.

Silvey, S. D. (1975) *Statistical Inference*, Chapman and Hall.

Smith, R. G. (1981) "The contract net protocol: high-level communication and control in a distributed problem solver," *IEEE Transactions on Computers*, C-29(12): 1104–13.

Tesauro, G. and Bredin, J. (2002) "Strategic sequential bidding in auctions using dynamic programming," in *Proceedings of the First International Joint Conference on Autonomous Agents and Multi-Agent Systems*, ACM Press, pp. 591–8.

_____ and Das, R. (2001) "High-performance bidding agents in the continuous double auction," in *Proceedings of IJCAI-01 Workshop on Economic Agents, Models and Mechanisms*, IJCAI Press, pp. 42–51.

Vishal, P. and Prashant, P. (2001) "Issues in bandwidth pricing using software agents," BTech Project, Indian Institute of Technology.

Vulkan, N. (1999) "Economic implications of agent technology and e-commerce," *Economic Journal*, 109: 67–90.

_____ (2000) "An economist's perspective on probability matching," *Journal of Economic Surveys*, 14(1): 101–18.

_____ and Jennings, N. (2000) "Efficient mechanisms for the supply of services in multi-agent environments," *Decision Support Systems*, 28 (1–2): 5–19.

_____ and Preist, C. (2003) "Automated trading in agents-based markets for communication bandwidth", *International Journal of Electronic Commerce*, 7(4): 119.

Walsh, W. E., Das, R., Tesauro, G. and Kephart, J. O. (2002) "Analyzing complex strategic interactions in multi-agent games," in *Proceedings of AAAI-02 Workshop on Game-Theoretic and Decision-Theoretic Agents*, AAAI Press, pp. 109–18.

Weiss, M. B. and Shin, S. (2002) "Internet interconnection economic model and its analysis: peering and settlement," *17th World Computer Congress*, pp. 215–31.

Wurman, P., Wellman, M. and Walsh, W. (1998) "The Michigan Internet AuctionBot: a configurable auction server for human and software agents," in *Proceedings of the Second Conference on Autonomous Agents*, ACM Press, pp. 301–8.

_____ Walsh, W. and Wellman, M. (1998) "Flexible double auctions for electronic commerce: theory and implementation," *Decision Support Systems*, 24(1): 17–27.

Yamaki, H., Yamauchi, Y. and Ishida, T. (1998) "Implementation issues on market-based QoS control," in *Proceedings of the Third International Conference on Multi-Agent Systems*, IEEE Press, pp. 357–64.

PART II
SECTION D

LAW DESIGN

CHAPTER 18

....

A MECHANISM DESIGN APPROACH TO LEGAL PROBLEMS

....

ALON KLEMENT AND ZVIKA NEEMAN[1]

INTRODUCTION

....

WE describe a mechanism design framework that could help identify a set of procedural mechanisms that would minimize the resources used to achieve one of the main goals of the court system, which is to differentiate between those who obeyed the law and those who did not. The proposed framework can also help to formulate and evaluate procedural rules, and to identify necessary and sufficient conditions for deciding disputes according to substantive law with minimal costs of litigation and delay. We illustrate our approach using three examples: fee-shifting rules, discovery rules, and third-party alternative dispute-resolution mechanisms.

The chapter proceeds as follows. We first identify a few inherent characteristics of the judicial process that make it apposite for modeling within a mechanism design framework, in which substantive law gives rise to a *social choice function*, and rules of procedure and evidence are captured by *game forms* or *mechanisms*. We illustrate our approach using four examples: the design of fee-shifting rules, the design of discovery rules, the design of fee structures for lawyers in class actions, and the use of third-party alternative dispute-resolution (ADR) mechanisms.[2]

---

[1] This chapter is based on our joint research that appears in Klement and Neeman (2004, 2005, 2008).

[2] The same framework may also be applied to analyze the rules of evidence. In fact, the approach advocated in this chapter closely resembles that of Sanchirico (1997).

CHARACTERISTICS OF THE CIVIL JUSTICE SYSTEM

Background

A civil justice system must provide just and efficient resolution of disputes. It must ensure the rule of law, offer redress to those whose rights were violated, and sanction those who infringed those rights. It must be accessible, accurate, and impartial. And it must consume as few social resources as possible.[3]

Most civil justice systems aspire to accomplish all these goals and more. Many civil justice systems (those of the USA, England, and Australia, to name but a few) have gone through a significant revision of their procedural rules in the past twenty years. These reforms were all fueled by a similar sense of crisis, and they all share a common set of principles that underlie the reformed rules. Although their relative weight and exact formulation varies, the following objectives can be found in most modern reformed rules of civil procedure: cost-effectiveness; proportionality; expeditiousness; and equality.

Cost-effectiveness means efficient use of judicial, as well as parties', resources. Proportionality addresses the need to distinguish and prioritize among cases based on their value (private and social) and complexity, due to the judicial system's limited resources. Expeditiousness requires that cases be resolved as quickly as possible, cutting down the time between filing and disposal. And equality commands that litigation be conducted on equal footing between the parties.

Each legal system establishes the measures it deems necessary to satisfy these objectives. Here, too, a comparative study demonstrates close similarities among the proposed, and often adopted, procedural mechanisms. They can be divided into two main categories. The first category includes measures that are intended to render the management of courts in general and litigation in particular more cost-effective. These measures include early judicial case management, timetabling, and alternative calendar systems.[4] These measures can be analyzed using methodologies from management science, and are outside the scope of this chapter.

The second category, which is of more interest to economists, includes procedural rules that affect litigants' incentives and decisions. Such decisions can be further divided into filing decisions, litigation investment decisions, and settlement decisions.[5] Filing decisions include the plaintiff's decision whether to file a lawsuit or not, and the defendant's decision whether to defend against it. Litigation investment decisions

---

[3] See for example Rule 1 of the American Federal Rules of Civil Procedure; Rule 1.1 of the English Civil Procedure Rules.

[4] See for example the reform proposals in England (Woolf, 1996), Hong Kong (Chief Justice's Working Party on Civil Justice Reform, 2004), and British Columbia (Civil Justice Reform Working Group, 2006).

[5] See Hay and Spier (1998), Spier (2005), and Daughety and Reinganum (2012).

include each party's decision regarding how much to spend on litigating the case. And settlement decisions include decisions regarding when to settle and for how much.

As fewer lawsuits are filed and defended, as litigants' investment in each case decreases, and as more of the lawsuits are settled, the justice system becomes less costly and delay is reduced. Yet, these three categories of decisions are interrelated, and may consequently interact with each other. For example, if litigation expenditures decrease, this may reduce incentives to settle, whereas the motivation to file and defend would increase. Moreover, these decisions affect the court's accuracy, and consequently attainment of its basic goal, which is to distinguish between liable and non-liable defendants.

A comparative study of civil justice reforms points to some procedural mechanisms that are often constructed to reduce cost and delay: dispute-resolution mechanisms based on third-party assistance; pretrial disclosure and discovery; fee-shifting rules; and pleadings procedures. Yet, there is little agreement about the effectiveness of these procedural mechanisms or their effect on the implementation of substantive law mandates. As we show, the mechanism design framework offers a fresh perspective on these issues.

The distinction between substance and procedure

One of the most fundamental distinctions in modern legal theory is the distinction between substance and procedure. Substantive law defines "rights, duties and powers of persons and institutions in their out-of-court relationships," whereas procedural law governs the "decision-making process by which substantive legal interests are maintained or redressed through courts." In its day-to-day application, the law of procedure implements substantive law.[6] Although the boundary between the two categories may be drawn differently depending on the context,[7] it is usually clear enough for practitioners to identify.

That such distinction exists does not imply, however, that procedural rules do not affect primary behavior, ex ante, before any dispute arises. Since procedural law imposes costs on litigants and because it influences the accuracy with which questions of rights and remedies are decided ex post, it also affects behavior ex ante.[8] Therefore, any measure of efficiency of the justice system must incorporate its ex ante effects.

The incorporation of ex ante deterrence effects and ex post costs of the judicial system into a single framework is a complex task. Whereas it is conceptually feasible to construct procedural mechanisms that would make litigants internalize all ex post litigation costs, it is much more difficult to do the same for deterrence. The deterrent effect of litigation is an ex ante effect, on behavior that pre-empts (and sometimes may even prevent) the dispute. By the time the dispute is brought into court, that behavior is already "sunk." Litigants, therefore, do not internalize the deterrent effects of their

---

[6] See James et al. (1992, p. 2). On the history of this distinction see, for example, Risinger (1982).
[7] See Cook (1933).
[8] See Scott (1975).

litigation decisions. This is referred to in the literature as the divergence between the social and the private incentive to use the legal system.[9]

The problem of civil justice reform has thus far been approached in two ways. One approach, which has been adopted by most reformers, was to ignore the ex ante deterrence effects and focus on ex post minimization of litigation costs and delay. From a social planning perspective this approach is at best incomplete. The other approach, which is sometimes used in the law and economics literature, is to ignore the inherent distinction between substance and procedure, and collapse all legal rules into one framework, in which the objective is to maximize ex ante efficiency. Because the distinction between substance and procedure is so fundamental in all legal systems, we believe that it should also be respected by the mechanism design analysis of legal problems, especially since this distinction is not merely a formalistic construct, and it may be explained on economic grounds as well.

First, the time gap between ex ante behavior and ex post litigation (ex ante and ex post relating to the time of dispute) makes it difficult to identify and quantify the deterrence effects of procedural rules. Consequently, the problem of constructing ex post procedural rules, which would be optimal from an ex ante perspective, may not be only conceptually difficult but also practically intractable.

Second, the ex ante deterrent effect of the same procedural mechanism may depend on the context in which it is applied. For example, the same discovery rule may influence behavior differently when the litigating parties are in a close relationship ex ante, as in a contractual setting, and when they are unaware of each other before the dispute, like in a typical tort case. Theoretically, then, it may be optimal to devise different discovery rules for different substantive contexts.

However, one inherent manifestation of the distinction between substance and procedure is that most modern procedural rules are trans-substantive. That is, they apply to all lawsuits, irrespective of their substantive cause of action. Therefore, associating an optimal procedural mechanism with a substantive context is usually unacceptable. A practicable framework for analysis must therefore allow for constraining the variability of procedural mechanisms across substantive contexts.

Finally, most people are unaware of procedural rules when conducting their out-of-court behavior. Procedural rules are usually in the realm of lawyers only. Hence, it may often be the case that the rules of procedure have no actual ex ante effect whatsoever. Distinguishing between rules that carry such effects, since individuals are aware of them ex ante, and those that do not, proves to be a difficult task.

To summarize, civil justice reforms have tended to ignore the ex ante effects of procedural rules, whereas the economic literature has often overlooked the inherent distinction between procedural rules and substantive law. We suggest a third alternative, which respects the distinction between substance and procedure, yet accounts for the influence of procedural rules on the implementability of substantive law. As explained in the next section, we do so by using a mechanism design framework, in which the social goal is to minimize litigation costs subject to the mandates of substantive law. Procedural

---

[9] See Shavell (1977, 1982a).

rules are used to determine game forms or mechanisms, for litigants to "play." We then look for procedural rules that would implement the social goal.

Private information and conflicting interests

The goal of the judicial process is to convey information to the court (judge or jury) so it can decide the dispute according to substantive law. Procedural rules regulate pretrial and trial activity, and consequently influence the sharing of information between the parties and its conveyance to the court. [10]

There are two types of information that the court does not hold. The first type is information shared by both litigants, but not by the court. This includes not only information regarding past events, but also information concerning specialized issues that require expert evidence in court. More generally, this is information which is observable by both parties, but is costly to verify in court.

The second type is information that is privately held by only one of the litigants, which the other litigant, as well as the court, does not know. Very often the defendant is privately informed about various aspects of her liability (what level of care she took, what information she had, etc.), whereas the plaintiff holds private information regarding her losses.

The lack of information makes the just and efficient implementation of procedural rules difficult. An uninformed court cannot apply such rules optimally without first learning the litigant's private information. Thus, for example, the decision whether to allow the plaintiff to use discovery measures against the defendant depends on the utility of such discovery and its costs, both unknown to the court. To take another example, a court contemplating whether to employ a provisional remedy against the defendant must weigh its costs against its utility in case the plaintiff prevails. Yet, the weights depend on the probability of plaintiff victory on trial, which the court does not know at the early stages after the lawsuit is commenced.

To overcome its lack of information the court relies on the adversarial nature of the lawsuit (even under inquisitorial systems[11]), which motivates litigants to reveal the relevant information and educate the court. Yet, it is exactly the adversarial behavior of the parties, or, more concretely, their conflicting interests, that requires innovative design of procedural mechanisms and active involvement of the court. Without such involvement, the litigants may engage in wasteful competition, spending more resources than socially optimal.

This combination of private information and conflicting interests complicates the design of an optimal procedural system. The more adversarial the system, the more information is uncovered, but the more costly the whole judicial process. On the other hand, if the court is endowed with greater powers to regulate and interfere in litigation

---

[10] The law of evidence, which is not analyzed here, determines which information can be brought to the court's attention, how, and what weight it should be given in the court's decision.

[11] On the adversarial nature of civil procedure in continental, usually perceived as inquisitorial, systems such as France, Italy, and Germany, see Davis (2002).

decisions, then total litigation costs may be reduced but less information may be conveyed to the court. Consequently, the court's decision would become less accurate, which often implies its decision is less just and efficient. The challenge is therefore to harness the litigants' private information in a way that would motivate them to educate the court about it, without increasing costs.

Mechanism design is a theoretical framework that is based on the two attributes described above: private information and conflicting interests. It allows the analyst to examine current and proposed mechanisms in situations that have these two attributes, and determine whether they are capable of implementing what is defined to be the social choice function. The next section presents the basic model for such analysis and demonstrates some of its possible applications.

THE MECHANISM DESIGN APPROACH

A mechanism design framework requires the analyst to define a social choice function or correspondence $f: \Theta \to C$ that maps every "state of the world" into an "outcome." Given the distinction between substance and procedure, and their above characteristics, we define the social choice function according to *substantive law* as follows:

> The set of states of the world, denoted Θ with typical elements $\theta \in \Theta$, describes everything that is relevant as far as the substantive law and the parties involved are concerned, including the involved parties' preferences and past actions. The states of the world are therefore not the ex ante states, before the dispute, but the *interim* states, after the dispute yet before litigation.

The set of outcomes, denoted C with typical elements $c \in C$, describes the set of all possible consequences, as conceived by substantive law. This set is independent of the procedural rule that is adopted to implement substantive law.

To take a simple example—the law of torts prescribes a remedy for the victim for any past action or omission of an alleged tort-feasor. Suppose that the alleged tort-feasor could have taken any one of n different actions, $\alpha 1, \ldots, \alpha n$. Suppose that, according to substantive law, if the alleged tort-feasor had taken any one of the actions $\alpha 1, \ldots, \alpha k$ then he is liable and should compensate the victim accordingly, and if he had taken any one of the actions $\alpha k + 1, \ldots, \alpha n$ then he is not liable and should not pay the victim anything. Hence, the state of the world consists of the action taken by the alleged tort-feasor, or $\Theta = \{\alpha 1, \ldots, \alpha n\}$, and the set of outcomes is given by the set of pairs (x, y) of non-negative real numbers, where x is the defendant's total liability, and y is the plaintiff's total recovery. If the damage to the victim is normalized to 1, then the social choice function under a negligence standard $f(\alpha_i)$ is $(1, 1)$ if i is between 1 and k, and $(0, 0)$ if i is between $k + 1$ and n. That is, the tort-feasor pays the victim's loss if the act or omission was negligent, and pays nothing otherwise.

Notice that the states of the world in this example are defined after the tort-feasor has already acted (and a loss was incurred). Thus, this formulation restricts attention to *liability rules* that determine the possible remedy after an action has been taken. [12]

Substantive law features twice in this model. First, it is instructive upon the court in its decision. Second, it describes the social choice function. The two are not the same because not all cases end in trial. The set of outcomes, C, describes the expected liabilities of the alleged tort-feasor and the expected recovery of the victim, which consists of cases that are litigated to judgment, as well as cases that are settled before or after they are brought to court. In particular, the case where the tort-feasor pays 100 in court with certainty, and the case where she pays 200 in settlement with probability 0.5, are treated as the same outcome for our analysis, assuming no litigation costs. The cases would be different if litigation costs were positive.

The problem of mechanism design is how to design a *game form* or a *mechanism*, M, whose solutions would belong to $f(\theta)$ for every state of the world $\theta \in \Theta$. [13] We interpret the choice of a mechanism as a choice of a procedural rule. Thus, the problem of finding a mechanism that accomplishes a certain goal or implements a certain social choice function becomes a problem of how to design a procedural mechanism that would implement substantive law.

More formally, a procedural mechanism consists of a pair $M = (A, m)$ where A is the set of actions that each party can take, and the mapping $m: A \times \Theta \to C$ describes the expected consequences of a profile of actions $a \in A$ when the state of the world is given by θ. The mechanism design problem can be conveniently described diagrammatically as shown in Figure 18.1. [14]

With reference to Figure 18.1, the mapping f describes the social choice function that maps states of the world in Θ into consequences in C. The mechanism M defines a set of rules that, together with the parties' preferences and relevant history as described in the relevant state of the world $\theta \in \Theta$, induces a game. The letter S denotes the "solution

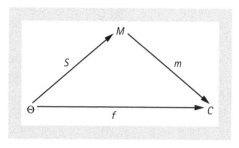

FIGURE 18.1. The mechanism design problem expressed as a Mount—Reiter triangle.

---

[12] This is opposed to *property rules*. The distinction between liability rules and property rules goes beyond the scope of this chapter. See, for example, Calabresy and Melamed (1972).

[13] Sometimes, the more stringent requirement that the solutions of M *coincide* with $f(\theta)$ for every state of the world $\theta \in \Theta$ is invoked.

[14] This diagram, which is known as a Mount–Reiter triangle, appeared in Mount and Reiter (1977).

concept" that is applied to this game. For example, in some situations, it may be reasonable to assume that the parties will play a Nash equilibrium; in other situations, a stronger solution concept such as dominant strategy equilibrium, or a weaker solution concept such as the sequential elimination of strictly dominated strategies, may be more appropriate. The point is that different states of the world will give rise to different games and different equilibrium outcomes, which will be mapped by the function m into different consequences. As explained earlier, the objective is that the outcome function, m, would map every relevant state of the world, $\theta \in \Theta s$, into $f(\theta)$.

Going back to the tort example, one procedural mechanism that can be examined is a pleadings rule. The defendant-alleged tort-feasor may be required to choose between acknowledging her liability or not. If she acknowledges her liability then she pays the plaintiff some amount. If she denies her liability then the plaintiff decides whether to pursue the case to trial or drop it. If he drops the lawsuit then each litigant gets some (possibly negative) payoff. If the plaintiff decides to proceed to trial then the court decides the case accurately, and, depending on its decision, awards each litigant some (possibly negative) payoff.

Different assumptions about the knowledge and beliefs of the parties, and about the appropriate S or "solution concept," translate into different mechanism design problems. In the tort example, the state of the world is privately known by the defendant, and so the appropriate solution concept is Bayesian Nash equilibrium.

Two remarks are in order. First, we assume that the outcome function, m: $A \times \Theta \to C$, depends on the parties' actions and on the state of the world. This formulation is more general than the common assumption in the implementation and mechanism design literature, where the outcome function depends only on the players' actions. The difference is due to the fact that a procedural mechanism typically involves a judge or an arbitrator, who may be able to observe the state of the world and to condition a decision on its realization.

Second, there are sometimes several procedural mechanisms that implement the same substantive rule. In such a case, we are interested in finding the mechanism that is optimal according to some other criterion of social welfare, such as the minimization of the sum of costs to the parties.

The next four sections illustrate the usefulness of the mechanism design approach through four examples: the design of fee-shifting rules, the design of discovery rules, the design of fee structures for lawyers in class actions, and the use of third-party alternative dispute-resolution (ADR) mechanisms.

THE DESIGN OF SETTLEMENT AND FEE-SHIFTING PROCEDURES

One important mechanism for inducing litigants to change their main litigation decisions is the shifting of litigation costs. Fee-shifting rules determine when and to what extent one litigant should reimburse another for her litigation costs. Fee-shifting rules can be divided into two main categories. First, there are outcome-based fee-shifting rules, which condition cost reimbursement on the outcome of trial. The two prominent fee-shifting rules in this category are the American rule, according to which each litigant bears her costs irrespective of the trial's outcome, and the English rule, in which the loser on trial fully reimburses the winner for her costs. Second, there are offer-of-settlement rules (sometimes called offer-of-judgment rules), which condition cost reimbursement on settlement offers that are rejected during litigation. One such rule is Rule 68 of the American Federal Rules of Civil Procedure, according to which "if the judgment finally obtained by the offeree is not more favorable than the offer, the offeree must pay the costs incurred after the making of the offer."

The law and economics literature has extensively analyzed the effect of both outcome-based fee-shifting rules, [15] and offer-of-settlement rules, [16] on the incentives to sue, settle, and invest in litigation. Yet, most of these studies have not attempted to identify the optimal settlement procedure and fee-shifting rule when the goal is to minimize the cost of litigation subject to the constraints imposed by substantive law such as maintaining deterrence.

An important exception is Spier (1994a), who offers a characterization of the fee-shifting rule that minimizes expected litigation costs or maximizes the likelihood of settlement but does not consider deterrence. In Klement and Neeman (2005) we extend her work by explicitly incorporating deterrence, and thus substantive law, into the analysis. We show that a settlement procedure that we call a *pleading mechanism* together with the English fee-shifting rule, according to which the loser in trial bears the legal costs of the winner, maximizes the likelihood of settlement, and maintains deterrence, as required by substantive law. We outline the main argument as follows.

Recall the tort example above. Suppose, for simplicity, that $n = 2$. That is, the tortfeasor is either liable or not, and the social choice function that one would like to implement, if possible, is $f(a_1) = (1, 1)$ and $f(a_2) = (0, 0)$. We show that this social choice function cannot be implemented. Intuitively, the reason for this is that implementation of this function requires the parties to go to trial with a positive probability, which imposes on the parties additional litigation costs that are not captured by the substantive liability rule.

We therefore examine a weaker substantive standard, which mandates only that the difference between the defendant's expected liability if she is liable and if she is not is 1. A liable defendant is thus still required to compensate the plaintiff for the entire damages caused, as under the original social choice function, but because of litigation costs, and the necessity of (ex post, inefficiently) going to trial in order to achieve justice, it is impossible to ensure that the expected payoff to the plaintiff be equal to the damage caused when the defendant is indeed liable, and be equal to zero otherwise.

We maintain the assumption that if the parties go to trial, then the court discovers whether or not the defendant is liable. And, we further assume that the court requires the

---

[15] See, for example, Shavell (1982b), Braeutigam et al. (1984), and Katz (1987).
[16] See, for example, Miller (1986), and Chung (1996).

defendant to compensate the plaintiff for her loss, 1, if she is found liable. We therefore assume a substantive law that does not allow decoupling, on the one hand, or punitive damages, on the other. In this example as well as in other contexts, substantive law imposes certain constraints, and procedural law provides the mechanism that satisfies these constraints.

However, since going to court is costly for both the plaintiff and the defendant, the objective is to design a settlement procedure that would compensate the plaintiff if and only if the defendant is indeed liable, and that would maximize the likelihood that the parties would settle outside of court and thus save the associated legal fees. The instrument that can be used for this purpose is fee-shifting rules.

As mentioned earlier, in Klement and Neeman (2005) we show that a *pleading mechanism*[17] together with the English fee-shifting rule maximizes the likelihood of settlement. The mechanism allows the plaintiff to make a take-it-or-leave-it settlement offer, which the defendant may either accept or reject. If the defendant rejects the offer then the plaintiff must decide whether to proceed to trial, and if she does then the court finds whether the defendant is liable or not, and allocates litigation costs according to the English fee-shifting rule.

The intuition for this result is the following. If it had been commonly known whether the defendant was truly liable or not, then, under the optimal mechanism, or the mechanism that maximizes the *interim* likelihood of settlement subject to a minimal deterrence constraint, the plaintiff and defendant would have settled with probability 1, and because of the deterrence constraint, the difference between the expected settlements of liable and non-liable defendants would have been equal to the extent of the damage caused to the plaintiff. Obviously, such a mechanism is not incentive compatible. In a world in which the defendant's true liability is not known to anyone but herself, a liable defendant has an incentive to pretend she is not liable so she can settle for less. It follows that an optimal mechanism must provide an incentive for liable defendants to admit their liability.

Because the defendant's true liability can be verified only in court, the only way to do this involves going to court with a positive probability. Because going to court is costly, the probability of going to court has to be minimized under the optimal mechanism. Conditional on the case going to trial, the English fee-shifting rule is the one that maximizes the difference between the expected payments of liable and non-liable defendants. Therefore, because the optimal mechanism should provide the "cheapest" possible incentives for being truthful, deterrence implies that it must rely on the English rule, because in this way it is possible to satisfy the deterrence constraint with the lowest possible probability of going to trial. The reason is similar to the well known argument that efficiency requires setting very large fines for those caught violating the law, but very small probabilities of detecting offenders (Becker, 1968).

---

[17] In Klement and Neeman's pleading mechanism the defendant is asked to plead whether she is liable or not. If she pleads liable, then she has to fully compensate the plaintiff for the damage caused. If she pleads not liable, then the plaintiff decides whether to litigate to trial or drop the case.

One interesting outcome of the model is that the maximum probability of settlement equals the probability that the defendant is liable, and it is independent of the litigant's litigation costs. This conclusion contradicts most theoretical and empirical findings, which identify a positive correlation between the probability of settlement and litigation costs. The intuition here is that under the English fee-shifting rule any increase in litigation costs renders litigation less profitable for the plaintiff, and she is therefore less willing to proceed to trial. But then, liable defendants have a stronger incentive to deny their liability and refuse to settle, hoping that the plaintiff will drop the suit. Under the optimal mechanism these two effects cancel out, and therefore the probability of settlement is kept constant.[18]

THE DESIGN OF DISCOVERY RULES

Historically, the common law has relegated all information transmission between the parties to the trial stage.[19] Modern civil justice systems have recognized, however, that pretrial discovery and disclosure devices are necessary. These may include: depositions, which are oral or written questioning of witnesses; interrogatories, which consist of written questions to a party; production of documents; physical or mental examinations of parties or persons under legal control of a party; and requests for admissions, which require a party to admit proposition of fact tendered in a written request.

Discovery can serve various objectives: it may eliminate fictitious controversies, upon which the parties would agree after discovery, and may even encourage pretrial settlement based on the information discovered; it can simplify the presentation of evidence at trial, by allowing the parties to exchange documents and review them before trial, and reduce the "gaming" effect of litigation; and it can secure the submission of accurate evidence, not deteriorated by the passage of time until trial. To put things more generally, pretrial discovery has two main goals: to encourage early settlement, and to raise the accuracy of trial.

Yet, pretrial discovery has its faults. Most significantly, discovery is costly. Since it is not limited to evidence which is admissible at trial, and since it is not constrained by the court's time, discovery usually increases total litigation costs for cases that are not settled.[20] Moreover, litigants can use discovery strategically, to force litigation costs upon their rivals. The question therefore stands, whether discovery's benefits outweigh its costs, and, if they do, whether and how discovery should be regulated by the court.

A large part of the literature has focused on the proper standard to implement in discovery disputes. That is, authors have attempted to draw up guidelines for

---

[18] For a similar conclusion see Nalebuff (1987), proposition 3.
[19] Equity cases were different, in that these facilitated the transmission of documentary evidence before trial. This was meant to overcome a party's privilege at common law trials not to testify against his own cause. See James et al. (1992, pp. 232–3).
[20] See Kakalik et al. (1998).

deciding whether a specific discovery application is justified and should therefore be allowed, and possible mechanisms to induce litigants to take mostly justified discovery measures. [21]

One exception is Mnookin and Wilson (1998), who used a mechanism design framework to show that under the optimal mechanism the initial allocation of information between the parties does not affect their expected ex ante gains from a joint discovery plan. In their model, therefore, the tension between the direct costs of discovery and its benefit in encouraging settlement and consequently saving litigation costs may be resolved through Coasian, pre-discovery, bargaining. Yet, their model does not account for the effect of discovery on the implementation of substantive standards.

In Klement and Neeman (2005) we have considered whether the likelihood of settlement under the optimal *pleading mechanism* and fee-shifting rule can be increased by the addition of a discovery phase. We showed that this is impossible, since under the assumptions of the model the probability of going to trial is a martingale. The expected posterior belief (i.e. following discovery) that the defendant is liable equals the prior belief (i.e. before discovery). Since the maximum probability of settlement equals the probability that the defendant is liable, discovery cannot increase the probability of settlement. Thus, any gain that discovery would produce in the probability of going to trial in some states of the world must be offset by corresponding losses in other states of the world.

THE DESIGN OF FEE STRUCTURES FOR LAWYERS IN CLASS ACTIONS

Class actions are private lawsuits in which the represented members of the plaintiff class are absent throughout the litigation, yet are bound by its outcome. It is not uncommon that in a single class action millions of plaintiffs may be represented, hundreds of millions of dollars may be at stake, and whole industries may be at risk of liability. [22] However, it is the opportunity for private profit, and not the concern for class members' interests, which motivates private attorneys to litigate class actions, invest their time and money, and bear the risk of no compensation if they fail to win a favorable judgment. Class actions thus provide a new paradigm for litigation—the private attorney general paradigm.

Courts have long been struggling with the challenges of managing class actions. Pursuing their own private profit often causes class attorneys to behave in an opportunistic manner, at the expense of the represented class. This tension, between the class action's

---

[21] See for example Sobel (1989), Cooter and Rubinfeld (1994), and Hay (1994).

[22] The most dramatic example is the asbestos industry, which has been exposed to numerous class actions since the 1970s, resulting in several defendants turning insolvent (Hensler et al., 1985).

social goals and the class attorneys' private profit, has generated much concern and debate with respect to the issues of how to select the class attorney, how to monitor her behavior, and how to compensate her. This section addresses the latter issue. It examines the way in which an attorney fee structure that maximizes the expected recovery for class members may be implemented in practice.

Unlike ordinary litigation, where courts do not usually intervene in the litigants' choice of attorney, in their attorney fee arrangements, or in their settlement decisions, in class actions courts are required to do all of these, in order to secure for class members proper compensation, given the merit of their case. Although it may seem that the court's problem in designing optimal fee structures for class attorneys is similar to the one faced by litigants in ordinary litigation, three important features of class actions render this problem more complicated.

First, whereas individual clients may choose to pay their lawyers a non-contingent fee, a class attorney's litigation fee must be contingent on winning the trial. Class members are dispersed and are very costly to identify, especially when the defendant wins an adverse judgment, because no individual class member has an incentive to step forward and identify herself just for the sake of bearing the class attorney's costs. Furthermore, as a matter of law and practice, absent class members are not liable for costs of litigation or attorneys' fees in the event of an adverse judgment against the class, so class attorneys are not compensated unless they create a common fund for the class by winning or settling the lawsuit.

Second, individual clients have strong incentives to take adequate measures to directly monitor their attorneys, which class members and their representatives lack. Most class actions are "lawyer driven" and the class attorney maintains all but absolute control over the lawsuit. She usually initiates the suit, selects the class representative, and controls both the litigation process and settlement decisions. The class representative, while supposedly in charge of the litigation as fiduciary for all those similarly situated, is in reality only a token figurehead, with no actual control over the lawsuit. Other class members' involvement is even less significant, as they are inclined to free ride on any litigation investment, sharing its proceeds without bearing the associated costs.

Finally, and as we show, most importantly, in ordinary litigation lawyers "compete" for individual clients, and are thus forced to offer optimal fee arrangements given the merits of individual clients' cases, in spite of the fact that the individual clients themselves may not always be aware of all the salient features of their cases. In contrast, in class actions the choice of attorney is usually made only indirectly. Typically, the court chooses the representative class member out of the class members who initiated the lawsuit, and the representative's attorney is then automatically appointed to represent the class. Although such a selection process is instrumental in motivating lawyers to search for worthy causes of action and appropriate class representatives, it nevertheless undermines the competitive forces in the selection of the class attorney. Moreover, the potentially large financial burden of the class action results in a limited and specialized class action bar, which further limits the possibility for a real market for class attorneys.

Using a mechanism design approach, it is possible to show that if the court can observe the class attorney's effort (the number of hours she spent on the case), then the optimal expected payment to the class may be realized using the *lodestar* method—a contingent hourly fee arrangement which is currently practiced in many class actions—but only if the hourly contingent fee is multiplied by a *declining*, as opposed to the practiced *constant* multiplier. That is, the optimal contingent fee to the class attorney should be concave in the number of hours worked. In some circumstances, the same optimal fee structure can be implemented even if the court cannot observe the class attorney's effort, and is therefore forced to use a *percentage* fee. In such cases the class attorney can optimally be offered a choice among a schedule of fees, each consisting of a fixed percentage and a threshold amount below which the class attorney earns no fee, with the threshold increasing with the chosen fixed percentage. The class attorney is paid the fixed percentage chosen only for amounts won above the threshold.

Both fee schedules allow the class attorney to capture a positive rent, over and above her reservation value. This positive rent is a direct consequence both of the court's inability to secure optimal effort by the class attorney (the moral hazard problem) and of the court's lack of information concerning the attorney's ability and the merit of the case (the adverse selection problem). The possible equivalence of the optimal percentage and lodestar methods suggests that the adverse selection problem should be of much concern to courts and regulators when considering how to reform class actions. This finding should be contrasted with the extensive attention given by the literature to lawyers' moral hazard problems, and the scant discussion, if any, devoted to adverse selection issues.

To gain some intuition for our results, suppose first that the court can perfectly observe and monitor the time the class attorney spends on the case but is not completely informed about either the attorney's ability, or the merits of the case. In other words, the court does not know the class attorney's production function—the way in which her effort would affect the expected judgment—which implies that the court faces the problem of determining the level of effort that should be optimally exerted by the attorney.

Clients in ordinary litigation do not usually face such a problem, for two reasons. First, the attorney can be paid her regular hourly fee independently of the outcome of trial. When paid the reservation value of her time, the attorney is likely to abide by both professional and ethical duties toward her client, and invest optimally in the case. Second, even assuming away professional and ethical considerations, competition among attorneys is likely to drive attorneys' fees toward their respective reservation values, leaving all the surplus to the client.

In contrast, in class actions the attorney's compensation must be contingent on winning, and therefore it must be adjusted to account for the risk of non-payment. The lower the probability of winning, the higher the likelihood of non-payment, and the higher should be the adjustment of the attorney's fee. In the absence of any competitive forces the attorney may therefore be tempted to pretend that the probability of not winning is higher than it actually is, in order to win a higher adjustment. Such behavior generates inefficiency, for two reasons. First, in order to reduce the rent a high-probability attorney can obtain from pretending to have a lower probability of winning, the court has to limit the number of hours paid to low-probability attorneys, thus having them exert less effort than their optimal level in the absence of asymmetric information. Second, this implies that it is impossible to prevent high-probability attorneys from obtaining a positive informational rent.

By pre-specifying different levels of effort and adjustments, the court should optimally screen among the different "types" of attorney, in order to have each attorney's investment in the case be as close as possible to the optimal investment, given her information. However, such optimal screening cannot avoid underinvestment of the attorney's effort on the one hand, and overpayment to the attorney on the other.

The main result of Klement and Neeman (2004) is that when the class attorney possesses private information about the probability of winning the class action, the rent that she extracts under the optimal fee schedule may be so large that, by using a percentage fee schedule, the same optimal pairs of effort and adjustments can be implemented even if the attorney's effort cannot be observed at all. Intuitively, a percentage fee induces the class attorney to work on the case up to the point where her marginal return equals her per-hour cost. Since the attorney's marginal return is increasing in her percentage, so is her choice of effort. Klement and Neeman (2004) show that to implement the optimal fee schedule the percentage that is chosen by the attorney must be increasing in her estimated probability of winning. At the same time, to extract at least part of the attorney's informational rent, each percentage must be coupled with a threshold amount below which the attorney earns no fee. They show that optimal screening among attorneys according to their estimated probabilities of winning requires coupling a higher percentage with a higher threshold, which still leaves the attorney an informational rent that increases in her probability of winning. As it turns out, the informational rent of the attorney under this payment scheme need not be higher than the rent she obtains under the optimal fee schedule when her effort is observable.

The design of third-party ADR mechanisms

Most proposals for reform of the judicial system include detailed plans to encourage litigants to use alternative dispute-resolution (ADR) mechanisms. Among those, two stand out: arbitration and mediation. Arbitration is an adjudicative procedure, in which a privately hired third party hears the evidence and then delivers a (potentially) binding decision. Mediation is a facilitative procedure, where the third party assists the litigants to reach an agreement and settle their dispute.

There is a vast amount of theoretical and empirical literature on the use of third-party ADR mechanisms.[23] Yet, none of this literature seems to answer a fundamental puzzle: how can third parties improve either the quality of decision-making or the efficiency of settlement negotiations? Indeed, a number of results in the mechanism design literature that describe how in some contexts it is possible to "decentralize" any social choice rule suggest that they cannot.[24]

Both mechanisms have many effects that go beyond the scope of a simple rational behavior model. Nevertheless, it would be interesting to examine their more limited implications within such a model. ADR mechanisms may figure into our framework in two possible variations – arbitration and mediation. In the first, one would replace the court with an arbitrator who could deliver an "intermediate" judgment. That is, the arbitrator, unlike the court, would not be required to decide the dispute on an all-or-nothing basis, and may therefore decide that the defendant should compensate the plaintiff for only part of his losses. More generally, an arbitrator would not be bound by substantive law. The question, thus, is whether relaxing this constraint can help implement substantive standards more efficiently.

Notice that the parties may opt for arbitration either before the dispute or after it. Signing an arbitration agreement before the dispute is not always possible. Yet, when it is possible, the parties would do so only if it would minimize their litigation costs, subject to the substantive law constraint (assuming, of course, that substantive law is efficient). That is, they will choose arbitration only if it is ex ante efficient. On the other hand, if the arbitration option is available only after the dispute, the parties would attribute no value to maintaining the substantive law constraint. They will, therefore, select arbitration only if its outcome is ex post efficient for both. It is interesting to examine the effect of the different timing of selection on the optimal structure of arbitration mechanisms.

In the second variation, we may want to allow the litigants to use a mediator who can transfer information between them before they decide whether to settle, and to help them coordinate on a specific correlated equilibrium. Referring back to Figure 18.1, the mediator can affect both the mechanism, M, and the appropriate solution concept, S. Some important work on these questions has already been done by Brown and Ayres (1994). Yet, they, too, have not accounted for the effects of mediation on the implementation of substantive standards.

A crucial first step toward addressing these issues hinges on the questions of when and how it is possible to find simple "practicable" game forms that would implement the same social choice function as some given abstract direct-revelation mechanism. These questions, which are still very much open questions in mechanism design theory, can be answered in some contexts (cf. auction theory). The challenge is to come up with a general answer that would shed light on the question of mediation versus arbitration versus abstract mechanism design.

---

[23] See for example Shavell (1994) and Bernstein (1992).

[24] For example, the second-price auction is a decentralized mechanism that implements the optimal allocation in a single good auction.

CONCLUDING REMARKS

This chapter draws a template for future research. It introduces a framework that respects the ingrained distinction between substance and procedure, yet does not undermine the substantive (or primary) effects of procedural rules. Using a mechanism design approach in which substantive law defines the social choice function and procedural rules describe possible game forms that may implement it, may prove useful in realizing the possible effects and limitations of various procedural mechanisms.

However, like any other model or approach, the mechanism design approach also has its weaknesses. It tends to abstract away from many complicating factors that often prove very important in practice. And it may prove to be sensitive to the allocation of information between the litigants, to their renegotiation opportunities, and to various sources of bounded rationality.

Mechanism design is therefore one more instrument in the policy-maker's toolkit. It may offer a fresh perspective over long-debated issues. Combining it with other theoretical and empirical methodologies would prove fruitful in the search for more efficiency and justice in legal systems.

References

Becker, G. S. (1968) "Crime and punishment: and economic approach," *Journal of Political Economy*, 76: 169–217.

Bernstein, L. (1992) "Opting out of the legal system: extralegal contractual relations in the diamond industry," *Journal of Legal Studies*, 21: 115–57.

Braeutigam, R., Owen, B. and Panzar, J. (1984) "An economic analysis of alternative fee shifting systems," *Law and Contemporary Problems*, 47: 173–204.

Brown, J. G. and Ayres, I. (1994) "Economic rationales for mediation," *Virginia Law Review*, 80: 323–402.

Calabresi, G. and Melamed, A. D. (1972) "Property rules, liability rules and inalienability: one view of the cathedralí," *Harvard Law Review*, 85: 1089–128.

Chief Justice"s Working Party on Civil Justice Reform, Hong Kong (2004) *Final Report*, available on <http://www.bcjusticereview.org/working_groups/civil_justice/cjrwg_report_11_06.pdf>.

Chung, T.-Y. (1996) "Settlement of litigation under Rule 68: an economic analysis," *Journal of Legal Studies*, 25: 261–86.

Civil Justice Reform Working Group to the Justice Review Task Force, British Columbia (2006) *Final Report: Effective and Affordable Civil Justice*, available on <http://www.bcjusticereview.org/working_groups/civil_justice/cjrwg_report_11_06.pdf>

Cook, W. W. (1933) "'Substance' and 'procedure' in the conflict of laws," *Yale Law Journal*, 42: 333–58.

Cooter, R. D. and Rubinfeld, D. L. (1994) "An economic model of legal discovery," *Journal of Legal Studies*, 23: 435–64.

Daughety, A. F. and Reinganum, J. (2012) "Settlement," in *The Encyclopedia of Law and Economics*, (Second ed.), Chapter 15, vol. 8 (Procedural Law and Economics, Sanchirico, C. W.), Edward Elgar Publishing Co Ltd., 386–471.

Davis, G. L. (2002) "The reality of civil justice reform: why we must abandon the essential elements of our system," *Journal of Judicial Administration*, 12(3): 155–171.

Hay, B. L. (1994) "Civil discovery: Its effects and optimal scope," *Journal of Legal Studies*, 23: 481–517.

_____ and Spier, K. E. (1998) "Settlement of litigation," in P. Newman (ed.), *The New Palgrave Dictionary of Economics and the Law*, Macmillan Reference Limited, vol. 3, pp. 442–51.

Hensler, D., Felstiner, W., Ebener, P., and Selvin, M. (1985) "Asbestos in the Courts: The Challenge of Mass Toxic Torts", *RAND*, R-3324-ICJ.

James, F., Hazard, G. C. and Leubsdorf, J. (1992) *Civil Procedure*, 4th edn, Little, Brown.

Kakalik, J., Hensler, D. R., McCaffrey, D., Oshiro, M., Pace, N. M. and Vaiana, M. E. (1998) *Discovery Management: Further Analysis of the Civil Justice Reform Act Evaluation Data*, RAND, The Institute for Civil Justice.

Katz, A. W. (1987) "Measuring the demand for litigation: is the English Rule really cheaper?" *Journal of Law, Economics, and Organization*, 3: 143–76.

Klement, A. and Neeman, Z. (2004) "Incentive structures for class action lawyers," *Journal of Law, Economics, and Organization*, 20: 102–24.

_____ _____ (2005) "Against compromise: a mechanism design approach," *Journal of Law, Economics, and Organization*, 21: 285–314.

_____ _____ (2008) "Civil justice reform: a mechanism design framework," *Journal of Institutional and Theoretical Economics*, 164: 52–67.

Miller, G. P. (1986) "An economic analysis of Rule 68," *Journal of Legal Studies*, 15: 93–125.

Mnookin, R. and Wilson, R. (1998) "A model of efficient discovery," *Games and Economic Behavior*, 25: 219–50.

Mount, K. and Reiter, S. (1977) "Economic environments for which there are Pareto satisfactory mechanisms," *Econometrica*, 45: 821–42.

Nalebuff, B. (1987) "Credible pretrial negotiation," *RAND Journal of Economics*, 18: 198–210.

Polinsky, A. M. and Rubinfeld, D. L. (1988) "The deterrent effects of settlements and trials," *International Review of Law and Economics*, 8: 109–16.

Risinger, D. M. (1982) " 'Substance' and 'procedure' revisited: with some afterthoughts on the constitutional problems of 'irrebuttable presumptions', " *UCLA Law Review*, 30: 189–216.

Sanchirico, C. W. (1997) "The burden of proof in civil litigation: a simple model of mechanism design," *International Review of Law and Economics*, 17: 431–47.

Scott, K. E. (1975) "Two models of the civil process," *Stanford Law Review*, 27: 937–50.

Shavell, S. (1982a) "The social versus the private incentive to bring suit in a costly legal system," *Journal of Legal Studies*, 11: 333–9.

Shavell, S. (1982b) "Suit, settlement and trial: a theoretical analysis under alternative methods for the allocation of legal costs," *Journal of Legal Studies*, 11: 55–82.

_____ (1994) "Alternative dispute resolution: an economic analysis," *Journal of Legal Studies*, 24: 1–28.

_____ (1997) "The fundamental divergence between the private and the social motive to use the legal system," *Journal of Legal Studies*, 26: 575–613.

Sobel, J. (1989) "An analysis of discovery rules," *Law and Contemporary Problems*, 52: 133–59.

Spier, K. E. (1994a) "Pretrial bargaining and the design of fee shifting rules," *RAND Journal of Economics*, 25: 197–214.

——— (1994b) "Settlement bargaining and the design of damage awards," *Journal of Law, Economics, and Organization*, 10: 84–95.

——— (2007) "Litigation," to appear in A. Mitchell Polinsky and Steven Shavell (eds), The Handbook of Law and Economics, North Holland.

Varano, V. and De Luca, A. (2007) Access to Justice in Italy, Global Jurist Vol. 7: Iss 1 (Advances), Article 6.

Woolf, Lord (1996) *Access to Justice: Final Report*, HMSO.

Zuckerman, A. A. (ed.) (1999) *Civil Justice in Crisis: Comparative Perspectives of Civil Procedure* (Oxford, 1999).

CHAPTER 19

LEGISLATION WITH ENDOGENOUS PREFERENCES

AVIAD HEIFETZ, ELLA SEGEV,
AND ERIC TALLEY[1]

INTRODUCTION

IN recent years, the rational actor model—pivotal to much modern economic theory—has fallen under renewed criticism from scholars both inside and outside economics proper. By at least some accounts, this scrutiny is long overdue. Indeed, there are scores of examples where observed behavior appears strikingly at odds with at least the most straightforward predictions of models with strong rationality assumptions. The growing literature in *behavioral economics* is largely dedicated to cataloguing and systematizing instances in which preferences are internally inconsistent, dynamically unstable, or actuarially biased. Examples of such phenomena include overconfidence, endowment effects, framing effects, self-serving biases, heuristics, cycling, and various forms of bounded rationality (see e.g. Rabin, 1998, for an overview).

Given the ascendancy of this literature, there is a natural urge to transcend the positive connections between decision-making problems and cognitive biases, and to explore the *normative* consequences that such phenomena imply. Not surprisingly, a number of recent efforts in the literature appear to do just that, using topics within behavioral economics as springboards for proposing market interventions or legal reforms that attempt to compensate for the existence of cognitive preference distortions.

[1] This chapter is based on Heifetz, A., Segev, E. and Talley, E. (2007) "Market design with endogenous preferences," *Games and Economic Behaviour*, 58, pp. 121–53, with permission from Elsevier. We thank Jennifer Arlen, Ian Ayres, Colin Camerer, Richard Craswell, and Ehud Kamar. All errors are ours. Support from the Hammer fund for Economic cooperation is also gratefully acknowledged.

Sunstein (2002), for example, considers how elimination of at-will employment doctrine may help address problems with endowment effects.[2]

In the main, these normative approaches tend to view cognitive biases as exogenous parameters within a behavioral model, and take preferences (distorted by biases) to be a primitive building block of equilibrium behavior. Such an assumption, however, stands in contrast with much of the existing experimental evidence, which suggests that many cognitive dispositions appear to be highly context specific, rising to first-order importance in certain settings, while curiously marginal in others (Camerer et al., 2003). Thus, without a more general theory of context, it is difficult to predict how (or whether) various biases occurring in the laboratory should translate to the real-world targets of policy reforms, and how these biases may be *affected* by such interventions.

In this chapter, we put forward a model for analyzing how context and cognition plausibly interact with one another, and a resulting framework for studying institutional design within such a setting. Our analysis reveals that the task of designing institutions in the presence of cognitive biases is somewhat more complicated than in the classical approach to design problems, for at least two reasons. First (and most centrally), regulatory interventions *themselves* are likely to distort context, and in so doing may affect the direction or magnitude of various cognitive dispositions. When contemplating issues of institutional design and policy, then, one must take care not only to identify the biases which cause inefficiencies, but also to anticipate the *feedback effects* induced by the very regulatory apparatus meant to compensate for them. Such feedback effects are frequently not incorporated into normative policy proposals, and their omission could very well lead to imprecise, inefficient, and ultimately ineffectual reforms.

Second, even if one could anticipate the feedback effects described above, a particularly thorny problem remains in specifying a reasonable definition of "optimality" in the presence of endogenous preferences. Indeed, conventional notions of economic welfare become more contested in environments where preferences themselves shift over time (see, e.g., Carmichael and MacLeod, 2002). As we demonstrate below, the nature of an optimal regulatory intervention may turn dramatically on whether one defines optimality in terms of context-specific or context-independent preferences.

The analytic model we propose is one in which cognitive dispositions—much like behavior itself—arise endogenously, through an equilibrium process. Thus this chapter adds to a recent literature on endogenous preferences which examines the relationships between endogenously changing preferences and economic institutions. We give here a brief overview of this literature.

In 1993, Stiglitz phrased the idea that the market structure influences preferences, and asked how should we design economic organizations. He analyzed examples in labor markets, financial markets, and others, in which changes in the market may change individuals' preferences such as altruism, cooperativeness, and willingness to take risks.

---

[2] Other examples include Jolls (1998), who argues that wealth redistribution is more efficiently accomplished through legal liability rules than through tax and trasnfer systems, since people systematically underestimate the likelihood of legal liability.

Following his footsteps, Aaron (1994) and Bowles (1998) also call for taking preference endogeneity into account when designing economic institutions.

The concept of endogenous preferences has been examined in different strategic environments. In his 1974 paper, Gintis was among the first to formalize the idea in a general equilibrium model that analyzes the market for labor. He compared the result of the neoclassical approach which treats the preferences as exogenous with his approach and derives conditions under which the two approaches lead to a different level of overall welfare. Palacios-Huertaa and Santos (2004) also develop a general equilibrium model in which the player's attitude toward risk (a parameter which influences her utility) changes over time as a result of the financial structure of the market, and discuss the formation of the long-run equilibrium of such a process. Bonatti (2007) analyzes optimal policies for the government in a general equilibrium model of a market with for-profit and non-profit organizations. Agents' willingness to devote effort to non-profit activities is indirectly influenced by the aggregate volume of non-profit activities. Thus public policy that supports such activities can also affect agents' preferences.

Another common strategic environment for analyzing the relationships between market institutions and preferences is that of a public good in which an individual decides how much, if at all, she will contribute to the public good. Her preferences then might change according to whether a subsidy or a different incentive is given, as in Bar-Gill and Fershtman (2000), and Bowles and Hwang (2008).

In political economy Gerber and Jackson (1993) examine empirical evidence suggesting that preferences are endogenous to the electoral process. Moreover, Fershtman and Heifetz (2006) show that a plausible outcome of the process of a change in policy variables followed by a change in voters', taste can be political instability. They emphasize that an elected politician should be sophisticated and adopt a policy that will be optimal to most voters *after* the opinions will adjust to that policy.

In bargaining theory Güth and Napel (2006) and Berninghaus, et al. (2007) show that if players play either an ultimatum game or a dictator game then preferences biases such as inequality aversion, reciprocity, and equity concerns change as the probability of playing any of these games (i.e. the environment) changes. Finally, in the context of law and economics there exists a string of works that examine the design of legal institutions in the presence of endogenously changing preferences. Among these are works by Sunstein (1993), Bar-Gill (2006), Bar-Gill and Fershtman (2004), and Güth and Ockenfels (2005).

The approach of endogenous preferences hinges on the *commitment value* of preferences: i.e., it may "pay" to be concerned with motives other than one's own wealth (in the form of biased preferences), since so doing can induce other actors to make favorable accommodations in their equilibrium behavior. A recent literature[3] whose origins trace

---

[3] Samuelson (2001) presents a brief overview of this literature, while Dekel et al. (2007) and Heifetz et al. (2007b) present general results in this vein. Examples include Güth and Yaari (1992), Huck and Oechssler (1998), Fershtman and Weiss (1997, 1998), Rotemberg (1994), Bester and Güth (1998), Possajennikov (2000), Bolle (2000), Bergman and Bergman (2000), Koçkesen et al. (2000a, 2000b), Guttmann (2000), Sethi and Somanathan (2001), Kyle and Wang (1997), Benos (1998), Heifetz and Segev (2004).

back (at least) to Becker (1976) has studied this concept of the commitment value of preferences. Ultimately, holding the economic environment constant, the interaction between individual biases and this responsive accommodation by others can generate a stable equilibrium both in preferences and in behavior. Accordingly, when the underlying economic environment governing the interaction changes, preferences and behavior will both adjust as well.

The linchpin of the equilibration process we posit is an assumption that those who adopt preference dispositions yielding larger material rewards (as measured by their context-independent preferences) also tend to become more prominent in the population of players. While this equilibrating process is certainly reminiscent of literal evolutionary equilibrium concepts, it is significantly broader than that. The same account, for example, would also apply to situations where individuals simply imitated and adopted the attitudes and norms of those who appear to be successful over the long term, thereby reducing the economic influence of others. This concept of preference equilibrium is a natural embarking point for an economic analysis of preference endogeneity, since (1) it is grounded in first principles rather than exogenous assumptions about biases; and (2) it ultimately subscribes to the notion that individuals adapt in a way which is beneficial to their own, genuine welfare (albeit indirectly and unconsciously).

In order to illustrate the application of our framework, we explore its consequences in what is perhaps the most fundamental arena of economic interaction: bilateral exchange. Using a familiar, canonical framework of non-cooperative bargaining with two-sided private information as a benchmark (e.g. Myerson and Satterthwaite, 1983), we characterize the emergence of preference distortions during bargaining that cause negotiators to skew their perceived private values away from those they would perceive outside the bargaining context. Such preference distortions are commonly observed in the experimental literature, often associated with the "endowment effect," the "self-serving bias," or both. We demonstrate how such cognitive dispositions can benefit private negotiators, effectively transforming them into "tougher" bargainers than they would be in the absence of bias, thereby augmenting the credibility of their threat to exit without an agreement. Moreover, based on the analysis in Heifetz and Segev (2004), we illustrate how such transitory preference distortions are a viable equilibrium trait within a population of parties who bargain in thin market settings, identifying the emerging preference-behavior equilibrium as a function of the bargaining scheme.

We then turn our attention to the question of optimal institutional design. Using the bilateral bargaining framework described above as a template, we demonstrate how various market interventions—either by the state or by a benevolent third party—can have profound effects on both the existence and the magnitude of transitory preference distortions during negotiation. Accounting for these effects can cause an optimal regulatory scheme to differ (sometimes dramatically) from that in which cognitive dispositions were either assumed away or treated as exogenous primitives.

Significantly, the market interventions we analyze are not merely fanciful figments of our collective imagination, but rather real-world mechanisms through which third parties can (and do) exercise regulatory power at various points in the bargaining process. In particular, we focus on three genres of actual market intervention (differentiated by the time at which regulation occurs) that are particularly salient:

- Ex post intervention. A number of institutional devices exist for rewarding traders upon reaching a negotiated outcome. For example, various elements of the tax code often act to subsidize the consummated bargains.[4] This approach is also increasingly common in the international arena, as funding sources (such as the World Bank) have begun to de-emphasize the importance of demonstrating economic need, basing their funding decisions more centrally on a model of rewarding the resolution of international conflicts and the implementation of internal agreements to distribute aid effectively.[5]
- Interim intervention. Other forms of regulatory intervention occur in the negotiation process itself, artificially constraining the types of bargains that are allowed. For example, numerous legal rules (such as the doctrines of consideration and unconscionability in contract law, and the doctrine of moieties in admiralty law) operate to narrow the range of enforceable bargaining outcomes relative to what the parties would find individually rational. In addition, in some circumstances price/wage ceilings and floors operate with a similar constraining effect.
- Ex ante intervention. Still other regulatory interventions take place before bargaining even begins, at the point at which initial property rights are assigned. A substantial portion of common-law doctrines and statutory provisions is dedicated to specifying the contours of individual property rights, ranging from strong monolithic entitlements protected by injunctive relief, to weak entitlements that are either protected solely with damages or are subject to other forms of divided ownership (e.g. Ayres and Talley, 1995). Individuals frequently negotiate transfers of title in the shadow of these entitlements.

Within each of these examples, we show how an optimal regulatory intervention would account not only for garden-variety market failures, but also for the endogenous cognitive shifts that regulatory interventions themselves can trigger. In so doing, we highlight how many of the now-accepted approaches for mitigating *strategic* barriers to trade might fare once *cognitive* barriers are also taken into account.

In some instances, the fit is a poor one. For example, while subsidizing successful trades (i.e. ex post intervention) has long been recognized as a method for counteracting the effects of strategic behavior by privately informed parties, we demonstrate that such subsidy schemes can themselves aggravate transitory preference shifts. As a result, optimal market design in the presence of endogenous cognitive biases may require a significantly greater amount of intervention in the market (i.e. a higher subsidy

[4] See, e.g., Cal. Rev. & Tax Code §17053.84 (paying a 15% tax credit for the purchase and installation of solar energy equipment by California residents).

[5] See, e.g., Dollar (2000).

rate) than would be necessary in an environment where agents were devoid of such distortions.[6]

Less pessimistically, we demonstrate that there are some forms of intervention in which strategic and cognitive concerns overlap. For instance, we show how "weak" property entitlements (such as joint ownership or "fuzzy" property rights) not only help to mitigate strategic misrepresentation (e.g. Cramton et al., 1987; Ayres and Talley, 1995), but they can also help to dissipate cognitive dispositions toward toughness.

In a similar vein, we find that interim interventions constraining the types of allowed bargains, typically devised to secure incentives in the process of information exchange, also help in mitigating transitory cognitive shifts. However, it turns out that *excessive* such intervention—*beyond* the extent prescribed in Myerson and Satterthwaite (1983)—will in fact make the traders happiest in expectation *given* the *endogenous* level of these shifts.

Our analysis illustrates also the second obstacle in conducting market design in the presence of cognitive biases: the elusive meaning of the term "optimal." Indeed, in a situation where individual preferences are prone to endogenous shifts, utilitarian notions such as efficiency become significantly more indeterminate than they are in traditional rational choice theory. In particular, one might justifiably choose to focus on a notion of welfare rooted in a-contextual preferences (what we shall intermittently refer to, perhaps with some inaccuracy, as "wealth"), corresponding to those preferences individuals manifest in the abstract, outside of an adversarial bargaining context. Alternatively, one might conceive of welfare rooted in their preferences during trade (what we shall intermittently refer to as "happiness"), corresponding to those that individuals would perceive through their transitory preference dispositions at the point of bargaining. As we demonstrate, this distinction is an important issue for policy design, as wealth- and happiness-maximizing approaches frequently point at divergent institutional structures.

The rest of this chapter is organized as follows. The following section introduces a general framework for institutional design problems in the presence of endogenous dispositions. The third section applies this framework to the case of bilateral exchange with private information, as in Heifetz and Segev (2004). We start by deriving the dependence of the equilibrium dispositions on the bargaining mechanism, and demonstrate how this idea is operationalized in a particular bargaining equilibrium. The fourth section then turns to our constructive enterprise, demonstrating (*ad seriatim*) how ex post, interim, and ex ante interventions into market structure can affect the existence and degree of preference distortions during bargaining. We characterize the optimal regulatory intervention under the alternative goals of maximizing actual gains from trade ("wealth") versus maximizing the gains from trade as the traders perceive them to be during trade (the traders' "happiness"). We compare not only these optimal interventions to one

---

[6] This particular result turns on the social planner attempting to maximize a-contextual preferences. However, as we show later, regardless of what the social planner attempts to maximize, the optimal subsidy will generally diverge (either above or below) that of a classical analysis, in which biases are assumed away.

another, but also against the baseline case in which cognitive dispositions were wholly absent. The fifth section concludes.

GENERAL FRAMEWORK: THE DESIGN
OF INSTITUTIONS WHEN DESIGN
AFFECTS PREFERENCES

In this section we describe a general framework to evaluate institutional design when preferences of individuals may be endogenously sensitive to this design. The next two sections apply these ideas to the case of bilateral exchange with private information.

Let O be a set of outcomes pertaining to the individuals $i \in I$. Let $\mathbb{U}_i$ be a set of utility functions $U_i : O \to R$ that individual i may have. We denote by $U = (U_i)_{i \in I}$ a utility profile of the individuals. The set of utility profiles is $\mathcal{U} = \prod_{i \in I} \mathbb{U}_i$.

Let there be a welfare aggregation function $\mathcal{W} : \mathcal{U} \to \mathbb{U}$, where $\mathbb{U}$ is also a set of utility functions $U : O \to R$. That is, for each utility profile U, $\mathcal{W}(U) : O \to R$ is a utility function itself.

The strategies available to individual $i \in I$ are S_i, and $S = \prod_{i \in I} S_i$ is the set of possible strategy profiles. A mechanism is a function $\mu : S \to O$, which specifies an outcome for each strategy profile of the individuals. When the individuals have the utility profile U, a strategy profile $s^* \in S$ is a Nash equilibrium of the mechanism μ if for each individual $i \in I$

$$U_i \left(\mu \left(s^* \right) \right) \geq U_i \left(\mu \left(s_i, s^*_{-i} \right) \right) \tag{1}$$

for every strategy $s_i \in S_i$. As usual, $\left(s_i, s^*_{-i} \right)$ denotes the strategy profile obtained from s^* by replacing only the strategy s^*_i of individual i by s_i.

Let $\mathcal{M}$ be the set of available mechanisms. We assume that for every utility profile $U \in \mathcal{U}$ there exists a mechanism $\mu \in \mathcal{M}$ which has a Nash equilibrium for U. For every mechanism $\mu \in \mathcal{M}$ for which this is the case, we assume that one Nash equilibrium for U is singled out, and denoted by $s^*(U, \mu)$.[7]

An institutional design is a map $\mathfrak{D} : \mathcal{U} \to \mathcal{M}$ such that for every utility profile $U \in \mathcal{U}$, the mechanism $\mathfrak{D}(U)$ has a Nash equilibrium for U. The induced outcome is then

$$o(U, \mathfrak{D}) = \mathfrak{D}(U) \left(s^* (U, \mathfrak{D}(U)) \right) \tag{2}$$

---

[7] With this simplifying assumption we abstract, of course, from the difficult issue of equilibrium selection in the case of multiple equilibria. We do so because our focus here is on market or mechanism design *given* the way individuals *actually* behave under each particular design, not on *why* this specific equilibrium behavior has emerged in lieu of other potential equilibria.

We denote by $\mathcal{D}$ the collection of available designs.[8]

For a given utility profile U, the classical problem of institutional design consists of choosing a design $\mathfrak{D} \in \mathcal{D}$ so as to maximize $W(U)(o(U, \mathfrak{D}))$ for every $U \in \mathcal{U}$. In words, the challenge is to find, for every utility profile $U \in \mathcal{U}$, a mechanism such that the outcome induced by that mechanism at its Nash equilibrium will maximize the aggregate welfare.

However, given the abundance of evidence on the sensitivity of preferences to context, it may very well be the case that the design itself also has an *unconscious effect* on individuals' preferences. We therefore assume that in the context induced by the design $\mathfrak{D}$, an individual with a utility function U_i will unconsciously try to maximize $U_i^{\mathfrak{D}}$ rather than U_i. Thus, the institution will ultimately impose the mechanism $\mathfrak{D}(U^{\mathfrak{D}})$ rather than $\mathfrak{D}(U)$, and the implemented outcome will be $o(U^{\mathfrak{D}}, \mathfrak{D})$ rather than $o(U, \mathfrak{D})$.

How does the utility profile adapt to the institutional design $\mathfrak{D}$? That is, how is the map $U \to U^{\mathfrak{D}}$ determined? If we are to assume an unconscious adaptation of preferences to context, it is first of all natural to assume that such an adaptation cannot be too "wild," but rather constrained to some "neighborhood" of specific distortions of the original preferences. Formally, we therefore assume that for each utility function $U_i \in \mathbb{U}_i$ there corresponds a set of utility distortions $\mathcal{N}(U_i) \subseteq \mathbb{U}_i$, which contains U_i and from which $U_i^{\mathfrak{D}}$ can emerge.

How is $U_i^{\mathfrak{D}}$ ultimately singled out of $\mathcal{N}(U_i)$? Here it is natural to assume that while the distortion from U_i to $U_i^{\mathfrak{D}}$ is not the result of a conscious process, $U_i^{\mathfrak{D}}$ eventually adjusts so as to maximize the base utility U_i given the institution $\mathfrak{D}$ and the emerging utility functions $U_j^{\mathfrak{D}}$ of the other individuals ($j \neq i$). Formally, denote by $(\tilde{U}_i, U_{-i}^{\mathfrak{D}})$ the utility profile one obtains from $U_i^{\mathfrak{D}}$ by replacing only the utility function $U_i^{\mathfrak{D}}$ of individual i by $\tilde{U}_i$. Then we say that $U^{\mathfrak{D}}$ is a *preference equilibrium* utility profile within the institutional design $\mathfrak{D}$ if for every individual $i \in I$

$$U_i^{\mathfrak{D}} \in \arg \max_{\tilde{U}_i \in \mathcal{N}(U_i)} U_i \left(o \left(\left(\tilde{U}_i, U_{-i}^{\mathfrak{D}} \right), \mathfrak{D} \right) \right) \tag{3}$$

In other words, $U^{\mathfrak{D}}$ is a Nash equilibrium of a meta-game with the strategy space $\mathcal{N}(U_i)$ for individual $i \in I$ and the payoff function $f_i : \prod_{i \in I} \mathcal{N}(U_i) \to R$, defined by

$$f_i(\tilde{U}) = U_i \left(o \left(\tilde{U}, \mathfrak{D} \right) \right) \tag{4}$$

Typically, the preference equilibrium utility profile $U^{\mathfrak{D}}$ indeed varies with $\mathfrak{D}$ and is different than U. Intuitively, this is because the distortion from U_i toward $U_i^{\mathfrak{D}}$ shifts the Nash equilibrium behavior of the other individuals from $s_{-i}^* \left((U_i, U_{-i}^{\mathfrak{D}}), \mathfrak{D}(U_i, U_{-i}^{\mathfrak{D}}) \right)$ to $s_{-i}^* \left(U^{\mathfrak{D}}, \mathfrak{D}(U^{\mathfrak{D}}) \right)$, both directly—through the first argument of s_{-i}^*, and also indirectly, via the effect on the mechanism implemented by the institution $\mathfrak{D}$—the second

---

[8] The definition of the institution as a map $\mathfrak{D} : \mathcal{U} \to \mathcal{M}$ assumes that the institution knows the utility functions of the individuals. Thus, our work is cast in the classical framework of mechanism design, and we have nothing to contribute here to the emerging literature on "robust mechanism design" (e.g. Bergemann and Morris, 2005), which aims at reducing the knowledge base that the institution is required to have.

argument of s^*_{-i}. This effect on others' equilibrium behavior may more than compensate the individual for the fact that her own equilibrium choice, $s^*_i\left(U^{\mathfrak{D}}, \mathfrak{D}\left(U^{\mathfrak{D}}\right)\right)$, maximizes $U^{\mathfrak{D}}_i$ rather than her genuine U_i.

This approach to endogenous preferences is now well established in the literature. It dates back at least to the seminal approach of Becker (1976) (e.g. the "rotten kid theorem," which exemplified the commitment value of altruism), and is elaborated further in the numerous contributions cited in footnote 3. In particular, many of these contributions analyzed the evolutionary viability of equilibrium preferences, and showed that they are either evolutionarily stable in the space of preferences represented by the utility functions in $\left(\mathcal{N}\left(U_i\right)\right)_{i\in I}$, or, even stronger, the sole survivors in any regular payoff-monotonic selection process in which preferences with higher fitness (with the fitness function f_i in equation 4) proliferate at the expense of less fit preferences (Heifetz et al., 2007c). The assumption that in equation (3) $U^{\mathfrak{D}}_i$, $i \in I$ constitute a Nash equilibrium of the meta-game (equation 4) does not necessarily hinge on the utility functions $U^{\mathfrak{D}}_i$ being mutually observed among the individuals. Indeed, Nash equilibrium—both in the game and in the meta-game – may very well emerge from various types of adaptive adjustment processes which do not rely on explicit observation and reasoning.[9]

The effect of the institutional design on the very preferences that the individuals try to maximize raises a new question regarding the object of social maximization. Should the design be chosen so as to maximize $\mathcal{W}\left(U\right)\left(o\left(U^{\mathfrak{D}}, \mathfrak{D}\right)\right)$, or rather $\mathcal{W}\left(U^{\mathfrak{D}}\right)\left(o\left(U^{\mathfrak{D}}, \mathfrak{D}\right)\right)$? The former approach is based on the assumption that the preference shifts from U_i to $U^{\mathfrak{D}}_i$ are short-lived, and welfare should be evaluated according to the base utility profile, U. The latter approach aims at maximizing the aggregate welfare of the individuals according to their utility profile $U^{\mathfrak{D}}$, i.e. according to their preferences when they interact within the institution. The analysis in this chapter will address both these approaches.

BILATERAL EXCHANGE WITH ENDOGENOUS PREFERENCES

We now proceed to analyze how the general framework of the previous section can be applied to the case of bilateral exchange with private information, as in Heifetz and Segev (2004).[10]

---

[9] See e.g. Al-Najjar et al. (2004), who postulate such an adaptive process for the emergence of biases. However, our own particular application of bilateral exchange is one with privately-known valuations, where Bayes–Nash play in the game itself is of doubtful meaning without assuming that the distribution of subjective valuations is commonly known. That's why we endorse this assumption later and spell it out explicitly.

[10] Huck et al. (2005) is a precursor analysis on the emergence of biases in Nash bargaining under *complete* information.

Consider a bilateral monopoly between a potential seller (denoted S) and buyer (denoted as B), who bargain over an undifferentiated good or legal entitlement. Both parties possess private information about their true valuations of the entitlement, but it is commonly known that these core valuations are drawn independently from uniform distributions on the same support $[\underline{a}, \bar{a}]$, which we normalize to be $[0, 1]$.[11] When we say that the core valuation of the seller is s, we mean that this is the minimum price for which she would be willing to sell the good were she to trade in some market as a price-taker. Similarly, the core valuation b of the buyer is the maximum price he would be willing to pay for the good as a price taker.

The bargaining scheme, however complicated, eventually gives rise to some probability of trade $p(s, b)$ for each pair of seller and buyer valuations $s, b \in [0, 1]$, and an average monetary transfer $t(s, b)$ from the buyer to the seller given trade. The ex ante probability of trade is therefore given by

$$P = \int_0^1 \int_0^1 p(s, b) ds db \tag{5}$$

and the ex ante expected gains from trade are

$$G = \int_0^1 \int_0^1 (b - s) p(s, b) ds db \tag{6}$$

which can be decomposed into

$$G = U + V$$

where

$$U = \int_0^1 \int_0^1 (t(s, b) - s) p(s, b) ds db$$

$$V = \int_0^1 \int_0^1 (b - t(s, b)) p(s, b) ds db$$

are the seller and buyer's expected payoffs, respectively.

Introducing cognitive dispositions during trade, we now assume that each of the parties may be subject to a preference drift in the course of bargaining, manifested by an additive distortion of its valuation. In particular, we assume that the seller's *perceived* valuation of the entitlement consists of $s + \varepsilon$, which represents the sum of her core valuation (s) and a distortion component (ε). Similarly, the buyer's *perceived* valuation of the entitlement consists of $b - \tau$, which represents the difference between his core valuation (b) and a distortion component (τ). Intuitively, ε and τ represent a type of emotional bargaining "toughness" exhibited by each side. Although the seller's "genuine" valuation of the entitlement is s, when bargaining over a sale she becomes

---

[11] To ease the exposition, we pursue the analysis with the uniform distribution, though a similar analysis can be carried out also with more general distributions—see Heifetz and Segev (2004) for details.

convinced that her true valuation is ε dollars higher still. Similarly, the buyer becomes convinced that his valuation is τ dollars lower than his true valuation b.

Such distortions have both empirical and theoretical justifications for coming about. Empirically, there is a vast and growing literature exploring the so-called "endowment effect" in bargaining, which operates much like the toughness distortion envisioned here; see Horowitz and McConnel (2002) or Arlen et al. (2002) for a literature survey. Theoretically, the above distortions may play a valuable role in enhancing each side's expected payoff during bargaining. Indeed, if each party to a negotiation perceives herself to possess a more "stingy" valuation than she would possess outside the bargaining context, and this perception is observed by the other bargaining party, then the preference distortion can, ironically, enhance her expected payoff (when viewed from the standpoint of her genuine, a-contextual preferences).

We therefore assume in what follows that the supports of the perceived valuations— $[\varepsilon, 1 + \varepsilon]$ for the seller, and $[-\tau, 1 - \tau]$ for the buyer—are commonly known, but we allow them to be endogenously determined over time as part of a preference equilibrium. How is this equilibrium determined?

For any *given* toughness dispositions ε, τ, trade takes place with a positive probability only when the true valuation s of the seller is in fact smaller than $1 - \tau - \varepsilon$,[12] and the true valuation b of the buyer is larger than $\tau + \varepsilon$.[13] We assume that the original bargaining mechanism, as characterized by the trade probability $p(s, b)$ and transfer $t(s, b)$ functions, is simply re-scaled to these new intervals of smaller length $1 - \tau - \varepsilon$. The overall probability of trade thus shrinks to

$$\int_0^1 \left(\int_0^1 p(s, b)(1 - \tau - \varepsilon)ds \right)(1 - \tau - \varepsilon)\, db = P(1 - \tau - \varepsilon)^2 \qquad (7)$$

and the ex ante gains from trade (as perceived by the bargaining parties) decrease to

$$\int_0^1 \left(\int_0^1 ((1 - \tau - \varepsilon)\,(b - s))\, p(s, b)\,(1 - \tau - \varepsilon)\, ds \right)(1 - \tau - \varepsilon)db = G(1 - \tau - \varepsilon)^3 \qquad (8)$$

Of this amount, $U(1 - \tau - \varepsilon)^3$ is enjoyed by the seller and the remaining $V(1 - \tau - \varepsilon)^3$ by the buyer.

Note, however, that the private payoffs characterized above are expressed in terms of each bargaining party's preferences she perceives herself to have from within the bargaining context. Of equal importance is how these distortions affect the parties, "genuine" payoffs away from the bargaining context. Under this metric, the "true" profit that the seller reaps increases by ε for every transaction she successfully consummates (reflecting the abandonment of her transitory cognitive attachment to the entitlement). As such, the seller's a-contextual payoff in the above game in expected value terms is given by:

---

[12] That is, when her perceived valuation S is smaller than $1 - \tau$, which is the maximum perceived valuation of the buyer.

[13] That is, when his perceived valuation B is higher than ε – the minimum perceived valuation of the seller.

$$f_{\text{seller}}(\varepsilon, \tau) = U(1 - \tau - \varepsilon)^3 + P(1 - \tau - \varepsilon)^2 \varepsilon \qquad (9)$$

Similarly, the "genuine" ex ante expected profit of the buyer increases by τ for every successfully consummated transaction, becoming:

$$f_{\text{buyer}}(\tau, \varepsilon) = V(1 - \tau - \varepsilon)^3 + P(1 - \tau - \varepsilon)^2 \tau \qquad (10)$$

Consequently, the a-contextual joint surplus of the parties is given by:

$$g(\varepsilon, \tau) = G(1 - \tau - \varepsilon)^3 + P(1 - \tau - \varepsilon)^2 (\varepsilon + \tau) \qquad (11)$$

Definition 1. *The bargainers' preferences are at equilibrium, if each bargainer's preferences confer the highest expected actual payoff given the preferences of the other bargainer, i.e. if the seller's "endowment effect" ε^* maximizes her expected payoff given the "toughness disposition" τ^* of the buyer, and vice versa.*

In other words, (ε^*, τ^*) are equilibrium dispositions if they constitute a Nash equilibrium of the meta-game with payoffs $f_{\text{seller}}, f_{\text{buyer}}$, which is straightforward to compute:

Proposition 1. *When $\frac{U}{P}, \frac{V}{P} < \frac{1}{3}$, the dispositions with the equilibrium preferences are*

$$\varepsilon^* = \frac{P - 3U}{P + 3(P - G)}$$

$$\tau^* = \frac{P - 3V}{P + 3(P - G)}$$

In fact, a sharper result obtains: In a population of individuals who are repeatedly matched at random to bargain, the preferences will indeed converge to having these levels (ε^*, τ^*) of the dispositions under any dynamic process that rewards material success with proliferation, for a genuinely wide range of initial distributions of preferences in the population. The proof of this theorem and the exact phrasing and proof of the sharper result requires a few more technical definitions, and can be found in Heifetz and Segev (2004).

To grasp the meaning of the conditions $\frac{U}{P}, \frac{V}{P} < \frac{1}{3}$ (and hence $\frac{G}{P} < \frac{2}{3}$), notice that, by equations (5) and (6), it is always the case that $P \geq G$, and the two quantities become closer to one another when the probability of trade $p(s, b)$ decreases when $b - s$ is small, and increases when $b - s$ is large.[14] Therefore, the condition $\frac{G}{P} < \frac{2}{3}$ means that the trade scheme allows for "a fair chance to strike even fairly profitable deals."[15]

---

[14] $P = G$ only in the limiting case when trade takes place with a positive probability exclusively when $b - s = 1$, i.e. $b = 1$ and $s = 0$.

[15] If we restrict attention to incentive-compatible (IC) and individually-rational (IR) trade mechanisms (i.e. budget-balanced (BB)), substituting the uniform distributions into inequality (2) of Myerson and Satterthwaite (1983) yields that such mechanisms must satisfy $\frac{G}{P} \geq \frac{1}{2}$. Thus, our condition is compatible with (IC),(IR), and (BB). Moreover, virtually all the particular equilibria of bargaining games we found in the literature satisfy the restriction $\frac{G}{P} < \frac{2}{3}$; see Heifetz and Segev (2004) for details.

Even though a tough spirit or character in the course of bargaining is unilaterally beneficial, and hence both bargainers adapt to such a tough mood during the bargaining process, these tendencies constitute a prisoners' dilemma type of inefficiency:

Proposition 2. *The parties might be better off without the distortion than with it, i.e. there are values for ε and τ such that $G \geq g(\varepsilon, \tau)$. In particular, they are better off without the equilibrium distortion than with it, i.e. $G \geq g(\varepsilon^*, \tau^*)$.*

The proof of this proposition, as well as the other propositions in this chapter, can be found in Heifetz et al. (2007a).

Thus, in terms of their a-contextual preferences (and a fortiori in terms of the traders' preferences in the heat of bargaining), cognitive biases might make both parties worse off than they would be if such biases were nonexistent. It is in precisely such instances that there may be a case for some form of measured paternalism, either by the state or by some other benevolent third party. Mitigating the cognitive shifts during trade is thus a new task for a social planner, on top of the classical task of mitigating strategic misrepresentation. In the section "Efficient market design" we shall explore how various measures of intervention perform in obtaining this duo of goals simultaneously.

An example: sealed-bid double auctions

In order to illustrate how the characterization of equilibrium preferences is operationalized, consider the canonical bargaining problem presented in Chatterjee and Samuelson (1983). Within their model there is a unique Bayes–Nash equilibrium profile in which strategies are smooth and strictly increasing in type. In the case of equal bargaining power and uniform distributions on $[\underline{S}, \overline{S}] \times [\underline{B}, \overline{B}]$, a seller with valuation s offers:

$$\sigma(s) = \frac{2}{3}s + \frac{1}{4}\overline{B} + \frac{1}{12}\underline{S} \tag{12}$$

and a buyer with valuation b bids:

$$\beta(b) = \frac{2}{3}b + \frac{1}{12}\overline{B} + \frac{1}{4}\underline{S} \tag{13}$$

Consequently, trade occurs only when

$$b \geq s + \frac{1}{4}(\overline{B} - \underline{S}) \tag{14}$$

When normalizing the intervals over the unit square, this condition has the familiar shape

$$b \geq s + \frac{1}{4}$$

and thus trade need not occur even when it is efficient. The total gains from trade in this case are given by:

$$G = \int_0^{\frac{3}{4}} \int_{s+\frac{1}{4}}^1 (b-s)\,db\,ds = \frac{9}{64}$$

$U = V = \frac{1}{2}G = \frac{9}{128}$; and finally, the probability of trade is given by:

$$P = \int_0^{\frac{3}{4}} \int_{s+\frac{1}{4}}^1 db\,ds = \frac{9}{32}$$

The equilibrium biases are therefore [16]

$$\varepsilon^* = \frac{P - 3U}{P + 3(P-G)} = \frac{1}{10} \tag{15}$$

$$\tau^* = \frac{P - 3V}{P + 3(P-G)} = \frac{1}{10} \tag{16}$$

It turns out that with these biases the ex ante probability of trade becomes: $P(1 - \tau^* - \varepsilon^*)^2 = \frac{9}{50}$, which is, of course, smaller than this probability without the biases (P), and the ex ante total gains from trade are also reduced to $G(1 - \tau^* - \varepsilon^*)^3 + P(1 - \tau^* - \varepsilon^*)^2 (\varepsilon^* + \tau^*) = \frac{27}{250}$. Thus we see in this case not only that the traders are better off without the distortions than with them (judged by their a-contextual preferences) but also the equilibrium strategies in the presence of the biases induce a less efficient mechanism.

EFFICIENT MARKET DESIGN

Because the existence of endogenous bargaining toughness creates a prima facie case for external intervention, we turn now to exploring the question of what form that intervention might take. As noted in the Introduction, we consider three possible candidates, differentiated by the time period in which the social planner enters: interven-

---

[16] To make the computations more explicit, observe that with biases ε, τ, trade can take place only when the *perceived* valuations S, B are in the interval $[\underline{S}, \overline{B}] = [\varepsilon, 1 - \tau]$. Hence, by equation (14), the probability of trade becomes

$$\int_{\underline{S}}^{\overline{B} - \frac{1}{4}(\overline{B} - \underline{S})} \int_{S + \frac{1}{4}(\overline{B} - \underline{S})}^{\overline{B}} db\,dS = \int_\varepsilon^{(1-\tau)-\frac{1}{4}(1-\tau-\varepsilon)} \int_{S+\frac{1}{4}(1-\tau-\varepsilon)}^{1-\tau} db\,dS = \frac{9}{32}(1 - \tau - \varepsilon)^2$$

and the *perceived* gains from trade are

$$\int_\varepsilon^{(1-\tau)-\frac{1}{4}(1-\tau-\varepsilon)} \int_{S+\frac{1}{4}(1-\tau-\varepsilon)}^{1-\tau} (B-S)\,db\,dS = \frac{9}{64}(1 - \tau - \varepsilon)^3$$

tion ex post, at the interim stage, or at the ex ante stage. For each case, moreover, we consider alternative efficiency definitions using, respectively, the players' a-contextual preferences (which we have labeled "wealth"), on the one hand, and their contextualized, "hot" preferences (labeled "happiness"), on the other. Although our analysis will focus on the example developed in the previous section, it is easily generalizable.

Ex post intervention: subsidizing trade

It has long been recognized in the bargaining literature that strategic barriers to trade can be reduced—and even eliminated—through an appropriately crafted ex post subsidy. Under such a scheme, a third-party insurer promises to pay a subsidy to the bargainers should they successfully consummate a transaction. If the subsidy is sufficiently large, it can counteract the incentives that players might otherwise have to extract information rents by threatening to walk out on the negotiations. The effect can be so pronounced as to eliminate completely the generic inefficiency that frequently attends bilateral bargaining (Myerson and Satterthwaite, 1983).

The attraction to trade subsidies, moreover, is more than a theoretical curiosity. Indeed, a number of legal and institutional mechanisms plausibly serve the very purpose of subsidizing successful bargaining outcomes. While a complete list of them is beyond the scope of this article, a few notable examples are as follows:

- *Tax incentives.* In state and federal tax law, there are typical deductions and credits that are allowed for certain categories of market purchases.[17]
- *Bankruptcy costs.* When a firm becomes financially distressed, it is generally agreed that the option of filing for bankruptcy adds considerable costs on the filing party and its creditors (Baird, et al., 2000). The significant costs due to bankruptcy have created substantial motivation for "private workouts" among debtors, their shareholders, and creditors. From a conceptual perspective, then, a successful workout allows the parties to forego a considerable cost, the savings of which can now be split among them. As such, the costs of bankruptcy effectively act as a type of subsidy for successful bargaining.
- *Conditionality in international aid.* As noted earlier, numerous donor institutions (e.g. the IMF, World Bank) condition their aid on the resolution of internal or international conflicts.[18] During the last decade of the 20th century, the amount of international assistance directly tied to the resolution of such conflicts amounted to more than $25 billion (Forman and Patrick, 2000, p. 10).
- *Anti-insurance.* Cooter and Porat (2002) have suggested greater use of "anti-insurance" to mitigate incentive problems in joint ventures. Under one example

---

[17] See footnote 4.

[18] Dollar (2000). There is some precedent for this change. Indeed, one of the benefits accrued to Egypt by signing the peace treaty with Israel in 1979 is sustained financial support from the US.

of such a scheme, business partners would execute a contingent debt contract with a third party that would bind the firm to pay off the principal when the firm's profits are low, but excuses the obligation when profits are high. (Because this type of insurance contract increases the volatility of the firm's cash flow, it has been dubbed "anti-insurance.") Although the idea behind anti-insurance is to provide efficient investment incentives on the margin, the same concept might be used to finance an insurance scheme that is triggered with contract negotiations, collective bargaining agreements, or other situations in which bargaining is successful.

Endogenous cognitive dispositions can significantly complicate the considerations underlying an optimal subsidy. Indeed, while continuing to dampen the parties' strategic incentives to misrepresent value, trade subsidies simultaneously raise the absolute size of the bargaining surplus available. This latter effect causes the parties to develop even tougher dispositions than they would have in the absence of subsidies, since a larger surplus enhances the returns that one derives from being committed to a tough mood. Consequently, the optimal subsidy policy will generally have to trade off desirable strategic repercussions with less desirable cognitive ones, and will therefore generally diverge from that implied in a wealth-maximizing actor model.

In order to make the appropriate comparisons, we first consider the optimal trade subsidies in the benchmark case, in which cognitive biases are wholly absent by definition. To focus on intuitions, we restrict attention to the special case in which the expected split of the surplus between the parties is symmetric ($U = V = \frac{G}{2}$). In such a situation, the optimal subsidy scheme generally awards an equal payment to each party upon the consummation of a transaction. Consider, then, the effects of a subsidy, α, paid to each trader when and only when a trade is consummated. To facilitate welfare comparisons, we shall assume that the cost of the subsidy is wholly internalized ex ante, financed by an ex ante head tax whose size is equal to the expected subsidy paid across all possible valuations. [19]

The inclusion of the subsidy causes the set of mutually advantageous trades to expand by α for each party. Thus, if the interval with gains from trade is originally of length z, its length increases by the total subsidy of 2α to become $z + 2\alpha$. In this case, the optimal subsidy is that which maximizes the expected total gains of the parties less the cost of the subsidy:[20]

$$\alpha^* = \arg \max_{\alpha} \left[G(1 + 2\alpha)^3 - 2\alpha P(1 + 2\alpha)^2 \right] = \frac{1}{6} \frac{3G - P}{P - G} \qquad (17)$$

---

[19] The assumption of self-finance ex ante is not critical. However, regardless of whether the subsidy is self-financed or financed by a social insurance scheme, the expected cost of the subsidy is a relevant component of social welfare.

[20] In the computations below, the first term is the sum of the ex ante profits of the seller and the buyer, and the second term is the tax.

and the eventual expected surplus is

$$G(1 + 2\alpha^*)^3 - 2\alpha^* P(1 + 2\alpha^*)^2 = \frac{4}{27} \frac{P^3}{(P - G)^2} \tag{18}$$

In the numerical example above (where $G = \frac{9}{64}$ and $P = \frac{9}{32}$), this implies that a subsidy of $\alpha^* = \frac{1}{6}$ is required in order to maximize the actual expected surplus. After accounting for their ex ante tax burden, the expected gains from trade are $G = \frac{1}{6}$, which represents an increase from $G = \frac{9}{64}$ in the absence of the subsidy.

Wealth-maximizing subsidies

With this benchmark in hand, we turn to analyze the effects of the endogenous cognitive dispositions. Consider first a social planner whose aim is to craft a subsidy to maximize the expected wealth of the parties. Under this approach, the social planner's problem must now account for the fact that the preference shifts ε, τ, will *generally depend on the subsidy level and eventually adjust to it.* Consequently, an optimal trade subsidy must take this endogeneity into account, as reflected by the following proposition, whose proof appears in Heifetz et al. (2007a).

Proposition 3. *With endogenous preferences, a wealth-maximizing social planner will choose a subsidy of*

$$\alpha^{**} = \frac{1}{6} \frac{P}{P - G}$$

*in order to maximize the expected gains from trade, which is **larger** than the optimal subsidy in the benchmark case. The equilibrium dispositions under this subsidy are*

$$\varepsilon^{**} = \tau^{**} = (1 + 2\alpha^{**}) \frac{2P - 3G}{2P + 6(P - G)},$$

*which are **larger** than those which would emerge without the subsidy. However, the eventual expected surplus will be the same as in the benchmark case with no dispositions –*

$$\frac{4}{27} \frac{P^3}{(P - G)^2}$$

The intuition behind this result is relatively straightforward. Because the subsidy marginally increases the aggregate bargaining surplus, there is more to be gained for each player from being credibly committed to a tough state of mind. Consequently, distortions in the direction of greater toughness are likely to be increasingly adaptive as the size of the subsidy increases, partially "cancelling out" the salubrious effects of the subsidy. A social planner must therefore ratchet the subsidy upward even further to eventually reach a state of first-best efficiency. Once this level of efficiency is attained, however, expected social welfare is identical to that which would emerge under the benchmark case.

In the above example, a subsidy of $\alpha^{**} = \frac{1}{3} > \frac{1}{6} = \alpha^*$ is required to maximize both players' actual ex ante wealth when we take into account their endogenous dispositions. The actual expected surplus upon introducing the subsidy will increase from $\frac{27}{250}$ without the subsidy, to $\frac{1}{6}$ (which is the same as in the benchmark case). The equilibrium dispositions induced by the optimal subsidy grow to $\varepsilon^{**} = \tau^{**} = \frac{1}{6}$, rather than the $\frac{1}{10}$ which would obtain without the external incentive.

Happiness-maximizing subsidies

Now consider the alternative case in which the social planner chooses a subsidy in order to maximize the expected sum of the parties' perceived level of happiness at the time of bargaining. This objective corresponds to the following expression:

$$U \left(1 - \varepsilon^{***} - \tau^{***}\right)^3 + V \left(1 - \varepsilon^{***} - \tau^{***}\right)^3$$

where $\varepsilon^{***}, \tau^{***}$ correspond to the equilibrium preferences under the subsidy policy, as in proposition 1. As before, we restrict attention to symmetric mechanisms in which $U = V$, so that each level of subsidy influences both players equally. Analysis of this problem generates proposition 4, whose proof appears in Heifetz et al (2007a):

Proposition 4. *With endogenous preferences, a happiness-maximizing social planner will choose a subsidy of*

$$\alpha^{***} = \frac{1}{6} \frac{9G - 4P}{4P - 5G}$$

*which is **smaller** than the optimal subsidy in the benchmark case. The equilibrium dispositions under this subsidy are*

$$\varepsilon^{***} = \tau^{***} = (1 + 2\alpha^{***}) \frac{2P - 3G}{2P + 6(P - G)}$$

*which are **larger** than those which would emerge without the subsidy.*

The intuition behind this proposition is as follows. Just as in the previous case, provision of a trade subsidy tends to exacerbate the equilibrium level of cognitive dispositions. Unlike that case, however, here the social planner's objective mutates along with the parties' perceived level of happiness at the time of bargaining. Since increasing the bargaining subsidy induces players to perceive that they are tougher bargainers (thereby reducing perceived gains from trade), bargaining failure imposes a smaller social cost on the parties. As a result, the optimal subsidy stops short of that in either the benchmark case or the wealth-maximizing case.

In the running numerical example, maximizing happiness requires imposing a subsidy of $\alpha^{***} = \frac{1}{18}$. This subsidy is clearly smaller than the optimal subsidy $\alpha^* = \frac{1}{6}$ in the benchmark case, and $\alpha^{**} = \frac{1}{3}$ in the wealth-maximization case. The equilibrium dispositions with the optimal subsidy α^{***} will adjust, and increase to $\varepsilon^{***} = \tau^{***} = \frac{1}{9}$

Table 19.1. Benchmark case vs. maximizing wealth case

	Benchmark case, no dispositions	Maximizing wealth
Endogenous disposition without subsidy	0	$\frac{1}{10}$
Expected surplus without subsidy	$\frac{9}{64}$	$\frac{27}{250}$
Optimal subsidy	$\frac{1}{6}$	$\frac{1}{3}$
Endogenous disposition with optimal subsidy	0	$\frac{1}{6}$
Expected surplus with optimal subsidy	$\frac{1}{6}$	$\frac{1}{6}$

Table 19.2. Benchmark case vs. maximizing happiness case

	Benchmark case, no dispositions	Maximizing happiness
Endogenous disposition without subsidy	0	$\frac{1}{10}$
Expected happiness without subsidy	$\frac{9}{64}$	$\frac{9}{125}$
Optimal subsidy	$\frac{1}{6}$	$\frac{1}{18}$
Endogenous disposition with optimal subsidy	0	$\frac{1}{9}$
Expected happiness with optimal subsidy	$\frac{1}{6}$	$\frac{2}{27}$

instead of $\frac{1}{10}$ without the subsidy. The expected happiness, given the maximizing subsidy, would rise from $G\left(1 - \tau^* - \varepsilon^*\right)^3 = \frac{9}{125}$ without the subsidy to $\frac{16}{27}\frac{P^3}{(4P-5G)^2} = \frac{2}{27}$ with it.

Tables 19.1 and 19.2 synthesize and compare the numerical example in the three cases studied above. Table 19.1 compares the benchmark case (no dispositions) and the wealth-maximizing case. Table 19.2 compares the benchmark case (no dispositions) and the happiness-maximizing case.

Note once again from the third row in the tables that the optimal subsidy in the benchmark case systematically diverges from that in either of the other two cases involving endogenous bias, falling short of the wealth-maximizing subsidy and exceeding the happiness-maximizing subsidy. Moreover, note from the fourth row in the tables that the equilibrium level of predicted toughness is not uniform across the two alternative objectives, and is significantly higher in the case of maximizing wealth. These respective differences exemplify our more general argument in this chapter: that accounting for the endogenous effects of regulation itself can lead to predictions that are distinct from those that would be rendered if one either ignored cognitive biases or treated them as an exogenous primitive. [21]

---

[21] Note also, by comparing the first and third entries of the last line of the tables, that if both bargainers were cold-blooded and incapable of developing cognitive distortions in the bargaining

Interim intervention: efficient bargaining mechanisms

Another important arena for regulatory intervention comes at the interim stage, in the design of bargaining procedures themselves. It is widely recognized in the literature that bargaining protocols matter, in that they can produce distinct trading probabilities and expected social payoffs. Consequently, the question of what constitutes an "optimal bargaining mechanism" in a given circumstance continues to receive a significant amount of attention. At the very least, the features of an optimal mechanism identify the limits of what can be accomplished in unmediated bargaining (Fudenberg and Tirole, 1991, p. 290). In this section, then, we explore optimal bargaining mechanisms, taking into account endogenous dispositions.

At core, all bargaining mechanisms specify both (1) the probability of trade for any pair of valuations, and (2) the distribution of gains from trade that ensue from a transaction. In conventional models, no such mechanism can be considered optimal if there is an alternative incentive compatible, individually rational mechanism that produces trade in strictly more situations. As we shall see later, however, the introduction of endogenous dispositions can sometimes cause the optimal mechanism to diverge from this general principle. In particular, under certain conditions, an optimal mechanism may be much more "draconian" than theory would otherwise predict, enforcing transactions in strictly *fewer* situations than other implementable mechanisms.

Interestingly, certain well known doctrines operate in much the same fashion, and can be interpreted (at least indirectly) as requiring some artificial lower bound in trade surplus before a court is willing to enforce a contract. Notably, most of the protections are limited to special cases, and contract law doctrine more generally is thought to implement the principle that courts should act to facilitate transactions in the most circumstances possible.

- *Unconscionability.* This doctrine instructs the court to refuse to enforce contracts in situations where the negotiation and resulting terms of a transaction are excessively one sided.[22] One interpretation of the unconscionability doctrine is that it has the effect of requiring each party to receive some minimum share of the joint surplus before a contract can be enforced. Under such an interpretation, the doctrine implies a requirement that the total amount of social surplus exceed some specified threshold before a contract is enforceable.
- *Moieties.* In admiralty law, the common-law doctrine of moieties dictated the division of rents from emergency salvage operations at sea. When, for instance,

phase, they would both be happier *even under the best-suited subsidy intervention.* Thus, even though these cognitive distortions may be unilaterally beneficial (when the preferences of the other bargainer are taken as given), their concurrent presence is socially destructive, just as in a prisoners' dilemma. The institutional intervention can amend this state of affairs, but to a limited extent.

[22] See, e.g., *Williams v. Walker Thomas Furniture Co. 350 F.2d* 445 (D.C. Cir. 1965) (refusing enforcement of a cross-collateralization clause in a consumer debt contract that would operate to preclude satisfaction on the debt of any single purchase until all purchases were paid off).

a vessel in distress off-loaded its cargo onto another ship that had come to its aid, the doctrine required that each party was to receive a "moiety" of a fixed fraction (usually either a third or a half) of the value of the cargo as computed by its trading price once sold on a commodities market.[23] Because the moieties doctrine constrains the feasible set of negotiated outcomes by fixing a price by reference to an external market, it has the likely effect of discouraging transactions in the bilateral monopoly setting, where the gains from trade are positive, but insubstantial.

- *Cooling off periods.* A number of state and federal laws in the United States require a specified period of time to pass before certain types of consumer contracts are enforceable. In federal law, for example, statutory cooling-off periods are required for door-to-door sales (16 C.F.R. Part 429), telemarketing, (16 C.F.R. Parts 308, 310, & 435), and sales of business franchises (16 C.F.R. Secs. 436.1 et seq.). As with the unconscionability doctrine, one interpretation of a cooling off period is as a doctrine that requires gains from trade to be relatively large. Indeed, where the gains are small at the time of negotiation, it is relatively likely that small post-transaction perturbations of the parties' respective valuations can cause recision of the agreement within the cooling-off period. Knowing this, rational parties might not find it in their interests to consummate deals that are likely to prove unenforceable.

Given these examples, we now turn to consider the characteristics of an optimal mechanism under endogenous biases. Such a mechanism is characterized by the pair of functions $(p(s, b), t(s, b))$ where $p(s, b)$ and $t(s, b)$ denote the equilibrium probability of trade and the equilibrium transfer payment, respectively, between a seller with a valuation s, and a buyer with a valuation b.

Once again, as a benchmark we first specify the mechanism which maximizes the players' payoff assuming away the emergence of biases. As demonstrated by Myerson and Satterthwaite (1983), the most efficient incentive-compatible (IC), individually-rational (IR), and budget-balanced (BB) mechanism is one in which trade is prohibited unless the seller's and buyer's stated valuations differ by at least $\frac{1}{4}$, i.e. a mechanism such that

$$p(s, b) = \begin{cases} 1 & b - s \geq \frac{1}{4} \\ 0 & b - s < \frac{1}{4} \end{cases}$$

The mechanism presented in the example of the sealed-bid double auction is thus an optimal one. The expected payoffs U, V of the bargainers might be different from one efficient mechanism to another. On the other hand, in every such optimal mechanism the total surplus and the expected probability of trade remain the same:

---

[23] See, e.g., *Post v. Jones*, 60 U.S. 150 (1856), in which a distressed vessel actually implemented a competitive bidding process among three aspiring salvagers. The lowest bidder later successfully challenged the terms of the contract under the doctrine of moieties, arguing that its accepted bid was too low to ensure it received its equitable share of the rents under the doctrine.

$$G = \int_0^1 \int_0^1 (b-s)\,p(s,b)dsdb = \int_{\frac{1}{4}}^1 \int_0^{b-\frac{1}{4}} (b-s)\,dsdb = \frac{9}{64}$$

$$P = \int_0^1 \int_0^1 p(s,b)dsdb = \int_{\frac{1}{4}}^1 \int_0^{b-\frac{1}{4}} dsdb = \frac{9}{32}$$

Wealth-maximizing mechanisms

As before, we now consider how the introduction of endogenous dispositions affects the analysis, focusing first on the social objective of maximizing the expected ex ante wealth of the bargainers. Also as before, to facilitate intuitions, we restrict attention to the symmetric case where $U = V = \frac{1}{2}G$. By equation (11), the expected sum of seller and buyer welfare is:

$$g\left(\varepsilon^*, \tau^*\right) = G\left(1 - \tau^* - \varepsilon^*\right)^3 + P\left(1 - \tau^* - \varepsilon^*\right)^2 \left(\varepsilon^* + \tau^*\right)$$

where ε^* and τ^* are the equilibrium biases from proposition 1. Analysis of the bargaining design problem gives rise to the following proposition whose proof appears in Heifetz et al., (2007a):

Proposition 5. *A Myerson–Satterthwaite (1983) mechanism with*

$$p(s,b) = \begin{cases} 1 & b-s \geq \frac{1}{4} \\ 0 & b-s < \frac{1}{4} \end{cases}$$

maximizes the expected wealth of the traders among all IC, IR, and BB mechanisms, even when the endogenous biases ε^*, τ^* from proposition 1 are taken into account. When the expected gains from trade are shared equally, the equilibrium dispositions are $\varepsilon^* = \tau^* = \frac{1}{10}$.

Interestingly, when the social planner is motivated by maximizing wealth, the efficient mechanism permits trade only when the seller's and buyer's reported valuations differ by at least $\frac{1}{4}$, exactly as in Myerson and Satterthwaite (1983). Consequently, every efficient mechanism as in Myerson and Satterthwaite is also efficient in order to maximize wealth paternalistically among distorted bargainers.

Happiness-maximizing mechanisms

Suppose instead that the social planner was motivated by a desire to maximize happiness rather than wealth. Under this alternative objective, the planner's maximand becomes:

$$G\left(1 - \varepsilon^{**} - \tau^{**}\right)^3$$

where ε^{**} and τ^{**} are the equilibrium biases from proposition 1. Analysis of this problem yields the following proposition, whose proof appears in Heifetz et al. (2007a):

Proposition 6. *The mechanism that maximizes the happiness of the players is character-ized by a threshold $h^{**} = \frac{1}{2}$ such that*

$$p(s, b) = \begin{cases} 1 & b - s \geq 1 - h^{**} = \frac{1}{2} \\ 0 & b - s < 1 - h^{**} = \frac{1}{2} \end{cases}$$

Such a mechanism induces trade in strictly fewer instances than the optimal mechanism in the baseline case. The equilibrium dispositions under this mechanism are $\varepsilon^{**} = \tau^{**} = 0$.

Note that the optimal happiness-maximizing mechanism is a significantly more "draconian" bargaining mechanism than that of either the baseline case or of the wealth-maximizing case, allowing trade if and only if the gains from trade $(b - s)$ exceed $\frac{1}{2}$. The intuition behind this result stems from a fundamental trade-off that a restrictive mechanism creates. On the one hand, more restrictive trading mechanisms impose a direct welfare loss, since they reduce the likelihood of any trades. On the other hand, this reduction in the likelihood of trade reduces the adaptiveness of a tough bargaining strategy, since the size of the expected bargaining surplus is smaller. The result reported in proposition 6 reflects the fact that the latter effect swamps the former one for all positive dispositions, so that an optimal trading rule coincides with the least restrictive mechanism that completely vitiates all biases. Consequently, under this mechanism, no biases ever evolve. It is easily verified that this mechanism yields a probability of trade of $P = \frac{1}{8}$, and an average payoff per player of $U = \frac{1}{24}$.

Note also that, unlike in the previous examples, the happiness-maximizing mecha-nism here is Pareto inferior to other candidates. For example, the Myerson and Satterth-waite (1983) optimal mechanism is clearly implementable in the case of zero biases, and both parties would prefer its implementation to the one given in the proposition. Allowing them to do so, however, would cause the parties to evolve increasingly tough dispositions, which in the long run would yield less trade and less ultimate happiness (as evaluated at the time of trade). Consequently, implementing the happiness-maximizing mechanism would require courts to actively prohibit trade except in situations where the surplus is sufficiently high. Many of the immutable legal doctrines discussed at the beginning of this section attempt to do just that.

Finally, note that just as with subsidies, the maximand favored by the social planner has a clear effect on the ultimate allocational rule. Maximizing wealth and maximizing happiness lead to very different solutions.

Ex ante intervention: property rights

For our final application, we consider regulatory interventions that occur before bargaining ever begins. Because transactions are little more than the transfer of property rights, it can be substantially affected by calculated manipulations to the content of those initial property rights. Indeed, the use of divided property rights has already been cited as a way to address problems of *strategic* barriers to trade (e.g. Cramton et al., 1987; Ayres and Talley, 1995).

Divided entitlements might, on first blush, appear exceptional within a capitalist economy. But on closer inspection, one can find dozens of areas where either courts or the parties themselves provide for divided property rights. Although a complete description of such partial entitlements is too lengthy to articulate here, the following represents a reasonable sampling:

- *Outright co-ownership.* In trade secret law, the "shop rights" doctrine provides for a type of divided ownership of inventions developed in the workplace. Explicitly, when a rank-and-file employee uses company time and/or resources in developing a new invention, the employee is awarded general ownership rights to the invention, while the employer receives a non-exclusive, zero-price license to use the invention. Employers and employees are generally free to contract around the shop rights doctrine, either prior to or after invention occurs (see Lester and Talley, 2000).
- *Temporal/subject matter divisions.* In patent law, patentees are generally awarded with a strong property right, but one that runs for only a prescribed, twenty-year statutory period after the effective filing date (or seventeen years from the date of the grant) (See 35 U.S.C. §§ 119, 120, 154(a)(2) (2002)). Viewed ex ante, this prospective temporal division can be thought to convey payoffs whose present value is divided between the patentee and its competitors that wish to use the patented technology. Once again, patent law allows for contracting around this statutory entitlement through licensing agreements.
- *Legal uncertainty.* In business law, corporate fiduciaries are prohibited from appropriating "corporate opportunities"—i.e. prospective business ventures that rightfully belong to the firm—for their own personal use. The standards that identify what exactly constitutes a bona fide corporate opportunity, however, are inherently casuistic, leaving an obscure doctrine that has been alternatively characterized by legal commentators as "vague," "in transition," "far from crystal clear," and "indecipherable" (see Talley, 1998). Although such randomness is generally perceived as undesirable, it has the effect of endowing both the fiduciary and the corporation with a probabilistic claim on the business opportunity, which has a number of characteristics resembling joint ownership. Moreover, corporate law generally allows for firms and their fiduciaries to allocate opportunities through bargaining.
- *Liability rules.* Even in the absence of a physical, temporal, or probabilistic division of property, legal rules can divide claims by altering the form of protection accorded one's entitlement. Much of modern nuisance law in the United States, for example, tends to award a successful plaintiff with money damages rather than injunctive relief for a defendant's incompatible activities. As with the above examples, the plaintiff and defendant are free to negotiate in the shadow of this liability rule (see Ayres and Talley, 1995).

In order to consider the effect of divided entitlements on bargaining with dispositions, suppose the parties bargained over ownership of an asset, but now assume that the initial property rights of the asset are given by $(q, 1 - q)$, where q represents the fractional ownership claimed by B, and $(1 - q)$ represent that claimed by S. Without loss of generality, suppose that $q \le \frac{1}{2}$. Following Ayres and Talley (1995) we explore what is the optimal allocation (i.e. what is the optimal q) that maximizes the expected surplus in case the partners wish to dissolve the partnership, but we now factor in the possibility of endogenous cognitive dispositions.

As a benchmark, we once again consider first the case in which no biases exist, and we explore a specific bargaining procedure, a sealed-bid double auction as the procedure to dissolve the partnership: Each of the partners submits a bid for the asset, and the partner with the higher bid buys her partner's share in the asset at the price (for the entire asset) which is the average of the two bids. Such a procedure with a linear-strategy equilibrium is known to be optimal in a Myerson-Satterthwaite framework with symmetric, uniform distributions of valuation.

As earlier, in the section presenting the example of sealed-bid double auction, we therefore explore an equilibrium with linear bidding strategies. These turn out to be

$$
r_B(b) = \begin{cases} \frac{2}{3}b + \frac{1}{12} + \frac{1}{6}q & \text{when } \frac{1}{4} - \frac{1}{2}q \le b \le 1 \\ \frac{1}{4} - \frac{1}{6}q & \text{when } 0 \le b < \frac{1}{4} - \frac{1}{2}q \end{cases} \tag{19}
$$

$$
r_S(s) = \begin{cases} \frac{2}{3}s + \frac{1}{4} - \frac{1}{6}q & \text{when } 0 \le s \le \frac{3}{4} + \frac{1}{2}q \\ \frac{3}{4} + \frac{1}{6}q & \text{when } \frac{3}{4} + \frac{1}{2}q < s \le 1 \end{cases} \tag{20}
$$

of which the Chatterjee and Samuelson (1983) equilibrium in the earlier section is the limiting case[24] for $q = 0$. Analysis of these expressions leads to proposition 7 for the benchmark case of no dispositions whose proof appears in Heifetz et al., (2007a):

Proposition 7. *In the absence of dispositions, the expected surplus in this double-auction equilibrium is maximized with the initial shares* $(q, 1 - q) = \left(\frac{1}{2}, \frac{1}{2}\right)$*. The asset will always end up in the hands of the partner who values it most.*

The fact that the optimal property rights scheme in the baseline case allocates equal ownership shares to each player should not be surprising. Indeed, in this case, the only impediment to a negotiated outcome is the parties' incentives to extract information rents by misstating their true valuations. Buyers tend to shade their private valuations downward, while sellers tend to shade theirs upward. A division of property rights tends to weaken these incentives, by making each player both a potential buyer and a potential

---

[24] The only difference is that in the Chatterjee and Samuelson (1983) case $(q = 0)$, no trade takes place when $0 \le b < \frac{1}{4} - \frac{1}{2}q$ or $\frac{3}{4} + \frac{1}{2}q < s \le 1$ with either the equilibrium strategies in equations (19), (20), or alternatively equations (12), (13). However, when $q > 0$ and $0 \le b < \frac{1}{4} - \frac{1}{2}q$, the partner with share q is certain to sell its part in the partnership at equilibrium, and will therefore bid the lowest equilibrium bid of its partner, and not below it. Similarly, when $\frac{3}{4} + \frac{1}{2}q < s \le 1$, the partner with share $1 - q$ is certain to buy its partner's part, and will therefore bid the highest equilibrium bid of its partner, and not above it

seller. The introduction of these dual roles causes the players to become more ambivalent about whether to shade their valuations (and in which direction to do so). When each side has a one-half initial ownership share of the asset, the incentives to overstate and understate exactly cancel one another out, thereby leading to first-best efficiency.

Wealth-maximizing property rights

We now consider how a social planner might maximize surplus in the presence of endogenous cognitive dispositions. To understand how biases alter the analysis, first fix q and consider only how dispositions are likely to evolve. As earlier, suppose that player S misperceives her valuation to be ε higher than it actually is, while B similarly misperceives his valuation to be τ lower than it actually is. Consequently, when the partners begin to negotiate, they observe each other's character (i.e. the supports $[\varepsilon, 1 + \varepsilon]$ and $[-\tau, 1 - \tau]$ of the distributions become common knowledge), but not the actual perceived valuations. Then they play the equilibrium where the bids r_B, r_S are linear in their perceived valuations. When translated back to their actual valuations, these equilibrium bids turn to be

$$
r_B(b) = \begin{cases} \frac{2}{3}b + \frac{1}{12} + \frac{1}{6}q + \frac{1}{4}\varepsilon - \frac{3}{4}\tau & \text{when } x \le b \le 1 \\ \frac{1}{4} - \frac{1}{6}q + \frac{3}{4}\varepsilon - \frac{1}{4}\tau & \text{when } 0 \le b < x \end{cases} \tag{21}
$$

$$
r_S(s) = \begin{cases} \frac{2}{3}s + \frac{1}{4} - \frac{1}{6}q + \frac{3}{4}\varepsilon - \frac{1}{4}\tau & \text{when } 0 \le s \le 1 - x \\ \frac{3}{4} + \frac{1}{6}q + \frac{1}{4}\varepsilon - \frac{3}{4}\tau & \text{when } 1 - x < s \le 1 \end{cases} \tag{22}
$$

where

$$
x = \frac{1}{4} - \frac{1}{2}q + \frac{3}{4}(\varepsilon + \tau) \tag{23}
$$

Analysis of these bid functions yields proposition 8, whose proof appears in Heifetz et al. (2007a):

Proposition 8. *In the double auction with a given value of q, the equilibrium dispositions of the parties are*

$$
\varepsilon^* = \tau^* = \frac{3}{10} + \frac{1}{3}q - \frac{1}{15}\sqrt{(9 + 80q)}
$$

Expected wealth is maximized when the initial entitlements are fixed at $(q, 1 - q) = (\frac{1}{2}, \frac{1}{2})$, so that

$$
\varepsilon^* = \tau^* = 0
$$

Proposition 8 illustrates that a wealth-maximizing property rights division under endogenous cognitive biases is identical to that in the baseline case: each party receives a one-half ownership share in the asset. This result suggests that there may be at least some forms of regulatory intervention that can address both strategic barriers to trade and cognitive barriers to trade simultaneously. Indeed, not only does divided ownership dampen information rents (as is well known in the literature), but proposition 7 demonstrates that it can also dampen the returns to developing a toughness disposition. When the parties could ultimately be either buyers or sellers of the asset, there is little to be gained from being committed to a set of preferences that biases one's valuation either upward or downward. Consequently, the allocation of equal shares $(q, 1 - q) = \left(\frac{1}{2}, \frac{1}{2}\right)$ both maximizes the players wealth, and also induces an equilibrium with no cognitive dispositions.

Happiness-maximizing property rights

The intuition developed in proposition 7 turns out to be quite general. In fact, not only does an initial property rights division maximize expected wealth, but it also turns out to be the optimal property rights allocation when the social planner's objective is to maximize happiness, as reflected in proposition 9, whose proof appears in Heifetz et al. (2007a):

Proposition 9. *In the double auction with a given value of q, expected happiness is maximized when the initial entitlements are fixed at $(q, 1 - q) = (\frac{1}{2}, \frac{1}{2})$, so that*

$$\varepsilon^* = \tau^* = 0$$

Recall that in the ex post and interim cases (studied earlier), the optimal regulatory intervention hinged crucially on the objective function of the social planner, and in particular whether she was motivated by maximizing "wealth" or "happiness." Here, in contrast, an ex ante intervention toward evenly divided property rights turns out to be optimal under both criteria as well as under the baseline case. Moreover, such a division has the effect of completely debiasing the players, so that they no longer develop cognitive distortions in the course of bargaining. This prediction squares nicely with some recent experimental work (e.g. Rachlinsky and Journden, 1998), in which endowment effects appear to dissipate when parties' interests are protected by weaker entitlements (such as liability rules).

DISCUSSION AND CONCLUSION

This chapter has presented a framework for designing optimal institutions in the presence of endogenous cognitive dispositions. This is a critically important problem if one wishes to draw meaningful policy implications from behavioral economics. At the same time, however, it is a problem that involves at least two unique complicating factors which are largely absent in conventional institutional design problems. First, the existence and size of cognitive biases may themselves be sensitive to the very institutional policies designed to address them. In such situations, policy makers must be keenly aware of the feedback effects that any candidate mechanism is likely to foster, and anticipate how cognition and regulatory context are likely to interact.

Second, the very definition of "optimality" may be even more contestable when preferences are endogenous. Policies designed to maximize wealth (i.e. welfare defined in terms of a-contextual preferences) need not coincide with those designed to maximize happiness (i.e. welfare defined in terms of the preferences induced within the institutional design). Consequently, policy makers may be forced to choose between these alternative objectives, since they generally will not produce the same policy prescriptions.

To illustrate our claims, we considered three families of regulatory interventions that have real-world institutional counterparts: ex post, interim, and ex-ante interventions in bilateral trade. Within these contexts, we have demonstrated how a failure to appreciate the complicating factors noted above can lead to unintended and undesirable institutional structures. In the context of ex post intervention, the optimal trade subsidy that incorporates evolutionary biases always diverges from the "baseline" case in which biases are ignored, regardless of the social objective adopted. In particular, wealth maximization requires a larger subsidy relative to the baseline, while happiness maximization requires a smaller one. Moreover, we have shown that an ex post intervention (of any size) induces the players to have even larger perception biases.

For interim interventions, the optimal trading mechanism in the baseline case turns out to be identical to that of a wealth-maximizing mechanism with biases. On the other hand, the trading mechanism that maximizes the players' happiness turns out to be relatively "draconian" in nature, prohibiting trade in strictly more circumstances than other implementable bargaining mechanisms would. Implementing such a Pareto-inferior mechanism would likely necessitate the implementation of immutable rules (such as that found in the doctrines of unconscionability or moieties).

More optimistically, we find that ex ante regulation through property rights allocations may be the most flexible and promising of all the interventions studied (at least within our framework). Here, the optimal allocation entails divided ownership, awarding half of the entitlement to each player. This allocation remains optimal regardless of whether the objective is to maximize wealth or happiness, and of whether we take the biases into account. Moreover, such a regulatory scheme completely de-biases the players, eventually eliminating their dispositions.

Two caveats to our analysis deserve specific mention. First, we do not attempt to offer in this chapter an all-encompassing explanation of preference dispositions within bargaining contexts. As noted in the Introduction, a number of such cognitive biases might manifest themselves in such contexts, and we explore but one. Second, the precise type of preference distortion we explore below—an endogenous disposition towards "toughness"—is assumed to be mutually observable. Although we posit that there are a number of real-world practices (such as personal affect, mannerisms, delegation practices, and so forth) which are manifestations of this form of observability, it likely does not carry over to all contexts. In those situations, the toughness dispositions we study are likely to be of little moment either to the parties or to a social planner. Nevertheless, our aim here is not to prove the ubiquity of the specific disposition we study, but rather to demonstrate how prudent market design should be mindful of both the existence and the endogeneity of preference distortions, be they of the species studied here or something else.

Although it is our motivating story, bilateral trade is likely not to be the sole arena in which the interplay between policy, context, and preferences is important. There are already several other attempts to address similar issues in other contexts.[25] Indeed, the approach suggested here may be relevant in virtually every instance of market or institutional design, and hence suggests a promising direction for further research.

References

Aaron, H. (1994) "Public policy, values and consciousness," *Journal of Economic Perspectives*, 8: 3–21.

Al-Najjar, N., Baliga, S. and Besanko, D. (2004) "The sunk cost bias and managerial pricing practices," Working Paper, available at SSRN <http://ssrn.com/abstract=825986>

Arlen, J., Spitzer, M. and Talley, E. (2002) "Endowment effects within corporate agency relationships," *Journal of Legal Studies*, 31(1): 1–37.

Ayres, I. and Talley, E. (1995) "Solomonic bargaining: dividing a legal entitlement to facilitate Coasean trade" *Yale Law Journal*, 104(5): 1027–117.

Baird, D. G., Jackson, T. H. and Adler, B. (2000) *Cases, Problems, and Materials on Bankruptcy*, Foundation Press.

Bar-Gill, O. (2006) "The evolution and persistence of optimism in litigation," *Journal of Law, Economics, and Organization*, 22: 490–507.

_____ and Fershtman, C. (2000) "The limit of public policy: endogenous preferences," CentER Working Paper No. 71. Available at SSRN: <http://ssrn.com/abstract=244676>.

_____ _____ (2004) "Law and preferences," *Journal of Law, Economics, and Organization*, 20: 331–52.

Becker, G. (1976) "Altruism, egoism and genetic fitness: economics and sociobiology," *Journal of Economic Literature*, 14: 817–26.

Benos, A. V. (1998) "Aggressiveness and survival of overconfident traders," *Journal of Financial Markets*, 1: 353–83.

Bergman N. and Bergman, Y. (2000) "Ecologies of preferences with envy as an antidote to risk-aversion in bargaining," Working Paper.

Bergemann, D. and Morrism, S. (2005) "Robust mechanism design," *Econometrica*, 73: 1771–813.

Berninghaus, S., Korth, C. and Napel, S. (2007) "Reciprocity—an indirect evolutionary analysis", *Journal of Evolutionary Economics*, 17: 579–603.

Bester, H. and Güth, W. (1998) "Is altruism evolutionary stable?" *Journal of Economic Behavior and Organization*, 34: 211–221.

Bolle, F. (2000) "Is altruism evolutionarily stable? And envy and malevolence? Remarks on Bester and Güth," *Journal of Economic Behavior and Organization*, 42(1): 131–3.

Bonatti, L. (2007) "Optimal public policy and endogenous preferences: an application to an economy with for-profit and non-profit enterprises," Working Paper.

Bowles, S. (1998) "Endogenous preferences: the cultural consequences of markets and other economic institutions," *Journal of Economic Literature*, 36: 75–111.

---

[25] Bar-Gill and Fershtman (2000), Fershtman and Heifetz (2006).

Bowles, S. and Hwang, S. (2008) "Social preferences and public economics: Mechanism design when social preferences depend on incentives," *Journal of Public Economics*, 92: 1811–20.

Camerer, C., Loewenstein, G. and Rabin, M. (eds) (2003) *Advances in Behavioral Economics,* Princeton University Press.

Carmichael, H. L. and MacLeod, W. B. (2002) "How should a behavioral economist do welfare economics?" USC CLEO Research Paper No. C02-18. Available at SSRN <http://ssrn.com/abstract=338280>.

Chatterjee, K. and Samuelson, W. (1983) "Bargaining under incomplete information," *Operations Research*, 31: 835–51.

Cooter, R. D. and Porat, A. (2002) "Anti-insurance," *Journal of Legal Studies*, 31: 203–32.

Cramton, P., Gibbons, R. and Klemperer, P. (1987) "Dissolving a partnership efficiently," *Econometrica*, 55(3): 615–32.

Dekel E., Ely, J. and Yilankaya, O. (2007) "Evolution of preferences," *Review of Economic Studies*, 74: 685–704.

Dollar, D. (2000) "What can research tell us about making lending aid more effective?" World Bank Working Paper.

Fershtman, C. and Heifetz, A. (2006) "Read my lips, watch for leaps: preference equilibrium and political instability," *Economic Journal*, 116: 246–65.

_____ and Weiss, Y. (1997) "Why do we care about what others think about us?," in A. Ben Ner and L. Putterman (eds), *Economics, Values and Organization*, Cambridge University Press.

_____ _____ (1998) "Social rewards, externalities and stable preferences," *Journal of Public Economics*, 70: 53–74.

Forman, S. and Patrick, S. (eds) (2000) *Good Intentions—Pledges of Aid for Postconflict Recovery*, Lynne Rienner Publishers.

Fudenberg, D. and Tirole, J. (1991) *Game Theory, MIT Press.*

Gerber, E. and Jackson, J. (1993) "Endogenous preferences and the study of institutions," *American Political Science Review*, 87: 639–56.

Gintis, H. (1974) "Welfare criteria with endogenous preferences: The economics of education," *International Economic Review*, 15: 415–30.

Güth W. and Napel, S. (2006) "Inequality Aversion in a Variety of Games—an indirect evolutionary analysis," *Economic Journal*, 116: 1037–56.

_____ and Ockenfels, A. (2005) "The coevolution of morality and legal institutions: an indirect evolutionary approach," *Journal of Institutional Economics*, 1: 155–74.

_____ and Yaari, M. (1992) "Explaining reciprocal behavior in simple strategic games: an evolutionary approach," in U. Witt (ed.), *Explaining Forces and Changes: Approaches to Evolutionary Economics,* University of Michigan Press.

Guttman, J. M. (2000) "On the evolutionary stability of preferences for Reciprocity," *European Journal of Political Economy*, 16: 31–50.

Heifetz, A. and Segev, E. (2004) "The evolutionary role of toughness in bargaining," *Games and Economic Behavior*, 49: 117–34.

_____ _____ and Talley, E. (2007a) "Market design with endogenous preferences," *Games and Economic Behavior*, 58: 121–53.

_____ Shannon, C. and Spiegel, Y. (2007b) "What to maximize if you must," *Journal of Economic Theory*, 133: 31–57.

_____ _____ _____ (2007c) "The dynamic evolution of preferences," *Economic Theory*, 32: 251–86.

Horowitz, J. K. and McConnel, K. E. (2002) "A review of WTA/WTP studies," *Journal of Environmental Economics and Management*, 44: 426–47.

Huck, S., Kirchsteiger, G. and Oechssler, J. (2005) "Learning to like what you have: explaining the endowment effect," *Economic Journal*, 115: 689–702.

―――― and Oechssler, J. (1998) "The indirect evolutionary approach to explaining fair allocations," *Games and Economic Behavior*, 28: 13–24.

Jolls, C. (1998) "Behavioral economic analysis of redistributive legal rules," *Vanderbilt Law Review*, 51: 1653.

Koçkesen, L., Ok, E. A. and Sethi, R. (2000a) "Evolution of interdependent preferences in aggregative games," *Games and Economic Behavior*, 31: 303–10.

―――― ―――― ―――― (2000b) "The strategic advantage of negatively interdependent preferences," *Journal of Economic Theory*, 92: 274–99.

Kyle, A.S. and Wang, A. (1997) "Speculation duopoly with agreement to disagree: can overconfidence survive the market test?" *Journal of Finance*, 52: 2073–90.

Lester, G. and Talley, E. (2000) "Trade secrets and mutual investments", USC Law School Working Paper # 00-15; Georgetown Law and Economics Research Paper No. 246406.

Myerson, R. and Satterthwaite, M. (1983) "Efficient mechanisms for bilateral trading," *Journal of Economic Theory*, 29: 265–81.

Palacios-Huertaa, I. and Santos, T. (2004) "A theory of markets, institutions, and endogenous preferences," *Journal of Public Economics*, 88: 601–27.

Possajennikov, A. (2000) "On the evolutionary stability of altruistic and spiteful preferences," *Journal of Economic Behavior and Organization*, 42(1): 125–9.

Rabin, M. (1998) "Psychology and economics," *Journal of Economic Literature*, 36(1): 11–46.

Rachlinski, J. and Jourden, F. (1998) "Remedies and the Psychology of Ownership," *Vanderbilt Law Review*, 51:1541.

Rotemberg, J. J. (1994) "Human relation in the workplace," *Journal of Political Economy*, 102: 684–717.

Samuelson, L. (2001) "Introduction to the evolution of preferences," *Journal of Economic Theory*, 97: 225–30.

Sethi, R. and Somanathan, E. (2001) "Preference evolution and reciprocity," *Journal of Economic Theory*, 97: 273–97.

Stiglitz, J. (1993) "Post Walrasian and post Marxian economics," *Journal of Economic Perspectives*, 7: 109–14.

Sunstein, C. R. (1993) "Endogenous preferences, environmental law," *Journal of Legal Studies*, 22: 217.

―――― (2002) "Switching the default rule," *N.Y.U. Law Review*, 77: 106.

Talley, E. (1998) "Turning servile opportunities: a strategic analysis of the corporate opportunities doctrine," *Yale Law Journal*, 108.

PART III

EXPERIMENTS

CHAPTER 20

....

COMMON-VALUE AUCTIONS WITH LIQUIDITY NEEDS

An Experimental Test of a Troubled-Assets Reverse Auction

....

LAWRENCE M. AUSUBEL,
PETER CRAMTON, EMEL FILIZ-OZBAY,
NATHANIEL HIGGINS, ERKUT Y. OZBAY,
AND ANDREW STOCKING[1]

INTRODUCTION

....

In the fall of 2008, US housing and financial markets were in the midst of severe adjustments. House prices were falling rapidly, and they were expected to continue to fall. Problems in the housing and mortgage markets had spread to a broader array of financial markets. The nation was facing serious disruption to the functioning of its financial markets that could substantially impair economic activity.[2] The adjustment began following the housing boom that ran from 2003 to early 2006, when delinquencies

[1] We thank Power Auctions LLC and its employees for customizing the auction software and making it available for this purpose. The analysis and conclusions expressed in this chapter are those of the authors and should not be interpreted as those of the Congressional Budget Office, the Economic Research Service, or the US Department of Agriculture.

[2] For an overview of the world financial crisis, see French et al. (2010). For a review of the conditions surrounding the credit markets specifically, see Mizen (2008), Congressional Budget Office (2008), and Brunnermeier, (2009). For a careful study of the conditions in the market for collateralized debt obligations (CDOs), which contributed substantially to the crisis, see Barnett-Hart (2009). Gorton (2008, 2009) and Swagel (2009) provide excellent reviews.

and foreclosures on mortgages rose, particularly on subprime adjustable-rate mort-gage loans (ARMs). Delinquencies also arose for prime ARMs and on so-called alt-A mortgage loans, which were made on the basis of little or no documentation of the borrower's income. Most mortgages were resold as mortgage-backed securities (MBSs), and the rise in delinquencies caused the value of MBSs to decline, in some cases quite sharply.

The problems in mortgage markets spread to the wider financial markets for several reasons. The number of bad mortgages and, consequently, losses on MBSs were expected to be large. The use of complex instruments to fund subprime lending, such as collateral-ized debt obligations (CDOs), also made it difficult for participants in financial markets to identify the magnitude of the exposure of other participants to losses. Moreover, a number of financial institutions borrowed heavily to finance their mortgage holdings, further increasing their risk exposure. Losses on mortgage assets, and the resulting contraction of the availability of credit to businesses and households, posed a significant threat to the pace of economic activity.

The US Department of Treasury and the Federal Reserve, led by Henry Paulson and Ben Bernanke, respectively, considered a host of policy responses to address the illiquid-ity triggered by market panic and the potential insolvency of many financial institutions. On October 3, 2008, the US Congress passed and the President signed the Emergency Economic Stabilization Act of 2008 (Public Law 110-343). The Act established the $700 billion Troubled Asset Relief Program (TARP), which authorized the Secretary of the Treasury to purchase, hold, and sell a wide variety of financial instruments, particularly those that are based on or related to residential or commercial mortgages issued prior to September 17, 2008. The authority to enter into agreements to purchase such financial instruments, which the proposal refers to as troubled assets, would expire two years after its enactment.

An immediate question was what auction designs were well suited to the task. Phillip Swagel, who served as Assistant Secretary for Economic Policy at Treasury from December 2006 to January 2009, recalls that, in September 2008, "we were already working hard to set up reverse auctions with which to buy structured financial products such as [mortgage-backed securities], focusing on mechanisms to elicit market prices. On this we received a huge amount of help from auction experts in academia—an outpouring of support that to us represented the economics profession at its finest" (Swagel, 2009, pp. 47–48). In a reverse auction, sellers compete with each other to sell a product to a single buyer.

Several potential mechanisms were suggested by market design economists. Ausubel and Cramton (2008a,b) suggested the use of a simultaneous descending clock auc-tion (with some particular features we describe below). Klemperer (2009) suggested a novel sealed-bid auction he dubbed the "product-mix" auction. The Treasury settled on a mixed approach: again according to Swagel, "We would have tried two auction approaches, one static and one dynamic—the latter approach is discussed by Lawrence Ausubel and Peter Cramton [2008a], who were among the academic experts providing enormous help to the Treasury in developing the reverse auctions" (Swagel, 2009, p. 56).

Regardless of the approach used, the Treasury had decided to use "reference prices" in order to purchase many different securities (i.e. securities with many different CUSIPs) in a single auction.[3]

The use of a reference price is necessary to hold a single auction during which bidders compete to sell a diverse mix of securities. Because ownership of many of the assets was highly concentrated (i.e. competition to sell a single CUSIP would have been relatively low) assets would be grouped together in a pooled-security reverse auction. Each asset is "scored" with a reference price so that the different types of asset can be compared on a single dimension and a single clearing price is determined in the auction.

Reference prices were to be based on the Treasury's best estimates—albeit imperfect estimates—of the value of each CUSIP. The Treasury, concerned that poor estimates would be taken advantage of by bidders, considered an auction format in which the reference prices would not be announced until after the auction. Armantier et al. (2010) conducted an experimental test of auctions where the reference price is announced only after bidding has taken place, and found that keeping reference prices secret reduced efficiency and did not save the government money.

The research experiments described in the present chapter were implemented in October 2008, and were designed to further test several auction mechanisms and design features considered for use by the Treasury. Both experiments include a comparison of sealed-bid and dynamic auctions for many assets (many securities with unique CUSIPs). Experiment 1 tests an auction appropriate for conditions in which ownership of the assets at auction are evenly distributed among banks. Experiment 2 tests an auction mechanism appropriate for conditions in which ownership is instead concentrated unevenly. The auctions in Experiment 2 are very similar to the auctions that would have been used by the Treasury to purchase toxic assets.

Several conclusions emerged from the experiments.

- The auctions were competitive. Owing to the bidders' liquidity needs, the Treasury paid less than the true common value of the securities under either format.
- The sealed-bid auction was more prone to bidder error.
- The dynamic clock auction enabled bidders to manage their liquidity needs better.
- The bidders attained higher payoffs (trading profits plus liquidity bonus) in the dynamic clock auctions than in the sealed-bid auctions.
- Nevertheless, the clock auctions resulted in equivalent aggregate expenditures, so that the benefit to the bidders did not come at the taxpayers' expense.
- The prices resulting from the clock auctions were a better indication of true values than those from the sealed-bid auctions. We conclude from this that, in the context of a troubled-asset crisis like the one facing the Treasury in 2008, the clock auction is apt to reduce risk for both banks and the Treasury, and to generate price information that may help to unfreeze secondary markets.

---

[3] The acronym "CUSIP" refers to the Committee on Uniform Security Identification Procedures. Each unique security has its own unique CUSIP number, or simply CUSIP, for short.

We conclude that the dynamic clock auction is more beneficial than the sealed-bid auction for both the banks and the taxpayer. The banks attain higher payoffs than in the sealed-bid auction, resulting from better liquidity management. Taxpayers are also better off, as the asset purchase program is better directed toward the liquidity needs of the banking sector without increasing the cost of the asset purchase program. The variability of outcomes is also reduced and the informativeness of prices is also increased with the clock format.

The experiments allowed us to do more than compare static and dynamic auctions. More broadly, the experimental format allowed us to create a market design test-bed. The test-bed helped us to do three things important for all applied market design: (1) demonstrate the feasibility for quick implementation of the auction design; (2) subject the auction design to testing for vulnerabilities; and (3) predict strategic behavior. The commercial auction platform was customized to handle both formats in one week, demonstrating feasibility. Both formats are easy to explain to bidders. Sophisticated subjects required only a three-hour training session to understand the setting, the auction rules, and to practice using the software. Since the auction design was novel and had not been field tested, a laboratory test was an important part of due diligence. Without the laboratory test-bed, we would not have discovered the special vulnerability of the sealed-bid auction to bidder errors. Finally, the test-bed was useful in helping to elicit probable strategies from bidders. Again, because the auction format was novel and further because the auction was too complex for equilibrium analysis, bidder behavior could not be predicted without a test-bed experiment.

Ultimately, on November 12, 2008, the Treasury decided to concentrate on negotiated equity purchases and to postpone the purchase of troubled assets via auction.[4] In March 2009, the Treasury proposed auctions to purchase pools of legacy loans from banks' balance sheets, but this time using a forward auction in which private investors competed to buy the pools of loans. Ausubel and Cramton (2009) describe the auction design issues in this new setting and argue for a two-sided auction in which the private investors compete to buy loan pools in a forward auction, and then banks compete in a reverse auction to determine which trades transact. The results we present here are fully applicable to the legacy-loan setting as well. The forward auction is analogous to the security-by-security auction, and the reverse auction is analogous to the reference-price auction.

The remainder of the chapter is organized as follows. In the next section we summarize the experimental literature with respect to dynamic and sealed-bid auctions. Our analysis builds on this literature. The following section briefly describes the experimental setup. The instructions and related materials are available in the appendices. The fourth section provides an econometric analysis of the results. The fifth section describes the implementation and results of a recombinant procedure, a procedure that explores the full range of outcomes in the sealed-bid auction. The sixth section presents an analysis and discussion, and the last section concludes.

---

[4] See <http://www.bloomberg.com/apps/news?pid=newsarchive&sid=aoD4QebIb_Fo&refer=home>.

Dynamic vs. static auction designs

There is a rich economic literature that points to the advantages of a competitive process over negotiation (see e.g., Bulow and Klemperer, 1996) and thus we focused exclusively on the use of auctions to accomplish the Treasury's objectives. It is within this context that we designed our auction experiment to help us understand the outcomes and relative advantages of alternate auction formats. One of the initial decisions facing the Treasury was whether to conduct a static (sealed-bid) or dynamic (descending-bid) auction.[5]

A frequent motivation for the use of dynamic auctions is reducing common-value uncertainty (Milgrom and Weber, 1982). In the troubled-asset setting there is a strong common-value element: a security's value is closely related to its "hold to maturity value," which is roughly the same for each bidder. Each bidder has an estimate of this value, but the true value is unknown. The dynamic auction, by revealing market supply as the price declines, lets the bidders condition their bids on the aggregate market information. As a result, common-value uncertainty is reduced and bidders will be comfortable bidding more aggressively without falling prey to the winner's curse—the tendency in a procurement setting of naïve sellers to sell at prices below true value.

In the context of many securities, the price discovery of a dynamic auction plays another important role. By seeing tentative price information, bidders are better able to make decisions about the quantity of each good to sell. This is particularly useful because the values of securities are related. Bidding in the absence of price information makes the problem much more difficult for bidders. Furthermore, with a dynamic auction, the bidder is better able to manage both liquidity needs and portfolio risk. In contrast, managing liquidity needs in a simultaneous sealed-bid auction is almost impossible.

Another advantage of a dynamic auction is transparency. Each bidder can see what it needs to do to win a particular quantity. If the bidder sells less, it is the result of the bidder's conscious decision to sell less at such a price. This transparency is a main reason for the high efficiency of the descending clock auction in practice.

Finally, as a practical matter, a clock auction allows for feedback between auction rounds, reducing the likelihood that a mistaken bid will go undetected. Bidders do make mistakes, entering bids incorrectly because of keystroke or other human error. An example occurred in the Mexican Central Bank's auction for US currency on May 19, 2009. A bank entered an erroneous bid that caused it to overpay by US$355,340. All other accepted bids in the auction were within 0.3% of the exchange rate traded that day, while the erroneous bid was 7.4% greater than the concurrent rate.[6] Reducing the

---

[5] See Ausubel and Cramton (2002, 2004, 2006), Cramton (1998), McAfee and McMillan (1987), and Milgrom (2004) for further discussion.

[6] See Bloomberg "Mexican Bank Pays $355,340 Above Cost in Dollar Sale", <http://www.bloomberg.com/apps/news?pid=newsarchive&sid=acbCBKTS6_As>.

likelihood of bidder error is important. We provide evidence in this chapter that bidder error is less likely under the clock format than under the sealed-bid format.

The experimental economics literature strongly supports the conclusion that dynamic auctions outperform sealed-bid auctions in terms of efficiency and price discovery. In sealed-bid auctions there is a tendency to consistently overbid (Kagel et al., 1987; McCabe et al., 1990), often resulting in inefficient outcomes. In contrast, many laboratory and field experiments have demonstrated that the clock auction format is simple enough that even inexperienced bidders can quickly learn to bid optimally (Kagel et al., 1987). Kagel (1995) finds that bidders readily transfer the experience gained in sealed-bid auctions to the clock auction format. Bidders in Levin et al. (1996) appear to adopt simple strategies that incorporate dynamically changing information from the clock auction, namely the prices at which other bidders drop out, and efficient outcomes are obtained.

A principal benefit of the clock auction is the inherent price-discovery mechanism that is absent in any sealed-bid auction. Specifically, as the auction progresses, participants learn how the aggregate demand changes with price, which allows bidders to update their own strategies and avoid the winner's curse (Kagel, 1995). Levin et al. (1996) show that bidders suffer from a more severe winner's curse in the sealed-bid format than in a clock auction. Kagel and Levin (2001, 2005) compare a clock auction and a sealed-bid auction when bidders demand multiple units, and confirm that outcomes are much closer to optimal in the clock auction. Efficiency in the clock auction always exceeded 97%. Moreover, in the Ausubel auction (a particular type of clock auction—see Ausubel, 2004, 2006) bidders achieve optimal outcomes 85.2% of the time, as compared with only 13.6% of the time in a sealed-bid auction. McCabe et al. (1990) found 100% efficient outcomes in forty-three of forty-four auctions using a clock auction. Kagel and Levin (2009) provide further evidence of more efficient outcomes with a clock format in the multi-unit setting. Alsemgeest et al. (1996) also find that clock auctions are efficient both in single-unit and multi-unit supply scenarios, achieving better than 99.5% efficiency and 98% efficiency.

The principal advantage of a sealed-bid auction is its apparent simplicity and relatively "inexpensive" setup. Some would argue that a sealed-bid auction is also less vulnerable to collusion. Some also fear that even a quick dynamic auction would expose participants to significant unhedged positions as a result of real-time interactions with financial markets. This latter complaint can be addressed by conducting the auction when the major financial markets are closed.

Experimental setup

During the period October 12, 2008, to November 11, 2008, using commercial auction software customized for our purpose, we tested two different auction environments at the University of Maryland's experimental economics laboratory.[7] The objective of the experimental setup was to mimic the environment faced by the Treasury Department. Specifically, the Treasury faced the challenge of purchasing assets so as to balance two competing criteria: (1) assuring that the taxpayer would not overpay for the assets; and (2) improving banking sector stability by purchasing assets from those banks most in need of liquidity. Ausubel and Cramton (2008a,b) discussed the design issues as they appeared in October 2008 and proposed a specific auction format. The experiments described in the present chapter were designed to provide insights into bidding behavior and performance of that format relative to alternate formats, as well as demonstrate the feasibility of quickly implementing either format of auction as part of the financial rescue. For each auction format and information setting analyzed, we compare sealed-bid uniform-price auctions with dynamic clock auctions, varying the level of competition, information, and banks' need for liquidity.

The experimental auction environments were closely tailored to the likely settings of the planned auctions for troubled assets. Specifically, to model the case where there was sufficient competition to conduct a competitive auction for individual securities, we ran an eight-security simultaneous reverse auction. Each security had a pure common value with unconditional expectation of 50 cents on the dollar, bidders had private information about the common value, and a fixed quantity of each security was purchased in the same reverse auction. This is what we refer to as a *security-by-security* auction. In the second auction environment, the ownership of the security was too concentrated to allow individual purchase. Securities of a similar quality were instead pooled together, thus mitigating the concentration-of-ownership problem. In this second auction environment, each security had a pure common value, bidders had asymmetric endowments, and bidders with larger holdings of a security had more private information about the common value. In order to implement an auction where dissimilar items are purchased together, bidders compete on the basis of a reference price, which reflects the government's best estimate of the security's value. Bidders then compete on a relative basis—a bid expresses willingness to tender a security at a stated percentage of the security's reference price. This is what we refer to as a *reference-price auction*.

The human subjects bidding in the auctions were experienced PhD students, highly motivated by the prospect of earning roughly $1,200 each—the actual amount depending on performance—for participating in twelve experimental sessions, each lasting two to three hours, over the three-week period. We chose to use experienced PhD students for these experiments, since the environment is considerably more complex than a typical economics experiment, and we believed that the PhD students' behavior would be more representative of the sophisticated financial firms that would be participating in the actual auctions.

In terms of scope, the experimental "banks" held roughly 8,000 distinct troubled securities, potentially available for purchase. For the purposes of this chapter, these

---

[7] See <http://www.econ.umd.edu/resources/computing/experimental>.

Table 20.1. Schedule of treatments

Order of Treatment:		First	Second	Third	Fourth
Session					
			Positive Liquidity Need		
1–4	Auction Type	Sealed-bid	Sealed-bid	Clock	Clock
	# of bidders	4	8	4	8
	reference prices	NA	NA	NA	NA
5,7	Auction Type	Sealed-bid	Clock	Sealed-bid	Clock
	# of bidders	8	8	8	8
	reference prices	More precise	More precise	Less precise	Less precise
6,8	Auction Type	Sealed-bid	Clock	Sealed-bid	Clock
	# of bidders	8	8	8	8
	reference prices	Less precise	Less precise	More precise	More precise
			Zero Liquidity Need		
9–10	Auction Type	Sealed-bid	Clock	Sealed-bid	Clock
	# of bidders	4	4	4	4
	reference prices	NA	NA	NA	NA
11–12	Auction Type	Sealed-bid	Clock	Sealed-bid	Clock
	# of bidders	8	8	8	8
	reference prices	Less precise	Less precise	Less precise	Less precise

assets fall into two general groups: (1) those securities with ownership concentrated among only a few firms; and (2) those securities with less concentrated ownership. By the nature of these troubled assets, both the banks and the government believed them to be worth less than face value. However, some securities are more "troubled" than others. Some are relatively high-valued securities (e.g. a market value of 75 cents on the dollar) and others are relatively low-valued securities (e.g. a market value of 25 cents on the dollar).

For purposes of exposition, we describe the two auction environments as Experiment 1, an eight-security simultaneous reverse auction, and Experiment 2, a pooled-security reverse auction. Experiments 1 and 2 were conducted over a total of twelve sessions. The schedule of treatments is given in Table 20.1. The individual and pooled auctions are described below, with more detail provided in the appendices. Each session involved four auctions in the order indicated.

Experiment 1: eight-security simultaneous reverse auction

In Experiment 1, bidders compete to sell their symmetric holdings of eight securities to the Treasury. Two formats are used:

- *Simultaneous uniform-price sealed-bid auction ("sealed-bid auction").* Bidders simultaneously submit supply curves for each of the eight securities. Supply curves are non-decreasing (i.e. upward-sloping) step functions. The auctioneer then forms the aggregate supply curve and crosses it with the Treasury's pre-announced and fixed demand. The clearing price is the lowest rejected offer. All quantity offered below the clearing price is sold at the clearing price. Quantity offered at the clearing price is rationed to balance supply and demand, using the proportionate rationing rule.[8]
- *Simultaneous descending clock auction ("clock auction").* The eight securities are auctioned simultaneously over multiple rounds. In each round, there is a price "clock" that indicates the start-of-round price and end–of-round price per unit of quantity. Bidders express the quantities they wish to supply at prices they select below the start-of-round price and above the end-of-round price. At the conclusion of each round, bidders learn the aggregate supply for each security. In subsequent rounds, the price is decremented for each security that has excess supply, and bidders again express the quantities they wish to supply at the new prices. This process repeats until supply is made equal to demand. The tentative prices and assignments then become final. Details of the design are presented in Ausubel and Cramton (2008a, b).

Six sessions in Experiment 1 were dedicated to testing the following three auction attributes: (1) the effect of sealed-bid vs. clock formats; (2) the effect of liquidity needs; and (3) the effect of increased competition. In sessions 1–4, we conducted paired sealed-bid and clock auctions with both low and high levels of competition (a total of four bidders competed in low-competition auctions, while eight competed in high-competition auctions). Sessions 9 and 10 were similar, except that bidders did not have liquidity needs. That is, subjects were not given a bonus based on the sale of securities during the auction. Instead, a subject's take-home pay was based entirely on the profits they made when they sold a security to the government for more than its true value. We focused on the low-competition case in sessions 9 and 10, substituting an extra pair of four-bidder auctions in place of the eight-bidder auctions. As a result of this schedule in sessions 9 and 10, we effectively gave players four auction pairs (sealed-bid and clock) of learning in two consecutive days, focused only on the four-bidder auction.

The experimental design was intended to facilitate a direct comparison of the sealed-bid auction and the clock auction. Before each sealed-bid auction, each bidder learned the realizations of one or more random variables that were relevant to the value of

---

[8] The proportionate rationing rule plays a role only in the event that multiple bidders make reductions at the clearing price. The rule then accepts the reductions at the clearing price in proportion to the size of each bidder's reduction at the clearing price. Thus, if a reduction of 300 is needed to clear the market and two bidders made reductions of 400 and 200 at the clearing price, then the reductions are rationed proportionately: the first is reduced by 200 and the second is reduced by 100. The actual reduction of the first bidder is twice as large as that of the second bidder, since the first bidder's reduction as bid is twice as large as the second bidder's reduction.

the securities that she owned. The same realizations of the random variables applied to the clock auction immediately following the sealed bid. Thus, in successive pairs of experimental auctions, the securities had the same values and the bidders had the same information. Bidders were not provided with any information about the outcome of a given sealed-bid auction before the following clock auction, in order to avoid influencing the behavior in the clock auction.[9]

The value of each security in cents on the dollar is the average of eight *iid* random variables uniformly distributed between 0 and 100:

$$v_s = \tfrac{1}{8} \sum_{i=1}^{8} u_{is}, \text{ where } u_{is} \sim_{iid} U[0, 100],$$

where a bidder's private information about security s is the realization u_{is}. This is true both for the eight-bidder and four-bidder auctions, so that only the first four draws are revealed in the four-bidder auction. This design allowed the true values to have the same distribution in both four-bidder and eight-bidder auctions, which caused the private information to have the same precision.

A bidder profits by selling securities to the Treasury at prices above the securities' true values. Profit (in million $) is defined as:

$$\pi_i(p, q_i, v) = \tfrac{1}{100} \sum_{s=1}^{8} (p_s - v_s) q_{is},$$

where the quantity sold is q_s of security s at price p_s.

In sessions 1–4, bidders also have a need for liquidity. The sale of securities to the Treasury is the source of a bidder's liquidity. The liquidity need, L_i, is drawn *iid* from the uniform distribution on the interval $[250, 750]$. Bidders know their own liquidity need, but not that of the other bidders. Bidders receive a bonus of $1 for every dollar of sales to the Treasury up to their liquidity need of L_i. Bidders do *not* get any bonus for sales to the Treasury above L_i. Thus, their bonus is:

$$\min \left[L_i, \tfrac{1}{100} \sum_{s=1}^{8} p_s q_{is} \right].$$

Given that bidders care about both profits and liquidity, their total payoff is the combination of the two:

$$U_i(p, q_i, v) = \begin{cases} \tfrac{1}{100} \sum_{s=1}^{8} (2p_s - v_s) q_{is} & \text{if } \tfrac{1}{100} \sum_{s=1}^{8} p_s q_{is} < L_i \\ L_i + \tfrac{1}{100} \sum_{s=1}^{8} (p_s - v_s) q_{is} & \text{otherwise} \end{cases}$$

---

[9] Observe that, inherently, information about a clock auction *must* be revealed, as bidders learn aggregate information about round 1 before the start of round 2, etc. Thus, it would have been impossible to run the sealed-bid auctions after the clock auctions without influencing the behavior in the sealed-bid auctions.

In each session, two auctions were selected at random (one from each pair of auctions) to determine bidders' take-home earnings. We used a conversion factor of $1 in take-home pay for every $10 million in experimental earnings.

Given the relatively tractable theoretical nature of the experimental setup without the liquidity constraint, we calculated a benchmark bid based on equilibrium bidding strategies in a common-value auction (Milgrom and Weber, 1982):

four-bidder sealed-bid strategy:

$$b_{is} = \frac{1}{8}\left(2u_{is} + 2\left(\frac{u_{is} + 100}{2}\right) + 4 \cdot 50\right) = \tfrac{3}{8}u_{is} + \tfrac{75}{2}.$$

eight-bidder sealed-bid strategy:

$$b_{is} = \frac{1}{8}\left(2u_{is} + 3\frac{u_{is}}{2} + 3\left(\frac{u_{is} + 100}{2}\right)\right) = \tfrac{5}{8}u_{is} + \tfrac{75}{4}.$$

These Bayesian Nash equilibrium strategies are based on a theoretical framework that differs from our experiment in two ways: (1) they ignore any behavioral adjustments resulting from the liquidity bonus; and (2) they assume that bidders sell their holdings to the Treasury as an indivisible block (i.e. either their entire endowment or nothing). Despite that, the benchmark strategies provide guidance in a static or dynamic setting. As a result, these strategies were explained to bidders and made operational in a bidding tool (i.e. the bidding tool facilitated updating of the strategy following a drop in supply by backwardly inducting the values of the bidder who reduced their supply). Assuming all players play the benchmark strategy, we simulated both the sealed-bid and clock auctions under the two competition levels. These simulations provide an expected clearing price for each of the eight securities as well as bidder-specific profits and payoffs. While in the experimental auction we anticipated that the liquidity bonus would be likely to cause players to bid more aggressively than predicted by the benchmark, this behavioral change was not included in our simulations.

Experiment 2: pooled-security reverse auction

In the reference-price auctions, the holdings of the eight individual securities are too concentrated for there to be competitive auctions on a security-by-security basis. Think of a reverse auction for apples and oranges. In a simultaneous auction all bidders would submit bids to sell apples and oranges at the same time—apple bids would compete against other apple bids and orange bids would compete against other orange bids. The result of the auction would be two separate clearing prices, one for apples and one for oranges. In a pooled-fruit auction, apple bids would compete against other apple bids *and* all the orange bids. Since apples and oranges are clearly different fruits, in order to consider the relative merit of apple bids and orange bids the auctioneer would state a price that she believes to be a fair price for apples (say $0.50), as well as a price that she believes to be a fair price for oranges (say $0.75). Bids are then ranked according to the discount on the estimated value, so that an apple bid of $0.50 (a discount ratio of 1) would rank as more *expensive* than an orange bid of $0.60 (a discount ratio of 0.8).

The defining features of the pooled auction are as follows:

- The clearing prices for different securities (i.e. securities with different CUSIP numbers) are determined within the same auction.
- Bidder endowment and thus price signals are asymmetric for each security.
- Before an auction, the Treasury determines and announces its estimate of the value of each security—these are referred to as *reference prices*.
- The prices in the sealed-bid auction, or in each round of the descending clock auction, are expressed as a percentage of the reference price for each security—these are referred to as *price points*.
- Clearing occurs when the cost of purchasing the securities offered at a given price point equals the budget allocated for the auction.

As in Experiment 1, two auction formats are considered:

- *Simultaneous uniform-price sealed-bid auction.* Bidders simultaneously submit supply curves for each of the securities within the pool. Supply curves are upward-sloping step functions, where prices are expressed as price points (a percentage of the reference price) and quantities are expressed in dollars of face value. The auctioneer then forms the aggregate supply curve and equates it with the Treasury's demand. The clearing price is the lowest rejected offer. All securities offered at price points below the clearing price point are purchased at the clearing price point. Securities offered at exactly the clearing price point are rationed by a proportional rationing rule.
- *Simultaneous descending clock auction.* There is a price "clock" indicating the current range of price points. For example, in round 1, bidders express the quantities that they wish to supply of each security at all price points from 106% to 102% of the respective reference price for securities within that auction. After round 1, the auctioneer aggregates the individual bids and informs bidders of the aggregate quantity that was offered at 102%. Assuming that supply exceeded demand, the price is decremented; for example, in round 2, bidders may express the quantities that they wish to supply of each security at all price points from 102% to 98% of the respective reference prices. The process is repeated, with the price decremented, bids submitted, and quantities aggregated, until supply is made equal to demand. Then, as in the sealed-bid auction, all securities offered at price points below the clearing price point are purchased at the clearing price point, and bids at exactly the clearing price point are rationed by a proportional rationing rule.

Details of the designs are described in Ausubel and Cramton (2008a, b).

Six sessions were dedicated to test the following three auction attributes: (1) the effect of sealed-bid vs. clock auction format; (2) the effect of the liquidity bonus; and (3) the effect of increasing precision with respect to the reference price. In sessions 5–8, we ran a low-precision sealed-bid and clock auction, and a high-precision sealed-bid and clock auction (four auctions total), in that order.[10] Thus bidders completed four auction pairs (sealed-bid and clock) for each of the low-precision and high-precision auctions (one pair of each precision level each day). In sessions 5 and 6, we removed the liquidity bonus and ran two low-precision sealed-bid and clock auctions per session for a total of four auctions in each session. As a result, we effectively gave players four auction pairs (sealed-bid and clock) of learning in two days, but only with the low-precision auction.

Bidder endowments for each security are described in Table 20.2. Each bidder had an endowment of $40 million of face value, divided differently across securities. Similarly, there were $40 million of face value for each security. The Treasury has a demand for 25% of the total face value within each pool of securities, which might involve the purchase of one or more individual securities.

The value of each high-quality security $s \in \{H1,H2,H3,H4\}$ in cents on the dollar is the average of n iid random variables uniformly distributed between 50 and 100:

$$v_s = \frac{1}{n} \sum_{j=1}^{n} u_{js}, \text{ where } u_{js} \sim_{iid} U[50, 100].$$

Table 20.2. Holdings of securities by bidder and security in million $ of face value

		High-quality securities				Low-quality securities				
		H1	H2	H3	H4	L1	L2	L3	L4	Total
Bidder	1	20	0	0	0	0	5	5	10	40
	2	0	20	0	0	10	0	5	5	40
	3	0	0	20	0	5	10	0	5	40
	4	0	0	0	20	5	5	10	0	40
	5	0	5	5	10	20	0	0	0	40
	6	10	0	5	5	0	20	0	0	40
	7	5	10	0	5	0	0	20	0	40
	8	5	5	10	0	0	0	0	20	40
	Total	40	40	40	40	40	40	40	40	
	Expected price	75	75	75	75	25	25	25	25	
	Expected value	30	30	30	30	10	10	10	10	
	Total value	120				40				

[10] In sessions 5 and 7, the more precise sealed-bid and clock auctions were conducted first. In sessions 6 and 8, the less precise sealed-bid and clock auctions were conducted first.

The value of each low-quality security $s \in \{L1,L2,L3,L4\}$ in cents on the dollar is the average of n *iid* random variables uniformly distributed between 0 and 50:

$$v_s = \frac{1}{n} \sum_{j=1}^{n} u_{js}, \text{ where } u_{js} \sim_{iid} U[0,50].$$

For auctions with more precise reference prices, $n = 16$; for auctions with less precise reference prices, $n = 12$. The reference price r_s for security s is given by

$$r_s = \frac{1}{n-8} \sum_{j=9}^{n} u_{js}.$$

Thus, the reference price is based on eight realizations in the more precise case (half of all realizations) and on four realizations in the less precise case (a third of all realizations). Reference prices are made public before each auction starts.

For each $5 million of security holdings, bidder i receives as private information one of the realizations u_{js}. Thus, bidder 1, who holds $20 million of security 1, gets four realizations (see Table 20.2). In this way, those with larger holdings have more precise information about the security's value. Observe that this specification requires the holders of each given security to receive collectively a total of eight realizations. Since there are eight realizations available (besides the ones that form the reference price), each of the realizations u_{js} ($i = 1, \ldots, 8$) can be observed by exactly one bidder.

Suppose that the auction clearing price point is p_H for the high-quality pool and p_L for the low-quality pool, where the price point in the auction is stated as a fraction of the reference price. Then $p_s = p_H r_s$ for $s \in \{H1,H2,H3,H4\}$ and $p_s = p_L r_s$ for $s \in \{L1,L2, L3,L4\}$.

If a bidder sells the quantity q_s of the security s at price p_s, then profit is

$$\pi_i(p, q_i, v) = \frac{1}{100} \sum_s (p_s - v_s) q_{is},$$

where the 1/100 factor converts cents into dollars. As with Experiment 1, when bidders have a liquidity need (sessions 5–8), it is drawn *iid* from the uniform distribution. In Experiment 2, however, the cash scale is increased, and thus liquidity needs are drawn from the interval $[2500, 7500]$. Each bidder knows his own liquidity need, but not that of the other bidders. The bidder receives a bonus of $1 for every dollar of sales to the Treasury up to his liquidity need:

$$\min\left[L_i, \frac{1}{100} \sum_s p_s q_{is}\right].$$

Combining the profit and the liquidity penalty results in the bidder's total payoff:

$$U_i(p, q_i, v) = \begin{cases} \frac{1}{100} \sum_s (2p_s - v_s) q_{is} & \text{if } \frac{1}{100} \sum_s p_s q_{is} < L_i \\ L_i + \frac{1}{100} \sum_s (p_s - v_s) q_{is} & \text{otherwise} \end{cases}$$

Thus, an additional $1 of cash is worth $2 when the bidder's liquidity need is not satisfied, but is worth $1 when the liquidity need is satisfied. To be roughly comparable to Experiment 1, bidders' take-home pay was calculated such that they received $1 in take-home pay for every $100,000 in experimental earnings.

Unlike Experiment 1, there is no Bayesian Nash equilibrium bidding strategy for a similar auction that we can use as a benchmark. The reference-price auction is beyond current theory. The pooling of securities combined with the use of reference prices violates monotonicity in signals, meaning that a higher signal does not necessarily translate to a higher bid. Monotonicity between signals and values exists within a particular security (i.e. *ceteris paribus*, a higher signal suggests a higher value); however, there is no monotonicity across securities within a particular pool.

Monotonicity holds when a higher signal implies a higher expected value to the bidder. This relationship is broken by the existence of the reference price. Consider a security with a higher reference price that has a higher expected value to each bidder, all else equal. Holding the common value of a security fixed, bidders prefer a higher reference price, since a high reference price makes the security more competitive in the pool. Thus, in determining her bid, a bidder must consider the countervailing forces of signals and reference prices. It is difficult to recommend how a bidder should respond to a high signal with a low reference price, a low signal with a high reference price, etc.

Experimental subjects

The training of subjects and all experimental sessions took place in the experimental lab of the University of Maryland's Economics Department. This is a new state-of-the-art facility for conducting economic experiments. Each subject has her own private cubical with computer and necessary software. The subject pool consisted of PhD students at the University of Maryland and George Mason University. The students had taken or were taking an advanced graduate course in game theory and auction theory, and were pursuing degrees in economics, business, computer science, or engineering.

In each session of approximately three hours, sixteen bidders, out of a total subject pool of nineteen, participated in four auctions. Each auction consisted of four or eight bidders (i.e. there were always multiple auctions conducted in parallel), and the bidders were randomly and anonymously matched.

Bidders' payoffs consisted of the sum of two terms. First, each bidder received trading profits according to the difference between the common value, v, of the security, and the price, p, at which the bidder's securities were purchased. Hence, if the bidder sold a quantity, q, of securities, the bidder's trading profits equaled: $q \cdot (p-v)$. Second, each bidder was randomly assigned a *liquidity need*, L, and received an additional dollar of payoff for each dollar in sales, $q \cdot p$, up to L that the bidder received in a given auction.

At the conclusion of all sessions, each subject received a check equal to a show-up fee of $22 per session plus an amount proportional to her total experimental payoff as described above. Average take-home pay was $100.43 per session.

The next section describes the results.

EXPERIMENTAL RESULTS

The primary results comparing sealed-bid and clock aspects of the two experiments are summarized in Tables 20.3–20.6. First, considering just the results from Experiment 1 with the liquidity bonus, we see that even though clearing price and profits are statistically indistinguishable between the two auction formats, the variability of profit is much higher in the sealed-bid auction compared with the clock. Thus the results from the clock auction would appear to be more stable and predictable. The Treasury would appear to best satisfy its first objective to consistently get the best possible price for the taxpayers using a clock auction, though the difference is small. This is particularly

Table 20.3. Comparison of mean outcomes by auction type in Experiment 1 with liquidity bonus

Variable	Auction Type		Result
	Sealed-Bid	Clock	
Clearing Price	47.79 (1.41)	49.57 (1.32)	The clearing price is statistically indistinguishable for the Clock and Sealed Bid auction (t-test p-value of 0.3621)
Profit	−39.54 (21.1)	−13.03 (15.5)	Profits are statistically indistinguishable between the two auction formats (t-test p-value of 0.3135)
Standard Deviation of Profit	239.3	175.8	Higher standard deviation of profit in sealed-bid than clock (variance ratio test p-value of 0.0006)
Liquidity Bonus	428 (16)	466 (13)	Clock liquidity bonus is significantly larger than sealed-bid liquidity bonus (t-test p-value of 0.0562)
Payoff	388 (25)	453 (20)	Clock payoff is significantly higher than sealed-bid payoff (t-test p-value of 0.0400)
Standard Deviation of Payoff	281.7	221.5	Higher standard deviation of payoff in sealed-bid than clock (variance ratio test p-value 0.0071)
Overshooting the liquidity need	692 (41)	605 (37)	Overshooting the liquidity need is almost significantly less in clock than in sealed-bid (t-test p-value of 0.1210)

Note: mean value is shown with standard error in parentheses

Table 20.4. Comparison of mean outcomes by auction type in Experiment 1 without liquidity bonus

Variable	Auction Type		Result
	Sealed-Bid	Clock	
Clearing Price	55.81 (0.66)	58.82 (0.51)	The clearing price is significantly higher for the Clock auction (t-test pvalue of 0.0004)
Profit = Payoff	118.07 (14.89)	178.20 (14.16)	Profits are significantly greater than zero in both cases, and are significantly higher in the Clock auction (t-test p-value of 0.0041)
Standard Deviation of Payoff	119.1	113.3	Standard deviation of payoff in sealed-bid is statistically identical to that of clock (variance ratio test p-value 0.6919)

Note: mean value is shown with standard error in parentheses

Table 20.5. Comparison of mean outcomes by auction type in Experiment 2 with liquidity bonus

Variable	Auction Type		Result
	Sealed-Bid	Clock	
Clearing Price	85.22 (0.81)	83.87 (1.18)	The clearing pricepoint is significantly indistinguishable between the two auction formats (t-test p-value of 0.3485)
Profit	−693.06 (51)	−798.60 (57)	Profits are significantly less than zero in both cases, but no significant difference in mean profits (t-test p-value of 0.1680)
Standard Deviation of Profit	574.1	645.2	No significant difference in the standard deviation on profit in clock compared to sealed-bid (variance ratio test p-value 0.1896)
Liquidity Bonus	3915.0 (172)	4517.2 (131)	Clock liquidity bonus is significantly larger than sealed-bid liquidity bonus (t-test p-value of 0.0059)
Payoff	3222.0 (146)	3718.6 (116)	Clock payoff is significantly higher than sealed-bid payoff (t-test p-value of 0.0083)
Standard Deviation of Payoff	1653.5	1311.9	Higher standard deviation of payoff in sealed-bid than clock (variance ratio test p-value 0.0095)
Overshooting the liquidity need	1984.0 (290)	904.8 (154)	Overshooting the liquidity need is less in clock than in sealed-bid (t-test p-value of 0.0014)

Note: mean value is shown with standard error in parentheses

Table 20.6. Comparison of mean outcomes by auction type in Experiment 2 without liquidity bonus

	Auction Type		
Variable	Sealed-Bid	Clock	Result
Clearing Price	93.7 (1.41)	94.6 (1.36)	The clearing pricepoint is significantly indistinguishable between the two auction formats (t-test p-value of 0.6716)
Profit=Payoff	160.1 (39)	244.6 (38)	Profits are significantly greater than zero, and are almost significantly higher in the Clock auction (t-test p-value of 0.1215)
Standard Deviation of Payoff	309.7	303.5	Standard deviation of payoff in sealed-bid is statistically identical to that of clock (variance ratio test p-value 0.8739)

Note: mean value is shown with standard error in parentheses

important when a liquidity bonus is in effect; without the liquidity bonus (Table 20.4) profits are statistically greater than zero and the clock profits are significantly higher for the clock auction (178) relative to the sealed-bid auction (118), with negligible differences between the standard deviation.

Turning to the Treasury's second objective, related to buying assets from those banks most in need of liquidity, we examine the payoffs from the two auction formats. Payoffs are significantly higher under the clock auction (453) than under the sealed-bid auction (388). We also see that the variability of total payoffs is higher under the sealed-bid auction than the clock, which supports the premise that the additional information provided by the clock auction format leads to more consistent, less variable outcomes. Once again, the Treasury is best served in achieving its second objective with a clock auction.

Turning to Experiment 2, with liquidity need, we see that there is no difference in the clearing price between the two auction formats and while the profits are lower in the clock auction (−799) than in the sealed-bid auction (−693), the difference is not significant (Table 20.5). In addition, there is not a significant variation in the standard deviation of the profit. This result is mimicked in Table 20.6, when the liquidity bonus is not present. In terms of achieving the Treasury's first objective, the two auction formats would seem indistinguishable.

However, when the liquidity bonus is included in the analysis (Table 20.5), we see that the mean payoff under the clock auction (3,719) is significantly higher than the payoff under the sealed-bid auction (3,222). Moreover, the standard deviation of payoff is higher under the sealed-bid auction and the magnitude by which experimental subjects overshot their liquidity need was higher in the sealed bid (1,984 sealed bid overshoot and 905 clock overshoot). Both of these results suggest that the clock auction is a more efficient and accurate means of helping the Treasury determine which banks are most in need of liquidity and allowing the banks to best manage their need for liquidity.

In the following two subsections we discuss in more detail the econometric analysis of the data.

Experiment 1: simultaneous descending clock

The baseline regression results demonstrating the effect of liquidity and learning across the six sessions of auctions in Experiment 1 are shown in Table 20.7. There are three striking results from this table. First, we see the results described in Tables 20.3–20.6: the profit between the sealed-bid and clock auction (regression 2) is statistically identical, whereas, when the liquidity bonus is zero, bidders earn a significantly higher profit in the clock auction ($60). Theory predicts that without the liquidity bonus, the expected payoff from the clock and sealed-bid auctions should be identical, though the sealed bid auction is likely to have higher profit variance. This is not borne out in the results and may be because additional information made available to bidders in the clock auction facilitated tacit collusion. In the auctions with liquidity, tacit collusion was more complicated to implement due to the multiple bidder objectives. When the liquidity

Table 20.7. Experiment 1: experimental subject fixed effects

	Liquidity>0		Liquidity=0
	(1)	(2)	(3)
dep var:	payoff	profit	profit=payoff
Liquidity	0.740***	−0.159*	
	[0.0855]	[0.0879]	
Session 2	183.4***	123.9***	
	[44.03]	[35.68]	
Session 3	283.9***	211.7***	
	[47.04]	[35.56]	
Session 4	250.0***	184.3***	
	[33.52]	[35.51]	
Session 9			58.82*
			[30.03]
Session 10			
Clock	65.39***	26.51	60.13***
	[16.36]	[24.31]	[16.08]
_cons	−147.2**	−93.30*	88.66***
	[54.83]	[51.58]	[16.64]
Subject FE	Yes	Yes	Yes
N	256	256	128
adj. R-sq	0.42	0.14	0.30
Robust standard errors in brackets; * p<0.1 ** p<0.05 *** p<0.01			

bonus is included in the payoff, the clock auction generates significantly higher payoffs ($65) relative to the sealed-bid auction.

The second observation from Table 20.7 is the large influence the liquidity bonus has on payoffs. For every $1 in liquidity bonus, payoffs are increased by a statistically significant $0.74, while profits are reduced by a marginally significant $0.15. The positive effect on payoffs and negative effect on profits are expected as a higher liquidity bonus should motivate players to bid more aggressively on some of their securities, driving the profits negative on those securities, but securing a positive payoff with the liquidity bonus. Given that liquidity bonus is directly added to a bidder's payoff, the coefficient on the liquidity bonus can be interpreted as the percentage of the bonus captured by bidders; overall, bidders captured 74% of their liquidity bonus over the four days.

Finally, Table 20.7 illustrates the effect of learning. Between session 1 and session 3, when there was a positive liquidity bonus, we see that payoffs and profits steadily increased. Specifically, session 2's payoffs were $183 greater than session 1's, and session 3's payoffs were an additional $100 greater than session 2's. With respect to profit, session 2's profits were $124 greater than session 1's, and session 3's profits were an additional $88 greater than session 2's. In session 4, however, the effect on learning appears to change. There is not a statistically significant difference between session 4 and session 3 in either the profit or payoff measure, which suggests that participants had learned all they could during the first three sessions. Alternately, it could be the case that in session 4 players were still learning, but because everyone was optimally responding to each other, there was no change in payoffs or profits.

The effect of learning is reversed when the liquidity bonus is set to zero. This is demonstrated in regression 3. We see that players in auction pair 1 and 2 during session 9 earned a statistically significant $59 more than during auction pair 1 and 2 of session 10. We consider this result in the discussion of Table 20.9.

These results are further reinforced in Table 20.8, where we explore the effect of competition and the expected payoff on actual payoffs. The most striking result in Table 20.8 is that increasing competition in both the sealed-bid and clock auctions results in a higher expected payoff for all players. Using the coefficients in Table 20.8, the incremental payoffs for the various auctions are as follows (assuming x payoff in the four-person sealed-bid auction):

- eight-person sealed-bid auction: $x +$ $239.70
- four-person clock auction: $x +$ $119.60
- eight-person clock auction: $x +$ $251.00

This competition benefit can be explained in the clock auction by the fact that players are learning information about all eight signals. That is, there are eight signals drawn for both auctions but in the four-person auction four of those signals are not represented by any players. Thus, it is impossible for players to learn anything about those four signals. As a result, the eight-player clock auction reveals much more information about the common value for each security than the four-player clock auction, which results in higher player payoff.

Table 20.8. Experiment 1: Effect of competition and expected payoff

	Liquidity>0				Liquidity=0
	(1) payoff	(2) payoff	(3) payoff	(4) payoff	(5) payoff
Liquidity	0.740***	0.810***	0.810***	0.875***	
	[0.0855]	[0.0594]	[0.0595]	[0.0609]	
Session 1	omitted	omitted	omitted	omitted	
Session 2	183.4***	186.2***	186.2***	188.8***	
	[44.03]	[43.97]	[44.06]	[43.30]	
Session 3	283.9***	285.4***	285.4***	290.9***	
	[47.04]	[47.14]	[47.23]	[47.74]	
Session 4	250.0***	249.7***	249.7***	252.4***	
	[33.52]	[34.87]	[34.94]	[34.70]	
Session 9	0	0	0	0	65.85*
	[0]	[0]	[0]	[0]	[29.60]
Session 10	0	0	0	0	omitted
	[0]	[0]	[0]	[0]	
Clock	65.39***	65.39***	119.6***	65.34***	59.09***
	[16.36]	[16.39]	[31.78]	[16.34]	[16.08]
8_Bidders		185.6***	239.7***	190.8***	
		[28.33]	[37.68]	[29.24]	
Clock*8_Bidders			−108.3**		
			[35.52]		
e_payoff				−0.0914	0.156
				[0.0648]	[0.116]
_cons	−147.2*	−274.2***	−301.3***	−265.9***	193.9***
	[54.83]	[43.63]	47.67	[42.81]	[30.30]
Subject FE	Yes	Yes	Yes		Yes
N	256	256	256		128
adj. R-sq	0.419	0.552	0.551		0.315

Standard errors in brackets clustered on subjects; * p<0.05 ** p<0.01 *** p<0.001

We also see that going from four to eight bidders in the sealed-bid auction increases payoffs. There is no theoretical support for this finding and thus we suggest that it is an experimental artifact. It is likely caused by the fact that four of the eight drawn signals were not observed by any player, and thus all of the possible outcomes (from the eight signals) were not represented in the outcomes.

Table 20.8 also illustrates that there is no correlation between actual payoff and expected payoff (i.e. simulated payoff), independent of liquidity. To further explore this result, we consider the interaction of expected payoff and session-specific effects for the zero-liquidity experiments in sessions 9 and 10. These results, shown in Table 20.9, illustrate that while there is no correlation between actual and expected payoff during

Table 20.9. Experiment 1: Effect of session–specific expected payoff

| Dep Var | Liquidity=0 | |
	(1) payoff Session=9	(2) payoff Session=10
Clock	65.93* [24.91]	50.47** [15.19]
e_payoff	0.0722 [0.301]	0.297 [0.164]
_cons	146.6*** [20.87]	64.34*** [10.77]
Subject FE	Yes	Yes
N	64	64
adj. R-sq	0.385	0.633

Standard errors in brackets clustered on subjects;
* p<0.05 ** p<0.01 *** p<0.001

session 9, there is a weakly significant correlation in session 10 (significant at the 9% threshold). Thus, it appears that players deviated from the benchmark during session 9, causing some players to experience larger profits, but reduced their deviation during session 10, lowering average profits.

Tables 20.10 and 20.11 illustrate the effects of adding various additional fixed effects to the regressions presented before for the auctions with liquidity and those without liquidity, respectively. Specifically, we include competitor fixed effects, which control for the effect of playing against specific opponents in the various auctions and subject*session fixed effects, which control for subject learning over the testing period. Table 20.10 demonstrates that during sessions 1–4, adding competitor fixed effects reduces the effect of liquidity in determining the payoff and dampens the effect of learning. Table 20.11 shows a similar phenomenon for sessions 9 and 10. Adding subject*session fixed effects appears to affect only the importance of liquidity in determining total payoff. This suggests that, over time, players got better at optimally managing their liquidity bonus.

1.1 Experiment 2: pooled securities

Table 20.12 reinforces the conclusions from Tables 20.5–20.7 for the pooled-security setting. The clock auction format creates no statistically significant change in profit, but does increase the payoff significantly. That is, the clock auction format is more efficient at helping the Treasury determine which banks are most in need of liquidity. Also in

Table 20.10. Experiment : Fixed effects regressions (fully interacted) with liquidity needs

	(1) payoff	(2) payoff	(3) payoff	(4) payoff	(5) payoff	(6) payoff	(7) payoff	(8) payoff	(9) payoff
liquidity	0.740*** [0.0855]	0.810*** [0.0594]	0.875*** [0.0609]	0.761*** [0.0679]	0.761*** [0.0679]	0.809*** [0.110]	0.791*** [0.127]	0.891*** [0.0930]	0.928*** [0.0795]
Session 1	omitted	omitted	omitted						
Session 2	183.4*** [44.03]	186.2*** [43.97]	188.8*** [43.30]	385.5* [151.2]	385.5* [151.2]	397.8* [150.2]			
Session 3	283.9*** [47.04]	285.4*** [47.14]	290.9*** [48.74]	522.0** [145.6]	522.0** [145.6]	528.1** [143.8]			
Session 4	250.0*** [33.52]	249.7*** [34.87]	252.4*** [34.70]	440.7** [142.5]	440.7** [142.5]	440.8** [141.9]			
Clock	65.39*** [16.36]	65.39*** [16.39]	65.34*** [16.34]	65.39** [17.01]	65.39** [17.01]	65.35** [16.99]	65.39** [17.93]	65.39** [17.97]	65.36** [17.98]
8_Bidders		185.6*** [28.33]	190.8*** [29.24]		935.7 [914.7]	777.4 [954.3]		187.9*** [32.52]	191.1*** [32.74]
e_payoff			−0.0914 [0.0648]			−0.0737 [0.134]			−0.0574 [0.0955]
_cons	−147.2* [54.83]	−274.2*** [43.63]	−265.9*** [42.81]	−545.2*** [135.5]	156.6 [716.1]	41.62 [741.0]	−222.1** [73.70]	−372.2*** [63.99]	−291.7*** [59.08]
Subject FE	Yes	Yes	Yes	Yes	Yes	Yes	Yes	Yes	Yes
Competitor FE				Yes	Yes	Yes			
Subject*Session FE							Yes	Yes	Yes
N	256	256	256	256	256	256	256	256	256
adj. R-sq	0.419	0.552	0.552	0.638	0.638	0.638	0.407	0.569	0.568

Standard errors in brackets clustered on subjects; * p<0.05 ** p<0.01 *** p<0.001

Table 20.11. Experiment : Fixed effects regressions (fully interacted) without liquidity needs

	(1) payoff	(2) payoff	(3) payoff	(4) payoff	(5) payoff	(6) payoff
Session 9	58.82	65.85*	26.71	35.74		
	[30.03]	[29.60]	[18.06]	[20.68]		
Session 10	omitted	omitted	omitted	omitted		
Clock	60.13**	59.09**	60.13**	59.34**	60.13**	58.65**
	[16.08]	[16.08]	[16.92]	[16.92]	[17.07]	[17.30]
e_payoff		0.156		0.119		0.223
		[0.116]		[0.0799]		[0.178]
_cons	88.66***	75.71***	76.67*	75.43*	94.37***	75.72***
	[16.64]	[17.51]	[34.71]	[35.26]	[8.534]	[15.74]
Subject FE	Yes	Yes	Yes	Yes	Yes	Yes
Competitor FE			Yes	Yes		
Subject*Session FE					Yes	Yes
N	128	128	128	128	128	128
adj. R-sq	0.299	0.315	0.649	0.657	0.447	0.471

Standard errors in brackets clustered on subjects; * p<0.05 ** p<0.01 *** p<0.001

Table 20.12 we see that the bidders are able to capture slightly less (61%) of their liquidity bonus on average than in Experiment 1 (74%). As in Experiment 1, Table 20.12 demonstrates that a larger liquidity bonus caused players to bid more aggressively, resulting in a statistically significant lower profit. This strategy is reflected in the individually reported strategies summarized in Appendix A.

The effect of learning over the first four sessions of Experiment 2 is somewhat more complicated than in Experiment 1. That is, the payoffs in Sessions 6–8 are statistically indistinguishable from those in session 5. However, we observe a statistically insignificant decline in profits between sessions 5, 6, and 7, and a dramatic and significant decline between session 7 and session 8 (–$566). This decline in profits between sessions 7 and 8 is matched by a statistically significant decline in payoffs (decline of $642, significant at the 99% level). This suggests that participants played significantly more aggressively during session 8, but to their own detriment. During the days with no liquidity bonus, it appears that learning may have played a role, albeit a weak one. The payoff during session 11 was $85 less than in session 12, significant at the 93% level.

Table 20.13 provides additional insights into the process of learning by looking at the interaction between the liquidity bonus and session, and clock and session. We see an upward trend in the percentage of the liquidity bonus captured by participants over the whole of Experiment 2 when liquidity was positive. This suggests that participants became more adept at managing their liquidity constraint over time. In addition, we see a negative trend in the benefit of clock auctions compared with sealed-bid auctions.

Table 20.12. Experiment 2: pooled-security reverse auction

dep var:	(1) payoff	(2) profit	(3) payoff=profit
Liquidity	0.606*** [0.0545]	−0.119*** [0.0217]	
Session 5	omitted	omitted	
Session 6	189.9 [220.0]	−120.8 [87.43]	
Session 7	326.4 [219.8]	−225.6* [87.37]	
Session 8	−315.4 [219.9]	−791.3*** [87.42]	
Session 11			−84.66 [46.74]
Session 12			omitted
Clock	496.7** [150.5]	105.5 [59.82]	−84.49 [46.74]
LessPrecise	−44.68 [150.6]	−43.82 [59.87]	
Subject FE	Yes	Yes	Yes
N	256	256	128
adj. R-sq	0.364	0.388	0.265

Standard errors in brackets; * p<0.05 ** p<0.01 *** p<0.001

Table 20.13. Experiment 2: Payoffs over time

dep var:	(1) Session 5 payoff	(2) Session 6 payoff	(3) Session 7 payoff	(4) Session 8 payoff	(5) Session 11 payoff	(6) Session 12 payoff
Liquidity	0.555* [0.197]	0.647*** [0.147]	0.637*** [0.0769]	0.727*** [0.201]	0 [0]	0 [0]
Clock	1030.1** [265.4]	846.9* [379.1]	224.3 [231.5]	−114.5 [278.0]	108.4** [29.05]	60.57* [24.91]
LessPrecise	200.9 [407.1]	−266.2 [320.5]	−139.4 [228.3]	−0.490 [267.6]	0 [0]	0 [0]
_cons	−39.38 [1118.1]	−28.80 [863.5]	466.0 [428.8]	−548.4 [965.0]	105.8*** [14.53]	214.4*** [12.45]
Subject FE	Yes	Yes	Yes	Yes	Yes	Yes
N	64	64	64	64	64	64
adj. R-sq	0.625	0.564	0.749	0.655	0.443	0.61

Standard errors in brackets clustered on subjects; * p<0.05 ** p<0.01 *** p<0.001

This suggests that participants determined a strategy in early rounds and played that strategy independent of other bidders' actions. That is, initially, when participants were unfamiliar with the pooled auction setting, the additional information revealed during the clock auction increased payoffs by a statistically significant $1,030. This benefit of the clock declined over time, independent of the presence of liquidity.

The final result from Table 20.12 is that we see no statistically significant effect with respect to the more precise and less precise cases. That is, providing bidders with reference prices that represent 25% or 50% of the total signals does not result in a significant change in payoff. As might be expected, the more precise cases appear to result in slightly higher payoffs ($45), but not statistically so. And when we look at the effect of the more or less precise case by day (Table 20.13), we again do not see a statistically significant affect or trend.

Given that there was not a tractable theoretical benchmark strategy to provide for the auction participants, they were forced to determine their own bidding strategies. At the conclusion of the auction, all participants provided a short synopsis of their strategies (see Appendix A). Participants described strategies that were heavily determined by their liquidity draws and ratio of private signals to reference prices. Using this information, we calculated an applied bidder strategy (ABS) ratio that appears to capture the substance of how bidders used their private information. This ABS ratio is calculated as follows:

$$b_{is} = \frac{(sig \cdot u_{is} + (8 - sig) \cdot E[v_{is}])}{8 \cdot ref_i} \qquad (20.1)$$

where sig is the number of private signals given to each player for security i, u_{is} is the average of those private signals, and $E[v_{is}]$ is the expected value of the unknown signals given the known uniform distribution for securities in that pool type (75 for high-quality securities, and 25 for low-quality securities). Finally, ref_i is the reference price for security i, which is given to all players.

Table 20.14 illustrates that when the lowest ABS ratio for participants was low (< 0.7) bidders did significantly better than when the lowest ABS ratio was higher, independent of the presence of the liquidity bonus. For example, there is not a statistically significant difference between an ABS ratio < 0.6 and an ABS ratio between 0.6 and 0.7. However, when the ABS ratio is between 0.7 and 0.8, payoffs fall by a statistically significant $906. When the ABS ratio rises above 0.8, payoffs are $674–$781 lower than when the ABS ratio is less than 0.6.

RECOMBINING THE SEALED-BID RESULTS

Although we held only four sessions of the eight-security simultaneous reverse auction, and thus a total of four low-competition and four high-competition auctions, we can

Table 20.14. Experiment 2: Effect of the lowest applied bidder strategy ratio on outcomes*

dep var:	(1) payoff	(2) profit	(3) payoff=profit
Liquidity	0.604*** [0.0753]	−0.121*** [0.0224]	0 [0]
Session 5	omitted	omitted	0 [0]
Session 6	150.7 [401.8]	−76.16 [114.3]	0 [0]
Session 7	357.7 [316.6]	−172.3 [130.4]	0 [0]
Session 8	−336.7 [276.8]	−729.9*** [145.4]	[0] 0
Session 11	0 [0]	0 [0]	0.893 [53.35]
Session 12	0 [0]	0 [0]	omitted
Clock	496.7* [199.4]	−105.5* [47.92]	84.49 [49.80]
LessPrecise	−73.33 [148.4]	42.52 [69.51]	0 [0]
ABS<0.6	omitted	omitted	omitted
0.6≤ ABS<0.7	−546.9 [377.6]	15.88 [139.3]	−135.1 [115.1]
0.7≤ABS <0.8	−905.8** [254.3]	−105.3 [96.79]	−267.7* [111.8]
0.8≤ABS<0.9	−674.4* [261.3]	−93.38 [102.3]	−329.5** [110.3]
0.9≤ABS<1	−780.8*** [189.4]	−340.0*** [63.61]	−502.3*** [120.2]
1≤ABS	−749.2** [258.7]	−528.6** [167.2]	−361.7 [223.1]
_cons	856.2 [579.2]	340.5 [180.3]	446.9*** [103.5]
Subject FE	Yes	Yes	No
N	256	256	258
adj. R-sq	0.410	0.476	0.165

Standard errors in brackets clustered on subjects when subject FE used;
* The ABS ratio was calculated for each security and the lowest was used

evaluate a rich set of data to determine the full range of possible outcomes from the bidding strategies employed by the subjects. To do this, we use a recombinant procedure.[11] The results of our recombinant analysis add strength to one of our major contentions—the sealed-bid format results in more varied outcomes than does the clock format. Further, we can also assert that a few anomalous bids—mistakes—can drive the outcomes of entire auctions using the sealed-bid format. In sum, the downside risk of poor price discovery and extremely low payoffs to the bidders is higher using the sealed-bid format.

The recombinant procedure is in principle very simple. To understand the theoretical justification, we present a simple example. Imagine a basic sealed-bid auction with two bidders. Suppose there are two of these basic auctions, auction A and auction B. Bidder 1 faces bidder 2 in auction A, while bidder 3 faces bidder 4 in auction B. A single outcome results from each auction, giving us a total of two price observations, for example. Now, consider that each bidder determined their bid using only their own private information, and no special knowledge of their opponents. Provided bidding is *anonymous*, we can expect that bidder 1 would have submitted the same bid if she had faced bidder 3 or bidder 4, just as she did when she faced bidder 2 in the auction we first observed. Exploiting this concept, we can compute the outcomes of several more auctions than those we actually observed. The total set of auctions is given by the set of bids $\{A \equiv \{b_1, b_2\}, \{b_1, b_3\}, \{b_1, b_4\}, \{b_2, b_3\}, \{b_2, b_4\}, B \equiv \{b_3, b_4\}\}$. We get a total of six outcomes to analyze instead of just two.[12]

We use this procedure to examine more closely the outcomes of the sealed-bid simultaneous reverse auctions.[13] We observe a total of 1,820 four-bidder auctions per session, rather than four, and a total of 12,870 eight-bidder auctions per session, rather than two. This gives us the ability to see a highly precise distribution of the outcomes that could have been produced by the strategies our subjects employed.[14]

The most important finding from this analysis is that the sealed-bid procedure is vulnerable to anomalous bids. Strong evidence is provided in Figure 20.1, which shows a histogram of the results in the four-bidder case. We display the difference between true common value and price, and point out the bimodality of the data in sessions 1 and 3. In both of these cases, small mistakes by bidders led to large differences between price and

---

[11] See Mullin and Reiley (2006) for details on the recombinant estimator.

[12] In this case, we get six outcomes, or four-choose-two. In general, we can calculate the number of outcomes equal to n-choose-k, where n is the total number of bidders and k is the number of bidders in each auction.

[13] Note that the theoretical justification for the recombinant procedure does not hold in a clock auction, since what one bidder does depends importantly on the signals she receives from her competitors. Thus the assumption of anonymity breaks down, and we cannot justify a recombinant analysis. Likewise, we cannot analyze reference-price auctions this way, since bidders in different auctions react to information that is specific to the auction, and thus combining bids from different auctions is not theoretically or statistically valid.

[14] Note carefully that we do not claim that all of these auctions are statistically independent. We cannot calculate multivariate test statistics with this data-set, as the distribution of the standard error is unknown. Instead, we use the procedure to examine in some detail what is possible, had our random matching of subjects turned out differently.

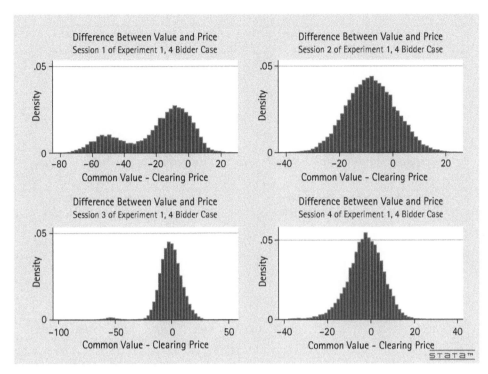

FIGURE 20.1. Difference between value and price in a four-bidder sealed-bid auction.

the true value of the securities. This clearly hampers price discovery, and certainly affects the profits of the winning bidders. Figure 20.2 shows the profit from the corresponding security sales. A heavy left tail is evident in the distribution of profits, driven again by the tendency of small mistakes by a few bidders to negatively influence the price of securities. Finally, note what happens to bidder payoffs under the same circumstances (Figure 20.3). Extremely negative outcomes are evident in session 1. Even in sessions 2–4, after learning has taken place, there is substantial mass below 200, substantially less than the average payoff of 388 we report in Table 20.2. The average payoff statistic from the sealed-bid auctions masks the possibility of very negative outcomes. The reason is simple—when a small number of bidders drive prices below their common value, due to poor strategy or low signals, they inhibit the ability of other bidders to satisfy their liquidity needs. When liquidity needs are an especially important goal of the auction, the sealed-bid format can result in an especially bad case of allocative inefficiency.

Such poor outcomes are extremely unlikely under the clock auction format. First, mistakes by bidders are simply less likely under the clock format. Since bidders have multiple chances to express their preferences and can continually update their strategies throughout the bidding process, anomalous bids are less likely. Second, as the results of our econometric analysis demonstrate, bidders are better able to manage their liquidity

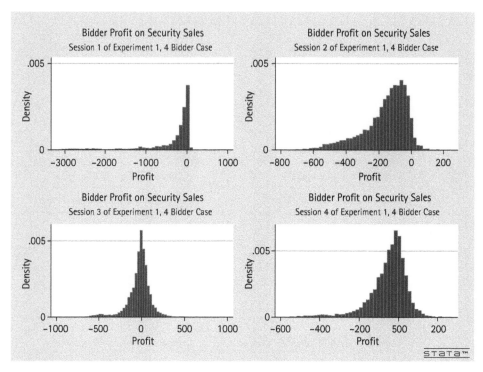

FIGURE 20.2. Bidder profit on security sales in a four-bidder sealed-bid auction.

needs under the clock format. Thus, should a small number of bidders drive the price of a handful of securities well below their common value, the clock format enables other bidders to respond by adjusting their bidding on other securities. Sensible bidders will bid more aggressively on other securities in their portfolio in order to meet their liquidity needs, allowing them to recover reasonable payoffs. The sealed-bid format gives bidders no such chance to adjust their strategies. When liquidity needs are a dominant concern of the Treasury, the sealed-bid format leaves gains from trade on the table.

ANALYSIS AND DISCUSSION

Feasibility of implementation

One of the points in conducting the experiments was to demonstrate the feasibility and practicality of conducting a computerized auction in which multiple securities are purchased simultaneously. In this regard, both the sealed-bid and descending clock auctions can be regarded as successfully implemented in a short time.

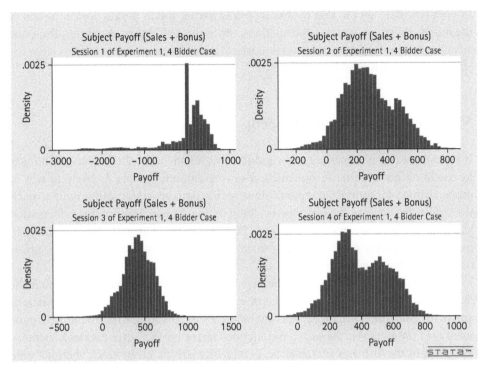

FIGURE 20.3. Subject Payoffs (sales plus liquidity bonus) in a four-bidder clock auction.

Competitiveness of price

Notwithstanding the presence of adverse selection, as a theoretical matter, the price in the auction can be driven below the security's fundamental value, to the extent that the bidders have liquidity needs above and beyond their objective of earning trading profits. This was demonstrated in the experiments. In both the sealed-bid and descending clock auctions, the prices were significantly below the fundamental values of the respective securities. This is seen in the second row (profit) in Tables 20.3–20.6. A bidder's mean trading profits in Experiment 1 were negative, though not significantly so under the sealed-bid (−39) and clock (−13). The mean trading profits in Experiment 2 were significantly negative under both formats: −693 with the sealed-bid format and −799 with the clock format.

Frequency of erroneous bids

In a relatively small but clearly noticeable subset of the auctions, bidders submitted what were almost certainly unintentionally low bids ("erroneous bids"). In descending clock auctions, the harm to a bidder from an erroneous bid was often fairly limited, as the extent of the damage was that the bidder found herself still in the auction at the end of the round. However, in sealed-bid auctions, the harm could be much greater. Thus, one advantage observed of the clock auction was that it helped to insulate bidders from the effects of their own mistakes.

Management of liquidity needs

If separate sealed-bid auctions are conducted simultaneously, the bidder has no way to condition her bidding in one auction on the amount that she is likely to win in other auctions. By contrast, if separate dynamic auctions are conducted simultaneously, the bidder can observe the outcome evolving in one auction and use that information to adjust her behavior in other auctions. Thus, the bidder has much greater ability to manage her liquidity need, without "overshooting" and selling more securities at prices below value than the bidder needs to sell.

This is demonstrated clearly in the experimental results. Despite the fact that the bidders had the same values and the same liquidity needs in the sealed-bid auctions and in the clock auctions, the bidders attained average payoffs in Experiment 1 (Experiment 2) of 388 (3,222) in the sealed-bid auctions and of 453 (3,719) in the clock auctions, as shown in the fifth row of Tables 20.3–20.6. The payoffs in the clock auctions were significantly higher (at the 1% level) than in the sealed-bid auction. This is entirely due to better management of liquidity with the clock auction. As shown in the fourth row of Tables 20.3–20.6, the mean liquidity bonus in Experiment 1 (Experiment 2) was significantly larger, 466 (4,517) with the clock format, compared with 428 (3,915) with the sealed bid format. Under sealed-bid, the bidder often overshoots her liquidity need. The mean overshoot in Experiment 1 (Experiment 2) with the sealed-bid auction was 692 (1,984), compared with 605 (905) for the clock format (see row 7 of Tables 3–6). This difference is significant at the 1% level (p-value of 0.0014).

Cost of purchasing securities

There appear to be three distinct effects determining the comparison of purchase prices between the sealed-bid and dynamic auction formats. First, a dynamic auction format is generally known to mitigate the winner's curse, leading to more aggressive bidding and to lower prices in a reverse auction. Second, as seen above, the bidders submit fewer erroneous bids in a dynamic format, leading to higher prices. Third, as seen above, a dynamic format allows bidders to manage better their liquidity needs; more than likely, this would lead to fewer desperation offers and cause higher prices. Combining these three effects, the price comparison between sealed-bid and dynamic auction formats is ambiguous.

In the experimental results, the price difference between the two formats is not statistically significant. The bidders, with the same values and the same liquidity needs in the two auction formats, produced mean clearing prices of 49.57 in the clock and 47.79 in the sealed-bid auction in Experiment 1. In Experiment 2, the two formats produced mean clearing price points of 85.22% in the sealed-bid auctions and of 83.87% in the clock auctions (see row 1 of Tables 20.3–20.6).

Combining the results on satisfying liquidity needs and on the cost of purchasing securities, the taxpayer would favor the descending clock auction. While the taxpayer's expenditure is approximately equal in the sealed-bid and the clock auctions, the latter gives "more bang for the buck"—for a given expenditure of money, the clock auction better directs resources toward satisfying the liquidity needs of the banking sector.

Variability of outcomes and informativeness of prices

Finally, there appears to be a fundamental difference in the experimental results between the sealed-bid auction and the corresponding dynamic auction. All other things being equal, the outcomes of the sealed-bid auction are more variable and random.

First, this means that the outcomes of the dynamic auction are more predictable, and thus more satisfying to risk-averse banks and regulators. The greater variability is seen in the variance of the bidder's payoff. The standard deviation of the bidder's payoff in Experiment 1 (Experiment 2) is 222 (1,312) with the clock format, compared with 285 (1,654) with sealed-bid (see row 6 of Tables 20.3–20.6). This difference is significant at the 1% level (p-value of 0.0095).

Second, one of the objectives of government purchases of troubled securities is to restart frozen secondary markets by providing relevant transaction prices. By doing so, the government can rely on the private market to accomplish some of the government's objectives, reducing the need for government resources. The experimental design limited the extent to which this can be seen in the data, as there were only two separate pools of securities and thus only two independent prices determined in the auctions. Nevertheless, it can be seen that the prices resulting from the dynamic auction are more accurate, an effect which can be expected to be enlarged when more independent prices are determined by an auction.

CONCLUSION

We present our findings from laboratory experiments of troubled-asset reverse auctions. The experiments demonstrate the feasibility of implementing a purchase program for troubled assets in a short period of time using either a sealed-bid or a dynamic auction format. The experiments suggest that the taxpayer cost of purchases using a well designed and well implemented auction program could be small using either auction format, to the extent that sellers have substantial liquidity needs. However, the dynamic auction format has significant advantages over the sealed-bid auction format for both the banks and taxpayers, because the informational feedback provided during the auction enables the seller to manage better its liquidity needs.

Our experiments focused on trading profits and liquidity needs as the bidder's principal objectives. In practice, a seller of troubled assets also cares about its portfolio risk. For reasons of simplicity, we ignored portfolio risk in the experiments. However, there is good reason to expect that what we learned about a bidder's challenges in managing liquidity would carry over to the issue of portfolio risk. It is the bidder's ability to see, while the auction is still running, which asset sales are likely to be successful that enables the bidder to better reach its revenue target in the dynamic auction. By the same token, the bidder's ability to see which asset sales are likely to be successful should enable the bidder to better manage its portfolio risk. Thus, explicitly including portfolio risk would likely strengthen the case for dynamic auctions over sealed-bid auctions for purchases of troubled assets.

APPENDIX A. ANALYSIS OF BIDDING STRATEGY IN EXPERIMENT 2

Each of the nineteen expert bidders provided a summary of their bidding strategy during Experiment 2. Although there are many variations of strategies employed, the following summarize the primary strategy components from the bidders.

Liquidity. All of the bidders described their strategy as heavily dependent on their liquidity draw. That is, they recognized that even though the securities were likely to close at a price below their respective common values, they could make money on the liquidity bonus. As a result, they selected certain securities on which to bid aggressively. Once determined, bidders would instantly reduce supply to a level that just allowed them to meet their liquidity bonus if the price were to fall to their estimated dropout ratio. The immediate reduction was an effort to signal to other bidders that they should be reducing supply also. They would hold that level of the security until the price went below their estimated dropout ratio, at which point they would rapidly drop out of the auction. Once the liquidity bonus was satisfied they would drop out of the auction completely.

Reference prices. Bidders were generally aware that securities with low reference prices were least likely to sell and thus they reduced their supply on those first. DI stated "I was more aggressive on the securities with a higher than average ... reference price." Many bidders compared their own signals to these reference prices as an indication of which securities to reduce the supply of first. Once determined, these ratios were often ranked from lowest (best) to highest (worst) to determine the appropriate aggressiveness to be used for each security. The lower the ratio on a security, the more aggressively the holder bid (i.e. the longer they held on to the security). If the ratio was above the reference price, they dropped out early.

High quality/low quality. Five bidders stated that they were going to focus their attention on the high-quality securities, as that would be the easiest way to achieve their liquidity bonus. However, three bidders stated that they spread their attention across both pools and one bidder focused on the lower-quality pool because he thought it "will have relatively higher or positive P-CV."

Learning. Four of the bidders stated that it took one to two days to optimize their strategy. The bidders were divided on whether observing the actions of the other bidders was helpful in optimizing their own behavior. Some stated that the actions of the other bidders were not a helpful signal in determining their values, because the other bidders were either making mistakes or misrepresenting. AP states that he could not get information from other bidders because "it was too easy to intentionally send out 'bad' information . . . make mistakes . . . and/or play irrationally." Others felt the opposite. BW said and DI concurred that they would watch what other bidders were selling and sell the same "because that specific security may have a lower common value." Still other bidders stated that while specific information was hard to determine from the actions of other bidders, they were able to learn the general behavior of the other bidders as the auctions progressed (i.e. that bidders were more aggressive than they thought initially). KD stated, "I learned that, broadly speaking, there were two types of players: those who appreciated the benefits of supply reduction and those who did not." KD continued that she would learn in the first few rounds the type of opponents she was up against and would adjust her strategy accordingly. PS stated, "I believed other bidders will [reduce supply]. I was wrong. . . . So in the remaining session, I just hung in the securities . . . without supply reducing."

Estimated dropout ratio. To achieve their respective liquidity goals, bidders would hold the necessary quantity of securities up to their calculated dropout ratio. These dropout ratios were estimated in two steps. First, they would target a ratio of the common value to reference price and then they would adjust it to determine the ratio at which they would drop out. Four of the bidders calculated the target ratio assuming the expected values for the unknown signals (75 and 25 for high and low, respectively). Seven bidders used only the ratio of their private signals to the reference price. Three bidders used the clearing ratio from previous similar experiments. Once these ratios were determined, seven of the bidders stated that they would drop out completely when the price got to 50% of the target ratio. Other bidders used different heuristics or downward adjustments to determine when they would drop out. For example, AV kept significant quantities of securities "until we got around 96–94% of the initial price." Several bidders stated that their level of confidence in their final dropout ratio was determined by the number of private signals they had: four private signals made them more confident with respect to the ratio; a single signal made them less confident. Lower confidence with respect to the ratio caused them to behave as if the ratio were slightly above their calculated value.

Supply reduction. Regardless of the reason, most bidders reduced their supply as quickly as possible, especially for those securities where their estimated true ratio was high, assuming these would not clear. Even in cases when they thought they could win, bidders often stated that they would immediately supply reduce 50% and then hold on that quantity for a while. Many of the bidders stated that they were often hoping for tacit collusion, even when it did not come. For example, AV complained that "the other players would not reduce quantities significantly from a round to another." JR stated, "What was more frustrating was the fact that people just wouldn't drop [their supply]." JB said, "I didn't see any way to elicit cooperation to get others to drop out sooner." KM said, "my aim was always to induce people to drop quantities early on by dropping myself . . . usually I assumed that players would take equal responsibility for reductions." Some bidders stated that they had a preference to supply reduce less for their larger holdings, because they knew more about those holdings.

Sealed bid vs. clock auctions. Many of the bidders stated that it was more difficult to play the sealed bid because they were not learning anything about the values of other bidders. As a result, they had to guess an optimal dropout ratio. BW and DI guessed an 80% ratio. Several players stated that they played more conservatively in the sealed-bid auction.

And there were other things that were conspicuously absent from their strategies:

Differentiation between less/more precise. Almost no one stated a strategy difference between the two auction types. In fact, two of the bidders explicitly stated that they did nothing different between the more precise and less precise scenarios. Anyone using the bidding tool calculated the estimated true ratio with different assumptions for each of these scenarios (i.e. twelve versus eight total bidders for the more and less precise scenarios, respectively), but many bidders stated that they did not use the bidding tool.

Activity points. Only one bidder discussed a strategy that involved shifting supply to high-signal/low-reference-price securities within the same pool in an effort to distract other bidders. Other than this, no one talked about using the activity point constraint as a means of passing information.

Appendix B. A common-value auction with liquidity needs: bidder instructions for Experiment 1

In this experiment, you are a bidder in a series of auctions conducted over four sessions. Each session is held on a different day and consists of four different auctions. Bidders are randomly assigned to each auction. Thus, you do not know who the other bidders are in any of your auctions. You will be bidding from your private cubical, which includes a computer and a bidder tool that is unique to you. We ask that you refrain from talking during the experiment. If you need assistance at any time, just raise your hand and one of the experimental staff will assist you.

You will be paid at the end of the experiment. Your payment is proportional to your total experimental payoff—the sum of your payoffs from each of your auctions. In particular, you will receive *$1 in take-home pay for every 10 million experimental dollars that you earn*. Throughout, dollar figures refer to your experimental payoff unless explicitly stated otherwise—and the "millions" are generally suppressed. When explicitly stated, your real dollar payment will be referred to as your *take-home payment*. We anticipate that each of you will earn a take-home payment of about $100 per experimental session on average. However, your actual take-home payment will depend on your bidding strategy, the bidding strategies of the other bidders you face, and the particular realizations of numerous random events.

In each auction, you will compete with other bidders to sell your holdings of eight securities to the Treasury. The eight different securities have different values. However, for each security, the bidders have a common value, which is unknown. Each bidder has an estimate of the common value. You profit by selling securities to the Treasury at prices above the securities' true values. You also have a need for liquidity (cash). The sale of securities to the Treasury is your source of liquidity. Thus, you care about both profits and liquidity (your sales to the Treasury).

Two formats are used:

- *Simultaneous uniform-price sealed-bid auction ("sealed-bid auction")*. Bidders simultaneously submit supply curves for each of the eight securities. Supply curves are non-decreasing (i.e. upward-sloping) step functions. The auctioneer then forms the aggregate supply curve and crosses it with the Treasury's demand. The clearing price is the lowest rejected offer. All quantity offered below the clearing price is sold at the clearing price. Quantity offered at the clearing price is rationed to balance supply and demand, using the proportionate rationing rule.
- *Simultaneous descending clock auction ("clock auction")*. The securities are auctioned simultaneously. There is a price "clock" for each security indicating its tentative price per unit of quantity. Bidders express the quantities they wish to supply at the current prices. The price is decremented for each security that has excess supply, and bidders again express the quantities they wish to supply at the new prices. This process repeats until supply is made equal to demand. The tentative prices and assignments then become final. Details of the design are presented in Ausubel and Cramton (2008a,b), which you received earlier.

In each session, you will participate in four different auctions in the following order:

1. Four-bidder sealed-bid. A sealed-bid auction with four bidders.
2. Eight-bidder sealed-bid. A sealed-bid auction with eight bidders.
3. Four-bidder clock. A clock auction with four bidders.
4. Eight-bidder clock. A clock auction with eight bidders.

Each session, one of your two four-bidder auctions and one of your two eight-bidder auctions will be selected at random. Your take-home payment from the session will be based on the number of (million) laboratory dollars that you earn in these two auctions only.

Securities

In each auction, eight securities are purchased by the Treasury. The bidders are symmetric before the draws of bidder-specific private information about security values and liquidity needs.

In each session, two sets of bidder-specific private information are drawn independently from the same probability distributions. The first set is used in the four-bidder auctions (auctions 1 and 3); the second set is used in the eight-bidder auctions (auctions 2 and 4). You are given no feedback following the sealed-bid auctions; thus, your behavior in the clock auctions cannot be influenced by outcomes in the sealed-bid auctions of a session.

In each four-bidder auction, the Treasury demand is 1,000 shares of each security, where each share corresponds to $1 million of face value. Each bidder has holdings of 1,000 shares of each security. Thus, it is possible for a single bidder to fully satisfy the Treasury demand for a particular security; that is, for each security there may be just a single winner or there may be multiple winners. One-quarter of the total available shares will be purchased by the Treasury.

In each eight-bidder auction, the Treasury demand is 2,000 shares of each security, where each share corresponds to $1 million of face value. Each bidder has holdings of 500 shares. Thus, at least four bidders are required to fully satisfy the Treasury demand—there must be at least four winners. One half of the total available shares will be purchased by the Treasury.

Your preferences

From each auction, your payoff depends on your profits and how well your liquidity needs are met.

Common-value auction

The value of each security in cents on the dollar is the average of eight *iid* random variables uniformly distributed between 0 and 100:

$$v_s = \tfrac{1}{8} \sum_{i=1}^{8} u_{is}, \text{ where } u_{is} \sim_{iid} U[0, 100].$$

Suppose you are bidder i.

Your private information about security s is the realization u_{is}. Notice that both for the eight-bidder and four-bidder auctions, the common value is the average of eight uniform draws, so that only the first four draws are revealed in the four-bidder auction. This means that the true values have the same distribution in both four-bidder and eight-bidder auctions, and your private information has the same precision.

If you sell the quantity q_s of the security s at the price p_s, then your profit (in million $) is:

$$\pi_i(p, q_i, v) = \tfrac{1}{100} \sum_{s=1}^{8} (p_s - v_s)q_{is}.$$

Liquidity need

You have a liquidity need, L_i, which is drawn *iid* from the uniform distribution on the interval $[250, 750]$. You know your own liquidity need, but not that of the other bidders. You get a bonus of $1 for every dollar of sales to the Treasury up to your liquidity need of L_i. You do *not* get any bonus for sales to the Treasury above L_i. Thus, your bonus is:

$$\min \left[L_i, \tfrac{1}{100} \sum_{s=1}^{8} p_s q_{is} \right].$$

Your payoff from an auction

Combining your profit and your liquidity bonus results in your total payoff:

$$U_i(p, q_i, v) = \begin{cases} \tfrac{1}{100} \sum_{s=1}^{8} (2p_s - v_s)q_{is} & \text{if } \tfrac{1}{100} \sum_{s=1}^{8} p_s q_{is} < L_i \\ L_i + \tfrac{1}{100} \sum_{s=1}^{8} (p_s - v_s)q_{is} & \text{otherwise} \end{cases}$$

Thus, an additional dollar of cash is worth two dollars when your liquidity need is not satisfied, but is worth one dollar when your liquidity need is satisfied.

Bidder tool and auction system

You have an Excel workbook that contains your private information for each auction. You will use the tool to submit bids in the sealed-bid auctions. In addition, the tool has features that will help your decision making in each of the auctions. Each auction has its own sheet in the tool. It is essential that you are working from the correct sheet for each auction. For example the four-bidder sealed-bid auction is the sheet named *Sealed Bid 4*.

For the clock auctions, bidding is done via a commercial auction system customized to this setting. You use the web browser to connect to the auction system. For each clock auction, you must go to a new auction site. The URL for the auction site is given in the bidder tool on the particular auction sheet, Clock 4 or Clock 8, for the four-bidder or eight-bidder clock auction. Once at the correct auction site, log in with the user name and password given on your auction sheet.

The auction system is easy to use. You will have an opportunity to use it in the training seminar.

An important feature of the tool is the calculation of expected security values conditional on information you specify. In a common-value auction, it is important for you to condition your estimate of value on your signal *and the information winning conveys.* Since your bid is only relevant in the event that you win, you should set your bid to maximize your payoff in that event. In this way, you avoid the winner's curse, which in this case is the tendency of a naïve bidder to lose money by selling shares at prices below what they are worth. In addition, in the clock auctions, the bidder also must condition on any information revealed through the bidding process. The tool provides one flexible method of calculating an appropriate conditional expected value for each security.

Bidding strategy

The auction environment has three complicating features:

- *Common-value auction.* You have an imperfect estimate of the good's common value.
- *Divisible-good auction.* Your bid is a supply curve, specifying the quantity you wish to sell at various prices.
- *Liquidity need.* You have a specific liquidity need that is met through selling shares from your portfolio of eight securities.

The combination of these factors makes a complete equilibrium analysis impossible. Nonetheless, equilibrium analysis is possible in a simplified environment. To aid your thinking about strategy, we discuss a particular strategy, which abstracts from the complications of a divisible good auction and the liquidity needs. In particular, we assume:

1. Each bidder submits a flat supply schedule; that is, the bidder offers to sell all of her holdings of a particular security at a specified price.
2. Each bidder ignores her liquidity need, bidding as if $L_i = 0$.

With these assumptions it is possible to calculate an equilibrium strategy, which we call the *benchmark strategy.* The analysis of this strategy will be helpful in thinking about the common value feature of the auction environment.

We wish to emphasize that the benchmark strategy focuses on only one element of the auction. Your challenge is to determine your own strategy to maximize your payoff that reflects all aspects of the auction environment.

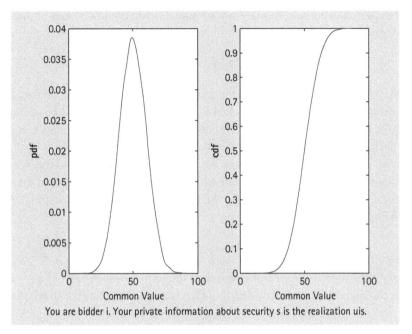

You are bidder i. Your private information about security s is the realization uis.

FIGURE 20B.1. Probability density and cumulative distribution of common value.

Common-value distribution

As mentioned earlier, the value of each security in cents on the dollar is the average of eight *iid* random variables uniformly distributed between 0 and 100:

$$v_s = \tfrac{1}{8} \sum_{i=1}^{8} u_{is}, \text{ where } u_{is} \sim_{iid} U[0, 100].$$

The probability density function (pdf) and cumulative distribution function (cdf) of the common value are shown in Figure 20B.1. The common value has a mean of 50 and a standard deviation of 10.2. Notice that the common value is approximately normally distributed, since it is the sum of many independent draws.

Sealed-bid uniform-price reverse auction

Under our strong simplifying assumptions, the auction is equivalent to a unit-supply common-value auction with uniform pricing. In this case, just as in a single-item second-price auction, your bid does not determine what you pay, only the likelihood of winning, thus the best strategy is to bid your true conditional expected value for the good. The trick, however, is to condition on your signal and the information that winning conveys.

In the four-bidder auction, under the benchmark assumptions, there is only a single winner, so the correct condition is as derived in Milgrom and Weber (1982). Your bid is your expected value conditional on your signal being the lowest and on the second-lowest signal being equal to yours:

$$b_{is} = E\left[v_s|u_1 = u_{is}, u_2 = u_1\right],$$

where u_1 is the lowest signal and u_2 is the second-lowest signal. The reason you condition on the second-lowest signal being as low as yours is that the bid must be optimal when it is on the margin and thus impacts whether you win. Your bid becomes marginal and hence decisive only in the event that you tie with the second lowest.

For the eight-bidder auction, there are exactly four winners. Hence, we need to generalize the above formula to multiple winners. The optimal rule is to bid the expected value conditional on your signal being the fourth-lowest signal and on the fifth-lowest signal being equal to yours:

$$b_{is} = E\left[v_s|u_4 = u_{is}, u_5 = u_4\right],$$

where u_4 is the fourth-lowest signal and u_5 is the fifth-lowest signal.

Since the signals are uniformly distributed, it is easy to calculate the above conditional values. In both cases, the conditional value is a linear function of your signal.

In the four-bidder auction, there is a single winner. The conditioning is that you win, the next lowest bidder has your value, and the remaining two bidders are evenly distributed above your value:

four-bidder sealed-bid strategy,

$$b_{is} = \frac{1}{8}\left(2u_{is} + 2\left(\frac{u_{is} + 100}{2}\right) + 4 \cdot 50\right) = \tfrac{3}{8}u_{is} + \tfrac{75}{2}.$$

In the eight-bidder auction, there are four winners. The conditioning is that you have the fourth-lowest signal, the fifth-lowest signal is the same, the three signals below you are evenly distributed below your signal and zero, and the remaining three bidders are evenly distributed above your signal:

eight-bidder sealed-bid strategy,

$$b_{is} = \frac{1}{8}\left(2u_{is} + 3\frac{u_{is}}{2} + 3\left(\frac{u_{is} + 100}{2}\right)\right) = \tfrac{5}{8}u_{is} + \tfrac{75}{4}.$$

Descending clock auction

In the clock auction, under the benchmark assumption, you will observe the price at which other bidders drop out. This provides additional information on which to condition your bid. Here we assume that the price clock is continuous. In the actual experiment, the price clock is discrete, and although bidders can make reductions at any price, you will learn the aggregate supply only at the end of round price. You may want to assume the quantity reduction occurred halfway between the prior price and the ending price.

When the clock starts, you calculate your dropout point in the same way as above. As the price clock falls, one of the other bidders may drop out. When the first bidder drops out, you can calculate this bidder's draw from the following equation.

$$u_{8s} = \frac{P_{\text{dropout}} - \text{intercept}}{\text{slope}}$$

where the slope and intercept are taken from the formulas above. With this new information on which to condition your bid, the revised optimal bid for the eight-bidder clock auction is straightforward to calculate.

Eight-bidder clock strategy

No one has dropped out: $b_{is} = \dfrac{1}{8}\left(2u_{is} + 3\dfrac{u_{is}}{2} + 3\left(\dfrac{u_{is} + 100}{2}\right)\right).$

One bidder has dropped out: $b_{is} = \dfrac{1}{8}\left(2u_{is} + 3\dfrac{u_{is}}{2} + u_8 + 2\left(\dfrac{u_{is} + u_8}{2}\right)\right).$

Two bidders have dropped out: $b_{is} = \dfrac{1}{8}\left(2u_{is} + 3\dfrac{u_{is}}{2} + u_8 + u_7 + 1\left(\dfrac{u_{is} + u_7}{2}\right)\right).$

Three bidders have dropped out: $b_{is} = \dfrac{1}{8}\left(2u_{is} + 3\dfrac{u_{is}}{2} + u_8 + u_7 + u_6\right).$

Four-bidder clock strategy
Similarly, we can calculate the analogous formulas for the four-bidder clock auction.

No one has dropped out: $b_{is} = \dfrac{1}{8}\left(2u_{is} + 2\left(\dfrac{u_{is} + 100}{2}\right) + 4\cdot 50\right)$

One bidder has dropped out: $b_{is} = \dfrac{1}{8}\left(2u_{is} + u_4 + 1\left(\dfrac{u_{is} + u_4}{2}\right) + 4\cdot 50\right).$

Two bidders have dropped out: $b_{is} = \dfrac{1}{8}(2u_{is} + u_4 + u_3 + 4\cdot 50).$

Note that the above formulas are all linear in the dropout price, so it is easy to invert to compute the bidder's signal.

Moving beyond the benchmark strategy

The bidding tool is set up to calculate the conditional expected values assuming the benchmark strategy. Of course, you (and others) may well deviate from the benchmark strategy as a result of liquidity needs or other reasons, since these other factors are ignored in the benchmark calculation.

The bidding tool allows for this variation in a number of ways. First, in the sealed-bid auctions your bid can be any upward sloping step function to account for liquidity and possible supply reduction by you or others. Second, in the clock auctions, the tool lets you interpret what a dropout is and where it occurs. This is useful and necessary when bidders make partial reductions of supply. In addition, although the tool will calculate a particular signal based on a dropout, you are free to type in any signal inference you like. Whatever you type as "my best guess" will be used in the calculation of the conditional expected value.

Further details of the tool will be explained in the training seminar.

If you have any questions during the experiment, please raise your hand and an assistant will help you.

Remember your overall goal is to maximize your experimental payoff in each auction. You should think carefully about what strategy is best apt to achieve this goal.

Many thanks for your participation.

Appendix C. A common-value reference-price auction with liquidity needs: bidder instructions for Experiment 2

In this experiment, you are a bidder in a series of auctions conducted over four sessions. Each session is held on a different day and consists of four different auctions. Bidders are randomly assigned to each auction. Thus, you do not know who the other bidders are in any of your auctions. You will be bidding from your private cubical, which includes a computer and a bidder tool that is unique to you. We ask that you refrain from talking during the experiment. If you need assistance at any time, just raise your hand and one of the experimental staff will assist you.

You will be paid at the end of the experiment. Your payment is proportional to your total experimental payoff—the sum of your payoffs from each of your auctions. In particular, you will receive *$1 in take-home pay for every one hundred thousand experimental dollars (100 $k) that you earn.* Throughout, dollar figures refer to your experimental payoff unless explicitly stated otherwise—and the "thousands" are generally suppressed. When explicitly stated, your real dollar payment will be referred to as your *take-home payment.* We anticipate that each of you will earn a take-home payment of about $100 per experimental session on average. However, your actual take-home payment will depend on your bidding strategy, the bidding strategies of the other bidders you face, and the particular realizations of numerous random events.

In each auction, you will compete with other bidders to sell your holdings of eight securities to the Treasury. The eight securities are split into two pools: four securities are low quality and four are high quality. The eight different securities have different values. However, for each security, the bidders have a common value, which is unknown. Each bidder has an estimate of the common value. You profit by selling securities to the Treasury at prices above the securities' true values. You also have a need for liquidity (cash). The sale of securities to the Treasury is your source of liquidity. Thus, you care about both profits and liquidity (your sales to the Treasury). The Treasury has allocated a particular budget for the purchase of each pool of securities within each auction. Its demand for high-quality securities is distinct from its demand for low-quality securities.

Before each auction, the auctioneer assigns each security a *reference price* (expressed in cents on the dollar of face value), which represents the Treasury's best estimate of what each security is worth. For example, a high-quality security might be assessed to be worth 75 cents on the dollar, while a low-quality security might be assessed to be worth 25 cents on the dollar. A *reference-price auction* determines the price point—a percentage of the reference price—for each pool of securities. A winning bidder is paid the pool's price-point × the security's reference price for each unit of the security sold.

Two formats are used:

- *Simultaneous uniform-price sealed-bid auction ("sealed-bid auction")*. Bidders simultaneously submit supply curves for each of the eight securities. Supply curves are non-decreasing (i.e. upward-sloping) step functions, offering a quantity at each price point. The auctioneer then forms the aggregate supply curve and crosses it with the Treasury's demand. This is done separately for each pool (i.e. for high- and low-quality securities, separately). The clearing price point is the lowest-rejected offer. All quantity offered at below the clearing price point is sold at the clearing price times the security's reference price. Quantity offered at exactly the clearing price is rationed to balance supply and demand, using the proportionate rationing rule.
- *Simultaneous descending clock auction ("clock auction")*. The securities are auctioned simultaneously. There are two descending clocks, one for high-quality securities and one for low-quality securities, indicating the tentative price point of each pool. Bidders express the quantities they wish to supply at the current price points. The price point is decremented for each pool of securities that has excess supply, and bidders again express the quantities they wish to supply at the new price points. This process repeats until supply is made equal to demand. The tentative price points and assignments then become final. Details of the design are presented in Ausubel and Cramton (2008a,b), which you received earlier.

The proportionate rationing rule plays a role only in the event that multiple bidders make reductions at the clearing price. The rule then accepts the reductions at the clearing price in proportion to the size of each bidder's reduction at the clearing price. Thus, if a reduction of 300 is needed to clear the market and two bidders made reductions of 400 and 200 at the clearing price, then the reductions are rationed proportionately: the first is reduced by 200 and the second is reduced by 100. The actual reduction of the first bidder is twice as large as that of the second bidder, since the first bidder's reduction as bid is twice as large as the second bidder's reduction. Ties can generally be avoided by refraining from bidding price points that are round numbers, instead specifying price points to odd one-hundredths of a percent (e.g. 98.42).

The clock auction has an activity rule to encourage price discovery. In particular, for each security pool, the quantities bid must be an upward-sloping supply curves as expressed in activity points. More precisely, *activity points*—equal to the reference price × quantity, summed over the four securities in the pool—are computed for each bid. The number of activity points is not permitted to increase as the price point descends. Thus, you are allowed to switch quantities across the four securities in a pool, but your total activity points for the pool cannot increase as the auction progresses.

The same activity rule applies to the sealed-bid auction, but then in a single round of bidding.

In each session, you will participate in four different auctions:

1. sealed-bid auction, with more precise reference prices
2. clock auction, with more precise reference prices
3. sealed-bid auction, with less precise reference prices
4. clock auction, with less precise reference prices.

On Tuesday and Thursday, the order of auctions will be as above. On Wednesday and Friday, the auctions with less precise reference prices will be done first.

In each session, one of your two auctions with more precise reference prices and one of your two auctions with less precise reference prices will be selected at random. Your take-home payment from the session will be based on the number of experimental dollars that you earn in these two auctions only.

Securities

In each auction, the Treasury has a demand for each pool of securities: high quality and low quality. The bidders have bidder-specific private information about security values and liquidity needs.

In each session, two sets of bidder-specific private information are drawn independently from the same probability distributions. The first set is used in the auctions with more precise reference prices (auctions 1 and 2); the second set is used in the auctions with less precise reference prices (auctions 3 and 4). You are given no feedback following each sealed-bid auction; thus, your behavior in the subsequent clock auction cannot be influenced by the outcome in the prior sealed-bid auction.

The bidders differ in their security holdings as shown in Table 20C.1.

Thus, there are four holders of each security: one large (50%), one medium (25%), and two small (12.5% each). Each bidder holds 20,000 ($k of face value) of high-quality securities and 20,000 ($k of face value) of low-quality securities.

The four high-quality securities are pooled together and sold as a pool; the four low-quality securities are pooled together and sold as a second pool. Whether done as a sealed-bid auction or done as a clock auction, the two pools are auctioned simul-

Table 20C.1. Holdings of securities by bidder and security in $thousands of face value

		High-Quality Securities				Low-Quality Securities				
		H1	H2	H3	H4	L1	L2	L3	L4	Total
	1	20,000					5,000	5,000	10,000	40,000
	2		20,000			10,000		5,000	5,000	40,000
	3			20,000		5,000	10,000		5,000	40,000
	4				20,000	5,000	5,000	10,000		40,000
Bidder	5		5,000	5,000	10,000	20,000				40,000
	6	10,000		5,000	5,000		20,000			40,000
	7	5,000	10,000		5,000			20,000		40,000
	8	5,000	5,000	10,000					20,000	40,000
	Total	40,000	40,000	40,000	40,000	40,000	40,000	40,000	40,000	
	Expected price	75	75	75	75	25	25	25	25	
	Expected value	30,000	30,000	30,000	30,000	10,000	10,000	10,000	10,000	
	Total value			120,000				40,000		

taneously. The Treasury has a budget of $30,000k for high-quality securities and a budget of $10,000k for low-quality securities. Thus, given the expected prices of 75 cents on the dollar for high-quality and 25 cents on the dollar for low-quality (see later), the Treasury can be expected to buy a quantity of about 40,000 ($k of face value), or 25% of face value, of each security pool. Between pools, there is no explicit interaction. However, the bidder's liquidity needs are based on sales from both pools together.

You are one of the eight bidders. You will have the same bidder number in auctions with more precise reference prices (auctions 1 and 2); you will have the same bidder number in auctions with less precise reference prices (auctions 3 and 4). Therefore, your holdings of securities and your signals will be the same in auctions 1 and 2, and they will also be the same in auctions 3 and 4. However, you will have different holdings of securities and signals in auctions 3 and 4, as compared with auctions 1 and 2.

Your preferences

From each auction, your payoff depends on your profits and how well your liquidity needs are met.

Common-value auction

The value of each high-quality security $s \in \{H1,H2,H3,H4\}$ in cents on the dollar is the average of n *iid* random variables uniformly distributed between 50 and 100:

$$v_s = \frac{1}{n} \sum_{j=1}^{n} u_{js}, \text{ where } u_{js} \sim_{iid} U[50, 100].$$

The value of each low-quality security $s \in \{L1,L2,L3,L4\}$ in cents on the dollar is the average of n iid random variables uniformly distributed between 0 and 50:

$$v_s = \frac{1}{n} \sum_{j=1}^{n} u_{js}, \text{ where } u_{js} \sim_{iid} U[0, 50].$$

For auctions with more precise reference prices, $n = 16$; for auctions with less precise reference prices $n = 12$. The reference price r_s for security s is given by:

$$r_s = \frac{1}{n-8} \sum_{j=9}^{n} u_{js}.$$

Thus, the reference price is based on eight realizations in the more precise case (half of all realizations) and on four realizations in the less precise case (a third of all realizations). Reference prices are made public before each auction starts.

For each 5,000 of security holdings, bidder i receives as private information one of the realizations u_{js}. Thus, bidder 1, who holds 20,000 of security 1, gets four realizations. In this way, those with larger holdings have more precise information about the security's value. Observe that this specification requires the holders of each given security to receive collectively a total of eight realizations. Since there are eight realizations available (besides the ones that form the reference price), each of the realizations u_{js} ($i = 1,\ldots, 8$) can be observed by exactly one bidder.

Suppose that the auction clearing price point is p_H for the high-quality pool and p_L for the low-quality pool, where the price point in the auction is stated as a fraction of the reference price. Then $p_s = p_H r_s$ for $s \in \{H1,H2,H3,H4\}$ and $p_s = p_L r_s$ for $s \in \{L1,L2,L3,L4\}$.

If a bidder sells the quantity q_s (in $thousands of face value) of the security s at the price p_s, then profit (in $thousands) is:

$$\pi_i(p, q_i, v) = \frac{1}{100} \sum_s (p_s - v_s) q_{is}.$$

The 1/100 factor in the formula above and other formulas involving price is to convert cents into dollars.

Liquidity

Each bidder has a liquidity need, L_i in thousands, which is drawn iid from the uniform distribution on the interval $[2500, 7500]$. Each bidder knows his own liquidity need, but not that of the other bidders. The bidder receives a bonus of $1 for every dollar of sales to the Treasury up to his liquidity need:

$$\min \left[L_i, \frac{1}{100} \sum_s p_s q_{is} \right].$$

Payoff of bidder from an auction

Combining the profit and the liquidity penalty results in the bidder's total payoff:

$$
U_i(p, q_i, v) = \begin{cases} \frac{1}{100} \sum_s (2p_s - v_s)q_{is} & \text{if } \frac{1}{100} \sum_s p_s q_{is} < L_i \\ L_i + \frac{1}{100} \sum_s (p_s - v_s)q_{is} & \text{otherwise} \end{cases}
$$

Thus, an additional dollar of cash is worth two dollars when the bidder's liquidity need is not satisfied, but is worth one dollar when the liquidity need is satisfied.

Bidder tool and auction system

You have an Excel workbook that contains your private information for each auction. The tool has features that will help your decision making in each of the auctions. Each auction has its own sheet in the tool. It is essential that you are working from the correct sheet for each auction. For example, the sealed-bid auction with more precise reference prices is the sheet named *Sealed Bid More.*

Bidding is done via a commercial auction system customized to this setting. You use the web browser to connect to the auction system. For each auction, you must go to a new auction site. The URL for the auction site is given in the bidder tool on the particular auction sheet. Once at the correct auction site, log in with the user name and password given on your auction sheet.

The auction system is easy to use. You will have an opportunity to use it in the training seminar.

An important feature of the tool is the calculation of expected security values conditional on information you specify. In a common-value auction, it is important for you to condition your estimate of value on your signal *and the information winning conveys.* Since your bid is relevant only in the event that you win, you should set your bid to maximize your payoff in that event. In this way, you avoid the winner's curse, which in this case is the tendency of a naïve bidder to lose money by selling shares at prices below what they are worth. In addition, in the clock auctions, the bidder also must condition on any information revealed through the bidding process.

The bidding tool provides one flexible method of calculating an appropriate conditional expected value for each security. In particular, the tool is set up to calculate the conditional expected values for each security, given the information that you know—the reference price and your own signals—and your best guesses for the relevant other signals. Making appropriate guesses for the other signals is an important element of your strategy. Once these guesses are made, the tool will calculate the common value of the security based on your entries. In the clock auction, you can adjust your estimates of other signals as you learn from the quantity drops of the other bidders.

The tool also lets you keep track of your liquidity bonus based on estimates that you enter for expected prices and expected quantities sold of each security.

Further details of the tool will be explained in the training seminar.

Bidding strategy

The auction environment has five complicating features:

- *Common-value auction.* You have an imperfect estimate of each security's common value.
- *Divisible-good auction.* Your bid is a supply curve, specifying the quantity you wish to sell at various price points.
- *Demand for pool of securities.* The Treasury does not have a demand for individual securities, but for pools of securities (high- and low-quality pools).
- *Asymmetric holdings.* Bidders have different holdings of securities.
- *Liquidity need.* You have a specific liquidity need that is met through selling shares from your portfolio of eight securities.

The combination of these factors makes a complete equilibrium analysis difficult or impossible, even when we make strong simplifying assumptions. For this reason we will refrain from providing any sort of benchmark strategy.

Your challenge is to determine your own strategy to maximize your payoff that reflects all aspects of the auction environment. The best response in this auction, as in any auction, is a best response to what the other bidders are actually doing.

It will be helpful to have an appreciation for the probability density of the common value in various circumstances.

Figure 20C.1 displays the pdf of the common value for low-quality securities in the more precise case by the size of the bidder's holdings, assuming that all the known signals take on the mean value of 25. Thus, when the bidder holds 20,000 of the security there are four unknown signals and the standard deviation is 1.8; when the bidder holds 10,000 there are six unknown signals and the standard deviation is 2.2; when the bidder holds 5,000 there are seven unknown signals and the standard deviation is 2.5.

Figure 20C.2 displays the pdf of the common value for low-quality securities in the less precise case by the size of the bidder's holdings, assuming that all the known signals take on the mean value of 25. Thus, when the bidder holds 20,000 of the security there are four unknown signals and the standard deviation is 2.4; when the bidder holds 10,000 there are six unknown signals and the standard deviation is 2.9; when the bidder holds 5,000 there are seven unknown signals and the standard deviation is 3.2.

If you have any questions during the experiment, please raise your hand and an assistant will help you.

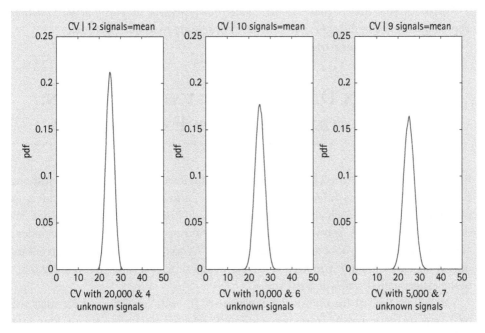

FIGURE 20C.1. Probability density of common value in the more precise case, by size of holdings

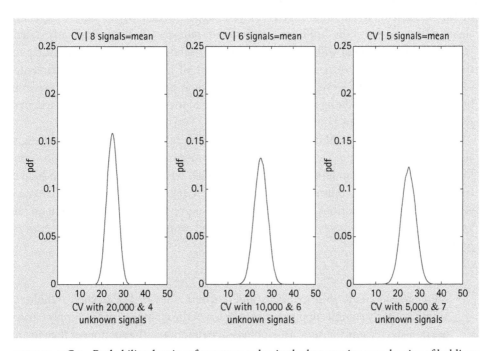

FIGURE 20C.2. Probability density of common value in the less precise case, by size of holdings

*Remember your overall goal is to maximize your experimental payoff in each auction.
You should think carefully about what strategy is best apt to achieve this goal.*

Many thanks for your participation.

APPENDIX D. A COMMON-VALUE AUCTION:
BIDDER INSTRUCTIONS FOR EXPERIMENT 3.1

In this experiment, you are a bidder in a series of auctions conducted over two sessions. Each session is held on a different day and consists of four different auctions. Bidders are randomly assigned to each auction. Thus, you do not know who the other bidders are in any of your auctions. You will be bidding from your private cubical, which includes a computer and a bidder tool that is unique to you. We ask that you refrain from talking during the experiment. If you need assistance at any time, just raise your hand and one of the experimental staff will assist you.

You will be paid at the end of the experiment. In each session, your earnings will be based on your payoff from two randomly selected auctions (out of four total auctions). Your take-home payment is then proportional to your total experimental earnings from sessions 1 and 2. In particular, you will receive *$0.40 in take-home pay for every one-million experimental dollars that you earn.* Throughout the remainder of the document, dollar figures refer to your experimental payoff unless explicitly stated otherwise—and the "millions" are generally suppressed. When explicitly stated, your real dollar payment will be referred to as your *take-home payment.* We anticipate that each of you will earn a take-home payment of about $100 per experimental session on average. However, your actual take-home payment will depend on your bidding strategy, the bidding strategies of the other bidders you face, and the particular realizations of numerous random events.

In each auction, you will compete with other bidders to sell your holdings of eight securities to the Treasury. The eight different securities have different values. However, for each security, the bidders have a common value, which is unknown. Each bidder has an estimate of the common value. You profit by selling securities to the Treasury at prices above the securities' true values. Unlike in previous experiments, you *do not* value liquidity. Thus your payoffs are based solely on profits from your sale of securities to the Treasury.

Two formats are used:

- *Simultaneous uniform-price sealed-bid auction ("sealed-bid auction").* Bidders simultaneously submit supply curves for each of the eight securities. Supply curves are non-decreasing (i.e. upward-sloping) step functions. The auctioneer then forms the aggregate supply curve and crosses it with the Treasury's demand. The clearing price is the lowest-rejected offer. All quantity offered below the clearing price is sold

at the clearing price. Quantity offered at the clearing price is rationed to balance supply and demand, using the proportionate rationing rule.
- *Simultaneous descending clock auction ("clock auction").* The securities are auctioned simultaneously. There is a price "clock" for each security indicating its tentative price per unit of quantity. Bidders express the quantities they wish to supply at the current prices. The price is decremented for each security that has excess supply, and bidders again express the quantities they wish to supply at the new prices. This process repeats until supply is made equal to demand. The tentative prices and assignments then become final. Details of the design are presented in Ausubel and Cramton (2008), which you received earlier.

The proportionate rationing rule plays a role only in the event that multiple bidders make reductions at the clearing price. The rule then accepts the reductions at the clearing price in proportion to the size of each bidder's reduction at the clearing price. Thus, if a reduction of 300 is needed to clear the market and two bidders made reductions of 400 and 200 at the clearing price, then the reductions are rationed proportionately: the first is reduced by 200 and the second is reduced by 100. The actual reduction of the first bidder is twice as large as that of the second bidder, since the first bidder's reduction as bid is twice as large as the second bidder's reduction. Ties can generally be avoided by refraining from bidding price points that are round numbers, instead specifying price points to odd one-hundredths of a percent (e.g. 98.42).

The clock auction has an activity rule to encourage price discovery. In particular, for each security, the quantities bid must be an upward-sloping supply curve. The quantity of a security is not permitted to increase as the price descends.

In each session, you will participate in *two pairs* of auctions in the following order:

1. Four-bidder sealed-bid, first pair. A sealed-bid auction with four bidders.
2. Four-bidder clock, first pair. A clock auction with four bidders. The values of securities and your signals will be identical to those in the sealed-bid above.
3. Four-bidder sealed-bid, second pair. A sealed-bid auction with four bidders.
4. Four-bidder clock, second pair. A clock auction with four bidders. The values of securities and your signals will be identical to those in the sealed-bid above.

Each session, one of your first pair of auctions and one of your second pair of auctions will be selected at random. Your take-home payment from the session will be based on the number of (million) laboratory dollars that you earn in these two auctions only.

Securities

In each auction, eight securities are purchased by the Treasury. The bidders are symmetric before the draws of bidder-specific private information about security values.

In each session, two sets of bidder-specific private information are drawn independently from the same probability distributions. The first set is used in the first pair of auctions (auctions 1 and 2); the second set is used in the second pair of auctions (auctions 3 and 4). You are given no feedback following the sealed-bid auctions; thus, your behavior in the clock auctions cannot be influenced by outcomes in the sealed-bid auctions of a session.

In each auction, the Treasury demand is 1,000 shares of each security, where each share corresponds to $1 million of face value. Each bidder has holdings of 1,000 shares of each security. Thus, it is possible for a single bidder to fully satisfy the Treasury demand for a particular security; that is, for each security there may be just a single winner or there may be multiple winners. One-quarter of the total available shares will be purchased by the Treasury.

Your preferences

From each auction, your payoff depends on your profits from the sale of securities to the Treasury.

Common-value auction

The value of each security in cents on the dollar is the average of eight *iid* random variables uniformly distributed between 0 and 100:

$$v_s = \tfrac{1}{8} \sum_{i=1}^{8} u_{is}, \text{ where } u_{is} \sim_{iid} U\,[0, 100]\,.$$

Suppose you are bidder i.

Your private information about security s is the realization u_{is}. Notice that the common value is the average of eight uniform draws, so that only the first four draws are revealed (as there are only four bidders in the auction).

If you sell the quantity q_s of the security s at the price p_s, then your profit (in million $) is:

$$\pi_i(p, q_i, v) = \tfrac{1}{100} \sum_{s=1}^{8} (p_s - v_s) q_{is}.$$

Bidder tool and auction system

You have an Excel workbook that contains your private information for each auction. The tool has features that will help your decision making in each of the auctions. Each auction has its own sheet in the tool. It is essential that you are working from the correct sheet for each auction. For example the sealed-bid, first-pair auction is the sheet named *Sealed Bid First Pair*.

For each of the auctions, bidding is done via a commercial auction system customized to this setting. You use the web browser to connect to the auction system. For each auction, you must go to a new auction site. The URL for the auction site is given in the bidder tool on the particular auction sheet, Sealed-Bid First Pair, Clock First Pair, Sealed-Bid Second Pair, or Clock Second Pair. Once at the correct auction site, log in with the user name and password given on your auction sheet.

The auction system is easy to use. It is identical to the system you used for bidding in all previous experiments.

An important feature of the tool is the calculation of expected security values conditional on information you specify. In a common-value auction, it is important for you to condition your estimate of value on your signal *and the information winning conveys*. Since your bid is only relevant in the event that you win, you should set your bid to maximize your payoff in that event. In this way, you avoid the winner's curse, which in this case is the tendency of a naïve bidder to lose money by selling shares at prices below what they are worth. In addition, in the clock auctions, the bidder also must condition on any information revealed through the bidding process. The tool provides one flexible method of calculating an appropriate conditional expected value for each security.

Bidding strategy

The auction environment has two complicating features:

- *Common-value auction.* You have an imperfect estimate of the good's common value.
- *Divisible-good auction.* Your bid is a supply curve, specifying the quantity you wish to sell at various prices.

The combination of these factors makes a complete equilibrium analysis difficult. Nonetheless, equilibrium analysis is possible in a simplified environment. To aid your thinking about strategy, we discuss a particular strategy which abstracts from the complications of a divisible-good auction. In particular, we assume that each bidder submits a flat supply schedule; that is, the bidder offers to sell all of her holdings of a particular security at a specified price.

With these assumptions it is possible to calculate an equilibrium strategy, which we call the *benchmark strategy*. The analysis of this strategy will be helpful in thinking about the common-value feature of the auction environment.

We wish to emphasize that the benchmark strategy focuses on only one element of the auction. Your challenge is to determine your own strategy to maximize your payoff that reflects all aspects of the auction environment.

Common-value distribution

As mentioned earlier, the value of each security in cents on the dollar is the average of eight *iid* random variables uniformly distributed between 0 and 100:

$$v_s = \frac{1}{8} \sum_{i=1}^{8} u_{is}, \text{ where } u_{is} \sim_{iid} U[0, 100].$$

The pdf and cdf of the common value are shown in Figure 20D.1. The common value has a mean of 50 and a standard deviation of 10.2. Notice that the common value is approximately normally distributed, since it is the sum of many independent draws.

Sealed-bid uniform-price reverse auction

Under our strong simplifying assumptions, the auction is equivalent to a unit-supply common-value auction with uniform pricing. In this case, just as in a single-item second-price auction, your bid does not determine what you pay, only the likelihood of winning, thus the best strategy is to bid your true conditional expected value for the good. The trick, however, is to condition on your signal and the information that winning conveys.

In the four-bidder auction, under the benchmark assumptions, there is only a single winner, so the correct condition is as derived in Milgrom and Weber (1982). Your bid

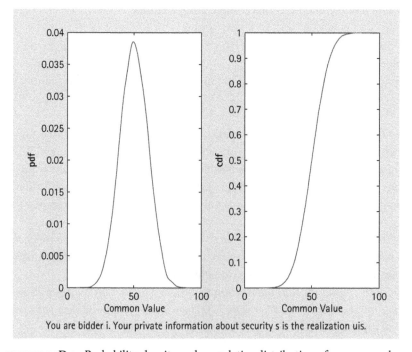

You are bidder i. Your private information about security s is the realization uis.

FIGURE 20D.1. Probability density and cumulative distribution of common value.

is your expected value conditional on your signal being the lowest, and on the second-lowest signal being equal to yours:

$$b_{is} = E[v_s | u_1 = u_{is}, u_2 = u_1],$$

where u_1 is the lowest signal and u_2 is the second-lowest signal. The reason you condition on the second-lowest signal being as low as yours is that the bid must be optimal when it is on the margin and thus impacts whether you win. Your bid becomes marginal and hence decisive only in the event that you tie with the second lowest.

Since the signals are uniformly distributed, it is easy to calculate the above conditional values. In both cases, the conditional value is a linear function of your signal.

In the four-bidder auction, there is a single winner. The conditioning is that you win, the next lowest bidder has your value, and the remaining two bidders are evenly distributed above your value:

four-bidder sealed-bid strategy:

$$b_{is} = \frac{1}{8}\left(2u_{is} + 2\left(\frac{u_{is} + 100}{2}\right) + 4 \cdot 50\right) = \tfrac{3}{8}u_{is} + \tfrac{75}{2}.$$

Descending clock auction

In the clock auction, under the benchmark assumption, you will observe the price at which other bidders drop out. This provides additional information on which to condition your bid. Here we assume that the price clock is continuous. In the actual experiment, the price clock is discrete, and although bidders can make reductions at any price, you will learn the aggregate supply only at the end-of-round price. You may want to assume the quantity reduction occurred halfway between the prior price and the ending price.

When the clock starts, you calculate your dropout point in the same way as above. As the price clock falls, one of the other bidders may drop out. When the first bidder drops out, you can calculate this bidder's draw from the following equation.

$$u_{8s} = \frac{P_{\text{dropout}} - \text{intercept}}{\text{slope}}$$

where the slope and intercept are taken from the formulas above. With this new information on which to condition your bid, the revised optimal bid for the four-bidder clock auction is straightforward to calculate.

Four-bidder clock strategy

$$\text{No one has dropped out: } b_{is} = \frac{1}{8}\left(2u_{is} + 2\left(\frac{u_{is} + 100}{2}\right) + 4 \cdot 50\right).$$

$$\text{One bidder has dropped out: } b_{is} = \frac{1}{8}\left(2u_{is} + u_4 + 1\left(\frac{u_{is} + u_4}{2}\right) + 4 \cdot 50\right).$$

$$\text{Two bidders have dropped out: } b_{is} = \frac{1}{8}(2u_{is} + u_4 + u_3 + 4 \cdot 50).$$

Note that the above formulas are all linear in the dropout price, so it is easy to invert to compute the bidder's signal.

Moving beyond the benchmark strategy

The bidding tool is set up to calculate the conditional expected values assuming the benchmark strategy. Of course, you (and others) may well deviate from the benchmark strategy.

The bidding tool allows for this variation in a number of ways. First, in the sealed-bid auctions your bid can be any upward-sloping step function to account for possible supply reduction by you or others. Second, in the clock auctions, the tool lets you interpret what a dropout is and where it occurs. This is useful and necessary when bidders make partial reductions of supply. In addition, although the tool will calculate a particular signal based on a dropout, you are free to type in any signal inference you like. Whatever you type as "my best guess" will be used in the calculation of the conditional expected value.

If you have any questions during the experiment, please raise your hand and an assistant will help you.

Remember your overall goal is to maximize your experimental payoff in each auction. You should think carefully about what strategy is best apt to achieve this goal.

Many thanks for your participation.

APPENDIX E. A COMMON-VALUE REFERENCE-PRICE AUCTION: BIDDER INSTRUCTIONS FOR EXPERIMENT 3.2

In this experiment, you are a bidder in a series of auctions conducted over two sessions. Each session is held on a different day and consists of four different auctions. Bidders are randomly assigned to each auction. Thus, you do not know who the other bidders are in any of your auctions. You will be bidding from your private cubical, which includes a computer and a bidder tool that is unique to you. We ask that you refrain from talking during the experiment. If you need assistance at any time, just raise your hand and one of the experimental staff will assist you.

You will be paid at the end of the experiment. In each session, your earnings will be based on your payoff from two randomly selected auctions (out of four total auctions). Your take-home payment is then proportional to your total experimental earnings from sessions 1 and 2. In particular, you will receive *$0.30 in take-home pay for every one thousand experimental dollars ($k) that you earn*. Throughout, dollar figures refer to your experimental payoff unless explicitly stated otherwise—and the "thousands" are generally suppressed. When explicitly stated, your real dollar payment will be referred to as your *take-home payment*. We anticipate that each of you will earn a take-home payment of about $100 per experimental session on average. However, your actual take-home payment will depend on your bidding strategy, the bidding strategies of the other bidders you face, and the particular realizations of numerous random events.

In each auction, you will compete with other bidders to sell your holdings of eight securities to the Treasury. The eight securities are split into two pools: four securities are low quality and four are high quality. The eight different securities have different values. However, for each security, the bidders have a common value, which is unknown. Each bidder has an estimate of the common value. You profit by selling securities to the Treasury at prices above the securities' true values. The Treasury has allocated a particular budget for the purchase of each pool of securities within each auction. Its demand for high-quality securities is distinct from its demand for low-quality securities.

Before each auction, the auctioneer assigns each security a *reference price* (expressed in cents on the dollar of face value), which represents the Treasury's best estimate of what each security is worth. For example, a high-quality security might be assessed to be worth 75 cents on the dollar, while a low-quality security might be assessed to be worth 25 cents on the dollar. A *reference-price auction* determines the price point—a percentage of the reference price—for each pool of securities. A winning bidder is paid the pool's price point × the security's reference price for each unit of the security sold.

Two formats are used:

- *Simultaneous uniform-price sealed-bid auction ("sealed-bid auction")*. Bidders simultaneously submit supply curves for each of the eight securities. Supply curves are non-decreasing (i.e. upward-sloping) step functions, offering a quantity at each price point. The auctioneer then forms the aggregate supply curve and crosses it with the Treasury's demand. This is done separately for each pool (i.e. for high- and low-quality securities, separately). The clearing price point is the lowest rejected offer. All quantity offered at below the clearing price is sold at the clearing price times the security's reference price. Quantity offered at exactly the clearing price is rationed to balance supply and demand, using the proportionate rationing rule.
- *Simultaneous descending clock auction ("clock auction")*. The securities are auctioned simultaneously. There are two descending clocks, one for high-quality securities and one for low-quality securities, indicating the tentative price point of each pool. Bidders express the quantities they wish to supply at the current price points. The price point is decremented for each pool of securities that has excess supply, and bidders again express the quantities they wish to supply at the new price points. This process repeats until supply is made equal to demand. The tentative price points and assignments then become final. Details of the design are presented in Ausubel and Cramton (2008a,b), which you received earlier.

The proportionate rationing rule plays a role only in the event that multiple bidders make reductions at the clearing price. The rule then accepts the reductions at the clearing price in proportion to the size of each bidder's reduction at the clearing price. Thus, if a reduction of 300 is needed to clear the market and two bidders made reductions of 400 and 200 at the clearing price, then the reductions are rationed proportionately: the first is reduced by 200 and the second is reduced by 100. The actual reduction of the first bidder is twice as large as that of the second bidder, since the first bidder's reduction as bid is twice as large as the second bidder's reduction. Ties can generally be avoided by refraining from bidding price points that are round numbers, instead specifying price points to odd one-hundredths of a percent (e.g. 98.42).

The clock auction has an activity rule to encourage price discovery. In particular, for each security pool, the quantities bid must be an upward-sloping supply curve as expressed in activity points. More precisely, *activity points*—equal to the reference price × quantity, summed over the four securities in the pool—are computed for each bid. The number of activity points is not permitted to increase as the price point descends. Thus, you are allowed to switch quantities across the four securities in a pool, but your total activity points for the pool cannot increase as the auction progresses.

The same activity rule applies to the sealed-bid auction, but then in a single round of bidding.

In each session, you will participate in *two pairs* of auctions in the following order:

1. sealed-bid auction, first pair
2. clock auction, first pair
3. sealed-bid auction, second pair
4. clock auction, second pair.

In all cases, the reference prices will be based on four signals. That is, the reference prices will be analogous to those used in the *less precise* auctions from Experiment 2.

Each session, one of your first pair of auctions and one of your second pair of auctions will be selected at random. Your take-home payment from the session will be based on the number of (million) laboratory dollars that you earn in these two auctions only.

Securities

In each auction, the Treasury has a demand for each pool of securities: high quality and low quality. The bidders have bidder-specific private information about security values.

In each session, two sets of bidder-specific private information are drawn independently from the same probability distributions. The first set is used in the first pair of auctions (auctions 1 and 2); the second set is used in the second pair of auctions (auctions 3 and 4). You are given no feedback following each sealed-bid auction; thus, your behavior in the subsequent clock auction cannot be influenced by the outcome in the prior sealed-bid auction.

Table 20E.1. Holdings of securities by bidder and security in $thousands of face value

		High-Quality Securities				Low-Quality Securities				
		H1	H2	H3	H4	L1	L2	L3	L4	Total
	1	20,000					5,000	5,000	10,000	40,000
	2		20,000			10,000		5,000	5,000	40,000
	3			20,000		5,000	10,000		5,000	40,000
	4				20,000	5,000	5,000	10,000		40,000
Bidder	5		5,000	5,000	10,000	20,000				40,000
	6	10,000		5,000	5,000		20,000			40,000
	7	5,000	10,000		5,000			20,000		40,000
	8	5,000	5,000	10,000					20,000	40,000
	Total	40,000	40,000	40,000	40,000	40,000	40,000	40,000	40,000	
Expected price		75	75	75	75	25	25	25	25	
Expected value		30,000	30,000	30,000	30,000	10,000	10,000	10,000	10,000	
Total value				120,000				40,000		

The bidders differ in their security holdings as shown in Table 20E.1.

Thus, there are four holders of each security: one large (50%), one medium (25%), and two small (12.5% each). Each bidder holds 20,000 ($k of face value) of high-quality securities and 20,000 ($k of face value) of low-quality securities.

The four high-quality securities are pooled together and sold as a pool; the four low-quality securities are pooled together and sold as a second pool. Whether done as a sealed-bid auction or done as a clock auction, the two pools are auctioned simultaneously. The Treasury has a budget of $30,000k for high-quality securities and a budget of $10,000k for low-quality securities. Thus, given the expected prices of 75 cents on the dollar for high-quality and 25 cents on the dollar for low-quality (see later), the Treasury can be expected to buy a quantity of about 40,000 ($k of face value), or 25% of face value, of each security pool. Between pools, there is no explicit interaction.

You are one of the eight bidders. You will have the same bidder number in the first pair of auctions (auctions 1 and 2); you will have the same bidder number in the second pair of auctions (auctions 3 and 4). Therefore, your holdings of securities and your signals will be the same in auctions 1 and 2, and they will also be the same in auctions 3 and 4. However, you will have different holdings of securities and signals in auctions 3 and 4, as compared with auctions 1 and 2.

Your preferences

From each auction, your payoff depends on your profits.

Common-value auction

The value of each high-quality security $s \in \{H1,H2,H3,H4\}$ in cents on the dollar is the average of twelve *iid* random variables uniformly distributed between 50 and 100:

$$v_s = \tfrac{1}{12} \sum_{j=1}^{12} u_{js}, \text{ where } u_{js} \sim_{iid} U[50, 100].$$

The value of each low-quality security $s \in \{L1,L2,L3,L4\}$ in cents on the dollar is the average of twelve *iid* random variables uniformly distributed between 0 and 50:

$$v_s = \tfrac{1}{12} \sum_{j=1}^{12} u_{js}, \text{ where } u_{js} \sim_{iid} U[0, 50].$$

The reference price r_s for security s is given by:

$$r_s = \tfrac{1}{4} \sum_{j=9}^{12} u_{js}.$$

Thus, the reference price is based on four realizations (one-third of all realizations). Reference prices are made public before each auction starts.

For each 5,000 of security holdings, bidder i receives as private information one of the realizations u_{js}. Thus, bidder 1, who holds 20,000 of security 1, gets four realizations. In this way, those with larger holdings have more precise information about the security's value. Observe that this specification requires the holders of each given security to receive collectively a total of eight realizations. Since there are eight realizations available (besides the ones that form the reference price), each of the realizations u_{js} ($i = 1,..., 8$) can be observed by exactly one bidder.

Suppose that the auction clearing price point is p_H for the high-quality pool and p_L for the low-quality pool, where the price point in the auction is stated as a fraction of the reference price. Then $p_s = p_H r_s$ for $s \in \{H1,H2,H3,H4\}$ and $p_s = p_L r_s$ for $s \in \{L1,L2,L3,L4\}$.

If a bidder sells the quantity q_s (in $thousands of face value) of the security s at the price p_s, then profit (in $thousands) is:

$$\pi_i(p, q_i, v) = \tfrac{1}{100} \sum_s (p_s - v_s)q_{is}.$$

The 1/100 factor in the formula above and other formulas involving price is to convert cents into dollars.

Bidder tool and auction system

You have an Excel workbook that contains your private information for each auction. The tool has features that will help your decision making in each of the auctions. Each auction has its own sheet in the tool. It is essential that you are working from the correct sheet for each auction. For example, the sealed-bid auction from the first pair is the sheet named *Sealed Bid First Pair*.

Bidding is done via a commercial auction system customized to this setting. You use the web browser to connect to the auction system. For each auction, you must go to a new auction site. The URL for the auction site is given in the bidder tool on the particular auction sheet. Once at the correct auction site, log in with the user name and password given on your auction sheet.

The auction system is easy to use. It is identical to the system you used for bidding in all previous experiments.

An important feature of the tool is the calculation of expected security values conditional on information you specify. In a common-value auction, it is important for you to condition your estimate of value on your signal *and the information winning conveys*. Since your bid is relevant only in the event that you win, you should set your bid to maximize your payoff in that event. In this way, you avoid the winner's curse, which in this case is the tendency of a naïve bidder to lose money by selling shares at prices below what they are worth. In addition, in the clock auctions, the bidder also must condition on any information revealed through the bidding process.

The bidding tool provides one flexible method of calculating an appropriate conditional expected value for each security. In particular, the tool is set up to calculate the conditional expected values for each security, given the information that you know— the reference price and your own signals—and your best guesses for the relevant other signals. Making appropriate guesses for the other signals is an important element of your strategy. Once these guesses are made, the tool will calculate the common value of the security based on your entries. In the clock auction, you can adjust your estimates of other signals as you learn from the quantity drops of the other bidders.

Bidding strategy

The auction environment has four complicating features:

- *Common-value auction.* You have an imperfect estimate of each security's common value.
- *Divisible-good auction.* Your bid is a supply curve, specifying the quantity you wish to sell at various price points.
- *Demand for pool of securities.* The Treasury does not have a demand for individual securities, but for pools of securities (high- and low-quality pools).
- *Asymmetric holdings.* Bidders have different holdings of securities.

The combination of these factors makes a complete equilibrium analysis difficult or impossible, even when we make strong simplifying assumptions. For this reason we will refrain from providing any sort of benchmark strategy.

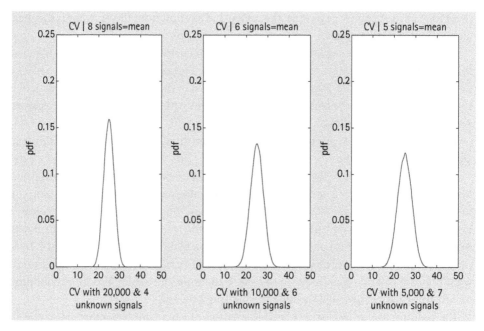

FIGURE 20E.2. Probability density of common value by size of holdings

Your challenge is to determine your own strategy to maximize your payoff that reflects all aspects of the auction environment. The best response in this auction, as in any auction is a best response to what the other bidders are actually doing.

It will be helpful to have an appreciation for the probability density of the common value in various circumstances.

Figure 20E.1 displays the pdf of the common value for low-quality securities by the size of the bidder's holdings, assuming that all the known signals take on the mean value of 25. Thus, when the bidder holds 20,000 of the security there are four unknown signals and the standard deviation is 2.4; when the bidder holds 10,000 there are six unknown signals and the standard deviation is 2.9; when the bidder holds 5,000 there are seven unknown signals and the standard deviation is 3.2.

If you have any questions during the experiment, please raise your hand and an assistant will help you.

Remember your overall goal is to maximize your experimental payoff in each auction. You should think carefully about what strategy is best apt to achieve this goal.

Many thanks for your participation.

REFERENCES

Alsemgeest, P., Noussair, C. and Olson, M. (1996) "Experimental comparisons of auctions under single- and multi-unit demand," Working Paper, Purdue University.

Armantier, O., Holt, C.A., Plott, C.R. (2010) "A reverse auction for Toxic Assets," *California Institute of Technology Social Science*, Working Paper No.1330.

Ausubel, L. M. (2004) "An efficient ascending-bid auction for multiple objects," *American Economic Review*, 94(5): 1452–75.

—— (2006) "An efficient dynamic auction for heterogeneous commodities," *American Economic Review*, 96(3): 602–29.

—— and Cramton, P. (2002) "Demand reduction and inefficiency in multi-unit auctions," University of Maryland Working Paper 96-07, revised July 2002.

—— —— (2004) "Auctioning many divisible goods," *Journal of the European Economic Association*, 2: 480–93.

—— —— (2006) "Dynamic auctions in procurement," in N. Dimitri, G. Piga, and G Spagnolo (eds), *Handbook of Procurement*, Cambridge University Press, pp. 220–45.

—— —— (2008a) "A troubled asset reverse auction," Working Paper, University of Maryland.

—— —— (2008b) "Auction design critical for rescue plan," *Economists' Voice*, 5:5, at <http://www.bepress.com/ev/vol5/iss5/art5>.

—— —— (2009) "A two-sided auction for legacy loans," University of Maryland.

Barnett-Hart, A. K. (2009) "The story of the CDO market meltdown: an empirical analysis," Harvard College Department of Economics.

Brunnermeier, M. K. (2009) "Deciphering the liquidity and credit crunch 2007–2008," *Journal of Economic Perspectives*, 23(1): 77–100.

Bulow, J. and Klemperer, D. (1996) "Auctions Versus Negotiations," *American Economic Review*, 86: 180–194.

Congressional Budget Office (2008) "Policy options for the housing and financial Markets."

Cramton, P. (1998) "Ascending auctions," *European Economic Review*, 42: 745–56.

French, K. R., Baily, M. N., Campbell, J. Y., Cochrane, J. H., Diamond, D. W., Duffie, D., Kshyap, A. K., Mishkin, F. S., Rajan, R. G., Shiller, R. J., Shin, H. S., Slaughter, M. J., Stein, J. C. and Stulz, R. M. (2010) *The Squam Lake Report: Fixing the Financial System*, Princeton University Press.

Gorton, G. B. (2008) "The panic of 2007," Yale ICF Working Paper No. 08–24.

—— (2009) "Information, liquidity, and the (ongoing) panic of 2007," *American Economic Review*, 99(2): 567–72.

Kagel, J. H. (1995) "Cross-game learning: experimental evidence from first-price and English common value auctions," *Economic Letters*, 49: 163–70.

—— and Levin, D. (2001) "Behavior in multi-unit demand auctions: experiments with uniform price and dynamic Vickrey auctions," *Econometrica*, 69: 413–54.

—— —— (2005) "Multi-unit demand auctions with synergies: some experimental results," *Games and Economic Behavior*, 53: 170–207.

—— —— (2009) "Implementing efficient multi-object auction institutions: an experimental study of the performance of boundedly rational agents," *Games and Economic Behavior*, 66:221–237.

—— Hartstad, R. M. and Levin, D. (1987) "Information impact and allocation rules in auctions with affiliated private values: a laboratory study," *Econometrica*, 55: 1275–304.

Klemperer, P. (2009) "A new auction for substitutes: central-bank liquidity auctions, 'toxic asset' auctions, and variable product-mix auctions," CEPR Discussion Paper No. DP7395.

Levin, D., Kagel, J. H. and Richard, J.-F. (1996) "Revenue effects and information processing in English common value auctions," *American Economic Review*, 86: 442–60.

McAfee, R. P. and McMillan, J. (1987) "Auctions and bidding," *Journal of Economic Literature,* 25: 699–738.

McCabe, K. A., Rassenti, S. J. and Smith, V. L. (1990) "Auction institutional design: theory and behavior of simultaneous multiple-unit generalizations of the Dutch and English auctions," *American Economic Review,* 80: 1276–83.

Milgrom, P. (2004) *Putting Auction Theory to Work,* Cambridge University Press.

—— and Weber, R. J. (1982) "A theory of auctions and competitive bidding," *Econometrica,* 5010: 1089–122.

Mizen, P. (2008) "The credit crunch of 2007–2008: a discussion of the background, market reactions, and policy responses," *Federal Reserve Bank of St. Louis Review,* 90(5): 531–68.

Mullin, C. H. and Reiley, D. H. (2006) "Recombinant estimation for normal-form games, with applications to auctions and bargaining," *Games and Economic Behavior,* 54: 159–82.

Swagel, P. (2009) "The financial crisis: an inside view," *Brookings Papers on Economic Activity,* spring: 1–63.

CHAPTER 21

....

INFORMATION
DISCLOSURE IN
AUCTIONS

An Experiment

....

MARTIN DUFWENBERG AND URI GNEEZY[1]

INTRODUCTION

....

DETAILS of market organization may influence economic performance. In many cases
these details are matters of choice for government procurement practices and government auctions (e.g. offshore oil leases, timber and grazing rights, or the broadcast
spectrum). Similarly, in the private sector the rise of electronic markets and auction
sites has focused attention on the specifics of market organization. In this chapter we
report experimental evidence concerning how one particular detail affects competition
in first-price auctions: the availability of information about historic bids submitted in
previous auctions. We consider three possible forms of information disclosure where,
in turn, all bids, all winning bids, or no bids at all are announced by the auctioneer at
the completion of an auction.

The core feature of the design is the following game. Each of two players simultaneously chooses a bid, which is an integer between 2 and 100. The player who chooses
the lowest bid gets a dollar amount times the number (s)he bids and the other player

[1] This article is reprinted from *Journal of Economic Behavior and Organization* (2002) 48: 431–44,
with permission from Elsevier, but with one new added paragraph in the Introduction, where we
discuss closely related work by Isaac and Walker (1985) (which was written before our paper) and by
Ockenfels and Selten (2005) (which was inspired by our paper and Isaac and Walker's). We thank Gary
Charness, Eric van Damme, David Grether, and Reinhard Selten for very helpful comments, and the
Swedish Competition Authority for financial support. We started this research while we were both at
the CentER for Economic Research at Tilburg University, and completed it during a visit of Uri Gneezy
to Stockholm University.

gets 0 (ties are split). This game may be interpreted as a first-price auction where the (common) value of the auctioned object is known. With this interpretation, a strategy measures the difference between the value of the object and the payment offered for it. For example, the bids could be profit levels or prices requested by two competing firms to perform some task desired by a government agency. This game has a unique Nash equilibrium in which each player submits a bid of 2 and gets a payoff of only 1 (times a dollar amount). This equilibrium is strict and it can be given a strong decision-theoretic justification, since a bid of 2 is the unique rationalizable strategy (Bernheim, 1984; Pearce, 1984) of the game.[2]

We wish to investigate whether this sharp prediction stands up in a laboratory test. We are primarily interested in the behavior of *experienced* participants, since in many first-price auctions the participants are experienced. Hence, we must let participants play the game several times. The most common method for inducing experience in experimental economics is to let a fixed group of participants interact over and over again. However, a drawback with this approach is that a confounding effect is introduced: Since *the same* participants interact repeatedly, opportunities for cooperation of the kind studied in the theory of repeated games may be created. See Pearce (1992) for a general overview. We wish to test the model described in the previous paragraph as it stands, and yet to allow for experience while avoiding repeated-game effects. To achieve this we use a random matching scheme such that participants play the game ten times, matched randomly with one out of eleven counterparts in each round. In the terminology of Jackson and Kalai (1997), our design approximates interaction in *recurring games*, as opposed to repeated ones.

The issue of information disclosure crops up when an auction is run many times. Since each bidder participates in more than one auction, a history of bids will exist. This history may or may not be public information. We use three treatments to investigate the importance of this issue. In the first treatment—*full information*—we publicly announce the entire vector of submitted bids at the end of each period. In the second treatment—*semi-information*—we announce only the winning bids, and in the third treatment—*no information*—we announce at the end of each period only which participant won. It is crucial to note that the theoretical prediction, described above, is invariant to the information condition.

Our study is closely related to work by Isaac and Walker (1985) and by Ockenfels and Selten (2005), who explore treatments similar to ours in various first-price auction games. All studies find that competition is most fierce when information about losing

---

[2] Two comments about this game and its solution: (1) In a finite two-player game, a strategy is rationalizable if it survives iterated elimination of strictly dominated strategies. If for each player a unique strategy does so, the corresponding profile must be the game's unique Nash equilibrium, which furthermore is strict. In our game, a bid of 100 is strictly dominated by a mixed strategy giving almost all weight to 99, and very low but positive weight to 2. Repeated analogous arguments reveal that 2 is the unique strategy surviving iterated elimination of strictly dominated strategies, and the desired conclusions follow. (2) We do not include (the per se reasonable) choices 0 and 1 in the strategy sets since this would eliminate the uniqueness of the theoretical prediction while (in terms of economic intuition) little would change (*all* equilibria entail *small* profits).

bids is not given. This prompts Ockenfels and Selten to talk about "a robust behavioral effect" (p. 156) (which they go on to explain using learning direction theory).

In this chapter we emphasize the first-price auction interpretations, but the game we study may also be thought of in terms of price competition, and our design may be compared to some price-competition experiments. Our study overlaps with Dufwenberg and Gneezy (2000) on the full-information treatment, and the results should be viewed as complementary. Both studies deal with the same game and use the random-matching set-up, but the investigation proceeds along different dimensions, as Dufwenberg and Gneezy (given full information) consider the case of more than two competitors. It is shown that bids come close to the Nash equilibrium when the number of competitors is three or four, but that bids remain much higher when only two competitors are matched. See Baye and Morgan (1999) for an analysis of how these results may be accounted for theoretically. Our old results on market concentration and our new results on the role of information feedback may be combined to yield insights about optimal auction design.

The random-matching set-up used here and in Dufwenberg and Gneezy distinguishes these studies crucially from most other experimental studies of price competition, because the usual approach is to consider repeated interaction among a fixed group of competitors. In such a setting, informational issues of various kinds (for example, information about cost structures or signals of future prices) have been investigated, but not the effect of information about historic strategic choices. The classic contribution is Fouraker and Siegel (1963); other relevant references include Dolbear et al. (1968), Selten and Berg (1970), Hoggatt et al. (1976), Friedman and Hoggatt (1980), Grether and Plott (1984), Holt and Davis (1990), Cason (1994, 1995), Cason and Davis (1995), Mason and Phillips (1997), and Gneezy and Nagel (1999). For overviews of some of the literature mentioned here, see Plott (1982, 1989) and Holt (1995). Three other studies of somewhat related games, which, however, are not conceptualized as price competition games, are Nagel (1995) (the guessing game), Capra et al. (1999) (the travelers' dilemma), and Rapoport and Amaldoss (2000) (an investment game framed in terms of research and development). The two last studies involve random matching.

In the next section we describe the experimental procedure. The following section reports the results. The fourth section contains a discussion, where we describe a signaling phenomenon that may be important for explaining the results, and make a recommendation concerning optimal auction design.

Experimental procedure

The experiment was conducted at Tilburg University. Students were recruited using an advertisement in the university newspaper as well as posters on campus. The experiment consisted of three treatments with two sessions per treatment. There were twelve bidders in each of the six sessions. In each period, six pairs of participants were formed according to a random-matching scheme. Each session had ten periods, or rounds.

In treatment F (full information feedback), participants were informed at the end of each period about the entire bid vector (that is, about all twelve bids). In treatment S (semi-full information feedback), only the vector of winning bids was communicated to the participants. In treatment N (no information feedback), participants were informed only about their personal payoff at the end of the period.

In each session, students received a standard-type introduction and were told that they would be paid 7.5 Dutch guilders ($f7.5$) for showing up.[3] Then, they took an envelope at random from a box which contained thirteen envelopes. Twelve of the envelopes contained numbers ($A1,\ldots,A12$). These numbers were called "registration numbers." One envelope was labeled "Monitor," and determined who was the person who assisted us and checked that we did not cheat.[4] We asked the participants not to show their registration number to the other students.

Each participant then received the instructions for the experiment (see the Appendix), and ten coupons numbered 1, 2,…, 10. After reading the instructions and asking questions (privately), each participant was asked to fill out the first coupon with her registration number and bid for period 1. The bids had to be between 2 and 100 "points," inclusive, with 100 points being worth $f5$. Participants were asked to fold the coupon, and put it in a box carried by the assistant. The assistant randomly took two coupons out of the box and gave them to the experimenter. In treatment F (sessions F1 and F2), the experimenter announced the registration number on each of the two coupons and the respective bids. If one bid was larger than the other, the experimenter announced that the low bid won the same amount of points as she bid, and the other bidder won 0 points. If the bids were equal the experimenter announced a tie, and said that each bidder won one-half of the bid. The assistant wrote this on a blackboard such that all the participants could see it for the rest of the experiment. Then the assistant took out another two coupons randomly, the experimenter announced their content, and the assistant wrote it on the blackboard. The same procedure was carried out for all the twelve coupons. All subsequent periods were conducted the same way; after period 10, payoffs were summed, and participants were paid privately.

Treatment S (sessions S1 and S2) was carried out the same way, except that the experimenter did not announce the losing bids. Treatment N (sessions N1 and N2) was carried out the same way, except that the experimenter did not announce bids at all (and hence communicated only the registration number(s) corresponding to the lowest bid for each matched pair).

EXPERIMENTAL RESULTS

The data from the sessions are presented in Tables 21.1–21.6, in which the average winning bids and the average bids are also presented. Correspondingly, the average

---

[3] At the time of the experiment, $\$1 = f1.7$.
[4] That person was paid the average of all other subjects participating in that session.

Table 21.1. The bids in session F1 (full information feedback)

	Round 1	Round 2	Round 3	Round 4	Round 5	Round 6	Round 7	Round 8	Round 9	Round 10
A1	49	34	24*	22*	16*	15*	100*	100	60	20*
A2	15*	20*	25*	20	19	19	14*	9*	19*	19*
A3	39	39	30	35	40	19	100*	99	99	99
A4	40*	29*	28*	26	18*	16*	13*	80*	40	28*
A5	10*	20*	29	24*	19*	15*	14	100	79	79
A6	40*	30*	26	20*	21	15*	14*	19*	50*	60*
A7	23*	29	31*	24*	28	20*	14*	17*	40*	50
A8	46	32	24*	26	18*	100	20	35	88	66
A9	40	38	25*	25	20*	20	15	40	100	40*
A10	40*	40	35	19*	19*	18	40	39	35*	60*
A11	20	25*	20*	19*	17*	15*	12*	12*	20*	39
A12	40*	35*	30	23*	25	16	14*	18*	39*	35
Average bid	33.5	30.9	27.3	23.6	21.7	24.0	30.8	47.3	55.8	49.6
Average winning bid	29.7	26.5	25.3	22.0	18.1	16.0	35.1	25.8	33.8	37.8

* A winning bid.

Table 21.2. The bids in session F2 (full information feedback)

	Round 1	Round 2	Round 3	Round 4	Round 5	Round 6	Round 7	Round 8	Round 9	Round 10
A1	66	50*	33*	66	44	85	98	96	50*	99
A2	30	24*	33*	22	30*	20	79	50*	54	40
A3	80	70	39	39	19	26	59*	69	67	46
A4	40*	50	40	20*	20*	80	79*	76	66	42
A5	85	85	85	20*	20	15*	20*	70	70	50
A6	22*	28*	18*	18*	28	20*	30*	49	48	39*
A7	98	40	84	85	99	99	99	99	99	99
A8	20*	30	28	20*	18*	80*	20	40*	40*	30*
A9	5	17*	20*	17*	17*	16*	13*	19*	35*	39*
A10	33*	29*	27*	26	17*	16	79*	49*	48*	38*
A11	21*	21	21	21	18*	16*	39	69*	48*	68*
A12	2*	2*	2*	2*	2*	2*	2*	2*	2*	2*
Average bid	41.8	37.2	35.8	29.7	27.7	39.6	51.4	57.3	52.3	49.3
Average winning bid	23.0	25.0	22.0	16.2	17.4	24.8	40.3	38.2	37.2	36.0

* A winning bid.

Table 21.3. The bids in session S1 (semi–information information feedback)

	Round 1	Round 2	Round 3	Round 4	Round 5	Round 6	Round 7	Round 8	Round 9	Round 10
A1	40*	35	25	15	9*	15	5	20	50	50
A2	47	27*	13*	12	9*	7*	3*	3*	3	2*
A3	22*	27*	22	12*	11	10	8	6*	4*	2*
A4	15*	24	15*	18	11	9*	7	15	2*	14*
A5	19*	19	17*	9*	9*	7*	5*	4*	2*	2*
A6	37*	18*	17	12*	9	7*	5*	3*	2*	2
A7	48	27*	25	14*	48	40	48	48	48	38
A8	21	15*	17*	15	11*	8*	8*	5*	5	4*
A9	25	25	13*	13	8*	8*	6*	5	4*	2*
A10	40	19*	18	10*	9*	8	6	5	3*	2*
A11	20*	15*	10*	10*	10	10	5*	2*	2*	2*
A12	44	28	25	16	11	8*	8	28	6*	20
Average bid	31.5	23.3	18.1	13.0	12.9	11.4	9.5	12.0	10.9	11.7
Average winning bid	25.5	21.1	14.2	11.2	9.2	7.7	5.3	3.8	3.1	3.8

* A winning bid.

Table 21.4. The bids in session S2 (semi-information feedback)

	Round 1	Round 2	Round 3	Round 4	Round 5	Round 6	Round 7	Round 8	Round 9	Round 10
A1	24*	39*	24*	19*	24	16	12*	9*	9	4
A2	75	50	75	25*	25	20	15	10*	10	99
A3	40	40	20	30	15*	15*	15	10*	8	4*
A4	28*	12*	21*	29	19	13*	11*	8*	10*	3
A5	49*	45	48	20*	19*	19*	13*	18	8	2*
A6	66	20*	35	20	16*	15	30	10	7*	20
A7	80	60	25*	28	22	14*	19	19	100	20*
A8	13*	45	19*	20*	22*	21	18*	10*	9	5*
A9	22	20	18*	23*	18*	15*	12*	11	7*	5
A10	25*	21*	30	20*	19	10*	14	10*	8*	5
A11	40	23*	24	24	15*	15*	14*	12*	7*	4*
A12	29*	39*	34*	23	17*	42	22	19	7*	2*
Average bid	40.9	34.5	31.1	23.4	19.3	17.9	16.3	12.2	15.8	14.4
Average winning bid	28.0	25.7	23.5	21.2	17.4	14.4	13.3	9.9	7.7	6.2

*A winning bid.

Table 21.5. The bids in session N1 (no information feedback)

	Round 1	Round 2	Round 3	Round 4	Round 5	Round 6	Round 7	Round 8	Round 9	Round 10
A1	100	100	100	100	100	100	100	100	100	100
A2	35	50	95	20*	30	20	90	30*	80	80
A3	100	50	50	25*	25	25	20	10	20*	10*
A4	35*	30*	35	25	20*	20*	20*	15*	25	15
A5	30	25	10*	14*	19*	24*	24*	24*	24	14*
A6	23*	23*	23*	23*	23	23	2	2*	2*	2*
A7	2*	100	2*	100	2*	100	2	2*	2*	2*
A8	29*	29*	29*	50	29*	29*	29*	29	15	39
A9	12*	14*	17*	22*	35	32	25	25*	20	10
A10	25*	25*	40	30	20*	20*	20*	20	5*	5*
A11	60	49	20	20*	20*	18*	18*	18*	10*	10*
A12	25	10*	20*	20	15*	15*	10*	10	10*	5*
Average bid	39.7	42.1	36.8	37.4	28.2	35.5	30.0	23.8	26.1	24.3
Average winning bid	21.0	21.8	16.8	20.7	17.9	21.0	20.2	16.6	8.2	6.9

*A winning bid.

Table 21.6. The bids in session N2 (no information feedback)

	Round 1	Round 2	Round 3	Round 4	Round 5	Round 6	Round 7	Round 8	Round 9	Round 10
A1	100	100	50	25	10*	10*	20*	50	20	20
A2	29*	19*	11*	9*	7	5*	4	9*	7*	9*
A3	21	98	99	2*	2*	2*	2*	2*	2*	2*
A4	45	25*	35	50	25	30	25	20	20	20
A5	39*	27*	18*	13*	11*	9	3*	4*	2*	2*
A6	25*	35*	40	40	30*	30	30	20	20	20
A7	46*	46	35*	40	33	24	15*	20	15*	20
A8	10*	10*	10*	10*	5*	10*	10*	10*	10*	10*
A9	49*	48	39	20	10*	6*	7	5*	5	5*
A10	49*	49	46*	42*	41	49	30	70	30	25
A11	29	20*	24*	27*	21	23	19	17*	22	30
A12	69	37	25	15	8*	10*	8*	10	8*	10*
Ave. bid	42.6	42.8	36.0	24.4	16.9	17.3	14.4	19.8	13.4	14.4
Ave. win bid	35.3	22.7	24.0	17.2	10.9	7.2	9.7	7.8	7.3	6.3

* A winning bid.

winning bids and the average bids are plotted in Figure 21.1. We start by describing the behavior in period 1, at which stage no elements of learning or experience exist. From observation of the data it is clear that the outcome predicted by theory was not achieved in this period.

The average bid (winning bid) was 33.5 (29.7) and 41.8 (23) in sessions F1 and F2, respectively, 31.5 (25.5) and 40.9 (28.0) in sessions S1 and S2, respectively, and 39.7 (21.0) and 42.6 (35.3) in sessions N1 and N2. We perform a statistical test of whether the bids in different sessions came from the same distribution. To this end, we consider each of the (fifteen) possible pairs of sessions, and investigate whether the two relevant sets of observed bids come from the same distribution. We use the non-parametric Mann–Whitney U-test based on ranks, and cannot, for any pair, reject (at a 5% significance level) the hypothesis that the observations come from the same distribution (see Table 21.7). In this sense, in period 1 the different rules in the different markets did not influence behavior.

When comparing the development of bids in later periods, however, we see a great difference between treatments. Figure 21.2 illustrates the average winning bids in the three treatments. In session F1, there was a slow decrease of the average winning bid from 29.7 in period 1 to 16.0 in period 6. From period 6 to period 7, a jump in the average winning bid from 16.0 to 35.1 is observed. From this point on the averages are 25.8 in period 8, 33.8 in period 9, and finally 37.8 in period 10. It is clear that no tendency of convergence towards bids of 2 is observed. In fact, the smallest bid in period 10 was 19. In session F2, the average winning bid decreased constantly, from 23.0 in period 1 to 16.2 in period 4. Then, however, the average winning bid started to rise, and in periods 8, 9, and 10 the average winning bids were 38.2, 37.2, and 36.0, respectively. An interesting observation is that participant number A12 in this session used a constant bid of 2 throughout the experiment. Of course, this bid of 2 was "strange" given the fact that the next-lowest bid in period 10 was 38. This bid was not enough to move the other bids to the neighborhood of 2. Furthermore, the bids in both sessions of treatment F were much alike in period 10; the average bids were 49.6 and 49.3 in sessions F1 and F2 respectively, and the average winning bids were 37.8 and 36.0 in the respective sessions.[5]

In session S1 there was a decrease in the average winning bid from 25.5 in period 1 to 3.8 in period 10. Bids decreased steadily, moving from period 1 to period 10. The lowest bid (as well as the median bid) in period 10 was 2. A similar behavior was observed in session S2, in which the average winning bid decreased from 28.0 in period 1 to 6.2 in period 10. The lowest bid in period 10 was also 2, with nine out of the twelve participants bidding 5 or less. When comparing the two sessions of treatment S we see that, as in the case of treatment F, the bids in both sessions were quite similar in period 10.

In session N1, the decrease in the average winning bid was not monotonic. The average winning bid fluctuated around its starting value (21) until the seventh period,

---

[5] Unlike the case of first-round behavior, it is not appropriate to use the Mann–Whitney U-test, because the assumption that all observations are independent is not justified.

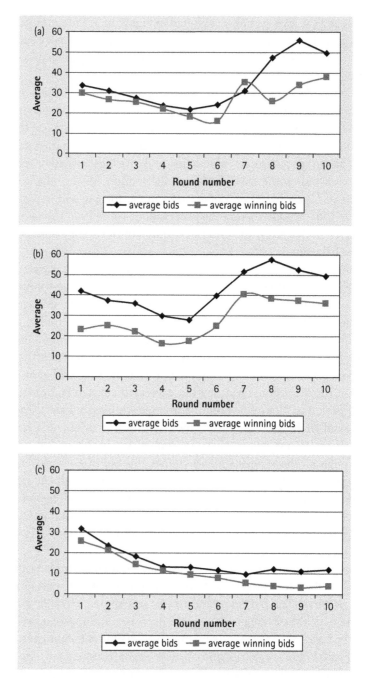

FIGURE 21.1. Average bids and winning bids, (a) session F1, (b) session F2, (c) session S1, (d) session S2, (e) session N1, (f) session N2.

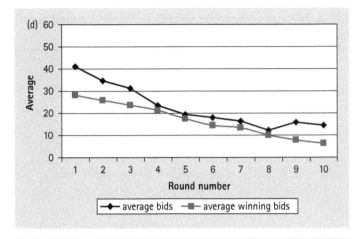

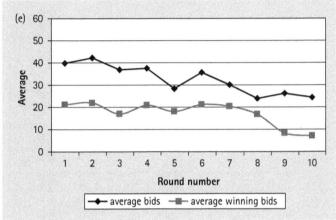

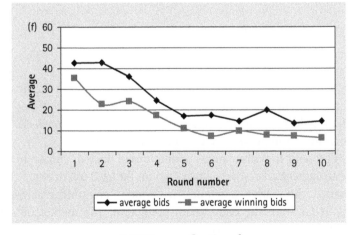

FIGURE 21.1. Continued

Table 21.7. A pairwise comparison of bids in the first period across sessions using Mann–Whitney U-test based on ranks

	Session F2	Session S1	Session S2	Session N1	Session N2
Session F1	0.00 (1.000)	0.29 (0.7728)	−0.61 (0.5444)	0.43 (0.6650)	−0.89 (0.3708)
Session F2		0.38 (0.7075)	−0.26 (0.7950)	−0.12 (0.9081)	−0.49 (0.6236)
Session S1			−1.10 (0.2727)	−0.23 (0.8179)	−1.39 (0.1659)
Session S2				0.38 (0.7075)	−0.26 (0.7950)
Session N1					−0.69 (0.4884)

The null hypothesis is that all bid vectors come from the same distribution. The numbers in the cells are the z-statistics. The Prob > $|z|$ is given in parentheses.

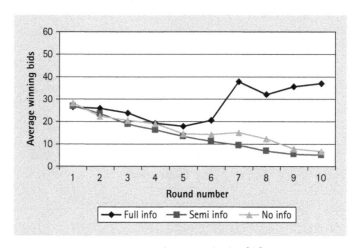

FIGURE 21.2. Average winning bids.

and only then did it start to decline. The average winning bid in the final period was 6.9, and the median bid was 10 (as compared with a median bid of 29.5 in period 1). In session N2 the decrease in the average winning bid was more steady (though not monotonic), from 35.3 in period 1 to 6.3 in the final period.

Figure 21.3 presents the cumulative distributions of the bids chosen in period 10 for each treatment, aggregated across the two sessions for each treatment.

To summarize, the market outcomes in period 1 are similar across treatments, and behavior is far from the equilibrium prediction. The market outcomes in the final periods differ dramatically between the full-information treatment and the other two treatments. Nearly all bids in the full-information sessions are far from equilibrium, while all winning bids in the other two treatments are relatively close to the equilibrium.

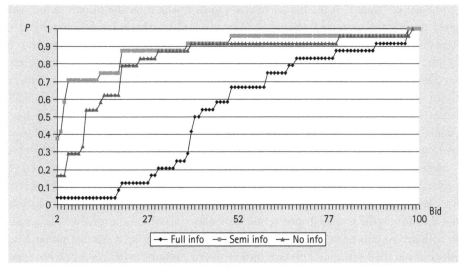

FIGURE 21.3. Cumulative distributions of the bids chosen in period 10 for each treatment, aggregated across the two sessions for each treatment.

DISCUSSION

In this chapter we consider the design of auctions by an auctioneer who may seem to have very little power: The auctioneer can decide only how much feedback to give to the bidders in the auction.[6] The bidders compete in an environment with recurrent competition. The theoretical prediction is unambiguous regarding all three degrees of information disclosure: each bidder should submit the lowest bid possible.

Yet when we test this model experimentally, bidders in the initial period choose bids higher than in the Nash equilibrium. However, bids rapidly moved in the direction of the theoretical prediction in two out of the three treatments. This occurred when each bidder either received information only about her own performance in previous periods or when each bidder received information only about her own performance and about the winning bids of the previous periods. However, in a third treatment in which the bidders were informed about the entire bid vector in previous periods, bids remained much higher than the theoretical prediction.

Apparently, the information about the *losing* bids is of great importance for the competitors. This result may be explained in terms of signaling behavior. The following intuitive argument is intended to be suggestive of the process. Assume each bidder has two possible actions at time t (when t is not the final period). The bidder can either

---

[6] By contrast, the theoretical literature on optimal auction design typically considers the effects of *different* mechanisms (e.g. Luton and McAfee, 1986; Laffont and Tirole, 1987; Piccione and Tan, 1996) or market structures (e.g. Dana and Spier, 1994; McGuire and Riordan, 1995), but does not address the issue of feedback in recurrent interaction.

"compete" or "signal." If the bidder chooses to compete, then she submits a bid which gives the highest expected reward at time t, based on the bidder's belief about the behavior of the competing bidders. Alternately, the bidder may choose to use her bid at time t to signal. Doing that, the bidder makes a conspicuously "high" bid at time t, sacrificing payoffs in that period in order to influence the beliefs of the other bidders at time $t+1$. If the bidder is successful in doing this, then she may expect a higher payoff at time $t+1$ than if she chooses to compete at time t.[7] Clearly, this kind of signaling may be profitable *only* when the other bidders can observe the signals. That is, the bidders will be aware of signals only in the treatment in which the entire bid vector can be observed.

Note that if this signaling story is relevant, the trade-off between profits in the current period and overall profits may favor signals when bids at time t are expected to be very low. Moreover, in the current random-matching context, this signaling explanation is not the same as the repeated-interaction explanation in which competitors are assumed to collude. To construct a formal model of signaling may be a feasible research task which could shed some light on how bids evolve in auctions over time, or on how prices evolve in markets in which firms compete on price. We hope that the findings we report in this study will serve to inspire such a line of inquiry. This, however, lies outside the scope of the present chapter.

What have we learned of relevance for optimal auction design? A fairly clear picture emerges if we first refer to the results of Dufwenberg and Gneezy (2000), which concern the same game as here except that more than two competitors may interact. That study may be interpreted as suggesting that auctioneers are well advised to try to have at least three bidders competing; with only two competitors, bids remained from the Nash equilibrium, while bids approached the equilibrium when the number of bidders was three or four. However, this result was derived under conditions of full information about historic bids. The present chapter shows that, even with two competitors, bids come close to the Nash equilibrium if information about losing bids is not disclosed. Based on this observation, we now venture upon the following piece of advice to auctioneers: *You may announce winning bids, but keep the losing bids secret!*

Appendix. Instructions for the full-information treatment

In the following game, which will be played for ten rounds, we use "points" to reward you. At the end of the experiment we will pay you 5 cents for each point you won (100

---

[7] In the context price competition experiments, observations of related kinds of signals are made by Fouraker and Siegel (1963, pp. 185–8), Hoggatt et al. (1976), and Friedman and Hoggatt (1980) for the case of repeated interaction among a fixed group of firms. See Plott (1982, pp. 1513–17) for a discussion. Surprisingly, these interesting observations seem to have been "forgotten"; we have seen no post-1982 discussion of the matter in the literature.

points equals 5 Dutch guilders). In each round your reward will depend on your choice, as well as the choice made by one other person in this room. However, in each round you will not know the identity of this person and you will not learn this subsequently.

At the beginning of round 1, you are asked to choose a number between 2 and 100, and then to write your choice on card number 1 (please note that the ten cards you have are numbered 1,2,...,10). Write also your registration number on this card. Then we will collect all the cards from round 1 from the students in the room and put them in a box.

The monitor will then randomly take two cards out of the box. The numbers on the two cards will be compared. If one student chose a lower number than the other student, then the student who chose the lowest number will win points equal to the number he/she chose. The other student will get no points for this round. If the two cards have the same number, then each student gets points equal to half the number chosen. The monitor will then announce (on a blackboard) the registration number of each student in the pair that was matched, and indicate which of these students chose the lower number and what his/her number was.

Then the monitor will take out of the box another two cards without looking, compare them, reward the students, and make an announcement, all as described above. This procedure will be repeated for all the cards in the box. That will end round 1, and then round 2 will begin. The same procedure will be used for all 10 rounds.

REFERENCES

Baye, M. and Morgan, J. (1999) "Bounded rationality in homogenous product pricing games," Mimeo, Indiana University and Princeton University.

Bernheim, D. (1984) "Rationalizable strategic behavior," *Econometrica*, 52: 1007–28.

Capra, M., Goeree, J., Gomez, R. and Holt, C. (1999) "Anomalous behavior in a travelers' dilemma," *American Economic Review*, 89: 678–90.

Cason, T. (1994) "The impact of information sharing opportunities on market outcomes: an experimental study," *Southern Economic Journal*, 61: 18–39.

—— (1995) "Cheap talk price signaling in laboratory markets," *Information Economics and Policy*, 7: 183–204.

—— and Davis, D. (1995) "Price communications in a multi-market context: an experimental investigation," *Review of Industrial Organization*, 10: 769–87.

Dana, J. and Spier, K. (1994) "Designing a private industry: government auctions with endogenous market structure," *Journal of Public Economics*, 53: 127–47.

Dolbear, F., Lave, L., Bowman, G., Lieberman, A., Prescott, E., Reuter, F. and Sherman, R. (1968) "Collusion in oligopoly: an experiment on the effect of numbers and information," *Quarterly Journal of Economics*, 82: 240–59.

Dufwenberg, M. and Gneezy, U. (2000) "Price competition and market concentration: an experimental study," *International Journal of Industrial Organization*, 18: 7–22.

Fouraker, L. and Siegel, S. (1963) *Bargaining Behavior*, McGraw-Hill.

Friedman, J. and Hoggatt, A. (1980) "An experiment in non-cooperative oligopoly," *Research in Experimental Economics*, 1 (Suppl. 1).

Gneezy, U. and Nagel, R. (1999) "Behavior in symmetric and asymmetric price-competition: an experimental study," Mimeo, Technion and Pompeu Fabra.

Grether, D. and Plott, C. (1984) "The effects of market practices in oligopolistic markets: an experimental investigation of the ethyl case," *Economic Inquiry*, 22: 479–507.

Hoggatt, A., Friedman, J. and Gill, S. (1976) "Price signaling in experimental oligopoly," *American Economic Review*, 66: 261–6.

Holt, C. (1995) "Industrial organization: a survey of laboratory research," in J. Kagel and A. Roth (eds), *Handbook of Experimental Economics*, Princeton University Press, pp. 349–444.

—— and Davis, D. (1990) "The effects of non-binding price announcements on posted offer markets," *Economics Letters*, 34: 307–10.

Isaac, M. and Walker, J. (1985) "Information and conspiracy in sealed-bid auctions," *Journal of Economic Behavior and Organization*, 6: 139–59.

Jackson, M. and Kalai, E. (1997) "Social learning in recurring games," *Games and Economic Behavior*, 21: 102–34.

Laffont, J.-J. and Tirole, J. (1987) "Auctioning incentive contracts," *Journal of Political Economy*, 95: 921–37.

Luton, R. and McAfee, P. (1986) "Sequential procurement auctions," *Journal of Public Economics*, 31: 181–95.

Mason, C. and Phillips, O. (1997) "Information and cost asymmetry in experimental duopoly markets," *Review of Economics and Statistics*, May: 290–9.

McGuire, T. and Riordan, M. (1995) "Incomplete information and optimal market structure: public purchases from private providers," *Journal of Public Economics*, 56: 125–41.

Nagel, R. (1995) "Unraveling in guessing games: an experimental study," *American Economic Review*, 85: 1313–26.

Ockenfels, A. and Selten, R. (2005) "Impulse balance equilibrium and feedback in first price auctions," *Games and Economic Behavior*, 51: 155–70.

Pearce, D. (1984) "Rationalizable strategic behavior and the problem of perfection," *Econometrica*, 52: 1029–50.

—— (1992) "Repeated games: cooperation and rationality," in *Advances in Economic Theory*, Cambridge University Press.

Piccione, M. and Tan, G. (1996) "Cost-reducing investment, optimal procurement, and implementation by auctions," *International Economic Review*, 37: 663–85.

Plott, C. (1982) "Industrial organization and experimental economics," *Journal of Economic Literature*, 20: 1485–527.

—— (1989) "An updated review of industrial organization: applications of experimental economics," in R. Schmalensee and R. Willig (eds), *Handbook of Industrial Organization*, North Holland, vol. 2, pp. 1109–76.

Rapoport, A. and Amaldoss, W. (2000) "Mixed strategies and iterative elimination of strongly dominated strategies: an experimental investigation," *Journal of Economy Behavior and Organization*, 42: 483–521.

Selten, R. and Berg, C. (1970) "Drei Experimentelle Oligopolspielserien mit Kontinuerlichem Zeitablauf," in H. Sauermann (ed.), *Beiträge zur Experimentellen Wirtschaftsforschung*, J. C. B. Mohr, vol. 2, pp. 162–221.

CHAPTER 22

EXPERIMENTS WITH BUYER-DETERMINED PROCUREMENT AUCTIONS

ELENA KATOK

INTRODUCTION

COMPETITIVE sourcing mechanisms, which include reverse auctions and requests for information, quotes, and proposals (RFI/Q/Ps), are fast becoming an essential part of the procurement toolkit. A large-scale study that surveyed close to 200 companies in a wide cross-section of industries (Center for Strategic Supply Research, 2006) reported that nearly 65% used electronic procurement mechanisms and over 60% regularly used online reverse auctions. The average total amount spent through reverse auctions was reported to be almost $9 billion (about 1% of gross sales), and growing at about 20% per year. Beall et al. (2003) report that while online reverse auctions account for less than 10% of the actual total amount spent, for some firms this figure can potentially increase to as much as 50%, indicating high growth potential. In addition to substantial cost savings, online reverse auctions deliver a number of other benefits, including an increase in the supplier base (Center for Strategic Supply Research, 2006) and accelerated transaction time (Shugan, 2005).

One of the most important factors that make procurement reverse auctions fundamentally different from forward auctions is that price is typically not the main attribute used to award contracts. Exogenous non-price attributes (e.g. distance from buyer, incumbency status, reputation) have a major effect on the expected buyer's surplus. While the bidding in most online reverse auctions is in terms of price (this is different from pure score auctions, in which bidders can submit bids in terms of price and quality—see for example Che, 1993; Branco, 1997), most awards are not based only on price. Jap (2002, p. 510) was the first to point out that "the vast majority of [online reverse] auctions...do not determine a winner."

Most procurement auctions are non-binding, because the buyer does not commit to awarding the contract to the lowest bidder, but instead reserves the right to award it on any basis, and the lag between the end of the auction and the announcement of the winner may be as long as six weeks (Jap, 2002). The long lag between the end of the auction and the award decision may indicate that sometimes buyers themselves are unable to quantify the value of quality until after the auction is over and they can see all the bids. This may be because some bidders have not been fully vetted before the auction (Wan and Beil, 2009; Wan et al., 2012). For the purpose of this chapter, I collectively group non-price attributes and label them *quality*—see also Tonca et al. (2008), and Zhong and Wu (2009), who use a similar modeling approach. Sometimes bidders have some private information about their own quality that is not fully known to other bidders, although it will eventually be revealed to the buyer. But sometimes the buyer may reveal this quality information, for example by setting individual reserves, or reporting feedback in the auction in terms of quality-adjusted prices. It is also possible that bidders may not even know their own quality. This may happen if the buyer does not announce the attributes of importance or their respective weights.

Because the best price bid does not necessarily win, if bidders do not know the qualities of their competitors, winning may seem random to them, but the probability of winning is affected by the price bid. In a dynamic (open-bid) buyer-determined auction, bidders do not know either their winning status, or by exactly how much they may be winning or losing. As a result, bidders in open-bid auctions do not have the dominant bidding strategy, and open-bid buyer-determined auctions have the feel of sealed-bid auctions, in the sense that bidders have to decide on their final bid without knowing their winning status with certainty.

COMPARING PRICE-BASED AND BUYER-DETERMINED FORMATS

Engelbrecht-Wiggans et al. (2007) model the bidders in the buyer-determined auction as having bidder-specific non-price attributes that are modeled using a parameter Q_i. The Qs are independent across bidders and can be arbitrarily related to bidder i's cost, C_i. In the Engelbrecht-Wiggans et al. (2007) model, bidder i knows his own Q_i and C_i but not the cost or quality of his competitors. Engelbrecht-Wiggans et al. (2007) also assume that the buyer knows the qualities of all the bidders, and awards the contract to the bidder whose bid results in the highest *buyer surplus* level $Q_i - B_i$, where B_i is bidder i's bid. The assumption that the buyer knows the Qs is reasonable because eventually the buyer has to determine the winner of the auction based in part on the qualities, so at some point in time he has to learn the Qs. Engelbrecht-Wiggans et al. (2007) compare the sealed-bid buyer-determined auction (RFP) to a standard (price-based) reverse sealed auction (RFQ),[1] and show analytically that buyer-determined auctions result in higher buyer surplus levels as long as there are enough suppliers competing and the positive relationship between Q and C is not too strong.

In the lab, the authors conducted an experiment to test the model.[2] In all treatments $C_i \sim U(0, 100)$ and $Q_i \sim U(C_i, C_i + \gamma)$. The model predicts that the buyer surplus level in the buyer-determined auction is below that of the price-based auction if and only if the number of bidders $N = 2$ and $\gamma > 200$. Therefore, the treatments in the experiment are: $N = 2$ and $\gamma = 100$; $N = 2$ and $\gamma = 300$; $N = 4$ and $\gamma = 300$.

Engelbrecht-Wiggans et al. (2007) conducted these three treatments with both mechanisms (price based and buyer determined) and also with experienced and inexperienced bidders. Experienced bidders came to the lab after having participated in a web-based session of the same auction as the one in the lab, but in which they bid against computerized rivals programmed to bid according to the risk-neutral Nash equilibrium.

Table 22.1 summarizes actual and theoretical buyer surplus levels in the Engelbrecht-Wiggans et al. (2007) study. The results are consistent with the model's qualitative predictions:

Table 22.1. Actual and predicted buyer surplus levels in the Engelbrecht-Wiggans et al. (2007) study

Mechanism		Experienced	Inexperienced				
		$N = 2$ $\gamma = 100$	$N = 2$ $\gamma = 300$	$N = 4$ $\gamma = 300$	$N = 2$ $\gamma = 100$	$N = 2$ $\gamma = 300$	$N = 4$ $\gamma = 300$
Buyer-determined	Actual	[49.00] 45.93 (12.32)	[123.00] 123.31 (30.12)	[235.00] 196.36 (29.00)	[42.00] 39.74 (12.72)	[132.00] 120.41 (37.16)	[202.50] 226.12 (40.56)
	Theoretical	[36.75] 34.30 (11.22)	[110.50] 102.88 (33.71)	[195.00] 184.25 (33.53)	[11.22] 34.30 (11.22)	[33.71] 102.88 (33.71)	[195.00] 184.25 (33.53)
Price-based	Actual	[27.50] 28.29 (32.66)	[136.50] 132.37 (88.37)	[123.00] 130.59 (89.08)	[31.50] 30.79 (33.08)	[137.00] 136.71 (88.37)	[150.00] 143.72 (90.18)
	Theoretical	[20.05] 19.51 (32.06)	[123.50] 123.54 (88.29)	[134.75] 133.80 (89.09)	[20.50] 19.51 (32.06)	[123.50] 123.54 (88.29)	[134.75] 133.80 (89.09)

Median buyer surplus levels are in square brackets, and standard deviations are in parentheses. The unit of analysis is a cohort; each treatment contains two cohorts.

---

[1] Usually the requests for proposals (RFPs) and requests for quotes (RFQs) differ in that the buyer commits to awarding the contract to the lowest bidder in the RFQ but not in the RFP.

[2] The experiments discussed in this chapter were implemented using zTree (Fischbacher, 2007).

1. When $N = 2$ and $\gamma = 100$ the buyer surplus is significantly higher under the buyer-determined than under the price-based format.
2. When $N = 2$ and $\gamma = 300$ the buyer surplus is significantly lower under the buyer-determined than under the price-based format.
3. When $N = 4$ and $\gamma = 300$ the buyer surplus is again significantly higher under the buyer-determined than under the price-based format.

Another notable regularity in this study is that the actual buyer surplus levels are above predicted in all treatments. Since all auctions in this study are in the sealed-bid format, these higher than predicted average buyer surplus levels are consistent with overly aggressive bidding, which has been reported in forward auction experiments. We will see this regularity persist in other auctions that are not in the sealed-bid format, but have a sealed-bid flavor. I term this overly aggressive bidding the "sealed-bid" effect.[3]

THE EFFECT OF INFORMATION

Haruvy and Katok (2013) consider the effect of information that bidders have in terms of price visibility during the auction, and in terms of their knowledge about the non-price attributes of the other bidders (Q). The study manipulates auction format (open bid vs. sealed bid) and whether or not all bidders know the non-price attributes of all other bidders, or whether they are the bidder's private information. In all treatments bidders continue to know their own Qs. In the open-bid format, bids are entered dynamically and the contract is awarded to the bidder who generates the highest buyer surplus, $Q_i - B_i$, with her final bid. In the sealed-bid auction (RFP), each bidder places a single bid B_i, and the contract is awarded to the bidder whose bid generates the highest buyer surplus. The open-bid format has full price visibility, and the sealed-bid format has no price visibility.

Quality transparency is the second factor that Haruvy and Katok (2013) manipulate. Bidders always know their own Qs, and in the full-information condition (F) they also know the Qs of their competitors, while in the private information condition (P) they do not.

There is an expected buyer surplus equivalence for risk-neutral bidders that holds between the sealed-bid auction with private information and the open-bid auction with information. This result follows from the expected-buyer surplus equivalence between the sealed-bid first- and second-price buyer-determined auctions (see Engelbrecht-Wiggans et al., 2007, for the proof), and the strategic equivalence between the sealed-bid second-price buyer-determined auction and the open-bid buyer-determined auction with full information (see Kostamis et al., 2009). So Haruvy and Katok (2013) have

---

[3] See Kagel (1995) for a review of studies documenting behavior in forward sealed-bid first-price auctions. The overly aggressive bidding has been attributed to risk aversion (Cox et al., 1982); it is also consistent with regret aversion (Engelbrecht-Wiggans and Katok, 2008; Feliz-Ozbay and Ozbay, 2007).

analytical benchmarks for two of the treatments in their study. They also show that as long as the score of bidder i, $Q_i - C_i$, and the quality of bidder i, Q_i, are not independent, bids in the sealed-bid buyer-determined auction with full information depend on the qualities of the competitors. For the open-bid buyer-determined auction with private information, Haruvy and Katok (2013) show that in equilibrium bids cannot fully reveal either the quality, Q_i, or the score, $Q_i - C_i$, of bidder i.

In their experiment, Haruvy and Katok (2013) use parameters similar to Engelbrecht-Wiggans et al. (2007), with $C_i \sim U(0, 100)$ and $Q_i \sim U(C_i, C_i + \gamma)$, with $\gamma = 300$, so the buyer-determined auction is advantageous over the price-based auction. In all of their treatments, auctions had four bidders ($N = 4$), and they used a 2×2 full factorial design in which they varied the auction mechanism (sealed or open bid) and the quality information (full or private). Each treatment included four cohorts of eight participants randomly rematched in groups of four for thirty auctions.

It is worthwhile at this point to understand the risk-neutral equilibrium bidding behavior. In the open-bid auction with full information, bidders know their winning status, so they have the weakly dominant strategy to bid down in the smallest allowable bid decrements as long as they are losing, and to stop bidding when they are winning. So, for losing bidders we have:

$$B_i = C_i \tag{1}$$

In the sealed-bid auction with private information, Engelbrecht-Wiggans et al. (2007) derived the risk-neutral Nash equilibrium to be:

$$B_i = C_i + \frac{1}{N}(Q_i - C_i) \tag{2}$$

In the other two treatments, however, equilibrium bidding strategies cannot be derived, due to the lack of bidder symmetry. Nonetheless, it is possible to compute a linear approximation of risk-neutral Nash equilibrium bidding strategies, using an iterative best-reply algorithm. The resulting linear approximations are as follows:

$$B_i = Intercept + \alpha C_i + \beta(Q_i - C_i)$$
$$+ \sum_{k=1}^{3} \chi_k Q_{(k)} + \sum_{k=1}^{3} \delta_k(Q_{(k)} \times HQ_i) \tag{3}$$
$$= 30 + C_i + 0.25(Q_i - C_i)$$
$$- 0.2Q_{(1)} + 0.05(Q_{(1)} \times HQ_1)$$

for the sealed-bid auction with full information, and

$$B_i = Intercept + \alpha C_i + \beta(Q_i - C_i)$$
$$+ \gamma(Avg(B_{-i})) + \varphi(Avg(B_{-i}) \times LB_i)$$
$$= -12.4 + C_i + 0.25(Q_i - C_i) \tag{4}$$
$$+ 0.41(Avg(B_{-i})) - 0.25(Avg(B_{-i} \times LB_i)$$

for the open-bid auction with private information. In expression (3), $Q_{(k)}$ is the kth highest quality, and HQ_i is 1 if bidder i is the bidder with the highest quality, and 0 otherwise. In expression (4), $Avg(B_{-i})$ is the average price bid of i's competitors, and LB_i is 1 if i is the lowest-price bidder, and 0 otherwise.

The authors then proceed to use either the equilibrium bid functions, or their linear approximations, to compute predicted average buyer surplus levels, and the predicted proportion of efficient allocations in their treatments, using the actual realizations of costs and qualities in their experiment.

Table 22.2 shows that the average surplus in the open-bid auctions with full information is in line with theory, while in the other formats, average buyer surplus levels are higher than predicted. The efficiency levels are slightly lower than predicted. A notable consequence of these deviations from equilibrium predictions is that the expected buyer surplus equivalence between the open-bid full and the sealed-bid private conditions fails to hold.

In terms of the individual bidding behavior, it is possible to fit linear regressions for bids as functions of the parameters in equations (1)–(4) and find that bidding behavior in the open-bid auction with full information is mostly in line with theoretical predictions. In the other three treatments, however, the coefficient on the score variable $Q_i - C_i$ is significantly lower than it should be, while the other coefficients are mostly in line with predictions. This result illustrates that in the three treatments in which bidders do not have the dominant bidding strategy, they bid too aggressively, primarily because they fail to mark up their bids sufficiently based on their high quality. The phenomenon is related to the "sealed-bid effect" because it is related to the overly aggressive bidding in sealed-bid first price auctions.

Table 22.2. Average buyer surplus levels, proportion of efficient allocations, and the comparison between actual and estimated theoretical buyer surplus levels and efficiency

	Open-bid, full	Sealed-bid, private	Open-bid, private	Sealed-bid, full
Actual buyer surplus (standard error)	186.11 (4.17)	224.60 (2.60)	211.85 (6.12)	205.88 (7.90)
Actual proportion of efficient allocations	86.88%	88.43%	85.94%	84.38%
Deviation of actual surplus from predicted (standard error)	−2.39 (2.11)	40.35** (1.04)	39.00** (1.87)	10.84* (3.91)
Deviation of actual efficiency from predicted.	−13.12%**	−11.57%**	−10.31%**	3.13%

$^*p < 0.05;\ ^{**}p < 0.01.$

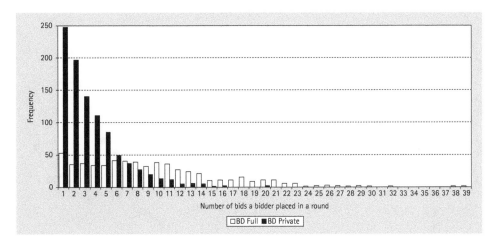

FIGURE 22.1. Average number of bids bidders place in the two open-bid treatments.

Figure 22.1 compares the average number of bids bidders place in the two open-bid treatments of the Haruvy and Katok (2013) study. We can see that while bidders place a large number of bids in open-bid auctions with full information, many place a very small number of bids in the open-bid auction with private information (the mode is 1). Many bidders do not use the price information from the auction because this information does not tell them their bidding status. Instead, they place what amounts to a sealed bid.

The main conclusion from the Haruvy and Katok (2013) study is that giving bidders less information (price information or quality information) appears to be better for the buyer. Katok and Wambach (2008) stress test this assertion by examining what happens when bidders do not know even their own quality. In procurement events, bidders often know the attributes that are important to the buyer, but often do not know the exact trade-offs between those attributes. In fact, sometimes buyers do not even know their own trade-offs until they evaluate the bids after the auction ends (Elmaghraby, 2007). In this setting, when bidder i does not know Q_i, winner determination looks random to the bidders.

Katok and Wambach (2008) show that in this setting there exists an equilibrium in which all bidders stop bidding at a point at which everyone has the same ex ante probability of winning (i.e. $1/N$). The reserve price has to be high enough relative to the differences in privately known parameters for this equilibrium to exist.

The Katok and Wambach (2008) experiment included three treatments. In all treatments, two bidders ($N=2$) whose cost is known to be $50(C_i = 50 \,\forall\, i)$ compete in auctions with a reserve price of 200. The bidders differ only in their quality ($Q_i \sim U(0,10)$), and compete in thirty open-bid descending auctions (with random rematching) that last for one minute and have a ten-second soft close. The three treatments are as follows:

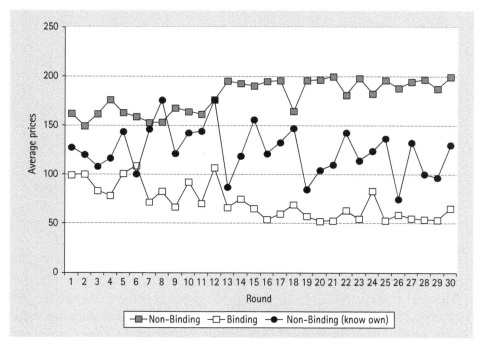

FIGURE 22.2. Average prices in the Katok and Wambach (2008) study.

1. *Binding*: all bidders know their own Q_i and Q_j of their competitor.
2. *Non-binding*: bidders do not know any Q_j.
3. *Non-binding (know own)*: bidders know their own Q_j but not their competitor's.

Figure 22.2 shows the average prices over time in the Katok and Wambach (2008) study. It is clear from the figure that after the initial thirteen rounds, bidders in the non-binding treatment learn to implicitly collude, driving prices essentially to the reserve level of 200. So giving bidders less information is not always better for the buyer, but the Katok and Wambach (2008) counterexample is quite extreme because their auctions have only two bidders (fewer players generally makes it easier to collude) and their cost is constant and known to all.

RANK-BASED FEEDBACK

Elmaghraby et al. (2012) examine the effect of price visibility, but unlike Haruvy and Katok (2013), who only examine the two extreme forms of price visibility, they look at the effect of rank-based feedback in buyer-determined auctions. With rank feedback, bidders in open-bid auctions are told the rank of their price bids, but they do not know the actual bids of their competitors. Elmaghraby et al. (2012) examine a simple setting with two quality types (high and low) and full information about the competitor's quality type. When a bidder is bidding against a competitor of the same type (called a *symmetric* market), he has the weakly dominant strategy of bidding down one bid decrement at a time as long as he is not at rank 1. When a bidder is bidding against a competitor of the opposite type (called an *asymmetric* market), in the simplified Elmaghraby et al. (2010) setting, he should simply bid down to the level of what his sealed bid would have been. Rank-based feedback is prevalent in practice. There is a perception that it leads to less adversarial relationships between buyers and suppliers (Jap, 2003, 2007), and suppliers prefer it because it reveals less information about their cost to their competitors. Buyers also prefer it because they believe that it leads to more competition.

All treatments include auctions with two bidders (N=2). There are two types of bidders, called *high* and *low* types). The quality for the high type is 200 and the quality for the low type is 100 ($Q^H = 200$, $Q^L = 100$). The costs of the two types come from different distributions: for the high type, $C_i^H \sim U(100, 200)$; and for the low type, $C_i^L \sim U(0, 100)$. The auctions are open bid with a one-minute duration and a ten-second soft close. The three treatments differ in their feedback: *full* feedback, *rank* feedback, and *sealed* bid.

Table 22.3 summarizes the average prices, bid decrements, and theoretical predictions in the Elmaghraby et al. (2012) study. Here we see again that sealed-bid prices are lower than open-bid prices. The consequence of this observation is that we observe the "sealed-bid effect" in asymmetric auctions with rank feedback, where average prices are very close to sealed-bid prices. Surprisingly, in symmetric auctions, average prices in auctions with rank feedback are lower than full feedback prices, which should not be the case, because symmetric bidders have the same dominant bidding strategy under both formats.

The explanation Elmaghraby et al. (2012) propose for overly aggressive bidding in symmetric auctions with rank feedback is bidder impatience. Even though bidders should be bidding down one bid decrement at a time, the average bid decrement in these auctions is 12.61. While significantly smaller than the average bid decrement of 21.86 in asymmetric auctions, it shows that jump bidding due to bidder impatience in these auctions is highly prevalent.

The Elmaghraby et al. (2012) results complement the findings of Isaac et al. (2005, 2007). Those authors study both real-world data from the wireless spectrum auctions (Isaac et al., 2007) as well as from the lab experiments (Isaac et al., 2005); they find that auction formats that allow for jump bids can help increase the bid-taker's revenue (or decrease procurement costs). In addition, their data suggest that jump bidding arises as a result of bidder impatience rather than effort by bidders to deter competition (signaling). Kwasnica and Katok (2007) report similar results in experiments in which bidder impatience was deliberately induced.[4]

---

[4] Impatience has also been used by Katok and Kwasnica (2008), who report that the speed of the clock in a "Dutch" (descending forward) auction affects the auction outcome, and reconcile the results with those reported by Cox et al. (1982) and Lucking-Reiley (1999).

Table 22.3. Summary of the average prices, bid decrements (standard deviations in parenthesis), and theoretical predictions (in square brackets) in the Elmaghraby et al. (2012) study

Treatment	Prices			Bid decrements	
	Overall	Symmetric	Asymmetric	Symmetric	Asymmetric
Full	67.11	69.03	65.65	6.35**	5.82**
	(3.06)	(3.93)	(3.07)	(1.26)	(1.14)
	[67.60]	[71.24]	[65.34]	[1]	[1]
Rank	58.73**	63.68**	55.66**	12.61**	21.86
	(3.59)	(3.03)	(4.41)	(3.59)	(4.65)
	[69.04]	[71.24]	[67.67]	[1]	[>>1]
Sealed bid	57.49*	58.21*	57.04*	N/A	N/A
	(2.20)	(2.86)	(1.93)		
	[68.26]	[69.20]	[67.67]		

Notes: Ho: Data = theoretical prediction; $*p < 0.10$, $**p < 0.05$.

Elmaghraby et al. (2012) conducted a robustness check of their results by running treatments in which bidders do not know the type of their competitor, as well as treatments in which the cost support of low- and high-type bidders overlaps. The results continue to hold.

QUALIFICATION SCREENING AND INCUMBENCY

Wan et al. (2012) focus on one particularly important quality attribute in buyer-determined auctions—incumbency status. In their model, an incumbent supplier competes against an entrant supplier whose probability of being able to meet the requirements to perform the contract (pass qualification screening) is $0 \le \beta < 1$. The buyer has a choice of: screening the entrant before the auction, and if he fails the qualification screening, which happens with probability $1 - \beta$, renewing the incumbent's contract at the current price of R; or waiting to screen him after the auction. In the latter case, the buyer will have to screen the entrant (which costs K to do) only in the event that the entrant wins the auction. But the auction between an incumbent and an entrant who may or may not be qualified is less competitive than an auction between two qualified bidders, because the incumbent may lose the auction but win the contract (with probability $1 - \beta$).

Whether the buyer is better off to screen the entrant before or after the auction is the central question that Wan et al. (2012) pose, and the answer hinges on the incumbent's bidding behavior when competing against an unscreened entrant (the entrant has the

weakly dominant strategy of bidding down to his cost). Wan et al. (2012) derive equilibrium bidding strategy for the (risk-neutral) incumbent supplier in this situation. In
equilibrium, a high-cost incumbent should bid the reserve, a very low-cost incumbent
should bid to win, and at intermediate cost levels incumbents should stop bidding at
some threshold above their costs.

Figure 22.3 shows the equilibrium bidding strategies for a risk-neutral incumbent
with cost $x_i \sim U(10, 110)$ bidding against an entrant with cost $e_i \sim U(0, 100)$. The graphs
represent the parameters in the Wan et al. (2012) experiment that sets β at 0.3 or 0.7,
and the cost of screening, K, at 2 or 20. The reserve price is set at 110 in all treatments.

When competing against an entrant who may or may not be qualified, the incumbent
often bids less aggressively than he would against a qualified competitor. Sometimes, the
incumbent may boycott the auction entirely (always places a bid of R), as should happen
when the qualification cost is low, and the entrant's probability of being qualified is also
low ($\beta = 0.3$, $K=2$ in the Wan et al. (2012) experiment).

The main lab finding is that in this dynamic auction, incumbents, bidding against
computerized entrants programmed to follow the weakly dominant strategy, bid with a
great deal of noise, and on average bid more aggressively than they should in equilibrium. Figure 22.4 summarizes the incumbent bidding data.

Each part of the figure displays behavior in one treatment. The top panel of each part
of the figure shows a scatterplot of actual bids as a function of x_i and compares them
to the equilibrium bid function. The bottom part of each panel shows the proportion
of bids (as a function of x_i) that are either boycotting ($Bid = R$) or bid all the way
down as low as needed to win the auction outright $Bid \leq \max(x_i, K)$. It turns out that
incumbents do not boycott enough when they should, and on average usually bid more
aggressively than they should (the "sealed-bid effect").

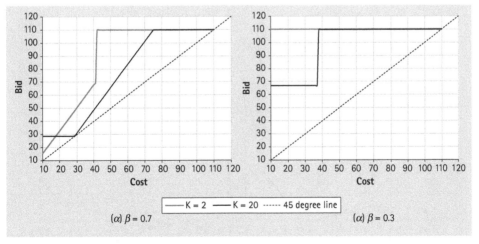

FIGURE 22.3. A risk-neutral incumbent's bidding functions when $K = 2$ and $K = 20$. Plots
assume $x_i \sim U[10, 110]$, $x_e \sim U[0, 100]$, and $\beta = 0.7$ (a) or $\beta = 0.3$ (b).

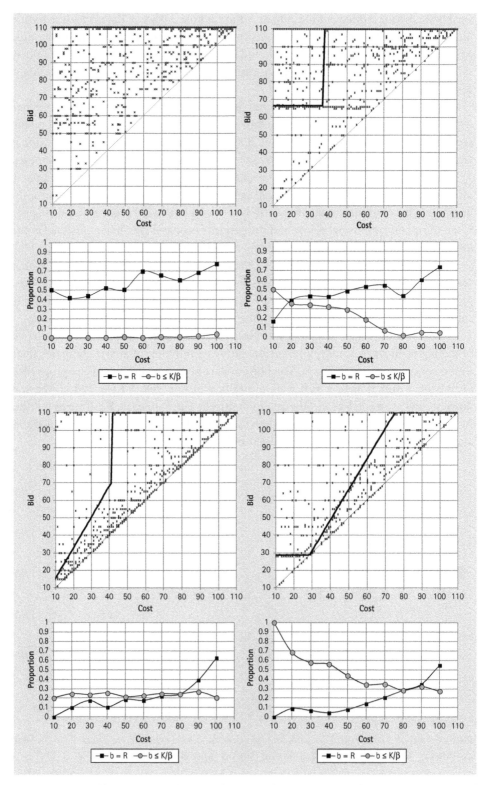

FIGURE 22.4. Bidding behavior: bids as a function of x_i, proportion of $Bid = R$, proportion of $Bid \geq \max(x_i, K)$.

As a result of the "sealed-bid effect," the buyer strategy of not evaluating the entrant supplier until after the auction is even more attractive than it should be in theory.

CONCLUSION

In summary, buyer-determined procurement auctions are a class of reverse auction that is prevalent in procurement. The bidding takes place during a dynamic auction, but winners do not know whether they are winning and losers do not know whether they are losing, or by how much. The resulting bidding behavior exhibits the "sealed-bid effect"— bidding is overly aggressive. Giving bidders less information appears to result in lower prices, unless there is so little information that bidders can profitably collude. Rank feedback results in lower prices than full price feedback, primarily because it promotes the sealed-bid effect, as well as bidder impatience.

REFERENCES

Beall, S., Carter, C., Carter, P. L., Germer, T., Hendrick, T., Jap, S. D., Kaufmann, L., Maciejewski, D., Monczka, R. and Petersen, K. (2003) "The role of reverse auctions in strategic sourcing," Center for Advanced Purchasing Studies (CAPS), Research Paper.

Branco, F. (1997) "The design of multi-dimensional auctions," *RAND Journal of Economics*, 28(1): 63–81.

Center for Strategic Supply Research (2006) "CAPS research: focus on eProcurement," April 19.

Che, Y.-K. (1993) "Design competition through multi-dimensional auctions," *RAND Journal of Economics*, 24(4): 668–80.

Cox, J. C., Roberson, B. and Smith, V. L. (1982) "Theory and behavior of single object auctions," in V. L. Smith (ed.), *Research in Experimental Economics*, JAI Press, pp. 1–43.

Elmaghraby, W. (2007) "Auctions within e-sourcing events," *Production and Operations Management*, 15(4): 409–22.

———— Katok, E. and Santamaria, N. (2012) "A laboratory investigation of rank feedback in procurement auctions," *Manufacturing and Services Operations Management*, 14(1): 128–44.

Engelbrecht-Wiggans, R. and Katok, E. (2008) "Regret and feedback information in first-price sealed-bid auctions," *Management Science*, 54(4): 808–19.

———— Haruvy, E. and Katok, E. (2007) "A comparison of buyer-determined and price-based multi-attribute mechanisms," *Marketing Science*, 26(5): 629–41.

Filiz-Ozbay, E. and Ozbay, E. Y. (2007) "Auctions with anticipated regret: theory and experiment," *American Economic Review*, 97(4): 1407–18.

Fischbacher, U. (2007) "z-Tree: Zurich toolbox for ready-made economic experiments," *Experimental Economics*, 10(2): 171–8.

Haruvy, E. and Katok, E. (2013) "Increasing revenue by decreasing information in procurement auctions," *Production and Operations Management*, 22(1): 19–35.

Isaac, R. M., Salmon, T. C. and Zillante, A. (2005) "An experimental test of alternative models of bidding in ascending auctions," *International Journal of Game Theory*, 33(2): 287–313.

———— ———— ———— (2007) "A theory of jump bidding in ascending auctions," *Journal of Economic Behavior and Organization*, 62(1): 144–14.

Jap, S. D. (2002) "Online reverse auctions: issues, themes and prospects for the future," *Journal of the Academy of Marketing Science*, 30(4): 506–25.

———— (2003) "An exploratory study of the introduction of online reverse auctions," *Journal of Marketing*, 67: 96–107.

———— (2007) "The impact of online reverse auction design on buyer–supplier relationships," *Journal of Marketing*, 71(1): 146–59.

Kagel, J. H. (1995) "Auctions: a survey of experimental research," in J. H. Kagel and A. E. Roth (eds), *The Handbook of Experimental Economics*, Princeton University Press, pp. 501–85.

Katok, E. and Kwasnica, A. M. (2008) "Time is money: the effect of clock speed on sellers revenue in Dutch auctions," *Experimental Economics*, 11(4): 344–57.

———— and Wambach, A. (2008) "Collusion in dynamic buyer-determined reverse auctions," Penn State Working Paper.

Kostamis, D., Beil, D. R. and Duenyas, I. (2009) "Total-cost procurement auctions: impact of suppliers' cost adjustments on auction format choice," *Management Science*, 55: 1985–99.

Kwasnica, A. M. and Katok, E. (2007) "The effect of timing on jump bidding in ascending auctions," *Production and Operations Management*, 16(4): 483–94.

Lucking-Reiley, D. (1999) "Using field experiments to test equivalence between auction formats: magic on the Internet," *American Economic Review*, 89(5): 1063–79.

Tonca, T. I., Wu, D. J. and Zhong, F. V. (2008) "An empirical analysis of price, quality, and incumbency in service procurement auctions," Georgia Tech Working Paper.

Shugan, S. M. (2005) "Marketing and designing transaction games," *Marketing Science*, 24(4): 525–30.

Wan, Z. and Beil, D. (2009) "RFQ auctions with supplier qualification screening," *Operations Research*, 57(4):, 934–49.

———— ———— and Katok, E. (2012) "When does it pay to delay supplier qualification? Theory and experiments," *Management Science*, 58(11): 2057–75.

Zhong, F. V. and Wu, D. J. (2009) "E-sourcing: impact of non-price attributes and bidding behavior," Georgia Tech Working Paper.

CHAPTER 23

....

THE INEFFICIENCY OF
SPLITTING THE BILL

....

URI GNEEZY, ERNAN HARUVY,
AND HADAS YAFE[1]

INTRODUCTION: A LESSON
IN INSTITUTION DESIGN

....

ROTH (2002) makes the case that experiments are a natural component of market design. This is because experiments allow designers to isolate particular designs and identify them as causes for any observed market effects.

The work we present here fits in a literature on the effect of economic incentives in markets, and in particular on the little-understood impact that economic incentives have on the perception of market participants regarding implicit social contracts. Economic incentives change the information that agents have on the environment, and therefore their effect on behavior may be the opposite of what would be expected. For example, Gneezy and Rustichini (2000) showed that in tasks ranging from answering test questions to volunteer work, monetary incentives produced improvements in the predicted direction but not monotonically. Incentives that were too small produced a decrease in effort relative to no payment at all. Thus, small incentives may have the opposite effect to what is intended.

In contrast to the reduction in effort resulting from small incentives, Charness and Gneezy (2009) showed that incentives to attend a gym will not only increase attendance but also lead to healthy habit formation in the longer term. The difference between this

[1] We thank Richard Thaler, Mark Walker, anonymous referees, the editor and seminar participants for comments. this chapter is reprinted from Gneezy, U., Haruvy, E. and Yafe, H. (2004) "The inefficiency of splitting the bill," *Economic Journal*, 114: 265–80, with permission from Wiley.

and the Gneezy and Rustichini (2000) result is striking, and shows the importance of understanding the context in which incentives are used.

A similar point can be made using the findings regarding the ability of individuals to adapt to incentive shocks. Theories like the fair-wage–effort theory and the theory of gift exchange in labor markets (Akerlof, 1982; Fehr et al., 1993) have been shown to be important in many experimental designs. Testing this theory in a field experiment using tasks ranging from library data entry to door-to-door fundraising, Gneezy and List (2006) provided wages that were roughly twice the wages advertised for the positions. They found that while initially these incentives resulted in a significant increase in effort, after a few hours the observed outcomes were indistinguishable. That is, the effect is short lasting. These results have implications for the applicability of these theories to labor market design.

The present study fits in this series on economic incentives. Like the cited studies, we find that economic incentives matter, but perhaps not in the direction one would expect. We study a restaurant setting in which groups of diners are faced with different ways of paying the bill. The manipulations are: splitting the bill, paying individually, having a portion of the bill picked up, and having the entire bill paid for. We find that people eat more food and spend more money when they pay a smaller portion of their bill. That is, they take advantage of others. This result is in line with economic theory, but in contrast to a body of laboratory evidence suggesting that individuals in more abstract laboratory environments would not free-ride to the same extent. Indeed, when we created an abstract version of this experiment and presented it in the lab to subjects from the same subject pool, we found results that were in line with the published literature. Individuals were sensitive to incentives but reluctant to free-ride at the expense of other participants. This might seem surprising, because the lab interaction was anonymous whereas the restaurant interaction was face to face, in a social setting, and the individuals were unaware that the bill set-up was part of an experiment.

This result goes counter to an implicit social contract. In other words, non-academic readers were often surprised that decision makers behaved according to selfish economic principles, ignoring the effect of negative externalities. As always, the implications of this study should be taken in the larger context of the effect of incentives on behavior, and the interaction between incentives and implicit social contracts.

Economic theory is unambiguous in its prediction that if externalities exist, outcomes are likely to be inefficient when agents selfishly maximize. The literature on externalities, as well as its derivatives in public goods, tragedy of the commons, and moral hazard studies, has shown that externalities lead to inefficient levels of production and consumption. This result depends crucially on the general assumption taken by such studies that human agents maximize selfish payoffs without regard for others.

With the emergence of behavioral economics, economists have come to question whether people actually ignore costs imposed on others when reaching economic decisions. If altruism is common, the various proposals in the literature to solve externality problems may be unnecessary or even harmful. For example, the government in a public-good setting may actually reduce voluntary contributions by interfering with the provision of a public good (Andreoni, 1993). Similarly, increased government monitoring for corruption may backfire by reducing intrinsic other-regarding behavior (Bohnet et al., 2001; Schulze and Frank, 2003), and the mere sanctioning of an activity may be counterproductive (Gneezy and Rustichini, 2001).

Experimental studies, with few exceptions, find evidence against theories based purely on selfish motives. The studies find that people free-ride, but not to the extent economic theory predicts (see Dawes and Thaler, 1988). Hence, despite the strong predictions generated by classical theory in externality settings, social scientists often question the truths provided by it.

To test economic predictions, we investigate a familiar environment. The *unscrupulous diner's dilemma* is a problem faced frequently in social settings. When a group of diners jointly enjoys a meal at a restaurant, often an unspoken agreement exists to divide the check equally. A selfish diner could thereby enjoy exceptional dinners at bargain prices. Whereas a naive approach would appear to suggest that this problem is not likely to be severe, it appears that even the best of friends can sometimes find it rather trying.[2] Furthermore, this dilemma typifies a class of serious social problems, from environmental protection and resource conservation to eliciting charity donations and slowing arms races (Glance and Huberman, 1994).

Here, we observe and manipulate conditions for several groups of six diners at a popular dining establishment. In one treatment the diners pay individually; in a second treatment they split the bill evenly between the six group members. In yet a third treatment, the meal is paid for entirely by the experimenter. Economic theory prescribes that consumption will be smallest when the payment is individually made, and largest when the meal is free, with the even-split treatment between the other two. The restaurant findings are consistent with these predictions. A fourth treatment, in which each participant pays only one-sixth of her own consumption costs and the experimenter pays the remainder, is introduced to control for possible unselfish and social considerations. The marginal cost imposed on the participants in this treatment is the same as in the even-split treatment. However, the externalities are removed: in the even-split case, increasing an individual's consumption by \$1 increases the individual's cost, as well as the cost of each of the other participants, by \$1/6. In the fourth treatment, this will increase only the individual's cost by \$1/6, but will have no effect on the payment of the other participants. In other words, the negative externality present in the even-split treatment is completely eliminated. If participants are completely selfish, the fourth treatment should not affect their consumption relative to the second treatment (the even split). On the other hand, if they care also for the well-being of the other participants (or for social efficiency), they can be expected to consume more in the last treatment than in the even-split treatment.

The efficiency implication of the different payment methods is straightforward. When splitting the bill, diners consume such that the marginal social cost they impose is larger than their own marginal utility, and as a result they over-consume relative to the social optimum. In fact, it is easy to show that the only efficient payment rule is the individual

---

[2] "Ross: ... plus tip, divided by six. Ok, everyone owes 28 bucks. Phoebe: No, uh uh, no way, I'm sorry, not gonna happen." (*Friends*, season 2, episode 5).

one. It turns out that subjects' preferences are consistent with increasing efficiency. When asked to choose, prior to ordering, whether to split the bill or pay individually, 80% choose the latter. That is, they prefer the environment without the externalities. However, in the presence of externalities, they nevertheless take advantage of others.

One example of an environment in which the selfishness hypothesis has been studied is public-goods games; for comprehensive reviews, see Davis and Holt (1993); and Ledyard (1995). Public-goods experiments in which non-contribution is a dominant strategy typically find that subjects are sensitive to free-riding incentives but nonetheless cooperate at a level that cannot be fully explained by mainstream economic theory.[3] However, as the typical public-goods game is repeated (regardless of whether opponents are the same or different), contributions fall substantially (Kim and Walker, 1984; Isaac and Walker, 1988; Andreoni, 1988; Asch et al., 1993; Weimann, 1994). In all these studies, subjects contribute less and less the longer they play. In other words, it seems that subjects may be contributing in part due to inexperience or confusion under lab conditions. Kim and Walker (1984) reviewed previous experiments that found little or no free-riding. They raised serious concerns about lab experiments, among which were misunderstanding and vagueness as well as insufficient economic incentives. In an experiment designed to overcome the criticisms raised, they indeed found that selfish behavior was in fact prevalent after only a few repetitions. Andreoni (1995) raised similar criticisms, which he labeled collectively "confusion." In order to explore this issue, he designed a zero-sum version of the public-goods game, in which the sets of strategies and corresponding token payoffs were the same as the public-goods game, but where token payoffs were mapped to monetary payoffs by the earnings ranks of the subjects. This mapping eliminated the monetary incentive to cooperate, and indeed cooperation dropped significantly, but not entirely. Andreoni concluded that "on average about 75 percent of the subjects are cooperative, and about half of these are confused about incentives, while about half understand free-riding but choose to cooperate out of some form of kindness" (Andreoni, 1995, p. 900).

The traditional lab environment could present some limitations when extrapolating to real-life settings. Such limitations may result from participants' lack of familiarity with the lab setting. It could be argued that subjects should be observed in settings with which they are familiar and experienced. For example, in a field experiment conducted during the orange-picking season in Israel (Erev et al., 1993), with different groups of four workers facing different payment schemes, it was found that, in line with the theoretical prediction, a collective payment resulted in substantial free-riding and 30% loss in production.

The current study proposes the restaurant setting as one with which subjects are expected to be familiar, thereby reducing the possibility of confusion. The idea of studying human economic behavior in a restaurant setting is not new. In a study discussed by Thaler (1980), consumers at an all-you-can-eat pizza restaurant were randomly given free lunches. These consumers ate less than the control group, who paid the $2.50

---

[3] In fact, even in public-goods games where some positive contribution is best response, subjects tend to substantially over-contribute relative to their best response (Keser, 1996).

normal bill. The main conclusion of that study was that, unlike the prescription of economic theory, people do not ignore sunk costs.

The paper is organized as follows. The next section sketches the theory as it pertains to the diner's dilemma and derives the appropriate hypotheses implied by the theory. The third section details the design and procedures for the restaurant setting. The fourth section lists and explains the results, and investigates possible implications of gender issues. The fifth section presents a related laboratory experiment. The sixth section concludes.

Theory

In this section we first introduce the mainstream assumptions and the resulting social inefficiency under the even-split and free-meal treatments. We then posit the hypotheses implied by the theory.

Mainstream assumptions

According to standard economic assumptions, consumers will find it optimal to increase consumption when marginal benefit exceeds marginal cost, and to lower consumption when the opposite holds. Therefore, at the utility-maximizing consumption level, marginal cost must equal marginal benefit. It is also a standard assumption that the marginal utility is decreasing (clearly the marginal utility reaches zero at some point, or else consumers in the free-meal treatment would consume at a level of infinity). Given these standard assumptions, economic theory predicts a negative relation between the marginal cost of the food and its consumption.

If the individuals do not internalize the negative externalities they impose on others, they will over-consume relative to the social best in all but the individual-pay treatment. In particular, if the six diners elected a social planner to "dictate" the allocations in the even-split treatment, this planner would be able to increase the value received by each diner. Similarly, under the free-meal treatment, the party financing the dinner could pay the diners to consume at individual-pay levels, such that all diners as well as the paying party would be better off. It follows that the individual-pay outcome is a Pareto improvement relative to the other treatments.

Hypotheses

Several hypotheses emerge from the theory of selfish utility-maximizing consumers. Since marginal benefit must equal marginal cost, it will take greater and greater consumption to equate marginal benefit to marginal cost, as we move from individual pay to even split and from even split to free meal. The following three hypotheses emerge:

- *Hypothesis 1.* Diners will eat more in the even-split treatment relative to the individual-pay treatment.
- *Hypothesis 2.* Diners will eat more in the free-meal treatment relative to the even-split treatment.
- *Hypothesis 3.* Diners will eat more in the free-meal treatment relative to the individual-pay treatment.

DESIGN OF THE RESTAURANT SETTING

Subjects were recruited through signs posted around the Technion campus (Israel Institute of Technology, Haifa), which promised a large amount for a one-hour experiment and invited them to call for information about the experiment. Upon calling, they were informed that the experiment would be conducted at a popular restaurant near the Technion campus. They were asked to show up at a specific time (during lunchtime). Six subjects, three males and three females, were invited for each time slot. A conscious effort was made not to invite to the same treatment students who were familiar with each other. Upon arrival, subjects received a show-up payment of 80 NIS (New Israeli Shekels) (roughly $20 at the time of the study) and brief instructions (see the Appendix A for the translation of the instructions from Hebrew). They were cautioned to maintain absolute silence for ten minutes, during which all participants were asked to complete the questionnaires in front of them. The questionnaires requested subjects to rate themselves on a wide range of emotions. They were told to expect the same questionnaire at the end of the meal.

In the instructions for the questionnaire, subjects were informed that they would be able to order from the restaurant menu following the completion of the questionnaire. They were asked to indicate their orders on a designated sheet of paper. Subjects wrote down their orders individually and separately, without any ability to communicate or coordinate with other participants. The intent of the questionnaire was to ensure the independence of observations, and the questionnaire was effective in keeping subjects silent.

Treatments differed only in the payment mechanism specified in the instructions: In the individual-pay treatment, subjects were told that they would pay for their own meal. In the even-split treatment, subjects were told that the bill would be evenly split between the six of them. In the free-meal treatment, subjects were told that the meal would be fully paid for by the experimenter. Four groups of six subjects participated in the individual-pay treatment, and four groups participated in the even-split treatment.

Two groups participated in the free-meal treatment. Two of the groups in each of the first two treatments were asked how they would prefer to pay—individually or by splitting the bill—prior to being informed of the actual payment mechanism.

The menu covered a broad international cuisine, with numerous delectable categories to encompass a wide range of tastes. Waiters were instructed not to communicate or otherwise interact with subjects before picking up the order sheet. That is, subjects had contact only with the experimenters before they ordered. The same two experimenters attended all treatments.

RESULTS

The three main treatments

Table 23.1 summarizes the results of the field study. For each of the three treatments, the first column reports the gender of the subject, the second column reports the number of items that subject ordered, and the third column reports the cost of the subject's meal. Subjects are ordered by the cost of their meals from highest to lowest.

Note the variability in subjects' costs for any given treatment. For example, the difference in cost between the least expensive subject and the most expensive subject in the free-meal treatment was 119 NIS. Normally, such heterogeneity could pose a problem for hypothesis testing. However, despite this enormous variability there was a fairly small overlap in meal costs between treatments. Treatment 3, for example, has only two observations out of twelve that fall below the highest observation of twenty-four observations in treatment 1. This surprisingly small overlap is clearly depicted in Figure 23.1.

The x-axis lists three values, corresponding to the three treatments. The y-axis represents the cost of the meals. Each point in the plot represents the meal cost in NIS for a particular subject in one of the three treatments. Different subjects are represented by different symbols. All in all, there are twenty-four values for each of the first two treatments and twelve for the third.

Recall hypotheses 1–3. These postulated differences between individual pay and even split, between even split and free meal, and between individual pay and free meal. To test hypotheses 1–3, we use two competing tests to determine if there are any reliable differences between any two independent groups—the parametric t-test and the nonparametric Mann–Whitney U-test. We find the "number of items ordered" not informative in the Mann–Whitney U-test due to the large number of ties. We expect the t-test similarly to produce a rough statistic at best, since it can hardly be assumed that the meal costs will be normally distributed, as required by the t-test. We nonetheless report p-values for this test. The cost of the meals, however, provided clear-cut evidence that the samples are significantly different under all three hypotheses. Table 23.2 shows

Table 23.1. Summary of the restaurant results

Subject	Individual pay			Even split			Free meal		
	Sex	No. of items	Cost (NIS)	Sex	No. of items	Cost (NIS)	Sex	No. of items	Cost (NIS)
1	F	2	59	F	2	81	M	4	168
2	M	2	54	M	2	73	M	5	123
3	M	2	50	M	2	71	M	3	101
4	M	2	49	F	1	66	F	4	94
5	M	2	47	F	2	64	M	3	81
6	F	1	46	F	3	62	M	3	75
7	F	2	45	F	2	60	F	3	69
8	M	2	45	M	2	59	F	2	61
9	F	2	43	F	2	59	F	2	57
10	M	2	43	M	3	56	M	3	59
11	M	2	40	M	2	52	F	2	51
12	F	2	40	F	2	47	F	2	49
13	F	2	39	M	2	46			
14	M	2	39	F	2	46			
15	F	1	35	M	2	45			
16	F	1	35	M	2	45			
17	M	1	35	M	2	44			
18	F	2	31	M	2	40			
19	F	2	31	F	2	40			
20	F	2	30	F	1	39			
21	M	1	16	F	1	37			
22	M	1	16	F	1	35			
23	F	1	15	M	2	33			
24	M	1	12	F	1	22			
Average		1.67	37.3		1.87	50.9		3	82.3

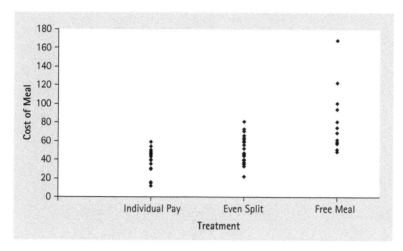

FIGURE 23.1. Summary of the restaurant results.

p-values for the three hypotheses, using the Mann–Whitney test on cost (column 1), the t-test on cost (column 2), and the t-test on number of items (column 3).

Finally, we use a graphical depiction of the population differences for the first three treatments to emphasize these results, using cumulative distribution plots in Figure 23.2.

We see from the plots that in the individual-pay treatment costs tend to be substantially lower in all percentiles of the distribution relative to the even-split and free-meal treatments. Similarly, the even-split treatment costs tend to be substantially lower in all percentiles of the distribution relative to the free-meal treatment.

Table 23.2. Hypothesis tests on the restaurant results: p-values for hypotheses 1–3

	Mann–Whitney U-test (one-sided) on cost of meal	Mann–Whitney U-test (one-sided) on number of items*	t-test (one-tailed) on cost of meal	t-test (one-tailed) on number of items ordered
Individual pay versus even split	0.0014	0.0948	<0.0001	0.0818
Individual pay versus free	<0.0001	<0.0001	<0.0001	<0.0001
Even split versus free	0.0008	<0.0001	0.0003	<0.0001

*The Mann–Whitney results on the number of items ordered may be unreliable due to the large number of ties.

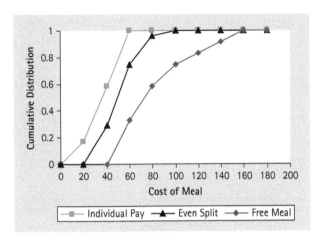

FIGURE 23.2. Cumulative distribution graphs for treatments 1–3 of the restaurant experiment.

Regard for others in the restaurant setting

Though it is clear from these results that an individual diner appears not fully to account for the cost her consumption imposes on her peers, the even-split treatment nonetheless leads to some other questions. In particular, does the individual ignore all of the cost she imposes on others, or does she account for some? The literature seems to present many approaches to answering questions of this kind. For example, we could suppose that the individual does not fully exploit her ability to consume at others' expense, since her utility is increasing in the consumption of others (for a review of the altruism and fairness literature, see Camerer, 2003).

The fourth restaurant treatment is introduced to examine the proposal that selfish considerations may not offer the best description of human agents. In that treatment, two groups of six diners, three males and three females, were recruited by signs around the Technion campus. The groups were summoned to the same restaurant used in the other treatments. Also, as in the other treatments, the groups were balanced between men and women. Unlike the other treatments, the instructions specified to the subjects that at the end of meal each would be asked to pay individually only one-sixth of his individual meal cost (see Appendix A for the translation of the instructions from Hebrew).

By the "selfish agent assumption," this treatment should not differ from the even-split treatment. The theories of altruism, equity, and reciprocity, however, would appear to suggest that agents are likely to consume more in this treatment than in the even-split treatment. This is because no costs are imposed on others in this "pay one-sixth" treatment and hence regard for others does not play a role in this treatment,[4] whereas

---

[4] This is assuming of course that regard for the experimenter is weaker than that for peers around the table. However, if we rely on the results reported thus far in the chapter, we should not be concerned about this possibility: in the free-meal treatment, subjects over-consumed relative to both other

Table 23.3. Summary of the results relevant to hypothesis 4

Even split			Pay one-sixth		
Sex	No. of items	Cost (NIS)	Sex	No. of items	Cost (NIS)
F	2	81	F	3	101
M	2	73	M	3	85
M	2	71	M	3	74
F	1	66	M	2	59
F	2	64	M	2	58
F	3	62	F	2	57
F	2	60	F	2	57
M	2	59	M	2	51
F	2	59	F	2	50
M	3	56	F	2	46
M	2	52	F	1	26
F	2	47	M	1	25
M	2	46			
F	2	46			
M	2	45			
M	2	45			
M	2	44			
M	2	40			
M	2	40			
F	1	39			
F	1	37			
F	1	35			
M	2	33			
F	1	22			
Avg.	1.87	50.9	Avg.	2.08	57.4

regard for others is expected to play some role in the even-split treatment. In other words, positive regard for others will raise the marginal cost of a meal, and will therefore lower the optimal spending under the assumption of decreasing marginal benefit. It is important to note that whereas altruism or utilitarian motives are unequivocal in this prediction, distributive and reciprocity concerns depend crucially on expectations, and could suggest predictions in either direction. We derive hypothesis 4:

- *Hypothesis 4.* Diners will exhibit the same levels of consumption in the even-split treatment and the "pay on-sixth" treatment.

treatments by a phenomenal amount. Hence it seems that any concern for the experimenter's welfare is miniscule at best.

Table 23.4. The *p*-values for hypothesis 4

	Mann–Whitney *U*-test (one sided) on cost of meal	Mann–Whitney *U*-test (one sided) on number of items ordered	*t*-test (one-tailed) on cost of meal	*t*-test (one-tailed) on number of items ordered
Even-split versus pay one-sixth	0.187	0.167	0.145	0.159

The last hypothesis addressed unselfish motives and postulated a difference between the even-split and the "pay one-sixth" treatments. Table 23.3 presents the relevant experimental results and Table 23.4 shows the *p*-values for this hypothesis, from the Mann–Whitney test on cost (column 1), the Mann–Whitney test on the number of items ordered (column 2), the *t*-test on cost (column 3), and the *t*-test on the number of items ordered (column 4).

Looking at both the cost of meals and at the number of items ordered, we find no significant differences between the even-split and the pay one-sixth treatments. This finding can be contrasted with the very significant differences between the even-split treatment and all other treatments we reported earlier. We should caution, however, that the lack of significance could be due to the smaller sample size of the pay one-sixth group.

Gender differences

Studies have shown that males and females have different propensities in relation to regard for others. In experiments, we find results in dictator games (e.g. Eckel and Grossman, 1998) and reward allocation games (e.g. Lane and Messe, 1971), which show more generosity in females than in males.[5] Such results would lead us to expect that women will not exploit the ability to impose cost on others to the same extent as men.

Another plausible gender difference has to do with different physical limitations as well as discriminatory cultural norms related to eating. Both physical capacity and discriminatory social norms would lead one to expect women to have a lower ceiling on food consumption. Hence we would expect women to under-utilize the ability to impose costs on others.

Surprisingly (or not, depending on one's prior assumptions), as Table 23.5 clearly shows, men and women did not differ in their consumption levels in three out of the

---

[5] There are, though, also studies that reject claims of gender differences (e.g. Bolton and Katok, 1995).

Table 23.5. The *t*-test *p*-values (two-tailed) for the null hypothesis of no gender effects in the restaurant results

	Gender effect
Individual pay	0.9623
Even split	0.8470
Free meal	0.0568
Pay one-sixth	0.8530

four treatments, under any reasonably acceptable level of significance. In the free-meal treatment, however, men tended to eat more than women in a manner (marginally) significant at the 5% level. However, given the lack of difference in the other three treatments, we tend to discount this finding.

A LABORATORY COMPARISON

Cross-country studies (e.g. Roth et al., 1991) have raised the possibility that subject pools in different countries may not share the same distribution of other-regarding preferences. Such cultural differences could affect the comparability of the present subject pool to other subject pools in the world. To exclude the possibility that our restaurant finding is driven by an odd subject pool, we briefly present the results of a simple negative externality experiment conducted in the lab with the same subject pool (Technion students) as in the restaurant study.

The laboratory setting

Subjects were recruited through signs around the Technion, as in the previous setting. Instead of meeting at a restaurant, however, subjects were summoned to the laboratory. The show-up fee was 80 NIS, the same as for the restaurant setting. As before, there were six subjects per session. All subjects were in the same room and could see each other. Subjects were shown a "production" table detailing the cost and revenue resulting from each production quantity, where the quantity of production could vary from 1 to 9. The production table is presented in Table 23.6.

Three different treatments were run. In the first treatment, subjects bore the full cost and reaped the full revenue from each unit of production. In the second treatment,

Table 23.6. Production table for a laboratory examination of the diner's dilemma

No. of units	Total cost	Total revenue
1	10	18
2	20	32
3	30	44
4	40	48
5	50	50
6	60	51
7	70	51.75
8	80	51.75
9	90	52

subjects reaped the full revenue, but the total cost of production was added up over subjects and then divided equally among them. In the third treatment, subjects incurred only one-sixth of the full cost of their production but the remainder was not imposed on any of the participants; instead, it just vanished.

The parallel between this production problem and the unscrupulous diner's dilemma in the restaurant setting is evident. Namely, the first production treatment corresponds to the restaurant's individual-pay treatment, the second production treatment corresponds to the restaurant's even-split treatment, and the third production treatment corresponds to the restaurant's pay one-sixth treatment.

Results

The full list of the thirty-six participants' choices is presented in Table 23.7.

In contrast to the restaurant results, the difference between the individual-pay and even-split treatments is not significant[6] (one-tailed p-value = 0.23), whereas the difference between the even-split and pay one-sixth treatments is highly significant (one-tailed p-value = 0.0007). This evidence of unselfish motives might lead to the conclusion that the bill-splitting convention would not result in any significant social detriment. This conclusion does not appear consistent with what was observed in the restaurant.

[6] It is interesting to note, however, that in the even-split treatment two subjects selected a quantity of two units, which is below the socially efficient level of production. Errors by subjects in the lab are not uncommon, nor are dominated choices unusual (as abundant evidence from second-price sealed-bid auctions shows). However, it is possible that the added level of complexity in understanding the even-split mechanism resulted in a higher chance of errors.

Table 23.7. Results of the laboratory experiment

Observation	Individual pay	Even split	Pay one-sixth
1	3	2	3
2	3	2	4
3	3	3	4
4	3	3	5
5	3	3	5
6	3	3	5
7	3	3	5
8	3	4	5
9	3	4	5
10	3	4	5
11	3	5	5
12	5	5	5
Average	3.17	3.42	4.67

This finding shows that the results typically reported in the experimental literature are easily replicated with the subject pool in the restaurant study. Clearly, in terms of design, many things are different between the restaurant study and the lab experiment.

Conclusions

The literature on negative externalities is based on the prediction that an economic agent who is able to impose some of the cost of his consumption on others will over-consume relative to the socially efficient level. This is a direct result of the assumption that economic agents equate individual marginal costs and marginal benefits with complete disregard for the costs imposed on others or for social efficiency. However, there is an emergent volume of evidence that places in doubt some of the assumptions of classical economic theory. Such studies often demonstrate that small groups in the laboratory are likely to secure voluntary cooperation.

The diner's dilemma gives us an opportunity to test this prediction in an environment close to real life. We find that the theoretical predictions work: people react to changes in incentives, and they seem to largely ignore negative externalities. These results have great importance in the design of institutions. Institutions and rules that ignore the effect of negative externalities are inefficient—not only in theory, but also in practice. This inefficiency is the result of people playing the equilibrium of the game, even if they all prefer to be in a "different game" (e.g. pay the bill individually). Interestingly, when asked which mechanism they would prefer, prior to informing them which mechanism they would face, nineteen out of the twenty-four subjects (80%) we asked indicated they would prefer individual pay over splitting the bill. However, when forced to play according to the less preferred set of rules (splitting the bill), subjects nevertheless minimized their losses by taking advantage of others.

Given the clear preferences of the diners in our study, we are left wondering why we ever observe splitting of the bill in restaurants and, more importantly, in economic institutions. We begin with the restaurant setting. Unlike our experiment, groups of diners eating together are generally not perfect strangers but rather friends or colleagues. Likewise, the custom of splitting the bill is generally prevalent among friends or colleagues, and not among strangers. This difference is critical, since with friends and colleagues the game is repeated, so punishment strategies in response to excessive waste are feasible. Nevertheless, one would expect some waste to result, even among friends, since monitoring and punishment are imperfect. In that case, is there any reason why one would prefer to split the bill?

Some cost is involved in paying individually. A part of it could be the mental cost of figuring out one's share of the bill, and calculating the portion of the tax and tip that apply to that share. Another part would be the social cost of appearing stingy or unfriendly. Given the cost of individually paying, and the ability to reduce the inefficiency of splitting the bill through repeated game strategies, it may in fact be individually and socially optimal to split the bill among friends. However, the danger in customs which are based on rational decision making is that once they become conventions they are resistant to change, even when circumstances change. For example, when you find yourself dining with distant acquaintances you are not likely to encounter any time soon, it may nevertheless be rude in some settings (e.g. conferences) to suggest paying individually.

An argument of socially inefficient conventions could be made for larger and less personal economic institutions. For example, until the 20th century, allocating fishing rights in coastal waters would have been a socially inefficient proposition. However, years of convention have produced the shibboleth of "freedom of the seas" advocated by maritime nations, which is most certainly socially inefficient, with large-scale fishing methods and inexhaustible demand from a growing human population. Similarly, the practice of common grazing areas in 14th-century English villages quickly became unsustainable once populations started growing. The practice of the commons is in fact not much different from the diner's dilemma. Though individual incentives for excess exist, in small communities the social mechanisms arising from repeated interaction and strong other-regarding preferences are in place to discourage excess consumption. Once these social mechanisms are eliminated, the tragedy of the commons results.

Finally, small groups in the laboratory, including the laboratory experiment presented here, have been shown to arrive closer to the socially efficient level than models of selfish behavior would. Given this common result, other-regarding preferences in many instances have been argued to be critical motives in decision making. Though the findings in the restaurant setting cannot preclude other-regarding considerations, they provide evidence in favor of other possible explanations for the results generally obtained in the laboratory. Such explanations include the concern of Kim and Walker (1984) that misunderstanding of the unfamiliar task could result in cooperation, and the concern of Andreoni (1995) that some cooperation could be due to confusion and lack of experience with the task. In contrast to unfamiliar laboratory tasks, the restaurant is a familiar setting, and ordering at a restaurant (as well as splitting the bill at a restaurant) is a familiar task. Another possibility is that the difference may be driven by the perception of the subjects regarding the task. In laboratory experiments, the subjects may perceive that they were brought to the lab in order to test their attitudes toward public goods, fairness, etc. This perception is less likely in our restaurant setting, where subjects may behave in a somewhat more natural manner. Though these explanations and others remain to be studied, we hope this study has provided food for thought.

Appendix A. Participant instruction sheet for the restaurant setting

Welcome to "Globes" Restaurant.

This experiment looks at emotions before and after eating. You therefore will be asked to eat.

Within the next ten minutes you must perform two tasks:

1. Fill out the questionnaire in front of you honestly and accurately.
2. Check the menu and write down your order on the empty sheet attached to the questionnaire. You will not have another opportunity to order. At the end of ten minutes, the waitress will pick up your order.

It is imperative that you **remain silent**. That is, do not communicate with the other participants at the table.

Following the ten minutes, before the meal, you will receive 80 NIS for your participation in the experiment.

[Treatment 1] At the end of the experiment you will receive a bill *for the food you order*. You will then have to pay the waitress. After that, you will be asked to fill out the same questionnaire.

[Treatment 2] At the end of the experiment you will receive a *bill for one-sixth of the entire bill of all participants at the table*. You will then have to pay the waitress. After that, you will be asked to fill out the same questionnaire.

[Treatment 3] At the end of the experiment you will be asked to fill out the same questionnaire. *You do not have to pay the bill. The meal is on us!*

[Treatment 4] At the end of the experiment you will receive a *bill for one-sixth of the cost of your individual order, which you will then have to pay the waitress*. After that, you will be asked to fill out the same questionnaire.

Bon appetit!

APPENDIX B. PARTICIPANT INSTRUCTION SHEET FOR THE LABORATORY SETTING

Welcome. This is an experiment in decision making. You will receive 80 NIS for showing up to the experiment, plus any amount that you earn in the course of the experiment. In the next ten minutes we ask that you read the instructions and make your choice of number of units to purchase. This is the only decision you will have to make in the experiment. You have only one chance to make a choice, after which the experiment ends. Hence, it is crucial that you make your choice carefully. If you have any questions, please raise your hand but do not exclaim out loud. We expect and appreciate your cooperation.

[Treatment 1] Your choice is in terms of quantity, or *number of units*, you wish to purchase. At the end of the experiment, we will pay you according to how many units you have purchased, but we will also charge you the *cost* for these units. So the earnings you take home at the end of the experiment, in addition to the show-up fee, are your revenue from the units you bought, minus the cost of the units you bought. The table below [Table 23B.1] specifies the revenue and cost from each quantity you choose. The amount you earn in this experiment is independent of the choices and earnings of other participants.

[Treatment 2] Your choice is in terms of quantity, or *number of units*, you wish to purchase. At the end of the experiment, we will pay you according to how many units have purchased, but we will also charge you the *cost* for these units as follows: The total cost of the quantity you choose will be added to the total costs of others' choices (there are five others in your group). You will then be asked to pay one-sixth of the total cost of everybody in your group. However, your revenue will be only the *revenue* corresponding

Table 23B.1. The individual revenue and cost from each quantity in the laboratory experiment

No. of units	Total cost	Total revenue
1	10	18
2	20	32
3	30	44
4	40	48
5	50	50
6	60	51
7	70	51.75
8	80	51.75
9	90	52

to your individual choice. So the earnings you take home at the end of the experiment, in addition to the show-up fee, are your individual revenue from the units you bought, minus one-sixth of the cost of the units *everybody in your group* bought. The table below [Table 23B.1] specifies the individual revenue and cost from each quantity you choose.

[Treatment 3] Your choice is in terms of quantity, or *number of units*, you wish to purchase. At the end of the experiment, we will pay you according to how many units have purchased, but we will also charge you one-sixth of the *cost*. So the earnings you take home at the end of the experiment, in addition to the show-up fee, are your individual revenue from the units you bought, minus the one-sixth of the cost of the units. The table below [Table 23B.1] specifies the individual revenue and cost from each quantity you choose. The amount you earn in this experiment is independent of the choices and earnings of other participants.

I choose to get a quantity of _____ units.

REFERENCES

Akerlof, G. A. (1982) "Labor contracts as partial gift exchange," *Quarterly Journal of Economics*, 97: 543–69.

Andreoni, J. (1988) "Why free ride? Strategies and learning in public goods experiments," *Journal of Public Economics*, 37(3): 291–304.

—— (1993) "An experimental test of the public-goods crowding-out hypothesis," *American Economic Review*, 83(5): 1317–27.

—— (1995) "Cooperation in public goods experiments: kindness or confusion?" *American Economic Review*, 85(4): 891–904.

Asch, P., Gigliotti, G. A. and Polito, J. (1993) "Free riding with discrete and continuous public goods: some experimental evidence," *Public Choice*, 77(2): 293–305.

Bohnet, I., Frey, B. and Huck, S. (2001) "More order with less law: on contract enforcement, trust and crowding," *American Political Science Review*, 95(1): 131–44.

Bolton G. and Katok. E. (1995) "An experimental test for gender differences in beneficent behavior," *Economics Letters*, 48: 287–92.

Camerer, C. (2003) *Behavioral Game Theory: Experiments in Strategic Interaction*, Princeton University Press.

Charness, G. and Gneezy, U. (2009) "Incentives to exercise," *Econometrica*, 77(3): 909–31.

Davis, D. and Holt, C. (1993) *Experimental Economics*, Princeton University Press.

Dawes, R. M. and Thaler, R. (1988) "Anomalies: cooperation," *Journal of Economic Perspectives*, 2(3): 187–97.

Eckel, C. and Grossman, P. (1998) "Are women less selfish than men? Evidence from dictator experiments," *Economic Journal*, 108: 726–35.

Erev, I., Bornstein, G. and Galili, R. (1993) "Constructive intergroup competition as a solution to the free rider problem: a field experiment," *Journal of Experimental Social Psychology*, 29: 463–78.

Fehr, E., Kirchsteiger, G. and Riedl, A. (1993) "Does fairness prevent market clearing? An experimental investigation," *Quarterly Journal of Economics*, 108: 437–60.

Glance, N. S. and Huberman, B. A. (1994) "The dynamics of social dilemmas," *Scientific American*, March: 76–81.

Gneezy, U. and List, J. (2006) "Putting behavioral economics to work: field evidence on gift exchange," *Econometrica*, 74(5): 1365–84.

―――― and Rustichini, A. (2000a) "A fine is a price," *Journal of Legal Studies*, 29: 1–17.

―――― ―――― (2000b) "Pay enough or don't pay at all," *Quarterly Journal of Economics*, August: 791–810.

Isaac, M. and Walker, J. (1988) "Group size and the voluntary provision of public goods: experimental evidence utilizing large groups," *Journal of Public Economics*, 54: 1–36.

Keser, C. (1996) "Voluntary contributions to a public good when partial contribution is a dominant strategy," *Economics Letters*, 50: 359–66.

Kim, O. and Walker, M. (1984) "The free rider problem: experimental evidence," *Public Choice*, 43: 3–24.

Lane, I. M. and Messe, L. A. (1971) "Equity and distribution of rewards," *Journal of Personality and Social Psychology*, 20: 1–17.

Ledyard, J. O. (1995) "Public goods: a survey of experimental research," in. J. Kagel and A. Roth (eds), *The Handbook of Experimental Economics*, Princeton University Press, pp. 111–94.

Roth, A. E. (2002) "The economist as engineer: game theory, experimentation, and computation as tools for design economics" (Fisher-Schultz Lecture), *Econometrica*, 70(4): 1341–78.

―――― Prasnikar., V., Okuno-Fujiwara, M. and Zamir, S. (1991) "Bargaining and market behavior in Jerusalem, Ljubljana, Pittsburgh, and Tokyo: an experimental study," *American Economic Review*, 81(5): 1068–95.

Schulze, G. and Frank, B. (2003) "Deterrence versus intrinsic motivation: experimental evidence on the determinants of corruptibility," *Economics of Governance*, 4(2): 143–60.

Thaler, R. (1980) "Toward a positive theory of consumer choice," *Journal of Economic Behavior and Organization*, 1: 39–60.

Weimann, J. (1994) "Individual behavior in a free riding experiment," *Journal of Public Economics*, 54(2): 185–200.

PART IV

....

COMPETING
DESIGNS

....

CHAPTER 24

....

COMPETING
MECHANISMS

....

MICHAEL PETERS

THERE are many ways to sell goods to people whose values you don't know. Auctions are one way, but there are many others. If you buy a car or house, the seller will often engage in a complicated negotiation process that typically involves auction-like tricks designed to sort the low- and high-value buyers. For example, a car dealer is happy to match lower prices you find at other dealers, but will warn you that his cars are in high demand, and the car you want may no longer be available when you get back. Whether you believe the assertion or not, the seller learns something about your value for the car when you are willing to take that chance. Houses are often sold at auction in Australia and New Zealand; however, in North America a very formalized offer–counter-offer process is more common.

In North America, fixed-price sales are common (for example in supermarkets). Yet outside North America, it isn't hard to find markets where haggling is the norm, even when the commodity being sold is of known quality and has a relatively low (and commonly known) value. A visit to the night market on Temple Street in Hong Kong gives an idea. The night market provides many alternatives to buyers, so there is little reason for a seller to hold out for a high price (their response to this is often to offer goods of relatively low quality).

One of the interesting things about the night market is that the ability to haggle is an attraction of the market itself. Beyond creating a tourist attraction for North Americans who aren't used to bargaining, the market provides an opportunity for very low-value buyers to buy stuff they otherwise might not want. Whether it is intended or not, bargaining provides a way of accommodating buyers with low values.

A more familiar selling technique is restaurant reservations. In principle, restaurants could auction their Saturday evening tables to the highest bidder. Instead, all they do is to require reservations, then uniformly raise the prices of all their meals. Presumably, the reservation system sorts out diners who particularly want to go to that restaurant from diners who simply want something to eat. Knowing that their reservation system is selecting buyers for them, they can set higher prices.

This proliferation of selling techniques can be partly explained by simple informational considerations. For example, if a good has a commonly known value which is the same for everyone, there is little point holding an auction in order to sell it. However, competitive considerations are also likely to determine how goods are sold. On eBay, sellers literally offer competing auctions. The impact of competition on eBay became very easy to see after the introduction of the "buy it now" option, which allows a bidder to circumvent what would otherwise be a fairly straightforward second-price auction by accepting a take-it-or-leave-it price offer. In data collected in 2005 on camera auctions at eBay, around 75% of all auctions were resolved using the "buy it now" option (there were dozens of simultaneous auctions for each model).

Many Internet sellers also experiment with selling techniques. For example, domain-name resellers like flippa.com resell domain names bundled with websites and software. Their auction site offers sellers a variety of different techniques, including eBay-like-second-price auctions, second-price auctions augmented with "buy it now," and a variant of the "buy it now" in which the seller essentially conducts a first-price sealed-bid auction. The site itself attributes some of its success to the fact that it allows sellers to auction domain names. This encourages sellers to bundle the names with websites which suggest ways that the domains can be used.

These examples suggest that selling methods may be as important as prices in attracting buyers. The theory of *mechanism design* was created to address exactly this issue, except that, as originally formulated, the theory allows only one mechanism designer. This chapter reviews a couple of models that explicitly model competition in mechanisms.

Competition makes it possible to address a perplexing theoretical issue as well. The theory of mechanism design with a single designer makes an implausible prediction. The theorem due to Cremer and McLean (1988) says that if buyers' valuations are even slightly correlated, then the seller can design a selling mechanism that provides the same expected revenue that he would have earned had he known the buyers' values. Of particular interest is the implication this theorem has for the properties of the seller's best selling mechanism. The technique that Cremer and McLean (1988) used to extract buyer surplus was to ask each buyer to commit to pay a fee after the auction finishes. This fee depends on the bids that were submitted by all of the other buyers. Since the fee depends only on the other buyers' bids, a buyer could not manipulate this fee at all. They then showed that the correlation in valuations could be used to design the fee so that the expected payment associated with the fee is exactly equal to the surplus the bidder receives by participating in a second-price auction.

Whether or not conditions in existing auction markets exactly mimic the conditions required for their theorem to hold, the result suggests that sellers' most profitable selling mechanisms should involve fees that depend on what other bidders do. No one has yet come up with a "real life" example in which sellers actually use these fees. The conclusion is that either sellers aren't designing selling mechanisms to maximize their revenues, or something else is going on.

In this chapter we discuss one possible resolution to this problem—competition. If buyers have better alternatives, they simply won't participate in a selling mechanism that takes away all their surplus. Yet it isn't the surplus extraction that presents a problem—it is the fees contingent on others' bids. Competition readily explains why surplus extraction doesn't happen, but it doesn't seem inconsistent with these fees.

DIRECTED SEARCH

Competition in prices is pretty straightforward, because every buyer is attracted to a low price. The night market example illustrates that selling mechanisms are quite different. Haggling is attractive to low-value buyers who can demonstrate their values by walking away, but unattractive to high-value buyers who need to find a good deal quickly.[1] A change in selling mechanism may not be attractive to all buyers.[2] One way to model competition in mechanisms is to borrow a technique originally designed to model competition in labor markets where search frictions were significant. We detour a bit to explain this method before returning to competing mechanisms.

Suppose there are two firms trying to attract workers. They offer wages w_1 and w_2, with w_1 being larger. There are two workers, each of whom applies to one and only one of the two firms. If a firm receives a single applicant, they hire him or her. If a firm receives two applicants, they hire one of them at random. If a firm doesn't receive an application, it does without a worker. If a worker applies and isn't hired, then that worker does without a job. A firm who hires a worker produces revenue of \$1. Workers who aren't hired and firms who don't hire both earn nothing.

The theory of directed search is based on the assumption that both workers apply to firm 1 with the same probability, say π. When firm 1's wage offer rises, this probability should increase. The firm trades off the higher wage that it offers against the higher probability that it will hire some worker in order to determine its wage. Eventually, we want to apply this idea to selling mechanisms. For example, it is reasonable to expect that if a firm lowers the reserve price that it sets in an auction, then all bidder types will be more likely to bid in that firm's auction.

---

[1] One of the interesting things about the Temple Street night market in Hong Kong is that the stalls where haggling occurs often obscure entrances to shops where more serious buyers can buy. Perhaps this is the device the night market uses to keep its high-value customers.

[2] One of the big changes that has occurred on eBay is the emergence of the "buy it now" feature. Rather than bidding in the auction, there is now a fixed (typically high) take-it-or-leave-it price that any buyer can agree to pay immediately, thus ending the auction. In data collected in 2005 for camera auctions, about three-quarters of all the auctions ended with some buyer clicking the buy-it-now price. Presumably, this new feature is great for high-value buyers who are anxious to get their cameras, but bad for low-value buyers, who have to work a lot harder to find a bargain.

In the job application problem, it is straightforward to tie down this bidding probability. Each worker is going to apply to the firm where it has the highest expected payoff. If we think a worker is going to apply to two different firms with positive probability, it better be the case that the worker receives the same expected payoff from both. In particular, if the other worker is applying to firm 1 with probability π, then the expected payoff to the worker if he applies there is

$$\pi \frac{w_1}{2} + (1 - \pi) \, w_1$$

The explanation is that if the other worker also applies to firm 1, then there is half a chance that the worker will be hired. If the other worker applies to firm 2, then the worker is hired for sure.

Using the same reasoning to compute the expected payoff associated with an application to firm 2, the probability with which the worker expects the other worker to apply to firm 1 had better satisfy

$$\pi \frac{w_1}{2} + (1 - \pi) \, w_1 = \pi w_2 + (1 - \pi) \frac{w_2}{2}$$

or

$$\pi = \frac{2w_1 - w_2}{w_1 + w_2}$$

What this algebra shows is that if worker 1 expects worker 2 to apply to firm 1 with probability $\frac{2w_1-w_2}{w_1+w_2}$, then worker 1 will be just indifferent about whether he applies to firm 1 or firm 2. Of course, if he is just indifferent, then it wouldn't be unreasonable to expect him to apply to firm 1 with probability $\frac{2w_1-w_2}{w_1+w_2}$, so that worker 2 would also be indifferent about which firm he applies to.

The application strategy $\pi = \frac{2w_1-w_2}{w_1+w_2}$ constitutes a Nash equilibrium for the application game that is played by the workers. This Nash equilibrium gives a very nice description of how workers go about choosing between different firms. As is apparent from the formula, as firm 1 raises its wage, both workers are more likely to apply to firm 1. It is exactly that logic that we want to apply when we think about competing mechanisms.

COMPETING MECHANISMS

The logic we want to develop is that when a firm alters a characteristic of its selling mechanism, this change will increase the probability with which bidders participate in the mechanism whenever this change increases the surplus they expect to earn. To illustrate, we can focus on auctions and assume that two firms compete in reserve prices. Then our logic suggests that raising the reserve price (which lowers all buyers' surplus *ceteris paribus*) will reduce participation probability.

To see the argument, suppose there are two firms, each of which possesses a single unit of output which they hope to sell to one of two buyers. We will imagine that each of the firms uses a second-price auction with a reserve price. Firms don't value their goods at all, apart from what they think they can sell them for. So we imagine the sellers' valuations are both 0. However, the sellers are explicitly concerned with how their reserve prices will affect buyer participation, since this affects the revenue they expect to earn from their auctions.

There are two buyers who both feel that the goods offered by the sellers are perfect substitutes for one another. However, the buyers differ in their valuations for the goods. We will suppose each buyer's valuation is independently drawn from a common probability distribution F.

As in the labor market story presented earlier, we imagine that the firms begin by describing their auctions, then each of the buyers chooses which of the two auctions he wants to participate in. If only one buyer bids in an auction, the good is sold for its reserve price; if two bidders bid, then the good is sold to the high bidder at a price equal to the second-highest bid. The probabilities with which the different types of bidder participate in seller 1's auction depend on the two reserve prices. Let $\pi(v)$ be the probability that a buyer with valuation v chooses to bid in seller 1's auction.

To see how to find the equilibrium, we use two ideas. One is a nice insight from McAfee (1993), the other a standard argument in mechanism design. Let's start with the mechanism design argument. When a bidder participates in an auction his expected payoff is equal to his probability of winning the auction when he participates, multiplied by his value, less the price he expects to pay to the firm. The winning probability and expected price both depend on his value as well as participation probabilities of the other bidder, and the reserve price he faces. For the moment, let's ignore the participation probabilities and reserve price, and write this out as

$$Q(v)v - P(v)$$

where $Q(v)$ is the probability that a seller of type v wins the auction, and $P(v)$ is the price he expects to pay. In the equilibrium of the second stage of the game, the bidders will adopt some participation strategies. We have no idea at the moment what they are, but the equilibrium participation strategy of a bidder with value v must be at least as good for him as the strategy that would be used by a bidder with a different value, say v'. If the bidder with value v were to adopt the participation strategy of the bidder with valuation v', then his payoff would be

$$Q(v')v - P(v')$$

In fact, since the payoff he gets by using his equilibrium strategy must be better than the payoff he could get by using *any* other bidder's strategy it must be that

$$Q'(v')v - P'(v') = 0$$

when $v' = v$. This tells us that in equilibrium $Q'(v)v = P'(v)$ for every v.

This is sort of helpful since the fundamental theorem of calculus tells us that

$$P(v) = \int_{\underline{v}}^{v} P'(v') \, dv' = \int_{\underline{v}}^{v} Q'(v') \, v' \, dv' \tag{1}$$

We can use this information to simplify the equilibrium payoff function. If we integrate the right-hand side of equation (1) by parts we get

$$P(v) = Q(v') \, v' \Big|_{\underline{v}}^{v} - \int_{\underline{v}}^{v} Q(v') \, dv'$$

If we assume that $\underline{v} = 0$ just to make things simple, then the equilibrium payoff to a buyer of type v is given by

$$Q(v) \, v - P(v) = \int_{\underline{v}}^{v} Q(v') \, dv' \tag{2}$$

It isn't particularly intuitive that the equilibrium payoff should be equal to the integral of the trading probability, but as you will see, it is an analytically very useful result. It is very general in the sense that the same result will hold no matter how many buyers there are. We use this fact later. It is very special in the sense that it relies heavily on the assumption that buyer types are independent. To see this, simply observe that in equation (1), we treat Q' as if it were the marginal impact of a change in buyer type on trading probability. In fact, it is the marginal impact on trading probability when a buyer pretends to have a higher-type, which is not the same. The reason is that a higher-type buyer will have a different belief about the types of the other buyers when types are correlated, whereas a buyer pretending to have a higher type won't.

Before going over how to simplify this, we should explain the seller's payoffs. Provided you recognize that the functions P and Q both depend on reserve prices the seller sets, the seller's payoffs can be written in a straightforward way using the information given so far. There are two potential bidders, each of whom makes expected payment $P(v)$ when their type is v. The distribution of types is given by $F(v)$, so the seller's payoff is just

$$2 \int_{\underline{v}}^{\bar{v}} P(v) \, dF(v)$$

From equation (2) this is equal to

$$2 \int_{\underline{v}}^{\bar{v}} \left\{ Q(v) \, v - \int_{\underline{v}}^{v} Q(v') \, dv' \right\} dF(v) \tag{3}$$

Now we want to simplify the buyer's and seller's payoffs as given by equations (2) and (3). We do this using the directed search logic explained earlier along with a very nice insight from McAfee (1993). A buyer with valuation v will win the auction in two circumstances: first, when the other bidder has a valuation below v; and second, when the other buyer chooses to participate in the other auction. This probability is very simple, and is given by:[3]

$$1 - \int_v^{\overline{v}} \pi\left(v'\right) F'\left(v'\right) dv'$$

Why might this simple expression be helpful? Well, suppose we have a couple of auctions and the bidders choose their participation strategies. Now suppose that bidder v bids in both auctions with positive probability. Then, as in the labor market example given earlier, the payoff that v gets in equilibrium from both auctions must be the same. This means that

$$\int_{\underline{v}}^v Q_1\left(v'\right) dv' = \int_{\underline{v}}^v Q_2\left(v'\right) dv'$$

where Q_1 and Q_2 are the equilibrium trading probabilities for the different buyer types at the two different auctions. The same equality should hold for all the higher types as well, so this expression is an identity. Then the derivatives of both sides with respect to v must also be equal that is:

$$Q_1\left(v\right) = Q_2\left(v\right)$$

or, using McAfee's idea,

$$\int_v^{\overline{v}} \pi_1\left(v'\right) F'\left(v'\right) dv' = \int_v^{\overline{v}} \pi_2\left(v'\right) F'\left(v'\right) dv'$$

Now since the original expression is an identity, we can differentiate again to get

$$\pi_1\left(v\right) F'\left(v\right) = \pi_2\left(v\right) F'\left(v\right)$$

or $\pi_1\left(v\right) = \pi_2\left(v\right) = \frac{1}{2}$ for every buyer type who participates in both auctions.

This is quite different from the labor market example. What it implies is that when a seller adjusts his reserve price, what he changes is not the participation probabilities, as in the labor market example, but the set of buyer valuations that apply. When a seller raises his reserve price, he chases away some of the lowest-valuation buyers completely. This means that the high-valuation buyers are less likely to face an opponent. It is this that keeps the higher-valuation buyers indifferent.

The main lesson from this argument is that when sellers compete in auctions, they aren't competing directly for the high-valuation buyers; it is only the low-valuation buyers who change their behavior in response to changes in their mechanisms.

It is easy enough to find the rest of the equilibrium in the second stage, where the buyer chooses where to bid. If the reserve prices are $r_1 < r_2$, then it is pretty obvious that

---

[3] This is just 1 minus the probability that the other buyer both has a higher valuation and comes to the same auction.

buyers whose valuations are below r_1 won't bother to bid at all. Buyers whose valuations are between r_1 and r_2 will bid with seller 1 for sure, since they can't afford (i.e don't want) to pay seller 2's reserve price. Even buyers whose valuation is slightly higher than r_2 are going to restrict their bidding to seller 1. Such buyer types aren't likely to win either auction. However, even if they are the only bidders at the auction, they only get a tiny surplus with seller 2. If they are the only bidder with seller 1, they get a much larger surplus because seller 1's reserve price is lower.

Suppose that v^* is the lowest bidder type who bids at seller 2's auction. Since buyers with lower valuations all go to seller 1 for sure, he will pay seller 2's reserve price if he wins, but he can win only if no other bidder participates. The probability that this happens, as described earlier, is $1 - \frac{1}{2} \int_{v^*}^{\bar{v}} F'(v') \, dv' = 1 - \frac{1 - F(v^*)}{2}$.

Putting all this together makes it possible to determine the value for v^*. The payoff that the marginal buyer of type v^* receives when she bids with seller 2 is

$$(v^* - r_2) \left(1 - \frac{1 - F(v^*)}{2} \right)$$

This should be just equal to the payoff she gets by bidding at seller 1 instead. As we explained, this is the integral of the trading probability with seller 1 up to the value v^*. Buyers whose values are below r_1 never trade with seller 1. Buyers whose valuations are below v^* (but above r_1) trade as long as the other bidder either has a lower valuation, or chooses to bid with seller 2. This probability is given by

$$1 - (F(v^*) - F(v)) - \frac{1 - F(v^*)}{2}$$

So v^* is determined by the condition that

$$(v^* - r_2) \left(1 - \frac{1 - F(v^*)}{2} \right) = \int_{r_1}^{v} \left\{ 1 - (F(v^*) - F(v')) - \frac{1 - F(v^*)}{2} \right\} \tag{4}$$

All that is left is to write down the profit functions for the two sellers. For seller 2, who charges the high reserve price, it is most straightforward since the marginal buyer type is highest. From equation (3), the high-reserve-price seller's profits are

$$2 \int_{\underline{v}}^{\bar{v}} \left\{ Q(v) v - \int_{\underline{v}}^{v} Q(v') \, dv' \right\} dF(v)$$

$$= 2 \int_{v_2^*}^{\bar{v}} \left\{ \left(1 - \frac{1 - F(v)}{2} \right) v - \int_{v_2^*}^{v} \left(1 - \frac{1 - F(v')}{2} \right) dv' \right\} dF(v).$$

Integrating the second term by parts gives

$$2 \int_{v_2^*}^{\bar{v}} \left(1 - \frac{1 - F(v)}{2}\right) v dF(v) -$$

$$2 \left\{ \int_{v_2^*}^{v} \left(1 - \frac{1 - F(v')}{2}\right) dv' F(v) \Big|_{v_2^*}^{\bar{v}} - \int_{v_2^*}^{\bar{v}} F(v) \left(1 - \frac{1 - F(v)}{2}\right) dv \right\}$$

$$= 2 \left\{ \int_{v_2^*}^{\bar{y}} \left(v - \frac{1 - F(v)}{F'(v)}\right) \left(1 - \frac{1 - F(v)}{2}\right) F'(v) \, dv \right\} \qquad (5)$$

This is a complicated function of the high-reserve-price seller's reserve price, r_2. Yet it is quite a simple function of the high-reserve-price seller's cut-off valuation, v_2^*. Once the cut-off valuation, is high enough such that $v^* > \frac{1 - F(v^*)}{F'(v^*)}$, the seller will have no interest in raising his cut-off valuation by raising his reserve price. To understand equilibrium, it is then necessary to understand what happens to the seller who sets the low reserve price.

Following the same logic as above, the low-reserve-price seller has payoff

$$2 \left\{ \int_{r_1}^{y_2^*} \left(v - \frac{1 - F(v)}{F'(v)}\right) F(y) F'(v) \, dv \right\}$$

$$+ 2 \left\{ \int_{v_2^*}^{\bar{y}} \left(v - \frac{1 - F(v)}{F'(v)}\right) \left(1 - \frac{1 - F(v)}{2}\right) F'(v) \, dv \right\} \qquad (6)$$

Once again, this is a fairly simple function of the cut-off valuations of the two sellers. However, there is one significant complication. As the low-reserve-price seller cuts his reserve price, r_1, he changes both the cut-off valuations v_2^* for the high-reserve-price seller. This is apparent from equation (4). In particular, when the low-reserve-price seller cuts his reserve price slightly, he raises the cut-off valuation at the high-price seller. In particular, this causes some buyer types who are bidding with equal probability at both sellers to decide to bid for sure at the low reserve price. So just as we wanted, cutting the reserve price draws customers away from the high-reserve-price seller.

The monopoly reserve price in a single-seller auction is the reserve price such that the buyer v^* whose type satisfies $v^* = \frac{1 - F(v^*)}{F'(v^*)}$ is just indifferent to bidding. Notably, this "optimal" reserve price is independent of the number of bidders. The argument shows why it cannot be an equilibrium for both sellers to set this monopoly reserve price when there is competition. Each of the sellers has an incentive to cut reserve price slightly to steal some of the customers away from the other seller. Once again, the customers who are attracted to this are not the high-value bidders. Instead, it is those bidders whose values are close to the reserve price anyway.

Calculating equilibrium reserve prices

A Nash equilibrium for the game just described can be defined as a pair of reserve prices that are jointly best replies to one another. In the story above, a seller attracts bidders from the other seller by lowering reserve prices. The cost of this is that the seller is now selling to some buyer types for whom $v < \frac{1-F(v)}{F'(v)}$. From equation (6), it is apparent that the seller lowers his profits slightly by selling to these buyers. In a Nash equilibrium, the marginal cost of selling to these buyers must be just offset by the marginal gain of selling to the higher-valuation buyers. Burguet and Sakovics (1999) analyze this game and show that equilibrium is in mixed strategies. Reserve prices in this mixed equilibrium are strictly larger than the seller's cost, but strictly less than the monopoly reserve prices.

Some auction markets are much more competitive than this example suggests. For example, a search for digital camera auctions on eBay will turn up thousands of opportunities to bid. Even if the auctions involve very different kinds of camera, this still provides hundreds of opportunities to bid on a new version of any particular model. In such large markets, sellers have a very small chance of attracting any of the high-value bidders. Instead of getting each high-value bidder with probability $\frac{1}{2}$, as occurs in the example, the seller gets each of them with probability $\frac{1}{100}$. By itself this isn't a problem since there are typically a lot more bidders to compete over.

However, this makes a big difference when the seller cuts his reserve price relative to the other sellers. Then, instead of raising the probability that higher-valuation bidders will bid in his auction from $\frac{1}{2}$ to 1, he raises it from $\frac{1}{100}$ to 1, and this has a big and positive impact on his profits. Peters and Severinov (1997) shows that as the number of bidders and sellers becomes very large, as it does in the eBay camera auction, it is profitable to cut reserve prices whenever they are positive. This result doesn't show that equilibrium reserve prices are zero when the number of buyers and sellers is large. When all sellers are identical and all reserve prices are zero, each seller has an incentive to raise his reserve price if there are enough buyers. A pair of papers establish the convergence of equilibrium in finite competing auction games. Hernando-Veciana (2005) shows that when there are a finite number of reserve prices, then very generally there will be an equilibrium in a large finite competing auction game in which each seller sets his reserve price equal to his value. More recently, Virag (2010) shows that in finite competing auction games with a continuum of feasible reserve prices, if all sellers are identical, mixed strategy equilibrium exists among sellers. As the number of buyers and sellers becomes large, all reserve prices converge in distribution to the sellers' value.

These are quite significant results. Recall from the earlier discussion that the *monopoly* reserve price ensures that a buyer whose type satisfies $v^* = \frac{1-F(v^*)}{F'(v^*)}$ is just indifferent about whether or not to bid. The odd thing about this result is that this "optimal" reserve price is very sensitive to what the seller thinks the distribution of bids is. In this sense the optimal auction is like the Cremer–McLean mechanism—to find the optimal reserve price requires a careful calculation involving information that is hard to get. The competitive results in Peters and Severinov (1997) and the convergence results in Hernando-Veciana (2005) and Virag (2010) explain how competition among sellers eliminates this counterintuitive result. In a large enough market, all the seller needs to know in order to set his reserve price is his own selling cost, more or less exactly what we assume he would do in a simple competitive market.

The competing auction game is complex because a seller has to explicitly calculate how a change in his own mechanism will affect the payoff that buyers get by going to some other mechanism—an effect that disappears as markets become large. It is possible to get around some of the complexity associated with the competing auction game by assuming that sellers behave "competitively" in relatively small markets. Usually, competitive sellers are price takers. Obviously, they can't literally be price takers, since price setting is an integral part of the mechanism that they offer. Instead, we might try to capture the competitive flavor of the limit results described earlier by assuming that sellers are "payoff takers." In other words, they believe that there is a market payoff that they have to provide buyers in order to attract them to their mechanisms.

The thing that makes this assumption nice is that, like price takers, they also believe that provided they offer buyers this market payoff, they can have any distribution of buyer types that they want. The trade-off that sellers have to work out when they design their mechanism is that the more buyers they plan to attract and the higher their types, the more competition buyers will face when they come. So sellers have to pick the types and participation probabilities that they want in such a way that all the types they expect to attract earn their market payoff. We turn now to this formulation.

COMPETITIVE EQUILIBRIUM IN MECHANISMS

In this section we return to the more general problem of equilibrium mechanisms. In particular, we want to add back correlation in valuations to create an environment like the one in Cremer–McLean in which a monopoly seller can extract all buyer surplus. We want to show that there is a unique symmetric "equilibrium" in which all sellers offer to run second-price auctions with reserve price equal to their cost. The surprising thing about this result is that competition will prevent sellers from using Cremer–McLean-type entry fees to extract surplus. Thus we have an argument that shows why competition keeps mechanisms simple.

There are a couple of remarks that need to be made before we proceed. First, we are going to restrict sellers to mechanisms that are "direct" in the naive sense that allocations depend on buyers' payoff types. Buyers' types are complex objects in competing mechanism games. These types include market information that a seller would want. For example, sellers might want to ask bidders to tell them whether some other seller has deviated from some convention that is usually used in the industry.[4]

---

[4] An appropriate formulation of bidder types is given in Epstein and Peters (1999).

Second, we use a competitive market payoff-taking assumption. Sellers will choose the distribution of types they want conditional on participants all receiving their market payoff when this distribution is realized. In equilibrium, the distributions that sellers choose will coincide with the true distribution of types that they face. However, as in all competitive models, the same will not be true outside equilibrium. If a seller deviates, he will typically anticipate a distribution that does not coincide with the distribution associated with the continuation equilibrium among buyers that follows his deviation. This is analogous to the idea that in a Walrasian equilibrium a buyer can deviate from his equilibrium demand and imagine the payoff he would get from buying more of some good, even though he wouldn't be able to find more of this good at prevailing prices.

This description of a large sub-market on eBay seems plausible. Camera sellers, for example, aren't likely to know much about all the different alternatives buyers consider before choosing to participate in their auctions. Certainly, other active sellers on eBay are observable. Yet buyers also purchase from standard retail outlets. Since sellers don't know where buyers live, they can't know much about these alternatives. On the other hand, many of the camera sellers have lots of experience selling on eBay. They are likely to know approximately what payoffs they need to offer buyers to keep them bidding. Generally, the eBay camera market is embedded in a much larger market that sellers may not fully understand, so that competitive assumptions about how this market works seem reasonable.

It should be mentioned at this point that a fully game theoretic treatment of the competing auction market runs up against a problem. Absent equilibrium refinements or explicit restrictions on feasible mechanisms, competing mechanisms can be used to support a large variety of equilibrium allocations.[5] Some stand has to be taken on how to restrict players in order for the model to have any predictive content at all. The "market payoff" assumption is the restriction adopted here.

In the market there are s sellers and n bidders. Each seller has a single unit of output to sell. He has no cost of offering this output for sale. Each bidder wants to acquire exactly one unit. Bidders have valuations x_i. A bidder who buys a unit of output at price p earns surplus $x_i - p$. The seller in this transaction earns p. Sellers offer direct mechanisms; bidders choose to participate in one and only one of these mechanisms.

A direct mechanism is a pair of functions $q : [0,1]^n \to [0,1]^n$, and $p : [0,1]^n \to \mathbb{R}^n$. The array $q(x)$ is a vector of probabilities with which objects are awarded to each of the different players, depending on their types. The sum of these probabilities should be less than or equal to one. The function $p(x)$ specifies a vector of payments to or from each bidder. These functions should specify probabilities and payments only for participating bidders. To compensate we will treat non-participants as if they had value $x_i = 0$. Feasible mechanisms are then required to assign $q_i(x_i, x_{-i}) = p_i(x_i, x_{-i}) = 0$ for any bidder i who has valuation 0.

---

[5] The most general folk theorem in this regard is Peters and Troncoso-Valverde (2009); however, the basic idea is due to Yamashita (2010).

The market is subject to an external shock, y, which is distributed $G(y)$ on some compact interval. Sellers face a joint distribution of types $Z(x_1, \ldots, x_n | y)$, which depends on the external shock. However, they also believe this distribution is related to the direct mechanism that they offer. Specifically, they believe the market provides a bidder of type x_i a payoff $\beta(x_i | y)$. Sellers don't think they have any impact at all on this market payoff. They also believe they can support any distribution of buyer types that they like, provided the mechanism they offer provides each buyer type with at least her expected payoff when buyers have the same belief about this distribution.

The surplus for a seller who offers mechanism (q, p) and faces distribution $Z(x|y)$ is given by

$$\int \int p(x) \, dZ(x|y) \, dG(y) \tag{7}$$

A bidder who participates in seller j's mechanism and shares the seller's beliefs earns surplus

$$\int \int \left[x_i q_i(x_i, x_{-i}) - p(x_i, x_{-i}) \right] dZ(x_{-i} | x_i, y) \, dG(y)$$

Given the market payoff function β, the seller chooses his mechanism (q, p) to maximize equation (7) subject to the constraint that

$$\int \int \left[x_i q_i(x_i, x_{-i}) - p(x_i, x_{-i}) \right] dZ(x_{-i} | x_i, y) \, dG(y) \geq \int \beta(x_i | y) \, dG(y) \tag{8}$$

for each x_i in the support of (the marginal distribution associated with) Z.

Bidders are more sophisticated. They share the belief that valuations are *conditionally independent*, with distribution $F(x_i | y)$ on the interval $[0, 1]$. Implicit in this is the assumption that valuations lie between 0 and 1. The joint distribution of valuations that buyers face is then given by

$$\int \prod_{i=1}^{n} F(x_i | y) \, dG(y)$$

Bidders adopt a symmetric participation strategy $\pi_j(x_i)$ which gives the probability with which they will participate in the mechanism offered by seller j (they can participate in only one mechanism). Given a participation strategy π_j for seller j, the true distribution of types faced by seller j is given by

$$z_j(x_1, \ldots, x_n) = \int \left[\prod_i 1 - \int_x^1 \pi_j(x') \, dF(x'|y) \right] dG(y)$$

A participation strategy $\{\pi_1(\cdot), \ldots, \pi_s(\cdot)\}$ is a *continuation equilibrium* if the sum of these participation strategies across sellers is less than or equal to one and

$$\int \left[x_i q_i^j \left(x_i, x_{-i} \right) - p_i^j \left(x_i, x_{-i} \right) \right] d \int \left[\prod_{i' \neq i} 1 - \int_{x_{i'}}^1 \pi_j \left(x' \right) dF \left(x' | y \right) \right] dG \left(y | x_i \right) \geq$$

$$\int \left[x_i q_i^{j'} \left(x_i, x_{-i} \right) - p_i^{j'} \left(x_i, x_{-i} \right) \right] d \int \left[\prod_{i' \neq i} 1 - \int_{x_{i'}}^1 \pi_{j'} \left(x' \right) dF \left(x' | y \right) \right] dG \left(y | x_i \right)$$

for each j for which $\pi_j \left(x_i \right) > 0$.

A *symmetric equilibrium in mechanisms* is a common mechanism, (q, p), to be used by each firm, a common conditional distribution function, $Z \left(x | y \right)$, and a market payoff function β having the property that (q, p) and Z jointly maximize equation (7) subject to equation (8), and such that

$$\int \left[x_i q_i \left(x_i, x_{-i} \right) - p_i \left(x_i, x_{-i} \right) \right] d \int \left[\prod_{i' \neq i} 1 - \int_{x_{i'}}^1 \pi_j \left(x' \right) dF \left(x' | y \right) \right] dG \left(y | x_i \right)$$

$$= \int \int \left[x_i q_i \left(x_i, x_{-i} \right) - p \left(x_i, x_{-i} \right) \right] dZ \left(x_{-i} | x_i, y \right) dG \left(y \right) = \int \beta \left(x_i | y \right) dG \left(y \right) \quad (9)$$

for each x_i in the support of Z, where $\pi_1 = \pi_2 = \cdots = \pi_s = \pi$ is a continuation equilibrium.

When all sellers offer an equilibrium mechanism, the distribution of types they expect to face is equal to the distribution of types associated with a continuation equilibrium. What makes this non-standard is what happens when there is a deviation. If a seller unilaterally chooses some other mechanism, he will expect a new distribution which provides each bidder the expected payoff they had before the deviation. In a sub-game perfect equilibrium, the payoff associated with the continuation equilibrium following a deviation would be different from what it was before the deviation. So, generally, sellers, expectations are incorrect out of equilibrium (as is true in every competitive model).

Efficient mechanisms

We now want to use the Cremer–McLean idea to show that there is a unique symmetric equilibrium. The main part of this argument shows that equilibrium mechanisms have to be efficient.

The approach that Cremer and McLean used was to imagine that some mechanism, say a second-price auction, is being used by a seller, and that this mechanism generates a payoff $\beta \left(x \right)$ to participants. They suggested the seller augment the auction with a menu of fees, $\{ \omega_\theta \left(x_2, \ldots, x_n \right) \}$. The variables x_2 through x_n are intended to represent the "bids" of the other participants. The variable θ simply indexes the fee schedule. In their story, there is a finite number of possible values for θ. In the new mechanism, participants would submit their bid as before, and, in addition, select one of the fee schedules. They provide a condition on the joint distribution of valuations such that for any continuous function $\beta : [0,1] \rightarrow [0,1]$, a menu of fees could be created such that

$$\beta (x) = \min_{\theta} \int \omega_{\theta} (x_2, \ldots, x_n) \prod_{j=2}^{n} dF (x_j|y) \, dG (y|x) \qquad (10)$$

for each x in the finite support of the distribution F. Each of a finite set of buyer types would choose one fee schedule (the one that minimized her expected fee conditional on her interim belief). This formulation has the property that the expected fee chosen by each bidder type is exactly the surplus they expect from the auction. Provided bidders bid their true values in the action, requiring them to select one of these fees ensures that their expected surplus from participation in the auction is zero. The second-price auction allocates the good to the bidder with the highest value, so that the expected surplus the seller earns is the same as his surplus under complete information.

The exact formulation is in the original article by Cremer and McLean (1988). The extension to the continuous case is difficult and is discussed in McAfee et al. (1989). They provide conditions under which equation (10) will hold approximately when the number of fees in the schedule is finite. As they assume that the distribution of type is absolutely continuous, their assumptions don't work in the competitive case since there is a strictly positive probability that buyers won't participate at all. Peters (2001) shows that their assumptions also ensure that fees can be designed to satisfy (10) in a competitive market assuming that fees are based on bids of participating bidders.

All we are interested in here is how to use the Cremer–McLean argument to understand the competitive case. So, rather than dealing with these issues here, we will simply assume that the joint distribution of types has enough correlation to support these fees when bidders choose among mechanisms with equal probability. In particular:

Definition 1. *The joint distribution of types has the Cremer–McLean property if for any continuous function β, there exists a family of fees $\{\omega_{\theta}\}$ mapping $[0, 1]^{n-1} \rightarrow \mathbb{R}$ such that*

$$\beta (x) = \inf_{\theta} \int \omega_{\theta} (x_2, \ldots, x_n) \prod_{j=2}^{n} d \left[1 - \int_{x_j}^{1} \frac{1}{s} dF (x'|y) \right] dG (y|x)$$

Our argument will then involve three parts. First, we are going to use the Cremer–McLean idea to show why competitive mechanisms must always allocate the object to the bidder with the highest value. The argument is to suppose that this isn't true and that sellers are using mechanisms that inefficiently allocate. Sellers have the option of replacing their mechanism with one that efficiently allocates. The complication of doing this is that such a mechanism may not provide the buyers the seller wants to attract with their market payoff. Here we will apply the Cremer–McLean idea, and augment existing fees with a new set that will just compensate all buyer types for their lost surplus, and extract any extra surplus that the new mechanism might create. This allows the seller to extract all the surplus gains from switching to an efficient mechanism.

This much establishes that every equilibrium involves mechanisms that allocate the good to the participant who has the highest valuation. Then, if all other sellers are offering second-price auctions with zero reserve price, no seller can improve his expected surplus by doing otherwise.

Finally, we show that the payoff function associated with second-price auctions is the only one that can satisfy our equilibrium conditions.

Theorem 2. *If the joint distribution of types has the Cremer–McLean property, then every competitive equilibrium in mechanisms has sellers using efficient mechanisms that award the good to the bidder with the highest value.*

Proof: Suppose $(q(\cdot), p(\cdot))$ is an equilibrium mechanism, and that it is not efficient in the sense that there is an event, E, having strictly positive probability for which $t_i > t_j$ for all j, and $q(t_i, t_{-i}) < 1$. Observe that since the bidders' payoff is always equal to the market payoff $\int \beta(x|y) \, dG(y)$, the seller's profit can be written as the total surplus, less what the seller expects to give to each participating bidder. That is:

$$\int \left\{ \int \cdots \int \sum_{i=1}^{n} q(x_i, x_{-i}) x_i dZ(x_1, \ldots, x_n | y) - n \int \beta(x|y) \, dZ(x_1, \ldots, x_n | y) \right\} dG(y)$$

$$(11)$$

In a symmetric competitive equilibrium, sellers' expectations about the distribution of types they face must be correct, so this must be equal to

$$\int \left\{ \int \cdots \int \sum_{i=1}^{n} q'(x_i, x_{-i}) x_i d \left[1 - \int_{x_j}^{1} \frac{1}{s} dF(x'|y) \right] \right.$$

$$\left. - n \int \beta(x|y) \, d \left[1 - \int_{x_j}^{1} \frac{1}{s} dF(x'|y) \right] \right\} dG(y)$$

Now replace the mechanism (q, p) with a simple second-price auction (q', p') (which, in particular, is an efficient and incentive compatible mechanism). Observe that

$$\int \left\{ \int \cdots \int \sum_{i=1}^{n} q'(x_i, x_{-i}) x_i d \left[1 - \int_{x_j}^{1} \frac{1}{s} dF(x'|y) \right] \right.$$

$$\left. - n \int \beta(x|y) \, d \left[1 - \int_{x_j}^{1} \frac{1}{s} dF(x'|y) \right] \right\} dG(y)$$

strictly exceeds equation (11). This comparison will be irrelevant if replacing the original mechanism with a second-price auction leaves some buyer types with a payoff that is less than their market payoff. To ensure this doesn't happen, we can augment the second-price auction with a fee. Let

$$\alpha(x) =$$

$$\int \left\{ \beta(x|y) - \int \int \cdots \int \sum_{i=1}^{n} \left\{ q'(x_i, x_{-i}) x_i - p'(x_i, x_{-i}) \right\} dZ(x_{-i}|x_i, y) \right\} dG(y)$$

By the Cremer–McLean property, there is a menu of fee schedules ω_θ such that

$$\alpha(x) = \min_\theta \int \int \cdots \int \omega_\theta(x_{-i}) d \left[1 - \int_{x_j}^1 \frac{1}{s} dF(x'|y) \right]^{n-1} dG(y)$$

Now, if we augment the second-price auction by requiring each player to choose any of the fee schedules that he likes, a bidder of type x will choose the fee indexed $\theta(x)$. If he does so, then he is participating in a mechanism $(q', p'(x_i, x_{-i}) + \omega_{\theta(x_i)}(x_{-i}))$. By construction, this scheme provides each participating player exactly his market payoff. No player has an incentive to misrepresent his type under this new mechanism since the fees don't depend on his type report, and nothing else in the mechanism depends on which fee the bidder chooses. However, the mechanism does give the seller strictly higher profits. This contradiction shows that equilibrium mechanisms must be second-price auctions augmented by Cremer–McLean like fees.

This makes the equilibrium slightly simpler. However, it still admits the possibility that sellers might extract full surplus in equilibrium. To rule this out, what we need is a restriction that allows sellers to use the fees outside of equilibrium. McAfee et al. (1989) provide a condition under which the Cremer–McLean property holds approximately when joint distributions are absolutely continuous with respect to the Lebesque measure. Their theorem doesn't apply to the competing mechanism case, since the distribution associated with the symmetric equilibrium is not absolutely continuous. Peters (2001) provides the extension of their theorem to allow for atoms in the joint distribution. We do not want to get into these mathematical issues here, so we provide a stronger set of restrictions.

Definition 3. *The joint distribution of types satisfies the extended Cremer–McLean property if there is some $\epsilon > 0$ such that for every family of conditional distribution functions $Z(x|y)$ satisfying*

$$\int_B \left[Z(x|y) - \left[1 - 1 - \int_{x_j}^1 \frac{1}{s} dF(x'|y) \right] \right] \le \epsilon$$

on each measurable subset B of $[0,1]$, and every continuous function β there is a family of fees $\{\omega_\theta^Z\}$ such that

$$\beta(x) = \inf_\theta \int \omega_\theta^Z(x_2, \ldots, x_n) \prod_{j=2}^{n} d[Z(x|y)] dG(y|x)$$

The extended Cremer–McLean property extends the surplus extraction property from the true distribution of types to a weakly open set of distributions around the true distribution. The implication of this property is that, at least for small changes in the distribution of types, the seller can always adjust fees so that bidders who participate earn their market payoff.

We can now finish the theorem.

Theorem 4. *If the distribution of types given by the family of conditional distributions $F(x|y)$ satisfies the extended Cremer–McLean property, then there is a unique competitive equilibrium in mechanisms in which all sellers offer second-price auctions with zero reserve price.*

Proof: In equilibrium, sellers choose a distribution of types that maximizes their expected payoff conditional on buyers receiving their market payoff when this distribution of types is realized. By the extended Cremer–McLean property, there is a weakly open neighborhood of the symmetric distribution in which sellers can design fees to compensate bidders for changes in their mechanisms. From the argument in theorem 2, if the seller wants one of these distributions, he might as well use a second-price auction augmented by fees to support it. So his payoff when he chooses one of these alternative distributions will be

$$\int \left\{ \int n x_i Z(x_i|y)^{n-1} dZ(x_i|y) - n \int \beta(x_i|y) dZ(x_i|y) \right\} dG(y)$$

The second term integrates by parts to

$$\int \left\{ \beta(1|y) - \int Z(x|y) \beta'(x|y) dx \right\} dG(y)$$

Similarly, the first term can be integrated by parts to

$$\int \left\{ 1 - \int Z(x|y)^n dx \right\} dG(y)$$

Putting them together gives the expression

$$\int \left\{ (1 - \beta(1|y)) - \int \{ Z(x|y)^n - nZ(x|y) \beta'(x|y) \} dx \right\} dG(y) \qquad (12)$$

Since the seller's payoff should be maximum in equilibrium, the conditional distribution $Z(y|y)$ should satisfy the necessary condition for point-wise maximization

$$Z(x|y)^{n-1} = \beta'(x|y)$$

If Z coincides with the distribution supported by the symmetric continuation equilibrium, this resolves to

$$\left[1 - \frac{1 - F(x|y)}{n}\right]^{n-1} = \beta'(x|y) \tag{13}$$

As we showed in equation (2) above, this means that, apart from a constant, the conditional payoff function $\beta(x|y)$ must coincide with the payoff function associated with a second-price auction in which each bidder participates in each auction with the same probability.

If $\int \beta(x|y)\, dG(x) = 0$ on some non-degenerate interval, then $\int \beta(x|y)\, dG(x) = 0$, which implies that $Z(x|y)^{n-1} = \beta'(x|y) = 0$ on this interval. Then $Z(x|y) = \left[1 - \frac{1-F(x|y)}{n}\right] = 0$, which can't be satisfied any by distribution $F(x|y)$. This ensures that $\int \beta(x|y)\, dG(y) > 0$ for all x.

Finally, $\int \beta(0|y)\, dG(y) = 0$, otherwise, by continuity, sellers would not want to attract buyers whose types are close enough to 0.

The argument that the second-price auction without reserve price is an equilibrium is straightforward. The seller will want every buyer whose type is at least his market payoff to participate. If all other sellers are offering second-price auctions, this will be so for all types.

CONCLUSION

We have reviewed the basic theory of competing mechanisms. A partial equilibrium analysis suggests that competition will ensure efficiency. In models in which types are assumed to be independent and sellers are restricted to auctions, it has been shown that the equilibrium in the partial equilibrium model approximates the equilibrium in large finite games. Hopefully the same property is true in the more general environment considered here, but a proof of this has not yet been established.

REFERENCES

Burguet, R., and Sakovics, J. (1999) "Imperfect competition in auction designs," *International Economic Review*, 40(1): 231–47.

Cremer, J., and McLean, R. (1988) "Full extraction of the surplus in Bayesian and dominant strategy auctions," *Econometrica*, 56(6): 1247–59.

Epstein, L., and Peters, M. (1999) "A revelation principle for competing mechanisms," *Journal of Economic Theory*, 88(1): 119–60.

Hernando-Veciana, A. (2005) "Competition among auctioneers in a large market," *Journal of Economic Theory*, 121(1): 107–27.

McAfee, P. (1993) "Mechanism design by competing sellers," *Econometrica*, 61(6): 1281–312.

McAfee, R. P., McMillan, J. and Reny, P. J. (1989) "Extracting the surplus in a common value auction," *Econometrica*, 57(6): 1451–59.

Peters, M. (2001) "Surplus extraction and competition," *Review of Economic Studies*, 68(3): 613–33.

——— and Severinov, S. (1997) "Competition among sellers who offer auctions instead of prices," *Journal of Economic Theory*, 75(1): 141–79.

——— and Troncoso-Valverde, C. (2013) "A folk theorem for competing mechanisms," *Journal of Economic Theory*, 148(3): 953–973.

Virag, G. (2010) "Competing auctions: finite markets and convergence," *Theoretical Economics*, 5: 241–74.

Yamashita, T. (2010) "Mechanism games with multiple principals and three or more agents," *Econometrica*, 78(2): 791–801.

CHAPTER 25

THREE CASE STUDIES OF COMPETING DESIGNS IN FINANCIAL MARKETS

NIR VULKAN AND ZVIKA NEEMAN[1]

INTRODUCTION

AT most major derivatives exchanges, traders face a choice between two parallel trading venues for the exchange of contracts. Theories of mechanism selection can help us better understand the growing trend of centralized and decentralized markets coexisting in leading stock exchanges. We begin by reviewing the literature on mechanism selection, which can be broadly classified into three main categories: cost-based, information-based, and strategy-based literature. Of course, these are by no means mutually exclusive.

The chapter is based on three Oxford theses of our students. The master's thesis by Felix Momsen (2006) tests three complementary hypotheses concerning the determinants of exchange mechanism selection at a European, a Japanese, and two North American derivatives exchanges. Firstly, there is evidence that transaction costs affect traders' choices between exchange mechanisms; at exchanges with higher contracting costs for block trades, bilateral negotiations are less important than centralized trades for the exchange of derivatives contracts. Secondly, findings concerning a strategic trade mechanism are mixed; except for the expansion of centralized trades at Eurex there is limited evidence of long-run shifts in an exchange's centralized market where prices are independent of individual reservation value. Lastly, there is support for a short-run, risk-induced explanation for trade mechanism selection. When liquidity falls and thus

[1] We are grateful to Felix Momsen, Simon Henry, and John Hutchins for their efforts. Our thanks go to Ammara Mahmood in helping us organize the materials in this chapter.

execution risk rises in the centralized market of a derivatives exchange, block trading increases in popularity relative to centralized trading.

An undergraduate thesis by Simon Henry (2003) examines the introduction of choice of traders in the London Stock Exchange, where an electronic exchange, SETS (Stock Exchange Electronic Trading Service) has been introduced in parallel to the existing dealership market. The main focus was on cost differences between the two trading systems, information and strategic factors that contribute to the routing decisions made by traders. The analysis further reveals that sector volatility has a decisive influence on the effects of the other factors.

And finally another undergraduate thesis, by John Hutchins (2004), assesses the long-term trends in choice of mechanism for trade at the New York Stock Exchange (NYSE), by comparing the share of orders held by the NYSE itself, a hybrid limit-order/dealer market, with a relatively new group of market centers called Electronic Communications Networks (ECNs). Market share in orders of the smallest size categories and most volatile industries is shifting from the NYSE to the ECNs, provided there is sufficient liquidity in the ECN order books, while the largest orders and in the least volatile sectors are shifting back to the NYSE. Evidence is also presented that shows that the NYSE is the market center of choice in periods of high volatility, but trade in the most volatile industries is shifting fastest to the ECNs in small orders. These three thesis projects form the basis of the case studies we look at in this chapter. Before we review their findings, we first review the finance literature, which can be divided into three main strands: explanation of choice of trading based on costs, information and strategic choice.

We now review the three strands of the literature.

THE LITERATURE ON MECHANISM SELECTION

Cost-based literature

When an investor wishes to trade a homogeneous commodity, such as a security, the principal objective will be to maximize profit. Investors seek to buy at low prices and then sell at higher prices, with minimal associated costs. There are, in fact, several costs associated with trading securities, some variable (bid–ask spread[2]), while others are fixed (subscription fees). According to the cost-based literature, mechanism selection is a balance between two different cost structures. In order for two mechanisms to run parallel to each other, there has to be a difference in cost structures between alternatives.

Several empirical studies test the impact of cost difference on choice of trading mechanism. DeJong et al. (1995) use a two-month tick data-set to estimate the effect of a specific type of transaction cost, namely bid–ask spreads, on the choices of stock traders

---

[2] The bid–ask spread essentially represents the cost of the trading through a dealer.

between the centralized auction market of the Paris Bourse and the SEAQ, the quote-driven, decentralized dealership system at the London Stock Exchange (LSE). They find that for French equity orders of small and normal market sizes, different measures of bid–ask spreads are lower at the Paris Bourse than at the LSE, while the opposite holds for very large orders. DeJong et al. (1995) observe that transaction costs determine where orders are traded, as the centralized market of the Paris Bourse attracts relatively more small orders, whereas the LSE's bilateral exchange system is more popular for the exchange of large orders.

In a similar study, Huang and Stoll (1996) compare transaction costs in the form of bid–ask spreads across rival trading routes. Over a period of one year they focus on two paired samples of stocks traded either via the decentralized dealer market NASDAQ (the National Association of Securities Dealers Automated Quotations) or the centralized auction market of the NYSE. In line with Christie and Schultz (1994), Huang and Stoll (1996) find that spreads are roughly twice as large on NASDAQ as on the NYSE. Unlike DeJong et al. (1995), Huang and Stoll (1996) rule out liquidity differences between the two exchange mechanisms as potential causes of the observed differences between spreads. Instead, they emphasize that transaction cost differentials persist because of other determinants of exchange mechanism selection, on which NASDAQ enjoys a competitive edge over the NYSE. These factors are related to differences in the trading rules that apply to the alternative exchange mechanisms.

Another stream of literature looks at the impact of transaction costs associated with bid reduction and zero quantity spread. A central theme of this literature is trade size. Bid reduction represents a transaction cost that varies according to the size of the order (Viswanathan and Wang, 1998). On the dealer market, the larger the order, the lower the bid will be, as the market maker is able to reduce the cost of trading per share, while still covering his costs. On the order book, the costs are inherent in the transaction cost, and therefore as the size increases so do the costs. Traders with large trades favor the dealer market, while smaller trades are more commonly routed through the order book (e.g. Bagliano et al., 1999; DeJong et al., 1995). This makes the hybrid structure more profitable for traders.

Another strand of the cost-based literature explores how trade mechanism selection is affected by *risk* differentials between a stock exchange's centralized auction market and its decentralized dealership system. In line with Haigh et al.'s (2005) assertion and Momsen's (2006) models, risk-averse equity traders prefer the "implicit insurance" (Pagano and Roell, 1996) of "firm prices" (Bagliano et al., 1999) quoted by dealers over exposure to price and execution risk inherent to trading via exchanges' auction markets. The literature has identified a further type of risk that traders of large equity orders may be able to hedge by means of bilateral negotiations in dealer markets. For instance, Seppi (1990) builds a model of adverse selection, under which block brokers of an exchange's dealer market protect traders (for reputational reasons) from being exploited by asymmetrically better-informed counterparties in the exchange's centralized market. Cheng and Madhavan (1997) fail to reject this theory empirically. However, since block trades in derivatives are negotiated directly between the participating sides without the intermediate screening of counterparties by brokers, an explanation for trade mechanism selection of this type does not apply to derivatives markets.

Ellul et al. (2005) consider liquidity as the foremost reason affecting mechanism selection. Traders may be prepared to pay a higher transaction cost so that orders are processed immediately, rather than waiting for orders on the other side of the market. Due to the discrete nature of the order book, as the trade size increases there is less chance of finding natural liquidity. This demand for immediacy and continuity, combined with the execution risk, justifies the dealer's power to demand the additional transaction costs (Grossman and Miller, 1998). Large trades are directed through the dealership system due to the guaranteed liquidity, the significantly greater depth provided, and the fact that the price impact of large trades is far smaller on the dealer market than on the order book (Cheng and Madhavan, 1997).

Information-based literature

The most widely explored area in information-based explanations of mechanism selection deal with the issue of transparency. Pagano and Roell (1996) define transparency as the extent to which market makers are able to observe the size and direction of current order flow. The idea that greater transparency lowers trading costs is popular, and central to information-based arguments. Transparency reduces the opportunities that informed traders have of taking advantage of those that are uninformed (Madhavan and Smidt, 1991). Auction markets are inherently more transparent than dealer markets; all information is available to all participants, whereas dealer markets display very limited information. Extant literature describes how trade disclosure through call auctions blends diverse information from the traders to achieve informational efficiency, while minimizing adverse selection problems, inherent with informational asymmetries (e.g. Admati and Pfleiderer, 1991; Economides and Schwartz, 1995; Pagano and Roell, 1996).

Strategy-based literature

Traders' *strategic motivation* for choosing between alternate trade mechanisms is empirically least developed. The strategy-based literature suggests that mechanism selection is part of a strategy that incorporates the costs, the information disseminated, the other investors and traders, and the timing of orders. Snell and Tonks (2003) conclude that inside information has a large influence on the stock price movements, so the value of sequential trading in revealing information (and reducing trading costs) is high. In contrast, where liquidity trading is the predominant source of stock price volatility, the value of competitive bidding in reducing trading costs is high, so the order book is preferred.

The games in this field of the literature are rooted in the previous cost-based and information-based literature, although the ideas are developed using game theory. Shin (1996) dynamically explores the price uncertainty associated with the order book. Similarly, Ellul et al. (2005) examine thick market externalities in which the gains from trading are dependent on the total number of traders in the market.

For the LSE, Ellul et al. (2005) observe that the probability that a trading day's first and last transactions are executed via a decentralized call auction instead of LSE's dealership system increases with a security's general liquidity properties (i.e. how much trading takes place in that security during the day). The authors explain this finding through the greater susceptibility of the call auction market to thick market externalities, which result more easily in strategic coordination failures among users of this exchange mechanism than among users of the dealership market, where market makers guarantee liquidity.

Kugler et al. (2006) experimentally study trade flows via the rival exchange mechanisms at the LSE. They find evidence in favor of the Neeman and Vulkan (2010) (NV hereafter) strategic model that predicts the unraveling of decentralized bargaining concurrently with a flourishing of centralized trading. In NV, we treat trader choice in a different manner, by making assumptions regarding the distribution of prices rather than on trader behavior. According to this theory, in perfect equilibrium all trades take place via a centralized market, such as an order book. Trade outside of this market cannot be mutually advantageous for both parties. Another equilibrium prediction is that traders trade only through a decentralized bargaining market. However, this equilibrium is unstable; if a small number of traders are forced into the centralized mechanism, then the rest will want to follow.

Centralized markets protect "weaker" traders against high prices on the buyer's side, and low costs on the seller's, through improved information and greater overall efficiency. A weak trader can be defined as a trader that is a price taker, rather than a price setter, a trader with little or no negotiation power. In order to prevent paying inflated prices through a dealership-style market, the weakest traders will opt for the order book. In doing so, the next weakest traders will be exposed to the same loss of control that the first group of traders had. They will therefore have to join the book as well. This continues, with the weakest traders leaving the dealership market and joining the order book until the "serious" traders are forced to change markets. When there are too many serious traders in the dealership market, compared with weak traders, it becomes more beneficial for the serious traders to join the alternate mechanism.

In NV, traders may prefer the ability to create long-term relationships with a small number of trading partners, where the prospect of future trade serves as a disciplinary device against opportunistic behavior. Anonymous centralized markets, such as an order book, offer little protection against opportunistic behavior (Kranton, 1996), and by negotiating with dealers at the expected centralized market price, risk-averse traders may reduce their exposure to the centralized market's volatility. These papers also highlight how strategic choice is often based on the impending future of a market. Where a trader can shift between markets that offer the least costly transactions, or the most advantageous place to use insider information, a mechanism that is not able to survive in equilibrium will start to have higher costs and worse conditions for trade as time goes on. This will happen until trade can no longer take place or until a new equilibrium is reached.

Conclusion

In conclusion, the three strands of the literature outline several explanations for trader mechanism choice; empirical studies corroborate theoretical findings by providing in-depth analysis of the interplay of factors influencing mechanism selection. The remainder of this chapter presents details of three case studies that test the above models of mechanism selection using actual data from leading stock exchanges.

CASE 1. DETERMINANTS OF TRADE MECHANISM SELECTION ON DERIVATIVES MARKETS

Since the addition of block trading facilities (BTF) to the exchanges' traditional markets, derivatives traders have faced a choice between two trade mechanisms. They can either execute orders via a centralized marketplace, or negotiate transactions bilaterally, away from the exchange. Momsen's (2006) study contributes to the literature on exchange mechanism selection in financial markets by focusing on the exchange of *derivatives contracts*, joint investigation of both short-term *and* long-run determinants of exchange mechanism selection, and analysis of their comparative character, employing data *across several* derivatives exchanges. The key findings of the study are summarized in Figure 25.1.

Momsen (2006) tests three complementary hypotheses concerning the determinants of exchange mechanism selection in derivatives markets. To this end, he employs data on financial derivatives contracts traded at Eurex, the Chicago Board of Trade (CBOT), the Chicago Mercantile Exchange (CME), and the Tokyo Financial Exchange (TFX, formerly Tokyo International Financial Futures Exchange, TIFFE). In addition, comprehensive intraday data-sets on prices, volumes, and frequencies of centralized and bilateral transactions from Eurex, and monthly trade volume data from all other exchanges were obtained. The sample periods during which trade flows and prices are analyzed range from January 2, 2002 until August 31, 2005 in the case of Eurex and CME. For the other exchanges, the sample is restricted to periods between the launch of the exchanges' BTF and August 2005.

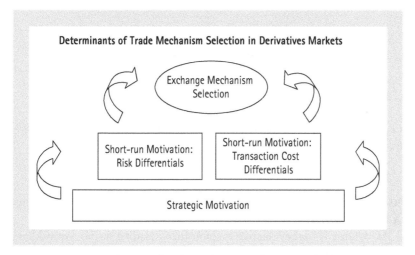

FIGURE 25.1. Determinants of trade mechanism selection in derivatives markets.

Impact of transaction costs differentials

Traders may opt for a specific exchange mechanism because of lower transaction costs associated with transferring contracts via this mechanism instead of the other trading route. Momsen (2006) focuses on a specific type of transaction cost, namely contracting costs (Coase, 1937; Williamson, 1975), which are defined as all the expenditures traders must incur in order to obtain clearance and settlement of a transaction through an exchange's clearing facilities. While the requirements imposed on centralized orders to be cleared and settled are essentially the same at most major exchanges, rules regarding the contracting costs of block trading (BT) diverge across individual exchanges. This allows estimation of the effect of differences in contracting costs on the relative popularity of the alternate trading routes at a sample of four major derivatives exchanges.

Building on the cost-based literature on mechanism design, empirical tests of the effect of transaction costs on traders' choices between alternate exchange mechanisms reveal that the importance of BT relative to centralized trading varies regionally between derivatives exchanges. While they play a significant role at exchanges outside the United States, BT is less important for the transfer of derivatives contracts at exchanges in the United States.

Strict BT rules asymmetrically impose contracting costs on BT participants, while leaving traders in an exchange's centralized market unaffected. Decentralized relative to centralized trading is considerably more popular at the two exchanges outside the United States. For instance, while Eurex's centralized market for options measures only 70% of the volume of bilateral options trades, at CME over 150 times as many options are exchanged centrally as they are bilaterally. Ignoring BT price restrictions and trading hours, BT rules are stricter (and thus BT contracting costs higher) at CBOT and CME,

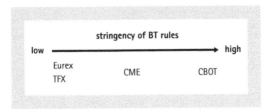

FIGURE 25.2. BT rules are stricter (and thus BT contracting costs higher) at North American derivatives exchanges than in European or Japan.

two North American derivatives exchanges, than at Eurex, a European exchange, and at the TFX (Figure 25.2).

Since BT volume is higher relative to centralized trading at the two exchanges outside the United States, we expect contracting costs imposed on BT to influence trade mechanism selection in derivatives markets in favor of exchanges' centralized markets.

Impact of strategic behavior

Apart from external rules constraining traders' freedom to engage in bilateral BT, a derivative trader's choice between exchange mechanisms may be subject to her own strategic considerations. If this were the case, we would expect traders to gravitate toward some sort of equilibrium behavior in the long run. In the long run (almost) all transactions should be conducted via the exchanges' centralized market, where prices are independent of individual traders' reservation values. According to NV, agents make endogenous decisions about whether to engage in bilateral negotiations, trade via a centralized marketplace, or postpone their desired trade to a future period. The model's equilibrium has almost all trades being executed via the centralized marketplace. NV stress (p. 24) that their model extends to derivatives markets where risk-averse traders engage in centralized trades and reach "private mutual insurance agreements," i.e. BT, to reduce their exposure to risk.

Since reservation values are an almost intrinsically unobserved measure of an agent's perceived worth of a good, the data available are unsuitable to test the theory's prediction regarding the dynamic form of the potential unraveling of an exchange's BTF. On the other hand, the data are rich enough to test for indications of a move toward the predicted equilibrium behavior. Momsen (2006) applies this model to the context of BT and centralized trades in derivatives markets, and observes that in Eurex's options contracts, and CME's Eurodollar future, the frequency and/ or volume of large centralized trades relative to large bilateral trades has increased over the years. However, in several other individual contracts and contract groups there is no trend. In the case of TFX, the centralized market has *contracted* relative to bilateral negotiations over the course of the sample period.

In addition to abstracting from traders' risk attitudes, NV assume that trade via the centralized and decentralized exchange mechanisms entails no or identical marginal transaction costs. This cost-equivalence assumption is difficult to maintain in the case of derivatives markets. It is conceivable, for instance, that asymmetric contracting costs might bolster a potential unraveling of an exchange's bargaining environment. Alternately, if the exchanges' rules favor BT over market transactions, the hypothesized unraveling might decelerate and potentially even fail to assume the extreme form that NV envision. However, regardless of which direction the effect of a potential failure of the cost-equivalence assumption takes, we should—under the theory—observe a movement *toward* equilibrium in the process of a gradual, though possibly incomplete, unraveling of decentralized bargaining.

While the empirical application of the NV model to alternate exchange mechanisms in derivatives markets should not suffer from the mere presence of asymmetric transaction costs, it may lose some of its thrust in the face of repeated, unobserved *changes* to the balance between the mechanisms' contracting costs. However, the currently observed differences between exchanges' BT rules are likely to result from a lack of "best practice" industry standards. Thus, with exchanges still in the process of adjusting the content and stringency of their BT rules, a *ceteris paribus* analysis of trade patterns resulting from strategic behavior may come into conflict with occasional exogenous changes to BT contracting costs.

Momsen (2006) studied the relationship between centralized transactions and BT in options and futures, and in selected individual contracts that were traded at three major derivatives exchanges during a period of either forty-four or twenty-seven months. While discrete, one-off rule adjustments may disrupt and delay a potential trend in the strategic competition between rival exchange mechanisms, such changes should only temporarily boost the popularity of one exchange mechanism at the expense of the other, while leaving traders' strategic behavior unaffected. Thus, it should still be possible to base an analysis of traders' strategic exchange mechanism selection on the quantification of long-term trade flows conducted via the centralized market and the BTF of derivatives exchanges. Potential trends can be interpreted as indicative of traders' strategic motivation for exchange mechanism selection, while sudden shifts in the popularity of decentralized bargaining relative to centralized trading will be attributed to exogenous rule changes. Formally, with $t = \{1, \ldots, 44\}$ for Eurex and CME, $t = \{1, \ldots, 27\}$ for TFX and $\varphi = \{options, futures\}$ we would expect the series

$$\alpha_{\varphi,t} = \frac{XL - Frequency_{\varphi,t}}{BT - Frequency_{\varphi,t}} \text{ and } \delta_{\varphi,t} = \frac{XL - TradeVolume_{\varphi,t}}{BT - TradeVolume_{\varphi,t}}$$

to increase over time.[3]

---

[3] *XL* represents extra large trades. *XL* trades are computed by splitting all trades via the central order book into two subgroups according to whether or not the centralized transactions meet the contract-specific volume requirements applicable to off-exchange BT. This generates a group of centralized transactions that are large enough that they could have been executed bilaterally away from Eurex's centralized market.

Unfortunately, only the intraday data-set of transactions at Eurex contains sufficient information to determine the frequency of trades and to distinguish centralized trades by size (i.e. α and δ). Across all exchanges, i.e. with $\chi = \{Eurex, TFX, CME\}$, we can only test if $\gamma_{\varphi,t}^{\chi} = \frac{XLX-TradeVolume_{\varphi,t}}{BT-TradeVolume_{\varphi,t}}$ increases over time.

Options

Turning to the analysis of the variables' time patterns, we observe a steady increase in $\alpha_{options,t}$ and $\delta_{options,t}$ over the sample period (Figure 25.3). This implies a positive trend both in the frequency and in the volume of *large* centralized options trades relative to large bilateral transactions in options. Thus, although still relatively small, Eurex' centralized market for large options orders has grown relative to bilateral negotiations.

However, the observed relationship across time between the volume of *all* centralized trades relative to that of BT at Eurex ($\gamma_{options,t}^{Eurex}$) and at CME ($\gamma_{options,t}^{CME}$) fails to show an increasing trend. While the time series seems to fluctuate around a fairly constant mean (0.7) in the case of Eurex, it fell sharply over the sample period's first six months in the case of CME and remained stable for the rest of the time covered in this study (i.e. to the end of 2005)—with the marked exception of four months in the summer of 2003. During these months, *total* BT volume was only half of the *average* monthly BT trade volume of the preceding seventeen months. This explains the observed spike in γ (Figure 25.3). This fleeting phenomenon cannot be ascribed to strategic motivations of exchange mechanism selection, but is instead related to a temporary rise in transaction costs for BT relative to centralized trading at CME during the summer of 2003.

Futures

Inspection of Figure 25.4 confirms that—contrary to options—Eurex's centralized market for futures dominates the exchange's BTF. Moreover, a striking co-movement of $\alpha_{futures,t}$ and $\delta_{futures,t}$ suggests that trade sizes of large on- and off-exchange trades in Eurex's futures contracts are similar. There seems to be a *negative* trend in the relationship between the volume of all centralized trades and BT at TFX ($\gamma_{futures,t}^{TFX}$) and at Eurex ($\gamma_{futures,t}^{Eurex}$) during the period before 2003. Thus, the centralized market in futures does not gain in importance relative to bilateral negotiations at any of the three derivatives exchanges.

Individual contracts

Inspection of Figure 25.5 does not reveal an overriding pattern across time in the relationship between centralized and decentralized trades in the exchanges' most heavily traded contracts. Nevertheless, it is worthwhile to consider some of the time series in more detail.

Furthermore, econometric analysis reveals a distinct seasonal pattern in several time series, which we attributed to the quarterly rolling over of derivatives contracts into

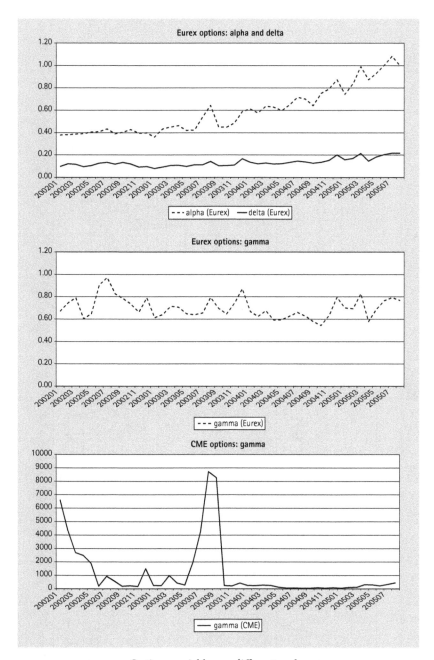

FIGURE 25.3. Options variables on different exchanges, 2002–05.

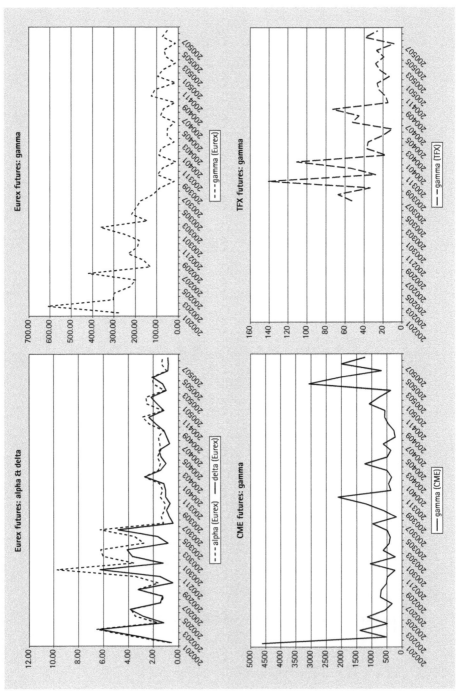

FIGURE 25.4. Futures variables on different exchanges, 2002–05.

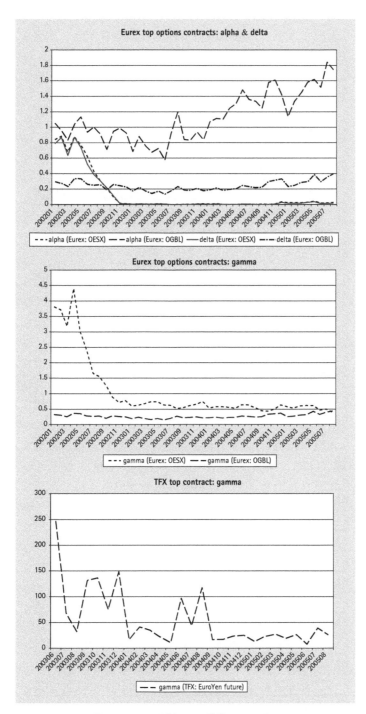

FIGURE 25.5. Contracts variables on different exchanges, 2002–05.

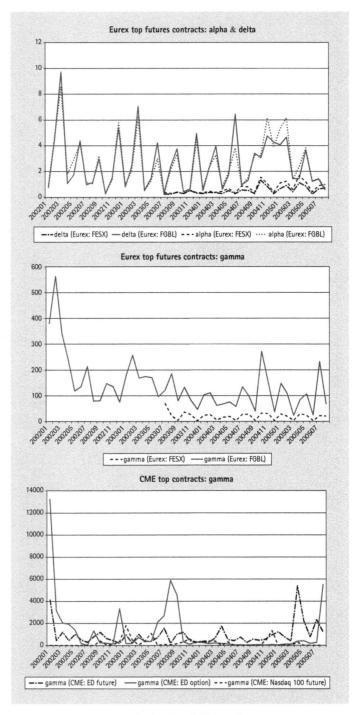

FIGURE 25.5. Continued

future maturity months. However, further research is required to formulate a robust theory regarding a trade-concept view of exchange mechanism selection in derivatives markets.

Impact of risk differentials

If BT protects risk-averse traders from price and execution risk, while large centralized trade does not, derivatives traders should prefer to negotiate the exchange of large-sized orders bilaterally instead of centrally when an exchange's centralized market becomes less liquid and the market price more volatile. The negative relationship between market liquidity and the probability of traders to prefer BT to centralized trading stems mostly from the estimation of primary effects.

Short-run trade mechanism selection is motivated by risks that arise under certain market conditions. Market illiquidity and price volatility are likely to determine the intensity of these risks. Momsen (2006) considers daily measures of liquidity (Figure 25.6) and price volatility to capture the information set available and employed by traders upon their choice between alternate trade mechanisms.

If derivatives traders engage in bilateral negotiations to avoid execution risk associated with the transaction of large orders in an exchange's centralized market, they should "be willing to pay to reduce their exposure to risk" (N V, p. 24). Thus, risk premia should be built into the prices of BT. Since BT shields both involved parties equally from the execution risks inherent to the exchange of large trades via the centralized market, we should not observe the prices of BT to systematically diverge from those of centralized trades. Thus, as long as the parties participating in BT are assumed to be equally risk averse, potential risk premia should cancel each other out.

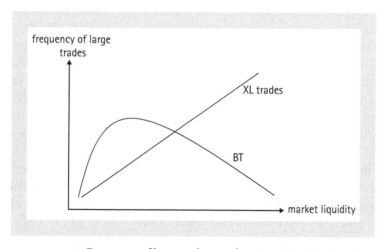

FIGURE 25.6. Frequency of large trades as a function of market liquidity.

Based on cross-sectional regression analysis[4] of information on daily transactions in a specific derivatives contract, *frequency*$_i$ and *volume*$_i$ are daily measures of market liquidity, and *volatility*$_i$ expresses the daily price volatility experienced in a contract's centralized market. The proposed argument holds that when liquidity and price stability fall in the centralized market, traders become increasingly exposed to two types of risk, which lead them to prefer riskless BT to risky centralized trades. However, there is no significant relationship between price volatility and traders' choices between alternate exchange mechanisms, which can be interpreted as evidence against the importance of price risk in traders' mechanism selections.

There is evidence in favor of the claim that BT becomes more attractive when liquidity decreases in an exchange's centralized market. Under such circumstances, traders risk increasingly long delays in the centralized execution of their orders, while BT offers traders the possibility of negotiating the complete transaction of an order and thus protects them from execution risk. This finding confirms that of Ozgit (2005), who observes that, as bet sizes increase, differences between prices charged to gamblers via the alternate betting mechanisms vanish. Second, Ozgit (2005) notes that pure price competition liquidity plays a critical role in determining punters' choices between betting venues, and "whenever exchange markets are thin, order flow migrates to the bookmaker" (p. 3). This again is consistent with the result that liquidity attracts derivatives traders to an exchange's centralized market, whereas a lack of market depth induces traders of large orders to bilaterally negotiate the exchange of derivatives contracts.

Furthermore, if traders are motivated in the short run to engage in BT in order to obtain execution security, prices of BT should be aligned with those of centralized trades. Using data on match prices, Momsen (2006) applies a method similar to Haigh et al.'s (2005), involving bid–ask prices. To test the hypothesis of the alignment of BT prices with prevailing market prices, four reference prices for every BT are constructed from the execution prices of centralized transactions that occurred in the same contract and on the same trading day prior to the BT.

If different price patterns were observed, traders might be motivated to engage in BT for reasons other than to avoid risk exposure in the centralized market, for example to negotiate more favorable price conditions away from an exchange's centralized market. However, there is no evidence to reject the hypothesis that BT prices are well aligned with market prices. This is consistent with the notion that, in the short run, traders engage in BT primarily for reasons related to fluctuating risk differentials between centralized trading and decentralized bargaining. There is no evidence to confirm that unusually high price volatility induces traders to go "off-exchange" and bargain over a BT instead of offering an order of a comparable size to Eurex's centralized market. We interpret these findings as indicative of the role that executions risk rather than price risk plays in short-run trade mechanism selection in derivatives markets. We find that the prices observed for BT in Eurex's options contracts are consistent with a risk-based explanation for short-run trade mechanism selection.

---

[4] Both linear and binary curve models were estimated.

CASE 2. IMPACT OF VOLATILITY ON MECHANISM SELECTION: EVIDENCE FROM THE LONDON STOCK EXCHANGE

Since 1997, the LSE has been a hybrid market structure. Traders have a choice between SEAQ, a quote-driven dealership market where trade is based around direct and bilateral negotiation between trader and market maker, and SETS, an electronic order-driven exchange with a limit-order book. SETS was initiated in October 1997, to complement the existing SEAQ by lowering trading costs and increasing trade volumes, as well as to counter the concentration of order flow on the sell side at SEAQ. With the limit-order book exchange, there is no market maker acting as the 'middleman' to set prices, so there is no bid–ask spread. Instead of a spread, there is simply a fixed price, determined by the exchange in order to induce maximum amounts of trade. During the first five years of its existence, SETS became the more popular mechanism, commanding upwards of 70% of all trades in SETS securities.

Henry (2003) investigates trader preference for trading mechanisms by comparing the differences between the two trading mechanisms at LSE. By focusing on differences in average trade size between the two mechanisms, Henry (2003) establishes the relationship of several price-related factors on the traffic of trade. The effects of factors such as volatility, price differences, spread size, and spread formation on mechanism selection have been considered at both a general level and a more detailed, sector-oriented level. Henry (2003) finds that sector volatility has a decisive influence on the effects of the other factors. This was evident in the difference in dealer behavior observed according to the sector of the security.

Impact of trade size on mechanism selection

Central to mechanism selection is the size of the trade; while the use of SETS is becoming more popular, it seems that large trades are still being routed through the dealer mechanism, and generally the trades that are switching from the dealer mechanism to SETS are smaller than the overall average trade size. Figure 25.7 shows that the average trade size (the percentage value of trades divided by the percentage number of trades) for the SETS mechanism is decreasing, while the dealer market trade size is increasing. Thus, the average trade sizes for the two mechanisms are diverging.

The argument for this switching is not solely based on just one cost factor, but on several: the liquidity of the dealer market, the operating costs, and the price uncertainty. Following is a brief overview of the factors considered by Henry (2003) and their impact on mechanism selection.

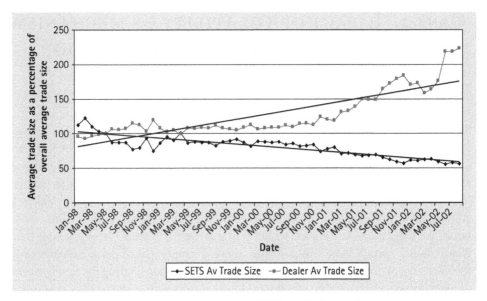

FIGURE 25.7. Standardized average trade size for SETS and dealer market trade, 1998–2002.

Volatility

In order to compare the volatility of the two mechanisms, the daily volatilities were compared.[5] For 60% of the days in the 420-day period from January 2001 to August 2002, the SETS mechanism was more volatile (Table 25.1). Although it appears that, overall, the SETS mechanism is more volatile than the dealer market, this result does not hold across sectors.[6]

Even though the ranking of the sectors was the same for both mechanisms, there is an actual difference between them. Table 25.1 shows that for some sectors the SETS mechanism is more volatile than SEAQ, and for others the opposite is true. As the dealer market becomes more volatile than SETS (the overall volatility is high), the dealer increases the ask volatility relative to the bid. When the overall volatility drops, so does the SEAQ's volatility (compared with SETS), and the dealer decreases the volatility of the ask price relative to the bid. The relevance of the volatility in the demand for dealership services was first suggested by Demsetz (1968), who believed that the demand for immediacy is dependent to a large extent on the degree of price uncertainty, which is a view also taken by Copeland and Galai (1983), who report that during periods of high volatility traders prefer trading through bilateral negotiation to leaving limit orders on the book. Friederich and Payne (2007) add that the demand for immediacy, and so the

---

[5] Volatility is a calculated using historical stock prices, meaning the SETS price, the mid-point of the SEAQ spread, and the actual bid and ask prices of the spread. These can all be used to calculate testable volatilities. The volatilities of the SETS price and the dealer mid-point were calculated for each sector.

[6] Over this period, information technology was more volatile in the dealer market than in SETS.

Table 25.1. Overall daily rank scores for volatility on both the dealer market and SETS

		Lower 95%	Upper 95%	Dealer Volatility Rank	Lower 95%	Upper 95%	SETS Volatility Rank
Industrials	Intercept	4.53	4.78	5	4.54	4.71	5
	Gradient	0.00	0.00		0.00	0.00	
Consumer Goods	Intercept	5.54	5.85	6	5.58	5.90	6
	Gradient	0.00	0.00		0.00	0.00	
Services	Intercept	2.97	3.31	3	3.12	3.49	3
	Gradient	0.00	0.00		0.00	0.00	
Utilities	Intercept	2.05	2.33	2	2.04	2.41	2
	Gradient	0.00	0.00		0.00	0.00	
Financials	Intercept	4.16	4.48	4	4.09	4.32	4
	Gradient	0.00	0.00		0.00	0.00	
Information Technology	Intercept	1.00	1.00	1	1.00	1.00	1
	Gradient	0.00	0.00		0.00	0.00	

demand for SEAQ, should be negatively related to a variable measuring trading activity in a stock.

Price formation effects

Since the introduction of SETS, the SETS price has trended toward the mid-point of the spread, suggesting that the SETS price is having a greater effect on dealer price formation. Until the introduction of the Central Counterparty (CCP) facility in 2001, the SETS price was, on average, higher than the mid-point of the spread. After March 2001, the opposite was true. The introduction of the CCP changed the absolute cost leadership mechanism. The effects of the CCP are evident in Figure 25.8, which demonstrates the relationship between the SETS price and the dealer price, showing the difference between the percentage value of trades through SETS and the percentage value of trades at the SETS price. This difference represents the value executed on the dealer market, at the SETS price.

As a consequence of the introduction of CCP, dealers no longer have as much negotiation power and are not able to justify such high prices. Dealers and traders negotiate SEAQ prices, and the establishment of the CCP has in fact increased trader negotiation power. Traders can demand lower prices from market makers, using the threat of switching to the order book as negotiation power. This power is dependent on the size

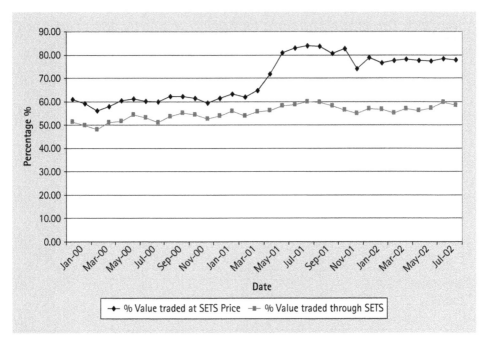

FIGURE 25.8. Percentage value through SETS vs. percentage value at SETS price.

of the trade being executed, with the larger traders commanding more power, but has allowed traders to demand more competitive prices. Dealership services are still able to offer immediacy services and are arguably a more liquid mechanism for exchange, although the launch of the CCP was aimed at improving the liquidity of the order book. Market makers are therefore not completely void of negotiation power and, in fact, when the market is strong, and the dealers are keen to trade, they command more negotiation power through the flexibility of the prices offered.

Furthermore, the spread width represents the pricing structure on the dealer market, and is a useful means of representing a dealer's willingness to trade. A wide spread indicates that the dealer is more reluctant to trade than if the spread were narrow. Dealers' pricing decisions are directly related to the volatility of the security. This willingness to trade obviously has an effect on trader mechanism selection. When the dealer is happy to trade, it is possible to make the dealer mechanism more attractive, by lowering the spread. It is no surprise that the spread size is positively related to the percentage of trade through the order book. The spread size is also related to the percentage at the SETS price. When the dealer is prepared to reduce the spread, the SETS price has less effect on price formation.

For high-volatility stocks the dealer market is preferred, whereas for less volatile stocks it is the SETS market that is preferred. This manifests itself in the spread widths. During times of high volatility, the width of spreads in highly volatile stocks increases, as dealers prefer not to trade. For less volatile stocks the opposite is true. Dealers service a larger percentage of the overall trade in these sectors, and can therefore set lower spreads, as they believe they are exposed to reduced risk, and are more eager to trade. The spread size is the dealer's reaction to the market situation, and is a way of comprehending the dealer's willingness to trade. The volatility is a factor that influences the dealer's price formation, which subsequently affects the trader's selection of mechanism.

Model of trade size

The observations from the aforementioned factors are linked to mechanism selection using a model that relates the average trade size to the traffic of trades. As the traffic of trade passing through SETS increases, the average trade size also increases. If all other factors are ignored there should exist an equilibrium average size associated with both the overall market and both mechanisms. So, while the sizes of the mechanisms are diverging, and have been diverging for all this period, there should eventually come a point when the average size of each sector is constant on both SETS and SEAQ. Assuming this happens, mechanism selection will resort to a simple decision concerning the size of the trade. Up to a certain value, the trade will be sent to SETS, and above that value SEAQ will be used.

The average overall trade size (for every sector and for the market as a whole) appears to be converging on a single sector-specific value. This does not mean that the number of bargains is converging on a single figure, nor the value of trades. The introduction of an order book reduced the overall average trade size steadily after 1998. Eventually equilibrium will exist where the average trade size of all trades is constant over time. In fact, the size of trades on the dealer market is growing, and the SETS average size is decreasing. These trends are not infinite; the SETS size is not tending to zero, but to a non-zero value, and the SEAQ size will not grow to infinity. Traders have a trade-off decision based around the cost structures of the two markets. Up to a certain cut-off point it is beneficial to send trades through SETS. Above this point the dealer market is the preferred mechanism.

Assuming that the overall average trade size is constant (denoted by the thicker line at $y = 1$ in Figure 25.9), there are several combinations of trade sizes that the SETS and the SEAQ mechanisms can select. Examples of these combinations can be seen in Figure 25.9, at the intersection of the vertical, thinner solid tie lines. Figure 25.9 shows the distribution of trade between the two mechanisms as well. The distance: $\frac{ab}{ac} \times 100\%$ represents the percentage of trade through the SETS mechanism, which has an average trade size of "SETS b."[7]

Figure 25.9 clarifies the effects of the other factors on the average trade size, and hence on the percentage of trade passing through either mechanism. As the average trade size on the SETS mechanism decreases, the percentage through the order book will also decrease, meaning that more trade is passing through the dealer market. Volatility,

---

[7] The SEAQ percentage is 1 – SETS percentage, and the trade size is SEAQ b.

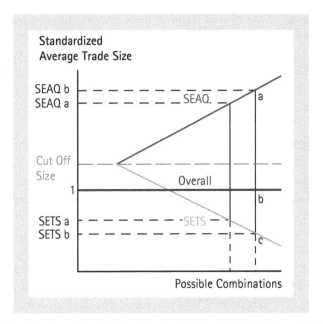

FIGURE 25.9. Combinations of trade sizes that the SETS and the SEAQ mechanisms can select.

price formation, price difference, and bid–ask spread width are all factors that influence mechanism selection, and in turn determine the combination of average trade sizes for the two mechanisms.

Extant literature discusses the existence of a cut-off point, which is determined primarily through a comparison of trading costs. The literature cites different types of costs; for instance, Bagliano et al. (1999) conclude that larger trades are executed through the dealer market, as the costs of price uncertainty outweigh dealer transaction costs. DeJong et al. (1995) conclude that the cost of trading on the dealer market is decreasing with increasing trade size. They cite the order processing costs as an important determinant of the bid–ask spread. Other dealership cost theories have been used to explain these types of observations (adverse selection, inventory control), but these all predict an increase in trading costs with trading size, which are not coherent with the model presented by Henry (2003).

Since the impact of these factors is not distributed uniformly across all industries, Henry (2003) applies his model to explain the differences he observed for different sectors. For instance, concerning the average SETS size, the information technology sector—the most volatile sector in his study (Figure 25.10)—seems to have a significantly larger decreasing rate, whereas the trade size in the much less volatile consumer goods sector falls significantly slower. Hence, for each sector the overall average size is tending toward a particular value, specific to the sector.

The relative volatility of the sectors plays a very large part in the investment decisions and strategies of traders, and these strategies clearly involve, and depend on, trading

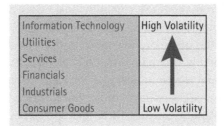

FIGURE 25.10. Volatility of the various sectors studied by Henry (2003).

mechanism selection. The key factor that influences the trader's mechanism selection is the volatility. The volatility induces changes in the dealer's bid–ask spread width, which in turn directly influence trader decisions. Dealers cope with volatility information, particularly mechanism comparison information, differently according to sector. Both dealers and traders have sector-specific strategies, and volatility is at the heart of these decisions.

CASE 3. LONG-TERM DETERMINANTS OF MECHANISM SELECTION: EVIDENCE FROM THE NYSE

Extant literature concentrates on the differences in the cost and information content of trade through different market centers. Hutchins (2004) establishes that long-term changes in market selection do not solely reflect cost or information benefits, but differences in actual mechanism efficiency and the environments in which different traders can survive depending on their bargaining power. Hutchins (2004) utilizes order execution quality disclosure data disseminated by all US securities market centers to analyze the market share of orders in all NYSE-listed stocks. In particular, Hutchins (2004) compares the share of orders held by the NYSE itself, a hybrid limit-order/dealer market, in comparison with a relatively new group of market centers called electronic communications networks (ECNs)—essentially web-based limit-order books.

The NYSE is an example of a continuous-session hybrid market. Trade via the NYSE mechanism is an order-driven/quote-driven hybrid. An increasing amount of NYSE stocks are traded via a relatively new group of market centers, ECNs. ECNs are networks of traders, connected to a central limit-order book; in this way they are continuous and order driven. However, in general, ECNs are thought to be more centralized than the NYSE, as all trade takes place in a single, observable order book, whereas NYSE is both quote and order driven. This combination offers a highly stylized environment for assessing whether traders prefer to trade (using Neeman and Vulkan's (2010) classification) in a decentralized bargaining market—a dealer market—or a centralized limit-order book.

Using data obtained from the NYSE and individual ECNs, cross-sectional analysis of market share of trade orders in different industries and size categories was performed. The data spanned a period of twenty-five months, from June 2001 to June 2003, and provided details of monthly order flow to each market center. The data were split into four size brackets and eleven industry sectors, and hence allowed detailed cross-sectional analysis of the characteristics of order flow to either the NYSE or the group of ECNs. Coupled with volatility statistics, these data serve as evidence to test the theory that trade is shifting from the NYSE to the ECNs in the smallest order categories, which are associated with traders who have less market power. In the largest order sizes, there is evidence for a market share recapture by the NYSE. Volatility correlation shows that the NYSE is the market of choice during periods of heightened fluctuation. In the longer term, trade in the most volatile markets is moving to the ECNs fastest.

A trader's strategic choice about market mechanisms depends on a much longer-term consideration of the optimum strategy for maximizing pay-offs. Extant literature in this field tends to model mechanisms within the same market in order to observe whether there are equilibria that allow mechanisms to exist side by side—as is the case in real markets. According to the literature, if one mechanism is more likely to survive, then rational traders will gradually shift to that mechanism over long periods of time, incorporating shorter-term factors, such as cost[8] and information,[9] within a longer-period trend. Glosten (1994) points out the ability of specialists and floor-brokers to recognize and sometimes penalize, in future trade, those who have access to information.

ECNs, on the other hand, operate completely anonymous order books, and so could harbor informed traders, increasing the chance of adverse selection in the order book. Reiss and Werner (2005) negate this possibility based on evidence from their study of brokered and direct interdealer trades. They find that trades that are information motivated tend to migrate to the direct (non-anonymous) public market. Short-term factors such as cost and information content at a market center will affect day-to-day trade, but in competitive markets the market center with lowest costs and fraction of informed traders can vary just as frequently. In the longer term, it is the characteristics of trading mechanisms that affect whether a market center attracts informed traders, has a larger average order size, or costs the least. The research to which Hutchins (2004) contributes endeavors to show long-term shifts in orders on each of these markets belying a movement of traders' mechanism preference. These trends would not solely reflect cost or information benefits, but differences in actual mechanism efficiency and the environments in which different traders can survive, depending on their bargaining power.

---

[8] Costs include transaction costs, the cost of giving the option to trade to others, and costs associated with risk of exposure to informed traders. Hutchins (2004) argues that speed of execution can also be thought of as a cost, and is intertwined with the risk factors within cost; if, for example, an order stays on a limit-order book for a long time, it is at risk of the market moving away from it, allowing execution at an unprofitable price (for the trader who provided the option to trade).

[9] Information factors refer to the amount of inside information, or just knowledge of companies and trading experience, that investors have and, particularly, where traders with differing knowledge choose to trade.

Long-term determinants of mechanism selection

Parlour and Seppi (2003) model a hybrid market, like the NYSE, alongside a pure limit-order market, like an ECN. They find that, given heterogeneous transaction costs, multiple equilibria are supported where both mechanisms can coexist, with different mechanisms dominant in different conditions, as well as equilibria with a single surviving mechanism. "Tie-breakers," such as payment for order flow and traders' behavioral habits, are key in breaking coexistence down to a single market. Similarly, NV examine centralized markets versus a direct negotiation market, i.e. a limit-order book and a telephone-based dealer market, respectively. Traders range in willingness to pay and willingness to sell: the weakest types are very willing to pay, or very willing to sell (have the lowest costs). When a centralized market is set up, the weakest traders find that their expected payoff is higher trading on this new market and thus shift their trade activity. In doing this, the next-strongest traders become the weakest in the marketplace, so they too shift their trade. The end-game scenario is that no trade can take place outside of the centralized market. Thus, if ECNs were to compete with a telephone-based dealer market, they would be expected to gain all order flow in the long run.

Data on order flow from the NYSE shows that market share in orders of the smallest size categories and the most volatile industries is shifting from the NYSE to the ECNs provided there is sufficient liquidity in the ECN order books, while the largest orders and those in the least volatile sectors are shifting back to the NYSE from the ECNs (see Figure 25.11 [10]).

This shift can be attributed to a process of the weakest traders moving in accordance with their optimal payoff strategy, as in NV. However, this trend could also be explained by a simple increase in the number of traders. ECNs have allowed an influx of new home traders using home investment programs and online broker services. This may mean that traders have not moved, causing a shift in market share, but that would suggest that the number of orders on the NYSE has remained constant.

A cost-based explanation of the trend would suggest that trading costs on ECNs have fallen way below costs on the NYSE. However, Bennett and Wei (2003) examined stocks that changed between listing on either the NASDAQ or the NYSE, and show that volatility reductions and better price discovery and improvement were consistently better on the NYSE. They noted that the proliferation of ECNs on the NASDAQ had improved market efficiency and thus improved trading costs, but that these costs still trailed those of the NYSE.

Impact of order size across industries

Figure 25.12 splits the size category statistics into industries. We can observe a range of market share growth and decline rates for each market center group. In the smallest two size brackets, the technology and telecommunications sectors had the fastest growth

---

[10] All of the plots shown in Figure 25.11 show market share of the total market, being the sum of orders at the NYSE and ECNs, which explains the mirror image effect between the two lines.

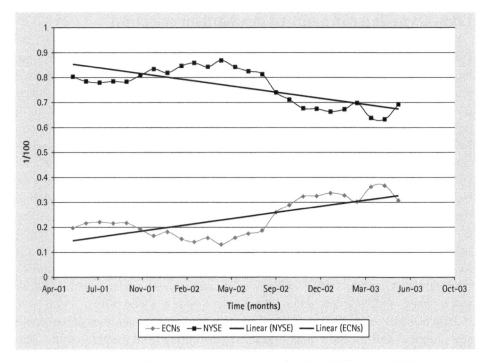

FIGURE 25.11. Percentage of market share of orders, NYSE versus ECNs.

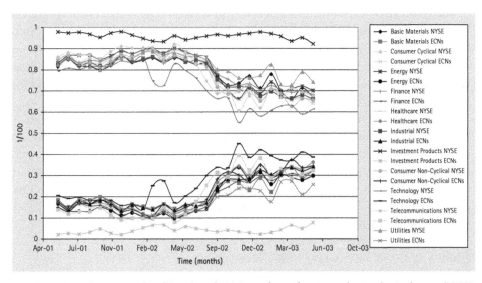

FIGURE 25.12. Price-correlated market share in orders of 100-49 shares, by industry (NYSE versus ECNs).

rates at the ECNs, and the utility and investment products the slowest, or showed little sign of growth. The former, fastest growth areas are linked to high volatility and activity. This leads to the conclusion that small traders in volatile stocks, whose trading position is weak, prefer to trade through ECNs, and the activity levels are such that ECNs have sufficient liquidity to support trade. The investment products and utility sectors have the least activity in these size categories and, on average, are the least volatile sectors. Low volatility is associated with a relatively stronger trading position, leading to a lower incentive to shift to an optimum market, while ECNs have little liquidity in these sectors due to lack of activity, explaining the slower rate of market share change.

Thus, the results of this study clearly show that ECNs are gaining market share in the smallest orders for trade. Without correcting for double reporting it is not possible to show that primary orders are sent to the ECNs with increasing frequency, but it is possible to state that ECNs are becoming more popular for trade in these sizes, as they are being "hit" more times. ECNs are finding ways around the trade-through rule by use of fleeting and hidden orders; this is not a logical step if the NYSE were the most popular choice for transactions. In the larger size brackets, there is even less likelihood of hidden depth, thus orders are routed directly to the NYSE for execution, so we see no shift, or a shift in favor of the NYSE. Meanwhile, traders in the largest orders who traded on the NYSE before they had a choice to move to the ECNs are thought to benefit much more from price improvement and the smart-order-book capability of floor brokers, as near-block orders will suffer from price shocks on a limit-order book with little liquidity.

In the larger size categories a slow-down in market-share change is observed (see Figure 25.13). Although relative activities in different sectors remain similar, the overall market activity in larger orders decreases, leaving ECNs with even less liquidity for a successful transaction where an order is filled. An overall change in growth direction is observed as the order sizes increase. This suggests that, after an initial trial period on the ECNs, larger-order traders have realized the lack of liquidity there and have subsequently shifted back to the NYSE.

Impact of liquidity

For the NYSE stocks, the hybrid market thus seems to make markets in the most illiquid of stocks, making it the market center of choice for sectors like investment products but essentially its market mechanism is a slower version of that which would take place in a liquid electronic order book. Traders can optimize their risk and speed by trading anonymously through an ECN, which increasingly offsets the monetary cost benefit of price improvement on the NYSE for many smaller traders. However, in the long term, ECNs must gain sufficient liquidity to be able to compete with the largest primary markets.

Impact of volatility

Finally, volatility is calculated on a standard deviation of the price, and is a measure of its dispersion around the twelve-month average. Based on correlation analysis between

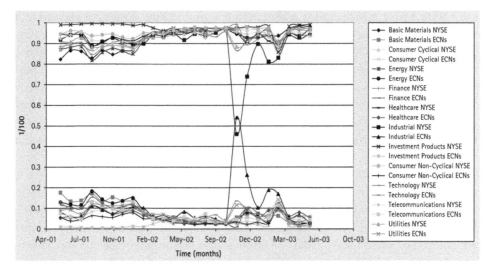

FIGURE 25.13. Price-correlated market share in orders of 5000–9999 shares, by industry (NYSE versus ECNs).

market share and volatility, the data illustrate that short-term periods of high volatility in a market are associated with increased market share at the NYSE. However, the most volatile markets on average are those with fastest growth of ECN market share in the smaller size categories. If the volatility of a market corresponds with the relative weakness of a trader, it is the weakest that move to the NYSE in times of heightened risk. This could be explained by the robustness of the NYSE; it is the oldest and most liquid market, and provides guaranteed execution with a chance of price improvement, whereas ECNs have a shorter, worse track record for liquidity, and do not guarantee execution. Guaranteed execution has particular importance in a volatile period where traders need to lay off positions. Hence, in the short term, weak traders favor the robustness of the NYSE in times of heightened risk. In the longer term, however, one can still observe a shift of weak traders to ECNs. Moreover, if ECNs were more liquid, we might see an increase in the chance of execution, which would encourage more orders during periods of volatility, as ECNs' ability to execute trades quickly would allow traders to lay off positions more quickly.

These trends corroborate previous work well, which showed that order flow had split on the LSE so that largest orders were more frequently sent to a decentralized alternative to a centralized market that received smaller orders more frequently. Although relative activities in different sectors remain similar, the overall market activity in larger orders decreases, leaving ECNs with even less liquidity for a successful transaction where an order is filled. An overall change in growth direction is observed as the order sizes increase, with more than half the sectors in the third size category insignificantly showing NYSE market share growth, and over half showing significant NYSE growth in the fourth. It is suggested that, after an initial trial period on the ECNs, larger-order traders have realized the lack of liquidity there and have subsequently shifted back to the NYSE. As the NYSE had most market share in many of the sectors to begin with, the ranking of the sectors' speeds of growth says less about the traders' relative strengths in each.

Conclusion

The three case studies presented in this chapter highlight the different factors that influence the choice of mechanism selection. While transaction costs are an important determinant of trading mechanism, in the long run strategic considerations play a key role. Momsen (2006) tested if derivatives traders are motivated in the short run by risk considerations when choosing between trade mechanisms. The proposed argument held that when liquidity and price stability fall in the centralized market, traders become increasingly exposed to two types of risk, which lead them to prefer riskless BT over risky centralized trades. Momsen (2006) failed to establish a relationship between price volatility and traders' choices between exchange mechanisms, which we interpreted as evidence against the importance of price risk in traders' mechanism selection decisions. However, there is strong evidence in favor of the claim that BT becomes more attractive when liquidity decreases in an exchange's centralized market. Under such circumstances, traders' risk increasingly long delays in the centralized execution of their orders, while BT offers traders the possibility of negotiating the complete transaction of an order and thus protects them from execution risk. While this study focuses on the determinants of trade mechanism selection in derivatives markets, a worthwhile extension might empirically examine to what degree the reservations of BT critics have materialized since the widespread launch of BTF at most derivative exchanges around the world.

By focusing on differences in average trade size between the two mechanisms, Henry (2003) attempted to confirm the relationship of several price-related factors on the traffic of trade. The effects of volatility, price differences, spread size, and spread formation on mechanism selection have been investigated at both a general level and at a more detailed, sector level. Based on a model relating trade size and trade volume, the aforementioned factors were linked to mechanism selection. According to Henry (2003), the key factor influencing mechanism selection is volatility. The volatility induces changes in the dealer's bid–ask spread width, which in turn directly influences trader decisions. Dealers cope with volatility information, particularly that of mechanism comparison, differently according to sector. It is apparent that both dealers and traders have sector-specific strategies, and volatility is at the heart of these decisions. Volatility induces changes in the dealer's bid–ask spread width, which in turn directly influence trader decisions.

Similarly, Hutchins (2004) also finds that volatility is a key factor influencing mechanism selection. Low volatility is associated with a relatively stronger trading position, leading to a lower incentive to shift to an optimum market. In the context of the NYSE, the largest orders are more frequently traded on a decentralized market compared with a centralized market that received smaller orders more frequently. Hutchins (2004) observes that, due to the lack of liquidity on the ECNs, large-order traders have shifted back to the NYSE after initial trading on the ECNs. Furthermore, evidence from correlation analysis suggests that high volatility in the short run is associated with increased market share at the NYSE. However, the most volatile markets on average have the fastest growth in trades through ECNs in the smaller size categories. This highlights the robustness of the NYSE at times of heightened risk for the weakest traders, who shift to the ECNs in the longer term.

Competition between exchanges is increasingly commonplace as more and more trade moves online. Lessons from financial exchanges where parallel trading has been ongoing for a number of years now are therefore of great value to all market designers, who must consider the incentives of their customers to either switch to using their exchange or to continue with it when competition is available. We hope that the cases presented here will be useful to future generations of market designers.

REFERENCES

Admati, A. and Pfleiderer, P. (1991) "Sunshine trading and financial market equilibrium," *Review of Financial Studies*, 4: 443–81.

Bagliano, F., Brandolini, A. and Dalmazzo, A. (1999) "Liquidity, trading size, and the coexistence of dealership and auction markets," *Economic Notes by Banca Monte dei Paschi di Siena SpA*, 29: 179–99.

Cheng, M. and Madhavan, A. (1997) "In search of liquidity: block trades in the upstairs and downstairs markets," *Review of Financial Studies*, 10: 175–203.

Christie, W. and Schultz, P. (1994) "Why do NASDAQ market makers avoid odd eighth quotes?" *Journal of Finance*, 49: 1813–40.

Coase, R. (1937) "The nature of the firm," *Economica*, 4: 386–405. Reprinted in Stigler, G. and Boulding, K. (eds), *Readings in Price Theory*, Irwin Publications, pp. 331–51.

Copeland, T. and Galai, D. (1983) "Informational effects on the bid ask spread," *Journal of Finance*, 38: 1457–69.

DeJong, F., Nijman, T. and Roell, A. (1995) "A comparison of the cost of trading French shares on the Paris Bourse and on Seaq International," *European Economic Review*, 39: 1277–1301.

Demsetz, H. (1968) "The cost of transacting," *Quarterly Journal of Economics*, 82: 33–53.

Economides, N. and Schwartz, R. (1995) "Electronic call market trading," *Journal of Portfolio Management*, 21(3): 10–18.

Ellul, A., Shin, H. and Tonks, I. (2005) "Opening and closing the market: evidence from the London Stock Exchange," *Journal of Financial and Quantitative Analysis*, 40(4): 779–801.

Friederich, S. and Payne, R. (2007) "Dealer liquidity in an auction market: evidence from the London Stock Exchange," *Economic Journal*, 117(552): 1168–91.

Glosten, L. (1994) "Equilibrium in an electronic open limit order book," *Journal of Financial Economics*, 21: 123–42.

Grossman, S. J. and Miller, M. H. (1998) "Liquidity and market structure," *Journal of Finance*, 43(3): 617–33.

Haigh, M., Overdahl, J. and Hranaiova, J. (2005) "Block trades in futures markets," CFTC Working Paper.

Henry, S. (2003) "Motivations for trading mechanism selection: hybrid markets vs. limit-order books. Evidence from the London Stock Exchange," Working Paper, University of Oxford.

Huang, R. and Stoll, H. (1996) "Dealer versus auction markets: a paired comparison of execution costs on NASDAQ and the NYSE," *Journal of Financial Economics*, 41: 313–57.

Hutchins, J. (2004) "Motivations for trading mechanism selection: hybrid markets vs. limit-order books. Evidence from the US securities market," Working Paper, University of Oxford.

Keim, D. and Madhavan, A. (1996) "The upstairs market for large-block transactions: analysis and measurement of price effects," *Review of Financial Studies*, 9: 1–36.

Kranton, R. (1996) "Reciprocal exchange: a self-sustaining system," *American Economic Review*, 86: 830–51.

Kugler, T., Neeman, Z. and Vulkan, N. (2006) "Markets versus negotiations: an experimental investigation," *Games and Economic Behavior*, 56(1): 121–34.

Madhavan, A. and Smidt, S. (1991) "A Bayesian model of intraday specialist pricing," *Journal of Financial Economics*, 30: 99–134.

Momsen, F. (2006) "Exchange mechanism selection in derivatives markets: an empirical investigation of trader behaviour at Eurex and other major derivatives exchanges," Working Paper, University of Oxford.

Neeman, Z. and Vulkan, N. (2010) "Markets versus negotiations: the predominance of centralized markets," *B.E. Journal of Theoretical Economics*, 10(1) (Advances): article 6.

Ozgit, A. (2005) "Auction versus dealer markets in online betting," Working Paper, UCLA Economics Department.

Pagano, M. and Roell, A. (1996) "Transparency and liquidity: a comparison of auction and dealer markets with informed trading," *Journal of Finance*, 51: 579–611.

Parlour, C. and Seppi, D. (2003) "Liquidity-based competition for order-flow," *Review of Financial Studies*, 16(2): 301–43.

Reiss, P. and Werner, I. (2002) "Anonymity, adverse selection, and the sorting of interdealer trades", *Review of Financial Studies*, 18(2): 599–636.

Roell, A. (1992) "Comparing the performance of stock exchange trading systems," in J. Fingleton and D. Schoenmaker (eds), *The Internationalisation of Capital Markets and the Regulatory Response*, Graham and Trotman, pp. 167–80.

Seppi, D. (1990) "Equilibrium block trading and asymmetric information," *Journal of Finance*, 45: 73–94.

Shin, H. S. (1996) "Comparing the robustness of trading systems to higher order uncertainty," *Review of Economic Studies*, 63: 39–59.

Snell, A. and Tonks, I. (2003) "A theoretical analysis of institutional investors' trading costs in auction and dealer markets", *Economic Journal*, 113: 576–98.

Viswanathan, S. and Wang, J. (1998) "Market architecture: limit-order books versus dealership markets," *Journal of Financial Markets*, 5: 127–67.

Williamson, O. (1975) *Markets and Hierarchies: Analysis and Antitrust Implications*, Free Press.

INDEX

·······················

Note: Bold entries refer to figures and tables.